T0185758

Aerodynamics

Peiqing Liu

Aerodynamics

 Science Press
Beijing

 Springer

Peiqing Liu
Lu Shijia Laboratory
Beihang University
Beijing, China

ISBN 978-981-19-4588-5   ISBN 978-981-19-4586-1   (eBook)
https://doi.org/10.1007/978-981-19-4586-1

Jointly published with Science Press, Beijing, China
The print edition is not for sale in China mainland. Customers from China mainland please order the print book from: Science Press.

Translation from the Chinese language edition: "Kong Qi Dong Li Xue" by Peiqing Liu, © Science Press 2021. Published by Science Press. All Rights Reserved.

This Springer imprint is published by the registered company Springer Nature Singapore Pte Ltd.
The registered company address is: 152 Beach Road, #21-01/04 Gateway East, Singapore 189721, Singapore

Foreword

This book is compiled on the basis of the lectures given by the author in their 20 years of teaching fluid mechanics and aerodynamics courses at Beihang University. In the compilation process, the author adhered to the style of the Göttingen School of Applied Mechanics and the idea of combining theory and application advocated by Dr. Lu Shijia. This book's characteristics are strengthening the foundation, facing engineering, and going from shallow to deep, guiding readers to obtain inspiration from natural phenomena and experimental observations in order to stimulate interest, condense key scientific issues, and improve independent innovation capabilities.

This book is divided into two parts: basic aerodynamics and applied aerodynamics. The basic part of aerodynamics includes the principles and basic equations of fluid kinematics and fluid dynamics. It clarifies the different characteristics of ideal and viscous flows, incompressible and compressible flows, and reveals waves, vortices, turbulence, and separation phenomena in flows. It focuses on analyzing the epoch-making achievements of modern mechanics—Prandtl boundary layer theory and its significance to the rapid development of aeronautical engineering. In the applied aerodynamics part, it mainly elaborates on the flow around the airfoil, wing, fuselage, and wing-body combination from low speed to high subsonic speed, transonic speed, and supersonic speed, focusing on the aerodynamic characteristics and flow control such as lift and drag, and principles, such as drag reduction, increased lift technology, and supercritical wing design methods for overcoming sound barriers. After the readers have mastered the above-mentioned basic principles of aerodynamics, they can draw inferences from one another and carry out research on low-speed industrial aerodynamics such as high-speed trains, large fans, high-rise buildings, and new bridges, and even become hypersonic researcher in the field of aerospace engineering by further learning aero-thermochemistry.

This book can be used as a reference textbook for undergraduates, postgraduates, teachers, scientific researchers, and engineering technicians in the fields of aircraft design and engineering, engineering mechanics, flight mechanics, missile design, etc.

This book is the crystallization of the author's many years of first-line teaching experience. I am convinced that readers will benefit a lot after studying it, so I am willing to write a preface for it, and I am willing to solemnly recommend this book to readers.

Beijing, China   Li Jiachun
June 2020

Preface

Aerodynamics is the core course of the aircraft design and engineering major at Beihang University. It is a professional basic course after advanced mathematics, theoretical mechanics, and material mechanics. Its overall positioning is through teaching the basic principles and methods of aerodynamics, focusing on cultivating students' aerospace awareness, the ability to pose and solve problems, and the ability to comprehensively use the knowledge they have learned to carry out innovative practices. This textbook is compiled based on the syllabus of aeronautical engineering majors in our school. It is divided into two parts: basic aerodynamics and applied aerodynamics. It focuses on the basic principles and methods of aerodynamics, as well as the aerodynamic characteristics of the low-speed, subsonic, and transonic and supersonic flows, which are mainly served by undergraduates majoring in aircraft design and universities. It can also be used as a reference book for students and technicians in adjacent majors.

This textbook is compiled based on the author's lectures on aerodynamics courses taught at Beihang University in the past 20 years. Beijing University of Aeronautics and Astronautics has offered aerodynamics courses since its establishment of the school in 1952. The textbook used in the 1980s was "Basic Aerodynamics" compiled by Xu Huafang of our school. The textbook used in the 1990s was Aerodynamics compiled by Beijing University of Aeronautics and Astronautics, Nanjing University of Aeronautics and Astronautics, and Northwestern Polytechnical University (edited by Chen Zaixin, Liu Fuchang, and Bao Guohua); Aerodynamics for aircraft was edited by Yang Zuosheng and Yu Shouqin from Nanjing University of Aeronautics and Astronautics and Aerodynamics by Qian Yiji from our school since the beginning of the twenty-first century. In order to adapt to the development of modern aerodynamics, under the impetus of teaching reform, the author has integrated and modified the teaching content in the course of teaching for many years to reflect the cutting-edge and contemporary nature of the teaching content. In the process of compiling, the author adhered to the academic thoughts of Applied Mechanics for Göttingen University and the principle of combining theory and application that Mr. Lu Shijia has always advocated, guiding readers to be good at getting inspiration from natural phenomena and experimental processes, condensing scientific ideas, stimulating learning interest, and raising the sense of innovation. For example, the classic potential flow theory in the original textbook has been deleted, and the contents of viscous flow, boundary layer theory and separation, and aerodynamic principles of high-lift devices have been added. This textbook has a total of 128 hours, including 80 hours for basic aerodynamics (including 64 hours for theoretical teaching and 16 hours for experimental teaching), and 48 hours for applied aerodynamics (including 32 hours for theoretical teaching and 16 hours for course design).

Part I of this textbook consists of seven chapters on the basis of aerodynamics. Chapter 1 explains the history of aerodynamics, research objects, classifications, methods, etc. Chapter 2 introduces the physical properties of fluid and air, the principle of hydrostatics, and the standard atmosphere. Chapter 3 introduces the principle of fluid kinematics and dynamics, the differential equations of ideal fluid motion (Eulerian equations) and vortex motion, etc. Chapter 4 introduces the ideal incompressible fluid plane potential flow theory, the superposition principle of singular points such as point vortices, point sources (sinks), and dipoles. Chapter 5 introduces the differential equations of viscous hydrodynamic motion (Navier–Stokes equations) and its characteristics. Chapter 6 introduces the boundary layer theory and its separation. Chapter 7 introduces the basics of compressible aerodynamics, one-dimensional compression flow equations, shock waves, and expansion waves.

Part II of this textbook consists of seven chapters on applied aerodynamics. Chapter 8 introduces the phenomenon of low-speed airfoil flows, aerodynamic characteristics, thin airfoil theory, etc. Chapter 9 introduces the phenomenon of low-speed wing flow, aerodynamic characteristics, and its lift line theory. Chapter 10 introduces the phenomenon of low-speed flow around the wing-body assembly, aerodynamic characteristics and interference mechanism, etc. Chapter 11 introduces the subsonic airfoil and wing flow and aerodynamic characteristics, as well as compression correction, etc. Chapter 12 introduces the supersonic airfoil and wing flow and air aerodynamic characteristics, as well as the impact of shock waves and expansion waves on aerodynamics, etc. Chapter 13 introduces the transonic airfoil and wing flow phenomenon and aerodynamic characteristics. Chapter 14 introduces the phenomenon of flow around high-lift devices of large aircraft and its aerodynamic characteristics.

Based on the theoretical teaching of this textbook, eight aerodynamic experiment items are configured, including Steady flow energy equation of incompressible fluid, Law of steady flow of incompressible fluid, Reynolds flow transition experiment, measurement of incompressible fluid in the boundary layer over the flat, Drawing Ma number distribution test along the valve nozzle, Display test of the flow around a cylinder and delta wing, and Airfoil pressure distribution and the wind tunnel test of the aerodynamic forces acting on wing.

The PPT that matches this textbook is compiled on the basis of the long-term teaching practice of all teachers in the course group, and has been revised twice. This set of teaching plans is based on theoretical teaching. The teaching content is appropriate, with illustrations and texts. The concept is clear and the structure is reasonable. It is very popular among students. The first edition was completed by

Peiqing Liu, Zhang Hua, Wu Zongcheng, and Chen Zemin; the second edition was a revised version of the first edition, completed by Qu Qiulin and Liu Peiqing; the third edition was a revised edition of the second edition, completed by Guo Hao and Liu Peiqing.

Approximately 90 WeChat textbooks supporting this textbook are distributed nationwide on the aerodynamics teaching platform and the Fengliu Zhiyin WeChat platform.

Academician Li Jiachun of the Institute of Mechanics from the Chinese Academy of Sciences was pleased to write a foreword for this book, and I would like to express my deep gratitude to him.

Professor Fu Song from Tsinghua University, Prof. Sun Mao from Beijing University of Aeronautics and Astronautics, and Prof. Zheng Yao from Zhejiang University reviewed the first draft of this book and provided valuable comments. I would like to express my sincere thanks to them.

Thanks to all the teachers in the Aerodynamics Course Group of Beihang University for their long-term support of the author's teaching work, especially the enlightenment and help in the years of teaching seminars. Thanks to Prof. Qu Qiulin and Assoc. Prof. Guo Hao for their great help in the writing process. Thanks to the Ph.D. student Liu Yuan for compiling the problem sets for each chapter.

Beijing, China
June 2020

Peiqing Liu

About This Book

This book is a textbook compiled for the major "aerodynamics" courses of my country's aerospace engineering. It is divided into two parts: basic aerodynamics and applied aerodynamics. It focuses on the basic principles and methods of aerodynamics, and the aerodynamic characteristics of low, subsonic, transonic, and supersonic flows around aircraft. The book is divided into fourteenth chapters, in which there are seven chapters on the basis of aerodynamics, including fluid kinematics and dynamics principles, ideal fluid motion differential equations (Eulerian equations) and vortex motion, the plane potential flow theory and singular point superposition principle for ideal incompressible fluid, the differential equations of mechanical motion (Navier–Stokes equations) and its characteristics for viscous fluid, boundary layer theory, and its separation, and the basis of compressible aerodynamics. There are seven chapters of applied aerodynamics, including the phenomenon around low-speed airfoil flows and thin wing airfoil theory, low-speed wing flows and lift line theory, low-speed flows and interference mechanism of wing-body assembly, aerodynamic characteristics of subsonic airfoil and wing flows, aerodynamic characteristics of supersonic airfoil and wing flows, aerodynamic characteristics of transonic airfoil and wing flows, and aerodynamic characteristics of high-lift devices of a large aircraft.

This book can be used as an undergraduate textbook for aircraft design and engineering, engineering mechanics, flight mechanics, missile design, and other majors. It can also be used as a reference by graduate students, teachers, researchers, and engineering technicians in aerospace-related majors.

Contents

Part II   Applied Aerodynamics

Part I
Fundamentals of Aerodynamics

Chapter 1
Introduction

This chapter mainly introduces the definition of aerodynamics, research objects and tasks, research methods and classifications, and the development history of aerodynamics, especially the leading role played in the development of modern aircraft.

Learning points:

(1) Definition, research object, research method, research content, and classification of aerodynamics.
(2) The development history of aerodynamics and its leading role in the development of modern aircraft.

1.1 Aerodynamics Research Tasks

There are five basic material forms in nature, including solid state, liquid state, gas state, plasma state, and ultra-dense state. Among them, solid corresponds to solid, liquid corresponds to liquid, and gas corresponds to gas. These three forms of matter are common and are jointly determined by the internal microstructure of matter, molecular thermal motion, and the force between molecules. The phase diagram of the object is shown in Fig. 1.1. From the perspective of macroscopic force, the force state of liquid and gas in a static state is the same (almost unable to withstand tensile and shear forces), so liquid and gas are called fluids.

According to the definition, mechanics is the subject of studying the laws of equilibrium and mechanical motion of objects and their applications. Therefore, solid mechanics is the subject of studying the laws of solids in equilibrium and mechanical motion and their applications, while fluid mechanics is the subject of studying the laws of fluids in equilibrium and mechanical motion and their applications and applied disciplines. Aerodynamics is a branch of fluid mechanics. It is a discipline that studies air in equilibrium and the laws of mechanical motion and its applications.

© Science Press 2022
P. Liu, *Aerodynamics*, https://doi.org/10.1007/978-981-19-4586-1_1

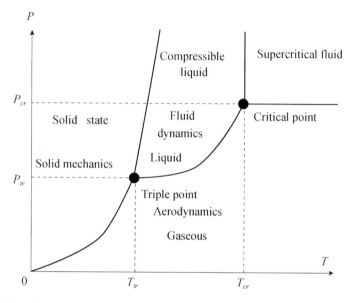

Fig. 1.1 Phase diagram of object

Aerodynamics is also a branch of physics, which mainly studies the laws of motion and force of air (or objects) when there is relative motion between objects and air. Traditional aerodynamics refers to the aerodynamics of aircraft, especially the aerodynamics of ordinary airplanes. Part of the force of the air on a moving aircraft is reflected in the lift (perpendicular to the flying direction of the aircraft), which acts on the aircraft; the other part is reflected in drag (opposite to the flying direction), which hinders the aircraft; and another part is caused by the resultant moment generated by the distributed pressure acting on the surface of the aircraft that controls the attitude of the aircraft. When people study aerodynamics, they often use the principle of relative flight to equate the movement of the aircraft through the air as the motion of the aircraft moving around the aircraft without moving the air. Using this principle, study the aerodynamic behavior of an aircraft when it moves in a uniform straight line at a certain speed in still air. The aerodynamic force experienced when bypassing the aircraft is equivalent, as shown in Fig. 1.2. Relative to the principle of flight, it provides convenience for the research of aerodynamics, which is the basis of aerodynamics experiments. During experimental research, people can fix the aircraft model, artificially create a straight and uniform airflow through the model, in order to observe the flow phenomenon, measure the aerodynamic force received by the model, conduct aerodynamic experimental research, and let the airflow in the wind tunnel test. Flow is easier to achieve than moving objects (as shown in Fig. 1.3). In flight mechanics, the speed relative to the aircraft is called airspeed (incoming flow speed), which is used to calculate aerodynamics; and the speed relative to the ground is called ground speed, which is used to calculate the flying distance of the aircraft.

When there is no crosswind, the airspeed and ground speed of the aircraft are equal and opposite; but when there is a crosswind, the two are different.

As a classic textbook, the main content of aerodynamics foundation includes aerodynamics and dynamics foundation. In low-speed flow, incompressible ideal fluid (inviscid fluid) two-dimensional and three-dimensional potential flow, thin wing theory, lift line and surface theory, flow around swept wings, etc. Viscous fluid dynamics equations, near-wall boundary layer theory, boundary layer separation, resistance of circumfluence objects, multi-section airfoil circumfluence, etc. In the subsonic flow, the full-velocity potential function of the compressible ideal potential flow is a nonlinear elliptic partial differential equation, the theory of perturbation linearization, the Prandtl–Glauert rule, and the Carmen–Qian Xuesen formula.

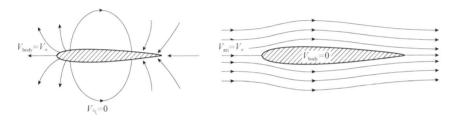

Fig. 1.2 Relative flight principle

Fig. 1.3 Wind tunnel test

In supersonic flow, small perturbation linear hyperbolic partial differential equations of compressible ideal potential flow velocity potential function, compression wave, expansion wave, shock wave, Prandtl–Meyer flow, etc. In transonic flow, the nonlinear mixed partial differential equation of ideal potential flow velocity potential function, critical Mach number, supercritical airfoil, and transonic airfoil flow around.

1.2 History of Aerodynamics

1.2.1 Qualitative Knowledge and Practice

For a long time, human beings have longed to fly freely in the air like small birds. There are many beautiful and mysterious myths and legends related to flying, such as the chariot flying in ancient Greece and Rome, the flying of the feather suit, and the flying of Cupid and the archery, the speeding vehicle of China, Chang'e, and so on. These wonderful legends all show mankind's reverie about flying in the sky and flying in the clouds. However, as a scientific record, the earliest description of airflow should belong to the ancient sage and ancient Greek scientist Aristotle (384–322 BC, as shown in Fig. 1.4). In 350 BC, Aristotle first gave a continuity model to describe air, and found that objects would experience resistance when moving in continuous air. In 250 BC, the ancient Greek scientist Archimedes (287–212 BC, as shown in Fig. 1.5) proposed the principle of hydrostatics (that is, the buoyancy theorem of an object in a static body of water) and found that when there is a pressure difference in a body of water, the body of water will move. After entering the Renaissance period in the dark Middle Ages, Italian all-round scientist Leonardo Da Vinci (1452–1519, as shown in Fig. 1.6), through a large number of observations and studies of bird flight, found that the wing surface exists under the bird's wing. The high-pressure air causes the bird's wings to receive upward force (lift). It is considered that the bird is an instrument that works according to the law of mathematics (the law of force balance). Objects can give a qualitative description such as drag reduction. According to records, in 1673, the French physicist Edme Mariotte (1620–1684, as shown in Fig. 1.7) discovered for the first time by measuring the impact force of a large number of water jets on a flat plate that the lift force of the flat plate is directly proportional to the square of the incoming flow velocity. In 1668, when the Dutch physicist Christiaan Huygens (1629–1695, as shown in Fig. 1.8) studied the falling characteristics of an object, he found that the resistance of an object was proportional to the square of the velocity, not the one-power relationship, given by Leonardo da Vinci. After the advent of calculus, scientific research entered the era of quantification. The classical continuum mechanics theory that combines continuous differentiable functions with particle mechanics constitutes the theoretical basis for the rapid development of mechanics. In 1687, the British scientist Isaac Newton (Isaac Newton, 1642–1726, as shown in Fig. 1.9) proposed the proportional relationship between lift and drag acting on the wings in his book "Mathematical Principles of Natural Philosophy", which is

$$L \propto \rho V_\infty^2 S, \quad D \propto \rho V_\infty^2 S$$

where L and D are lift and drag, V_∞ is the incoming flow velocity, S is the wing area, and ρ is the air density. But Newton only paid attention to the air lifting force of the lower wing surface, and did not consider the suction effect of the upper wing surface, and concluded that the lift force is proportional to the square of the angle of attack, rather than the first power of the angle of attack found in the experiment. The lift predicted by Newton at a small angle of attack is significantly smaller than the actual lift value. The British aerodynamicist George Cayley (1773–1857, as shown in Fig. 1.10) is known as the father of classical aerodynamics. He has done a lot of research on the principles of bird flight. Observing the flight speed, we estimated the relationship between speed, wing area, and lift, and found that the lift of the bird's wing changes with the angle of attack of the bird's wing in addition to the square of the flight speed and the bird's wing area. At the same time, it is suggested that the propulsion power and lift surface of artificial aircraft should be considered separately.

$$L \propto V_\infty^2 S C_L(\alpha)$$

American scientist Langley Samuel Pierpont (1834–1906, as shown in Fig. 1.11) proposed a formula for calculating wing lift. The German engineer and gliding pilot Otto Lilienthal (1848–1896, as shown in Fig. 1.12) began to manufacture gliders. He was one of the aviation pioneers in the manufacture and practice of fixed-wing gliders,

Fig. 1.4 Ancient sage, ancient Greek scientist Aristotle (384–322 BC)

Fig. 1.5 Archimedes,
ancient Greek scientist
(287–212 BC)

and he flew more than 2000 tests near Berlin. Secondly, a wealth of information has
been accumulated, providing valuable experience for the Wright brothers in the
United States to realize powered flight in the future. The British aerodynamicist
F. W. Lanchester (1868–1946, as shown in Fig. 1.13), in a paper in 1891, pointed
out the principle of flight of a vehicle that is heavier than air, and discovered the
wingtip vortices of the wings (as shown in Fig. 1.14); in 1894, the principle of
lift generation by the wing was first explained, and the correct calculation method
was proposed. The American aircraft inventors Wright brothers are two imaginative
and visionary engineers with both practical experience and theoretical knowledge
(Wilbur Wright, 1867–1912 and Orville Wright, 1871–1948, as shown in Fig. 1.15).
On December 27, 1903, Orville Wright drove them to design and manufacture the
"Aviator One" for the first test flight. This was the first powered, manned, continuous,
and stable and steerable aircraft. Since then, a new era of powered flights has been
opened. Thereafter, the development of aircraft promoted the rapid development of
aerodynamics. As shown in Fig. 1.16, the qualitative study from "Changer to the
moon" to a qualitative understanding process by Vinci in 1945. From the bird-like
model of Da Vinci to the power flight of the Wright brothers, humans went through
430 years of quantitative experiments and scientific cognitive processes.

Fig. 1.6   Italian all-round
scientist Leonardo da Vinci
(1452–1519 AD)

Fig. 1.7   French physicist
Edme Mariotte (1620–1684)

Fig. 1.8 Dutch physicist Christiaan Huygens (1629–1695)

1.2.2 Low Speed Flow Theory

In the late seventeenth century, British scientist Newton (shown in Fig. 1.9) and German scientist Gottfried Wilhelm Leibniz (1646–1716, shown in Fig. 1.17). After the invention of calculus, mathematicians and mechanics quickly combined the continuous differentiable function of calculus with the theory of particle mechanics. In 1727, Swiss scientist Johann Bernoulli (1667–1748) proposed classical continuum mechanics, which laid a solid theoretical foundation for the development of fluid mechanics and aerodynamics. In 1738, the Swiss scientist Daniel Bernoulli (1700–1782, as shown in Fig. 1.18) applied the kinetic energy theorem of a particle to the same ideal fluid micro-element flow tube, and derived the equation for the conservation of mechanical energy of the univariate flow, which is the famous steady flow

Fig. 1.9 British physicist
Isaac Newton (1642–1726)

Fig. 1.10 British
aerodynamicist George
Cayley (1773–1857)

of ideal fluid energy equation (hereinafter referred to as Bernoulli equation). For the
steady flow of an ideal incompressible fluid, without considering the mass force,
the Bernoulli equation shows that the sum of the pressure potential energy and the
kinetic energy of a unit mass fluid particle along the same streamline is a constant.

$$\frac{p}{\rho} + \frac{V^2}{2} = C$$

The discovery of Bernoulli's equation correctly answers the reason why the
suction of the upper wing surface contributes to the lift. Later, wind tunnel tests

Fig. 1.11 Samuel Pierpont
Langley, American scientist
(1834–1906)

Fig. 1.12 German engineer
and glider Otto Lilienthal
(1848–1896)

showed that for the airfoil, the contribution of the upper airfoil suction accounted for about 60–70% of the total lift of the airfoil. In 1755, Swiss mathematician and fluid dynamicist Leonhard Euler (1707–1783, as shown in Fig. 1.19) proposed the Euler method to describe fluid motion. Based on the continuum hypothesis and the ideal fluid model, using Newton's second theorem, the ideal fluid motion differential equations, namely Euler equations, are established.

$$\frac{\mathrm{d}\,\vec{V}}{\mathrm{d}t} = \vec{f} - \frac{1}{\rho}\nabla p$$

Fig. 1.13 British fluid mechanics F. W. Lanchester (1868–1946)

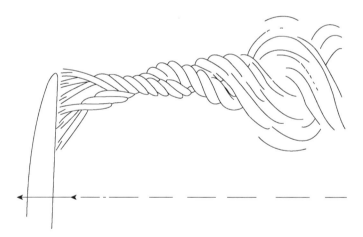

Fig. 1.14 Wingtip vortex given by Lanchester

For the steady flow of an ideal incompressible fluid with potential mass force, the Bernoulli equation can be obtained by integrating Euler's equations along the streamline. Further research shows that not only the Bernoulli equation is satisfied along the same streamline, but it is also satisfied along the same vortex line, potential flow field, and spiral flow. In the nineteenth century, fluid mechanics focused on the development of ideal fluid non-rotational motion solutions, and established ideal fluid

Fig. 1.15 The Wright Brothers, American aircraft inventors (left: Wilbur Wright, 1867–1912; right: Orville Wright, 1871–1948)

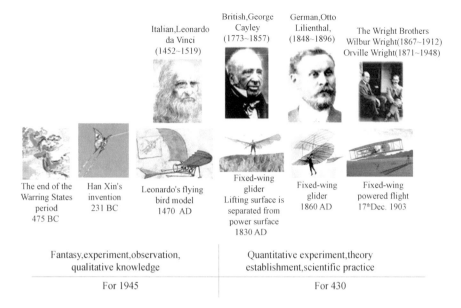

Fig. 1.16 The cognitive process of aviation flight

Fig. 1.17 German
mathematician Gottfried
Wilhelm Leibniz
(1646–1716)

vortex motion theory and viscous fluid motion differential equations. In 1858, the German fluid dynamicist Hermann Ludwig Ferdinand von Helmholtz (1821–1894, as shown in Fig. 1.20) proposed the velocity decomposition theorem of fluid clusters, and at the same time studied the rotational motion of an ideal incompressible fluid under the action of a powerful force. The Helmholtz vortex motion conservation law is proposed, that is, the vortex intensity invariance law along the vortex tube, the vortex tube retention law, and the vortex intensity conservation law, and the ideal fluid vortex motion theory is established. Choose a closed curve in the flow field. The

Fig. 1.18 Daniel Bernoulli, mathematician and fluid dynamicist (1700–1782)

Fig. 1.19 Leonhard Euler, Swiss mathematician and fluid dynamicist (1707–1783)

linear integral of the speed along the closed curve is called the velocity circulation of the closed curve. The sign of the velocity circulation is not only determined by the velocity direction of the flow field, but also related to the detour direction of the closed curve. Relevantly, it is stipulated that the counterclockwise detour direction is positive during integration. According to the Stokes line integral and area fraction formula, in the velocity vector field, the velocity loop along any closed curve is equal to the vortex flux of any curved surface stretched by the closed curve, which is

Fig. 1.20 German fluid
dynamicist Hermann Ludwig
Ferdinand von Helmholtz
(1821–1894)

$$\Gamma = \oint_{L} \vec{V} \cdot d\vec{s} = \iint_{S} 2\vec{\omega} \cdot d\vec{S} = \iint_{S} \nabla \times \vec{V} \cdot d\vec{S}$$

where Γ is the vortex intensity (velocity circulation) in the region passing through the envelope L. $\vec{V}$ is the velocity field. $\vec{\omega}$ is the rotational angular velocity of the fluid particle. $\nabla \times \vec{V} = 2\vec{\omega}$ is the vorticity of the fluid particle. In order to determine the lift of the airfoil, in 1902, the German mathematician Martin Wilhelm Kutta (1867–1944, as shown in Fig. 1.21) and the Russian physicist N. Joukowski (1847–1921, as shown in Fig. 1.22) in 1906. The formula for calculating the lift force of the flow around a circular cylinder is extended to the flow around an object of any shape, and it is proposed that for the flow around an object of any shape, as long as there is a velocity circulation, lift will be generated, and the direction of the lift will follow the direction of the incoming flow according to the reverse circulation. Rotate 90°. Later, it is called Kutta and Joukowski's law of lift circulation (as shown in Fig. 1.23), which is

$$L = \rho V_{\infty} \Gamma$$

where L is the lift force acting on the circumfluence object, ρ is the incoming air density, V_{∞} is the incoming flow velocity, and Γ is the velocity circulation of the circumfluence object. In 1909, Joukowski used the conformal transformation method of complex variable functions to study the steady flow around an ideal fluid airfoil, and proposed the famous Joukowski airfoil theory (as shown in Fig. 1.24). During

Fig. 1.21 German
mathematician Martin
Wilhelm Kutta (1867–1944)

Fig. 1.22 Nikolai
Yegorovich Joukowski,
Russian scientist (January
1847–1921)

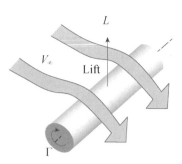

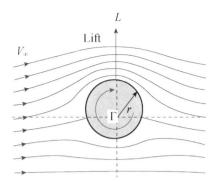

Fig. 1.23 Joukowski's law of lift circulation

the First World War, the warring countries all explored some better performance airfoils in practice, such as the Joukowski airfoil, the German Gottingen airfoil, the British RAF airfoil (Royal Air Force; later changed to the RAE airfoil–Royal Aircraft Establishment), the United States Clark–Y, and so on. After the 1930s, the American NACA airfoil (National Advisory Committee for Aeronautics, changed to NASA, National Aeronautics and Space Administration after 1958) and the ЦАГИ airfoil (Central Air Fluid Research Institute) of the former Soviet Union appeared. The National Aviation Advisory Committee (abbreviated as NACA, now NASA) in the late 1930s conducted a systematic study on the performance of airfoils (as shown in Fig. 1.25), and proposed NACA four-digit wing families and five-digit wing families, Wing family, etc., as shown in Fig. 1.26. At the same time, in view of the fact that the conclusion that there is no resistance to flow around an ideal fluid cylinder is inconsistent with reality, people began to study the movement of viscous fluids, starting with the French engineer Claude–Louis Navier (1785–1836, as shown in Fig. 1.27) in 1822, and finally by 1845 British scientist George Gabriel Stokes (1819–1903, as shown in Fig. 1.28) completed the differential equations of Newtonian fluid motion at Trinity College, Cambridge University, the famous Navier–Stokes equations, referred to as NS equations. For the differential equations of incompressible viscous fluid motion, the vector form is

$$\frac{\mathrm{d}\overrightarrow{V}}{\mathrm{d}t} = \overrightarrow{f} - \frac{1}{\rho}\nabla p + \nu\Delta\overrightarrow{V}$$

This system of equations shows that the mass force, differential pressure (surface normal force), and viscous force acting on the fluid micelles are caused to change the acceleration of the fluid micelles. In 1904, the world master of fluid mechanics, German mechanic Ludwig Prandtl (1875–1953, as shown in Fig. 1.29), published a paper on the motion of small viscous fluids at the Third International Annual Conference of Mathematics in Heidelberg, Germany, and proposed the well-known boundary layer concept. In the case of a large Reynolds number, the boundary layer flow characteristics and control equations of the boundary layer affected by viscosity

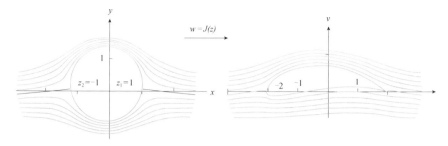

Fig. 1.24 Joukowski airfoil

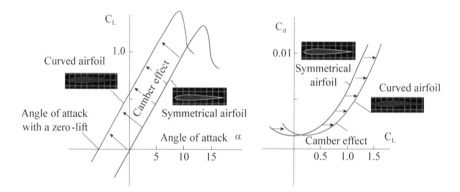

Fig. 1.25 Lift coefficient and drag coefficient of the airfoil

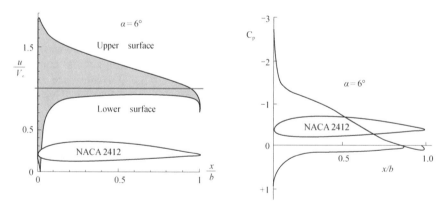

Fig. 1.26 Velocity and pressure distribution on the upper and lower airfoils of NACA2412 airfoil

on the surface of the flowing object are profoundly explained, and the relationship between the overall flow and the local flow is cleverly solved. Thus, a new way has been found to solve the problem of the resistance of the viscous fluid to bypass the object (as shown in Fig. 1.30), which has played an epoch-making role.

Fig. 1.27 French mechanic
Claude-Louis Navier
(1785–1836)

Fig. 1.28 George Gabriel
Stokes, British mechanic and
mathematician (1819–1903)

Fig. 1.29 Ludwig Prandtl, German mechanic and world master of fluid mechanics (1875–1953)

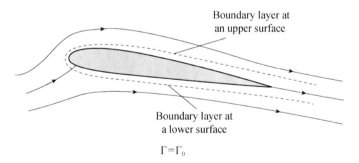

Fig. 1.30 Boundary layer development around airfoil

From 1911 to 1918, Prandtl conducted wind tunnel tests and found that straight wings with a large span (the leading-edge sweep angle of the wing is less than 20°, and the presentation ratio is greater than 5) flowed around because of the flow in the span direction. As shown in Fig. 1.31, the flow around the wing can be replaced by a straight uniform flow superimposed on the attached vortex (line) and the free vortex surface model (as shown in Fig. 1.31). The attached vortex and the free vortex surface are connected by countless Π-shaped horseshoe vortices. It is called the lifting surface model. The aerodynamic model only conforms to the actual flow around, for the following reasons: (1) The model conforms to the invariance theorem of the ideal fluid vortex strength along a vortex line and cannot be interrupted in the fluid. (2) The vertical flow of the Π-shaped horseshoe vortex is the attached vortex, which can replace the lift effect of the wing. The number of vortex lines passing through each section in the spanwise direction is different, and the strength of the attached vortex is also different. Among them, the vortex line passing through the middle section is the most, and the loop volume is the largest. There is no vortex line passing through the wingtip section, and the annular volume is zero, which simulates the spanwise distribution of the annular volume and lift (the elliptical distribution is the best). It can be seen that the strength of the attached vortex varies in the span direction, which is the same as the profile lift distribution, which is zero at the wing tip and maximum at the wing root. (3) The Π-shaped horseshoe vortex system flows in parallel and drags infinitely downstream, simulating the free vortex surface. Since the strength of the free vortex drawn between two adjacent sections in the spanwise direction is equal to the difference in the amount of the attached vortices in these two sections, the relationship between the strength of the free vortex line in the spanwise direction and the strength of the attached vortex on the wing is established. (4) For a straight wing with a large aspect ratio, since the chord length is much smaller than the extension length, the attached vortex system on the wing can be approximately merged into an attached vortex line with variable strength in the spanwise direction. Acting on this line is called the lift line hypothesis. Because the lift increment of the low-speed airfoil is at the focal point, approximately at the 1/4 chord point, the attached vortex line can be placed on the line connecting the 1/4 chord point of each profile in the spanwise direction. This line is the lift line. The lift line theory is an approximate potential flow theory for solving a straight wing with a large aspect ratio. After knowing the plane shape of the wing and the aerodynamic data of the airfoil, the circulation volume distribution, the profile lift coefficient distribution and the lift coefficient of the entire wing, the slope of the lift line, and the induced drag coefficient can be obtained. Its outstanding advantage is that the influence of the wing plane parameters on the aerodynamic characteristics of the wing can be clearly given.

As shown in Fig. 1.32, the development of classic low-speed aerodynamics has gone from Newton's boulder theory to Joukowski's law of lift circulation, and the theory of lift lines from airfoil flow to three-dimensional airfoil flow is established. Has promoted the development of multi-layer wing to single-layer wing.

In summary, the establishment of any theory is developed on the basis of a large number of experimental studies, and the establishment of fluid mechanics theory is no exception. From a historical perspective, without a large number of experiments, without calculus and continuum mechanics, there would be no establishment of Bernoulli equations. It is no exaggeration to say that Bernoulli's equation is a ground-breaking achievement in people's study of fluid motion and has played a milestone role. If there is no Bernoulli equation, it is impossible to accurately express some seemingly irrelevant flow phenomena with unified theoretical formulas. If there is no modeling idea of Bernoulli equations, it is impossible to have a later Euler equation

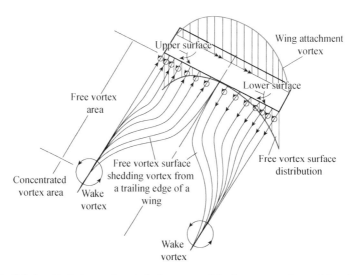

Fig. 1.31 Relationship between the attached vortex and the free vortex of a straight wing with a large aspect ratio

Fig. 1.32 Development of classic aerodynamics

system that characterizes the motion of ideal fluid clusters. If there are no Euler equations, it will not be generalized to the Navier–Stokes equations (N–S equations) that characterize the movement of viscous fluid clusters. Of course, without these, there would be no basic theories of fluid mechanics, nor would there be the establishment of later theories of boundary layer theory, turbulence, flow control, aerodynamic noise, etc. In summary, you can use Fig. 1.33 to express the development history of fluid mechanics. As the core equation of fluid mechanics, Bernoulli's equation plays the role of the soul (as shown in Fig. 1.34).

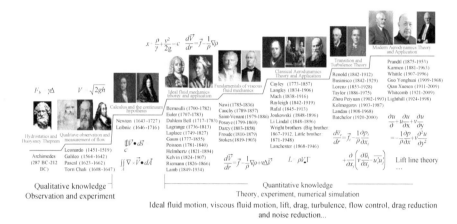

Fig. 1.33 The development history of fluid mechanics

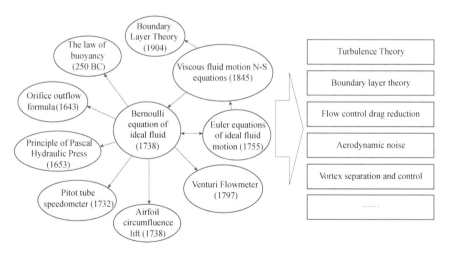

Fig. 1.34 Establishment and development of basic theories of fluid mechanics

1.2.3   High-Speed Flow Theory

With the development of jet technology, the flying speed of aircraft has increased rapidly. Experiments have found that when the incoming flow Mach number is greater than 0.3, the influence of air density on the flow cannot be ignored. At this time, fluid mechanics and thermodynamic theories must be combined to establish a correct high-speed aerodynamic theory. From 1887 to 1896, the Austrian scientist Ernst Mach (1836–1916, as shown in Fig. 1.35), when studying the propagation of projectile motion disturbance, pointed out that in the flow less than or greater than the speed of sound, the characteristics, and area of the disturbance propagation caused by the projectile are different. This introduces the ratio of flow velocity to local sound velocity as a dimensionless parameter to characterize. In 1929, German aerodynamicist J. Ackeret (1898–1981) first called this dimensionless parameter the Mach number, which was later widely cited in aerodynamics. The flow is usually divided according to the magnitude of the undisturbed incoming flow Mach number Ma far ahead. When Ma is less than 0.3, the compressibility of the air is not considered, and it is called low-speed flow; when Ma is between 0.3 and 0.8, it is flow at subsonic speed. In this range, the effect of compressibility on aerodynamic characteristics can be achieved through the Modification of results of low-speed flows. When Ma is between 0.8 and 1.2, it is transonic flow. In this range, there will be local supersonic or local subsonic regions in the flow field, and shock waves will generally appear. Within this range, the aerodynamic coefficient will change greatly with the increase of Ma. When Ma is between 1.2 and 5, the flow is supersonic. When Ma exceeds 5, the flow is hypersonic.

1. **For subsonic flow,** the governing equation of the ideal compressible potential flow is a nonlinear second-order elliptic partial differential equation. The main

Fig. 1.35 Ernst Mach, Austrian physicist (1836–1916)

approximate method for studying this type of flow is the theory of small disturbance linearization. The international fluid mechanics master Prandtl (1922) and British aerodynamics H. Glauert (1928) established the subsonic flow compressibility correction law, the Prandtl–Glauert law. According to this law, the influence of compressibility on aerodynamic characteristics can be corrected from the result of low-speed flow. It is not necessary to solve the compressible flow equation separately. In 1939, the American aerodynamicist Von Kármán (1881–1963, as shown in Fig. 1.36) and the Chinese scientist Qian Xuesen (1911–2009, as shown in Fig. 1.37) further revised Prandtl–Glauert Law, the famous Carmen–Money Law (Carmen proposed ideas, Qian Xuesen derives the result). This law better establishes the correction relationship between the compressibility of air in subsonic airflow and the surface pressure, and the range of adaptation is significantly wider than the Prandtl–Glauert law, especially for the correction of the airfoil leeward surface pressure coefficient reasonable.

2. **For supersonic flow,** the governing equation of the ideal compressible potential flow is a nonlinear second-order hyperbolic partial differential equation. Similarly, based on the theory of supersonic small perturbation linearization, a linear second-order hyperbolic partial differential equation is established and solved by the characteristic line method. In supersonic flow, the main research is on the influence of compression wave, expansion wave, shock wave, etc., on the flow. The shock wave of an ideal gas has no thickness and is a discontinuity in the mathematical sense. In 1870, the British scientist Rankine (1820–1872, as shown in Fig. 1.38) and the French scientist Hugoniot in 1887 independently derived

Fig. 1.36 American aerodynamicist Von Kármán (1881–1963)

Fig. 1.37 Chinese scientist
Qian Xuesen (1911–2009)

the Rankine–Hugoniot relationship from the continuous equation, momentum equation, and energy equation before and after the shock. Prandtl established the relationship between the Mach number before and after the positive shock. Regarding the small disturbance problem of the thin wing, Ackeret proposed the two-dimensional linear airfoil theory in 1925, and later extended it to the linear theory of three-dimensional airfoil. For two-dimensional and three-dimensional steady supersonic gas flows, the interface between the disturbed and undisturbed areas is the Mach wave. If the supersonic gas flows through a series of Mach wave expansion and acceleration, it is called an expansion wave. Prandtl and his student T. Meyer (1907–1908) established the relationship between expansion waves. Figure 1.39 shows the oblique shock wave and the normal shock wave at the head of the circling object.

3. **For transonic flow,** in the surrounding flow field, there will be part of the supersonic region (with the appearance of shock waves, as shown in Fig. 1.40), the flow changes are complex, and the governing equation of the flow is a second-order nonlinear mixed partial differential equation, which makes a theoretical solution more difficult. Especially when the flight speed or the flow speed is close to the speed of sound, the aerodynamic performance of the aircraft will change sharply, the resistance will increase suddenly, and the lift will drop sharply, which will seriously affect the maneuverability and stability of the aircraft. This is the famous sound barrier. The high-thrust jet engine broke through the sound barrier, but it did not solve the complex transonic flow problem well. Until the 1960s, due to the demand for transonic cruise flight, the study of transonic flow received great attention. American aerodynamicist Richard Whitcomb (1921–2009, as shown in Fig. 1.41) proposed the transonic area of aircraft in 1952, Law theory

Fig. 1.38 British scientist W. J. M. Rankine (1820–1872)

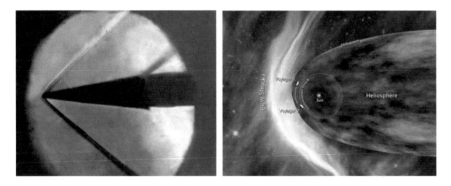

Fig. 1.39 Oblique shock wave and normal shock wave (bow shock wave)

(Area Rule). In 1967, the Supercritical Airfoil was proposed. This type of airfoil delays the generation of local shock waves and greatly increases the resistance divergence Mach number of the airfoil, thereby increasing the cruising speed of subsonic aircraft.

4. **For hypersonic flow,** it was proposed by Chinese scientist Qian Xuesen in 1946, mainly studying the theory, calculation method, and experimental technology of hypersonic flow. Hypersonic flow generally refers to a flow with an airflow velocity of more than five times the speed of sound. The main problems are aerodynamics (lift, drag, torque, pressure distribution, etc.), aerodynamics (heat flow

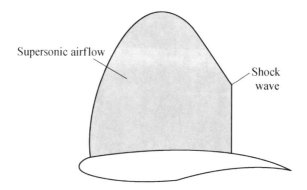

Supersonic airflow

Shock
wave

Fig. 1.40 Flow around a transonic airfoil

Fig. 1.41 Richard T. Whitcomb, American aerodynamicist (1921–2009)

calculation, preventive measures, etc.), and aerodynamics (flow field photoelec-
tric characteristics). In terms of experimental technology, the main parameters
of hypersonic flow ground simulation include free-flow Mach number, Reynolds
number, total flow enthalpy, density ratio before and after shock wave, test gas,
ratio of wall temperature to total temperature, and thermochemical properties of

the flow field. Common ground simulation equipment includes a shock tube, arc heating wind tunnel, hypersonic wind tunnel, and free ballistic target.

1.3 The Leading Role of Aerodynamics in the Development of Modern Aircraft

As a basic and forward-looking discipline of aerospace technology, aerodynamics has always played a leading and key role in the development of various aircraft. Therefore, its level of development has played a decisive role in the advancement of aircraft. For example, in the development of fighter jets (as shown in Fig. 1.42), in the 1950s, the emergence of jet engines led to the development of the first generation of supersonic fighters (MiG-15, F-86, etc.). In the 1960s, the breakthrough of large swept wings and area law aerodynamic problems led to the development of the second generation of fighter jets, which increased the speed of the aircraft to twice the speed of sound (MiG-21, F-4, French Mirage III, etc.). In the 1980s, breakthroughs in new aerodynamic technologies such as nonlinear lift technology and side strip and wing layout led to the development of third-generation fighter jets (Su-27, F-15, etc.), which improved weapon performance, airborne equipment, and maneuverability. After the 1990s, breakthroughs in wing-body integration design, new materials, electronics, and other new technologies have developed the fifth-generation fighter jet represented by the American F-22, which has stealth, supersonic cruise, super vision, distance combat capability, high mobility, and agility (as shown in Fig. 1.43). China's fifth-generation fighter J-20 (as shown in Fig. 1.44) is a stealth fifth-generation fighter with high stealth, high situational awareness, and high maneuverability. It can be seen that in the development of fighter jets, in addition to the application of high-tech achievements such as propulsion technology, electronic technology, new material technology, and stealth technology, aircraft pose more severe challenges to aerodynamics. How to expand the use range of the angle of attack as much as possible, how to improve the agility of the aircraft, how to achieve the smallest detectability while satisfying the flight performance, and how to exert the high efficiency of the propulsion system, etc. These are all aerodynamics that require breakthroughs, a difficult problem. For this reason, NASA has listed aerodynamics as one of the key technologies in its future research strategy in recent years. The goal of aerodynamics is to develop new concepts, put forward physical understanding and theory, test and CFD calculation verification, etc., and ultimately ensure the effective design and safe operation of the aircraft.

In addition, from the perspective of the development trend of civil aircraft, high-performance power plants and excellent aerodynamic characteristics are the guarantee for civil aircraft to obtain excellent cruise performance, take-off and landing performance, and economy. Since the British Comet in 1952, the development of large civil aircraft has been almost inseparable from the progress of aerodynamics, the introduction and introduction of technologies such as supercritical wing and lifting device design, winglet, flow control, and deformable wing. The breakthrough

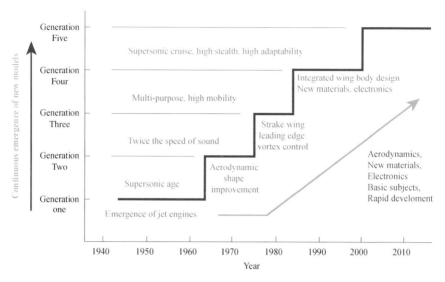

Fig. 1.42 The important role of aerodynamics in promoting the development of fighter jets

Fig. 1.43 U.S. single-seat dual-engine fifth-generation high-stealth fighter (F-22)

has greatly promoted the development of civil aircraft (as shown in Fig. 1.45), in terms of large passenger aircraft, such as the new 300-seat aircraft B787 produced by Boeing in the United States (as shown in Fig. 1.46), the 550-seat A380 produced by Airbus (as shown in Fig. 1.47), and the C919 produced in China (as shown in Fig. 1.48); and in terms of large transport aircraft, such as the large strategic military transport aircraft C-5 (shown in Fig. 1.49) produced by the United States Lockheed,

Fig. 1.44 China's fifth-generation single-seater dual-engine high-stealth fighter (J-20)

and China's Y20 (shown in Fig. 1.50). In order to solve the various challenging problems faced by aerodynamics today, it is generally believed that it is very important to strengthen the basic research work of the subject, and it is necessary to continuously explore the mechanisms and laws of various complex flow phenomena, such as the turbulent structure, the transition process of laminar flow, the leading-edge vortex and its rupture, the cause and control of asymmetric vortices at high angles of attack, effective control of laminar flow and turbulent drag reduction, shock wave, and boundary layer interference. These will be the main research directions of the subject in the future.

1.4 Aerodynamics Research Methods and Classification

Aerodynamics is the subject of studying the flow field and aerodynamic laws of aircraft (aircraft, missiles, etc.) under different flight conditions. The main research methods include three methods: theory, experiment, and calculation. Among them, theoretical research methods mainly use basic concepts, laws, and mathematical tools, grasp the main factors of the problem, and make quantitative analysis through some abstract models to reveal the laws. The experimental research method is mainly based on the principle of relative flight, with the help of wind tunnel and water tunnel equipment, experimental measurement of the physical quantity changes that bypass the model, in addition to free flight experiments and high-speed rail car experiments. Numerical calculation methods mainly use numerical discrete methods (finite difference method, finite element method, etc.) to numerically simulate flow phenomena. The three methods of theoretical analysis, experimental research, and

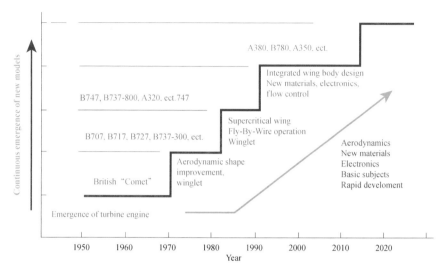

Fig. 1.45 The important role of aerodynamics in promoting the development of large passenger aircraft

Fig. 1.46 New 300-seat aircraft B787 produced by Boeing in the United States

numerical calculation have their own advantages and disadvantages, complement each other, and promote each other.

According to the research method, aerodynamics is divided into theoretical aerodynamics, experimental aerodynamics, and computational aerodynamics. According

Fig. 1.47 A 550-seat large wide-body airliner A380 produced by Airbus

Fig. 1.48 A 150-seat large narrow-body passenger aircraft C919 produced in China

to the speed of the aircraft, it is divided into low-speed aerodynamics, subsonic aerodynamics, supersonic aerodynamics, and transonic aerodynamics. According to the

Fig. 1.49 Large-scale strategic military transport aircraft C-5 produced by Lockheed in the United States

Fig. 1.50 Large transport aircraft Y20 produced in China

service object, it is divided into aircraft aerodynamics, missile aerodynamics, automobile aerodynamics, etc. According to whether viscosity is considered, it is divided into ideal aerodynamics, viscous aerodynamics, and so on.

1.5 Dimension and Unit

Physical quantity (abbreviated as quantity) is an attribute that specifies the distinction and quantitative determination of phenomena and substances. In nature, the quantities involved in physical phenomena can be divided into two categories according to their attributes: the size of one type of physical quantity is related to the unit selected for measurement, which is called a dimensional quantity, common ones being length, time, mass, speed, acceleration, force, kinetic energy, work, etc. The size of another type of physical quantity has nothing to do with the unit selected when measuring; it is called a dimensionless quantity, such as angle, ratio of two lengths, ratio of two times, ratio of two forces, and ratio of two energies. The easiest way to understand any physical phenomenon is to establish an inherent connection that reflects the nature of the phenomenon through dimensional analysis.

In a mechanical system, only three quantities are independent, called basic physical quantities, and other physical quantities are derived from physical quantities (according to definitions, laws, and relationships). In the International System of Units, people agree that length L, mass M, and time T are the basic physical quantities. The same type of quantity refers to the quantity that can be added, subtracted, and compared physically. The measurement unit is also a physical quantity, and belongs to the same type of quantity as the measured physical quantity. The size of a physical quantity represents the product of the value and the unit, and the same type of unit with different sizes is used to express a quantity without changing the type and value of the quantity. For example, 1 m = 100 cm = 1000 mm. Dimension refers to the category of a physical quantity, and the same type of quantity has the same dimension. The relationship between the unit and the dimension is the dimension represents the category of the physical quantity, and the unit represents the category and size of the physical quantity.

Assuming that the mass dimension M, the length dimension L, and the time dimension T are three independent physical dimensions, the expression of the physical dimension is derived as

$$[q] = M^x L^y T^z \tag{1.1}$$

where x, y, and z are dimensional exponents, which can be determined by physical theorems or definitions (in dimensional expressions, only the power product of basic quantities can be used instead of exponents, logarithms, trigonometric functions, and addition and subtraction operations). A dimensionless quantity means that in the dimensional expression of a quantity, all dimension indices are zero, otherwise it is a dimensioned quantity. Dimensionless quantities are different from pure numbers and have specific physical meanings and quantitative properties. The value of a dimensional quantity changes with different units, and the value of a non-dimensional quantity does not change with different units. From the principle of dimensional harmony, it can be known that in the physical equation that correctly reflects the objective law, the dimensions of addition and subtraction are the same. The dimensions in the physical equations are consistent, and have nothing to do with the unit system used by each physical quantity. Any dimensionless physical equation can be expressed by a dimensionless equation. The regularity in the physical equation does not change due to the difference in the basic physical quantities. The similarity principle and dimensional analysis will be given in Chapter 5, Basics of Viscous Fluid Dynamics.

Exercises

A. **Thinking questions**

1. Why did double-wing and triple-wing aircraft appear in the early days, but they are all gone today. Explain the main measures to increase the lift of the aircraft.
2. Qualitatively explain the reasons for the difference in the aerodynamic force of the flat plate, curved plate, and airfoil.
3. If the landing gear of the aircraft is not retracted, please explain how it affects the flight performance of the aircraft.
4. When an airplane is in flight, the balance requires that the total torque around the center of gravity of the airplane is zero. If the torque is unbalanced, what parts of the airplane are used to adjust the torque?
5. Please give the dimension and unit of force, moment, and acceleration.
6. What is the relative principle of flight? The difference between ground speed and airspeed?
7. What is the difference in aerodynamics between flying upwind and flying downwind?
8. What is the scientific idea of aerodynamics of British physicist George Kelly?
9. Explain the velocities of the upper and lower wing surfaces in the airfoil flow, and indicate the force characteristics of the wing surfaces.
10. Explain the theorem (magnitude and direction of lift) of Russian physicist Joukowski.
11. In a straight and uniform flow, the lift of the cylinder is determined by the rotation speed of the cylinder. For a non-rotating wing, what can be used to change the lift?
12. During the flight of an airplane, the thrust and drag balance are generated by the engine. With what force and drag balance do birds fly?

13. Why is the fuselage of an airplane slender, and the slenderness ratio of an airplane generally taken? (The ratio of the length of the fuselage to the maximum diameter.)

14. Why does a fixed-wing aircraft roll during take-off, and explain the main factors affecting the take-off speed of the aircraft?

15. What is the main function of the winglet? Briefly describe the principle of pneumatics.

16. Briefly describe the reasons for the use of lift-increasing devices on the aircraft.

B. Calculation problems

1. The Stokes–Oseen formula for the resistance on the sphere at low-speed V is

$$F = 3\pi\mu DV + \frac{9\pi}{16}\rho V^2 D^2$$

where D = sphere diameter, μ = viscosity, ρ = density. Prove that the dimensions of both sides of the formula equal sign are the same.

2. If p is the pressure and y is the coordinate, using {MLT} as the benchmark, give the dimensional expressions of the following quantities.

$$(a)\ \partial p/\partial y; \quad (b)\ \int p\,dy; \quad (c)\ \partial^2 p/\partial y^2; \quad (d)\ \nabla p$$

3. It is proved that the dimensions are the same, and the volume flow Q through the orifice with diameter D satisfies the following formula, and the hole is located on the side of h below the liquid surface.

$$Q = 0.68 D^2 \sqrt{gh}$$

where g is the acceleration due to gravity. What is the dimension of the constant 0.68?

4. Prove that the dimensions of the boundary layer x-momentum equation are consistent.

$$u\frac{\partial u}{\partial x} + v\frac{\partial u}{\partial y} = f_x - \frac{1}{\rho}\frac{\partial p}{\partial x} + \frac{1}{\rho}\frac{\partial \tau}{\partial y}$$

Among them, u and v are the velocity components of the fluid particle movement, f_x is the mass force per unit mass, p is the pressure, τ is the shear stress, and ρ is the fluid particle density.

Chapter 2
Basic Properties of Fluids and Hydrostatics

This chapter mainly introduces the continuum assumption, the flow of fluid, the basic mechanical properties (such as liquidity, compressibility and elasticity, viscosity, and Newton's laws of internal friction), the stress of the fluid of classification, the static fluid inside of the isotropic characteristics of the pressure, export Euler equilibrium differential equations, and the integral form; this paper introduces the characteristics of standard atmosphere.

Learning points:

(1) Become familiar with the fluid continuum hypothesis, the basic mechanical properties of fluid, and the mechanical characteristics of fluid micro-elements;
(2) Grasp the isotropy proof process of any point in the static fluid;
(3) Master the derivation process of Euler balanced differential equations and their integral forms;
(4) Understand the standard atmospheric characteristics.

2.1 Basic Properties of Fluids

From the perspective of micromechanics, the basic properties of fluids include fluid density, fluidity, compressibility and elasticity, viscosity, etc.

2.1.1 Continuum Hypothesis

The fluid is composed of a large number of molecules, and the distance scale between the molecules is much larger than the scale of the molecules themselves. All molecules are in endless and irregular movement, exchanging energy and momentum. Therefore, the movement of the fluid is microscopically uneven, discrete, and

© Science Press 2022
P. Liu, *Aerodynamics*, https://doi.org/10.1007/978-981-19-4586-1_2

random. However, if people use the naked eye or instruments to observe that the macroscopic motion behavior of the fluid is uniform, continuous, and definite. Thus, two methods for studying fluid mechanics have been formed: one method is statistical mechanics starting from the motion of microscopic molecules and atoms, and the method of statistical average is used to establish the equations to be satisfied by macroscopic physical quantities, so as to determine the macroscopic properties and motion behavior of the fluid. The application of this method is limited, and it is currently unable to provide sufficient theoretical support for fluid mechanics; another method is continuum mechanics based on the assumption of a continuum. From a macro perspective, the fluid is assumed to be composed of countless particles, and the physical properties of the macro fluid and the laws of motion of objects (diffusion, viscosity, heat conduction and other transport properties, mass, momentum and the law of conservation of energy, and the law of thermodynamics) are used to study the regularity of physical quantities (such as mass, velocity, pressure, and temperature) of fluid particles. This method is widely used in the field of fluid mechanics and is the basis of fluid mechanics research. Of course, the statistical mechanics method tries to explore the macroscopic laws from the micro level and reveal the relationship between the macro and the micro, which is of great benefit to the understanding of the basic properties and concepts of the macro in fluid mechanics.

The law of fluid motion and force is studied from the macro point, and the object of study is the medium composed of fluid particles. Fluid particle is defined as follows: a particle in fluid mechanics is a space point with mass (material point, as shown in Fig. 2.1) that is large enough (composed of a large number of molecules) on the micro level and small enough to ignore the volume size on the macro level. It is the smallest unit to express the macroscopic mechanical behavior of a fluid. The hypothesis of fluid continuum is that fluid is composed of countless mass points, which in any case fill the occupied space without gaps. That is to say, from a macro perspective, fluid particles and spatial points must meet a one-to-one relationship under any circumstances (moving and static), that is, each fluid particle can only occupy one spatial point at any time, but not More than two spatial points to ensure that the physical quantity of the fluid particle is not interrupted in space; each spatial point can only be occupied by one fluid particle at any time, and cannot be occupied by more than two particles, ensuring that the physical quantity of the fluid particle is in space and No multi-value appears on the table. In this way, people naturally introduce the calculus knowledge of the single-valued continuous differentiable function into the analysis of the physical quantity change of the fluid particle movement.

Macroscopically, the volume occupied by the fluid particle is zero, but microscopically, the particle represents the molecular group composed of a large number of molecules, and the position of the particle represents the centroid of the molecular group. As shown in Fig. 2.2, take a small volume $\Delta\tau$ in the fluid arbitrarily, and the mass contained in it is Δm. According to the definition, the average density of the fluid medium is expressed as

$$\overline{\rho} = \frac{\Delta m}{\Delta \tau} \tag{2.1}$$

Fig. 2.1 The relationship between the average density of fluid ρ and the size of molecular clusters

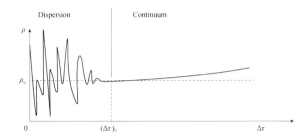

From the microscopic point of view, when $\Delta \tau \rightarrow \Delta \tau_0$, the average density of fluid clusters is independent of the number of molecules and is a function of spatial position. If viewed from a macroscopic point of view, it can be considered that the density of fluid particles at $\Delta \tau \rightarrow 0$ is

$$\rho = \lim_{\Delta \tau \rightarrow 0} \frac{\Delta m}{\Delta \tau} = f(x, y, z, t) \qquad (2.2)$$

Now take air as an example to illustrate the relationship between mass points and molecules. At sea level, the standard pressure at room temperature is 101325 Pa and the temperature is 288.15 K. Currently, every 1 cm^3 of space contains 2.7×10^{19} air molecules, the effective diameter of the molecules is 3.1×10^{-10} m, and the mean free path of the molecules is 8.71×0^{-8} m; under what circumstances does the flow of molecules meet the definition of a continuous flow of particles, rather than the discrete motion problem, which involves the relationship between the molecular mean free path and the characteristic scale of the object? Danish physicist Knudsen (1871–1949, as shown in Fig. 2.3) proposed to use the Knudsen number to judge the relative dispersion degree of molecules when studying molecular motility theory and low-pressure phenomena in airflow, that is, Kn number is defined as the ratio of the mean free path of molecules to the characteristic scale of the object under study. In 1946, Qian Xuesen, a Chinese scientist, proposed to use the number Kn

Fig. 2.2 Fluid micro-elements in fluid space

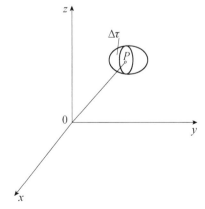

Fig. 2.3 Martin Knudsen
(1871–1949), Danish
physicist

to judge the continuity condition of fluid motion when studying the dynamics of rarefied gases. Generally, the average free path of air molecules is about 10 nm, and the Knudsen number less than 0.01 is called continuous flow, that is, the macro scale should be more than 1000 nm to be considered as continuous flow, then the fluid dynamics macro equation can be used to describe the fluid motion. While the Kn number between 0.01 and 1 is the slip flow, the viscous fluid motion equation with slip boundary condition can be used to describe the fluid motion. The Kn number between 1.0 and 10 is a transition flow. When Kn is greater than 10, it is the molecular flow (the macro scale is less than 1 nm). In this case, the assumption of molecular discrete flow is adopted, and the Boltzmann equation of molecular motion can be used to describe the fluid motion. In other words, the macroscopic scale of less than 1 nm is the complete molecular flow motion (discrete motion). Once in a state of continuous flow, the impact of the collisions and interludes of individual molecules on the mainstream is almost negligible, just as the behavior of an elephant (object size) is immovable by the random movement of individual ants (molecules) on the elephant.

2.1.2   Fluidity of Fluid

The essential difference between fluid and solid in macroscopic force that lies in that fluid is easy to flow, that is, they cannot withstand shear stress in a static state, while solid can withstand a certain intensity in a static state. As shown in Fig. 2.4, a solid can withstand tension, pressure, and shear forces in a static state because of its fixed shape and volume. For fluids (including gases and liquids), which have a fixed volume but no fixed shape, they can only withstand pressure but hardly tensile and shear forces at static, as shown in Fig. 2.5.

Fig. 2.4 Force on solid
micro-elements at static

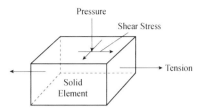

Fig. 2.5 Force on fluid
micro-elements at static

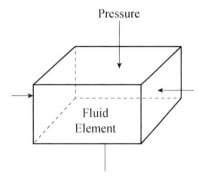

Solid micro blocks can bear a certain shear force at static, as shown in Fig. 2.6. According to Hooke's law (Robert Hooke, British scientist, 1635–1703), the shear force on a solid micro block is proportional to the shear angular deformation θ of the micro block, i.e.,

$$F = G\Delta A\theta, \tau = \frac{F}{\Delta A} = G\theta \tag{2.3}$$

where G is the shear modulus (unit: N/m)^2), ΔA is the bottom area of the micro block, and τ is the shear stress (unit: N/m^2).

However, for the fluid, if a similar experiment is conducted for any fluid micro-elements, it is found that under any shear force, the fluid micro-elements will undergo continuous deformation movement, and the angular deformation of the fluid micro-elements is not only related to the shear stress τ, but is also proportional to the shear stress τ duration, as shown in Fig. 2.7. It is shown that no matter how small the added shear stress τ is, as long as it is not equal to zero, the fluid will continue to deform

Fig. 2.6 Shear forces on
solids

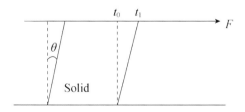

Fig. 2.7 Shear forces on fluid micro-elements

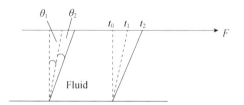

under the action of shear stress, which is called the fluidity of fluid. In other words, the fluidity of the fluid means that the fluid cannot withstand any shear force at static.

The difference between fluid and solid is not absolute. There are some substances whose properties are intermediate between solids and fluids, and have the properties of both. Substances such as gelatinous materials and paint that behave like solids after being placed for a period of time lose their elasticity when shaken or painted, undergo a lot of deformable motion, and behave exactly like fluids. This textbook focuses on pure fluids such as water or air.

2.1.3   Compressibility and Elasticity of Fluid

When a fluid is in motion, its volume changes due to changes in pressure and temperature called compressibility of the fluid. All real fluids have compressibility, and the ability of the fluid to resist compression deformation is called the elasticity of the fluid. Similar to the mechanics of materials, the elastic modulus of the volume is used to measure the elasticity of a fluid. By definition, the volumetric elastic modulus of the fluid represents the increase in pressure required to cause a change in unit relative volume, which is

$$E = -\frac{\mathrm{d}p}{\mathrm{d}\forall/\forall} \tag{2.4}$$

where E is the volumetric elastic modulus of the fluid (unit: N/m^2); $\forall$ is fluid micro-elements volume; the negative sign indicates that the volume is decreased (i.e., the volume increment is negative $\mathrm{d}\forall < 0$) when the fluid is under pressure ($\mathrm{d}p > 0$). For a given mass of fluid, $m = \rho\forall$ is a constant, obtained by differentiation

$$\mathrm{d}m = 0, \mathrm{d}(\rho\forall) = 0, \rho\mathrm{d}\forall + \forall\mathrm{d}\rho = 0 \tag{2.5}$$

Substitute Eq. (2.5) into Eq. (2.4), the volumetric elastic modulus can be written as

$$E = \frac{\mathrm{d}p}{\mathrm{d}\rho/\rho} = \rho\frac{\mathrm{d}p}{\mathrm{d}\rho} \tag{2.6}$$

Table 2.1 Density and elastic modulus of common fluids

Name of the fluid	Temperature (K)	Density (kg/m^3)	Modulus of elasticity (N/m^2)
Alcohol	288.15	789.5	0.9×10^9
Glycerin	288.15	1259.9	4.8×10^9
Mercury	288.15	13,600.0	20.3×10^9
Water	288.15	1000.0	2.1×10^9
Carbon dioxide (liquid)	288.15	1177.0	1.6×10^9
Air	288.15	1.225	1.4×10^5

It can be seen that when E is large, the fluid is not easily compressed; on the contrary, when E is small, the fluid is easily compressed. The bulk elastic modulus of liquids is generally large and can usually be regarded as an incompressible fluid. The bulk elastic modulus of gas is usually small, and is related to the thermal process, so the gas has compressibility. Whether compressibility of air should be considered for specific flow problems depends on the influence of density changes caused by pressure changes generated by flow on flow. In general, when the airflow velocity is low, the density change caused by pressure change is very small, regardless of the effect of air compressibility on the flow. Table 2.1 shows the density and elastic modulus of common fluids.

Water under normal temperature and pressure: $\rho_w = 1000 \, kg/m^3$, $E_w = 2.1 \times 10^9 \, N/m^2$. For air, under normal temperature $T = 288.15 \, K$ and standard atmospheric pressure: $\rho_a = 1.225 \, kg/m^3$, $E_a = \rho \frac{dp}{d\rho} = 1.42 \times 10^5 \, N/m^2$.

2.1.4 Viscosity of Fluid (Momentum Transport of Fluid)

The viscosity of a fluid reflects the momentum transport of a fluid. Although the fluid cannot withstand the shear stress in a static state, when relative movement occurs between adjacent fluid layers in a moving state (shear deformation movement), the fluid has the ability to resist the relative movement of the fluid layer. This resistance is called the shear stress of the flow layer. In fact, all fluids are viscous, but some are large and some are small, and the viscosity of air and water are very small (belonging to small viscous fluids), which makes people's daily activities such as drinking water and breathing not feel strenuous. However, it is found that although the fluid viscosity is small, when the relative motion between the fluid layers is large, the resistance of the fluid viscosity to the relative motion of the fluid layers cannot be ignored, which is reflected in the shear force at the flow layer.

In 1686, In the Mathematical Principles of Natural Philosophy, Newton formulated the law of internal friction against the relative motion of layers of flow. For different fluids placed between two parallel plates with a distance of h, Newton found through experiments that when the upper plate is moved, the fluid particles close to the upper plate adhere to the plate and move together with the plate, and the velocity of the fluid gradually decreases when leaving the moving plate until the speed of the stationary plate is zero (adhered to the stationary plate); the shear force F of the dragging moving plate is proportional to the moving speed U of the upper plate and the area A of the plate, and inversely proportional to the distance h between the plates, which is

$$F \propto \frac{U}{h} A, \ \tau = \frac{F}{A} \propto \frac{U}{h} \tag{2.7}$$

By writing Eq. (2.7) into an equation, we get

$$\tau = \mu \frac{U}{h} \tag{2.8}$$

In the formula, the proportional coefficient μ is related to the properties of the fluid, and its value varies with different fluids, which is called the hydrodynamic viscosity coefficient (unit: N/m)^2 S, Pa.s); $\frac{U}{h}$ is the increment of velocity per unit height, called the velocity gradient. As shown in Fig. 2.8, the direction of shear stress τ can promote the stationary plate and hinder the moving plate.

As shown in Fig. 2.7, in the period Δt, the horizontal distance of the moving plate is $\Delta x = U \Delta t$, and the distance between the plates is h. According to the geometric relationship, the shear deformation Angle $\Delta \theta$ of the fluid mass between the plates is

$$\text{tg}(\Delta \theta) \approx \Delta \theta \approx \frac{\Delta x}{h} = \frac{U \Delta t}{h}, \frac{d\theta}{dt} = \lim_{\Delta t \to 0} \frac{\Delta \theta}{\Delta t} = \frac{U}{h} \tag{2.9}$$

Equation (2.9) indicates that the relative velocity gradient of the flow layer is equal to the shear deformation rate of the flow layer. Therefore, Newton's law of internal friction is also called the constitutive relationship between shear stress and shear deformation rate of the flow layer. The dynamic viscosity coefficient μ is determined by the physical properties of the fluid, and its dimension is

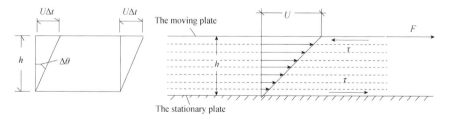

Fig. 2.8 Newton's law of internal friction

$$[\mu] = \frac{[\tau]}{[U/h]} = \frac{M}{LT} \tag{2.10}$$

where M is the mass dimension, L is the length dimension, and T is the time dimension.

From the perspective of molecular dynamics, the viscosity of the fluid is the result of momentum exchange caused by molecular thermal motion and collisions between the flow layers. The velocity of fluid molecules is composed of average velocity and molecular thermal velocity. The former reflects the macroscopic velocity of fluid particles, while the latter represents the random thermal velocity of fluid microscopic molecules. For adjacent fluid layers, when they move at their respective macroscopic speed u, due to the random thermal motion and collision of molecules, momentum exchange between fluid layers occurs. The fluid layer with high macro-velocity reduces its momentum through an exchange, and the fluid layer with low macro-velocity increases its momentum through an exchange. According to the momentum theorem, a pair of shear forces parallel to the direction of motion are generated between the two layers of fluid, as shown in Fig. 2.9. The shear force is opposite to the velocity direction of the fast layer, and hinders the fast layer. For the slow layer, it is in the same direction as the motion speed and plays a promoting role. Therefore, this shear force occurs in pairs inside the flow layer and is equal in size and opposite in direction, so it is called the fluid friction force, or viscous shear force. According to the molecular dynamics theory and Newton's law of internal friction, the dynamic viscosity coefficient of the fluid can be approximated as

$$\mu = 0.499 \rho c \lambda \tag{2.11}$$

where C is the molecular thermal velocity; λ is the mean free path of the molecule. The dynamic viscosity coefficients of different fluids are shown in Table 2.2. It can

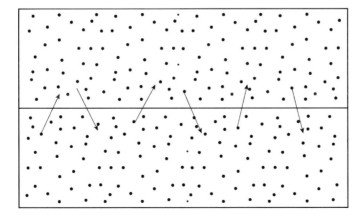

Fig. 2.9 Thermal motion of molecules

Table 2.2 Dynamic viscosity coefficient of common fluids

Name of the fluid	Temperature (K)	Density (kg/m^3)	Dynamic viscosity coefficient (N s/m^2, Pa s)
Alcohol	300.00	789.5	1.57×10^{-2}
Glycerin	300.00	1259.9	79.9×10^{-2}
Mercury	300.00	13,600.0	1.56×10^{-3}
Water	300.00	1000.0	1.0×10^{-3}
Carbon dioxide (liquid)	300.00	1177.0	1.49×10^{-5}
Air	293.15	1.225	1.79×10^{-5}

be seen from Newton's law of internal friction that fluid and solid are completely different in the law of friction. The friction force in a fluid depends on the relative motion between the fluids, that is, its magnitude is positive with the velocity gradient; the friction between the solids is independent of the velocity, but is proportional to the positive pressure exerted between the two solids. Under normal temperature and pressure, the dynamic viscosity coefficients of air and water are $\mu_a = 1.7894 \times 10^{-5}$ Pa s, $\mu_w = 1.139 \times 10^{-3}$ Pa s.

If we observe the viscous straight and uniform flow around the plate, we will find that the flow velocity distribution above the plate is similar to the parabolic distribution. The reason is fluid viscous effect: a layer of fluid mass points close to the surface of the plate adheres to the plate, and there is no relative movement with the surface of the plate, and the velocity of the fluid layer drops to zero. This condition is called the non-slip condition of viscous fluid bypassing the surface. The fluid particles in the outer layer are hampered by the underlying fluid speed which also dropped to close to zero. However, since it has left the surface flow layer to some extent, the viscosity effect of the fluid will decrease layer by layer as the distance from the plate increases. The speed keeps increasing. After a certain distance from the board surface, the viscous effect disappears and the flow layer velocity becomes uniform, as shown in Fig. 2.10. In this case, Newton's law of internal friction needs to be extended to flow with velocity gradient. Assuming that the coordinate y is perpendicular to the plate, the velocity u parallel to the plate is a continuous differentiable function $u(y)$ of y. Now, at the distance y from the plate, take the fluid layer of thickness dy, and the fluid particle velocity under the fluid layer is $u(y)$, the upper fluid particle velocity is $u(y + dy)$, and the upper and lower fluid particle velocity increment is du; according to Newton's internal friction Eq. (2.8), the shear stress between the fluid layers can be obtained for

$$\tau = \mu \frac{U}{h} = \mu \frac{U(h) - U(0)}{h} = \mu \frac{u(y + dy) - u(y)}{dy} = \mu \frac{du}{dy} \qquad (2.12)$$

Equation (2.12) is Newton's law of internal friction at any point in the velocity changing flow field. The equation shows that, for flow around a plate, the velocity gradient and shear stress on the plate are the largest, and the resistance of the plate

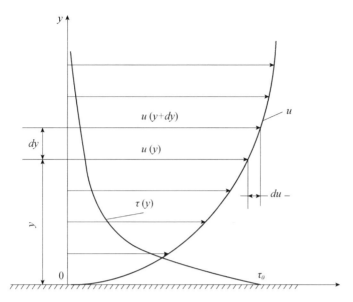

Fig. 2.10 Velocity distribution of straight and uniform flow of viscous fluid around a plate

with such shear stress to the fluid is called wall friction resistance. As the distance from the plate increases, the plate's obstructive effect on the fluid weakens, and the shear stress decreases. Until a certain distance away from the plate, the velocity gradient tends to zero, the shear stress also tends to zero, and the plate's blocking effect disappears.

According to Newton's law of internal friction:

(1) The shear stress of the fluid is independent of the pressure P (note that the friction force of solids is related to the positive pressure).

(2) When $\tau \neq 0$, $\frac{du}{dy} \neq 0$, it indicates that no matter how small the shear stress is, as long as it is not equal to zero, the fluid will continue to deform. Therefore, Newton's law of internal friction can also be regarded as a mathematical expression of fluidity.

(3) When $\left(\frac{du}{dy}\right) = 0$, $\tau = 0$, which means that as long as the fluid is stationary or has no relative motion (deformation motion), there is no shear stress, in other words, there is no static friction in the fluid.

(4) Since there is no slip between the fluid and the solid surface, the velocity gradient at the wall surface is finite, so the shear stress τ_0 at the wall is also a finite value.

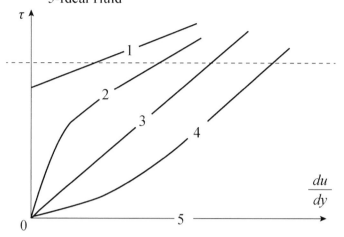

Fig. 2.11 Constitutive relationships of different fluids

In fluid mechanics, the fluid whose shear stress satisfies Newton's law of internal friction and whose dynamic viscosity coefficient is constant is called Newtonian fluid. Otherwise, it is a non-Newtonian fluid. The air studied in this book belongs to Newtonian fluid. For non-Newtonian fluids (as shown in Fig. 2.11), the general constitutive relationship can be written as

$$\tau = \tau_i + \mu \left(\frac{du}{dy} \right)^n \tag{2.13}$$

Among them, τ_i is initial stress and n is exponential. For example,

(1) For Bingham fluids such as mud, plasma, and toothpaste, $\tau_i \neq 0, n = 1.0$.
(2) For pseudo-plastic fluids such as nylon, rubber, paint, and insulation, $\tau_i = 0, n = 0.5$.
(3) For Newtonian fluids such as water, air, gasoline, and alcohol, $\tau_i = 0, n = 1.0$.
(4) For swelling plastic fluids such as dough and thick starch paste, $\tau_i = 0, n = 2.0$.
(5) If the viscosity of the fluid is not considered, it is called an ideal fluid, $\tau_i = 0$, $\mu = 0.0$.

It is found that the dynamic viscosity of the fluid changes with the temperature of the fluid, but does not change much with the pressure of the fluid (except in the case of ultra-high pressure, the viscosity of the fluid increases with the increase of pressure). When the temperature increases, the dynamic viscosity of gas increases.

This is because the viscous force of gas is mainly the result of transverse momentum exchange caused by molecular thermal motion in adjacent flow layers. Obviously, the higher the temperature is, the momentum exchange caused by molecular thermal motion will be intensified, and the viscosity coefficient will increase. However, it is different for liquids. As the temperature increases, the viscosity coefficient of the liquid decreases. Because the viscous force of liquid mainly comes from the cohesion of molecules between adjacent flow layers, as the temperature increases, the thermal motion of liquid molecules intensifies and the distance between liquid molecules becomes larger, so the cohesion of molecules decreases and the viscosity coefficient decreases.

The relationship between the dynamic viscosity coefficient of liquids and gases as a function of temperature can be found in the corresponding manuals or approximate formulas, the most commonly used of which is Sutherland's formula

$$\frac{\mu}{\mu_0} = \left(\frac{T}{288.15}\right)^{1.5} \frac{288.15 + C}{T + C} \tag{2.14}$$

where μ_0 is the corresponding value at $T = T_0 = 288.15\ °K$. For air, $\mu_0 = 1.7894 \times 10^{-5}$ Pa s. C is a constant, and is equal to $110.4\ °K$.

In many aerodynamic problems, the influence of flow is not the dynamic viscosity coefficient, but the unit mass of the dynamic viscosity coefficient, known as the kinetic viscosity coefficient, expressed by ν.

$$\nu = \frac{\mu}{\rho} \tag{2.15}$$

The unit of kinetic viscosity is m²/s, and the dimension is $[\nu] = \left[\frac{L^2}{T}\right]$. Under normal temperature and pressure, the kinematic viscosity coefficients of water and air are $\nu_a = 1.461 \times 10^{-5}$ m²/s $\nu_w = 1.139 \times 10^{-6}$ m²/s, respectively.

Because the dimension of ν only contains the length and time, for the kinematic quantity, it is called the kinematic viscosity coefficient. For a fluid with small viscosity, the viscosity effect can be ignored in some flows. We call fluids that do not consider viscosity as ideal fluids.

2.1.5 The Thermal Conductivity of the Fluid (The Heat Transport of the Fluid)

Thermal conductivity is the phenomenon of heat transfer when there is no macroscopic movement in the medium. It can occur in solids, liquids, and gases, but strictly speaking, only in solids is pure heat conduction. Even if the fluid is in a static state,

Fig. 2.12 French
mathematician and physicist
J. B. J. Fourier (1768–1830)

natural convection will occur due to the density difference caused by the temperature gradient. Therefore, the heat convection and heat transfer in the fluid occur at the same time. The essence of heat conduction is the process in which a large number of molecules in the fluid collide with each other, and the energy is transferred from the high-temperature part of the fluid to the low-temperature part. Assuming that the temperature in the fluid is a function of space coordinates and time $T(x, y, z, t)$, the process of heat transfer from high temperature to low temperature satisfies Fourier's law (first proposed by French physicist Fourier in 1822, as shown in Fig. 2.12); the heat transfer quantity per unit area q (also known as heat flux density) perpendicular to the heat transfer direction is proportional to the temperature gradient in this direction, and is written as an equation

$$q_x = -k\frac{\partial T}{\partial x}, q_y = -k\frac{\partial T}{\partial y}, q_z = -k\frac{\partial T}{\partial z} \qquad (2.16)$$

where, k is the thermal conductivity (unit: W/(K m)); q is the heat flux density (unit: W/m^2), the subscript of q is the direction of heat conduction, and the negative sign means that heat is always transferred from high temperature to low temperature.

2.1.6   *Diffusivity of Fluid (Mass Transport of Fluid)*

Diffusion phenomenon refers to the phenomenon in which the substance molecules transfer from the high-concentration area to the low-concentration area until uniform distribution. The diffusion rate is proportional to the concentration gradient of the substance. Diffusion is a mass transfer phenomenon caused by the thermal motion of molecules, which is mainly caused by density difference. It is believed that molecular thermal motion does not occur at absolute zero. A large number of facts, such as the diffusion phenomenon, show that the molecules of all substances are constantly doing an irregular movement. As early as 1855, the German physiologist Fick (as shown in Fig. 2.13) proposed that the amount of material diffusion per unit time through a unit area perpendicular to the diffusion direction is proportional to the material concentration gradient at the cross section. In other words, the larger the concentration gradient, the greater the material diffusion flux. And Fick established a mathematical expression describing the diffusion of substances from a high-concentration area to a low-concentration area by referring to Fourier's law. Assuming that $C(x, y, z, t)$ is the distribution of the material concentration field, the material diffusion flux is

$$J_x = -D\frac{\partial C}{\partial x}, \ J_y = -D\frac{\partial C}{\partial y}, \ J_z = -D\frac{\partial C}{\partial z} \qquad (2.17)$$

where D is the material diffusion coefficient (unit: m^2/s), C is the volume concentration of the diffused substance (unit: $kg/(m)^3$ J is the material diffusion flux (unit: $kg/(m)^2 \cdot S$), J subscript is the direction of substance diffusion, and the negative sign indicates that substance concentration always diffuses from high concentration to low concentration.

The above three types of transport phenomena all belong to the gradient transport process, which is a type of physical phenomenon that exists universally in nature. In fact, the transport properties of momentum, heat, and mass of fluids are similar in their formation mechanisms from a microscopic point of view. They are all transport caused by molecular thermal motion and molecular collision. From a macroscopic point of view, they are all related to their respective transport physical quantities. The gradient is proportional, which is

Fig. 2.13 Adolf Fick, German physiologist (1829 −1901)

Momentum transport (Newton's internal friction law): $\tau = \mu \frac{du}{dy}$

Heat transport (Fourier's law): $q_x = -k \frac{\partial T}{\partial x}$

Mass transport (Fick's law):. $J_x = -D \frac{\partial C}{\partial x}$

2.2 Classification of Forces Acting on a Differential Fluid Element

In the fluid, take any volume τ with a closed surface S as the interface (as shown in Fig. 2.14), and the force acting on the fluid micro-elements can be classified into mass force and surface force according to the action mode. Among them, mass force is the external force field acting on each fluid particle, and its size is proportional to the mass of the fluid micro-element. It belongs to non-contact force, such as gravity, inertial force, and electromagnetic force of magnetic fluid. Because mass is proportional to volume, mass force is also called volume force or total physical force. Assuming the mass force acting on the fluid micro-element is Δm, the mass force per unit mass can be expressed as $\Delta \vec{F}$.

$$\vec{f} = \lim_{\Delta m \to 0} \frac{\Delta \vec{F}}{\Delta m} = f_x \vec{i} + f_y \vec{j} + f_z \vec{k} \tag{2.18}$$

Fig. 2.14 Forces in the fluid

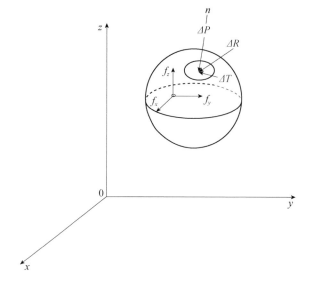

where Δm is the mass of the fluid micro-elements, $\Delta \vec{F}$ is the mass force acting on the fluid micro-elements, $\vec{i}, \vec{j}, \vec{k}$, respectively, represents the unit vector in three coordinate directions, and f_x, f_y, and f_z is the component of the unit mass force in three directions.

On the closed surface S, there is the force of adjacent fluids or objects on the taken fluid micro-element. The force is proportional to the surface area of the acting surface, which belongs to the contact force and is called the surface force. For example, pressure and friction are both surface forces. Since the surface force is distributed by area and generally does not coincide with the normal direction of the surface, the surface force can be decomposed into normal force and tangential force, as shown in Fig. 2.13. Assuming an arbitrary micro-element area ΔS is taken on S, the positive force on the unit area pointing to the normal direction in the action plane is called pressure in fluid mechanics, which is defined as

$$p = \lim_{\Delta S \to 0} \frac{\Delta P}{\Delta S} \qquad (2.19)$$

The concept of pressure and vacuum was introduced by the French physicist Blaise Pascal (1623–1662).

The tangential force on a unit area tangent to the acting surface is called the shear stress and is defined as

$$\tau = \lim_{\Delta S \to 0} \frac{\Delta T}{\Delta S} \qquad (2.20)$$

In fluid mechanics, the research objects are divided into ideal fluids and viscous fluids. For an ideal fluid, no matter in motion or static state, the force on any micro-element surface in the fluid is only normal force and no tangential force, and only pressure P exists. However, for viscous fluid (also called real fluid), the force behavior on the surface of the micro-element is different under static and moving states. In the static state, only the normal force (pressure) is applied to the surface of the micro-element without tangential force. But under the motion state, the normal force and shear stress are acting on the surface of the micro-element, the tangential force is completely caused by the shear deformation of viscous resistance to fluid micro-elements, and the normal force includes the normal force caused by inertia and viscosity; but the normal force caused by viscosity is generally much smaller than the normal force caused by inertia, so the internal normal stress is often called pressure in the moving fluid.

Pressure p is a characteristic quantity in the fluid, and its dimension and unit are expressed as follows:

(1) pressure dimension: $[p] = \left[\frac{\Delta P}{\Delta S}\right] = \frac{N}{L^2} = \frac{ML}{L^2 T} = \frac{M}{LT}$
(2) Force per unit area, pressure unit N/m², Pa or kPa;
(3) Expressed by the liquid column height ($h = p/\gamma$), the pressure unit can be expressed as m, cm, mm;
(4) Expressed by atmospheric pressure (a common barometer), the unit of pressure is Pa (one atmospheric pressure);
(5) Expressed by the units bar and mbar in meteorology, 1 ba = 100,000 Pa = 1000 Mbar.

Atmospheric pressure is divided into standard atmospheric pressure (atm) and engineering atmospheric pressure (at) where

$$p_{atm} = 101,300\,\text{Pa} = 101.3\,\text{kPa} = 1.013\,\text{bar} = 1013\,\text{mbar}$$
$$p_{at} = 98,000\,\text{Pa} = 98\,\text{kPa} = 980\,\text{mbar}$$

2.3 Isotropic Characteristics of Pressure at Any Point in Static Fluid

It has been pointed out in Sect. 2.1 that a distinguishing feature of fluids from solids is that static fluids cannot withstand any shear stress, that is, only normal stresses and no shear stresses exist in a fluid at static. The fluid mentioned here is in a static state, which means that the fluid is stationary relative to a certain coordinate system. This coordinate system can be an inertial coordinate system or a non-inertial coordinate system. No matter which coordinate system is used, as long as the fluid is stationary relative to the coordinate system, there is only normal stress (pressure p) in the fluid, and no tangential stress. Now in the static fluid, an arbitrary differential tetrahedron is taken around point O, and the rectangular coordinate system is taken over point O (as shown in Fig. 2.15). The lengths of the micro-element body in the three coordinate directions are Δx, Δy, Δz, respectively, and the four faces of the micro-element body are $\triangle BCO$, $\triangle ACO$, $\triangle ACO$, and $\triangle ABO$, and the slant $\triangle ABC$, respectively. The pressure at the centroid of each surface is p_x, p_y, p_z and p_n, respectively, and the unit mass force acting on the micro-element body is f_x, f_y, f_z. According to the force balance principle, the force balance equation along each direction is established as follows:

$$p_x \frac{1}{2}\Delta y \Delta z - p_n \Delta S \cos(n, x) + \Delta F_x = 0 \tag{2.21}$$

$$p_y \frac{1}{2}\Delta z \Delta x - p_n \Delta S \cos(n, y) + \Delta F_y = 0 \tag{2.22}$$

Fig. 2.15 Force balance of
micro tetrahedron

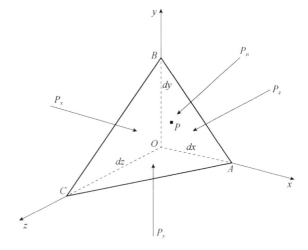

$$p_z \frac{1}{2}\Delta x \Delta y - p_n \Delta S \cos(n, z) + \Delta F_z = 0 \qquad (2.23)$$

From the geometric relation, it can be obtained as

$$\Delta S \cos(n, x) = \frac{1}{2}\Delta y \Delta z, \ \Delta S \cos(n, y) = \frac{1}{2}\Delta z \Delta x, \ \Delta S \cos(n, z) = \frac{1}{2}\Delta x \Delta y \qquad (2.24)$$

If the density of the fluid is ρ, the mass force can be expressed as

$$\Delta F_x = \rho f_x \frac{1}{6}\Delta x \Delta y \Delta z, \ \Delta F_y = \rho f_y \frac{1}{6}\Delta x \Delta y \Delta z, \ \Delta F_z = \rho f_z \frac{1}{6}\Delta x \Delta y \Delta z \quad (2.25)$$

By substituting Eqs. (2.24) and (2.25) into Eqs. (2.21)–(2.23), we can get

$$p_x - p_n + \rho f_x \frac{1}{3}\Delta x = 0 \qquad (2.26)$$

$$p_y - p_n + \rho f_y \frac{1}{3}\Delta y = 0 \qquad (2.27)$$

$$p_z - p_n + \rho f_z \frac{1}{3}\Delta z = 0 \qquad (2.28)$$

Let the micro-element tetrahedron shrink to zero, and set $\Delta x \to 0, \Delta y \to 0, \Delta z \to 0$, then from the formula (2.26)–formula (2.28),

$$p_x = p_y = p_z = p_n \qquad (2.29)$$

Since the normal n of the slant passing O is arbitrary, Eq. (2.29) indicates that the pressure acting on the slant in any direction at a point in the static fluid is equal (independent of the direction of the acting plane), and the direction is perpendicular to the acting plane. Thus, in the fluid, the pressure at any point can be expressed as a continuous function of the coordinate position, which is

$$p = p(x, y, z, t) \tag{2.30}$$

In the same way, it can be easily proved that for an ideal fluid, since there is no shear stress between the flow layers, the pressure at any point inside the flow field satisfies the isotropic condition at rest and in motion.

2.4   Euler Equilibrium Differential Equations

In a fluid of static equilibrium (absolutely static or relatively static), the Cartesian coordinate system O-xyz is adopted, and the coordinate axis is in any direction. Take any point $M(x, y, z)$ in the fluid, and then take the M point as the center, take three lengths Δx, Δy, and Δz along the three directions of the coordinate axis, and draw a micro-element hexahedron as the analysis object, as shown in Fig. 2.16. Let the pressure p at the point $M(x, y, z)$ be a function $p(x, y, z)$ of the coordinate position, and the force acting on the hexahedron be composed of surface force and mass force. Using the force balance principle, a system of fluid static equilibrium can be established. As shown in Fig. 2.16, the volume of the hexahedron is Δx, Δy, Δz, the density at the point $M(x, y, z)$ is $\rho(x, y, z)$, and the unit mass force acting on the point M is f_x, f_y, f_z. The surface force of the infinitesimal hexahedron can be expressed by the first-order Taylor series expansion of the pressure p at the center point, and the equilibrium equation is established by the force in the x-direction. The resultant force of the surface force of the infinite hexahedron in the x-direction is

$$\left(p - \frac{\partial p}{\partial x}\frac{\Delta x}{2}\right)\Delta y \Delta z - \left(p + \frac{\partial p}{\partial x}\frac{\Delta x}{2}\right)\Delta y \Delta z = -\frac{\partial p}{\partial x}\Delta x \Delta y \Delta z \tag{2.31}$$

The mass force in the x-direction is

$$f_x \rho \Delta x \Delta y \Delta z \tag{2.32}$$

Since the fluid is in a static state, the resultant external force in the x-direction is zero, and $\sum F_x = 0$, thus

$$-\frac{\partial p}{\partial x}\Delta x \Delta y \Delta z + f_x \rho \Delta x \Delta y \Delta z = 0 \tag{2.33}$$

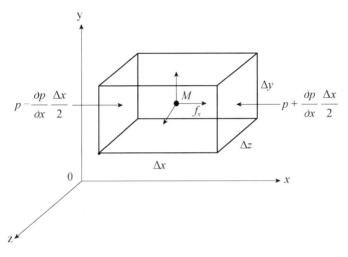

Fig. 2.16 Forces on a differential hexahedron

Divide the above equation by $\Delta x \Delta y \Delta z$ and let the element approach zero to obtain an equilibrium differential equation in the x-direction, which is

$$f_x - \frac{1}{\rho}\frac{\partial p}{\partial x} = 0 \tag{2.34}$$

Similarly, the equilibrium differential equation in the y- and z-directions can be obtained as

$$f_y - \frac{1}{\rho}\frac{\partial p}{\partial y} = 0 \tag{2.35}$$

$$f_z - \frac{1}{\rho}\frac{\partial p}{\partial z} = 0 \tag{2.36}$$

These three equations show that when the fluid is in static equilibrium, the component of the unit mass force in a certain direction is equal to the partial derivative of the pressure in that direction divided by the density. This system of equilibrium equations was derived by the Swiss scientist Euler in 1755, called Euler equilibrium differential equations.

According to Eqs. (2.34)–(2.36), the vector expression equation can be obtained, which is

$$\vec{f} - \frac{1}{\rho}\nabla p = 0 \tag{2.37}$$

where $\nabla = \frac{\partial}{\partial x}\vec{i} + \frac{\partial}{\partial y}\vec{j} + \frac{\partial}{\partial z}\vec{k}$ is a differential vector operator, also known as the Hamiltonian operator. Multiply Eqs. (2.34)–(2.36) by dx, dy, and dz, respectively, and then add them up to get

$$\frac{\partial p}{\partial x}dx + \frac{\partial p}{\partial y}dy + \frac{\partial p}{\partial z}dz = \rho\left(f_x dx + f_y dy + f_z dz\right) \qquad (2.38)$$

The left-hand side of this equation is the total differential of the pressure

$$dp = \frac{\partial p}{\partial x}dx + \frac{\partial p}{\partial y}dy + \frac{\partial p}{\partial z}dz \qquad (2.39)$$

The integral formula (2.38) along any closed curve in the fluid, as shown in Fig. 2.17, is obtained as

$$\oint_C \frac{\partial p}{\partial x}dx + \frac{\partial p}{\partial y}dy + \frac{\partial p}{\partial z}dz = \oint_C \rho\left(f_x dx + f_y dy + f_z dz\right) \qquad (2.40)$$

Since pressure is a single-valued continuous differentiable function, the integral on the left side of the above equation is zero, that is,

$$\oint_C \frac{\partial p}{\partial x}dx + \frac{\partial p}{\partial y}dy + \frac{\partial p}{\partial z}dz = \oint_C dp = 0$$

Thus, the integration on the right side of Eq. (2.40) is also zero

$$\oint_C \rho\left(f_x dx + f_y dy + f_z dz\right) = 0 \qquad (2.41)$$

Fig. 2.17 Curvilinear integral of mass force

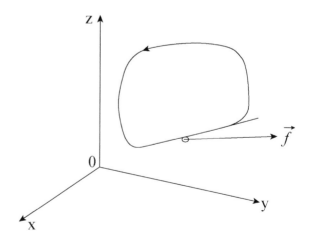

Since the integral curve is arbitrary and the fluid density is not zero, the fractional component of the mass force in parentheses on the left of Eq. (2.41) must be the total differential of the function Π, which is called the force potential function of the mass force, namely

$$f_x = -\frac{\partial \Pi}{\partial x}, f_y = -\frac{\partial \Pi}{\partial y}, f_z = -\frac{\partial \Pi}{\partial z}$$

$$f_x dx + f_y dy + f_z dz = -d\Pi \qquad (2.42)$$

It indicates that the mass force acting on the fluid particle must be powerful when the fluid is in a static state. But under force, the fluid can be stationary or moving.

The sufficient and necessary condition for the mass force to be a potential force is

$$\frac{\partial f_x}{\partial y} = \frac{\partial f_y}{\partial x}, \frac{\partial f_y}{\partial z} = \frac{\partial f_z}{\partial y}, \frac{\partial f_z}{\partial x} = \frac{\partial f_x}{\partial z} \qquad (2.43)$$

According to Eq. (2.38), the force balance equation can be written in full differential form, namely

$$dp = -\rho d\Pi \qquad (2.44)$$

When ρ is a constant, the above equation is obtained as

$$p = -\rho\Pi + C \qquad (2.45)$$

In the formula, C is the integral constant, which is determined by the value of Π. If the pressure value P_a and the force potential function Π_a at a point are known, the relationship between the pressure and the potential function at any other point is

$$p = p_a - \rho(\Pi - \Pi_a) \qquad (2.46)$$

In fluid, a geometric curve or surface consisting of space points with equal pressure is called an isobar or an isobaric surface. From Eq. (2.44), the isobaric surface ($P =$ constant) in the fluid (including the free liquid surface) must be the isobaric surface of the mass force ($\Pi =$ constant). As shown in Fig. 2.18, on the isobaric surface, $dp = 0$,

$$-d\Pi = f_x dx + f_z dy + f_z dz = \overleftarrow{f} \cdot d\vec{r} = 0 \qquad (2.47)$$

$d\vec{r} = dx\vec{i} + dy\vec{j} + dz\vec{k}$ is the radial of the isobaric surface. The above equation indicates that the mass force is orthogonal to the isobaric surface.

For example, the isobaric surface of a stationary liquid under gravity is horizontal. In addition to gravity, the liquid in the accelerated ascending or descending elevator is also subjected to downward or upward inertial force. The mass force combined

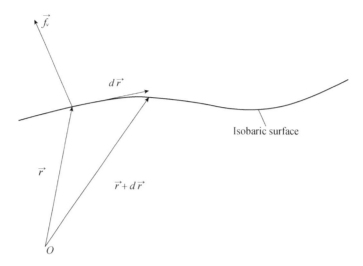

Fig. 2.18 Isobars

by the two is parallel to gravity, so the isobaric surface is also a horizontal plane. In addition to the vertical gravity, the liquid in the horizontal right accelerating vessel is also subjected to the left inertial force, and the combined mass force of the two deviates from the left of the vertical line, so the isobaric surface is a slanting face, as shown in Fig. 2.19.

Fig. 2.19 Horizontally accelerating to the right on the inclined water surface of the trolley

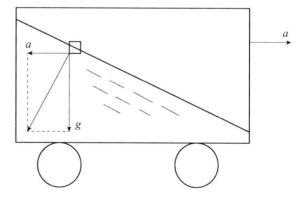

2.5 Pressure Distribution Law in Static Liquid in Gravitational Field

Now consider the pressure distribution in a stationary liquid in a gravitational field at the earth's surface. In the gravity field, let's say we have a liquid in a closed container, and the pressure at the free surface is p_0, establish the coordinate system as shown in Fig. 2.20, and consider a fluid particle at a height z from the horizontal axis; its unit mass force is only gravity, which can be expressed as

$$f_x = 0, f_y = 0, f_z = -g$$

The differential of the force potential function is

$$d\Pi = -(f_x dx + f_y dy + f_z dz) = -(-g dz) = g dz \tag{2.48}$$

where g is the gravity, which is substituted into the equilibrium total differential Eq. (2.40)

$$dp = -\rho d\Pi = -\rho g dz$$

Take $\gamma = \rho g$, and integrate along the height direction

$$\frac{p}{\gamma} + z = H \tag{2.49}$$

Fig. 2.20 Static liquid pressure under gravity

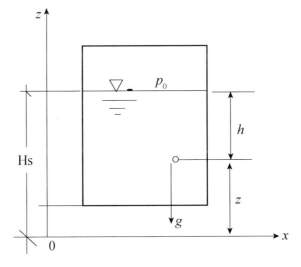

This formula is the equilibrium integral equation of static liquid under gravity. The formula shows that the sum of p/γ and z at any point in j static liquid is constant under gravity. Obviously, the isobaric surface in the static liquid is the horizontal surface $z = C$.

The geometric and physical meanings of Eq. (2.49) are explained below, where z denotes the height of the position of the liquid particle in the coordinate system, which is called positional head in hydraulics; p/γ *represents the height of the liquid column under pressure of the liquid particle. Hydraulics is called pressure head. H represents the sum of the above positions and the pressure height, which is constant for any point in the liquid and hydraulically called the pressure tube head (also known as the total head in still fluids).* As shown in Fig. 2.21, for points 1, 2, and 3 at different heights in the same container, it can be known from the equilibrium integral Eq. (2.49)

$$\frac{p_1}{\gamma} + z_1 = \frac{p_2}{\gamma} + z_2 = \frac{p_3}{\gamma} + z_3 = H \tag{2.50}$$

This shows that the position head and the pressure head are interchangeable in a static fluid, but the total head remains the same.

From the physical point of view, the physical meanings of each item in Eq. (2.49) are as follows: z represents the gravitational potential energy of the liquid per unit weight, because for the liquid mass with mass m, the position height is z, then the potential energy of the liquid mass relative to the origin of the reference coordinate is mgz, and the gravitational potential energy per unit weight is $mgz/(mg) = z$; p/γ *is the pressure potential energy per unit weight of the liquid; H is the total potential energy per unit weight of the liquid;* the equilibrium equation $\frac{p}{\gamma} + z = H$ indicates.

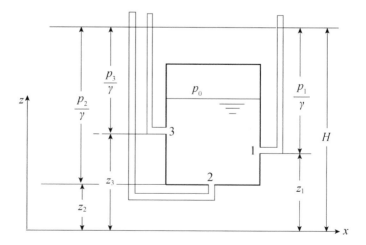

Fig. 2.21 The sum of position and pressure height is constant

Fig. 2.22 French scientist
B. Pascal (1623–1662)

The sum of potential energy and pressure energy per unit weight of liquid in static liquid is constant, but they can be converted into each other.

Assuming that the distance between the free liquid surface and the horizontal axis (reference axis) is H, then the balance equation between the free surface and the liquid particle at z is

$$\frac{p}{\gamma} + z = \frac{p_0}{\gamma} + H_s$$

And we get

$$p = p_0 + \gamma (H_s - z) \quad \text{or} \quad p = p_0 + \gamma h \tag{2.51}$$

where $h = H_s$, and $-z$ is the depth below the free surface at point z, also known as the underwater depth. Equation (2.51) indicates that in the static liquid, the pressure at h from the depth of the free surface comes from the contributions of two parts: one is the weight of the liquid column in unit area γh above the point z, which is proportional to the submerged depth under water; the other is the pressure P_0, at the free surface, which is transferred to any point inside the liquid with the same value, regardless of the depth. This equivalent transfer characteristic is the famous Pascal principle (as shown in Fig. 2.22, proposed by the French physicist Pascal in 1653). By using this principle, Pascal invented the hydraulic press. This principle indicates that when the external pressure changes for the static liquid in a closed container, the pressure at any point in the liquid will change by the same magnitude, that is, the pressure applied to the static liquid is transferred to all points in the liquid equivalent, as shown in Fig. 2.23.

If the free surface of the liquid is atmospheric pressure pa, the pressure at the underwater depth h from the free surface is

Fig. 2.23 Principle of
hydraulic press

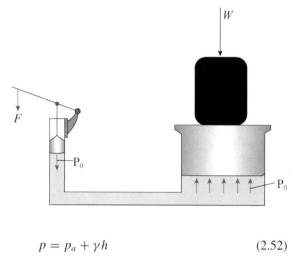

$$p = p_a + \gamma h \tag{2.52}$$

In hydraulics, the water pressure of the retaining wall can be calculated by this formula, as shown in Fig. 2.24.

For pressure, different reference values can be used, as shown in Fig. 2.25. If the pressure measured with the absolute vacuum as the reference value is called the absolute pressure, such as p in Eq. (2.52), the absolute pressure can only be greater than or equal to zero without negative values; if the pressure measured with atmospheric pressure pa as the reference pressure is called relative pressure, namely $p_r = p - p_a$, the relative pressure can be positive or negative; in gas mechanics, the pressure gauge is called surface pressure. Taking atmospheric pressure pa as the reference pressure, the pressure of insufficient atmospheric pressure is called vacuum

Fig. 2.24 Distribution of
water pressure on the
retaining wall

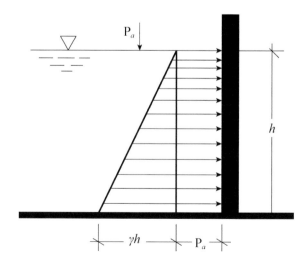

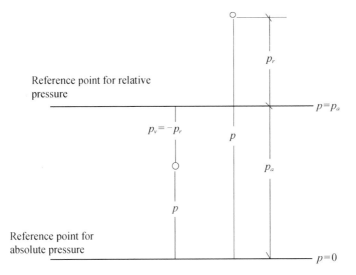

Fig. 2.25 Pressure measurement method

degree $p_V = p_a - p$. For the same pressure value p, the relationship between the relative pressure pr and the vacuum degree pv is $p_r = -p_v$.

Example 1 The working principle of wet atmospheric pressure gauge. There is an atmospheric pressure gauge that uses the height of the mercury column to express the value of the atmospheric pressure, as shown in Fig. 2.26. A long glass tube is closed at the top and a bottom box is containing mercury, with the glass tube standing upright. The mercury in the glass tube is connected to the mercury in the bottom box. The top of the mercury column in the glass tube is completely vacuumed. As shown in Fig. 2.26, place the coordinate plane xz on the upper surface of the mercury column in the tube, where $p_0 = 0$, $z_0 = 0$. According to the formula $p = p_0 + \gamma h$, the pressure p_A below the glass tube is $p_A = \rho_{Hg}gh$ at A (the depth from the upper surface is h) equal to the mercury surface in the box. p_A is equal to atmospheric pressure p_a, namely $p_a = \rho_{Hg}gh = \gamma_{Hg}h$. The density ratio of mercury to water is 13.6, so $\gamma_{Hg} = 13.6\gamma_w$. In this way, to calculate the atmospheric pressure, just multiply the height of the mercury in meters read on the barometer by the bulk density of the mercury; the atmospheric pressure readings often only refer to the height of the mercury column, and a standard pressure is 760 mm mercury column.

Example 2 The buoyancy of an underwater closed object. As early as 250 BC, Archimedes, an ancient Greek scientist, put forward the buoyancy theorem that the buoyancy of any object is equal to the weight of the object's volume discharged from the water. As shown in Fig. 2.27, for any object submerged under water, the water pressure on the surface of the object is calculated according to Eq. (2.52), and then the component of the water pressure in the vertical direction is integrated around the closed surface, so that the buoyancy can be easily obtained, namely $F_w = \gamma_w \forall$.

Fig. 2.26 Wet atmospheric
pressure gauge

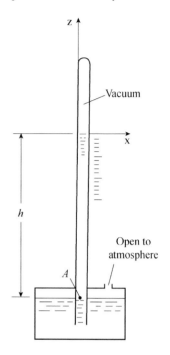

Fig. 2.27 Buoyancy
theorem

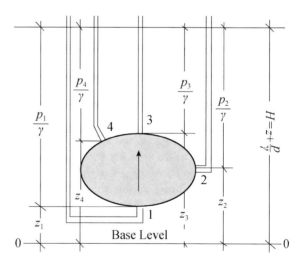

2.6 Equilibrium Law of Relative Static Liquid

Euler's equations of equilibrium differential equations are not only applicable to absolute static fluids, but also to relatively static fluids, such as homogeneous fluids in which the liquid moves uniformly in a straight line like a rigid body, and also to relatively static homogeneous fluids after being stabilized in a rotating container. As shown in Fig. 2.28, it is assumed that a container containing part of the water body rotates at an angular velocity ω about the central axis. When the water body inside the container is stable, the water body is in a relatively static state. Euler equilibrium differential equations are used to solve the pressure distribution and the shape of the free surface of the water body in the container.

(1) Water pressure distribution

As shown in Fig. 2.28, the unit mass force (including centrifugal inertial force) received at any point in the body of water is

$$f_x = x\omega^2, \; f_y = y\omega^2, \; f_z = -g \tag{2.53}$$

By substituting the equilibrium differential equation $dp = \rho(f_x dx + f_y dy + f_z dz)$, we can get

Fig. 2.28 The water body in the rotating container is relatively balanced

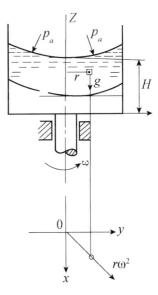

$$dp = \rho(\omega^2 x dx + \omega^2 y dy - g dz) \tag{2.54}$$

Integral to

$$p = \frac{\rho \omega^2}{2}(x^2 + y^2) - \rho g z + c \tag{2.55}$$

From the free surface conditions, the integral constant c is determined. As shown in Fig. 2.28, when $x = y = 0$, $z = H$, and $p = p_a$, the integral constant $C = P_a + \rho g H$ can be obtained from Eq. (2.55). And we get

$$p = p_a + \rho g \left[\frac{\omega^2}{2g}(x^2 + y^2) + (H - z) \right] \tag{2.56}$$

(2) Free surface shape

According to the definition of isobaric surface, on the isobaric surface, it can be obtained by Eq. (2.54)

$$\omega^2 x dx + \omega^2 y dy - g dz = 0 \tag{2.57}$$

Integral to

$$z = \frac{\omega^2}{2g}(x^2 + y^2) + c \quad \text{Or} \quad z = \frac{\omega^2 r^2}{2g} + c \tag{2.58}$$

The above equation is the paraboloid of rotation. On the free surface, the lowest points is $r = 0$, $c = z = H$.
So the free surface equation is

$$z_s = \frac{\omega^2}{2g}(x^2 + y^2) + H \tag{2.59}$$

where $z_s = \frac{\omega^2 r^2}{2g}$ is called the superelevation, that is, the free surface is higher than the parabola vertex.

2.7 Standard Atmosphere

It is well known that the air surrounding the earth is always called the atmosphere. The total mass of the atmosphere is about one-millionth of the mass of the Earth. The main components of the air are nitrogen (78%) and oxygen (21%), and 99.9% of the air mass is concentrated within 50 km above the ground. In the atmosphere, temperature, pressure, etc. change with altitude. According to its changing characteristics, the atmosphere can be divided into several layers.

(1) Troposphere: The lowest layer of the atmosphere from sea level. The height varies with latitude. It is about 16–18 km at the equator, 10–12 km at mid-latitudes, and 7–10 km at the poles. The density is highest in this layer, and the air quality contained accounts for about 3/4 of the entire air quality. There is up and down movement in the air. Complex meteorological changes such as thunderstorms and storms all occur in this layer, and the temperature decreases linearly with height. There is a transition layer between the troposphere and the stratosphere with a thickness of only a few hundred meters to one or two kilometers, which is called the convective top layer.

(2) Stratosphere: Above the top layer of convection, the range is about 32 km from the top layer of convection, and the air quality contained in it accounts for 1/4 of the entire atmosphere. The atmosphere only moves in the horizontal direction, and there are no meteorological changes such as thunderstorms. The altitude is 12–20 km, and the temperature is maintained at $T = 216.65$ K. This layer is called the stratosphere. The altitude is 20–32 km, and the temperature rises with altitude.

(3) Intermediate atmosphere: the altitude is 32–80 km. The temperature of this layer first rises with altitude, reaches 282.66 K at 53 km, then drops, and drops to 196.86 K at 80 km. The air quality in this layer accounts for about 1/3000 of the total mass.

(4) High-temperature layer: The altitude is 80–400 km, and the temperature rises with altitude, reaching 1500–1600 K at 400 km, which is due to the direct shortwave radiation of the sun. In this area, air molecules are decomposed into ions by shortwave radiation, so there are several ionospheres: the lowest layer is called the D layer, at 60–80 km; the second layer is the E layer, at 100–120 km; the third layer is the F_1 layer, at 180–220 km; the highest layer is called the F_2 layer, at 300–350 km. The high-altitude air above 100 km is a good electrical conductor. Above 150 km, because the air is too thin, there is no audible sound.

(5) Outer atmosphere: At an altitude of 400–1600 km, air molecules have the opportunity to escape into space without colliding with other molecules. The air quality accounts for 10^{-11} of the total mass.

Ordinary aircraft mainly operate in the troposphere and stratosphere. According to records, the maximum altitude of the aircraft is 39 km, the detection balloon is 44 km, the perigee of the artificial satellite is more than 100 km, and the apogee can reach several thousand kilometers. Most meteorites are destroyed at an altitude of 40–60 km. The weather conditions change from day to day, and the weather varies in different regions. Whether for aircraft design or experimental research, atmospheric conditions must be used. In order to facilitate comparison, engineering needs to specify a standard atmosphere. This standard is based on the average weather conditions in the mid-latitude area. When doing this calculation, all calculations are carried out according to this standard; when doing experiments, they are all converted into data under standard conditions.

(1) Standard atmospheric temperature distribution

The standard atmosphere is set at sea level, with an atmospheric temperature of 15 °C or $T_0 = 288.15$ K, pressure $p_0 = 760$ mm Hg, column $= 101{,}325$ N/m^2, and density $\rho_0 = 1.225$ kg/m^3. As shown in Fig. 2.29, the upper air from the base level to 11 km is called the troposphere. In the troposphere, the atmospheric density and temperature change significantly with height, and the temperature decreases with the increase of height. For every 1 km increase in height, the temperature decreases by 6.5 K, which is $T = 288.15 - 0.0065z$. The high-altitude atmospheric temperature from 11 to 21 km is basically unchanged, and the temperature remains at 216.5 K; when the altitude is greater than 21 km, the change of atmospheric temperature with altitude is complicated. The main factors include the following: the surface absorbs solar heat, and ozone absorbs heat and ionizes it. Air or cosmic dust is heated by shortwave radiation, etc.

(2) Standard atmospheric pressure distribution

As shown in Fig. 2.30, the coordinate system was established and the law of pressure variation with height was derived by a static equilibrium differential equation. Since atmospheric density ρ is a variable, atmospheric pressure at a certain height can be seen as the result of the weight of an unbounded column of air per unit area, as shown in Fig. 2.30. Considering a unit mass air element at a certain height, its mass force component is

$$f_x = 0,\ f_y = 0,\ f_z = -g$$

If I plug it into the equilibrium differential equation, I get

$$\mathrm{d}p = -g\rho\mathrm{d}z \tag{2.60}$$

According to the complete gas equation of state, density can be written as an expression of pressure and temperature. Namely

Fig. 2.29 Atmospheric temperature distribution

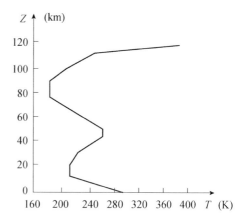

Fig. 2.30 Balance equation of atmospheric column

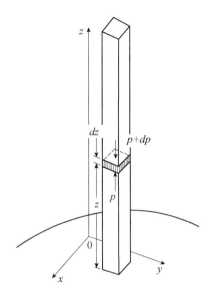

$$\rho = p/RT \tag{2.61}$$

Substitute it into the equilibrium differential equation, and get

$$\frac{\mathrm{d}p}{\mathrm{d}z} = -\frac{gp}{RT} \tag{2.62}$$

where temperature T is a known function of height y, and strictly speaking, g also varies with y, but within the troposphere, its influence being minimal, which is considered a constant here, with a value of 9.80665 m/s^2. The variable is separated by substituting the expression for T.

In the troposphere, $T = 288.15 - 0.0065z$, substitute the equilibrium differential equation, and get

$$\int_{p_0}^{P_h} \frac{\mathrm{d}p}{p} = -\frac{g}{R} \int_0^h \frac{\mathrm{d}z}{288.15 - 0.00065z}$$

If we integrate it, we get

$$\frac{p_h}{p_0} = \left(\frac{288.15 - 0.0065h}{288.15} \right)^{5.25588} = \left(\frac{T_h}{T_0} \right)^{5.25588} \tag{2.63}$$

In the formula, the subscript h represents the atmospheric parameters at an altitude of h meters. The corresponding air density is

$$\frac{\rho_h}{\rho_0} = \left(\frac{T_h}{T_0}\right)^{4.25588} \tag{2.64}$$

Based on the standard atmospheric parameters on the ground, the pressure and density distribution at a certain height h in the troposphere can be obtained.

In the stratosphere, the height ranges from 11 to 20 km, $T = 216.65K$, and substituting into the differential equation, the integral is

$$\int_{p_{11}}^{p_H} \frac{\mathrm{d}p}{p} = -\frac{g}{RT}\int_{11}^{h} \mathrm{d}z \tag{2.65}$$

The results have

$$\frac{p_h}{p_{11}} = \frac{\rho_h}{\rho_{11}} = \mathrm{e}^{-\frac{h-11000}{6341.62}} \tag{2.66}$$

The subscript "11" represents the atmospheric parameters at $h = 11000$ m. The pressure and density parameters at other heights can be obtained by integrating the relationship between temperature and height into the equilibrium differential equation. The atmospheric parameters (pressure, density, temperature, etc.) calculated in this way are listed in the standard atmospheric table for reference. The graph below (2.30) shows the curve of temperature T, pressure p, and density ρ with height h (Fig. 2.31).

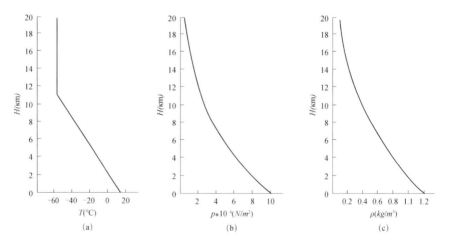

Fig. 2.31 Variation curve of atmospheric parameters

Exercises

A. Answer the following questions

1. What is the fluid continuum hypothesis?
2. How to describe the fluid state from a microscopic view? How to describe the fluid state from a macroscopic view?
3. Why can the Knudsen number be used to characterize the continuity of the fluid? Combining the standard atmosphere table, please indicate the height above which the continuum hypothesis is not satisfied for an object 1000 mm long.
4. What is the fluidity of a fluid? Please explain the fluidity of the fluid from the microscopic view.
5. For air and water, if the volume is compressed by 1%, what is the increase in pressure?
6. What is the viscosity of the fluid? Why is the viscous shear stress in the fluid proportional to the rate of shear deformation, but not the amount of shear deformation?
7. Please explain the relationship between shear stress and deformation rate of Bingham fluid, pseudo-plastic fluid, swelling fluid, and Newton fluid.
8. Given that the barometer reads 2.5 atmospheres at a certain point, what is the relative pressure and the absolute pressure at that point?
9. What is the isotropy of static pressure at any point in a static fluid?

B. Calculate the questions

1. The distance between one plate and the other fixed plate is 0.5 mm. The space between the two plates is filled with liquid. The upper plate moves at a speed of 0.25 m/s under the force of 2 N per square meter. Please calculate the viscosity of the liquid.
2. There is a flat plate with a base area of 5 kg, sliding along the slant at an angle of 20° with the horizontal plane 60 cm × 40 cm.
   The thickness of the oil layer between the plane and the slant is 0.6 mm. If the sliding speed is 0.84 m/s, the dynamic viscosity μ of oil.

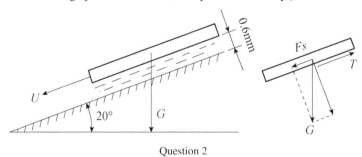

Question 2

3. In order to conduct insulation treatment, the wire is pulled through the middle of the mold filled with insulation coating. The diameter of the wire is known to be 0.8 mm and the viscosity of the coating $\mu = 0.02$ Pa s, the diameter of the mold is 0.9 mm and the length is 20 mm, and the pulling speed of the wire is 50 m/s. Try to calculate the required pulling force.

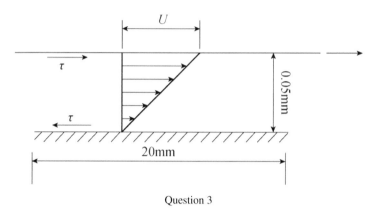

Question 3

4. The two parallel disks have a diameter of D, the distance between them is h, the lower disk is fixed, the upper disk rotates at a uniform angular velocity ω, and there is a liquid with a viscosity of μ between the disks. Assuming that the distance h between the two disks is small compared with the diameter D, the velocity distribution of the liquid between the two disks is linear. Try to derive the relationship between viscosity μ and torque T and angular velocity ω.
5. The upper and lower parallel disks have a diameter of d, the distance between the two disks is δ, and the dynamic viscosity of the liquid in the gap is μ. If the lower disk is fixed and the upper disk rotates at an angular velocity ω, find the expression of the required torque M.

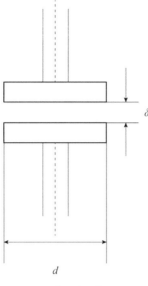

Question 5

6. Given that $\omega = 16\,\mathrm{rad/s}$, $\delta = 1\,\mathrm{mm}$, $R = 0.3\,\mathrm{m}$, $H = 0.5\,\mathrm{m}$, and $\mu = 0.1\,\mathrm{Pa\ s}$, calculate the resistance moment acting on the cone.

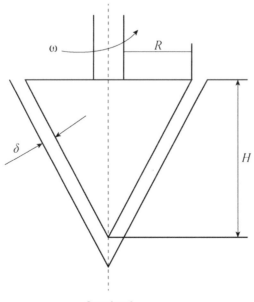

Question 6

7. For a wide channel with a rectangular section, the flow velocity distribution is $u = 0.002\frac{\gamma}{\mu}\left(hy - \frac{y^2}{2}\right)$, where γ is the bulk density of water, μ is the dynamic viscosity coefficient of water, y is the water depth, and $h = 0.5$ m. Calculate the shear stress $\tau0$ at the bottom of the canal at $y = 0$.

8. As shown in the picture, two plates are filled with two kinds of unmixed liquids, the viscosity coefficient of which is liquid dynamic viscosity $m_1 = 0.14$ Pa s, $m_2 = 0.24$ Pa s, and the liquid thickness is $\delta_1 = 0.8$ mm, $\delta_2 = 12$ mm, respectively. Assuming that the velocity distribution is a linear law, try to calculate the force required to push the upper plate with the bottom area $A = 0.1\text{m}^2$ to move at a uniform velocity of 0.4 m/s.

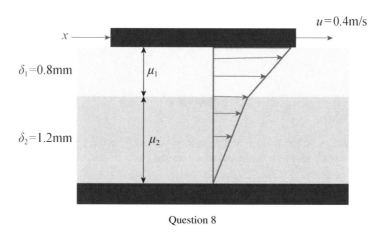

Question 8

9. The height of the Mariana Mountains in the Pacific Ocean is 11,034 m. At this height $\gamma_{\text{seawater}} = 10520\text{N/m}^3$. Estimate the pressure at this altitude.

10. Assuming that the density of the atmosphere is a constant of 1.225 kg/m3, what is the upper bound of the atmosphere in m? (Assume that the pressure at sea level is the same as the international standard atmospheric value.)

11. A vertical clean glass manometer tube has an inside diameter of 1 mm. When pressure is applied, the water at 20 °C rises to a height of 25 cm. After correcting the surface tension, estimate the pressure.

12. The compressor compresses the air, the absolute pressure increases from 9.8067 $\times\ 10^4$ Pa to 5.8840 $\times\ 10^5$ Pa, and the temperature increases from 20 to 78 °C, how much is the air volume reduced?

13. An open glass tube with an inner diameter of 10 mm is inserted into water with a temperature of 20 °C, and the known contact angle $\theta = 10°$ between water and glass. Try to calculate the height of water rising in the pipe.

14. A closed container contains 1.5 m of oil, 1 m of water, 20 cm of mercury, and a top air space, all of which are in 20 °C. If $p_{\text{bottom}} = 60\text{kPa}$, then what is the pressure in the air?

15. In the figure, the temperature of the two fluids is 20 °C. If the effect of surface tension is negligible, what is the density of the oil?

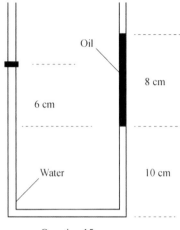

Question 15

16. The design height of an engine is 1000 m; try to find the atmospheric pressure density and temperature at this height, and compare them with the parameters given on the international standard atmospheric table.

17. In the figure, sensor A reading is 1.5 kPa. All the liquids are at 20 °C. Determine the liquid level height $Z(m)$ of the open manometer tubes B and C.

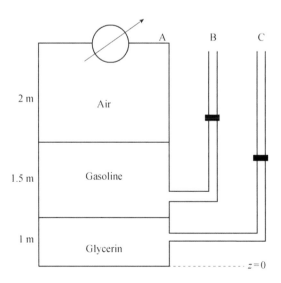

18. The volume of a gas storage tank is 6 m³, which stores 48.1 kg of air. Try to determine the density of the air in the air storage tank.

19. A 5 mm diameter capillary tube is used as a viscometer for oil. When the flow rate is 0.071 m³/h, the measured pressure drop per unit length is 375 kPa/m. Estimate the viscosity of the fluid. Is it laminar flow? Please estimate the density of the fluid.

20. As shown in the figure, it is a water container with first-order acceleration moving downward. The water depth $h = 2$ m and the acceleration $A = 4.9$ m/s². Try to calculate (1) the absolute static pressure of the fluid at the bottom of the container; (2) what is the acceleration when the pressure on the bottom of the container is atmospheric pressure? (3) what is the value of acceleration when the absolute static pressure at the bottom of the container is equal to zero?

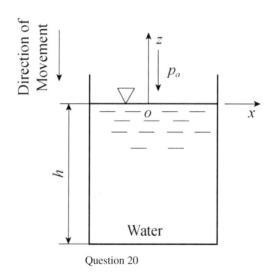

Question 20

21. The reading of the barometer on a certain day is 101.672 kPa. Try to calculate the force in N per 1 m² of the atmospheric pressure.

22. The volume of a gas tank is 27.1 m³, stored with compressed air. Given that the temperature of the air in the tank is 303 K and the pressure is 127.825 kPa, what is the mass of the compressed air in the tank in kilograms?

23. Use a metal tank with a volume of 1000 m³ for a water pressure test. First, fill the container with water with a pressure of $1.013 \times 10°$Pa, and then pressurize the water to increase the pressure in the container to 7×10^5 Pa. How much more water is needed?

24. Assuming that the temperature of the atmosphere is a constant of 288.15 K, try to calculate the pressure at 5000 m altitude. Please compare the pressure value with the corresponding value of the standard atmosphere at the same altitude, and explain the main reason for this difference.

Chapter 3
Foundation of Fluid Kinematics and Dynamics

This chapter mainly introduces the Lagrange method and Euler method to describe the fluid motion, the motion characteristics of fluid particles (streamline, trace, etc.), and the change rate of physical quantities of fluid particles (including velocity, acceleration, material derivative of physical quantities, etc.). The decomposition theorem of the basic motion forms of fluid elements, the divergence and curl of the flow field, the continuity equation and the differential equations of motion of fluid elements are derived. Bernoulli integral equation and its application, Reynold's transport equation and integral equations of fluid motion are introduced. The vortex motion and its characteristics of fluid micro-elements are introduced.

Learning points:

(1) Familiar with the two description methods of fluid particle motion and the characterization of the elements of fluid particle motion (including velocity, acceleration, material derivative of physical quantities, etc.);
(2) Master the basic decomposition of fluid micro-elements motion and the mathematical expressions of each motion component, the divergence and curl of the flow field and the vortex motion and its characteristics;
(3) Master the derivation process of Euler motion differential equations and their integral forms, Reynold's transport equations and fluid motion integral equations;
(4) Master the applicable conditions and applications of the Bernoulli equation.

© Science Press 2022
P. Liu, *Aerodynamics*, https://doi.org/10.1007/978-981-19-4586-1_3

3.1   Methods for Describing Fluid Motion

Two basic questions must be answered before correctly characterizing the motion characteristics of each fluid particle: (1) How to track and distinguish each fluid particle. (2) How to describe the motion characteristics and changes of each fluid particle, which is also the basic problem of fluid kinematics. Depending on the observer, the motion of the convective mass points can be described by the Lagrange method and the Euler method.

3.1.1   Lagrange Method (Particle Method or Particle System Method)

This method, called the particle system of fluids method, was first proposed by The Swiss mathematician and fluid mechanics Euler, and then developed by the French mathematician and physicist Lagrange (1736–1813, Fig. 3.1) in 1781. In this method, all fluid particles are identified (spatial points are not identified), and then the position coordinates of each particle at different times are recorded, so as to understand the overall flow behavior. Obviously, this method requires the observer to track each fluid particle at any time and place, record the motion history of the particle (direct measurement data is the position of the particle at different times, leading to the concept of particle track line), so as to obtain the motion law of the whole flow. Let's say we're at rest or at some initial time t_0, take the initial position coordinates (a, b, c) of the fluid particle as the identifier of the particle (as shown in Fig. 3.2), then at any time t, the spatial position of the particle can be expressed as

$$\begin{cases} x = x(a, b, c, t) \\ y = y(a, b, c, t) \\ z = z(a, b, c, t) \end{cases} \tag{3.1}$$

In the formula, a, b, c and t are called Lagrange variables. a, b and c are identifiers of fluid particles, which are used to distinguish and identify each particle, and are also followability conditions of fluid particles. t represents time. If t is given, Eq. (3.1) represents the spatial positions of different particles at a given time. If (a, b, c) is given, Eq. (3.1) represents the position of a given particle at a different time and gives the trajectory of the particle. By tracking the entire process of all the particles, a complete picture of the flow can be obtained. Among them, the position records of any particle at different times are direct measurement data, and the velocity and acceleration data obtained through definitions and laws are indirect measurement data.

If the position of a given particle is known, the velocity of a particle can be defined, namely:

Fig. 3.1 French mathematician and physicist Joseph-Louis Lagrange (1736–1813)

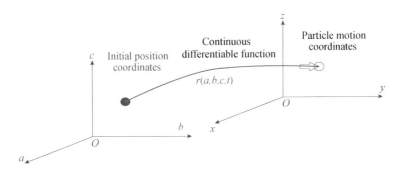

Fig. 3.2 Particle identification method

$$u = \lim_{\Delta t \to 0} \frac{x(a, b, c, t + \Delta t) - x(a, b, c, t)}{\Delta t} = \frac{\partial x(a, b, c, t)}{\partial t}$$

$$v = \lim_{\Delta t \to 0} \frac{y(a, b, c, t + \Delta t) - y(a, b, c, t)}{\Delta t} = \frac{\partial y(a, b, c, t)}{\partial t}$$

$$w = \lim_{\Delta t \to 0} \frac{z(a, b, c, t + \Delta t) - z(a, b, c, t)}{\Delta t} = \frac{\partial z(a, b, c, t)}{\partial t} \tag{3.2}$$

The time partial derivative is applied here because the particle of the fluid is given. Similarly, the acceleration expression of a given particle can be obtained by definition, namely

$$a_x = \lim_{\Delta t \to 0} \frac{u(a, b, c, t + \Delta t) - u(a, b, c, t)}{\Delta t} = \frac{\partial u(a, b, c, t)}{\partial t} = \frac{\partial^2 x}{\partial t^2}$$

$$a_y = \lim_{\Delta t \to 0} \frac{v(a, b, c, t + \Delta t) - v(a, b, c, t)}{\Delta t} = \frac{\partial v(a, b, c, t)}{\partial t} = \frac{\partial^2 y}{\partial t^2}$$

$$a_z = \lim_{\Delta t \to 0} \frac{w(a, b, c, t + \Delta t) - w(a, b, c, t)}{\Delta t} = \frac{\partial w(a, b, c, t)}{\partial t} = \frac{\partial^2 z}{\partial t^2} \tag{3.3}$$

where, u, v and w respectively represent the velocity components of the fluid particle along x, y and z directions. a_x, a_y and a_z represent the acceleration components of fluid particle acceleration along x, y and z directions, respectively. For any given fluid particle, the lines of its spatial positions at different times are called the trace lines of the particle (as shown in Fig. 3.3). This method of tracking particles (vividly seen as the working mode of "police tracking thieves") is a direct extension of the particle system method in theoretical mechanics. This method has a clear concept and is convenient for the direct generalization of physical laws. However, the disadvantage is that there are too much recorded data, especially for the flow characteristics of only a local area is very inconvenient. For example, during the flood season every year, we are only concerned about the water situation of the Yangtze River in the Wuhan section (Wuhan Pass water level). When described by this method, it is necessary to make clear the origin and context of all water quality points in the Yangtze River through the Wuhan section, and track and record the flow process of each particle in the whole process to depict the flow characteristics in the Yangtze River in Wuhan section. Many records are actually useless for water quality points not in the Wuhan section of the Yangtze.

3.1.2   Euler Method (Space Point Method or Flow Field Method)

This method was proposed by The Swiss mathematician and physicist Leonhard Euler (1707–1783) in 1752–1755. This method is also known as the spatial point method or the flow field method. In order to avoid unnecessary data record in the method of Lagrange, Euler proposed the method which avoids identifying fluid particle Instead, it identifies the flow area of the space point (the relationship between the particle in space still satisfy the continuity assumption). The observer is relatively static to the space point. The observer directly records the particle velocity at a given

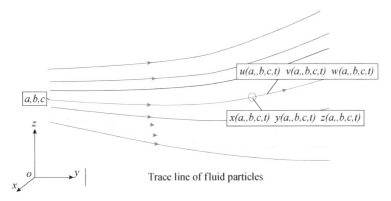

Fig. 3.3 Trace lines of different fluid particles characterized by Lagrange method

space point. By placing observers at every space point, a comprehensive under-standing of regional fluid characteristics can be obtained. Note that although this method identifies spatial points, it still deals with fluid particles, so it can be said to be an unidentified particle system method. As shown in Fig. 3.4, it is assumed that in the flow region under investigation, the position coordinate of any spatial point is (x, y, z), and the velocity of fluid particle passing through this spatial point is directly recorded by the observer at this point at time t

$$\begin{cases} u = u(x, y, z, t) \\ v = v(x, y, z, t) \\ w = w(x, y, z, t) \end{cases} \qquad \vec{V} = u\vec{i} + v\vec{j} + w\vec{k} \qquad (3.4)$$

In the formula, x, y and z are space coordinates, t is time, and x, y, z and t are called Euler variables. If x, y and z are given, and t changes, the above equation represents the velocity (fast or slow) of different fluid particles passing through the same space

Fig. 3.4 Flow field described by Euler method

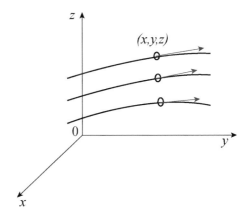

point at different times. If t is given, x, y and z change, it represents the velocity of a fluid particle occupying different space points at a given time and the velocity field is given. It should be noted that velocity at a space point is essentially the velocity of a fluid particle occupying that space point at time t. This point is particularly important in the derivation of the basic equations of fluid dynamics. This method doesn't need to identify fluid particle information but needs to record any fluid particle by fixed space point the speed of information. Therefore, it is called as a flow field method. In physics, a space full of certain physical quantities is called a field. The space occupied by fluid flow is the flow field. If the quantity is velocity, it describes the velocity field. If it's pressure, it's called a pressure field. At high speed, the density and temperature of the flow also change with the flow, so there are density fields and temperature fields. They are all included in the concept of the flow field. If the field is only a function of space coordinates and independent of time it is called a steady field, otherwise, it is an unsteady field.

It is pointed out that the velocity of the fluid particle passing through any space point at any time recorded by the observer is a direct measurement value, and the physical values such as acceleration obtained through the definition and law are indirect measurements. Figuratively speaking, this approach can also be viewed as a way of working as "waiting for something to happen."

If the velocity is a function of first order continuous partial derivatives, then the acceleration of a particle can be determined. Although the Euler method does not need to track the fluid particle in the whole process, when studying the specific physical process, it needs to track the object of study locally, otherwise, it is meaningless. In Euler's method, the following condition for locally tracking the motion of a fluid particle is

$$\frac{dx}{dt} = u, \quad \frac{dy}{dt} = v, \quad \frac{dz}{dt} = w \tag{3.5}$$

In the study of fluid particle acceleration, it is necessary to locally track the particle velocity change. As shown in Fig. 3.5, suppose that a mass point is located at point $M\,(x, y, z)$ in the flow field at time t. After a finite time Δt, this particle point moves from point M to point $N\,(x + \Delta x, y + \Delta y, z + \Delta z)$. According to the definition of particle acceleration Δt

$$\vec{a} = \frac{d\vec{V}}{dt} = \lim_{\Delta t \to 0} \frac{\Delta \vec{V}}{\Delta t} = \lim_{\Delta t \to 0} \frac{\vec{V}(N, t + \Delta t) - \vec{V}(M, t)}{\Delta t}$$

$$\frac{d\vec{V}}{dt} = \lim_{\Delta t \to 0} \frac{\vec{V}(N, t + \Delta t) - \vec{V}(N, t)}{\Delta t} + \lim_{\Delta t \to 0} \frac{\vec{V}(N, t) - \vec{V}(M, t)}{\Delta t} \tag{3.6}$$

Equation (3.6) indicates that the acceleration of a particle consists of two parts. The first part is the acceleration caused by the velocity change of a particle through a fixed space point during a period, and the second part is the acceleration caused by the velocity change of a particle through an adjacent space point at a given time Δt. According to the Taylor series expansion, for the first part, it is equivalent to the

Fig. 3.5 Acceleration of a
fluid particle

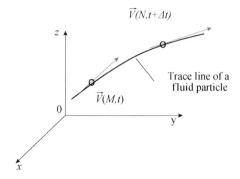

Trace line of a
fluid particle

acceleration caused by the change of fluid particle velocity with time at a fixed space
position:

$$\vec{V}(N, t + \Delta t) = \vec{V}(N, t) + \frac{\partial \vec{V}(N, t)}{\partial t} \Delta t + O(\Delta t^2) \quad (3.7)$$

And it can be obtained:

$$\lim_{\Delta t \to 0} \frac{\vec{V}(N, t + \Delta t) - \vec{V}(N, t)}{\Delta t}$$

$$= \lim_{\Delta t \to 0} \frac{\frac{\partial \vec{V}(N, t)}{\partial t} \Delta t + O(\Delta t^2)}{\Delta t} = \frac{\partial \vec{V}(M, t)}{\partial t} = \frac{\partial \vec{V}}{\partial t} \quad (3.8)$$

For the second part, it is equivalent to the acceleration caused by the velocity
change of adjacent fluid particles in different spatial positions at a fixed time, which
can also be obtained by the Taylor series expansion.

$$\vec{V}(N, t) = \vec{V}(x + \Delta x, y + \Delta y, z + \Delta z, t)$$

$$= \vec{V}(x, y, z, t) + \frac{\partial \vec{V}(x, y, z, t)}{\partial x} \Delta x + \frac{\partial \vec{V}(x, y, z, t)}{\partial y} \Delta y$$

$$+ \frac{\partial \vec{V}(x, y, z, t)}{\partial z} \Delta z + O(\Delta x^2, \ldots,)$$

$$\vec{V}(N, t) - \vec{V}(M, t) = \frac{\partial \vec{V}(M, t)}{\partial x} \Delta x + \frac{\partial \vec{V}(M, t)}{\partial y} \Delta y + \frac{\partial \vec{V}(M, t)}{\partial z} \Delta z + O(\Delta x^2, \ldots,)$$

$$(3.9)$$

By substituting the second part on the right side of Eq. (3.6), we can get

$$\lim_{\Delta t \to 0} \frac{\vec{V}(N, t) - \vec{V}(M, t)}{\Delta t}$$

$$= \lim_{\Delta t \to 0} \frac{\Delta x}{\Delta t} \frac{\partial \vec{V}(M,t)}{\partial x} + \lim_{\Delta t \to 0} \frac{\Delta y}{\Delta t} \frac{\partial \vec{V}(M,t)}{\partial y} + \lim_{\Delta t \to 0} \frac{\Delta z}{\Delta t} \frac{\partial \vec{V}(M,t)}{\partial z}$$

$$= u \frac{\partial \vec{V}(M,t)}{\partial x} + v \frac{\partial \vec{V}(M,t)}{\partial y} + w \frac{\partial \vec{V}(M,t)}{\partial z} \tag{3.10}$$

Note that since the same fluid particle is always being tracked, $\Delta x = u \Delta t$, $\Delta y = v \Delta t$, $\Delta z = w \Delta t$ are satisfied. This is the followability condition for a given fluid particle. In the Euler frame, the total velocity of the fluid particle is

$$\vec{a} = \frac{d\vec{V}}{dt} = \frac{\partial \vec{V}}{\partial t} + u \frac{\partial \vec{V}}{\partial x} + v \frac{\partial \vec{V}}{\partial y} + w \frac{\partial \vec{V}}{\partial z}$$

$$\vec{a} = \frac{d\vec{V}}{dt} = \frac{\partial \vec{V}}{\partial t} + (\vec{V} \cdot \nabla)\vec{V} \tag{3.11}$$

Written in terms of components:

$$a_x = \frac{du}{dt} = \frac{\partial u}{\partial t} + u \frac{\partial u}{\partial x} + v \frac{\partial u}{\partial y} + w \frac{\partial u}{\partial z}$$

$$a_y = \frac{dv}{dt} = \frac{\partial v}{\partial t} + u \frac{\partial v}{\partial x} + v \frac{\partial v}{\partial y} + w \frac{\partial v}{\partial z}$$

$$a_z = \frac{dw}{dt} = \frac{\partial w}{\partial t} + u \frac{\partial w}{\partial x} + v \frac{\partial w}{\partial y} + w \frac{\partial w}{\partial z} \tag{3.12}$$

In the expression of total acceleration, the first term on the right side of the equation represents the partial derivative of velocity with respect to time, which is caused by the unsteadiness of the flow field and is called local acceleration or local acceleration. The second term on the right represents the acceleration due to the change in the position of the fluid particle. It is called the migration acceleration or the locational acceleration or the convective acceleration. The combination of the two is called the total acceleration or the acceleration with the body (the acceleration following the fluid particle).

To understand the physical meaning of acceleration expression, a simple water tank and pipe outlet system are presented. As shown in Fig. 3.6. The first picture in Fig. 3.6 shows the outflow of an equal-diameter pipe with a constant water level in the tank. The second picture shows an equal-diameter pipe outflow with the water level dropping. The third picture shows the flow out of the shrinking pipe with a constant water level. The fourth picture shows a shrinking pipe outflow with the water level falling. Assume that the water flow in the pipeline is approximately regarded as a one-dimensional flow and the central axis of the pipeline is the x-axis, the velocity in the pipeline is $U(x, t)$ and the acceleration is expressed as

$$\frac{du}{dt} = \frac{\partial u}{\partial t} + u \frac{\partial u}{\partial x}$$

Fig. 3.6 Body point
acceleration under Euler
frame

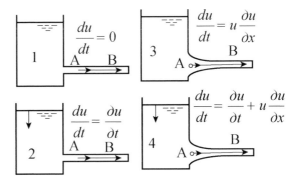

Now, the flow in the first picture of Fig. 3.6 is steady and uniform, which means
acceleration $\frac{du}{dt} = 0$. In the second picture, the pipeline water acceleration is $\frac{du}{dt} =$
$\frac{\partial u}{\partial t} < 0$ and the flow is unsteady and uniform. The pipeline velocity decreases with
the decrease in water level. In the third picture, the acceleration in the pipeline
is $\frac{du}{dt} = u\frac{\partial u}{\partial x} > 0$ and the flow is steady and non-uniform. The pipeline velocity
increases along the pipeline; In the fourth picture, the acceleration in the pipeline
is $\frac{du}{dt} = \frac{\partial u}{\partial t} + u\frac{\partial u}{\partial x}$. The flow is unsteady and non-uniform and the pipeline velocity
increases along the pipeline but decreases with time.

It shows that the acceleration of the flow field is caused by the unsteadiness and
non-uniformity of the flow field.

The above method of locally tracking fluid particles to derive particle acceleration
can also be applied to other physical quantities. If the physical quantity is pressure
P, then the derivative with the body of pressure is

$$\frac{dp}{dt} = \frac{\partial p}{\partial t} + u\frac{\partial p}{\partial x} + v\frac{\partial p}{\partial y} + w\frac{\partial p}{\partial z} \tag{3.13}$$

Similarly, if the physical quantity is temperature T, then the derivative with respect
to the body of temperature is

$$\frac{dT}{dt} = \frac{\partial T}{\partial t} + u\frac{\partial T}{\partial x} + v\frac{\partial T}{\partial y} + w\frac{\partial T}{\partial z} \tag{3.14}$$

In fluid mechanics, the derivative of the random body is defined as

$$\frac{d}{dt} = \frac{\partial}{\partial t} + u\frac{\partial}{\partial x} + v\frac{\partial}{\partial y} + w\frac{\partial}{\partial z} \tag{3.15}$$

Notice that the material derivative here is different from the total derivative in
field theory. In field theory, the total derivative of any function is ϕ

$$\frac{d\phi}{dt} = \frac{\partial \phi}{\partial t} + \frac{dx}{dt}\frac{\partial \phi}{\partial x} + \frac{dy}{dt}\frac{\partial \phi}{\partial y} + \frac{dz}{dt}\frac{\partial \phi}{\partial z} \tag{3.16}$$

If it's a material derivative, the following conditions must be satisfied: $dx = udt$, $dy = vdt$, $dz = zdt$, namely

$$\frac{d\phi}{dt} = \frac{\partial \phi}{\partial t} + u\frac{\partial \phi}{\partial x} + v\frac{\partial \phi}{\partial y} + w\frac{\partial \phi}{\partial z} \tag{3.17}$$

It can be seen from the derivative expressions above that, in Euler coordinate system, the acceleration of any fluid particle is composed of local acceleration and migration acceleration. The former is determined by the unsteadiness of the velocity field, while the latter is determined by the inhomogeneity of the velocity field. The derivative of the physical quantity of fluid particles in the Euler coordinate system refers to the material derivative.

In summary, the feature of the Lagrange method is: "**Whole process tracking, whole area recording**". The feature of the Euler method describing fluid movement is: "**Partial process tracking, local area recording**". In fact, in the analysis of fluid mechanics, the Euler method is generally used for the whole and the Lagrange method for the local.

3.2 Basic Concepts of Flow Field

In fluid mechanics, the flow field refers to the space occupied by the fluid flow. The physical quantity of the flow field is the physical measurements of a fluid particle occupying a different space. The physical quantities include velocity field, pressure field, temperature field, density field and so on. These quantities are also known as key elements of the moving fluid field. The change of these quantities with time and space constitutes a characteristic of the flow field. According to the continuum hypothesis of fluid, these physical quantities can be regarded as continuously differentiable functions of space and time in analysis.

3.2.1 Steady and Unsteady Fields

In the flow field, if the motion elements of the upper flow points of each spatial point do not change with time. such a flow field is called a steady flow field, and the corresponding flow is called a steady flow, such as the pipe flow in Fig. 3.6. In this case, the physical quantity of fluid particles is only a function of spatial position and independent of time, namely

$$\begin{cases} u = u(x, y, z) \\ v = v(x, y, z) \\ w = w(x, y, z) \\ p = p(x, y, z) \\ \cdots \end{cases} \tag{3.18}$$

If the time partial derivatives of the above physical quantities are zero, i.e.

$$\frac{\partial u}{\partial t} = 0, \ \frac{\partial v}{\partial t} = 0, \ \frac{\partial w}{\partial t} = 0, \ \frac{\partial p}{\partial t} = 0, \ \frac{\partial T}{\partial t} = 0, \dots \tag{3.19}$$

For the unsteady flow field, the physical quantity of fluid particle is not only the continuous differentiable function of spatial coordinates, but also the continuous differentiable function of time, and the partial derivative of any physical quantity (which can represent the velocity component, pressure, temperature, etc.) with respect to time is not zero, namely

$$\phi = \phi(x, y, z, t), \ \frac{\partial \phi}{\partial t} \ne 0 \tag{3.20}$$

3.2.2 Streamline and Path Line

In fluid mechanics, the Lagrangian method and Euler method are used to describe the motion of fluid particles. The Lagrange method is also known as the method of the particle system. The investigator tracks each particle and records its position in space at different times. This method naturally leads to the path line of fluid particle, and the corresponding mathematical expression is the path line equation of fluid particle, namely

$$\frac{dx}{u} = \frac{dy}{v} = \frac{dz}{w} = dt \tag{3.21}$$

Since this equation describes the line of the space occupied by the same fluid particle in the continuous time process, in the trace line equation, the independent variable is time t, the particle identifier a, b and c are parameters, and different values of a, b and c represent different particle traces. Then the above equation can also be expressed as

$$\frac{dx(a, b, c, t)}{u(a, b, c, t)} = \frac{dy(a, b, c, t)}{v(a, b, c, t)} = \frac{dz(a, b, c, t)}{w(a, b, c, t)} = dt \tag{3.22}$$

Euler's method is also known as flow field method (spatial point method), in which the researcher records the change of velocity of fluid particles passing through a fixed space position, and field theory knowledge can be naturally introduced. Because the flow field method can investigate the flow of fluid particles through different spatial points at the same time, this method can lead to the concept of streamline. Streamline refers to a virtual space curve that can be drawn in the flow field at a given moment, and the velocity direction of fluid particles at each space point on the curve is parallel to the tangent direction of the curve, as shown in Fig. 3.7. At a certain moment, according to the streamline definition, the streamline equation passing any point in the flow field is

$$\frac{dx}{u(x, y, z, t)} = \frac{dy}{v(x, y, z, t)} = \frac{dz}{w(x, y, z, t)} \quad (3.23)$$

In the above equation, x, y and z are independent variables and t is a parameter. Different t represents different streamlines at different times, as shown in Fig. 3.8.

Streamline is a curve reflecting the direction of the instantaneous flow field, which is composed of different particles at the same time. Trace lines are the trace lines of the same particle at different times. According to the definition of a streamline, a streamline has the following properties:

Fig. 3.7 Streamline and velocity direction

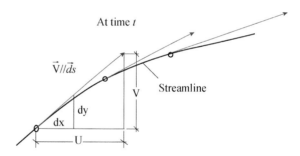

Fig. 3.8 Streamlines passing through different points in the flow field at the same time

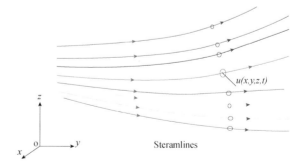

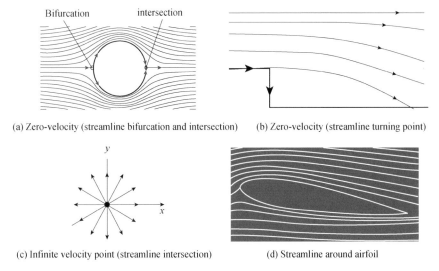

(a) Zero-velocity (streamline bifurcation and intersection)   (b) Zero-velocity (streamline turning point)

(c) Infinite velocity point (streamline intersection)   (d) Streamline around airfoil

Fig. 3.9 Features of streamline around different points

(1) In steady flow, trace lines of fluid particles coincide with streamline lines. However, in unsteady flow, streamline and trace lines are generally not coincident. If the flow path is fixed, the trace and streamline are coincident even for unsteady flow.

(2) Streamline cannot intersect, fork, converge or turn. Streamline can only be a smooth curve. That is, only one streamline can pass through a point at the same time, as shown in Fig. 3.9.

(3) The singularity (the point with infinite velocity) and the zero-velocity point are exceptions, which do not satisfy the property (2), as shown in Fig. 3.9.

In the flow field, the flow diagram of different spatial points can be drawn at the same time to fully understand the flow situation in the investigated area. In the experiment, in order to obtain flow pictures bypassing the flow area, some "indicators" that can show the flow direction (such as dyeing liquid in the flume experiment, smoke line in the wind tunnel experiment, etc.) are often spread in the observed flow field, and the flow conditions in the flow field are recorded by a camera, as shown in Fig. 3.9d.

At a certain point in the flow field, an arbitrary closed curve (ABCD in Fig. 3.10) without a streamline is taken and a streamline is made through each point of the closed curve to form a surface composed of streamline lines. The region surrounded by this surface is called a flow tube. According to the definition of streamline, the fluid in the pipe will not flow out through the pipe, and the fluid outside the pipe will not flow through the wall of the pipe. If taking any non-closed curve from a non-streamline in the flow field, a series of streamline lines can also be obtained through the curve to form a surface, and the surface is non-closed, and the fluid cannot cross the surface.

Fig. 3.10 Flow tube and
flow surface

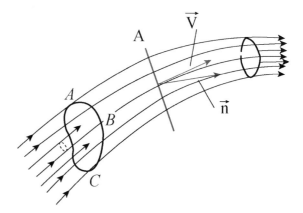

According to the characteristics of the flow pipe, the flow rate through the flow
pipe can be calculated. Flow rate is the amount of fluid (volume, mass, or weight)
passing through a given section per unit of time. As shown in Fig. 3.10, the volume
flow rate, mass flow rate and weight flow rate passing through any section A in the
flow tube are respectively defined as

$$Q = \int_A (\vec{V} \cdot \vec{n}) \mathrm{d}A \tag{3.24}$$

$$m' = \int_A \rho(\vec{V} \cdot \vec{n}) \mathrm{d}A \tag{3.25}$$

$$G' = \int_A \rho g(\vec{V} \cdot \vec{n}) \mathrm{d}A \tag{3.26}$$

where $\vec{V}$ is the local velocity vector, ρ is the fluid density, $\vec{n}$ is the normal vector of
the microelement area $\mathrm{d}A$ of the section.

3.2.3 One-Dimensional, Two-Dimensional and Three-Dimensional Flows

Flow field refers to the distribution of physical quantities of fluid particles in space,
so it can be divided into one-dimensional flow, two-dimensional flow and three-
dimensional flow according to the relationship between physical and spatial coordinates. One-dimensional flow means that the physical quantity of the flow field is
related to only one spatial coordinate, such as $u(x, t)$. Introduce the flow coordinate
s, and the flow velocity can be stated as $u = u\,(s, t)$. In this case, the acceleration is
expressed as

$$\frac{du(x,t)}{dt} = \frac{\partial u}{\partial t} + u\frac{\partial u}{\partial x}, \quad \frac{du(s,t)}{dt} = \frac{\partial u}{\partial t} + u\frac{\partial u}{\partial s} \tag{3.27}$$

If the physical quantity of the flow field is a function of two spatial coordinate variables, that is, $u = u(x, y, t)$, then the flow is called a two-dimensional flow. In this case, the acceleration is expressed as

$$\frac{du(x,y,t)}{dt} = \frac{\partial u}{\partial t} + u\frac{\partial u}{\partial x} + v\frac{\partial u}{\partial y} \tag{3.28}$$

If the physical quantity of the flow field is a function of three spatial coordinate variables, that is, $u = u(x, y, z, t)$, then the flow is called a three-dimensional flow.

3.3 Motion Decomposition of a Differential Fluid Element

3.3.1 Basic Motion Forms of a Differential Fluid Element

In theoretical mechanics, the research objects are particles and rigid bodies (without deformable bodies), whose basic motion forms can be expressed as:

(1) A particle (a material point with no volume) has only translational motion;
(2) The motion of a rigid body (an object with a certain volume but no deformation) includes translation and rotation.

For the translational motion of a particle, only the coordinates of the particle at a different time need to be determined. In addition to the translational motion of the centroid, there is also the rotational motion of the rigid body. However, since there is no relative motion between the particles of the rigid body, the rotational motion of the rigid body is the overall rotational motion.

In fluid mechanics, in addition to the particles, the research objects also include fluid groups (with the size and shape of particle group). Determining the position and posture of fluid mass effect on the flow is crucial, because there is relative motion between fluid mass of internal particles. Therefore, translation, rotation and deformation must be considered when studying the motion of micro fluid elements, as shown in Fig. 3.11. There are two kinds of deformation movement. One is side line extension/shorten which causes the change of volume. The other is the angular deformation which causes the change of shape. Thus, the motion of micro fluid elements can be decomposed into translation, rotation, length deformation and angular deformation, as shown in Fig. 3.12.

For the convenience of analysis, at the given time t, any plane micro-element is taken from the flow field, as shown in Fig. 3.13. Let the side lengths of the tiny group be Δx and Δy respectively. The velocity of point A is $u(x, y)$ and $v(x, y)$. According to the Taylor series expansion, the velocity components of the other three vertices of the tiny group are respectively as follows:

The velocity of point B is:

$$
\begin{cases}
u_{\mathrm{B}} = u(x + \Delta x, \, y) = u(x, \, y) + \dfrac{\partial u}{\partial x}\Delta x \\[2mm]
v_{\mathrm{B}} = v(x + \Delta x, \, y) = v(x, \, y) + \dfrac{\partial v}{\partial x}\Delta x
\end{cases}
\tag{3.29}
$$

The velocity of point C is:

$$
\begin{cases}
u_{\mathrm{C}} = u(x, \, y + \Delta y) = u(x, \, y) + \dfrac{\partial u}{\partial y}\Delta y \\[2mm]
v_{\mathrm{C}} = v(x, \, y + \Delta y) = v(x, \, y) + \dfrac{\partial v}{\partial y}\Delta y
\end{cases}
\tag{3.30}
$$

The velocity of point D is:

Fig. 3.11 Motion diagram of micro fluid elements

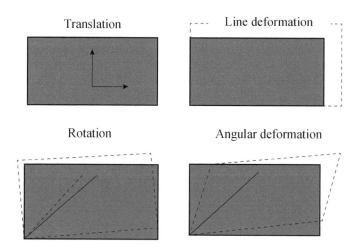

Fig. 3.12 The motion of fluid elements can be decomposed into different forms

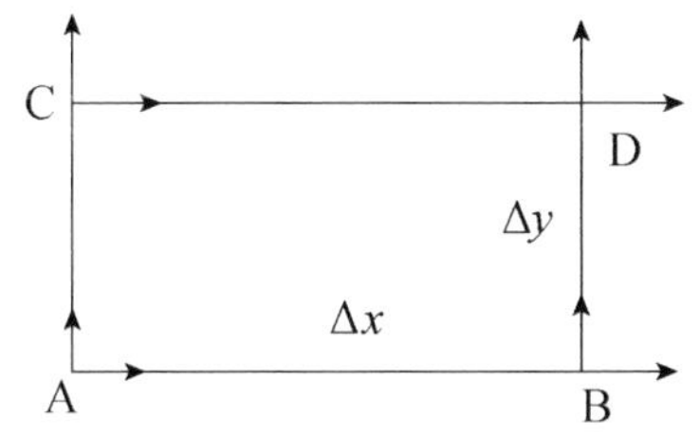

	A:	u		v
	B:	$u + \dfrac{\partial u}{\partial x}\Delta x$		$v + \dfrac{\partial v}{\partial x}\Delta x$
	C:	$u + \dfrac{\partial u}{\partial y}\Delta y$		$v + \dfrac{\partial v}{\partial y}\Delta y$
	D:	$u + \dfrac{\partial u}{\partial y}\Delta x + \dfrac{\partial u}{\partial y}\Delta y$		$v + \dfrac{\partial v}{\partial x}\Delta x + \dfrac{\partial v}{\partial y}\Delta y$

Fig. 3.13 Taylor expansion of the velocity for a micro-element

$$\begin{cases} u_D = u(x + \Delta x, y + \Delta y) = u(x, y) + \dfrac{\partial u}{\partial x}\Delta x + \dfrac{\partial u}{\partial y}\Delta y \\[3mm] v_D = v(x + \Delta x, y + \Delta y) = v(x, y) + \dfrac{\partial v}{\partial x}\Delta x + \dfrac{\partial v}{\partial y}\Delta y \end{cases} \tag{3.31}$$

(1) If the velocities of all vertices are the same, the translational motion of the micro-element will be represented. Therefore, the translational velocity of the micro-element is called $u\,(x, y)$ and $v\,(x, y)$;

(2) The line deformation motion refers to the stretching motion of each side length of the micro-element. Line deformation rate is defined as the amount of line deformation per unit length per unit time. For the side length AB, the increment of the side length in the finite period Δt is

$$\Delta(\text{AB}) = \left[u + \frac{\partial u}{\partial x}\Delta x - u \right]\Delta t = \frac{\partial u}{\partial x}\Delta x\,\Delta t \tag{3.32}$$

Thus, the line deformation rate of the micro-element in the x direction is

$$\varepsilon_x = \lim_{\Delta t \to 0} \frac{\Delta(\text{AB})}{\Delta t\,\Delta x} = \frac{\partial u}{\partial x} \tag{3.33}$$

Similarly, the line deformation rate of micro-elements in the y direction is

$$\varepsilon_y = \lim_{\Delta t \to 0} \frac{\Delta(\text{AC})}{\Delta t\,\Delta y} = \frac{\partial v}{\partial y} \tag{3.34}$$

The area change rate (the change rate per unit area per unit time) of the micro group is

$$\lim_{\Delta t \to 0} \frac{\Delta(\text{AB} \times \text{AC})}{\Delta x\,\Delta y\,\Delta t} = \lim_{\Delta t \to 0} \frac{\left(\Delta x + \frac{\partial u}{\partial x}\Delta x\,\Delta t \right)\left(\Delta y + \frac{\partial v}{\partial y}\Delta y\,\Delta t \right) - \Delta x\,\Delta y}{\Delta x\,\Delta y\,\Delta t}$$

$$= \lim_{\Delta t \to 0} \frac{\left(\frac{\partial u}{\partial x} + \frac{\partial v}{\partial y} \right)\Delta x\,\Delta y\,\Delta t + \frac{\partial u}{\partial x}\frac{\partial v}{\partial y}\Delta x\,\Delta y\,\Delta t^2}{\Delta x\,\Delta y\,\Delta t}$$

Fig. 3.14 Rotation and
angular deformation of
micro fluid elements

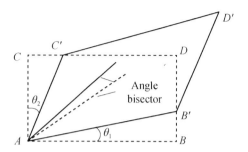

$$= \frac{\partial u}{\partial x} + \frac{\partial v}{\partial y} = \varepsilon_x + \varepsilon_y \tag{3.35}$$

(3) Angular deformation rate and angular velocity. In the finite period Δt, the change of the angle between two orthogonal edges AB and AC is related to the angular deformation and rotation of the micro-element, as shown in Fig. 3.14. In the period Δt, the deflection angle of AB side (positive when rotated counterclockwise) is

$$\theta_1 = \frac{BB'}{\Delta x} = \frac{\left(v + \frac{\partial v}{\partial x}\Delta x - v\right)\Delta t}{\Delta x} = \frac{\partial v}{\partial x}\Delta t \tag{3.36}$$

Similarly, In the period Δt, the deflection Angle of AC side (negative when rotated clockwise) is

$$\theta_2 = -\frac{CC'}{\Delta y} = -\frac{\left(u + \frac{\partial u}{\partial y}\Delta y - u\right)\Delta t}{\Delta y} = -\frac{\partial u}{\partial y}\Delta t \tag{3.37}$$

As shown in Fig. 3.14, the total angle variation of AC and AB can be decomposed into the rotation and pure angular deformation, as shown in Fig. 3.15.

In the period Δt, assuming that the rotation angle of the angular bisector of the micro-element is α and the pure angular deformation of the edge line is β, then, according to the geometric relationship of Eq. (3.15), it can be obtained

$$\theta_1 = \alpha + \beta, \theta_2 = \alpha - \beta \tag{3.38}$$

Solve the above equation to obtain:

$$\alpha = \frac{\theta_1 + \theta_2}{2} \quad \beta = \frac{\theta_1 - \theta_2}{2} \tag{3.39}$$

The angular velocity of rotation (rotation angle per unit time) of the micro-element is defined as

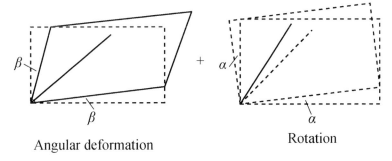

Angular deformation   Rotation

Fig. 3.15 Angular deformation and rotational motion of micro-elements

$$\omega_z = \lim_{\Delta t \to 0} \frac{\alpha}{\Delta t} = \frac{1}{2}\left(\frac{\partial v}{\partial x} - \frac{\partial u}{\partial y}\right) \tag{3.40}$$

The angular deformation rate (single side angular deformation per unit time) of the micro-element is

$$\varepsilon_{xy} = \lim_{\Delta t \to 0} \frac{\beta}{\Delta t} = \frac{1}{2}\left(\frac{\partial v}{\partial x} + \frac{\partial u}{\partial y}\right) \tag{3.41}$$

For three-dimensional hexahedral micro-groups, their motion forms can also be divided into translation, rotation and deformation. Like plane micro-groups, it is easy to derive related formulas.

(1) The translational velocity of the micro-element: $u(x, y, z, t)$, $v(x, y, z, t)$, $w(x, y, z, t)$.
(2) Deformation rate of the edges of micro-elements:

$$\varepsilon_x = \frac{\partial u}{\partial x}, \quad \varepsilon_y = \frac{\partial v}{\partial y}, \quad \varepsilon_z = \frac{\partial w}{\partial z} \tag{3.42}$$

(3) Angular deformation rate (shear deformation rate) of micro-elements:

$$\varepsilon_{xy} = \frac{1}{2}\left(\frac{\partial v}{\partial x} + \frac{\partial u}{\partial y}\right), \quad \varepsilon_{yz} = \frac{1}{2}\left(\frac{\partial w}{\partial y} + \frac{\partial v}{\partial z}\right), \quad \varepsilon_{zx} = \frac{1}{2}\left(\frac{\partial u}{\partial z} + \frac{\partial w}{\partial x}\right) \tag{3.43}$$

(4) Angular velocity of rotation:

$$\omega_x = \frac{1}{2}\left(\frac{\partial w}{\partial y} - \frac{\partial v}{\partial z}\right), \quad \omega_y = \frac{1}{2}\left(\frac{\partial u}{\partial z} - \frac{\partial w}{\partial x}\right), \quad \omega_z = \frac{1}{2}\left(\frac{\partial v}{\partial x} - \frac{\partial u}{\partial y}\right) \tag{3.44}$$

3.3.2   Velocity Decomposition Theorem of Fluid Elements

Helmholtz (1821–1894), a German physicist, put forward the velocity decomposition theorem of the flow field in 1858, and clarified the effect of the motion forms of micro fluid elements on the spatial variation of velocity. In the flow field at a given time, take any two points with a small distance (as shown in Fig. 3.16) and expand the decomposition according to the Taylor series.

Set the speed as $M_0(x, y, z, t)$

$$\begin{cases} u = u(x, y, z, t) \\ v = v(x, y, z, t) \\ w = w(x, y, z, t) \end{cases} \tag{3.45}$$

At the neighboring point, the velocity is $M_1(x + \Delta x, y + \Delta y, z + \Delta z, t)$

$$\begin{cases} u = u(x + \Delta x, y + \Delta y, z + \Delta z, t) \\ v = v(x + \Delta x, y + \Delta y, z + \Delta z, t) \\ w = w(x + \Delta x, y + \Delta y, z + \Delta z, t) \end{cases} \tag{3.46}$$

According to the Taylor series expansion

$$u(x + \Delta x, y + \Delta y, z + \Delta z, t) = u(x, y, z, t) + \frac{\partial u}{\partial x}\Delta x + \frac{\partial u}{\partial y}\Delta y + \frac{\partial u}{\partial z}\Delta z$$

$$v(x + \Delta x, y + \Delta y, z + \Delta z, t) = v(x, y, z, t) + \frac{\partial v}{\partial x}\Delta x + \frac{\partial v}{\partial y}\Delta y + \frac{\partial v}{\partial z}\Delta z$$

$$w(x + \Delta x, y + \Delta y, z + \Delta z, t) = w(x, y, z, t) + \frac{\partial w}{\partial x}\Delta x + \frac{\partial w}{\partial y}\Delta y + \frac{\partial w}{\partial z}\Delta z$$

$$\tag{3.47}$$

The linear incremental part in the above equation is decomposed and combined according to the corresponding motion. Take the u direction velocity component as

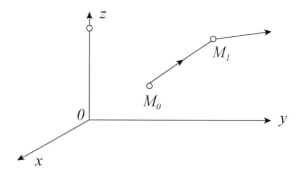

Fig. 3.16 Velocity of two adjacent points

an example, namely

$$u(x + \Delta x, y + \Delta y, z + \Delta z, t)$$

$$= u(x, y, z, t) + \frac{\partial u}{\partial x}\Delta x + \frac{\partial u}{\partial y}\Delta y + \frac{\partial u}{\partial z}\Delta z$$

$$= u(x, y, z, t) + \frac{1}{2}\left(\frac{\partial u}{\partial z} - \frac{\partial w}{\partial x}\right)\Delta z - \frac{1}{2}\left(\frac{\partial v}{\partial x} - \frac{\partial u}{\partial y}\right)\Delta y + \frac{\partial u}{\partial x}\Delta x$$

$$+ \frac{1}{2}\left(\frac{\partial v}{\partial x} + \frac{\partial u}{\partial y}\right)\Delta y + \frac{1}{2}\left(\frac{\partial u}{\partial z} + \frac{\partial w}{\partial x}\right)\Delta z$$

$$= u(x, y, z, t) + (\omega_y \Delta z - \omega_z \Delta y) + \varepsilon_x \Delta x + \varepsilon_{xy} \Delta y + \varepsilon_{zx} \Delta z \qquad (3.48)$$

According to formula (3.48), the velocity components v and w can also be decomposed into

$$u(x + \Delta x, y + \Delta y, z + \Delta z, t) = u(x, y, z, t)$$
$$+ (\omega_y \Delta z - \omega_z \Delta y) + \varepsilon_x \Delta x + \varepsilon_{xy} \Delta y + \varepsilon_{xz} \Delta z$$
$$v(x + \Delta x, y + \Delta y, z + \Delta z, t) = v(x, y, z, t)$$
$$+ (\omega_z \Delta x - \omega_x \Delta z) + \varepsilon_{yx} \Delta x + \varepsilon_y \Delta y + \varepsilon_{yz} \Delta z$$
$$w(x + \Delta x, y + \Delta y, z + \Delta z, t) = w(x, y, z, t)$$
$$+ (\omega_x \Delta y - \omega_y \Delta x) + \varepsilon_{zx} \Delta x + \varepsilon_{zy} \Delta y + \varepsilon_z \Delta z \qquad (3.49)$$

Equation (3.49) is the Helmholtz velocity decomposition theorem, in which the influence of translation, rotation and deformation motion is included.

It should be noted that the motion of an actual fluid pellet can be one or a combination of several forms of motion, such as:

(1) If the fluid particle has only translational motion, then the velocity field decomposition theorem can be simplified as

$$u(x + \Delta x, y + \Delta y, z + \Delta z, t) = u(x, y, z, t)$$
$$v(x + \Delta x, y + \Delta y, z + \Delta z, t) = v(x, y, z, t)$$
$$w(x + \Delta x, y + \Delta y, z + \Delta z, t) = w(x, y, z, t) \qquad (3.50)$$

(2) If the fluid particle has only translation and deformation. Then the velocity field decomposition theorem can be simplified as

$$u(x + \Delta x, y + \Delta y, z + \Delta z, t) = u(x, y, z, t) + \varepsilon_x \Delta x + \varepsilon_{xy} \Delta y + \varepsilon_{xz} \Delta z$$
$$v(x + \Delta x, y + \Delta y, z + \Delta z, t) = v(x, y, z, t) + \varepsilon_{yx} \Delta x + \varepsilon_y \Delta y + \varepsilon_{yz} \Delta z$$
$$w(x + \Delta x, y + \Delta y, z + \Delta z, t) = w(x, y, z, t) + \varepsilon_{zx} \Delta x + \varepsilon_{zy} \Delta y + \varepsilon_z \Delta z \quad (3.51)$$

(3) If the fluid particle has only translation and rotation, the velocity decomposition theorem can be simplified as

$$u(x + \Delta x, y + \Delta y, z + \Delta z, t) = u(x, y, z, t) + (\omega_y \Delta z - \omega_z \Delta y)$$
$$v(x + \Delta x, y + \Delta y, z + \Delta z, t) = v(x, y, z, t) + (\omega_z \Delta x - \omega_x \Delta z)$$
$$w(x + \Delta x, y + \Delta y, z + \Delta z, t) = w(x, y, z, t) + (\omega_x \Delta y - \omega_y \Delta x) \qquad (3.52)$$

If the angle of rotation is $\omega_x = 0$, $\omega_y = 0$, $\omega_z = \omega$, and put in the above equation:

$$u(x + \Delta x, y + \Delta y, z + \Delta z, t) = u(x, y, z, t) - \omega \Delta y$$
$$v(x + \Delta x, y + \Delta y, z + \Delta z, t) = v(x, y, z, t) + \omega \Delta x \qquad (3.53)$$

Obviously, this is the velocity field of a rigid body rotating around the z axis, like the velocity field of a fluid in a rotating container.

In essence, there is an important difference between the velocity decomposition theorems of rigid bodies and those of fluid elements in addition to the deformation motion. The decomposition theorem of rigid body velocity is for the whole rigid body, so it belongs to the wholeness theorem. The velocity decomposition theorem of fluid elements is only valid for fluid elements, so it is a locality theorem. For example, the angular velocity of a rigid body is a characteristic quantity describing the rotation of the whole rigid body, which is constant at any point on the rigid body. The rotational angular velocity of a fluid is a local characteristic quantity to describe the rotation of local fluid elements about their own axes. The rotational angular velocity of fluid elements at different points is generally different.

3.4   Divergence and Curl of Velocity Field

3.4.1   *Divergence of Velocity Field and Its Physical Significance*

For any velocity field, according to the knowledge of field theory, the divergence of velocity field is defined $\vec{V} = u\vec{i} + v\vec{j} + w\vec{k}$

$$\operatorname{div} \vec{V} = \nabla \cdot \vec{V} = \frac{\partial u}{\partial x} + \frac{\partial v}{\partial y} + \frac{\partial w}{\partial z} \qquad (3.54)$$

By analyzing the above expression, the three items on the right of the above equation are actually the linear deformation rates of the micro fluid elements in three mutually perpendicular directions. Therefore, the divergence of the velocity field can be characterized as the sum of the linear deformation rates of the three vertical axis

Fig. 3.17 Volume expansion of micro fluid elements

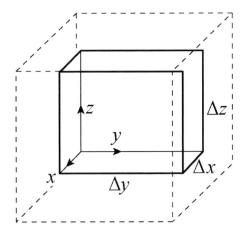

directions of the micro-elements. In fact, from another point of view, velocity field divergence also represents the relative volume expansion rate (expansion amount per unit volume per unit time) of micro fluid elements. As shown in Fig. 3.17, it is assumed that the length of each side of the fluid pellet before deformation is Δx, Δy and Δz respectively, and the original volume is $\Delta x \Delta y \Delta z$. After Δt period, the length of the three sides becomes

$$\Delta x_1 = \left(1 + \frac{\partial u}{\partial x} \Delta t\right) \Delta x$$

$$\Delta y_1 = \left(1 + \frac{\partial v}{\partial y} \Delta t\right) \Delta y \qquad (3.55)$$

$$\Delta z_1 = \left(1 + \frac{\partial w}{\partial z} \Delta t\right) \Delta z$$

Then the relative volume expansion rate (the increase per unit volume per unit time) is:

$$\text{div } \vec{V} = \lim_{\Delta t \to 0} \frac{1}{\Delta x \Delta y \Delta z \Delta t}$$

$$\left[\left(1 + \frac{\partial u}{\partial x} \Delta t\right) \Delta x \left(1 + \frac{\partial v}{\partial y} \Delta t\right) \Delta y \left(1 + \frac{\partial w}{\partial z} \Delta t\right) \Delta z - \Delta x \Delta y \Delta z\right]$$

$$= \frac{\partial u}{\partial x} + \frac{\partial v}{\partial y} + \frac{\partial w}{\partial z} = \nabla \cdot \vec{V} \qquad (3.56)$$

No matter how the shape and volume change, the mass of the actual fluid group is always the same in motion. Considering that mass is equal to volume times density, for fluid motion with constant density, the volume of fluid micro-element is also constant, and the divergence of its velocity field must be zero.

$$\text{div}\ \vec{V} = \nabla \cdot \vec{V} = \frac{\partial u}{\partial x} + \frac{\partial v}{\partial y} + \frac{\partial w}{\partial z} = 0 \tag{3.57}$$

Therefore, for a fluid with constant density, the volume of the fluid in motion is also constant, so it is called an incompressible fluid. The continuity equation of the incompressible fluid is the divergence equal to zero.

For the flow field whose velocity field divergence is not equal to zero, the velocity field divergence represents the distribution size of the source (sink) and the deformation rate of the micro fluid elements in the flow field. If the divergence is greater than zero, the source is positive. If the divergence is less than zero, the source is negative (sink). According to the field theory, for any closed volume and closed surface area S in the flow field, the Relation between the surface integral and the volume fraction can be represented by the Gaussian equation (Gauss, German scientist, 1777–1855, as shown in Fig. 3.18)

$$\iiint\limits_{V} (\nabla \cdot \vec{V})\mathrm{d}x\mathrm{d}y\mathrm{d}z = \oiint\limits_{S} \vec{V} \cdot \mathrm{d}\vec{S} \tag{3.58}$$

If the volume of the fluid goes to zero, the divergence of the velocity field can also be written as

Fig. 3.18 Johann Carl Friedrich Gauss (1777–1855)

$$\nabla \cdot \vec{V} = \lim_{\forall \to 0} \frac{\oint_S \vec{V} \cdot d\vec{S}}{\forall} \tag{3.59}$$

The above equation represents the volume of fluid scattered from a unit volume per unit time or the volume flux scattered from a unit volume.

3.4.2 Curl and Velocity Potential Function of Velocity Field

It can be seen from the motion form decomposition of micro fluid elements that the three components of the rotational angular velocity of micro fluid elements around its own axis are ω_x, ω_y and ω_z, the angular velocity vector of Eq. (3.45) can be expressed as

$$\vec{\omega} = \omega_x \vec{i} + \omega_y \vec{j} + \omega_z \vec{k} = \frac{1}{2} \text{rot } \vec{V} = \frac{1}{2} \nabla \times \vec{V} \tag{3.60}$$

In the formula, rot $\vec{V}$ or $\nabla \times \vec{V}$ represents the curl of the velocity field, so it can be said that the curl of the velocity field equals 2 times of the angular velocity of the micro-element about its own axis. In the flow field, if the curl in a certain region is large, it means that the fluid mass in this region rotates quickly, and vice versa. The determinant of curl is

$$\text{rot } \vec{V} = \nabla \times \vec{V} = \begin{vmatrix} \vec{i} & \vec{j} & \vec{k} \\ \frac{\partial}{\partial x} & \frac{\partial}{\partial y} & \frac{\partial}{\partial z} \\ u & v & w \end{vmatrix} \tag{3.61}$$

If ω is equal to zero everywhere in a flow field, such a flow field is called a non-swirling flow field and the corresponding flow is called a non-swirling flow. Otherwise, it is a swirling flow field, and the corresponding flow is called a swirling flow.

In a non-swirling flow field, there is a velocity potential function, which is only a function of coordinate position and time.

$$\varphi = \varphi(x, y, z, t) \tag{3.62}$$

A sufficient and necessary condition for the existence of the velocity potential function is that the curl of the velocity field is zero (no eddy flow), namely

$$\frac{\partial w}{\partial y} - \frac{\partial v}{\partial z} = 0, \quad \frac{\partial u}{\partial z} - \frac{\partial w}{\partial x} = 0, \quad \frac{\partial v}{\partial x} - \frac{\partial u}{\partial y} = 0 \tag{3.63}$$

In this case, the relationship between the velocity potential function and the velocity component is

$$u = \frac{\partial \varphi}{\partial x}, v = \frac{\partial \varphi}{\partial y}, w = \frac{\partial \varphi}{\partial z}, \ \vec{V} = \nabla \varphi \tag{3.64}$$

It says that the partial derivative of the velocity potential in one direction is equal to the velocity component in that direction. In A non-swirling (potential) field, the result of velocity line integration along any curve connecting two points A and B is only related to the difference of the velocity potential function values at the two ends and independent of the integration path, namely

$$\int_A^B (u dx + v dy + w dz) = \int_A^B d\varphi = \varphi_B - \varphi_A \tag{3.65}$$

Example Given a two-dimensional flow field, the velocity distribution is $u = 2ax$, $v = -2ay$. Is the flow non-swirling flow or swirling flow? Is there a velocity potential function? What is the streamline equation and deformation rate?

Solution:

According to the velocity component of the flow field, the rotational angular velocity of the fluid micro-element around the z axis is

$$\omega_z = \frac{1}{2} \left(\frac{\partial v}{\partial x} - \frac{\partial u}{\partial y} \right) = \frac{1}{2} (0 - 0) = 0$$

It shows that the flow is irrotational and has a velocity potential function. The Differential of the velocity potential function is

$$d\varphi = u dx + v dy = 2ax dx - 2ay dy$$

Integral to get

$$\varphi = a(x^2 - y^2) + C \tag{3.66}$$

where C is the integral constant. By definition, the streamline equation is

$$\frac{dx}{u} = \frac{dy}{v}, \frac{dx}{2ax} = \frac{dy}{-2ay}$$

So integrate it to get

$$xy = C \tag{3.67}$$

Take a series of values for constant C, and a series of streamlines can be drawn, as shown in Fig. 3.19.

The linear deformation rate of micro fluid elements is

Fig. 3.19 Flow around the
angle

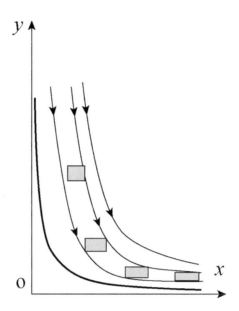

$$\varepsilon_x = \frac{\partial u}{\partial x} = 2a, \, \varepsilon_y = \frac{\partial v}{\partial y} = -2a$$

The area expansion rate is

$$\text{div } \vec{V} = \varepsilon_x + \varepsilon_y = 0$$

The angular deformation rate is

$$\varepsilon_{xy} = \frac{1}{2}\left(\frac{\partial v}{\partial x} + \frac{\partial u}{\partial y}\right) = 0 \tag{3.68}$$

The rectangular micro-element ABCD is investigated. In the flow field shown in
Fig. 3.19, it flows from the top left to the bottom right. Since the flow is non-rotating,
the micro-element does not rotate during the flow process. Since the area expansion
rate is zero, the x direction line segment of the fluid micro-element will elongate,
the y direction line segment will shorten in the process of motion, and the area of
the fluid micro-element will remain unchanged. The fluid particles have no angular
deformation and their shapes remain unchanged.

3.5   Continuous Differential Equation

Fluid mechanics is to study the macroscopic motion behavior of fluid particles, which must satisfy the general laws of material motion. The law of conservation of mass is one of them, which is used to characterize the continuity conditions of fluid motion. Based on Lagrange's view, the mass conservation law requires that the mass of a fluid mass remains unchanged in the process of flow when any fluid mass is taken from a passive (no sink) flow field. If a fixed control body is arbitrarily taken in a passive flow field based on Euler's point of view, the law of conservation of mass requires that the mass difference between the inflow and outflow of the control body (net inflow) is equal to the mass increment caused by the change of fluid mass density in the control body in a differential period.

3.5.1   Continuity Differential Equation Based on Lagrange View

In the flow field, take any fluid particle Δx, Δy and Δz, the fluid density is ρ, and the mass of the fluid particle is Δm, i.e.

$$\Delta m = \rho \Delta x \Delta y \Delta z \tag{3.69}$$

The law of mass conservation requires that the mass of the micro-element remains unchanged during the flow, i.e., $\Delta m = $ Const. Using the continuum hypothesis, the law of conservation of mass can be written as

$$\frac{d(\Delta m)}{dt} = 0 \tag{3.70}$$

Substituting Eq. (3.69) into Eq. (3.70), we can get

$$\frac{d(\rho \Delta x \Delta y \Delta z)}{dt} = 0 \tag{3.71}$$

By differentiating by parts, we get

$$\frac{d\rho}{dt} + \rho \frac{d(\Delta x \Delta y \Delta z)}{\Delta x \Delta y \Delta z dt} = 0 \tag{3.72}$$

In this expression, the second term on the left represents the relative volume expansion rate (the increase per unit volume per unit time). According to Eq. (3.56), this term represents the divergence of the velocity field and is substituted into Eq. (3.72).

$$\frac{d\rho}{dt} + \rho \nabla \cdot \vec{V} = 0 \tag{3.73}$$

It can be seen from Eqs. (3.72) and (3.73)

$$\nabla \cdot \vec{V} = \frac{d(\Delta x \Delta y \Delta z)}{\Delta x \Delta y \Delta z dt} = -\frac{1}{\rho} \frac{d\rho}{dt} \tag{3.74}$$

Use the derivative of the density

$$\frac{d\rho}{dt} = \frac{\partial \rho}{\partial t} + u \frac{\partial \rho}{\partial x} + v \frac{\partial \rho}{\partial y} + w \frac{\partial \rho}{\partial z} \tag{3.75}$$

Equation (3.73) can be obtained

$$\frac{\partial \rho}{\partial t} + u \frac{\partial \rho}{\partial x} + v \frac{\partial \rho}{\partial y} + w \frac{\partial \rho}{\partial z} + \rho \left(\frac{\partial u}{\partial x} + \frac{\partial v}{\partial y} + \frac{\partial w}{\partial z} \right) = 0 \tag{3.76}$$

By arranging Eq. (3.76), we can get

$$\frac{\partial \rho}{\partial t} + \frac{\partial(\rho u)}{\partial x} + \frac{\partial(\rho v)}{\partial y} + \frac{\partial(\rho w)}{\partial z} = 0 \tag{3.77}$$

In vector form

$$\frac{\partial \rho}{\partial t} + \nabla \cdot (\rho \vec{V}) = 0 \tag{3.78}$$

3.5.2 Continuity Differential Equation Based on Euler's Viewpoint

Now, in the flow field, take any rectangular hexahedron (control body) whose side length is Δx, Δy and Δz respectively. The microelement body is fixed and invariable relative to the coordinate system and is passed by the fluid. Based on the continuity hypothesis, the continuity differential equation controlled by the mass conservation law can be derived. Suppose at time t, the coordinate of the center point of the differential hexahedron is (x, y, z), the velocity is u, v, w, and the density is ρ. As shown in Fig. 3.20, the static inflow of fluid passing through the control body in the x direction is deduced. In the period Δt, the mass inflow from the left side of the differential control body is

$$m_{ix} = \rho(x - \frac{\Delta x}{2}, y, z, t)u(x - \frac{\Delta x}{2}, y, z, t)\Delta y \Delta z \Delta t \tag{3.79}$$

Fig. 3.20 Fluid mass variation of control volume

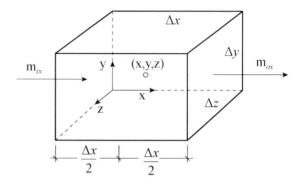

Using the Taylor series expansion and taking the small quantity of the first order in brackets, the above formula can be written as

$$m_{ix} = \rho(x - \frac{\Delta x}{2}, y, z, t)u(x - \frac{\Delta x}{2}, y, z, t)\Delta y\Delta z\Delta t$$

$$= \left(\rho - \frac{\Delta x}{2}\frac{\partial \rho}{\partial x}\right)\left(u - \frac{\Delta x}{2}\frac{\partial u}{\partial x}\right)\Delta y\Delta z\Delta t$$

$$= \left(\rho u - \frac{\Delta x}{2}\frac{\partial(\rho u)}{\partial x}\right)\Delta y\Delta z\Delta t \qquad (3.80)$$

Similarly, the mass flowing out of the right side of the differential control body is

$$m_{ox} = \rho(x + \frac{\Delta x}{2}, y, z, t)u(x + \frac{\Delta x}{2}, y, z, t)\Delta y\Delta z\Delta t$$

$$= \left(\rho u + \frac{\Delta x}{2}\frac{\partial(\rho u)}{\partial x}\right)\Delta y\Delta z\Delta t \qquad (3.81)$$

The static inflow mass flowing into the differential control body along the x direction is

$$\Delta m_x = m_{ix} - m_{ox} = \left(\rho u - \frac{\Delta x}{2}\frac{\partial(\rho u)}{\partial x}\right)\Delta y\Delta z\Delta t$$

$$- \left(\rho u + \frac{\Delta x}{2}\frac{\partial(\rho u)}{\partial x}\right)\Delta y\Delta z\Delta t = -\frac{\partial(\rho u)}{\partial x}\Delta x\Delta y\Delta z\Delta t \qquad (3.82)$$

The mass in and out of the y and z directions can be written as

$$m_{iy} = \left(\rho v - \frac{\Delta y}{2}\frac{\partial(\rho v)}{\partial y}\right)\Delta z\Delta x\Delta t$$

$$m_{oy} = \left(\rho v + \frac{\Delta y}{2}\frac{\partial(\rho v)}{\partial y}\right)\Delta z\Delta x\Delta t$$

$$\Delta m_y = -\frac{\partial(\rho v)}{\partial y}\Delta x\Delta y\Delta z\Delta t \tag{3.83}$$

$$m_{iz} = \left(\rho w - \frac{\Delta z}{2}\frac{\partial(\rho w)}{\partial z}\right)\Delta x\Delta y\Delta t$$

$$m_{oz} = \left(\rho w + \frac{\Delta z}{2}\frac{\partial(\rho w)}{\partial z}\right)\Delta x\Delta y\Delta t$$

$$\Delta m_z = -\frac{\partial(\rho w)}{\partial z}\Delta x\Delta y\Delta z\Delta t \tag{3.84}$$

In the period Δt, the mass of the fluid flowing in through the differential control body is

$$\Delta m_{xyz} = \Delta m_x + \Delta m_y + \Delta m_z = -\left(\frac{\partial(\rho u)}{\partial x} + \frac{\partial(\rho v)}{\partial y} + \frac{\partial(\rho w)}{\partial z}\right)\Delta x\Delta y\Delta z\Delta t \tag{3.85}$$

Meanwhile, in the period Δt, the mass increment caused by density change in the differential control body is

$$\Delta m_t = \left(\rho + \frac{\partial\rho}{\partial t}\Delta t - \rho\right)\Delta x\Delta y\Delta z = \frac{\partial\rho}{\partial t}\Delta x\Delta y\Delta z\Delta t \tag{3.86}$$

According to the law of conservation of mass, for the flow field with no source (no sink), the static inflow mass of the control body through differential is equal to the mass increment caused by the change of density in the control body. That is:

$$\frac{\partial\rho}{\partial t} + \frac{\partial(\rho u)}{\partial x} + \frac{\partial(\rho v)}{\partial y} + \frac{\partial(\rho w)}{\partial z} = 0 \tag{3.87}$$

This formula is consistent with Formula (3.78).
Write Eq. (3.87) in the form of divergence as

$$\frac{\partial\rho}{\partial t} + \nabla\cdot(\rho\vec{V}) = 0, \quad \frac{d\rho}{dt} + \rho\nabla\cdot\vec{V} = 0 \tag{3.88}$$

For incompressible fluids, the continuity equation becomes $\frac{d\rho}{dt} = 0$

$$\nabla\cdot\vec{V} = 0, \quad \frac{\partial u}{\partial x} + \frac{\partial v}{\partial y} + \frac{\partial w}{\partial z} = 0 \tag{3.89}$$

The incompressible condition of fluid $\frac{d\rho}{dt} = 0$ means that the density of fluid particle remains constant in the motion, but the density of this fluid particle and that fluid particle can be different. That is, the fluid can be non-mean. Therefore, the density of the incompressible fluid is not necessarily constant everywhere. For example, parallel flow with variable density. The mean fluid is defined as $\nabla \rho = 0$. That is, the density is uniform in space, but it cannot be guaranteed that it does not change with time. That is, $\frac{\partial \rho}{\partial t}$ is not zero. The density of a fluid is constant everywhere only if it is both incompressible and mean. Substitute the mean condition into the incompressible condition $\nabla \rho = 0$

$$\frac{d\rho}{dt} = \frac{\partial \rho}{\partial t} + u \frac{\partial \rho}{\partial x} + v \frac{\partial \rho}{\partial y} + w \frac{\partial \rho}{\partial z} = \frac{\partial \rho}{\partial t} + \vec{V} \cdot \nabla \rho = 0$$

The following equation can be obtained:

$$\frac{\partial \rho}{\partial t} = 0 \tag{3.90}$$

Since $\rho = $ constant, it means fluid density does not change with time, nor with the position. Therefore, it is constant throughout the flow field.

The continuity differential equations derived above are only the kinematic behavior of micro fluid elements, independent of dynamics, so they are suitable for both ideal and viscous fluids. The above continuous differential equation reflecting the conservation of mass was derived by The Swiss mathematician Euler in 1753.

3.6 Differential Equations of Ideal Fluid Motion (Euler Equations)

In 1755, based on the continuum hypothesis and ideal fluid model, based on the view of local tracking fluid micro group, and by applying Newton's second theorem (momentum conservation law), Swiss mathematician and fluid mechanics Leonhard Euler (1707–1783) established the ideal fluid motion differential equations, referred to as Euler's differential equations. Now, in the flow field, take any fluid micro-element at time t, the three sides of the micro-element are Δt, Δy and Δz, the coordinates of the center point of the micro-element are (x, y, z), the velocity is u, v, w, the density is ρ, the pressure is P, and the unit mass force is f_x, f_y, f_z. For an ideal fluid, only pressure and are applied on the surface of the fluid element, and shear stress is not considered. As shown in Fig. 3.21, at time t, the pressure at the shape point is P (x, y, z, t), then the Taylor series is used and the higher-order terms are ignored. In the x direction, the pressure acting on the left side of the micro-element is $\left(p - \frac{\partial p}{\partial x} \frac{\Delta x}{2}\right) \Delta y \Delta z$, and the pressure on the right side of the micro-element is $\left(p + \frac{\partial p}{\partial x} \frac{\Delta x}{2}\right) \Delta y \Delta z$. The component of the mass force in the x direction of the micro-element is $\rho \Delta x \Delta y \Delta z f_x$. According to Newton's second law, the force acting on the x direction of the micro-element is equal to the mass times the acceleration of the x direction of the micro-element, i.e.

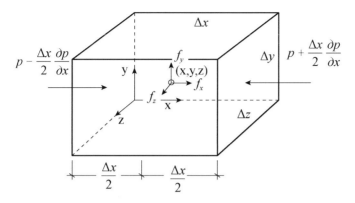

Fig. 3.21 Forces on an ideal fluid element in the x direction

$$\left(p - \frac{\partial p}{\partial x}\frac{\Delta x}{2}\right)\Delta y\Delta z - \left(p + \frac{\partial p}{\partial x}\frac{\Delta x}{2}\right)\Delta y\Delta z + \rho f_x \Delta x\Delta y\Delta z = \frac{du}{dt}\rho\Delta x\Delta y\Delta z$$

$$(3.91)$$

Divide both sides of Eq. (3.91) by the mass of the micro-element $\rho\Delta x\Delta y\Delta z$, and take the limit to obtain the differential equation of motion in the x direction

$$\frac{du}{dt} = f_x - \frac{1}{\rho}\frac{\partial p}{\partial x} \tag{3.92}$$

The form of total acceleration velocity is used here because it follows the micro fluid elements when establishing the above equation. Equation (3.92) can also be written as

$$\frac{\partial u}{\partial t} + u\frac{\partial u}{\partial x} + v\frac{\partial u}{\partial y} + w\frac{\partial u}{\partial z} = f_x - \frac{1}{\rho}\frac{\partial p}{\partial x} \tag{3.93}$$

Similarly, differential equations of motion in the other two directions can be established. In sum, the Euler equations are

$$\frac{du}{dt} = f_x - \frac{1}{\rho}\frac{\partial p}{\partial x}$$
$$\frac{dv}{dt} = f_y - \frac{1}{\rho}\frac{\partial p}{\partial y}$$
$$\frac{dw}{dt} = f_z - \frac{1}{\rho}\frac{\partial p}{\partial z} \tag{3.94}$$

Its vector form is

$$\frac{d\vec{V}}{dt} = \vec{f} - \frac{1}{\rho}\nabla p \tag{3.95}$$

The above three equations are the differential equations of ideal fluid motion in the Cartesian coordinate system. The Euler differential equation shows that the acceleration of a fluid element along a certain direction is equal to the mass force minus the pressure gradient force. If the mass force is zero, the acceleration of the fluid element in one direction is equal to the negative pressure gradient force. Take the differential equation in the x direction as an example. If $f_x = 0$, then the equation in the x direction is:

$$\frac{du}{dt} = -\frac{1}{\rho}\frac{\partial p}{\partial x} \tag{3.96}$$

For a favourable pressure gradient flow $\frac{\partial p}{\partial x} < 0$ (the pressure decreases along the flow direction), the fluid particles move in an accelerated motion along the flow direction. For a flow with a adverse pressure gradient $\frac{\partial p}{\partial x} > 0$ (the pressure increases along the flow direction), the fluid particles move in an accelerated motion along the flow direction. For zero-pressure gradient flow $\frac{\partial p}{\partial x} = 0$, the velocity of the fluid particle is constant along the flow direction.

The other form of Eq. (3.95) is

$$\frac{\partial u}{\partial t} + u\frac{\partial u}{\partial x} + v\frac{\partial u}{\partial y} + w\frac{\partial u}{\partial z} = f_x - \frac{1}{\rho}\frac{\partial p}{\partial x}$$

$$\frac{\partial v}{\partial t} + u\frac{\partial v}{\partial x} + v\frac{\partial v}{\partial y} + w\frac{\partial v}{\partial z} = f_y - \frac{1}{\rho}\frac{\partial p}{\partial y}$$

$$\frac{\partial w}{\partial t} + u\frac{\partial w}{\partial x} + v\frac{\partial w}{\partial y} + w\frac{\partial w}{\partial z} = f_z - \frac{1}{\rho}\frac{\partial p}{\partial z} \tag{3.97}$$

The vector form is

$$\frac{\partial \vec{V}}{\partial t} + (\vec{V} \cdot \nabla)\vec{V} = \vec{f} - \frac{1}{\rho}\nabla p \tag{3.98}$$

For one-dimensional flows, the equation of motion can be written as (Fig. 3.22)

$$\frac{\partial u}{\partial t} + u\frac{\partial u}{\partial s} = f_s - \frac{1}{\rho}\frac{\partial p}{\partial s} \tag{3.99}$$

If the acceleration terms are reassembled and the rotation angle component can be shown in the acceleration term, equations called Gromik-Lamb (British scientist, as shown in Fig. 3.23) type equations can be obtained. For example, in the x direction, we have

Fig. 3.22 One-dimensional flow

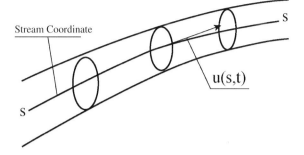

Fig. 3.23 Horace Lamb, British mathematician and mechanist (1849–1934)

$$u\frac{\partial u}{\partial x} + v\frac{\partial u}{\partial y} + w\frac{\partial u}{\partial z}$$

$$= u\frac{\partial u}{\partial x} + v\frac{\partial v}{\partial x} + w\frac{\partial w}{\partial x} - v\left(\frac{\partial v}{\partial x} - \frac{\partial u}{\partial y}\right) + w\left(\frac{\partial u}{\partial z} - \frac{\partial w}{\partial x}\right)$$

$$= \frac{\partial}{\partial x}\left(\frac{V^2}{2}\right) - 2v\omega_z + 2w\omega_y \tag{3.100}$$

where $\frac{V^2}{2} = \frac{u^2+v^2+w^2}{2}$ represents the kinetic energy per unit mass of micro fluid elements. Thus, the equations of the Gromyko-Lamb form can be obtained as

$$f_x - \frac{1}{\rho}\frac{\partial p}{\partial x} = \frac{\partial u}{\partial t} + \frac{\partial}{\partial x}\left(\frac{V^2}{2}\right) - 2(v\omega_z - w\omega_y)$$

$$f_y - \frac{1}{\rho}\frac{\partial p}{\partial y} = \frac{\partial v}{\partial t} + \frac{\partial}{\partial y}\left(\frac{V^2}{2}\right) - 2(w\omega_x - u\omega_z)$$

$$f_z - \frac{1}{\rho}\frac{\partial p}{\partial z} = \frac{\partial w}{\partial t} + \frac{\partial}{\partial z}\left(\frac{V^2}{2}\right) - 2(u\omega_y - v\omega_x) \tag{3.101}$$

Write it as a vector

$$\vec{f} - \frac{1}{\rho}\nabla p - \nabla\left(\frac{V^2}{2}\right) = \frac{\partial \vec{V}}{\partial t} - 2\vec{V} \times \vec{\omega} \tag{3.102}$$

Gromyko-Lamb type equations are still equations of differential in an ideal fluid motion. The advantage is that the rotational angular velocity term is shown in the equation, which facilitates the analysis of irrotational flows. For an ideal fluid, the fluid micro-element will not be affected by tangential force (i.e., viscous shear force) in the process of motion, so the fluid micro-element will not change its curl in the process of motion. For example, if the original curl is zero (i.e., there is no cyclone), it will also remain cyclone-free in motion. The ones that have swirls continue to be swirls, and their curl stays the same.

3.7 Bernoulli's Equation and Its Physical Significance

3.7.1 Bernoulli Equation

In 1738, Swiss scientist Daniel Bernoulli (1700–1782) established the one-variable fluid mechanical energy conservation equation, which is the famous ideal fluid steady flow energy equation (later called Bernoulli equation). In 1755, Euler derived the differential equations of ideal fluid motion. Then for the ideal fluid steady flow with gravitational potential, by integrating Euler equations, Bernoulli equation can also be obtained. The derivation shows that Bernoulli's equation is not only true along the same streamline, but also for the same vortex line, potential flow field and spiral flow. The following is an integral derivation of Bernoulli's equation based on the Gromyko-Lamb type equation.

For an ideal positive pressure fluid (density is only a function of pressure ρ (P), if the density ρ (t, P) is a function of temperature and pressure, it is called a baroclinic fluid. Under the potential condition of the mass force, the flow is steady. Now examine the terms in Eq. (3.102). namely

$$\vec{f} - \frac{1}{\rho}\nabla p - \nabla\left(\frac{V^2}{2}\right) = \frac{\partial \vec{V}}{\partial t} - 2\vec{V} \times \vec{\omega}$$

(1) The mass force $\vec{f}$ belongs to the potential field, and there is a potential function of the mass force

$$\vec{f} = -\nabla\Pi$$

$$f_x = -\frac{\partial \Pi}{\partial x}, f_y = -\frac{\partial \Pi}{\partial y}, f_z = -\frac{\partial \Pi}{\partial z} \tag{3.103}$$

where the negative sign indicates that the gradient of the force potential function is opposite to the direction of the mass force.

(2) For an ideal positive pressure fluid whose density is a constant (incompressible fluid) or only a function of pressure (such as adiabatic or isentropic flows), define a function $P(p)$ such that

$$\nabla P = \frac{1}{\rho}\nabla p \tag{3.104}$$

(3) For steady flow, there is

$$\frac{\partial \vec{V}}{\partial t} = 0 \tag{3.105}$$

By substituting Eqs. (3.103)–(3.105) into Gromyko-Lamb Equation (3.102), we can get

$$-\nabla\Pi - \nabla P - \nabla\left(\frac{V^2}{2}\right) = -2\vec{V} \times \vec{\omega}$$

$$\nabla(\Pi + P + \frac{V^2}{2}) = 2\vec{V} \times \vec{\omega} \tag{3.106}$$

Write the above equation in component form as

$$\frac{\partial}{\partial x}\left(\Pi + P + \frac{V^2}{2}\right) = 2(\vec{V} \times \vec{\omega})_x$$

$$\frac{\partial}{\partial y}\left(\Pi + P + \frac{V^2}{2}\right) = 2(\vec{V} \times \vec{\omega})_y$$

Fig. 3.24 Integral projected
along a curve

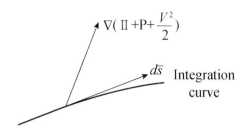

$$\frac{\partial}{\partial z}\left(\Pi + P + \frac{V^2}{2}\right) = 2(\vec{V} \times \vec{\omega})_z \tag{3.107}$$

Now, take any smooth curve in the flow field and project the above equation onto the curve (as shown in Fig. 3.24). In other words, multiply dx, dy and dz by each component of Eq. (3.107), and get

$$\frac{\partial}{\partial x}\left(\Pi + P + \frac{V^2}{2}\right)dx = 2(\vec{V} \times \vec{\omega})_x dx$$

$$\frac{\partial}{\partial y}\left(\Pi + P + \frac{V^2}{2}\right)dy = 2(\vec{V} \times \vec{\omega})_y dy$$

$$\frac{\partial}{\partial z}\left(\Pi + P + \frac{V^2}{2}\right)dz = 2(\vec{V} \times \vec{\omega})_z dz \tag{3.108}$$

By adding up the components of Eq. (3.108) and writing them in vector form, we can get

$$\nabla\left(\Pi + P + \frac{V^2}{2}\right) \cdot d\vec{s} = 2\vec{V} \times \vec{\omega} \cdot d\vec{s}$$

$$\frac{\partial}{\partial s}\left(\Pi + P + \frac{V^2}{2}\right)ds = 2\vec{V} \times \vec{\omega} \cdot d\vec{s} \tag{3.109}$$

If the right-hand side of Eq. (3.109) is zero, then

$$\vec{V} \times \vec{\omega} \cdot d\vec{s} = 0 \tag{3.110}$$

Substituted into Eq. (3.107), it can be concluded that the equation holds up and down the curve.

$$\frac{\partial}{\partial s}\left(\Pi + P + \frac{V^2}{2}\right) = 0 \tag{3.111}$$

Integrate it along the curve to get

$$\Pi + P + \frac{V^2}{2} = C(s) \tag{3.112}$$

This integral is known as the Bernoulli equation or Bernoulli integral. The integral constant $C(s)$ is constant along the integral curve S. The above equation indicates that for the steady flow of an ideal positive pressure fluid, the sum of potential energy (Π), pressure energy (P) and kinetic energy $\left(\frac{V^2}{2}\right)$ of a unit volume fluid element along a particular integral curve S is invariable under the condition of potential mass force, which is the conservation of total mechanical energy.

The sufficient and necessary conditions for Bernoulli equation to be valid are: $\vec{V} \times \vec{\omega} \cdot d\vec{s} = 0$ The following analysis meets this condition.

(1) Bernoulli's equation is satisfied along any streamline, because, in this case, the differential segment ds on the streamline is parallel to the velocity of the fluid particle, and the velocity is perpendicular to the angular velocity of rotation, and so its dot product with the vector ds is zero. On the streamline there exist $d\vec{s} = \vec{V}dt$. That is

$$d\vec{s}//\vec{V}, \quad \vec{V} \times \vec{\omega} \perp \vec{V}, \quad \vec{V} \times \vec{\omega} \cdot d\vec{s} = 0 \tag{3.113}$$

$$\vec{V} \times \vec{\omega} \cdot d\vec{s} = \begin{vmatrix} dx & dy & dz \\ u & v & w \\ \omega_x & \omega_y & \omega_z \end{vmatrix} = \begin{vmatrix} u & v & w \\ u & v & w \\ \omega_x & \omega_y & \omega_z \end{vmatrix} dt = 0 \tag{3.114}$$

(2) Along any vortex line, Bernoulli's equation is established. This is because, in this case, on the vortex line, the differential segment vector ds is parallel to the rotational angular velocity vector ω. That is, on the vortex line, let $d\vec{s} = k\vec{\omega}|ds|$ (k is the scaling coefficient), and then get

$$d\vec{s}//\vec{\omega}, \quad \vec{V} \times \vec{\omega} \perp \vec{\omega}, \quad \vec{V} \times \vec{\omega} \cdot d\vec{s} = 0 \tag{3.115}$$

$$\vec{V} \times \vec{\omega} \cdot d\vec{s} = \begin{vmatrix} dx & dy & dz \\ u & v & w \\ \omega_x & \omega_y & \omega_z \end{vmatrix} = \begin{vmatrix} \omega_x & \omega_y & \omega_z \\ u & v & w \\ \omega_x & \omega_y & \omega_z \end{vmatrix} k|ds| = 0 \tag{3.116}$$

(3) Under the following conditions, Bernoulli equation is independent of the curve taken, and the integral constant remains unchanged in the whole flow field and is equal to the same constant, i.e., $\vec{V} \times \vec{\omega} = 0$ namely

(a) Static flow field, $\vec{V} = 0$;
(b) There is no swirling flow field and potential flow, $\vec{\omega} = 0$;

(c) The streamline line coincides with the vortex line, that is, the spiral flow $\vec{V}//\vec{\omega}$.

For an incompressible fluid, the Bernoulli equation along the streamline, regardless of the mass force, is

$$\frac{p}{\rho} + \frac{V^2}{2} = C(s), \; p + \rho\frac{V^2}{2} = C(s) \tag{3.117}$$

Equation (3.117) answers the contribution of suction on the upper wing surface to the lift force, and the mechanism of the lift force generated by the flow around the airfoil. Later wind tunnel tests show that for the airfoil, the contribution of upper wing suction to lift is about 70% of the total lift. This was a great advance on the "Skipping Stone Theory" proposed by Newton in the Mathematical Principles of Natural Philosophy in 1686. Newton believed that the lift of an airfoil was the result of the impact of the lower wing of the airfoil on the airflow, and had nothing to do with the upper wing.

If the mass force is only gravity, Bernoulli's equation along the streamline is

$$gz + \frac{p}{\rho} + \frac{V^2}{2} = C(s) \tag{3.118}$$

If divide both sides by g, Bernoulli's equation can be written as the following form

$$z + \frac{p}{\gamma} + \frac{V^2}{2g} = H(s) \tag{3.119}$$

The above equation is Bernoulli's equation for the streamline integral of an ideal incompressible fluid under the mass force of gravity. The physical meaning of each item can be expressed as the sum of potential energy, pressure energy and kinetic energy of the fluid per unit weight remains unchanged. That is

(1) z represents the positional potential energy of a fluid micro-element per unit time and per unit weight. This height relative to the datum surface is also called the positional head. For example, a fluid micro-element with mass m has potential energy mgz at z, and the potential energy per unit weight is $mgz/(mg) = z$.

(2) P/γ represents the pressure potential energy of micro fluid elements per unit time and per unit weight. This potential energy is stored in the fluid through the pressure. This formula is expressed in terms of the height of the liquid column, that is, $mg(P/\gamma)/(mg) = P/\gamma$, which is called the pressure height. If the absolute vacuum point is selected as the pressure reference, the height of the liquid column is the height of the absolute pressure. If the pressure reference is atmospheric pressure, the liquid column height is the relative pressure height.

Fig. 3.25 Meanings of Bernoulli equation

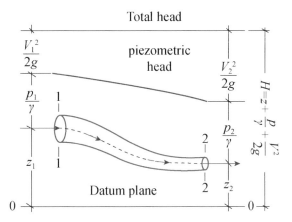

(3) $\frac{V^2}{2g}$ represents the kinetic energy per unit time and per unit weight of the fluid micro-element, namely $mg\left(\frac{V^2}{2g}\right)/(mg) = \frac{V^2}{2g}$, which is equivalent to the height that can be reached when thrown vertically with the initial velocity V, and is called kinetic energy height or velocity head.

(4) $H(s)$ indicates that the total mechanical energy of a fluid micro-element per unit time and per unit weight remains unchanged along the streamline and is called the total head. As shown in Fig. 3.25, it can be expressed graphically according to Bernoulli's equation. For example, the total energy line is the total head, pressure pipe line or pressure pipe head line $\left(= z + \frac{p}{\gamma}\right)$.

3.7.2 Application of Bernoulli Equation

Example 1: The formula of the steady outflow In 1643, Italian scientist Evangelista Torricelli (1608–1647) put forward the basic formula of steady orifice outflow through a large number of orifice outflow experiments, showing that the orifice outflow velocity is proportional to the square root of water depth h over the orifice. As shown in Fig. 3.26, the outflow velocity of the small hole in the smooth container is V, and the height between the center of the small hole and the free surface is H. Ignoring the viscosity of the fluid and assuming that the flow is steady, Bernoulli equation of the streamline at 1–2 points is established, i.e.

$$\frac{p_a}{\gamma} + h + 0 = \frac{p_a}{\gamma} + 0 + \frac{V^2}{2g} \tag{3.120}$$

Transform the above equation to obtain

$$V = \sqrt{2gh} \tag{3.121}$$

Fig. 3.26 Orifice outflow at
atmospheric pressure

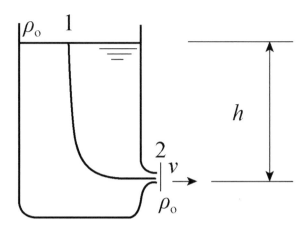

where h is the height of the liquid level and V is the velocity of water flow at the outlet. This is called the Torricelli formula. The physical meaning of Eq. (3.121) is as follows: when a fluid particle per unit weight moves from the liquid level position at point 1 to the outlet position at point 2, its gravitational potential energy becomes the corresponding kinetic energy. For a viscous fluid, there is a certain loss of mechanical energy in the process of cavitation outflow, so the actual cavitation outflow velocity will be lower than the theoretical value, which is usually written as $v = c_v\sqrt{2gh}$, where c_v is called the velocity coefficient, and the experimental results show that $c_v = 0.97$).

Example 2: Pitot tube speedometer In 1732, French hydraulic engineer Henri Pitot (1695–1771, as shown in Fig. 3.27) invented a device to measure the total pressure in the flow field, namely the Pitot tube (as shown in Fig. 3.28, also called Pitot tube). Pitot found that the height of the Pitot tube column in the river was proportional to the square of the Pitot tube inlet velocity, and that the velocity at any point in the flow could be measured by the difference between the total pressure tube and the static pressure tube measurements at the same point. Later in 1905, Ludwig Prandtl (1875–1953), a master of world fluid mechanics, developed a device to simultaneously measure the total pressure and static pressure of fluid particles with this method. This device is called a Prandtl anemometer, or Pitot speedometer (as shown in Fig. 3.29). The Pitot tube velocity measurement principle is established according to Bernoulli equation, indicating that the dynamic pressure of a fluid particle is equal to the difference between the total pressure and the static pressure of the fluid particle at the same point.

$$\rho\frac{V_0^2}{2} = p_o - p_s$$

$$V_0 = \sqrt{\frac{2(p_0 - p_s)}{\rho}} \tag{3.122}$$

where, p_o is the total pressure, p_s is the static pressure at the velocity measuring position, and V_0 is the measured velocity. In the actual velocity measurement, due to the influence of structure and loss, Eq. (3.122) can be written as

$$V_0 = \sqrt{\frac{2\xi(p_0 - p_s)}{\rho}}$$

(3.123)

Fig. 3.27 French hydraulic engineer Henri Pitot (1695–1771)—http://baike. baidu.com/view/64741.htm

Fig. 3.28 Pitot total pressure tube (Bernoulli equation application)

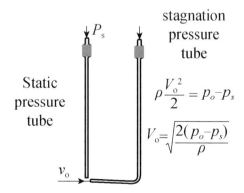

Static pressure tube

P_s

stagnation pressure tube

$$\rho\frac{V_o^2}{2} = p_o - p_s$$

$$V_o = \sqrt{\frac{2(p_o - p_s)}{\rho}}$$

v_o

Fig. 3.29 Prandtl
anemometer (Pitot tube
speedometer)

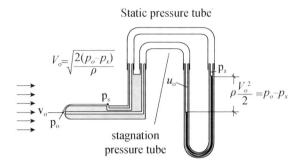

where, ξ is the correction coefficient, about 0.98–1.05.

Example 3: Venturi flowmeter In 1797, The Italian physicist Giovanni Battista
Venturi (1746–1822) found that the velocity in a small section was large and the
pressure was small (Venturi effect) through experiments on the pipeline with variable
section. He proposed to use this effect and continuous conditions to measure the
fluid flow in the pipeline of shrinking-expansion pipe, which is called Venturi pipe
(as shown in Fig. 3.30). The principle is as follows: for a horizontal pipe passing
through an ideal incompressible fluid, if a pipe segment is inserted into the pipe that
first shrinks and then expands, Bernoulli equation between the first section of the
pipe before and the second section after shrinkage is established according to the
Venturi effect, and the volume flow through the pipe can be obtained by using the
continuity condition, namely.

$$p_1 + \rho\frac{V_1^2}{2} = p_2 + \rho\frac{V_2^2}{2},$$
$$Q = V_1 A_1 = V_2 A_2 \tag{3.124}$$

$$Q = V_1 A_1 = A_1 \sqrt{\frac{2}{\rho}\frac{(p_1 - p_2)}{\frac{A_1^2}{A_2^2} - 1}} \tag{3.125}$$

where, Q is the flow rate, V_1, p_1, A_1 are the velocity, pressure and cross-sectional
area at section 1, and V_2, p_2, A_2 are the velocity, pressure and cross-sectional area
at section 2, respectively.

Example 4 At sea level, there is a straight uniform flow around the airfoil. far ahead
of the static pressure of the straight uniform flow is $p = p_\infty = 101200$ N/m². The
velocity of the flow is $v_\infty = 100$m/s. It is known that the velocities of A, B and C
on the airfoil are $V_A = 0$, $V_B = 150$ m/s and $V_C = 50$ m/s respectively. The density
of air at sea level is $\rho = 1.255$ kg/m³. Assuming that the flow around the airfoil is
irrotational, try to calculate the pressure at A, B and C (Fig. 3.31).

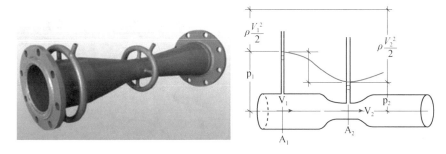

Fig. 3.30 Principle of Venturi flow tube and flowmeter

Fig. 3.31 Flow around airfoil

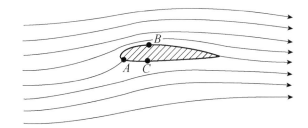

Solution:

Because the flow is irrotational, the Bernoulli constant is the same in the whole flow field. Bernoulli's equation without considering the mass force is

$$p + \rho \frac{V^2}{2} = p_\infty + \rho \frac{V_\infty^2}{2} = C \qquad (3.126)$$

Based on the conditions in the far front, the constant C is zero

$$C = p_\infty + \rho \frac{V_\infty^2}{2} = 101,200 + \frac{1.225}{2} \times (100)^2 = 107,325 \text{ N/m}^2$$

Using Eq. (3.124), the pressure at each point can be calculated as

$$p_A = C - \frac{\rho}{2} V_A^2 = 107,325 \text{ N/m}^2;$$

$$p_B = C - \frac{\rho}{2} V_B^2 = 107325 - 0.6125 \times 22,500 = 93,825 \text{ N/m}^2;$$

$$p_C = C - \frac{\rho}{2} V_C^2 = 107,325 - 1531 = 105,794 \text{ N/m}^2.$$

The calculation results show that: A is the stagnation point $p_A > p_\infty$, p_A is the maximum; Point C is the point on the lower wing surface, the velocity is less than

Fig. 3.32 Rotating container
flow

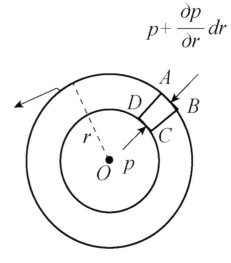

$$p + \frac{\partial p}{\partial r} dr$$

the incoming flow velocity, $p_C > p_\infty$; Point B is the point on the upper wing, and the velocity is greater than the incoming flow velocity, $p_B < p_\infty$.

Example 5: Container rotation In a rotating container, it is assumed that the flow field only has circumcircular velocity, which is proportional to the radius r. That is $V_\theta = kr$, as shown in Fig. 3.32. Verify that Bernoulli's constant C is a function of r.

Proof: Since the flow field has rotational motion, Bernoulli equation along the streamline is written as

$$C = p + \frac{\rho}{2} V_\theta^2 \tag{3.127}$$

Take the derivative with respect to the radius r

$$\frac{\partial C}{\partial r} = \frac{\partial p}{\partial r} + \rho V_\theta \frac{\partial V_\theta}{\partial r} \tag{3.128}$$

Radial pressure difference must balance the centrifugal force of micro-element $ABCD$, so there is

$$\left(p + \frac{\partial p}{\partial r} dr\right)(r + dr) d\theta - \frac{1}{2}\left(p + p + \frac{\partial p}{\partial r} dr\right)(r + dr - r) d\theta - pr d\theta$$
$$= \frac{V_\theta^2}{r} \rho \left(\frac{r + r + dr}{2}\right) d\theta dr \tag{3.129}$$

The second term on the left of the above equation is the projection in the r direction of the pressure on the planes AD and BC. Omit the higher-order to obtain

$$\frac{\partial p}{\partial r} = \rho \frac{V_\theta^2}{r} \tag{3.130}$$

Substituting the above equation into Eq. (3.128), and substituting $V_\theta = kr$. It can be obtained

$$\frac{\partial C}{\partial r} = 2\rho k^2 r, \quad C = \rho k^2 r^2 \tag{3.131}$$

If the velocity field is zero

$$V_\theta = \frac{k}{r} \tag{3.132}$$

By substituting it into Eq. (3.128), we can get

$$\frac{\partial C}{\partial r} = \frac{\partial p}{\partial r} + \rho v_\theta \frac{\partial v_\theta}{\partial r} = \rho \frac{1}{r}\left(\frac{k}{r}\right)^2 + \rho \frac{k}{r}\left(-\frac{k}{r^2}\right) = 0 \tag{3.133}$$

It shows that the constants in Bernoulli's equation are independent of r, and are the same for the entire flow field. It can be proved that the flow field is eddy free and the integral constant of the energy equation is constant, namely

$$v_\theta = \frac{K}{r} \quad u = -K\frac{y}{x^2 + y^2} \quad v = K\frac{x}{x^2 + y^2}$$
$$\omega_z = \frac{1}{2}\left(\frac{\partial v}{\partial x} - \frac{\partial u}{\partial y}\right) = \frac{K}{2}\left(\frac{y^2 - x^2}{(x^2 + y^2)^2} - \frac{y^2 - x^2}{(x^2 + y^2)^2}\right) = 0 \tag{3.134}$$

For a concentrated vortex in the flow field, the induced flow field outside the vortex core and the vortex core are divided. In the vortex core, the fluid particle rotates around the vortex axis like a rigid body, and its circumferential velocity is proportional to r. In the induced flow field outside the vortex core, there is no vortex movement, and its circumferential velocity is inversely proportional to r.

Example 6 As shown in Fig. 3.33, the working principle of the jet pump is that the vacuum is generated by the high-speed jet from the nozzle at section 1–1 of the upstream water tank pipe of the studio. The water in the external downstream pool is sucked into the tank. Then it flows out through the jet into the outlet pipe. We know: $H = 2$ m, $h = 1$ m, $d_3 = 15$ cm, $D_2 = 10$ cm, $D_1 = 10$ cm, $D_0 = 15$ cm, $z = 2$ m, water density $\rho = 1000$ kg/m³. The static pressure in the tank stays the same and ignores the viscosity loss:

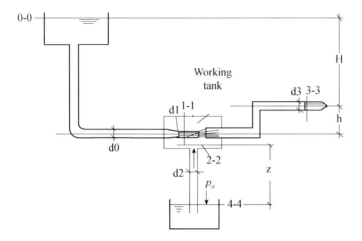

Fig. 3.33 Working principle of jet pump

(1)   The flow rate Q_3 in the outlet pipe;
(2)   The flow Q_1 through the jet pump;
(3)   The flow rate Q_2 through the suction pipe;
(4)   The vacuum degree h_v in the tank;
(5)   The maximum vacuum degree in the tank h_{vm} ($Q_2 = 0$).

Solution:

When the jet pump works steadily, the water body is mixed in the tank. With the help of jet flow V_1, the water in the ejector pool enters the outlet pipe and flows out. Now, the energy equation between 0–0 and the 3–3 section of the outlet pipe is established

$$H = \frac{V_3^2}{2g}, \quad V_3 = \sqrt{2gH} = \sqrt{2 \times 9.8 \times 2} = 6.261 \text{ m/s}$$

By the continuity equation

$$A_1 V_1 + A_2 V_2 = A_3 V_3, \quad V_1 = \frac{A_3}{A_1} V_3 - \frac{A_2}{A_1} V_2 = 2.25 V_3 - V_2$$

The energy equation between the 0–0 cross-section and the 1–1 cross-section is

$$H + h = \frac{p_1}{\gamma} + \frac{V_1^2}{2g}$$

The energy equation between sections 4–4 and 2–2 of the downstream pool is

$$\frac{p_a}{\gamma} = \frac{p_a + p_1}{\gamma} + z + \frac{V_2^2}{2g}$$

By using the above three formulas, we can obtain

$$V_2 = \frac{2.25^2 V_3^2 - 2g(H + h + z)}{4.5 V_3} = 3.565 \, \text{m/s}$$

The flow through the suction pipe is

$$Q_2 = V_2 A_2 = 3.565 \times 0.7854 \times 0.1^2 = 0.0280 \, \text{m}^3/\text{s}^3$$

The flow through the jet pump is

$$Q_1 = V_1 A_1 = 10.522 \times 0.7854 \times 0.1^2 = 0.0826 \, \text{m}^3/\text{s}^3$$

The flow through the outlet pipe is

$$Q_3 = V_3 A_3 = 6.261 \times 0.7854 \times 0.15^2 = 0.111 \, \text{m}^3/\text{s}^3$$

The vacuum in the studio is

$$h_V = -\frac{p_1}{\gamma} = z + \frac{V_2^2}{2g} = 2 + \frac{3.565^2}{2g} = 2.648 \, \text{m}$$

The maximum vacuum in the tank is

$$h_{Vm} = -\frac{p_1}{\gamma} = -(H + h) + \frac{V_1^2}{2g} = -3 + \frac{2.25^2 \times 6.261^2}{2g} = 7.125 \, \text{m}$$

3.8 Integral Equation of Fluid Motion

3.8.1 Basic Concepts of Control Volume and System

In fluid dynamics, there are always three aspects that need to be solved, namely.

(1) Fluid kinematics;
(2) Characteristics of the forces acting on the fluid;
(3) The general laws governing fluid motion (mass conservation, momentum conservation, energy conservation, etc.).

The task of fluid mechanics is to describe the general law of the motion of substances, and apply it to the physical phenomenon of fluid motion, so as to obtain the relations between the physical quantities of fluid motion. These relations are the basic equations of fluid dynamics. If the relationship is given in integral form, it is called a system of fluid dynamics integral equations. If given in differential form, it is called a system of differential equations. The previous section established a system of differential equations for fluid motion. In this section, fluid dynamics integral equations are derived and established, including: continuity equation; momentum equation; moment of momentum equation; Energy equation.

1. System

A system is any aggregate containing a determinedly invariant substance. In fluid mechanics, a system is a group consisting of any given fluid particle. The basic characteristics of the system are:

(1) The boundary of the system moves with the fluid;
(2) There is no mass exchange at the boundary of the system;
(3) There exists external surface force on the boundary of the system;
(4) There is an exchange of energy at the boundary of the system.

For example, for a system, Newton's second law $F = ma$ says: F is the net force of all external forces acting on the system. A is the acceleration of the center of mass of the system. The fluid system corresponds to the Lagrangian point of view. That is, to determine the aggregate of fluid particles as the research object, to study the relationship between the physical quantities of the system. Such equations are called the Lagrangian integral equations.

2. Control Volume

A control volume is any volume that is fixed and invariable relative to the coordinate system. The boundary of the control volume is called the control surface. In a fluid motion, the control volume is fixed, but is passed by the fluid. That is, the fluid particles occupy the control volume change with time. It should be noted that the control volume can also move and deform, but this is not discussed in this book. The characteristics of the control volume are:

(1) The boundary of the control volume is fixed relative to the coordinate system;
(2) Mass exchange can occur on the control surface, that is, the fluid can flow into and out of the control surface;
(3) the control surface is subjected to external forces acting on the fluid in the control volume;
(4) The fluid in the control volume will be affected by the mass force;
(5) There is energy exchange on the control plane.

For example, for the control volume, Newton's second law $F = ma$ says: F is the net force of all external forces acting on the control volume and the fluid at its boundary. The control volume corresponds to Euler's point of view. That is, the fluid particle system through the control volume is taken as the research object to study the relationship between the physical quantities of the fluid in the control volume. Such equations are called Euler integral equations.

3.8.2 *Lagrangian Integral Equations*

In the flow field, take any fluid system as the object of investigation, the volume of the system is τ_0, and the boundary surface area of the system is S_0. Lagrangian integral equations are established according to the general law of motion.

(1) Continuity equation (mass conservation)

$$\frac{dM}{dt} = \frac{d}{dt} \iiint\limits_{\tau_0} \rho \, d\tau_0 = 0 \tag{3.135}$$

Its physical meaning is that the mass of a fluid system does not change with time in the absence of sources and sinks.

(2) Momentum equation

$$\frac{d\vec{K}}{dt} = \frac{d}{dt} \iiint\limits_{\tau_0} \rho \vec{V} \, d\tau_0 = \iiint\limits_{\tau_0} \rho \vec{f} \, d\tau_0 + \oiint\limits_{S_0} \vec{p}_n \, dS_0 \tag{3.136}$$

The physical implication is that the rate of change of momentum in a fluid system with respect to time is equal to the vector sum of all external forces acting on the system.

(3) Moment of momentum equation

$$\frac{d\vec{M}_r}{dt} = \frac{d}{dt} \iiint\limits_{\tau_0} \rho \vec{r} \times \vec{V} \, d\tau_0 = \iiint\limits_{\tau_0} \rho (\vec{r} \times \vec{f}) d\tau_0 + \oiint\limits_{S_0} \vec{r} \times \vec{p}_n \, dS_0 \tag{3.137}$$

The physical meaning is that the rate of change of momentum of a fluid system at a point is equal to the sum of the moments of all external forces acting on the system at the same point.

(4) Energy equation

$$Q + W = \frac{dE}{dt} = \frac{d}{dt} \iiint\limits_{\tau_0} \rho \left(e + \frac{V^2}{2} \right) d\tau_0 \tag{3.138}$$

where e is the internal energy of unit mass fluid, and the total energy of unit mass fluid is $E = e + \frac{V^2}{2}$. The physical meaning of Eq. (3.138) is that the sum of the heat transmitted Q from the outside to the fluid system in unit time and the work W done by the force acting on the fluid system in unit time is equal to the rate of change of the total energy E of the system with time. The heat transferred from the outside to the system includes heat conduction Q_h and heat radiation Q_r.

Heat generated by head conduction refers to the total heat conduction transmitted from the system surface per unit time Q_h

$$Q_h = \oiint_{S_0} q_\lambda dS_0 = \oiint_{S_0} k\frac{\partial T}{\partial n}dS_0 = \oiint_{S_0} k\nabla T \cdot \vec{n}\,dS_0 \tag{3.139}$$

where k is the heat conduction coefficient and T is the temperature.

The heat radiated by heat per unit time can be expressed as Q_r

$$Q_r = \iiint_{\tau_0} \rho q_r d\tau_0 \tag{3.140}$$

Among them, the q_r is the heat transmitted by radiation per unit mass.

The forces acting on a fluid system include mass forces and surface forces. The work W done to the system by force per unit time includes mass force work W_b and surface force work W_s. For an ideal fluid, the surface force is pressure work W_s, namely

$$W_s = \oiint_{S_0} \vec{p}_n \cdot \vec{V} dS_0 = -\oiint_{S_0} p\vec{n}\cdot\vec{V} dS_0 \tag{3.141}$$

The mass force work is W_b

$$W_b = \iiint_{\tau_0} \rho\vec{f} \cdot \vec{V} d\tau_0 \tag{3.142}$$

In unit time, the power by mass force and surface force is

$$W = \iiint_{\tau_0} \rho\vec{f} \cdot \vec{V} d\tau_0 + \oiint_{S_0} \vec{p}_n \cdot \vec{V} dS_0 \tag{3.143}$$

The resulting energy equation is

$$\oiint_{A_0} q_\lambda dS_0 + \iiint_{\tau_0} \rho q_R d\tau_0 + \iiint_{\tau_0} \rho\vec{f} \cdot \vec{V} d\tau_0$$
$$+ \oiint_{S_0} \vec{p}_n \cdot \vec{V} dS_0 = \frac{d}{dt} \iiint_{\tau_0} \rho\left(e + \frac{V^2}{2}\right) d\tau_0 \tag{3.144}$$

So far, the Lagrangian integral equations based on fluid system are given. In practical problems, the motion characteristics of the fluid in a given region are often concerned, rather than the overall picture of the fluid system. Therefore, it is necessary to transform Lagrangian integral equations into Eulerian integral equations. The main problem is to convert the integral of the system into the integral of the control volume.

3.8.3 Reynolds Transport Equation

For any function σ, the integral over the system is

$$N = \iiint_{\tau_0} \rho\sigma\,d\tau_0 \tag{3.145}$$

With different values of σ, N represents the integral of different physical quantities. For example:

When $= \sigma 1$, $N = M$ is the mass of the system.
When $\sigma = \vec{V}$, $N = K$ is the momentum of the system;
When $\sigma = \vec{r} \times \vec{V}$, $N = M_r$ represents the moment of momentum of the system;
When $\sigma = e + \frac{V^2}{2}$, $N = E$ is the energy of the system.

In order to distinguish between the system and the control volume, the subscript 0 in the volume and area is for the system, and the variables without subscript are for the control volume. Let the fluid system and the control weight at time t, and then follow the fluid system. The volume and position of the system change at time $t + \Delta t$. Suppose at time t, the volume of the system is $\tau_0(t)$. At time $t + \Delta t$, the system moves to the next position, the volume of the system becomes $\tau_0(t + \Delta t)$. τ_{01} represents the common part of the two, τ_{03} equals $\tau_0(t)$ subtracting τ_{01}, τ_{02} equals $\tau_0(t + \Delta t)$ subtracting τ_{01}. S_{01} is the boundary between τ_{01} and τ_{02} (for the control volume, it is the outflow surface). S_{02} is the boundary not in common with τ_{03} and τ_{01} (for the control volume, it is the inflow surface). Geometrically, as shown in Fig. 3.34:

$$\tau_0(t) = \tau_{01} + \tau_{03}, \quad \tau_0(t + \Delta t) = \tau_{01} + \tau_{02} \tag{3.146}$$

It can be obtained $d\tau_0 = \left(\vec{V} \cdot \vec{n}\right)dS\Delta t$. Then the integral $\iiint_{\tau_{02}} \rho\sigma\,d\tau_0$ can be transformed as

$$\iiint_{\tau_{02}} \rho\sigma\,d\tau_0 = \iint_{S_{01}} \rho\sigma\left(\vec{V} \cdot \vec{n}\right)dS\Delta t \tag{3.147}$$

Fig. 3.34 Relationship between system and control volume

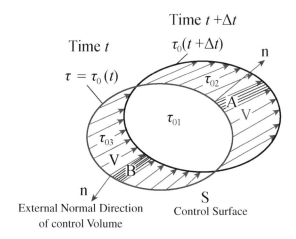

Similarly, we can obtain $d\tau_0 = -\left(\vec{V} \cdot \vec{n}\right) dS \Delta t$. Then the integral $\iiint\limits_{\tau_{03}} \rho\sigma d\tau_0$ can be transformed as

$$\iiint\limits_{\tau_{03}} \rho\sigma d\tau_0 = -\iint\limits_{S_{02}} \rho\sigma \left(\vec{V} \cdot \vec{n}\right) dS \,\Delta t \qquad (3.148)$$

The increase of function over time Δt can be stated as:

$$\Delta N = N(t + \Delta t) - N(t) = \iiint\limits_{\tau_0(t+\Delta t)} \rho\sigma d\tau_0 - \iiint\limits_{\tau_0(t)} \rho\sigma d\tau_0$$

$$= \iiint\limits_{\tau_{01}+\tau_{02}} \rho\sigma(t + \Delta t) d\tau_0 - \iiint\limits_{\tau_{01}+\tau_{03}} \rho\sigma(t) d\tau_0$$

$$= \iiint\limits_{\tau_{01}} \rho[\sigma(t + \Delta t) - \sigma(t)] d\tau_0 + \iiint\limits_{\tau_{02}} \rho\sigma(t + \Delta t) d\tau_0 - \iiint\limits_{\tau_{03}} \rho\sigma(t) d\tau_0$$

$$(3.149)$$

Now analyze the physical meaning of the above terms.

(1) ΔN is the increment of system physical quantity;

(2) $\iiint\limits_{\tau_{01}} \rho[\sigma(t + \Delta t) - \sigma(t)] d\tau_0$ is the increment of physical quantity with constant volume over time;

(3) $\iiint\limits_{\tau_{02}} \rho\sigma(t + \Delta t) d\tau_0 - \iiint\limits_{\tau_{03}} \rho\sigma(t) d\tau_0$ is the physical quantity increment caused by the volume change.

Divide both sides of Eq. (3.149) with Δt and take the limit:

$$\lim_{\Delta t \to 0} \frac{N(t + \Delta t) - N(t)}{\Delta t} = \frac{dN}{dt} \tag{3.150}$$

$$\lim_{\Delta t \to 0} \frac{1}{\Delta t} \iiint_{\tau_{01}} \rho[\sigma(t + \Delta t) - \sigma(t)]d\tau_0 = \iiint_{\tau} \frac{\partial \rho \sigma}{\partial t} d\tau \tag{3.151}$$

$$\lim_{\Delta t \to 0} \frac{1}{\Delta t} \iiint_{\tau_{02}} \rho\sigma(t + \Delta t)d\tau_0 = \iint_{S_{01}} \rho\sigma\left(\overrightarrow{V} \cdot \overrightarrow{n}\right)dS \tag{3.152}$$

$$\lim_{\Delta t \to 0} \frac{1}{\Delta t} \iiint_{\tau_{03}} \rho\sigma(t)d\tau_0 = -\iint_{S_{02}} \rho\sigma\left(\overrightarrow{V} \cdot \overrightarrow{n}\right)dS \tag{3.153}$$

As shown in Fig. 3.34, Eq. (3.152) is subtracted from Eq. (3.153), and the net outflow through the control plane S (the boundary plane of the control volume) is

$$\iint_{S_{01}} \rho\sigma\left(\overrightarrow{V} \cdot \overrightarrow{n}\right)dS + \iint_{S_{02}} \rho\sigma\left(\overrightarrow{V} \cdot \overrightarrow{n}\right)dS = \oiint_{S} \rho\sigma\left(\overrightarrow{V} \cdot \overrightarrow{n}\right)dS \tag{3.154}$$

Finally, Reynolds (British scientist Reynolds, 1842–1912) transport equation is obtained

$$\frac{dN}{dt} = \frac{d}{dt} \iiint_{\tau_0} \rho\sigma d\tau_0 = \iiint_{\tau} \frac{\partial \rho \sigma}{\partial t} d\tau + \oiint_{S} \rho\sigma\left(\overrightarrow{V} \cdot \overrightarrow{n}\right)dS$$

$$= \frac{\partial}{\partial t} \iiint_{\tau} \rho\sigma d\tau + \oiint_{S} \rho\sigma\left(\overrightarrow{V} \cdot \overrightarrow{n}\right)dS \tag{3.155}$$

Formula (3.155) is the expression that the derivative of the fluid system is transformed into the control volume, namely Reynold's transport equation. The physical meanings of the items on the right side of the equation are: $\frac{\partial}{\partial t} \iiint_{\tau} \rho\sigma d\tau$ is the rate of change of physical quantities in the control volume with time, and represents the increment caused by the unsteadiness of the flow field $\oiint_{S} \rho\sigma\left(\overrightarrow{V} \cdot \overrightarrow{n}\right)dS$ is the net increment of physical quantity of the control volume output through the control surface per unit time, which represents the net increment caused by the inhomogeneity of the flow field.

Using the Gaussian integral and the continuity equation, Eq. (3.155) can also be derived directly. The matter derivative of the control volume of Eq. (3.145) is obtained

$$\frac{dN}{dt} = \frac{d}{dt} \iiint_{\tau_0} \rho\sigma \delta\tau_0 = \iiint_{\tau} \frac{d\sigma}{dt} \rho\delta\tau + \iiint_{\tau_0} \sigma \frac{d(\rho\delta\tau_0)}{dt} \tag{3.156}$$

From the continuity equation

$$\frac{d(\rho\delta\tau_0)}{dt} = \frac{d(\delta m)}{dt} = 0 \tag{3.157}$$

The above equation indicates that the fluid mass remains unchanged when moving with the fluid micro-element. Substitute it into Eq. (3.156) to get

$$\frac{dN}{dt} = \frac{d}{dt}\iiint_{\tau_0}\rho\sigma\,\delta\tau_0 = \iiint_{\tau}\frac{d\sigma}{dt}\rho\delta\tau \tag{3.158}$$

Due to the

$$\frac{d\sigma}{dt} = \frac{\partial\sigma}{\partial t} + \vec{V}\cdot\nabla\sigma \tag{3.159}$$

Multiply both sides of this equation by ρ, and you get

$$\rho\frac{d\sigma}{dt} = \rho\frac{\partial\sigma}{\partial t} + \rho\vec{V}\cdot\nabla\sigma \tag{3.160}$$

It's given by the continuity equation

$$\sigma\frac{d\rho}{dt} = \sigma\frac{\partial\rho}{\partial t} + \sigma\nabla\cdot(\rho\vec{V}) = 0 \tag{3.161}$$

By summing Eqs. (3.160) and (3.161), we can get

$$\rho\frac{d\sigma}{dt} = \rho\frac{\partial(\rho\sigma)}{\partial t} + \nabla\cdot(\rho\sigma\vec{V}) \tag{3.162}$$

Substituting Eq. (3.162) into Eq. (3.158):

$$\begin{aligned}
\frac{dN}{dt} &= \iiint_{\tau}\left[\frac{\partial(\rho\sigma)}{\partial t} + \nabla\cdot(\rho\sigma\vec{V})\right]\delta\tau \\
&= \frac{\partial}{\partial t}\iiint_{\tau}\rho\sigma\delta\tau + \iiint_{\tau}\nabla\cdot(\rho\sigma\vec{V})\delta\tau \\
&= \frac{\partial}{\partial t}\iiint_{\tau}\rho\sigma\delta\tau + \oiint_{S}\rho\sigma(\vec{V}\cdot\vec{n})\delta S
\end{aligned} \tag{3.163}$$

In summary, the physical meaning of the Reynolds transport equation can be expressed as follows: the derivative with the body of the physical quantity of the fluid system is equal to the time change rate of the physical quantity of the fluid system in the control volume and the net outflow of the physical quantity of the fluid through the control surface in unit time.

3.8.4 Eulerian Integral Equations

Euler's integral system is integral equations established for the control volume, based on Reynold's transport equation, which is easy to obtain.

(1) Continuity equation (mass conservation)

Let $= \sigma 1$. it can be obtained from Eq. (3.155)

$$\frac{dM}{dt} = \frac{d}{dt} \iiint_{\tau_0} \rho d\tau = \frac{\partial}{\partial t} \iiint_{\tau} \rho d\tau + \oiint_{S} \rho\left(\overrightarrow{V} \cdot \overrightarrow{n}\right) dS = 0 \qquad (3.164)$$

Without source and sink in the control volume, the net mass outflow from the control volume per unit time is equal to the reduction of the control volume mass.

(2) The momentum equation

Let $\sigma = \overrightarrow{V}$, it can be obtained from Eq. (3.155)

$$\begin{aligned}
\frac{d\overrightarrow{K}}{dt} &= \frac{d}{dt} \iiint_{\tau_0} \rho \overrightarrow{V} d\tau_0 \\
&= \frac{\partial}{\partial t} \iiint_{\tau} \rho\overrightarrow{V} d\tau + \oiint_{S} \rho\overrightarrow{V}\left(\overrightarrow{V} \cdot \overrightarrow{n}\right) dS \\
&= \iiint_{\tau} \rho \overrightarrow{f} d\tau + \oiint_{S} \overrightarrow{p}_n dS \qquad (3.165)
\end{aligned}$$

In unit time, the increment of momentum in the control volume plus the net momentum flowing through the control surface is equal to the vector sum of all external forces acting on the control volume.

(3) Moment of momentum equation

Let $\sigma = \overrightarrow{r} \times \overrightarrow{V}$, it can be obtained from Eq. (3.155)

$$\begin{aligned}
\frac{d\overrightarrow{M}_r}{dt} &= \frac{d}{dt} \iiint_{\tau_0} \rho \overrightarrow{r} \times \overrightarrow{V} d\tau_0 = \frac{\partial}{\partial t} \iiint_{\tau} \rho \overrightarrow{r} \times \overrightarrow{V} d\tau \\
&+ \oiint_{S} \rho \overrightarrow{r} \times \overrightarrow{V}\left(\overrightarrow{V} \cdot \overrightarrow{n}\right) dS
\end{aligned}$$

$$= \iiint_{\tau} \rho(\vec{r} \times \vec{f}) \mathrm{d}\tau + \oiint_{S} \vec{r} \times \vec{p_n} \mathrm{d}S \qquad (3.166)$$

In unit time, the increment of momentum moment in the control volume plus the net momentum moment flowing through the control surface is equal to the sum of all external torques acting on the control volume.

(4) Energy equation

Let $\sigma = e + \frac{V^2}{2}$, it can be obtained from Eq. (3.155)

$$\frac{\mathrm{d}}{\mathrm{d}t} \iiint_{\tau_0} \rho \left(e + \frac{V^2}{2} \right) \mathrm{d}\tau_0$$

$$= \frac{\partial}{\partial t} \iiint_{\tau} \rho \left(e + \frac{V^2}{2} \right) \mathrm{d}\tau + \oiint_{S} \rho \left(e + \frac{V^2}{2} \right) \left(\vec{V} \cdot \vec{n} \right) \mathrm{d}S$$

$$= \oiint_{S} q_\lambda \mathrm{d}S + \iiint_{\tau} \rho q_R \mathrm{d}\tau + \iiint_{\tau} \rho \vec{f} \cdot \vec{V} \mathrm{d}\tau + \oiint_{S} \vec{p_n} \cdot \vec{V} \mathrm{d}S \qquad (3.167)$$

The increment of total energy in the control volume plus the net total energy flowing through the control surface per unit time is equal to the heat transferred to the control volume fluid plus the work done by the forces on the control volume fluid.

For the ideal and adiabatic steady flow with potential mass forces, the energy equation can be simplified as follows. For an adiabatic flow, where there is no heat conduction or radiation, there is

$$\oiint_{S} q_\lambda \mathrm{d}S + \iiint_{\tau} \rho q_R \mathrm{d}\tau = 0 \qquad (3.168)$$

In the case that the mass force has potential $\vec{f} = -\nabla \Pi$, the continuity equation of steady flow $\nabla \cdot (\rho \vec{V}) = 0$ is applied

$$\iiint_{\tau} \rho \vec{f} \cdot \vec{V} \mathrm{d}\tau = - \iiint_{\tau} \rho \nabla \Pi \cdot \vec{V} \mathrm{d}\tau = - \iiint_{\tau} \nabla \cdot (\Pi \rho \vec{V}) \mathrm{d}\tau$$

$$+ \iiint_{\tau} \Pi \nabla \cdot (\rho \vec{V}) \mathrm{d}\tau$$

$$= - \iiint_{\tau} \nabla \cdot (\Pi \rho \vec{V}) \mathrm{d}\tau = - \oiint_{S} \vec{n} \cdot (\Pi \rho \vec{V}) \mathrm{d}S \qquad (3.169)$$

For steady flow, it can be obtained from the continuity equation

$$\oint\!\!\!\!\!\oint_S \left(\vec{n} \cdot \rho \vec{V}\right) dS = \iiint_\tau \nabla \cdot (\rho \vec{V}) d\tau = 0 \tag{3.170}$$

For an ideal fluid, we have

$$\oint\!\!\!\!\!\oint_S \vec{p_n} \cdot \vec{V} dS = -\oint\!\!\!\!\!\oint_S \vec{n} \cdot \vec{V} p dS, \quad \vec{p_n} = \vec{n} p \tag{3.171}$$

For steady flow, there is

$$\frac{\partial}{\partial t} \iiint_\tau \rho \left(e + \frac{V^2}{2}\right) d\tau = 0 \tag{3.172}$$

By substituting Eqs. (3.168)–(3.172) into the energy Eq. (3.167), we can get

$$\oint\!\!\!\!\!\oint_S \rho \left(e + \frac{V^2}{2} + \frac{p}{\rho} + \Pi\right)\left(\vec{V} \cdot \vec{n}\right) dS = 0 \tag{3.173}$$

For the adiabatic steady flow of an incompressible fluid, there is

$$\oint\!\!\!\!\!\oint_S \rho e(\vec{n} \cdot \vec{V}) dS = 0 \tag{3.174}$$

$$\oint\!\!\!\!\!\oint_S \rho \left(\frac{V^2}{2} + \frac{p}{\rho} + \Pi\right)\left(\vec{V} \cdot \vec{n}\right) dS = 0 \tag{3.175}$$

Considering the arbitrariness of integral volume, Eq. (3.175) can derive

$$\frac{V^2}{2} + \frac{p}{\rho} + \Pi = C \tag{3.176}$$

This formula is known as Bernoulli's equation.

3.8.5 Reynolds Transport Equation of the Control Volume with Arbitrary Movement Relative to the Fixed Coordinate System

It is assumed that the control volume is moving relative to the coordinate system, and the moving speed of the boundary surface of the control volume relative to the coordinate system is $\vec{V}_b$, and the static control volume is τ. The boundary surface of

Fig. 3.35 Relative positions
of fluid system, motion
control volume and static
control volume at $t + \Delta t$

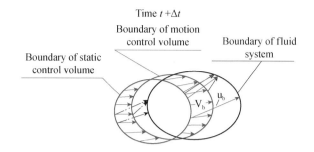

the static control volume is S. The motion control volume is τ_V. The boundary surface
of the motion control volume is S_0. The fluid system volume is τ_0. The boundary
surface of the fluid system is S_0. As shown in Fig. 3.35, at time t, the fluid system,
the motion control volume, and the static control volume are coincided, namely τ_0
$= \tau_V = \tau$ and $S_0 = S_V = S$, the boundary surface S_0 of the fluid system is $\vec{u}_b$. On
the motion control boundary surface S_V, the velocity is $\vec{V}_b$. At the time $t + \Delta t$, the
relative positions of the fluid system, the motion control volume and the static control
volume are taken, as shown in Fig. 3.35. If the motion control is regarded as a fluid
system, then Reynold's transport Eq. (3.155) is used. For any function $f(x, y, z, t)$,
there is

$$\frac{\mathrm{d}}{\mathrm{d}t} \int_{\tau_V} f(x, y, z, t)\delta\tau = \int_{\tau} \frac{\partial f(x, y, z, t)}{\partial t}\delta\tau + \oint_S \vec{n} \cdot (\vec{V}_b f)\delta A \qquad (3.177)$$

If the fluid system is applied to the Reynolds transport Eq. (3.155), there is

$$\frac{\mathrm{d}}{\mathrm{d}t} \int_{\tau_0} f(x, y, z, t)\delta\tau_0 = \int_{\tau} \frac{\partial f(x, y, z, t)}{\partial t}\delta\tau + \oint_S \vec{n} \cdot (\vec{u}_b f)\delta A \qquad (3.178)$$

Formula (3.178) is used to replace the first term on the right of formula (3.177),
then formula (3.177) becomes

$$\frac{\mathrm{d}}{\mathrm{d}t} \int_{\tau_V} f(x, y, z, t)\delta\tau = \frac{\mathrm{d}}{\mathrm{d}t} \int_{\tau_0} f(x, y, z, t)\delta\tau_0$$

$$- \oint_S \vec{n} \cdot (\vec{u}_b f)\delta A + \oint_S \vec{n} \cdot (\vec{V}_b f)\delta A$$

Rearrange the above equation to obtain:

$$\frac{\mathrm{d}}{\mathrm{d}t} \int_{\tau_V} f(x, y, z, t)\delta\tau = \frac{\mathrm{d}}{\mathrm{d}t} \int_{\tau_0} f(x, y, z, t)\delta\tau_0 - \oint_S \vec{n} \cdot (\vec{u}_b - \vec{V}_b) f\delta A \qquad (3.179)$$

According to Eq. (3.179), the derivative with respect to the motion control volume is equal to the derivative with respect to the fluid system minus the increment of the net outflow motion control volume.

By using Eq. (3.179), it can also be written as

$$\frac{d}{dt} \int_{\tau_0} f(x, y, z, t)\delta\tau_0 = \frac{d}{dt} \int_{\tau_V} f(x, y, z, t)\delta\tau + \oint_S \vec{n} \cdot (\vec{u}_b - \vec{V}_b)f\delta A \quad (3.180)$$

Equation (3.180) indicates that the derivative with respect to the fluid system is equal to the derivative with respect to the motion control volume fraction plus the increment of the net outflow of the motion control volume.

If $f(x, y, z, t) = \rho\vec{u}$, it can be written as

$$\frac{d}{dt} \int_{\tau_0} \rho\vec{u}\delta\tau_0 = \frac{d}{dt} \int_{\tau_V} \rho\vec{u}\delta\tau + \oint_{S_c} \vec{n} \cdot (\vec{u}_b - \vec{V}_b)\rho\vec{u}_b\delta A \quad (3.181)$$

Equation (3.181) indicates that the derivative of the momentum integral with the body of the fluid system is equal to the momentum integral with the body of the fluid system through the motion control volume and the momentum increment of the net outflow of the motion control volume.

3.9 Vortex Motion and Its Characteristics

3.9.1 Vortex Motion

There are two kinds of fluid motion: non-rotational motion and rotational motion. Non-rotational motion refers to the motion in which the angular velocity of the fluid group is zero, while rotational motion refers to the motion in which the angular velocity of the fluid group is not zero. The fluid rotating motion discussed in this section is a kind of motion prevalent in nature. It is found that once the vortex is formed, it will play a controlling role in the flow and the force on the object, such as the tornado in nature, the eddy around the island, and the tail vortex of the aircraft, as shown in Fig. 3.36. The German aerodynamics Kuchemann (1911–1976, Prandtl's student, shown in Fig. 3.37) once said that vortices are tendons of fluid motion. Chinese fluid mechanics Lu Shijia (1911–1986, Prandtl's student, as shown in Fig. 3.38) also pointed out that: the fluid cannot be rubbed, rubbed out of a vortex. Thus, vortex is the essence of fluid movement.

(a) tornado (b) whirlpool around ocean reef flow (c) aircraft tail vortex

Fig. 3.36 Some typical vortices in nature

Fig. 3.37 German
aerodynamicist Kuchemann
(1911–1976)

Fig. 3.38 Chinese
hydrodynamicist Lu Shijia
(1911–1986)

Vortex motion can be expressed in a variety of ways. This section mainly discusses vortex, which is defined as a group of mass motions rotating about a common central axis in the flow field, as shown in Fig. 3.39. The structure of concentrated vortex mainly includes cylindrical vortex, spiral vortex and disk vortex (ring), as shown in Fig. 3.40.

In the second half of the nineteenth century, an ideal fluid model was used to establish the theory of fluid vortex motion. In 1858, based on the decomposition theorem of an ideal incompressible fluid, Germany Fluid mechanics Helmholtz (1821–1894) put forward the Helmholtz vortex of the three laws of motion, namely, the laws of the constant vortex density along the vortex tube, vortex tube maintain law and vortex strength conservation law. He also established the ideal fluid vortex movement theory.

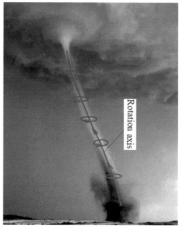

Fig. 3.39 Vortex

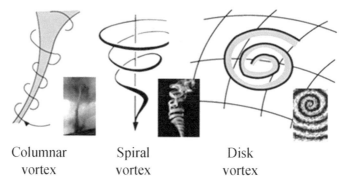

| Columnar | Spiral | Disk |
| vortex | vortex | vortex |

Fig. 3.40 Structure of concentrated vortices

3.9.2 *Vorticity, Vorticity Flux and Circulation*

(1) Vorticity

Vorticity equals twice rotation angular velocity. For example, in a plane problem (for the *xy* plane) the vorticity is $2\omega_z$. Vorticity is a purely kinematic concept. In the three-dimensional flow field, there are three angular velocity components of any fluid micro-element, namely, ω_x, ω_y and ω_z. The angular velocity vector can be stated as:

$$\vec{\omega} = \omega_x \vec{i} + \omega_y \vec{j} + \omega_z \vec{k}, \quad |\vec{\omega}| = \omega = \sqrt{\omega_x^2 + \omega_y^2 + \omega_z^2} \tag{3.182}$$

Vorticity is also the curl of the velocity field, i.e.

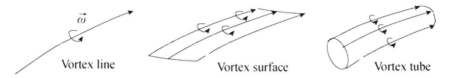

Fig. 3.41 Vortex line, vortex surface and vortex tube

$$\vec{\Omega} = \text{rot}\,\vec{V} = 2\vec{\omega} = \nabla \times \vec{V} \tag{3.183}$$

$$\vec{\Omega} = \text{rot}\,\vec{V} = \begin{vmatrix} \vec{i} & \vec{j} & \vec{k} \\ \frac{\partial}{\partial x} & \frac{\partial}{\partial y} & \frac{\partial}{\partial z} \\ u & v & w \end{vmatrix} \tag{3.184}$$

The axis of rotation of vorticity is determined by the right hand rule. Just like streamline, there exists a smooth curve in an instantaneous vorticity field, and the vorticity at each point on the curve is tangent to the curve, and this curve is called vortex line. The differential equation of the vortex line is (given time, t is the parameter).

$$\frac{dx}{\omega_x} = \frac{dy}{\omega_y} = \frac{dz}{\omega_z} \tag{3.185}$$

In the vorticity field, an arbitrary non-vorticity line curve (as shown in Fig. 3.41) is taken and a vortex line is made through each point on the curve to form a surface composed of vortex lines, and this surface is called the vortex surface. If it is a closed curve, the tubular surface formed by the vortex line on the closed curve is called a vortex tube.

(2) Vorticity flux

According to the nature of the vortex line, the vorticity inside the vortex tube will not pass through the wall to leave the vortex tube, and the vorticity outside the vortex tube will not cross the wall to enter the vortex tube. Vortex flux refers to the sum of vorticity passing through a certain area, which also represents the intensity of vorticity flux in this region. For the plane problem (as shown in Fig. 3.42), the eddy flux through any area A can be expressed as

$$I = \iint\limits_{A} 2\omega_z dA \tag{3.186}$$

For the three-dimensional problem, the vorticity flux through any space surface can be expressed as

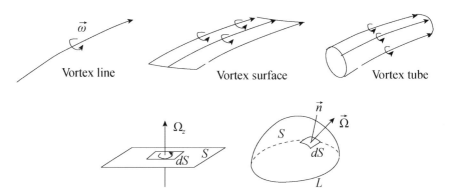

Fig. 3.42 Plane and surface vorticity flux

$$I = \iint\limits_{S} \vec{\Omega} \cdot \mathrm{d}\vec{S} = \iint\limits_{S} \mathrm{rot}\vec{V} \cdot \mathrm{d}\vec{S} \tag{3.187}$$

where S is a surface of any shape, $\mathrm{d}S$ is the area of the microelement on the surface, and n is the normal vector of the microelement.

(3) Circulation

In any closed curve in the flow field, the line integral of the velocity vector along the closed curve is called the velocity circulation of the closed curve. Velocity circulation was first proposed by British scientist Lord Kelvin in 1869 (1824–1907, as shown in Fig. 3.43). In the same way work done by force can be calculated, the velocity circulation can also be figuratively called the velocity work around a closed curve. The sign of the velocity circulation is not only determined by the velocity direction of the flow field, but also related to the direction of the closed curve. It is stipulated that the counterclockwise direction of the integration is positive, that is, the area surrounded by the closed curve is always on the left of the traveler. According to the above definition, as shown in Fig. 3.44, the velocity loop Γ can be expressed in the form of line integral, namely

$$\Gamma = \oint\limits_{L} \vec{V} \cdot \mathrm{d}\vec{s} = \oint\limits_{L} V \cos\alpha \mathrm{d}s \tag{3.188}$$

If the velocity is represented by its three components u, v, w, and the line segment ds is represented by its three components $\mathrm{d}x$, $\mathrm{d}y$, $\mathrm{d}z$, then we have

$$\vec{V} \cdot \mathrm{d}\vec{s} = u\mathrm{d}x + v\mathrm{d}y + w\mathrm{d}z \tag{3.189}$$

So the expression for the circulation is

Fig. 3.43 British scientist
Lord Kelvin (1824–1907)

Fig. 3.44 Velocity
circulation

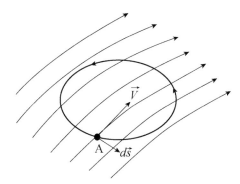

$$\Gamma = \oint_L (u\mathrm{d}x + v\mathrm{d}y + w\mathrm{d}z) \tag{3.190}$$

If the flow is irrotational, there must be a velocity potential function φ. Then the relationship between the velocity component and the velocity potential function is

$$u = \frac{\partial \varphi}{\partial x}, \quad v = \frac{\partial \varphi}{\partial y}, \quad w = \frac{\partial \varphi}{\partial z} \tag{3.191}$$

$$\Gamma = \oint_L (\vec{V} \cdot d\vec{s}) = \oint_L \left(\frac{\partial \varphi}{\partial x} dx + \frac{\partial \varphi}{\partial y} dy + \frac{\partial \varphi}{\partial z} dz \right) = \oint_L d\varphi = 0 \tag{3.192}$$

It shows that the velocity circulation along any closed curve is equal to zero in a non-swirling flow field. But in a swirling flow, the velocity circulation around any closed curve is generally not equal to zero.

3.9.2.1 Relationship Between Vorticity Flux and Circulation

In a swirling flow, velocity circulation and vortex flux are physical quantities representing the overall rotational strength of the vortex from different angles, and there must be a certain relationship between them.

Firstly, a two-dimensional flow field with radius R is investigated, where the vorticity distribution is uniform (as shown in Fig. 3.45), and the vorticity flux passing through a circle with radius R is

$$I = \Omega_z A = 2\pi R^2 \omega_z \tag{3.193}$$

The velocity circulation around the circle is

$$\Gamma = \oint_{2\pi R} \vec{V} \cdot d\vec{s} = \oint_{2\pi R} V_s ds = \int_0^{2\pi} V_s R d\theta = 2\pi R V_s \tag{3.194}$$

where V_s is the circumcircle velocity $(= \omega_z R)$. Substitute the V_s into the above equation to obtain

Fig. 3.45 Vortexes of equal vorticity distribution

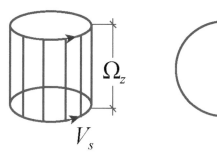

$$\Gamma = \oint_{2\pi R} \vec{V} \cdot d\vec{s} = \oint_{2\pi R} V_s ds = \int_0^{2\pi} V_s R d\theta = 2\pi R^2 \omega_z = I \qquad (3.195)$$

Equations (3.193) and (3.195) indicate that the velocity circulation in the circular region with radius R is equal to the eddy flux, which represents the intensity of the vortex.

Now study the relation between the vortex flux and the velocity circulation in any region in the plane flow field. As shown in Fig. 3.46, any closed curve L is taken from the plane flow field, and the area enclosed by the closed curve is divided into a series of small areas by using parallel lines of two sets of coordinates. The velocity circulation of each small area is made and summed, so that the total velocity circulation of the contour can be obtained. For the microelement ABCD, the velocity circulation is

$$\begin{aligned}
d\Gamma &= \int_{ABCDA} \vec{V} \cdot d\vec{s} \\
&= \left(u + \frac{\partial u}{\partial x} \frac{dx}{2} \right) dx + \left(v + \frac{\partial v}{\partial x} dx + \frac{\partial v}{\partial y} \frac{dy}{2} \right) dy \\
&\quad - \left(u + \frac{\partial u}{\partial y} dy + \frac{\partial u}{\partial x} \frac{dx}{2} \right) dx - \left(v + \frac{\partial v}{\partial y} \frac{dy}{2} \right) dy \\
&= \left(\frac{\partial v}{\partial x} - \frac{\partial u}{\partial y} \right) dx dy = 2\omega_z dx dy \qquad (3.196)
\end{aligned}$$

By summing the velocity circulation increments in all mesh regions, the velocity circulation values around the entire closed curve can be obtained, namely

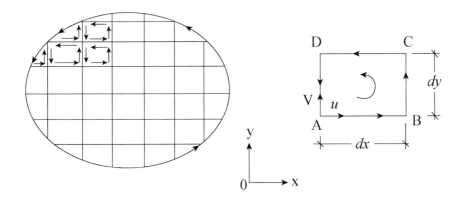

Fig. 3.46 Relationship between velocity circulation and vortex flux in plane region

$$\Gamma = \oint_L \vec{V} \cdot d\vec{s} = \oint_L (u dx + v dy) = \iint_A \left(\frac{\partial v}{\partial x} - \frac{\partial u}{\partial y} \right) dS = \iint_A 2\omega_z dS \quad (3.197)$$

This formula indicates that, in a plane flow, if a region A is taken at any time and its closed contour is L, then the velocity circulation around L is equal to the eddy flux through region A. This formula is called Green's formula in field theory.

For any three-dimensional surface S, its bottom closed contour line is L. The relation between line integral and surface integral obtained by British mechanic and mathematician George Gabriel Stokes (1819–1903) shows that the vortex flux passing through the surface S is equal to the velocity circulation around the closed contour line L, as shown in Fig. 3.47, namely

$$\Gamma = \oint_L \vec{V} \cdot d\vec{r} = \iint_S \text{rot}\vec{V} \cdot d\vec{S} = \iint_S \vec{\Omega} \cdot d\vec{S} = I \quad (3.198)$$

$$\Gamma = \oint_L \vec{V} \cdot d\vec{s}$$
$$= \int_S \left[\left(\frac{\partial w}{\partial y} - \frac{\partial v}{\partial z} \right) \cos(n, x) + \left(\frac{\partial u}{\partial z} - \frac{\partial w}{\partial x} \right) \right.$$
$$\left. \cos(n, y) + \left(\frac{\partial v}{\partial x} - \frac{\partial u}{\partial y} \right) \cos(n, z) \right] dS \quad (3.199)$$

If it is non-swirling field, its curl is zero everywhere, which can be obtained from Eq. (3.198)

$$\oint_L \vec{V} \cdot d\vec{r} = 0 \quad (3.200)$$

Considering the arbitrariness of the integral curve, the above equation indicates that the curve integral of the velocity field is independent of the path and is only a function of the coordinate position. Mathematically, the differential components in

Fig. 3.47 Relationship between velocity circulation and flux in three-dimensional flow field

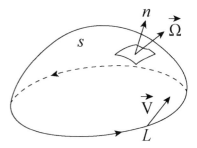

Eq. (3.200) can be expressed as the total differential of a function, i.e.

$$d\varphi = \vec{V} \cdot d\vec{r} = udx + vdy + wdz \tag{3.201}$$

Substitute it into Eq. (3.200), and get

$$\oint_L \vec{V} \cdot d\vec{r} = \oint_L d\varphi = 0 \tag{3.202}$$

3.9.2.2 Induction Velocity of Vortex

Take a microsegment ds on any vortex line with intensity of Γ. It produces an induced velocity dv for any point P outside the line. The induction of vortex is the same as that of electromagnetic induction, and the velocity formula dimension induced by microsegment vortices can be obtained by the Biot-Savart formula

$$d\vec{v} = \frac{\Gamma}{4\pi} \frac{d\vec{s} \times \vec{r}}{r^3} \tag{3.203}$$

where dv is the velocity perpendicular to the plane composed of line segment ds and disturbed point P (as shown in Fig. 3.48), which is proportional to the vortex strength Γ and the vortex segment length ds, but inversely proportional to the distance r squared. For the finite line segment AB, the induced velocity is

$$\vec{v} = \int_A^B \frac{\Gamma}{4\pi} \frac{d\vec{s} \times \vec{r}}{r^3} \tag{3.204}$$

Now for a linear vortex of intensity Γ, find the velocity induced by any point outside it. As shown in Fig. 3.49, let AB be vortex line. P is a point outside the line, and the distance between P and AB is h. Let the angle between the line of any

Fig. 3.48 Vortex line and induced velocity

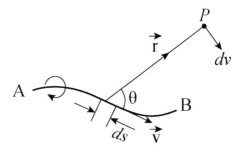

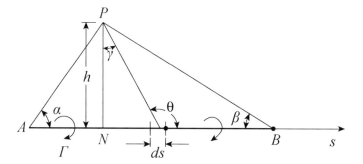

Fig. 3.49 Induction velocity of linear vortex

microsegment ds and P and the perpendicular line of AB PN be γ, and the geometric relation is

$$\mathrm{d}s = \mathrm{d}(h t g \gamma) = h \sec^2 \gamma \, \mathrm{d}\gamma, \ \mathrm{d}v = \frac{\Gamma}{4\pi h} \cos \gamma \, \mathrm{d}\gamma \tag{3.205}$$

Let the angle between PA and AB be α, and the angle between PB and BA be β. In the integral formula (3.209), γ changes from $-\left(\frac{\pi}{2} - \alpha\right)$ to $+\left(\frac{\pi}{2} - \beta\right)$

$$v = \frac{\Gamma}{4\pi h}(\cos \alpha + \cos \beta) \tag{3.206}$$

The induced velocity is perpendicular to the paper surface, and according to the γ direction shown in Fig. 3.49, the induced velocity refers to the outer. If one end of the vortex line is infinite, there exist

$$v = \frac{\Gamma}{4\pi h}(1 + \cos \alpha) \tag{3.207}$$

If the vortex line is semi-infinite in length and point P is located at the bottom of the vortex line, then there exist

$$v = \frac{\Gamma}{4\pi h} \tag{3.208}$$

If both ends of the vortex line extend to infinity, there exist

$$v = \frac{\Gamma}{2\pi h} \tag{3.209}$$

3.9.2.3   Flow Field of Rankine Vortex Model

In 1872, W. J. M. Rankine, a British scientist (1820–1872, as shown in Fig. 3.50), proposed mathematical models of free vortex, forced vortex and combined vortex based on the ideal fluid model, namely the famous Rankine eddy model.

The Rankine vortex model assumes that the concentrated eddy field is composed of vortex core and its induced flow field, in which the vortex core is the equal vorticity rotating flow field, and outside the vortex core is the vortex core induced flow field. The velocity field and pressure field obtained are:

(1)   Flow field in vortex core (forced vortex field)

There is a vortex field with equal vortexes in the vortex core, and its circumferential velocity satisfies the rotation law of the rigid body about the axis, namely

$$u_\theta = \frac{\Gamma}{2\pi R^2} r \tag{3.210}$$

Among them, u_θ is the velocity in the circumferential direction at the radius r. R is the radius of the vortex core, Γ is the vortex intensity (velocity circulation). By using Eq. (3.130), the corresponding static pressure strength at any radius r can be obtained by integral as

$$p = p_c + \frac{1}{2}\rho u_\theta^2 \tag{3.211}$$

Among them, the p_c is the static pressure strength at the center of the vortex core.

Fig. 3.50 W. J. M. Rankine (1820–1872)

Fig. 3.51 Flow field of
Rankine vortex model

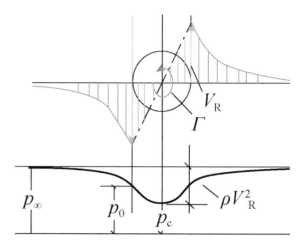

(2) Flow field outside the vortex core (free vortex field)

Outside the vortex core, there is the flow field induced by the point vortex (the field without eddy), and the circumferential velocity at the radius r is

$$u_\theta = \frac{\Gamma}{2\pi r} \tag{3.212}$$

Because it is a potential flow field, the static pressure strength is obtained by the Bernoulli equation

$$p = p_\infty - \frac{1}{2}\rho u_\theta^2 \tag{3.213}$$

Among them, the p_∞ is the pressure at infinity. The difference between the outflow pressure and the pressure at the vortex center (pressure funnel) is

$$\Delta p = p_\infty - p_c = \rho u_\theta^2(R) = \rho V_R^2 \tag{3.214}$$

The Rankine eddy model provides theoretical basis for people to understand the formation mechanism of tornadoes, as shown in Figs. 3.51, 3.52 and 3.53. From the perspective of fluid mechanics, tornado is actually a development process of spatial concentrated vortex.

3.9.2.4 Helmholtz Vortex Theorem

In 1858, Helmholtz (1821–1894), a German fluid dynamicist, proposed three theorems on vorticity field for ideal positive pressure fluid under the condition of potential mass force, which were later called Helmholtz Theorems.

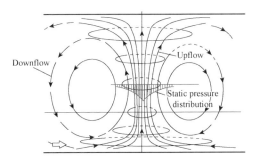

Fig. 3.52 Tornado structure generated by rising thermals

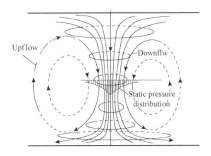

Fig. 3.53 Tornado structure generated by descending cold air flow

Theorem 1 *The vortex intensity along the vortex line or vortex tube remains constant at the same instant (vortex intensity preservation theorem).*

As shown in Fig. 3.54, The circulation is calculated along the enclosing line $PQRR'Q'P'P$. Since the stretched surface is located on the wall of the vortex tube and no vortex line passes through, the total circulation is zero.

$$\Gamma_{PQRR'Q'P'P} = \Gamma_{PQR} + \Gamma_{RR'} + \Gamma_{R'Q'P'} + \Gamma_{P'P} = 0$$

Due to the

$$\Gamma_{P'P} = -\Gamma_{RR'} \quad \Gamma_{R'Q'P'} = -\Gamma_{P'Q'R'}$$

Then obtain

$$\Gamma_{PQR} = \Gamma_{P'Q'R'} \tag{3.215}$$

That is to say, wherever the circularity (vorticity) is calculated along the vortex tube, its value remains the same. This theorem is called Helmholtz's first theorem, or simply the first vortex theorem.

Fig. 3.54 Strength of vortex
tube

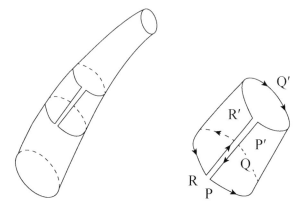

Fig. 3.55 Conservation of
vortex tube strength (left)
and possible existence forms
of vortex tube

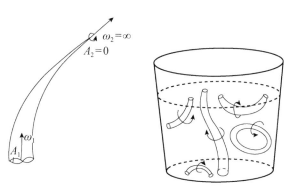

Generalizations of the first theorem: a vortex tube is impossible to break in a fluid.
It can be extended to an infinite distance. It can be self-connected into a vortex ring
(not necessarily a ring), and it can stop at a boundary, a solid boundary or a free
boundary (such as a free liquid surface), as shown in Fig. 3.55.

Theorem 2 *A fluid particle forming a vortex line or vortex tube at a certain moment
will still form a vortex line or vortex tube in the later motion (vortex line or vortex
surface retention theorem).*

For an ideal positive pressure fluid, under the condition of potential mass force,
the vortex line and vortex tube move together with the fluid particles constituting
it. That is to say, the fluid particles constituting the vortex surface, vortex tube and
vortex line at a certain moment still constitute the vortex surface, vortex tube and
vortex line before or after this point.

Theorem 3 *The strength of a vortex tube does not change with time, nor does it
increase, weaken or disappear (conservation law of vortex strength).*

The above three theorems show that for an ideal positive pressure fluid, the vortex
motion of the fluid can neither be generated nor die out under the condition of

potential mass force. That is, a rotational motion will always be a rotational motion, and non-rotational motion will always be non-rotational motion. Then the cause of vortex motion is: viscous fluid, non-barotropic fluid (baroclinic fluid), and mass force is not potential.

3.9.2.5 Differential Equations of Lagrangian Motion for Ideal Fluid

(1) Lagrangian continuous differential equations

In t_0, the coordinate of the center point of the fluid micro-element is

$$\begin{cases} x_0 = x_0(a, b, c, t_0) \\ y_0 = y_0(a, b, c, t_0) \\ z_0 = z_0(a, b, c, t_0) \end{cases} \tag{3.216}$$

The density is $\rho_0 = \rho_0(a, b, c, t_0)$. At time t, the coordinate of the center point of the fluid micro-element is

$$\begin{cases} x = x(a, b, c, t) \\ y = y(a, b, c, t) \\ z = z(a, b, c, t) \end{cases} \tag{3.217}$$

The density is $\rho = \rho(a, b, c, t)$. According to the law of conservation of mass, the mass of the micro-element remains unchanged in motion, then

$$\rho_0 dx_0 dy_0 dz_0 = \rho dx dy dz \tag{3.218}$$

Using the substitution rule of integral variables, by

$$\rho_0 \frac{\partial(x_0, y_0, z_0)}{\partial(a, b, c)} = \rho \frac{\partial(x, y, z)}{\partial(a, b, c)} \tag{3.219}$$

Among them

$$\frac{\partial(x_0, y_0, z_0)}{\partial(a, b, c)} = \begin{vmatrix} \dfrac{\partial x_0}{\partial a}, & \dfrac{\partial y_0}{\partial a}, & \dfrac{\partial z_0}{\partial a} \\[2mm] \dfrac{\partial x_0}{\partial b}, & \dfrac{\partial y_0}{\partial b}, & \dfrac{\partial z_0}{\partial b} \\[2mm] \dfrac{\partial x_0}{\partial c}, & \dfrac{\partial y_0}{\partial c}, & \dfrac{\partial z_0}{\partial c} \end{vmatrix} \tag{3.220}$$

$$\frac{\partial(x, y, z)}{\partial(a, b, c)} = \begin{vmatrix} \dfrac{\partial x}{\partial a}, & \dfrac{\partial y}{\partial a}, & \dfrac{\partial z}{\partial a} \\[2mm] \dfrac{\partial x}{\partial b}, & \dfrac{\partial y}{\partial b}, & \dfrac{\partial z}{\partial b} \\[2mm] \dfrac{\partial x}{\partial c}, & \dfrac{\partial y}{\partial c}, & \dfrac{\partial z}{\partial c} \end{vmatrix} \qquad (3.221)$$

For an incompressible fluid, $\rho = \rho_0$, and the initial time is $x_0 = a$, $y_0 = b$, $z_0 = c$. Then

$$\frac{\partial(x, y, z)}{\partial(a, b, c)} = 1 \qquad (3.222)$$

(2)   Lagrangian differential equations of motion

The Lagrangian equations can be directly transformed from Euler's equations of motion. For an ideal fluid, the Euler motion differential equation (Eq. 3.94) is

$$\begin{cases} \frac{du}{dt} = f_x - \frac{1}{\rho}\frac{\partial p}{\partial x} \\ \frac{dv}{dt} = f_y - \frac{1}{\rho}\frac{\partial p}{\partial y} \\ \frac{dw}{dt} = f_z - \frac{1}{\rho}\frac{\partial p}{\partial z} \end{cases}$$

In the Lagrange method, the components of the fluid particle acceleration can be written as

$$a_x = \frac{\partial^2 x}{\partial t^2}, a_y = \frac{\partial^2 y}{\partial t^2}, a_z = \frac{\partial^2 z}{\partial t^2} \qquad (3.223)$$

By substituting the above equation into Euler's equations, we get

$$\begin{cases} f_x - \dfrac{\partial^2 x}{\partial t^2} = \dfrac{1}{\rho}\dfrac{\partial p}{\partial x} \\[3mm] f_y - \dfrac{\partial^2 y}{\partial t^2} = \dfrac{1}{\rho}\dfrac{\partial p}{\partial y} \\[3mm] f_z - \dfrac{\partial^2 z}{\partial t^2} = \dfrac{1}{\rho}\dfrac{\partial p}{\partial z} \end{cases} \qquad (3.224)$$

According to the coordinate transformation relation

$$\begin{cases} \dfrac{\partial p}{\partial a} = \dfrac{\partial p}{\partial x}\dfrac{\partial x}{\partial a} + \dfrac{\partial p}{\partial y}\dfrac{\partial y}{\partial a} + \dfrac{\partial p}{\partial z}\dfrac{\partial z}{\partial a} \\[3mm] \dfrac{\partial p}{\partial b} = \dfrac{\partial p}{\partial x}\dfrac{\partial x}{\partial b} + \dfrac{\partial p}{\partial y}\dfrac{\partial y}{\partial b} + \dfrac{\partial p}{\partial z}\dfrac{\partial z}{\partial b} \\[3mm] \dfrac{\partial p}{\partial c} = \dfrac{\partial p}{\partial x}\dfrac{\partial x}{\partial c} + \dfrac{\partial p}{\partial y}\dfrac{\partial y}{\partial c} + \dfrac{\partial p}{\partial z}\dfrac{\partial z}{\partial c} \end{cases} \qquad (3.225)$$

The Lagrangian differential equations of motion can be obtained as

$$
\begin{cases}
\dfrac{1}{\rho}\dfrac{\partial p}{\partial a} = \left(f_x - \dfrac{\partial^2 x}{\partial t^2}\right)\dfrac{\partial x}{\partial a} + \left(f_y - \dfrac{\partial^2 y}{\partial t^2}\right)\dfrac{\partial y}{\partial a} + \left(f_z - \dfrac{\partial^2 z}{\partial t^2}\right)\dfrac{\partial z}{\partial a} \\[3mm]
\dfrac{1}{\rho}\dfrac{\partial p}{\partial b} = \left(f_x - \dfrac{\partial^2 x}{\partial t^2}\right)\dfrac{\partial x}{\partial b} + \left(f_y - \dfrac{\partial^2 y}{\partial t^2}\right)\dfrac{\partial y}{\partial b} + \left(f_z - \dfrac{\partial^2 z}{\partial t^2}\right)\dfrac{\partial z}{\partial b} \\[3mm]
\dfrac{1}{\rho}\dfrac{\partial p}{\partial c} = \left(f_x - \dfrac{\partial^2 x}{\partial t^2}\right)\dfrac{\partial x}{\partial c} + \left(f_y - \dfrac{\partial^2 y}{\partial t^2}\right)\dfrac{\partial y}{\partial c} + \left(f_z - \dfrac{\partial^2 z}{\partial t^2}\right)\dfrac{\partial z}{\partial c}
\end{cases}
\tag{3.226}
$$

3.9.2.6   Typical Examples of Euler Integral Equations

Example 1 In wind tunnel experiments, the wake flow field is commonly used to measure the drag coefficient of flow around an airfoil, called the wake method. To determine the drag coefficient C_D of the airfoil, the airfoil was placed in a two-dimensional steady incompressible flow field and the flow field was measured. The height of the flow field h_0. The velocity distribution measured in sections 1–3 (inflow section with uniform velocity distribution) and 2–4 (outflow section) is shown in Fig. 3.56, and the upper flow field pressure of these two sections is set as uniform P_∞. Try to calculate the drag coefficient of airfoil $C_D = \dfrac{D}{\frac{1}{2}\rho V_\infty^2 b}$, where the density is ρ and the chord length of airfoil is b.

Solution:

For steady incompressible flow, the mass m flowing out 1–2 and 3–4 per unit time can be obtained from the continuous equation

$$
\int_0^{h_0} \rho u_2 dy - \int_0^{h_0} \rho V_\infty dy + m = 0, \quad m = \int_0^{h_0} \rho V_\infty dy - \int_0^{h_0} \rho u_2 dy
$$

Fig. 3.56 Airfoil drag coefficient measured by wake method

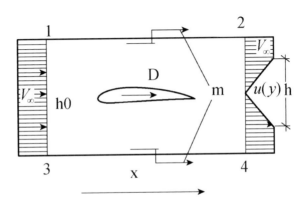

Among them, the u_2 is the distribution of outflow velocity at sections 3–4. Assuming the force exerted by the airfoil on the air flow is F, which is equal to and opposite to the force exerted by the air flow on the airfoil. Now the momentum integral equation (the component of massless force) is established along the x direction

$$-D = \int_0^{h_0} \rho u_2^2 dy - \int_0^{h_0} \rho V_\infty^2 dy + m V_\infty$$

$$D = \int_0^{h_0} \rho V_\infty^2 dy - \int_0^{h_0} \rho u_2^2 dy - V_\infty \int_0^{h_0} \rho V_\infty dy + V_\infty \int_0^{h_0} \rho u_2 dy$$

$$= \int_0^{h_0} \rho u_2 (V_\infty - u_2) dy \tag{3.227}$$

Drag coefficient is

$$C_D = \frac{D}{\frac{1}{2}\rho V_\infty^2 b} = \frac{2}{b} \int_0^{h_0} \frac{u_2}{V_\infty} \left(1 - \frac{u_2}{V_\infty}\right) dy = \frac{1}{3}\frac{h}{b}$$

Example 2 as shown in Fig. 3.57, there is plane cascade with space h. There is an ideal incompressible adiabatic stationary flow. Assume that the distribution of flow velocity and pressure is uniform far away from the cascade. The velocity components and pressure of sections 1–1 (before cascade) are u_1, v_1, p_1. The velocity components and pressure of sections 2–2 (after cascade) are u_2, v_2, p_2. Try to find the fluid force on a single blade.

Solution:

Suppose the horizontal component of the fluid acting on a single blade is F_x and the vertical component is F_y. This force is equal to and opposite to the force exerted by the blade on the fluid. The control volume as shown in the figure is taken around the

Fig. 3.57 Flow around plane cascade

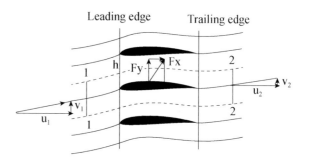

blade. The controlled inflow section is 1–1, the outflow section is 2–2, and the side is 1–2, as shown in Fig. 3.57. The continuity equation can be obtained

$$\rho u_1 h = \rho u_2 h, \quad u_1 = u_2 = u$$

Using the energy equation, we get

$$p_1 + \frac{1}{2}\rho V_1^2 = p_2 + \frac{1}{2}\rho V_2^2$$

$$p_2 - p_1 = \frac{1}{2}\rho\left(u^2 + v_1^2\right) - \frac{1}{2}\rho\left(u^2 + v_2^2\right) = \frac{1}{2}\rho v_1^2 - \frac{1}{2}\rho v_2^2 \qquad (3.228)$$

Using the momentum integral equation, we get

$$\rho u h(u_2 - u_1) = (p_1 - p_2)h - F_x$$
$$\rho u h(v_2 - v_1) = -F_y \qquad (3.229)$$

$$F_x = (p_1 - p_2)h = \left(\frac{1}{2}\rho v_2^2 - \frac{1}{2}\rho v_1^2\right)h$$

$$F_y = \rho u h(v_1 - v_2) \qquad (3.230)$$

Introduce the velocity circulation, and take the clockwise rotation of 1–2–2–1–1, the velocity circulation (clockwise rotation) is

$$\Gamma = h(v_1 - v_2)$$

$$F_x = -\rho\frac{v_2 + v_1}{2}\Gamma$$

$$F_y = \rho u \Gamma$$

$$F = \rho\Gamma\sqrt{u^2 + \left(\frac{v_2 + v_1}{2}\right)^2} = \rho\Gamma V_m \qquad (3.231)$$

$$v_m = \frac{v_1 + v_2}{2}, u_m = u, V_m = \sqrt{u_m^2 + v_m^2}$$

$$\frac{F_x}{F_y}\frac{u}{v_m} = -1$$

The above equation is the Kutta-Joukowski lifting circulation law for a single blade of flow around a planar cascade. The direction of lift is to rotate 90 against the direction of incoming flow. If I take $h \to \infty$, and the circulation Γ stays the same, then we can obtain $(v_1 - v_2) \to 0$, $V_m = V_\infty$, and

Fig. 3.58 Thrust of an
aspirated jet engine

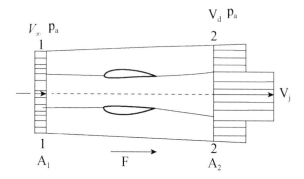

$$F = \rho \Gamma V_\infty \tag{3.232}$$

This is known as the Kutta-Joukowski lifting circulation law for flow around a single airfoil.

Example 3 As shown in Fig. 3.58, there is an aspirated jet engine. In the cruising state, it is assumed that the engine is moving steadily and uniformly in a straight line with the aircraft at the speed of V_∞. Now the coordinate system is taken to consolidate the engine, and the 1–1 section in the far front and 2–2 section in the far back are taken to form the control volume. The inlet section is 1–1, the velocity is V_∞, and the pressure is atmospheric pressure P_a. The flow area is A_1. after engine combustion, pressurization and expansion, the high temperature air flow velocity at outlet sections 2–2 is V_j. The cold air flow of ejection is V_d. The cross-sectional area is A_2. The pressure is atmospheric pressure P_a. Regardless of mass force, try to find the thrust of air flow on the engine.

Solution:

It is assumed that the mass of gas drawn from the engine inlet section per unit time is m_1. The mass that comes out of the engine is m_3. The engine burns with a mass of m_f. According to the law of conservation of mass

$$m_1 + m_f = m_3$$

The mass flow of cold air outside the engine is

$$m_1' = m_2' = m'$$

If F is the thrust from the engine to the airflow, then the momentum theorem can be obtained

$$F = m_3 V_j - m_1 V_\infty + m'(V_d - V_\infty) \tag{3.233}$$

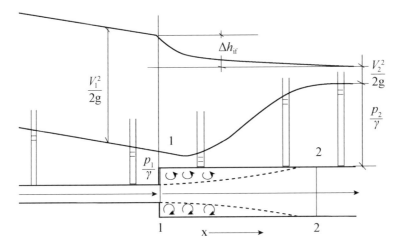

Fig. 3.59 Sudden expansion separation flow in pipeline

The thrust of the airflow on the engine is equal to and opposite to F. If, $V_d \approx V_\infty$

$$F = m_3 V_j - m_1 V_\infty \tag{3.234}$$

Example 4 As shown in Fig. 3.59, there is an incompressible steady flow with a sudden expansion of a horizontally placed pipe (regardless of the fluid mass force). The French physicist J. C. Borda (1733–1799) obtained the famous formula of local energy loss of the pipeline sudden expansion separated flow by using the total flow momentum, which is called The Borda formula for short. Try to derive Borda's formula.

Solution:

Set the fluid density as ρ, and take the control volume as shown in Fig. 3.59. The fluid velocity at the inlet section is 1–1, and the outlet section is 2–2. The flow velocity of sections 1–1 is V_1, and the pressure is p_1, cross-section area is A_1. The flow velocity of sections 2–2 is V_2, and the pressure is p_2, cross-section area is A_2. Using the continuity equation, can be obtained

$$\rho V_1 A_1 = \rho V_2 A_2 \tag{3.235}$$

By applying the momentum equation along the x direction and ignoring the friction force of the pipe wall, we can get

$$p_1 A_1 + p_s(A_2 - A_1) - p_2 A_2 = \rho V_1 A_1(V_2 - V_1) \tag{3.236}$$

Among them, the p_s is the fluid pressure in the section 1–1 with a sudden expansion. It is found in the experiment that $p_s \approx p_1$, substituted into the above

Fig. 3.60 Steady orifice
outflow

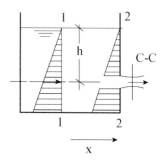

equation

$$p_1 A_2 - p_2 A_2 = \rho V_2^2 A_2 - \rho V_1^2 A_1$$

$$(p_1 + \rho V_1^2) - (p_2 + \rho V_2^2) = \rho V_1^2 \left(1 - \frac{A_1}{A_2}\right) \tag{3.237}$$

The energy loss of fluid per unit weight (γ is fluid bulk density) between 1–1 and 2–2 cross-sections is

$$h_w = \left(\frac{p_1}{\gamma} + \frac{V_1^2}{2g}\right) - \left(\frac{p_2}{\gamma} + \frac{V_2^2}{2g}\right) = \frac{(V_1 - V_2)^2}{2g} \tag{3.238}$$

Example 5 As shown in Fig. 3.60, A constant water tank is investigated. A small hole is opened on the side wall of the water tank. The area of the hole is A (Ignore the energy loss).

Answer:

As shown in Fig. 3.60, the volume enclosed between sections 1–1, 2–2 and C–C is taken as the control volume. The height from the center of the hole to the water surface is set as h, and the outlet velocity of the hole is set as u_c. Based on Bernoulli's equation

$$h = \frac{u_c^2}{2g}$$

Suppose F represents the thrust of the side wall on the water flow. The contraction section of the outlet flow is A_c. The shrinkage coefficient is

$$\varepsilon = \frac{A_c}{A}$$

The momentum equation is given by

$$p_c A + F = \rho u_c A_c (u_c - 0) = \rho u_c^2 A \varepsilon \tag{3.239}$$

The hydrostatic pressure is zero

$$p_c = \rho g h$$

$$F = \rho g h A (2\varepsilon - 1)$$

The experiments shows that the shrinkage coefficient is 0.62, then

$$F = \rho g h A (2\varepsilon - 1) = 0.24 \rho g h A \tag{3.240}$$

The thrust of outlet flow on the side wall is equal to and opposite to F.

Example 6 There is a jet mixer. The high-speed fluid is ejected from the nozzle and the surrounding fluid is ejected. It is assumed that the two are the same fluid with the same density and the same diameter of the mixing chamber. The jet section is known as A_1; the section of the mixing chamber is $A_3 (= 3A_1)$, and the jet velocity is V_1, the velocity of the ejection fluid is V_2. The fluid density is ρ. Try to find the outflow velocity V_3 of the mixing chamber and $p_3 - p_1 = ?$, as shown in Fig. 3.61.

Answer:

The control volume as shown in Fig. 3.61 is taken. The inflow cross-section of the control volume is 1–3 and the outflow cross-section is 2–4. It is assumed that the uniform pressure on the section 1–3 is P_1. From the continuity equation, there is

$$\rho V_1 A_1 + \rho V_2 (A_3 - A_1) = \rho V_3 A_3$$

$$V_3 = V_2 + (V_1 - V_2) \frac{A_1}{A_3} \tag{3.241}$$

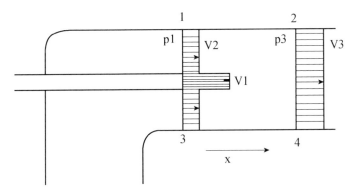

Fig. 3.61 Jet mixer

Establishing the momentum equation in the x direction (ignoring friction) gives

$$p_1 A_3 - p_3 A_3 = \rho V_3^2 A_3 - \rho V_1^2 A_1 - \rho V_2^2 (A_3 - A_1)$$

$$p_3 - p_1 = \rho V_1^2 \frac{A_1}{A_3} + \rho V_2^2 \left(1 - \frac{A_1}{A_3}\right) - \rho V_3^2 = \rho(V_1 - V_2)^2 \frac{A_1}{A_3}\left[1 - \frac{A_1}{A_3}\right]$$

$$(3.242)$$

The mechanical energy consumed per unit weight of fluid per unit time in a mixer

$$\Delta h_w = \frac{\Delta P_w}{\gamma Q_3} = \left[Q_1\left(p_1 + \frac{1}{2}\rho V_1^2\right) + Q_2\left(p_1 + \frac{1}{2}\rho V_2^2\right) - Q_3\left(p_3 + \frac{1}{2}\rho V_3^2\right)\right]$$

$$= \frac{Q_1}{Q_3}\left(\frac{p_1}{\gamma} + \frac{V_1^2}{2g}\right) + \frac{Q_2}{Q_3}\left(\frac{p_1}{\gamma} + \frac{V_2^2}{2g}\right) - \left(\frac{p_3}{\gamma} + \frac{V_3^2}{2g}\right) \quad (3.243)$$

$$\Delta h_w = \frac{\Delta P_w}{\gamma Q_3} = \frac{4}{9}\frac{Q_1}{Q_1 + Q_2}\frac{V_1^2}{2g}\left(1 + \frac{V_2}{V_1} - 2\frac{V_2^2}{V_1^2}\right)$$

$$- \frac{4}{9}\frac{Q_2}{Q_1 + Q_2}\frac{V_1^2}{2g}\left(-\frac{5}{4} + \frac{V_2}{V_1} + \frac{1}{4}\frac{V_2^2}{V_1^2}\right)$$

$$(3.244)$$

If $V_1 \gg V_2$, there are

$$\Delta h_w = \frac{4}{9}\frac{V_1^2}{2g} \quad (3.245)$$

Example 7 As shown in Fig. 3.62, on a horizontal floor, the water flow suddenly jumps up, and the flow is incompressible steady flow (water density is ρ). The incoming velocity and the depth are V_1 and h_1. The water depth after jump is h_2 and the velocity is V_2. The distribution law of hydrostatic pressure is approximately satisfied along the vertical line at sections 1–1 and 2–2. Try to solve: (1) h_2/h_1; (2) Mechanical energy loss of water body per unit weight.

Answer:

The volume between sections 1–1 and 2–2 is taken as the control volume, which can be obtained from the continuity equation

$$\rho V_1 h_1 = \rho V_2 h_2$$

Ignoring the frictional force on the wall between 1–1 and 2–2, the momentum equation is established along the direction of horizontal flow

$$\frac{1}{2}\gamma h_1^2 - \frac{1}{2}\gamma h_2^2 = \rho V_2^2 h_2 - \rho V_1^2 h_1$$

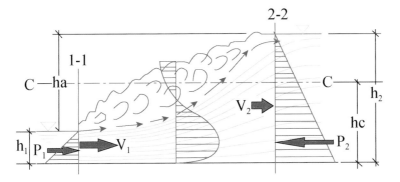

Fig. 3.62 Hydrualic jump

Solve the above two equations simultaneously, and get

$$2\text{Fr}_1^2 = \frac{h_2}{h_1}\left(1 + \frac{h_2}{h_1}\right)$$

$$\frac{h_2}{h_1} = \frac{1}{2}\left[-1 + \sqrt{1 + 8Fr_1^2}\right], \text{Fr}_1 = \frac{V_1}{\sqrt{gh_1}} \tag{3.246}$$

The energy loss between 1–1 and 2–2 cross-sections is

$$h_w = h_1 + \frac{V_1^2}{2g} - \left(h_2 + \frac{V_2^2}{2g}\right) = \frac{(h_2 - h_1)^3}{4h_2h_1} \tag{3.247}$$

Example 8 As shown in Fig. 3.63, there is a contraction pipe in which the ideal incompressible fluid flows steadily. The velocity of inlet section 1–1 is V_1, and the area is A_1. The velocity of inlet section 1–1 is V_1, and the area is A_1. Try to find the pressure p_1 at the inlet section and the flow thrust on the contraction segment.

Fig. 3.63 Flow in
contraction segment

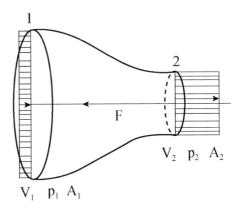

Answer:

Take the control volume as shown in Fig. 3.63. The inflow section of the control volume is 1–1 and the outflow section is 2–2. Assuming that F is the force exerted by the contraction section on the fluid, it can be obtained from the continuity equation

$$\rho V_1 A_1 = \rho V_2 A_2, \quad V_2 = V_1 \frac{A_1}{A_2}$$

From the energy equation, we have

$$p_1 + \frac{1}{2}\rho V_1^2 = p_2 + \frac{1}{2}\rho V_2^2$$

Solve the above equation

$$p_1 = p_2 + \frac{1}{2}\rho V_2^2 - \frac{1}{2}\rho V_1^2 = p_2 + \frac{1}{2}\rho V_1^2 \left[\frac{A_1^2}{A_2^2} - 1 \right] \qquad (3.248)$$

Establish the momentum equation along the direction of flow

$$p_1 A_1 - p_2 A_2 - F = \rho V_2^2 A_2 - \rho V_1^2 A_1$$

Solve the above equation

$$F = \frac{1}{2}\rho V_1^2 A_1 \left(\frac{A_1}{A_2} - 1 \right)^2 + p_2 A_1 \left(1 - \frac{A_2}{A_1} \right)$$

$$F = \frac{1}{2}\rho V_2^2 A_1 \left(1 - \frac{A_2}{A_1} \right)^2 + p_2 A_1 \left(1 - \frac{A_2}{A_1} \right) \qquad (3.249)$$

The thrust of the airflow on the contraction section is equal to F and opposite (to the right).

Example 9 Find the impact force of the two-dimensional incompressible steady jet on the fixed inclined plate with width b_0 (Angle with the horizontal line is θ), the jet width ratio b_1/b_2, and the point at which the impact force is applied. Gravity and flow losses are ignored.

Solution:

Take the control volume enclosed by the dotted line as shown in Fig. 3.64. The inlet section 1, outlet section 2 and the pressure at the boundary surface contacting with the surrounding atmosphere are all atmospheric pressure. Establish Bernoulli's equation regardless of gravity along the upper and lower streamline respectively (or think that the flow is uniform without rotation, Bernoulli constant is valid in the whole field), and get

Fig. 3.64 Impact of
horizontal jet on plate

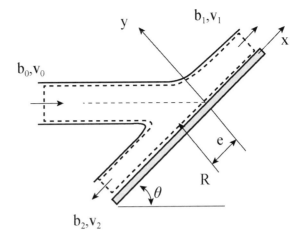

$$p_0 + \frac{1}{2}\rho V_0^2 = p_0 + \frac{1}{2}\rho V_1^2 = p_0 + \frac{1}{2}\rho V_2^2$$

get

$$V_1 = V_2 = V_0$$

From the continuity equation

$$V_2 b_2 + V_1 b_1 - V_0 b_0 = 0, \quad b_0 = b_2 + b_1 \qquad (3.250)$$

(1) Impact force R of jet on plate

Assume that the force exerted by the plate on the jet is R (as shown in Fig. 3.64),
and the 0–0, 1–1 and 2–2 velocities of the three sections are evenly distributed. The
momentum equation along the normal direction of the plate (regardless of gravity)
is established, namely

$$0 - \iint\limits_{0-0} \rho V_0^2 dS \sin\theta = -R$$
$$R = \rho V_0^2 b_0 \sin\theta \qquad (3.251)$$

The impact force of jet on plate is equal to R and opposite to R. So, the impact
force R is related to θ. When the jet is vertically incident, $\theta = 90°$, the impact R is
maximum.

(2) Jet width ratio b_1/b_2

Establish the momentum equation along the direction x of the plate, and get

$$\rho V_1^2 b_1 - \rho V_2^2 b_2 - \rho V_0^2 b_0 \cos\theta = 0$$

Considering the $V_1 = V_2 = V$, and this becomes

$$b_0 \cos\theta = b_1 - b_2$$

Consider $b_0 = b_1 + b_2$ and solve it all at once to get

$$b_1 = \frac{1 + \cos\theta}{2} b_0, \quad b_2 = \frac{1 - \cos\theta}{2} b_0 \tag{3.252}$$

Therefore, the jet width ratio is

$$\frac{b_1}{b_2} = \frac{1 + \cos\theta}{1 - \cos\theta} \tag{3.253}$$

Since the velocities are equal, this is also the flow ratio Q_1/Q_2.

(3) The applied point e of impact force R

As shown in Fig. 3.64, the distance between the applied point of the impact force and the y-axis is set as e. The clockwise moment is taken as positive, which can be obtained from the moment of momentum equation

$$R \cdot e = 0 + \left(V_1 \frac{b_1}{2}\right) \rho V_1 b_1 - V_2 \frac{b_2}{2} \rho V_2 b_2$$

Arrange the above equation to get

$$e = \frac{\frac{1}{2}\rho V_0^2 (b_1^2 - b_2^2)}{\rho V_0^2 b_0 \sin\theta} = \frac{1}{2} \frac{(b_1 + b_2)(b_1 - b_2)}{b_0 \sin\theta} = \frac{b_0 \cos\theta}{2 \sin\theta} \tag{3.254}$$

Only when $\theta = 90°$, the applied point of the resultant force passes through the center of the jet.

Example 10 using the law of moment of momentum, deduce the turbine mechanical Euler equation. Suppose that the rotating angular velocity of the driven wheel of the turbine machine is ω. When the fluid passes through the passage of the driven wheel, mechanical energy exchange occurs with the driven wheel. The machinery that the driven wheel works on the fluid is called the turbine compressor (compressor, fan and pump, etc.), and the machinery that the driven wheel accepts the mechanical energy of the fluid is called the turbine engine (hydroelectric generator, wind generator, etc.). As shown in Fig. 3.65, the channel between two blades is taken as the control

Fig. 3.65 Moment of
momentum of rotating
machinery (centrifuge)

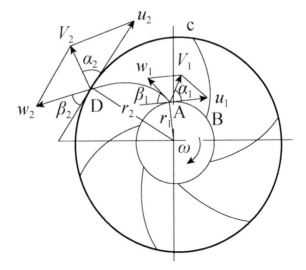

volume ABCD. The relative velocity of fluid entering AB plane is w_1. The absolute velocity is V_1. The rotational speed is u_1. The relative velocity of fluid out of CD plane is w_2. The absolute velocity is V_2. The rotational speed is u_2. The density of the fluid is ρ.

Solution:

For steady flow, using the moment of momentum equation about the axis of rotation

$$\oiint_S \rho \left(\vec{r} \times \vec{V} \right)_z \left(\vec{V} \cdot \vec{n} \right) dS = M_z$$

where M_z is the moment of the blade on the passing fluid. α_1 and α_2 are the angle between the absolute velocity and the tangent line. A_1 and A_2 are the inlet and outlet area of the impeller respectively. The following equation can be obtained

$$\iint_{A_2} \rho V_2 r_2 \cos \alpha_2 V_{2n} dA - \iint_{A_1} \rho V_1 r_1 \cos \alpha_1 V_{1n} dA = M_z \qquad (3.255)$$

The continuity equation in integral form is

$$Q = \iint_{A_2} V_{2n} dA = \iint_{A_1} V_{1n} dA$$

Substitute the above equation as

$$\rho V_2 r_2 \cos \alpha_2 \iint_{A_2} V_{2n} dA - \rho V_1 r_1 \cos \alpha_1 \iint_{A_1} V_{1n} dA = M_z$$

$$\rho Q(V_{2\tau} r_2 - V_{1\tau} r_1) = M_z \qquad (3.256)$$

The power transferred to the fluid per unit time is

$$P = M_z \omega = \rho Q(V_{2\tau} r_2 \omega - \rho V_{1\tau} r_1 \omega) = \rho Q(V_{2\tau} u_2 - V_{1\tau} u_1) \qquad (3.257)$$

Among them, the $u_2 = r_2 \omega$ and $u_1 = r_1 \omega$ are the circumferential velocity of the impeller inlet and outlet respectively. The mechanical energy gained by a fluid per unit weight per unit time is

$$H = \frac{P}{\gamma Q} = \frac{1}{g}(V_{2\tau} u_2 - V_{1\tau} u_1) \qquad (3.258)$$

Among them, the $V_{1\tau}$ and $V_{2\tau}$ are the absolute velocity component of the impeller inlet and outlet along the circumferential tangent direction. β_1 and β_2 are the angles between the relative velocity and the circumferential direction, as shown in Fig. 3.65.

$$V_{2\tau} = u_2 - w_2 \cos \beta_2, \quad V_{1\tau} = u_1 - w_1 \cos \beta_1$$

$$V_2^2 = u_2^2 + w_2^2 - 2u_2 w_2 \cos \beta_2, \quad V_1^2 = u_1^2 + w_1^2 - 2u_1 w_1 \cos \beta_1$$

Substitute it into the above equation to get

$$H = \frac{P}{\gamma Q} = \frac{1}{g}\left(\frac{V_2^2 - V_1^2}{2} + \frac{u_2^2 - u_1^2}{2} + \frac{W_1^2 - W_2^2}{2}\right) \qquad (3.259)$$

For an axial flow turbine ($u_1 = u_2$, as shown in Fig. 3.66), the formula can be simplified as

$$H = \frac{P}{\gamma Q} = \frac{1}{g}\left(\frac{V_2^2 - V_1^2}{2} + \frac{W_1^2 - W_2^2}{2}\right) \qquad (3.260)$$

Example 11 The speed of the fan is 1500 rpm and the inner diameter is $D_1 = 480$ mm, inlet angle $\beta_1 = 60°$, the entrance width $b_1 = 105$ mm, outside diameter $d_2 = 600$ mm, exit angle $\beta_2 = 120°$, exit width $b_2 = 84$ mm, flow $Q = 12,000$ m³/h, air bulk density $= 11.8$ n/m γ^3. Try to find the circumferential velocity, relative velocity, absolute velocity, the theoretical pressure height and effective power of the impeller inlet and outlet.

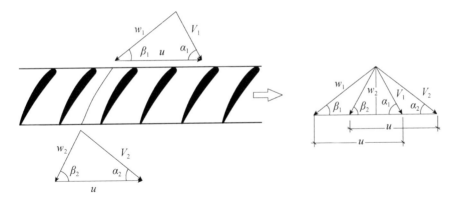

Fig. 3.66 Axial flow turbine blade

Solution:

$$u_1 = \frac{\pi d_1 n}{60} = \frac{3.14159 \times 0.48 \times 1500}{60} = 37.7 \, \text{m/s}$$

$$V_{1n} = \frac{Q}{\pi d_1 b_1} = \frac{12,000}{3600 \times 3.14159 \times 0.48 \times 0.105} = 21 \, \text{m/s}$$

$$w_1 = \frac{V_{1n}}{\sin \beta_1} = \frac{21}{\sin(60)} = 24.3 \, \text{m/s}$$

$$V_{1\tau} = u_1 - w_1 \cos \beta_1 = 37.7 - 24.3 \times \cos(60) = 25.5 \, \text{m/s}$$

$$V_1 = \sqrt{V_{1n}^2 + V_{1\tau}^2} = \sqrt{21^2 + 25.5^2} = 33 \, \text{m/s}$$

$$u_2 = \frac{\pi d_2 n}{60} = \frac{3.14159 \times 0.6 \times 1500}{60} = 47.1 \, \text{m/s}$$

$$V_{2n} = \frac{Q}{\pi d_2 b_2} = \frac{12,000}{3600 \times 3.14159 \times 0.6 \times 0.084} = 21 \, \text{m/s}$$

$$w_2 = \frac{V_{2n}}{\sin \beta_2} = \frac{21}{\sin(120)} = 24.3 \, \text{m/s}$$

$$V_{2\tau} = u_2 - w_2 \cos \beta_2 = 47.1 - 24.3 \times \cos(120) = 59.3 \, \text{m/s}$$

$$V_2 = \sqrt{V_{2n}^2 + V_{2\tau}^2} = \sqrt{21^2 + 59.3^2} = 63 \, \text{m/s}$$

The theoretical pressure height of the impeller is

$$H = \frac{59.3 \times 47.1 - 25.5 \times 37.7}{9.8} = 187\,\text{m}$$

The effective power of the impeller is

$$P = \gamma QH = 11.8 \times \frac{12,000}{3600} \times 187 = 7.4\,\text{kW}$$

Example 12 There is a rocket whose initial total mass is M_0, flying straight up with velocity $V_0(t)$. The exhaust velocity relative to the rocket is V_j. The exhaust mass per unit time is Q_m. The exhaust pressure is p_j. The exhaust area is A_j. Assume that V_j, Q_m and p_j are all constants. Assume that the resistance of the rocket is D, try to solve the differential equation of the rocket motion.

Solution:

As shown in Fig. 3.67, assume that the upward acceleration of the rocket is

$$a_0 = \frac{dV_0}{dt}$$

The mass outflow is Q_m, and there are

$$\oiint_A \rho\left(\vec{V} \cdot \vec{n}\right) dA = \rho_j V_j A_j = Q_m$$

Fig. 3.67 Rocket movement

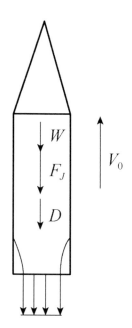

If the total mass in the control volume is $m(t)$, then

$$\iiint_\tau \frac{\partial \rho}{\partial t} d\tau = \frac{\partial}{\partial t} \iiint_\tau \rho d\tau = \frac{\partial m(t)}{\partial t} = -Q_m \tag{3.261}$$

Integral above equation

$$m(t) = m_0 - Q_m t \tag{3.262}$$

Set up the momentum equation in the vertical direction

$$m(t) \frac{dV_0}{dt} - Q_m V_j = -m(t)g - D + (p_j - p_a)A_j$$

Solve above equation

$$\frac{dV_0}{dt} = \frac{-D + (p_j - p_a)A_j + Q_m V_j}{m_0 - Q_m t} - g \tag{3.263}$$

Question 13: As shown in Fig. 3.68, there is a horizontal pipe placed horizontally with length L. At the initial moment, the length of the pipe filled with water is x and the inlet jet velocity is V_1. The area is A_1 (diameter D_1). The pipeline area is A_2 (diameter D_2), and the velocity is V_2. The inlet and outlet of the pipe are in the atmosphere, the friction shear stress of the horizontal pipe is $\tau_0 = \frac{1}{8}\rho\lambda V_2^2$, and the density of water is a constant. Try to solve:

The change rate of x over time $\frac{dx}{dt} =$?
The change rate V_2 over time $\frac{dV_2}{dt} =$?

Solution:

Take the length of the control volume as L. At any time t, the full length of the pipe is x. The continuity equation in integral form is

$$\frac{\partial}{\partial t} \iiint_\tau \rho d\tau + \oiint_S \rho \vec{V} \cdot d\vec{S} = 0$$

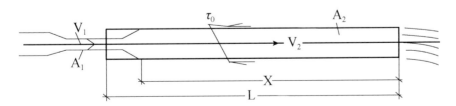

Fig. 3.68 Fluid flow in horizontal pipe

As shown in Fig. 3.68

$$\frac{\partial}{\partial t}[\rho(L - x)A_1 + \rho x A_2] + \rho(V_2 A_2 - V_1 A_1) = 0$$

Simplify it to get

$$\frac{dx}{dt} = -\frac{V_2 A_2 - V_1 A_1}{A_2 - A_1} \tag{3.264}$$

The momentum integral equation along the x direction is established as

$$\frac{\partial}{\partial t} \iiint_\tau \rho V_x d\tau + \oiint_S \rho V_x (\vec{V} \cdot d\vec{S}) = \sum F_x$$

As shown in Fig. 3.68

$$\frac{\partial}{\partial t}[\rho A_2 x V_2 + \rho A_1 (L - x)V_1]$$

$$+ \rho A_2 V_2^2 - \rho A_1 V_1^2 = -\rho\lambda\frac{1}{8}V_2^2 \pi D_2 x$$

Substitute the expression dx/dt to get

$$\frac{dV_2}{dt} = \frac{1}{A_2 x}\left[\frac{(A_2 V_2 - A_1 V_1)^2}{A_2 - A_1} + A_1 V_1^2 - A_2 V_2^2 - \lambda\frac{1}{8}V_2^2 \pi D_2 x\right] \tag{3.265}$$

Exercises

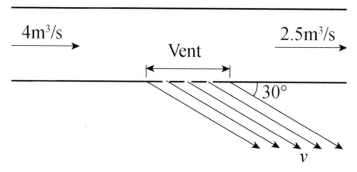

Question 17

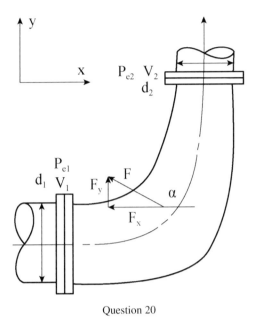

Question 20

Exercises

A. Answer the following questions

1. Explain the physical nature of the Lagrangian method for describing the motion of a fluid particle?

2. Explain the physical nature of Euler's method for describing the motion of fluid particles?

3. Try to explain the relationship between Lagrange variables a, b, c, t and Euler variables x, y, z, t.

4. In The Lagrange method, the pressure of the particle is $p(a, b, c, t)$. Please explain the physical meaning of this expression?

5. In Lagrange method, the following particle motion is characterized by (a, b, c) invariance. How to characterize the particle following in Euler's method? Explain the difference between the following two expressions.

$$\frac{dp}{dt} = \frac{\partial p}{\partial t} + \frac{\partial p}{\partial x}\frac{dx}{dt} + \frac{\partial p}{\partial y}\frac{dy}{dt} + \frac{\partial p}{\partial z}\frac{dz}{dt}$$

$$\frac{dp}{dt} = \frac{\partial p}{\partial t} + u\frac{\partial p}{\partial x} + v\frac{dp}{dt} + w\frac{dp}{dt}$$

6. Explain the physical meaning of the components of acceleration in Euler's method? For steady flow, why can't we say that the acceleration is zero?

7. What are the physical meanings of trace and streamline lines of fluid particles? At what point do they overlap? For unsteady flows, why are they generally not coincident?

8. Explain the physical meaning of heat conduction and heat convection by using the matter derivative of temperature.

9. What are the basic motion forms of fluid particles? What are the basic modes of motion of fluid elements? Write the velocity expression (translational velocity, rotational velocity, angular deformation rate, linear deformation rate) that characterizes the motion of the fluid group.

10. What are the similarities and differences between the rotation motion of micro fluid elements and the rotation motion of rigid bodies?

11. Derive an expression for Helmholtz's velocity decomposition theorem.

12. What is the physical meaning of divergence of velocity field? In the process of fluid motion, the divergence can be greater than zero, less than zero and equal to zero, please explain the physical meaning of each expression?

13. What is the expression of the law of conservation of mass in fluid motion? If the fluid particles do not have continuity conditions, how to express the conservation of mass?

14. If there is a mass source in the control volume, how can we express the law of conservation of mass?

15. Use the continuity equation to prove that the density decreases when the volume of the micro fluid elements expands. The density increases when the volume is compressed. The volume doesn't change when the density doesn't change with time.

16. In an ideal fluid, what are the main factors that change the motion of a fluid particle?

17. How to change particle velocity in an isobaric flow field? How to change the velocity of a fluid particle regardless of mass force?

18. Combined with Euler's differential equations of motion, explain how the pressure changes when the velocity increases along the pipeline with horizontal contraction. If the pipe is expanded horizontally, and the velocity decreases along the way, how does the pressure change?

19. Try to derive the Lagrangian differential equations of fluid motion?

20. Please explain the physical meaning of curl of velocity field? How does it relate to the angular velocity of the fluid mass about its own axis? And write the curl expression.

21. The flow field where the rotational angular velocity is equal to zero is called a non-swirling flow field. Please use the Stokes formula to prove that the line integral of the velocity is independent of the integration path and there is a velocity potential function.

22. At the same spacing, what's the relationship between the magnitude of the velocity potential change and the magnitude of the velocity?

23. The fluid flows along the horizontal expansion pipe and the contraction pipe. Describe the relationship between acceleration and pressure from Euler's equation. What causes the acceleration of a fluid particle?

24. Can you write Euler's equation in vector form? The projection form along the flow direction?

25. Write down the expressions for the components of the Euler equation along the x direction.

26. Please write down the Gromyko-Lamb equation in the x direction. Write down the Gromyko-Lamb equation in vector form.

27. What are the conditions for the Existence of Bernoulli integrals? Write the ideal incompressible fluid, with gravity as mass force, steady flow, energy equation per unit mass, and explain the physical meaning of each?

28. Write the energy equation for an ideal incompressible fluid, without mass force, steady flow, unit mass, and explain the physical meaning of each.

29. If we assume that the relationship between pressure and density is $p = C\rho^\gamma$, where C is a constant. What is the energy equation per unit mass for an ideal compressible steady flow without mass force?

30. What is the velocity change of flow with and without mass force? Please use specific examples to illustrate this.

31. What is a system? What is the control volume? Explain the differences between the system and the control volume.

32. The physical meaning of Reynolds transport equation? What are the conditions for the Reynolds transport equation?

33. Write down the momentum integral equation and explain the physical meaning of the terms?

34. Write down the energy integral equation, and explain the physical meaning of the terms?

35. What are ideal positive pressure fluid, steady flow and mass force potential?

36. For pipe flow, can you give a simplified form of the Reynolds transport equation?

37. Under the condition of ideal positive pressure fluid and potential mass force, vorticity is strongly conserved. Explain how a viscous fluid flows around a vortex.

38. What is a vortex? What are vorticity, eddy flux and velocity circulation?

39. Explain the physical meaning of Stokes integral? In any closed curve, the velocity circulation is equal to zero, why can't it be judged that there is no vortex field in this region?

40. How to characterize the rotation speed of a group of fluid masses?

41. What is Helmholtz's conservation of three vortices?

42. Why a vortex tube cannot be interrupted in a fluid.

43. Why can tornadoes pull up big trees? How to characterize the strength of the tornado?
44. How does the wingtip vortex in flight affect the lift of the aircraft?
45. How does the leading-edge vortex of modern fighter wings affect the lift?
46. How to calculate the vorticity flux in a vortex tube if the vorticity distribution is not uniform?
47. Describe the expression of the induced velocity of an infinitely long cylindrical vortex perpendicular to the plane of symmetry of the vortex axis.

B. Calculate the following questions

1. The velocity component of the flow field $u = y/(x^2 + y^2)$ and $v = -x/(x^2 + y^2)$. Calculate the equation of streamline through the point (0, 5).

2. The velocity field is $u = y/(x^2 + y^2)$ and $v = -x/(x^2 + y^2)$. Try to calculate the circulation around a circular path with a radius of 5 m.

3. The following velocity fields

$$\begin{cases} u = 2t + 2x + 2y \\ v = t - y + z \\ w = t + x - z \end{cases}$$

   Find the acceleration at the point (2, 2, 1) at $t = 3$.

4. The two-dimensional velocity field is made up of

$$V = (x^2 - y^2 + x)i - (2xy + y)j$$

   At that time, the calculation $(x, y) = (1, 2)$.
   (a) the accelerations a_x and a_y,
   (b) the velocity component in the direction $\theta = 40°$,
       i. Maximum velocity direction, and
       ii. The direction of maximum acceleration.

5. The following velocity field

$$u = xy^2, v = -\frac{1}{3}y^3, w = xy$$

   Try to find out: (1) the acceleration of point (1, 2, 3); (2) whether it is a multi-dimensional flow; (3) whether it is a steady flow or an unsteady flow; (4) whether it is a uniform flow or an inhomogeneous flow.

6. A section of U-shaped pipe. The inner diameter of the pipe is 0.5 m. Air enters one section of the pipe at an average speed of 100 m/s and flows out the other section at the same speed, but in the opposite direction. The

flow pressure at the inlet and outlet is the ambi■nt pressure. Calculate the magnitude and direction of the force exerted by ¬e air flow on the pipe. Air density is 1.23 kg/m³.

7. Through the flow, the adjusting valve is gradua y closed, so that the flow is linearly reduced, the flow is reduced to zer in 20 s. Try to find the acceleration at point A on the axis of the pip⊂when the valve is closed for 10 s (assuming that the velocity is evenly dis−ibuted on the section).$l = 60$ cm, $D = 20$ cm, $d = 10$ cm, $Q = 0.2$ m³/s

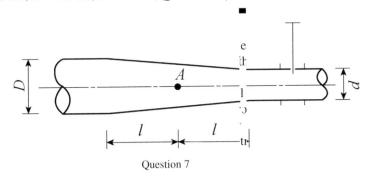

Question 7

8. The radial and tangential components of the ve⊃city are $V_r = 0$ and $V_\theta = C_r$, respectively, where C is a constant. Try to ⊏lculate the equation of the streamline.

9. The velocity field of plane flow is $u = -\frac{cy}{x^2+y}$, $v = \frac{cx}{x^2+y^2}$, where C is a constant, try to find the streamline equation.

10. Consider an object of arbitrary shape. If the ┏essure distribution on the surface of the object is constant, prove that th net force on the object is zero.

11. Consider an airfoil in a wind tunnel. It is prove⊣ that the lift per unit span can be obtained from the pressure distribution ⌐ the top and bottom walls of the wind tunnel (i.e., the pressure distribut₁ρn on the top and bottom walls of the airfoil).

12. Fluid motion has partial velocity

$$
\begin{cases}
u = \dfrac{x}{(x^2+y^2+z^2)^{3/2}} \\
v = \dfrac{y}{(x^2+y^2+z^2)^{3/2}} \\
w = \dfrac{z}{(x^2+y^2+z^2)^{3/2}}
\end{cases}
$$

Is there rotation in the flow field? If there is no rotation, find the velocity potential function.

13. The velocity field of an incompressible fluid in steady motion is

$$u = ax$$
$$v = ay$$
$$w = -2az$$

where a is a constant:
(1) Line deformation rate, angular deformation rate;
(2) Whether there is rotation in the flow field;
(3) Whether there is a velocity potential function.

14. Let the following functions represent the three fractional velocities of the flow:

(a) $u = x^2yz$, $v = -y^2x$
(b) $u = x^2 + 3z^2x$, $w = -z^3 + y^2$

Which case represents incompressible flow?

15. Given that an incompressible fluid is flowing in a plane, the velocity component in the y direction is $u_y = y^2 = 2x + 2yu_x$. Try to solve the velocity component in the x direction

16. Let the following functions represent the three fractional velocities of the flow:

(1) $u = kx$, $v = -ky$, $w = 0$
(2) $u = kx$, $v = -ky$, $w = kx$
(3) $u = kx.v = -ky$, $w = kz$
(4) $u = kx.v = -ky$, $w = -2kz$
(5) $u = kx$, $v = ky$, $w = kx$

where k is a constant. Which cases represent incompressible flow.

17. There is a vent with an area 0.4 m² on the wall of the air supply duct. Try to find the average velocity v of the outlet flow

18. Find the flow rate of the fluid in a single width between two parallel plates. The velocity distribution is known as

$$u = u_{max}\left[1 - \left(\frac{y}{b}\right)^2\right]$$

where $y = 0$ is the center line, $y = \pm b$ is the position of the plate, and m is a constant.

19. Mass flow $q_m = 8000$ kg/h, the temperature is 400 °C, the pipe section size is 400 × 600 mm. The density of the air in standard state (0 °C, 101,325 Pa) is $\rho = 1.29$ kg/m³. Please calculate the average flow rate of the air in the gas pipeline.

20. A pipe system is placed on a horizontal plane. pipe diameters are $d_1 = $ 15 cm, $d_2 = 7.5$ cm. Mean flow rate at the inlet is $v_1 = 2.5$ m/s. Static pressure (measured pressure) is $p_{e1} = 6.86 \times 10^4$ Pa. Without considering energy loss, try to find the horizontal force required to support the bend in its position.

Chapter 4
Plane Potential Flow of Ideal Incompressible Fluid

This chapter introduces the governing equations of the plane potential flow of ideal incompressible fluids and their solutions. The concepts of velocity potential function and stream function are introduced, the superposition principle of singularities and their solutions are explained, the basic solutions of sources (sink), dipoles and point vortices are introduced, and the characteristics and potential solutions of flows around blunt bodies and cylinders (with and without circulars) are introduced.

Learning points:

(1) Familiar with the governing equation of ideal incompressible potential flow and its basic solution.
(2) Master the principle of singularities superposition, the basic solutions of sources (sinks), dipoles and point vortices, and the characteristics and potential solutions of flows around blunt bodies and cylinders.
(3) Master the derivation process and physical meaning of Kutta–Joukowski theorem.

4.1 Basic Equations of Plane Potential Flow of Ideal Incompressible Fluid

For an ideal incompressible fluid, the basic equations are the continuous differential equations and Euler's differential equations of motion. These equations have been derived in Chap. 3 of this book, and this chapter focuses on their simplification and solutions. Euler equations belong to the first-order nonlinear partial differential equations. It is not easy to solve this set of partial differential equations mathematically, especially for the flow around the complex shape of aircraft, there is almost no exact solution, only an approximate solution. In order to simplify the problem, the plane potential flow of ideal fluid studied in this chapter belongs to the simplest type of flow problem.

© Science Press 2022
P. Liu, *Aerodynamics*, https://doi.org/10.1007/978-981-19-4586-1_4

4.1.1   Basic Equations of Irrotational Motion of an Ideal Incompressible Fluid

The basic governing equations include continuity differential equations and Euler differential equations. which is Vector form

$$
\begin{cases}
\nabla \cdot \vec{V} = 0 \\[2mm]
\dfrac{\partial \vec{V}}{\partial t} + (\vec{V} \cdot \nabla)\vec{V} = \vec{f} - \dfrac{1}{\rho}\nabla p
\end{cases}
\tag{4.1}
$$

Component form

$$
\begin{cases}
\dfrac{\partial u}{\partial x} + \dfrac{\partial v}{\partial y} + \dfrac{\partial w}{\partial z} = 0 \\[2mm]
\dfrac{\partial u}{\partial t} + u\dfrac{\partial u}{\partial x} + v\dfrac{\partial u}{\partial y} + w\dfrac{\partial u}{\partial z} = f_x - \dfrac{1}{\rho}\dfrac{\partial p}{\partial x} \\[2mm]
\dfrac{\partial v}{\partial t} + u\dfrac{\partial v}{\partial x} + v\dfrac{\partial v}{\partial y} + w\dfrac{\partial v}{\partial z} = f_y - \dfrac{1}{\rho}\dfrac{\partial p}{\partial y} \\[2mm]
\dfrac{\partial w}{\partial t} + u\dfrac{\partial w}{\partial x} + v\dfrac{\partial w}{\partial y} + w\dfrac{\partial w}{\partial z} = f_z - \dfrac{1}{\rho}\dfrac{\partial p}{\partial z}
\end{cases}
\tag{4.2}
$$

This is a set of first-order nonlinear partial differential equations, four equations, and four unknowns, respectively, u, v, w, p.

The initial conditions are

$$
t = t_0 \text{ when } \vec{V} = \vec{V}(x, y, z), \ p = p(x, y, z)
\tag{4.3}
$$

The boundary conditions are

On the boundary of the object, $V_n = 0$(non-penetrating conditions)   (4.4)

At infinity, $V = V_\infty$.

It can be seen from Eq. (4.2) that if there is an irrotational flow condition to further simplify the equations, it is quite difficult to solve. This is because the system is not only nonlinear, but also requires the velocity V and the pressure P to be coupled together. However, in the case of irrotational flows, the complexity of the problem can be simplified, especially the velocity and pressure can be solved separately. This is because, for irrotational motion, the velocity curl of the flow field is zero, i.e.,

$$
\text{rot } \vec{V} = \nabla \times \vec{V} = 2\vec{\omega} = 0
$$
$$
\omega_x = 0, \ \omega_y = 0, \ \omega_z = 0
\tag{4.5}
$$

Then there exists velocity potential function φ in the flow field, and there is

$$\vec{V} = \nabla\varphi, \ u = \frac{\partial\varphi}{\partial x}, \ v = \frac{\partial\varphi}{\partial y}, \ w = \frac{\partial\varphi}{\partial z} \qquad (4.6)$$

Substitute Eq. (4.4) into the continuity equation of incompressible fluid, and obtain

$$\left\{ \frac{\partial u}{\partial x} + \frac{\partial v}{\partial y} + \frac{\partial w}{\partial z} = 0, \ \frac{\partial^2\varphi}{\partial x^2} + \frac{\partial^2\varphi}{\partial y^2} + \frac{\partial^2\varphi}{\partial z^2} = 0 \right. \qquad (4.7)$$

It can be seen from Eq. (4.7) that the well-known Laplace equation about the velocity potential function can be obtained by the combination of the irrotational flow and the continuity equation (as shown in Fig. 4.1, the Laplace mathematician and the French fluid mechanics). This is a second-order linear homogeneous partial differential equation, which is a kinematic equation. If this equation is given the definite solution, the velocity potential function can be solved separately, and then the velocity value can be obtained. And the pressure p is not coupled to solve, so how to determine the pressure? In this case, the velocity value can be substituted as a known quantity into the equation of motion to solve the p-value. The actual solution is not directly substituted into the equation of motion, but by Bernoulli integral. Therefore, the solving steps of the whole problem can be summarized as follows:

(1) Calculate the velocity potential function and velocity component according to the kinematics equation;
(2) Bernoulli equation is used to determine the pressure at each point in the flow field.

Fig. 4.1 Laplace (1749–1827), French mathematician and fluid mechanic

This solution simplifies the problem in two aspects: one is the linearization of the original nonlinear problem, and the other is that the coupling problem of the original velocity and pressure becomes an independent solution.

In summary, the governing equation and its initial boundary conditions for an ideal incompressible irrotational flow are

$$
\begin{cases}
\dfrac{\partial^2 \varphi}{\partial x^2} + \dfrac{\partial^2 \varphi}{\partial y^2} + \dfrac{\partial^2 \varphi}{\partial z^2} = 0 \\[2mm]
\dfrac{\partial \varphi}{\partial t} + \dfrac{V^2}{2} + \dfrac{p}{\rho} + \Pi = C(t)
\end{cases}
\tag{4.8}
$$

where V is the fluid particle resultant velocity, $V = \sqrt{u^2 + v^2 + w^2}$.

The initial conditions $t = t_0$ $\vec{V} = \vec{V}_0(x, y, z)$ $p = p_0(x, y, z)$.

The boundary conditions are

$$
\begin{cases}
\dfrac{\partial \varphi}{\partial n} = 0 \text{ solid wall condition} \\[2mm]
p = p_s \text{ free surface condition} \\[2mm]
\vec{V} = \vec{V}_\infty \text{ infinite distance}
\end{cases}
\tag{4.9}
$$

Equation (4.9) shows that the wall conditions of flow around an object belong to the second type of boundary conditions, namely, the partial derivative of a given velocity potential function on the boundary.

4.1.2 Properties of Velocity Potential Function

The velocity potential function is the key unknown quantity to solve the potential flow problem. Because of its special relationship with the velocity of the flow field, understanding the nature of the function is beneficial to the solution process.

(1) The partial derivative of the velocity potential function along a certain direction and the velocity component in that direction, and the increment direction of the velocity potential function is along the streamline direction.
(2) The velocity potential function allows an arbitrary constant difference without affecting the flow velocity.
(3) The velocity potential function satisfies the Laplace equation and is a harmonic function. Because it is a second-order linear homogeneous partial differential equation, its solution satisfies the principle of linear superposition, that is, the linear combination of the velocity potential function is also full of the Laplace equation.

$$\varphi = \sum_{i=1}^{n} C_i \varphi_i \tag{4.10}$$

$$\frac{\partial^2 \varphi}{\partial x^2} + \frac{\partial^2 \varphi}{\partial y^2} + \frac{\partial^2 \varphi}{\partial z^2} = \sum_{i=1}^{n} C_i \left(\frac{\partial^2 \varphi_i}{\partial x^2} + \frac{\partial^2 \varphi_i}{\partial y^2} + \frac{\partial^2 \varphi_i}{\partial z^2} \right) = 0 \tag{4.11}$$

(4) The lines connected by points with equal velocity potential functions are called equipotential lines, which are perpendicular to the velocity direction (as shown in Fig. 4.2). The velocity direction is the increment direction of the velocity potential function.

$$d\varphi = 0, \ d\varphi = \vec{V} \cdot d\vec{s} = udx + vdy + wdz = 0 \tag{4.12}$$

$$\vec{V} \perp d\vec{s}$$

(5) The integral of the velocity curve on any curve connecting two points in the flow field is equal to the difference of the velocity potential functions of the two points. The velocity line integral is independent of the path, it just depends on the position of the two points. If it's a closed curve, the velocity circulation is zero.

$$\int_{A}^{B} \vec{V} \cdot d\vec{s} = \int_{A}^{B} (udx + vdy + wdz)$$

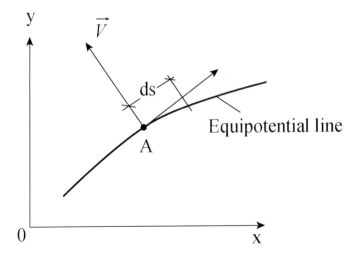

Fig. 4.2 The equipotential line is perpendicular to the velocity vector

$$= \int_A^B \left(\frac{\partial \varphi}{\partial x} dx + \frac{\partial \varphi}{\partial y} dy + \frac{\partial \varphi}{\partial z} dz \right) = \int_A^B d\varphi = \varphi_B - \varphi_A \qquad (4.13)$$

4.1.3   Stream Functions and Their Properties

According to Green's formula

$$\oint_L P dx + Q dy = \iint_\sigma \left(\frac{\partial Q}{\partial x} - \frac{\partial P}{\partial y} \right) dx dy \qquad (4.14)$$

where $P(x, y)$ and $Q(x, y)$ are smooth and continuous differentiable functions. If the

$$P = -v \quad Q = u \qquad (4.15)$$

Substituted into Eq. (4.14), it can be obtained

$$\oint_L P dx + Q dy = \oint_L -v dx + u dy = \iint_\sigma \left(\frac{\partial u}{\partial x} + \frac{\partial v}{\partial y} \right) dx dy \qquad (4.16)$$

From this equation, we can see the continuity differential equation for the plane flow of incompressible fluid

$$\frac{\partial u}{\partial x} + \frac{\partial v}{\partial y} = 0 \qquad (4.17)$$

It is a sufficient and necessary condition for (4.16) that the integral is path independent. In this way, in the plane flow of an incompressible fluid, there must be a total differential of a certain function ψ, which satisfies

$$d\psi = -v dx + u dy \qquad (4.18)$$

get

$$u = \frac{\partial \psi}{\partial y}, \quad v = -\frac{\partial \psi}{\partial x} \qquad (4.19)$$

This function is called a stream function. It can be seen that for the plane flow of incompressible fluid, whether it is an ideal fluid or a viscous fluid, whether it is a rotational flow or non-rotational flow, there is a stream function. Because the continuity differential equation of incompressible fluid is a necessary and sufficient condition for the existence of a stream function. The concept of stream function was first introduced by French mathematician and physicist Lagrange (Lagrange, 1736–1813) in 1781. The basic properties of flow function:

(1) The value of the stream function can be changed by any constant without affecting the velocity of the flow field.
(2) The line between the points with the same value of stream function is a streamline. That is, the tangent direction of the current function line coincides with the direction of the velocity vector. For incompressible fluid plane flow, on the stream function line $d\psi = 0$, namely,

$$d\psi = -vdx + udy = 0 \tag{4.20}$$

It can be obtained from this formula

$$\frac{dx}{u} = \frac{dy}{v} \tag{4.21}$$

It can be seen that Eq. (4.21) is the streamline equation of plane flow. It shows that the current line is a streamline, that is, the value of the function does not change over the same streamline.

(3) In the field of fluid, the partial derivative of the stream function at any point in a certain direction is equal to the velocity component rotated 90° clockwise in that direction, as shown in Fig. 4.3.

$$V_s = \frac{\partial\psi}{\partial n} = \frac{\partial\psi}{\partial x}\frac{\partial x}{\partial n} + \frac{\partial\psi}{\partial y}\frac{\partial y}{\partial n} = -v\cos(n, x) + u\cos(n, y) \quad \vec{n}\perp\vec{s} \tag{4.22}$$

According to the property of the stream function, if we take S along the streamline and rotate 90° counterclockwise in the direction of n, we have (4.23).

$$V_s = \frac{\partial\psi}{\partial n}, \quad V_n = -\frac{\partial\psi}{\partial s} = 0 \tag{4.23}$$

The formula (4.23) shows that the direction of the stream function increment is to rotate 90° counterclockwise along the streamline direction.

(4) The ideal incompressible fluid plane potential flow, and the stream function satisfies the Laplace equation. Namely,

$$\omega_z = \frac{1}{2}\left(\frac{\partial v}{\partial x} - \frac{\partial u}{\partial y}\right) = \frac{1}{2}\left(\frac{\partial}{\partial x}\left(-\frac{\partial\psi}{\partial x}\right) - \frac{\partial}{\partial y}\left(\frac{\partial\psi}{\partial y}\right)\right)$$
$$= -\frac{1}{2}\left(\frac{\partial^2\psi}{\partial x^2} + \frac{\partial^2\psi}{\partial y^2}\right) = 0 \tag{4.24}$$

Fig. 4.3 Stream function
derivative and velocity
component

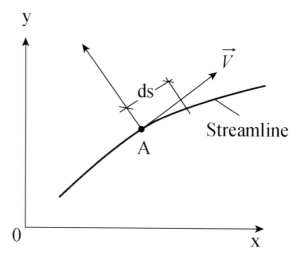

(5) The uniform velocity potential function line passing through the same point is
orthogonal to the iso-current function line (the equipotential line is orthogonal
to the stream line), as shown in Fig. 4.4. The line of the iso-current function is
a streamline, then

$$d\psi = -v dx + u dy = 0$$

$$K_1 = \frac{dy}{dx} = \frac{v}{u} \tag{4.25}$$

On the other hand, the equation of the equipotential correspondence line through
this point is

Fig. 4.4 Streamlines are
perpendicular to
equipotential lines

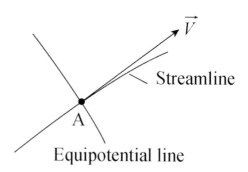

$$d\varphi = \frac{\partial \varphi}{\partial x}dx + \frac{\partial \varphi}{\partial y}dy = udx + vdy = 0$$

$$K_2 = \frac{dy}{dx} = -\frac{u}{v} \tag{4.26}$$

Then at the same point, the product of the slope of the streamline and the equipotential line is zero

$$K_1 K_2 = \frac{v}{u}\left(-\frac{u}{v}\right) = -1 \tag{4.27}$$

(6) Flow net and its characteristics

For the steady plane potential flow of an ideal incompressible fluid, the velocity potential function and flow function exist at every point in the flow field. In this way, there are two families of curves in the flow field, one is streamline, the other is equipotential, and they are orthogonal to each other. In fluid mechanics, this grid of orthogonal curves is called a flow net, as shown in Fig. 4.5. In a flow net, the ratio of the side length of each grid is equal to the ratio of the potential function to the increment of the flow function.

$$d\psi = V_s dn, \quad d\varphi = V_s ds$$

$$\frac{dn}{ds} = \frac{d\psi}{d\varphi} \tag{4.28}$$

The flow net can not only show the direction of the speed, but also reflect the magnitude of the speed. For example, where the streamline is dense, the velocity is large, and where the streamline is sparse, the velocity is small. If the flow function difference between adjacent streamlines is constant, it is equal to the increment of flow per unit width. Namely,

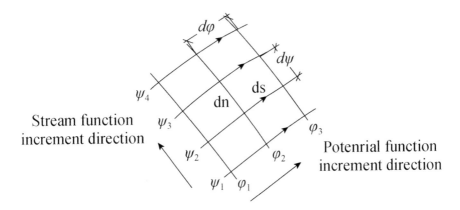

Fig. 4.5 Flow network

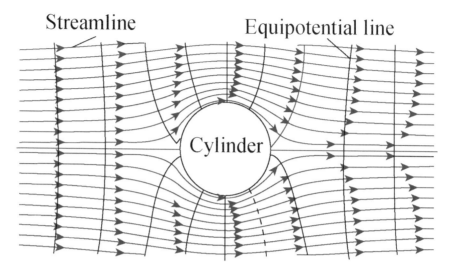

Fig. 4.6 Flow network around a cylinder (streamline and equipotential lines)

$$V_s = \frac{d\psi}{dn} = \frac{dq}{dn} \frac{V_{s1}}{V_{s2}} = \frac{dn_2}{dn_1} \tag{4.29}$$

Indicates that the velocity is inversely proportional to the grid spacing, so the density of streamline reflects the velocity. As shown in Fig. 4.6.

Example Assuming the plane potential flow of an ideal incompressible fluid, its potential function is

$$\varphi(x, y) = \frac{1}{2}a(x^2 - y^2) \tag{4.30}$$

Calculate the velocity and pressure in the flow field, and make the flow network diagram of streamline and equipotential lines.

Solution: According to the definition of velocity potential function, the two velocity components are

$$u = \frac{\partial \varphi}{\partial x} = ax, \quad v = \frac{\partial \varphi}{\partial y} = -ay$$

As shown in Fig. 4.7, the streamline equation is

$$\frac{dx}{u} = \frac{dy}{v}, \quad \frac{dx}{x} = -\frac{dy}{y}$$

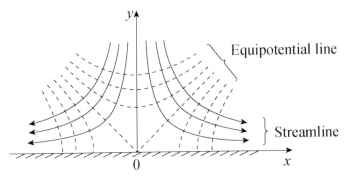

Fig. 4.7 Flow line and equipotential line of the vertical jet impingement flow field on plate

$$\frac{dx}{x} = -\frac{dy}{y}$$

The integral is $xy = C$ (4.31)

Streamline is a family of hyperbolas, with x- and y-axes as asymptotes. The family of equipotential lines is

$$a(x^2 - y^2) = C \qquad (4.32)$$

Isopotential lines are also hyperbolic families, with two straight lines $x = y$ an $x = -y$ as its asymptotic lines. The flow network is shown in Fig. 4.7.

4.1.4 Formulation of the Mathematical Problem of Steady Plane Potential Flow of Ideal Incompressible Fluid

There are three mathematical formulations for the steady plane potential flow problem of an ideal incompressible fluid. Basic problem: For a given plane object C, the flow from infinity is straight uniform flow, solve the problem of flow around a plane object.

(1) Taking the velocity potential function as an unknown function

Find the velocity potential function φ in the unbounded region outside object C, which satisfies the Laplace equation.

$$\frac{\partial^2 \varphi}{\partial x^2} + \frac{\partial^2 \varphi}{\partial y^2} = 0 \qquad (4.33)$$

And the following boundary conditions are

$$
\begin{cases}
\dfrac{\partial \varphi}{\partial n}\bigg|_C = 0 & \text{On the object,} \\[3mm]
\dfrac{\partial \varphi}{\partial x} = u_\infty, \dfrac{\partial \varphi}{\partial y} = v_\infty & \text{At infinity}
\end{cases}
\tag{4.34}
$$

This is a typical Neumann problem (given the value of the partial derivative of an unknown function on the boundary, Neumann condition).

(2)   Taking the stream function as an unknown function

Find the stream function ψ in the unbounded region outside the object C, which satisfies the Laplace equation.

$$
\frac{\partial^2 \psi}{\partial x^2} + \frac{\partial^2 \psi}{\partial y^2} = 0
\tag{4.35}
$$

And the following boundary conditions are

$$
\begin{cases}
\psi = \text{constant} & \text{On the object} \\[3mm]
\dfrac{\partial \psi}{\partial x} = -v_\infty, \dfrac{\partial \psi}{\partial y} = u_\infty & \text{At infinity}
\end{cases}
\tag{4.36}
$$

This is a typical Dirichlet problem (given an unknown function value on the boundary, the Dirichlet condition).

(3)   Take the reset potential $W(z)$ as an unknown function

For an ideal incompressible plane potential flow, both the velocity potential function and the stream function satisfy the Laplace equation, so they are harmonic functions and can be studied as complex velocity potential (reset potential), namely,

$$
w(z) = \varphi + i\psi
\tag{4.37}
$$

We need to solve for analytic functions that satisfy the condition of solution in the region outside C.

In the above three mathematical formulations, the first formulation belongs to the Riemann problem of the Laplace equation; The second formulation belongs to the Dirichlet problem of the Laplace equation; The third formulation belongs to the category of solving functions of complex variables.

4.2   Typical Singularity Potential Flow Solutions

This section mainly discusses some common typical singularity potential flows. The solutions of these potential flows are the basic solutions of some complex flows such as the flow around a cylinder.

4.2.1 *Uniform Flow*

Uniform flow is a kind of parallel flow with constant velocity, uniform flow in a straight line. The flow direction is arbitrary in the coordinate system, and the velocity field (as shown in Fig. 4.8) is

$$u = a, \ v = b \tag{4.38}$$

The velocity potential function is zero

$$u = \frac{\partial \varphi}{\partial x} = a, \ v = \frac{\partial \varphi}{\partial y} = b \ d\varphi = \frac{\partial \varphi}{\partial x} dx + \frac{\partial \varphi}{\partial y} dy = a dx + b dy$$

After integration, get

$$\varphi = ax + by + c \tag{4.39}$$

The commonly used is the straight and uniform flow parallel to the *X*-axis. The velocity of the flow from the far left is V_∞, and the corresponding stream function and potential function are

$$\varphi = V_\infty x + c$$
$$d\psi = \frac{\partial \psi}{\partial x} dx + \frac{\partial \psi}{\partial y} dy = -v dx + u dy$$
$$\psi = V_\infty y + c \tag{4.40}$$

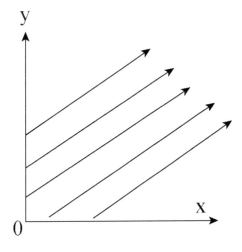

Fig. 4.8 Straight and uniform flow

4.2.2 Point Source (Sink)

Point source refers to the flow into the flow field, whereas point sink refers to the flow out of the flow field. For the convenience of discussion, point sink can also be called a negative source. For a positive point source, the flow emanates from a point in all directions, while for a negative source, the flow converges from a point in all directions.

If a point source is placed at the origin of coordinates (as shown in Fig. 4.9), considering that the point source is a flow diverging in all directions, there is only a radial velocity v_r in the flow field, but no circumferential v_θ. Suppose the radial velocity at the radius r is v_r, and the total flow rate of the point source is Q, which is obtained by the law of conservation of mass

$$Q = 2\pi r v_r, \quad v_r = \frac{Q}{2\pi} \frac{1}{r} \tag{4.41}$$

Since the flow Q is constant, the radial velocity is inversely proportional to the radius. v_r It can be proved that the velocity field described in Eq. (4.41) is a potential flow field.

As shown in Fig. 4.10, the radial velocity v_r can be expressed as cartesian coordinate components,

$$u = v_r \cos\theta = \frac{Q}{2\pi} \frac{x}{x^2 + y^2}, \quad v = v_r \sin\theta = \frac{Q}{2\pi} \frac{y}{x^2 + y^2} \tag{4.42}$$

Fig. 4.9 Point source induced velocity field

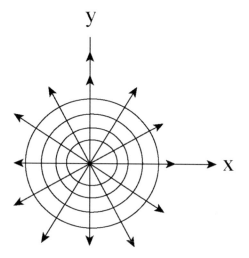

Fig. 4.10 Relationship
between cartesian
coordinates and polar
coordinates

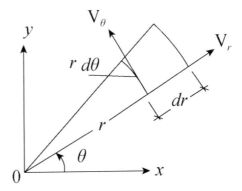

The rotation angle around the z-axis is

$$\omega_z = \frac{1}{2}\left(\frac{\partial v}{\partial x} - \frac{\partial u}{\partial y}\right) = \frac{Q}{4\pi}\left(\frac{-2xy}{(x^2+y^2)^2} - \frac{-2xy}{(x^2+y^2)^2}\right) = 0 \qquad (4.43)$$

The velocity field induced by this point source is potential flow field, so there are velocity potential function and flow function. In order to solve the problem conveniently, it is necessary to establish the velocity component expression in the polar coordinate system. As shown in Fig. 4.10, the velocity component in polar coordinates is

$$v_r = \frac{\partial \varphi}{\partial r}, \quad v_\theta = \frac{1}{r}\frac{\partial \varphi}{\partial \theta}$$
$$v_r = \frac{1}{r}\frac{\partial \psi}{\partial \theta}, \quad v_\theta = -\frac{\partial \psi}{\partial r} \qquad (4.44)$$

By using Eqs. (4.41) and (4.44), it can be obtained.
The velocity potential function is

$$\varphi = \frac{Q}{2\pi}\ln r, \quad r\sqrt{x^2+y^2} \qquad (4.45)$$

The stream function is

$$\psi = \frac{Q}{2\pi}\theta, \quad \psi = \frac{Q}{2\pi}\text{arctg}\frac{y}{x} \qquad (4.46)$$

If the position of the point source is not at the origin of coordinates but at $A(\xi, \eta)$, then the velocity, potential, and stream functions are

$$\varphi = \frac{Q}{2\pi}\ln\sqrt{(x-\xi)^2 + (y-\eta)^2} \qquad (4.47)$$

$$\psi = \frac{Q}{2\pi} \text{arctg} \frac{y - \eta}{x - \xi} \tag{4.48}$$

$$u = \frac{\partial \varphi}{\partial x} = \frac{Q}{2\pi} \frac{(x - \xi)}{(x - \xi)^2 + (y - \eta)^2}$$

$$v = \frac{\partial \varphi}{\partial y} = \frac{Q}{2\pi} \frac{(y - \eta)}{(x - \xi)^2 + (y - \eta)^2} \tag{4.49}$$

4.2.3 Dipole

A point source and a sink of equal strength are placed on the X-axis. The point source is placed at $(-h, 0)$ and the sink is placed at $(0, 0)$. All flows from the source go into the sink. By using the superposition principle of potential flow, the velocity potential function and stream function after superposition can be obtained. According to Eqs. (4.47) and (4.48), it can be obtained.

The velocity potential function is

$$\varphi = \frac{Q}{2\pi} \left[\ln \sqrt{(x + h)^2 + y^2} - \ln \sqrt{x^2 + y^2} \right] \tag{4.50}$$

The stream function for

$$\psi = \frac{Q}{2\pi} (\theta_1 - \theta_2) \tag{4.51}$$

where $\theta_1 = \text{arctg} \frac{y}{x+h}$, $\theta_2 = \text{arctg} \frac{y}{x}$ represents the angle between the flow field point P and the line between the source and sink and the x-axis. The superimposed streamlines and equipotential lines are shown in Fig. 4.11.

Now we consider a limit, when $h \to 0$, Q increases at the same time, but $\frac{Qh}{2\pi} = M$ remains the same limit. At this time, the velocity potential function becomes

$$\phi(x, y) = \lim_{h \to 0} \frac{Q}{4\pi} \left[\ln \frac{x^2 + y^2 + 2xh + h^2}{x^2 + y^2} \right]$$

$$= \lim_{h \to 0} \frac{Qh}{2\pi} \frac{x}{x^2 + y^2} = M \frac{x}{x^2 + y^2} \tag{4.52}$$

The equipotential line is the circle with its center on the X-axis, and it passes through the origin.

$$\frac{x}{x^2 + y^2} = C', \ (x - c)^2 + y^2 = c^2 \tag{4.53}$$

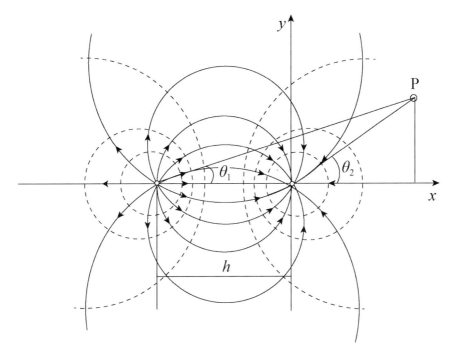

Fig. 4.11 Synthesis of point source and sink

From the expression of the stream function, it can be seen that taking $h \to 0$ and the limit that $Qh/2\pi = M$ remains unchanged, there is

$$\psi = -M\frac{y}{x^2 + y^2} \tag{4.54}$$

$$\frac{y}{x^2 + y^2} = C', \ x^2 + (y - c)^2 = c^2 \tag{4.55}$$

The streamline is also a circle, but the center of the circle is on the Y-axis, and they both pass the origin O. The velocity component is expressed as

$$u = \frac{\partial \varphi}{\partial x} = \frac{M(y^2 - x^2)}{(x^2 + y^2)^2} = -M\frac{\cos 2\theta}{r^2}$$

$$v = \frac{\partial \varphi}{\partial y} = -\frac{M2xy}{(x^2 + y^2)^2} = -M\frac{\sin 2\theta}{r^2} \tag{4.56}$$

And the resultant velocity of

$$V = \sqrt{u^2 + v^2} = \frac{M}{r^2} \tag{4.57}$$

Note that the dipole is the limit flow field where the source and sink are infinitely close. The above arrangement of dipoles takes the x-axis as the axis, and its positive direction points to the negative x-direction, as shown in Fig. 4.12. If the dipole axis and the x-axis form an angle θ, and the positive direction points to the third quadrant, the velocity potential function is

$$\varphi = \frac{M}{x^2 + y^2}(x\cos\theta + y\sin\theta) \tag{4.58}$$

The corresponding stream function is

$$\psi = -\frac{M}{x^2 + y^2}(y\cos\theta - x\sin\theta) \tag{4.59}$$

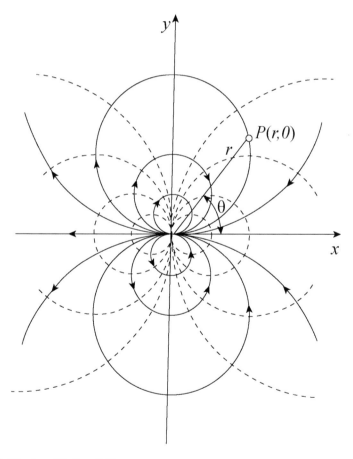

Fig. 4.12 Dipole and its flow field

If the dipole is located at (ξ, η), the axis and the x-axis form an angle θ, and the positive direction points to the third quadrant, then the potential function and the stream function are, respectively:

$$\phi = M \frac{(x - \xi) \cos \theta + (y - \eta) \sin \theta}{(x - \xi)^2 + (y - \eta)^2} \tag{4.60}$$

$$\psi = -M \frac{(y - \eta) \cos \theta - (x - \xi) \sin \theta}{(x - \xi)^2 + (y - \eta)^2} \tag{4.61}$$

4.2.4 Point Vortex

There is a point vortex with a strength of Γ at the origin, the streamline is concentric circle and the radial ray is an equipotential line, as shown in Fig. 4.13. For the flow field induced by the point vortex, there is only axial velocity component v_θ, but no radial velocity component v_r namely,

$$\Gamma = v_\theta (2\pi r), \quad v_\theta = \frac{\Gamma}{2\pi} \frac{1}{r} \tag{4.62}$$

where Γ is the point vortex strength (is constant), and the counterclockwise direction is positive. The velocity component v_θ is inversely proportional to the distance r from the center point and points counterclockwise, as shown in Fig. 4.13. It shows that the circumferential velocity of the flow field outside the point vortex is inversely proportional to the distance to the center point, and the further away from the point vortex, the smaller the induced velocity is, which is consistent with the reality.

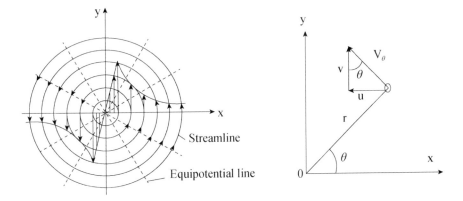

Fig. 4.13 Point vortex and its flow field

According to the velocity field formula (4.62), in the rectangular coordinate system, the velocity component is

$$u = -\frac{\Gamma}{2\pi}\frac{y}{x^2 + y^2}, \quad v = \frac{\Gamma}{2\pi}\frac{x}{x^2 + y^2} \tag{4.63}$$

The Angle of rotation about the z-axis is

$$\omega_z = \frac{1}{2}\left(\frac{\partial v}{\partial x} - \frac{\partial u}{\partial y}\right) = \frac{\Gamma}{4\pi}\left(\frac{y^2 - x^2}{(x^2 + y^2)^2} - \frac{-(x^2 - y^2)}{(x^2 + y^2)^2}\right) = 0 \tag{4.64}$$

It shows that the velocity field induced by this point vortex is a potential flow field, so there are velocity potential function (equipotential line is ray) and stream function (streamline line is circle), namely,

$$\varphi = \frac{\Gamma}{2\pi}\theta + C \tag{4.65}$$

$$\psi = -\frac{\Gamma}{2\pi}\ln r + C \tag{4.66}$$

If the position of the point vortex is not at the origin but at (ξ, η), then the velocity potential function and stream function of the point vortex are, respectively,

$$\phi = \frac{\Gamma}{2\pi}\operatorname{arctg}\frac{y - \eta}{x - \xi} \tag{4.67}$$

$$\psi = -\frac{\Gamma}{2\pi}\ln\sqrt{(x - \xi)^2 + (y - \eta)^2}. \tag{4.68}$$

The calculated value of the circulation of the closed line around the point vortex of any shape is Γ, but the circulation of the closed line excluding the point vortex is equal to zero. According to Helmholtz vortex theorem, as shown in Fig. 4.14, the ARCDEA circumferential velocity of the circulation around the enveloping point vortex is Γ, and the IGFHI circumferential velocity of the circulation around the non-enveloping point vortex is zero.

This point vortex is equivalent to an infinite straight vortex line in the z-direction, and the induced flow field in the symmetric section. Vortex originally is rotational flow, but the flow field induced by a single vortex line like this, except for the vortex center (a point on the symmetry plane perpendicular to the vortex line), the rest of the area is the potential flow field induced by the point vortex. Call such a flow field a potential vortex. When $r \to 0$, the velocity tends to infinity, and the corresponding pressure also tends to negative infinity, which is unrealistic, indicating that the induction formula is invalid at the vortex center. According to the induced velocity formula (4.62), the rate of change of velocity along the radius direction is

Fig. 4.14 The circulation of point vortices

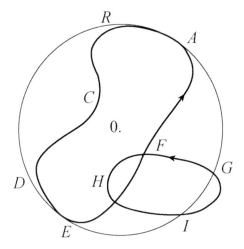

$$\frac{\partial v_\theta}{\partial r} = -\frac{\Gamma_0}{2\pi} \frac{1}{r^2} \tag{4.69}$$

When r is very small, the rate of change is very large, and the viscous force is very strong, the potential flow theory fails. The actual vortex always has a vortex core with a great viscous effect. The velocity of the fluid particles in the core is proportional to r, and there is vortex movement in the core. However, the induced velocity outside the vortex core is inversely proportional to r, as shown in Fig. 4.15. If the area with vortex inside the core is A_c, and the area outside the core with irrotational flow is A, Since the ratio between the area outside the vortex core and the area outside the vortex core is a small amount of $A_c/A \ll 1$ much less than 1. Therefore, for convenience, the influence of the vortex core area can be ignored when studying the external flow field, and it can be regarded as a point vortex with intensity Γ. The actual vortex structure is like the Rankine vortex model.

Fig. 4.15 Point vortex structure

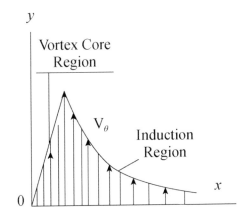

4.3 Singularity Superposition Solution of Flow Around Some Simple Objects

In this section, we discuss some potential flow solutions of singularities by using the principle of singularities superposition. Since the potential flow satisfies the principle of superposition, the composite flow superimposed by each singularity is also a potential flow. According to Eqs. (4.10) and (4.11), it is assumed that the velocity potential functions generated by the singularities are $\varphi_1, \varphi_2, \varphi_3, \ldots$ respectively, then the function after their linear combination is also the velocity potential function

$$\varphi = C_1\varphi_1 + C_2\varphi_2 + C_3\varphi_3 + \cdots \tag{4.70}$$

In the formula, the coefficient C_i can be any constant that is not zero, and the combination of different coefficients can be understood as the contribution of each singularity potential function to the composite potential function.

4.3.1 Flow Around a Blunt Semi-infinite Body

The flow around a blunt semi-infinite body can be obtained by the superposition of a straight uniform flow and a point source. Suppose a straight and uniform flow flowing parallel to the x-axis to the right, a source of strength Q is arranged at the origin of the coordinate, and the velocity potential function of the resultant potential flow superposition is

$$\varphi(x, y) = C_1 V_\infty x + C_2 \frac{Q}{2\pi} \ln r = C_1 V_\infty x + C_2 \frac{Q}{4\pi} \ln(x^2 + y^2) \tag{4.71}$$

The velocity component of the resultant potential flow is

$$\begin{cases} u = \dfrac{\partial \varphi}{\partial x} = C_1 V_\infty + C_2 \dfrac{Q}{2\pi} \dfrac{x}{x^2 + y^2} \\ v = \dfrac{\partial \varphi}{\partial y} = C_2 \dfrac{Q}{2\pi} \dfrac{y}{x^2 + y^2} \end{cases} \tag{4.72}$$

For this synthetic flow field, physical conditions need to be satisfied. Such as

(1) It can be seen from Eq. (4.72) that, when $x \to -\infty$ $u = C_1 V_\infty$, $v = 0$. In order to make the incoming flow condition of the composite flow field consistent with the straight and uniform flow V_∞ it is obvious that only the coefficient $C_1 = 1.0$ can be taken here.

(2) When $r \to 0$, take the contour around the point source, the flow through the contour obtained from the point source is Q, and the flow through the same

contour calculated by the formula (4.72) is $C_2 Q$, which is physically required for the two should be equal, so the coefficient $C_2 = 1.0$.

It can be seen that in the solution of singularities superposition, the coefficient C_i is 1. This style (4.71) and formula (4.72) can be written as

$$\varphi(x, y) = V_\infty x + \frac{Q}{2\pi} \ln r = V_\infty x + \frac{Q}{4\pi} \ln(x^2 + y^2) \tag{4.73}$$

The velocity component of the resultant potential flow is

$$\begin{cases} u = \dfrac{\partial \varphi}{\partial x} = V_\infty + \dfrac{Q}{2\pi} \dfrac{x}{x^2 + y^2} = V_\infty + \dfrac{Q}{2\pi} \dfrac{\cos \theta}{r} \\ v = \dfrac{\partial \varphi}{\partial y} = \dfrac{Q}{2\pi} \dfrac{y}{x^2 + y^2} = \dfrac{Q}{2\pi} \dfrac{\sin \theta}{r} \end{cases} \tag{4.74}$$

Firstly, the flow characteristics of the synthetic velocity Eq. (4.74) are analyzed. The resultant flow has a point on the x-axis where the resultant velocity is zero, which is the stagnation point A. Requirement at stagnation point, $y_A = 0$, $u_A = 0$, $v_A = 0$. From the formula (4.74)

$$V_\infty + \frac{Q}{2\pi x_A} = 0 \tag{4.75}$$

The coordinates of the stagnation point A are

$$x_A = -\frac{Q}{2\pi V_\infty}, \quad y_A = 0 \tag{4.76}$$

This formula shows that the distance between the stagnation point and the origin is directly proportional to the intensity of the point source and inversely proportional to the speed of a uniform direct current. Physically, the stagnation point is the point at which the velocity of the fluid particles from the point source cancels out the velocity of the straight uniform flow there (Fig. 4.16).

The stream function for the resultant flow is

$$\psi = V_\infty r \sin \theta + \frac{Q}{2\pi} \theta \tag{4.77}$$

For zero streamline

$$\psi = 0, \quad \theta = 0 \tag{4.78}$$

It's a horizontal line through the origin of the coordinates.

If $\psi = \frac{Q}{2}$, solve the formula (4.77) to get $\theta = \pi$, which is a horizontal streamline through the stagnation point. For another boundary non-horizontal streamline, take $\psi = \frac{Q}{2}$, and the radius r solved by Eq. (4.77) is

$$r = \frac{Q}{2V_\infty \sin \theta} \left(1 - \frac{\theta}{\pi} \right) \tag{4.79}$$

For $\theta = \frac{\pi}{2}, \theta = \frac{3\pi}{2}$, the corresponding radius r is

$$r_E = \frac{Q}{4V_\infty}, r_F = \frac{Q}{4V_\infty} \tag{4.80}$$

On the boundary streamline, the distance between E and F is

$$D_{EF} = r_E + r_F = \frac{Q}{2V_\infty} \tag{4.81}$$

Because the streamline of $\psi = \frac{Q}{2}$ is a boundary streamline BAB_1 passing through the stagnation point A. This streamline divides the flow field into two parts. The flow on the outside of the streamline is a straight and uniform flow around the streamline, and the flow inside is the flow of a point source in the streamline. In this way, the external flow can be regarded as a turbulent flow field where a straight and uniform flow passes around a blunt semi-infinite body shaped like BAB_1. Because the back of the object is unsealed, it is called a semi-infinite body. This semi-infinite body is at $+x$ infinity, and its width (y-direction size) tends to an asymptotic value D (the thickness of the circumfluence object)

$$D = \frac{Q}{V_\infty} \tag{4.82}$$

For the dimensionless pressure coefficient C_p, its definition is the dynamic pressure head of the current after subtracting the current static pressure from the local static pressure. On the boundary streamline, the pressure at any point is obtained by Bernoulli's equation. which is

$$p_s = p_\infty + \frac{\rho}{2} V_\infty^2 - \frac{\rho}{2} V_s^2$$

Defined by C_p as

$$C_p = \frac{p_s - p_\infty}{\frac{1}{2}\rho V_\infty^2} = 1 - \frac{V_s^2}{V_\infty^2} = 1 - \frac{u_s^2 + v_s^2}{V_\infty^2} \tag{4.83}$$

On the boundary streamline of $\psi = \frac{Q}{2}$, substituting the radius $r_s = \frac{Q}{2V_\infty \sin \theta} \left(1 - \frac{\theta}{\pi} \right)$ into the Eq. (4.74) can be obtained

$$\begin{cases} u_s = V_\infty + \dfrac{Q}{2\pi}\dfrac{\cos\theta}{r_P} = V_\infty\left(1 + \dfrac{\sin\theta\cos\theta}{\pi - \theta}\right) \\[3mm] v_s = \dfrac{\partial\varphi}{\partial y} = \dfrac{Q}{2\pi}\dfrac{\sin\theta}{r_P} = V_\infty\dfrac{\sin^2\theta}{\pi - \theta} \end{cases} \tag{4.84}$$

Substituting Eq. (4.83), the pressure coefficient along the outer surface of the semi-infinite body is

$$C_p = -\frac{\sin 2\theta}{\pi - \theta} - \left(\frac{\sin\theta}{\pi - \theta}\right)^2 \tag{4.85}$$

At the stagnation point A, take the $\theta_A = \pi$ from the above formula and obtain the pressure coefficient at the stagnation point A according to the Rubida rule

$$C_{pA} = \lim_{\theta\to\pi}\left[-\frac{\sin 2\theta}{\pi - \theta} - \left(\frac{\sin\theta}{\pi - \theta}\right)^2\right] = 1 \tag{4.86}$$

After leaving the stagnation point, C_p decreases rapidly. Not far from point A, C_p drops to zero, and the flow velocity at this point reaches the incoming flow velocity far ahead. After that, the airflow accelerates along the object surface. When it reaches a certain position, the velocity reaches the maximum and C_p decreases to the minimum. This is called the point of maximum velocity, or the point of minimum pressure, and after that point the flow begins to slow down, to the right of infinity, to the same velocity as the incoming flow far ahead. This is characteristic of most blunt objects flowing at low velocities. A low-speed and high-pressure zone is formed near the head, and then the speed increases rapidly, and the pressure drops sharply, which belongs to the pressure gradient flow.

According to formula (4.85), find the derivative of θ for C_p and set it to zero to obtain the minimum value of C_p, which is

$$\frac{dC_p}{d\theta} = 0, \ \theta_m = \mathrm{tg}^{-1}\left(\frac{\pi - \theta_m}{\pi - \theta_m - 1}\right), \ \theta_m = 62.96^0 \tag{4.87}$$

The minimum pressure coefficient is $C_{p\min} = -0.587$, and the corresponding maximum velocity around the wall is $V_{\max} = 1.26\,V_\infty$ (Fig. 4.17).

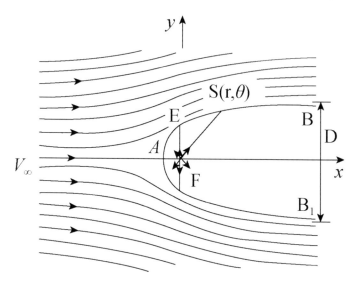

Fig. 4.16 Flow around a semi-infinite body with a blunt head

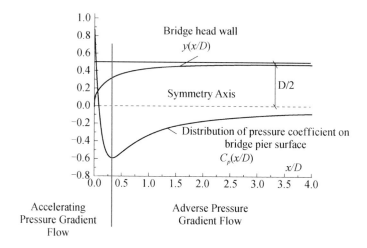

Fig. 4.17 Pressure distribution around the wall of a semi-infinite body at the pier head

4.3.2 Flow Around Rankine Pebbles

A straight uniform flow flowing to the right is arranged along the X-axis, a point source with strength Q is placed at $H/2$ to the left of the origin of the coordinate, and a point sink with strength Q at $h/2$ to the right of the origin of the coordinate. The resultant potential flow after superposition can simulate the flow field around Rankine pebbles. The resultant velocity potential function is zero

$$\varphi(x, y) = V_\infty x + \frac{Q}{4\pi} \ln\left[(x + h/2)^2 + y^2\right]$$
$$- \frac{Q}{4\pi} \ln\left[(x - h/2)^2 + y^2\right] \tag{4.88}$$

The velocity component of the resultant potential flow is

$$\begin{cases} u = \dfrac{\partial \varphi}{\partial x} = V_\infty + \dfrac{Q}{2\pi} \dfrac{x + h/2}{(x + h/2)^2 + y^2} - \dfrac{Q}{2\pi} \dfrac{x - h/2}{(x - h/2)^2 + y^2} \\ v = \dfrac{\partial \varphi}{\partial y} = \dfrac{Q}{2\pi} \dfrac{y}{(x + h/2)^2 + y^2} - \dfrac{Q}{2\pi} \dfrac{y}{(x - h/2)^2 + y^2} \end{cases} \tag{4.89}$$

The stream function for the resultant flow is

$$\psi = V_\infty y + \frac{Q}{2\pi} \mathrm{tg}^{-1}\left(\frac{y}{x + h/2}\right) - \frac{Q}{2\pi} \mathrm{tg}^{-1}\left(\frac{y}{x - h/2}\right) \tag{4.90}$$

Use the stream function formula (4.90) to take different stream function values to draw streamline diagrams; use the velocity potential function formula (4.88) to take different potential function values to draw equipotential lines. Figure 4.18 shows the streamline diagram.

If the half length of the long axis of the Rankine pebble is taken as a and the half length of the short axis is taken as b, then the shape of the different slenderness than the Rankine pebble is given in Fig. 4.19 (in the case of flow function $\psi = 0$, by Eq. (4.90) Calculated.), Fig. 4.20 shows the ratio of the surface tangential velocity to the incoming flow velocity of the Rankine pebbles with different slender ratios, and Fig. 4.21 shows the distribution curve of the surface pressure coefficient of the Rankine pebbles with different slender ratios. It can be seen from 4.12 that

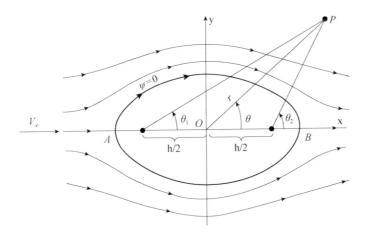

Fig. 4.18 Rankine pebble flow

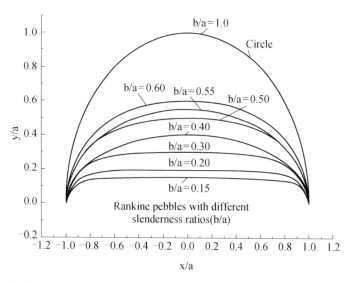

Fig. 4.19 Rankine pebble shape with different slenderness ratio *b/a*

different head shapes directly affect the distribution of head pressure coefficient. For the NACA66-012 laminar airfoil with a sharp head, the minimum pressure coefficient on the object surface appears downstream of the maximum thickness point, as shown in Fig. 4.22. For the NACA0012 airfoil with a blunt head, the minimum pressure coefficient on the object surface appears upstream of the maximum thickness point, as shown in Fig. 4.23.

4.3.3   Flow Around a Circular Cylinder Without Circulation

The flow around a cylinder without circulation can be obtained by superimposing a straight uniform flow with a dipole. Add a point source in a straight and uniform flow, and a flow around a semi-infinite body can appear, and the shape of the object will not be closed. In order to finish, it is necessary to add a point sink (negative point source). Only when the total intensity of the positive point source and the negative point source is equal to zero, the object shape is closed. Set a straight uniform flow parallel to the x axial right flow, and then an axis point to the negative x dipole is arranged at the coordinate origin. At this time, the velocity potential function of the resultant potential flow is

$$\varphi(x, y) = V_\infty x + M \frac{x}{x^2 + y^2} = V_\infty r \cos \theta + M \frac{\cos \theta}{r} \tag{4.91}$$

where M is the strength of the dipole.

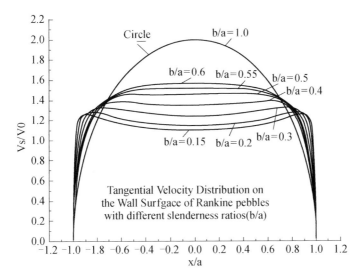

Fig. 4.20 The ratio of the surface tangential velocity to the incoming flow velocity of Rankine pebbles with different slenderness ratio *b/a*

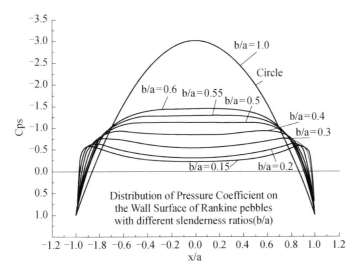

Fig. 4.21 The distribution curve of the surface pressure coefficient of Rankine pebbles with different slenderness ratio *b/a*

The stream function of the resultant potential flow field is

$$\psi(x, y) = V_\infty y - M \frac{y}{x^2 + y^2} = V_\infty r \sin \theta - M \frac{\sin \theta}{r} \qquad (4.92)$$

Fig. 4.22 Surface pressure coefficient distribution of laminar flow airfoil NACA66-012

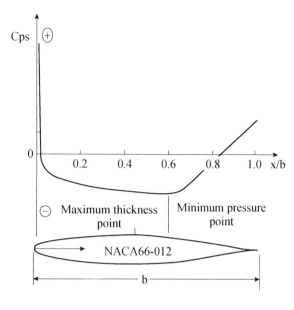

Fig. 4.23 Pressure coefficient distribution of NACA0012 airfoil

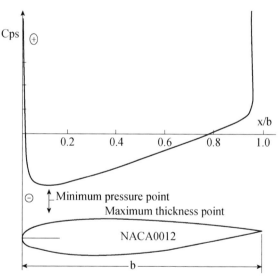

The resultant velocity field is

$$u = \frac{\partial \varphi}{\partial x} = V_\infty + M \frac{y^2 - x^2}{(x^2 + y^2)^2}$$

$$v = \frac{\partial \varphi}{\partial y} = M \frac{-2xy}{(x^2 + y^2)^2}$$

(4.93)

Using Eq. (4.92), the stagnation point on the X-axis can be determined. At the stagnation point, $y_A = 0, u_A = 0, v_A = 0$ get

$$V_\infty - \frac{M}{x_A^2} = 0 \tag{4.94}$$

get

$$x_A^2 = M/V_\infty, a^2 = M/V_\infty \tag{4.95}$$

where a is the radius of the cylinder. So the velocity potential function can be written as

$$\varphi(x, y) = V_\infty \left(x + \frac{a^2 x}{r} \right) = V_\infty \left(r + \frac{a^2}{r} \right) \cos \theta \tag{4.96}$$

The stream function equation is

$$\psi(x, y) = V_\infty \left(r - \frac{a^2}{r} \right) \sin \theta \tag{4.97}$$

For a streamline with $\psi = 0$, it is easy to prove that there is a circle on the streamline besides the x-axis passing through the stagnation point, there is also a $r = a$ circle on the streamline, as shown in Fig. 4.24. According to the expression for the velocity component

$$V_r = \frac{\partial \varphi}{\partial r} = V_\infty \left(1 - \frac{a^2}{r^2} \right) \cos \theta$$

$$V_\theta = \frac{1}{r} \frac{\partial \varphi}{\partial \theta} = -V_\infty \left(1 + \frac{a^2}{r^2} \right) \sin \theta \tag{4.98}$$

On the circumference, $r = a$, the velocity component is

$$V_{rs} = \frac{\partial \varphi}{\partial r} = V_\infty \left(1 - \frac{a^2}{a^2} \right) \cos \theta = 0$$

$$V_{\theta s} = \frac{1}{r} \frac{\partial \varphi}{\partial \theta} = -V_\infty \left(1 + \frac{a^2}{a^2} \right) \sin \theta = -2 V_\infty \sin \theta \tag{4.99}$$

According to the Bernoulli equation and the definition of pressure coefficient, the distribution of pressure coefficient on the cylindrical surface is

$$C_p = 1 - \frac{V_{\theta s}^2}{V_\infty^2} = 1 - 4 \sin^2 \theta \tag{4.100}$$

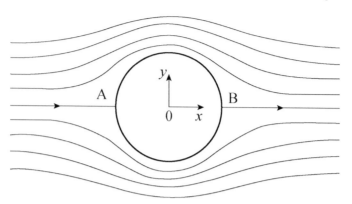

Fig. 4.24 Flow around a cylinder

The pressure coefficient distribution on the cylinder surface is shown in Fig. 4.25.
At the stagnation point before and after the circle, $\theta = 0°$, $\theta = \pi$, the pressure coefficient is equal to 1.0. Flowing from the front stagnation point to the back, the flow velocity at $\theta = \pm 150°$ accelerates to the same size as the flow velocity of the incoming flow. Continue to accelerate afterward and reach the maximum speed at $\theta = \pi/2$, which is twice the incoming flow speed, and C_p is -3.0. Starting from the front stagnation point, the pressure of the fluid particle along the object surface decreases and the velocity increases. After the maximum velocity point, the fluid particle velocity decreases and drops to zero at $\theta = 0°$. This point is called the post-stagnation point. Therefore, on the cylindrical object surface, from the front stagnation point to the rear stagnation point, the flow undergoes an acceleration

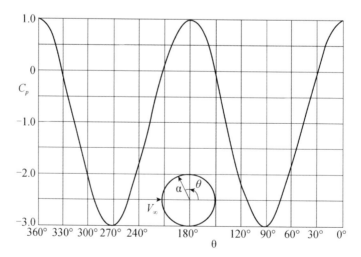

Fig. 4.25 Distribution of pressure coefficient on cylindrical surface

Fig. 4.26 Force of flow
around a cylinder

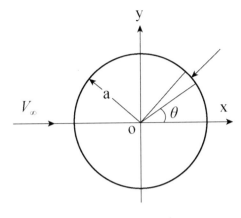

process and a deceleration process. During the acceleration process, the pressure
gradient along the cylindrical surface is negative, and this area is called the zone
of the favorable pressure gradient. During the deceleration of the flow, the pressure
gradient along the cylinder surface is positive, and this area is called the zone of the
adverse pressure gradient.

As shown in Fig. 4.26, by integrating the pressure on the cylinder surface, the
resultant force acting on the cylinder can be obtained. The component along the
x-axis is called resistance, which is

$$D = \oint_{2\pi a} -p_s \cos\theta \, ds = \int_0^{2\pi} -p_s a \cos\theta \, d\theta \tag{4.101}$$

The component along the direction perpendicular to the incoming flow is called
lift, which is

$$L = \oint_{2\pi a} -p_s \sin\theta \, ds = \int_0^{2\pi} -p_s a \sin\theta \, d\theta \tag{4.102}$$

Using Eq. (4.99), the pressure on the cylindrical surface can be obtained from
Bernoulli's equation as

$$p_s = p_\infty + \frac{1}{2}\rho V_\infty^2 - \frac{1}{2}\rho(-2V_\infty \sin\theta)^2 \tag{4.103}$$

By substituting Eq. (4.103) into Eqs. (4.101) and (4.102), it can be obtained

$$D = 0, \ L = 0 \tag{4.104}$$

Fig. 4.27 D 'Alembert,
French mathematician, and
mechanist (1717–1783)

It can be seen that the resultant force of an ideal incompressible fluid flowing around a cylinder is zero. Because the pressure distribution on the surface of the flow around the cylinder is not only symmetrical up and down, but also symmetrical left and right, the resultant force is zero. However, the actual flow is asymmetric on the left and right, because the actual fluid is viscous. When the fluid particle bypasses the maximum velocity point, the flow enters the inverse pressure gradient zone, and the fluid particle decelerates and pressurizes along the flow direction, due to the loss of viscosity and the reverse pressure. The dual effect of the gradient makes it impossible to always flow against the cylindrical surface to the back stagnant point, but to leave the object surface at a certain point on the back of the cylinder, resulting in flow separation, which causes the pressure distribution on the cylindrical surface to be asymmetric before and after, resulting in resistance. Regarding the problem of no resistance in the flow of an ideal incompressible fluid cylinder, as early as 1752, the French mathematics and mechanics D'Alembert (D'Alembert, 1717–1783, as shown in Fig. 4.27), in his paper "A New Theory of Fluid Damping," published, proved for the first time that ideal incompressible fluid flows around any three-dimensional object without resistance by means of mathematical mechanics, later known as the D 'Alembert paradox. These people study viscous fluid movement to promote.

4.3.4 Flow Around a Cylinder with Circulation

On the basis of the superposition of the straight uniform flow and the dipole, a point vortex of strength A is added at the center of the circle (clockwise to negative).

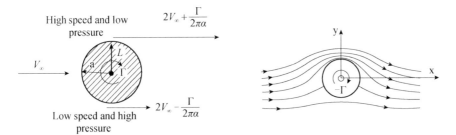

Fig. 4.28 Flow around a cylinder with circulation

Since the flow caused by the point vortex is a circular motion around the point vortex. Therefore, after the point vortex is superimposed on the flow around without circulation, the original symmetrical flow field will not exist, as shown in Fig. 4.28.

According to the singularity arrangement in Fig. 4.28, the velocity potential function and flow function of the resultant potential flow superposed are respectively.

The resultant potential flow velocity potential function is

$$\varphi(x, y) = V_\infty \left(r + \frac{a^2}{r} \right) \cos \theta - \frac{\Gamma}{2\pi} \theta \tag{4.105}$$

The stream function of the resultant potential flow is

$$\psi(x, y) = V_\infty \left(r - \frac{a^2}{r} \right) \sin \theta + \frac{\Gamma}{2\pi} \ln r \tag{4.106}$$

In polar coordinates, the velocity component of the resultant potential flow is

$$\begin{cases} v_r = \dfrac{\partial \varphi}{\partial r} = V_\infty \left(1 - \dfrac{a^2}{r^2} \right) \cos \theta \\[3mm] v_\theta = \dfrac{1}{r} \dfrac{\partial \varphi}{\partial \theta} = -V_\infty \left(1 + \dfrac{a^2}{r^2} \right) \sin \theta - \dfrac{\Gamma}{2\pi r} \end{cases} \tag{4.107}$$

The cylindrical surface of $r = a$ is a streamline. On this circle $V_r = 0$, the circumferential velocity (shown in Fig. 4.28) is

$$V_{\theta s} = -2V_\infty \sin \theta - \frac{\Gamma}{2\pi a} \tag{4.108}$$

It can be seen that the stagnation point is not in $\theta = \pi$ and $\theta = 0$, let $V_\theta = 0$, a new stagnation point can be determined, which is

$$V_{\theta s} = -2V_\infty \sin \theta - \frac{\Gamma}{2\pi a} = 0, \ \sin \theta_0 = -\frac{\Gamma}{4\pi a V_\infty} \tag{4.109}$$

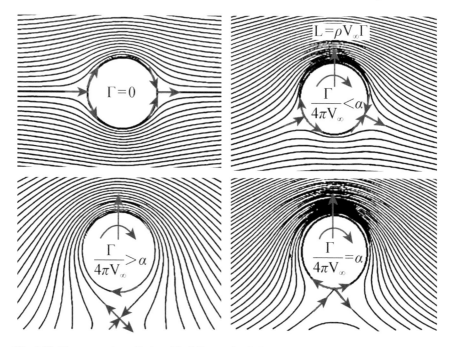

Fig. 4.29 Flow around a cylinder with different circulation

The front and back stagnation points are in the third and fourth quadrants. They are symmetrical to the y-axis. The deviation of the new stagnation points from π and $0°$ is determined by the ratio of the added circulation and the product of the incoming flow velocity and the radius of the cylinder. When the incoming flow velocity and cylinder radius are constant, the surrounding flow field generated by different Γ values are shown in Fig. 4.29

It can be seen from Fig. 4.29 that as the amount of circulation increases, stagnation points occur at different positions, which is

(1) In the case of the circulation $\Gamma = 0$, the flow around the cylinder is symmetrical, and the front and rear stagnation points are on the x-axis.

(2) For the case of $\frac{\Gamma}{4\pi a V_\infty} < 1$, the flow around the cylinder is asymmetric, and the stagnation points are located in the third and fourth quadrants. But as the amount of circulation increases, the deflection angle of the stagnation point also increases.

(3) For the case of $\frac{\Gamma}{4\pi a V_\infty} = 1$, the stagnation points before and after on the cylinder are synthesized into a point, which is the maximum value of the circulation of the stagnation point on the object surface.

(4) In the case of $\frac{\Gamma}{4\pi a V_\infty} > 1$, there is no stagnation point on the cylinder because the velocity circulation is too large, and the stagnation point appears in the flow field leaving the cylinder surface.

According to Bernoulli's equation, the pressure at any point on the cylinder is

$$p_s = p_\infty + \frac{\rho}{2}V_\infty^2 - \frac{\rho}{2}V_{\theta s}^2 \tag{4.110}$$

By substituting Eq. (4.108) into the above equation, get

$$p_s = p_\infty + \frac{\rho}{2}V_\infty^2 - \frac{\rho}{2}\left(-2V_\infty \sin\theta - \frac{\Gamma}{2\pi a}\right)^2 \tag{4.111}$$

The pressure coefficient on the cylinder is

$$C_p = 1 - \frac{V_{\theta s}^2}{V_\infty^2} = 1 - 4\left(\sin\theta + \frac{\Gamma}{4\pi a V_\infty}\right)^2 \tag{4.112}$$

The pressure coefficient distribution on the cylinder surface drawn by Eq. (4.112) is shown in Fig. 4.30. It can be seen from the figure that at the stagnation point, $\sin\theta + \frac{\Gamma}{4\pi a V_\infty}$, and the pressure coefficient is 1.0; at $\theta = 90°$, where the velocity on the cylindrical surface is the largest and the pressure coefficient is the smallest, its value is

$$C_{p\min} = 1 - 4\left(1 + \frac{\Gamma}{4\pi a V_\infty}\right)^2 \tag{4.113}$$

For the limit case where the stagnation point is on the cylinder surface, $\frac{\Gamma}{4\pi a V_\infty} = 1$ the minimum pressure coefficient $C_{p\min} = -15$.
If taken $\frac{\Gamma}{4\pi a V_\infty} = \frac{1}{2}$, the minimum pressure coefficient $C_{p\min} = -8$.
As shown in Fig. 4.30, by using the formula (4.111), the pressure integral on the cylindrical surface is projected along the x-axis, and the resistance is obtained.

$$D = \int_0^{2\pi} -p_s a \cos\theta d\theta$$

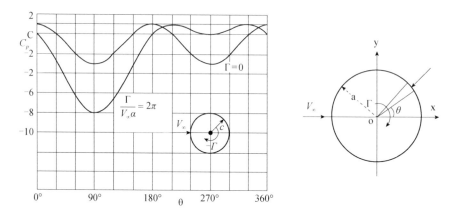

Fig. 4.30 Pressure coefficient distribution of flow around a cylinder

$$= -\int_0^{2\pi} \left[p_\infty + \frac{\rho}{2} V_\infty^2 - \frac{\rho}{2} \left(-2V_\infty \sin\theta - \frac{\Gamma}{2\pi a} \right)^2 \right] a\cos\theta \, d\theta$$

$$= \frac{1}{2}\rho V_\infty^2 \int_0^{2\pi} \left(2\sin\theta + \frac{\Gamma}{2\pi V_\infty a} \right)^2 a\cos\theta \, d\theta = 0 \qquad (4.114)$$

Projection along the direction perpendicular to the incoming flow, get lift, which is

$$L = \int_0^{2\pi} -p_s a \sin\theta \, d\theta$$

$$= \int_0^{2\pi} - \left[p_\infty + \frac{\rho}{2} V_\infty^2 - \frac{\rho}{2} \left(-2V_\infty \sin\theta - \frac{\Gamma}{2\pi a} \right)^2 \right] a\sin\theta \, d\theta$$

$$= \int_0^{2\pi} - \left[p_\infty + \frac{\rho}{2} V_\infty^2 \right] a\sin\theta \, d\theta + \int_0^{2\pi} \left[\frac{\rho}{2} V_\infty^2 \left(2\sin\theta + \frac{\Gamma}{2\pi V_\infty a} \right)^2 \right] a\sin\theta \, d\theta$$

$$= \frac{\rho}{2} V_\infty^2 \int_0^{2\pi} \left[4\sin^2\theta + \frac{2\Gamma \sin\theta}{\pi V_\infty a} + \left(\frac{\Gamma}{2\pi V_\infty a} \right)^2 \right] a\sin\theta \, d\theta$$

$$= \frac{\rho V_\infty \Gamma}{\pi} \int_0^{2\pi} \sin^2\theta \, d\theta = \rho V_\infty \Gamma \qquad (4.115)$$

Equation (4.115) is called the Kutta–Joukowski theorem flowing through the object with circulation. The German mathematician Kutta (1867–1944) proposed it in 1902, and the Russian physicist Joukowski (1847–1921) proposed it in 1906. The Kutta–Joukowski theorem states that: for an ideal incompressible fluid to bypass any enclosed object, the lift is equal to the product of the incoming flow velocity, the incoming flow density, and the circulation, and the direction of lift is that the direction of the incoming flow rotates 90° in the opposite direction of Γ, as shown in Fig. 4.31, which is

$$\vec{L} = \rho \vec{V}_\infty \times \vec{\Gamma} \tag{4.116}$$

In fact, as early as 1852, the German scientist Magnus (1802–1870, as shown in Fig. 4.32) experimentally found that a lateral force (lift) appeared in the flow around a rotating cylinder, causing the cylinder to move laterally. This phenomenon is called Magnus effect. In this regard, the arc trajectory of the rotating ball is a typical example.

4.4 Numerical Method for Steady Flow Around Two-Dimensional Symmetrical Objects

By superposing a straight uniform flow with a distributed dipole (or a distributed point source and sink of zero total intensity), the resultant flow can represent the flow around a symmetric closed body. As shown in Fig. 4.33, the straight and uniform flow flows forward along the x-axis, and the incoming flow velocity is V_∞, continuously distributed dipoles are arranged in the range $x = a$ and $x = b$ on the x-axis, and the dipole intensity distribution function is $m(x)$ (dipole density) in unit length. If the dipole density is distributed in the region discussed, then in the tiny region with the

Fig. 4.31 Direction of lift and circulation

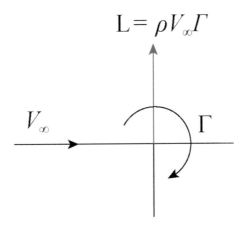

Fig. 4.32 German scientist
Magner (1802–11,870)

distance of ζ from the origin, the stream function $d\zeta$ generated by the tiny segment $d\zeta$ dipole can be obtained by using Eq. (4.54)

$$d\psi = -\frac{m(\zeta)y\,d\zeta}{(x-\zeta)^2 + y^2} \tag{4.117}$$

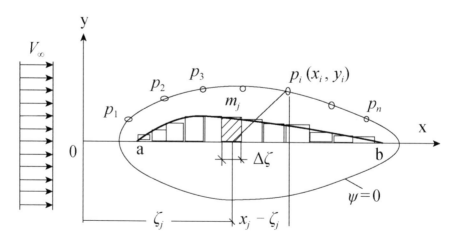

Fig. 4.33 Numerical method of flow around an arbitrary object

The total stream function of the resultant potential flow is

$$\psi = V_\infty y - \int_a^b \frac{m(\zeta)y}{(x - \zeta)^2 + y^2} d\zeta \tag{4.118}$$

The shape of the object is represented by zero-streamline lines. The shape of the object can be determined by the value of the zero-flow function, and the pressure distribution on the surface of the object can be determined by the velocity on the zero-streamline line.

For practical problems, it is often the shape of an object to determine its flow characteristics. In this case, the dipole density distribution function needs to be solved by the stream function. For the dipole density, the stream function is an integral equation, and it is difficult to find the exact solution, which can be solved by numerical methods.

The steps of numerical solution are as follows:

(1) Divide the distribution region of dipole density function into n segments of equal width, set the width of each segment as $\Delta\zeta$, and the number of segments n can be determined according to the accuracy requirements.

In the flow field, the stream function at a certain point P is

$$\psi = V_\infty y - \sum_{j=1}^n \frac{m_j \Delta\zeta y}{(x - \zeta_j)^2 + y^2} \tag{4.119}$$

where ζ_j is the distance from the origin of the midpoint in segment j; m_j is the mean value of dipole density in segment j; $m_j \Delta\zeta$ represents the strength of the dipole in segment j.

(2) The distribution of dipole density function is determined by the boundary condition of object surface

For n known points (x_i, y_i) on the surface of a given object, an n-element simultaneous algebraic system for an unknown function can be obtained as

$$\psi_i = 0, \quad V_\infty y_i - \sum_{j=1}^n \lambda_{ij} m_j = 0, \quad i = 1, 2, \cdots, n \tag{4.120}$$

Among them, the influence coefficient λ_{ij}

$$\lambda_{ij} = \frac{y_i \Delta \zeta}{\left(x_i - \zeta_j\right)^2 + y_i^2} \tag{4.121}$$

Indicates the contribution of the unit dipole density to the stream function of the point p_i (x_i, y_i) on the surface of the object at ζ_j on the x-axis.

Expansion (4.118), we can get

$$
\begin{aligned}
\lambda_{11} m_1 + \lambda_{12} m_2 + \cdots + \lambda_{1n} m_n &= V_\infty y_1 \\
\lambda_{21} m_1 + \lambda_{22} m_2 + \cdots + \lambda_{2n} m_n &= V_\infty y_2 \\
&\vdots \\
\lambda_{n1} m_1 + \lambda_{n2} m_2 + \cdots + \lambda_{nn} m_n &= V_\infty y_n
\end{aligned}
\tag{4.122}
$$

(3)   To solve the numerical solution of the dipole density distribution

Equation (4.122) is a system of linear equations of order n, which can be solved numerically to determine the unknown dipole density distribution. Once the dipole density distribution of the given object shape is solved, the stream function at any point in the flow field can be determined, and then the velocity value on the surface can be determined by the stream function. Finally, the velocity and pressure distribution at each point in the flow field and the surface can be determined by the Bernoulli equation.

In the numerical solution above, the dipole distributed in segment j is replaced by an equal-strength dipole concentrated at the midpoint of the segment. Obviously, this approximation can only be accurate if the number of sections is large. Theoretically, as the number of segments n tends to infinity, the numerical results of the dipole density distribution tend to the exact solution. In practical application, the number of segments is always finite. Of course, it can also be solved by the velocity potential function, but when the potential function is used, the corresponding boundary condition of the object surface is the non-penetration condition, that is, the normal deviation of the object surface is zero. These two methods are equivalent.

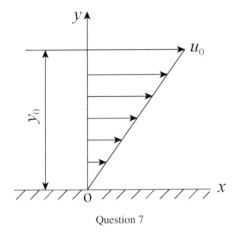

Question 7

8. Try to prove that the following two flow fields are equivalent

$$\varphi = x^2 + x - y^2$$
$$\psi = 2xy + y^2$$

9. There is a uniform linear incoming flow with velocity u_0 at infinity and a confluence with intensity $-q$ at the origin of coordinates. Try to find the flow function after the superposition of the two flows, the position of stagnation point and the boundary equation of fluid inflow and flow through the confluence.

10. There are two equal-strength point vortices with opposite rotation directions at $(0, a)$ and $(0, -a)$. When $a \rightarrow 0$, $2\pi a\Gamma$ is a constant. Prove that the corresponding flow is exactly the same as the dipole whose axis is on the x-axis.

11. There Are Two-Point Sources Located at $(1,0)$ and $(-1,0)$ with the Same Intensity of 4π, Calculate the Speed at $(0,0)$, $(0,1)$, $(0, -1)$, and $(1,1)$.

12. The point source and sink of equal intensity with a distance of $2a$ and the intensity of Q are located on a straight line at an angle of $45°$ to the x-axis, and the point source and sink are symmetrical with respect to the origin. Prove that when $a \rightarrow 0$ and keep $2\pi a Q$ equal to the constant M, the flow function of the resulting dipole is

$$\varphi = -\frac{M}{2\pi} \frac{\sqrt{2}}{2} \frac{y - x}{x^2 + y^2}$$

13. Verify that in a straight and uniform flow, the pressure coefficient on the surface of a cylinder with radius a is

$$C_p = 1 - 4\sin^2\theta \left(1 + \frac{\Gamma}{4\pi a V_\infty \sin\theta}\right)^2$$

The amount of circulation around the cylinder is r.

14. A cylinder with a diameter of 1.2 m and a length of 50 m rotates around its axis at an angular velocity of $90r/\min$. and the air flows around the cylinder at a speed of 80 km/h in a direction perpendicular to the axis of the cylinder. Try to calculate the amount of velocity circulation, lift and the position of the stagnation point. Assuming that there is no sliding between the circulation and the cylinder $\rho = 1.205$ kg/m^3.

15. The flow function of a two-dimensional flow field can be written as follows

$$\psi = 100y\left(1 - \frac{25}{r^2}\right) + \frac{628}{2\pi}\ln\left(\frac{r}{5}\right)$$

Try to calculate the shape of the zero streamline, the position of the stagnation point, the amount of circulation around the object, the speed at infinity, and the force acting on the object.

Exercises

A. Answer the following questions

1. Ideal incompressible potential flow, why the velocity potential function satisfies the Laplace equation.

2. For compressible potential flows, derive the differential equation that the velocity potential function satisfies.

3. Please explain the conditions for the ideal incompressible potential flow velocity potential function to satisfy the Laplace equation?

4. Please explain the conditions for the ideal incompressible potential flow function to satisfy the Laplace equation?

5. For incompressible plane flows, what is the relationship between flow function and vorticity?

6. For an ideal incompressible potential flow, prove that the streamline is orthogonal to the equipotential line.

7. Please give the formulation of the definite solution of the velocity potential function and the flow function for the potential flow problem around an object.

8. Please analyze the variation of pressure distribution along the surface in the flow around a blunt object to find the minimum pressure coefficient.

9. In the flow around the cylinder, please explain the variation of velocity and pressure along the cylinder surface from the front stagnation point to the rear stagnation point.

10. Please give the Angle of the stationary point on the cylinder surface corresponding to the point vortex strength Γ = $\pi a V_\infty$, $2\pi a V_\infty$, $3\pi a V_\infty$, $3.464\pi a V_\infty$, $4\pi a V_\infty$.

11. For the flow around a circular cylinder, integrate the pressure on the cylinder surface to find the lift of the cylinder. The pressure coefficient on the cylinder is

$$C_p = 1 - \left(2 \sin \theta + \frac{\Gamma}{2\pi a V_\infty}\right)^2$$

B. Calculate the following questions

1. Let $G(x, y)$ be the solution of the two-dimensional Laplace equation. Please prove that $G(x, y)$ can represent the potential function or flow function of the two-dimensional inviscid and incompressible flow.

2. Try to prove that the incompressible fluid plane flow: $v_x = 2xy + x$, $v_y = x^2 - y^2$, which satisfies the continuity equation, is a potential flow, and finds the velocity potential.

3. Given the velocity potential $\varphi = xy$, find the velocity component and the flow function, and draw the equipotential lines with φ being 1, 2, and 3. Prove that equipotential lines and streamlines are orthogonal to each other.

4. The flow function of incompressible fluid in plane flow is $\varphi = xy + 2x - 3y + 10$, try to find its velocity potential.

5. Superposition point vortices and point sources with the center at the origin, and test that the resultant flow is a rotational flow. In this flow, the angle between the velocity and the polar radius is equal everywhere, and its value is equal to $\arctan(-\Gamma/Q)$.

6. Given that the velocity of plane irrotational flow is velocity potential $\varphi = \frac{2x}{x^2 - y^2}$, try to find the flow function and velocity field.

7. Given that the velocity of plane flow is a linear distribution, if $y_0 = 4\text{m}$, $u_0 = 80\text{m/s}$, try to calculate: (1) flow function ψ: (2) whether the flow is a potential flow.

Chapter 5
Fundamentals of Viscous Fluid Dynamics

This chapter introduces the basics of viscous fluid mechanics, including the stress state of viscous fluid motion, generalized Newton's internal friction theorem (constitutive relationship), the derivation of viscous flow differential equations (Navier–Stokes equations), the basic properties of viscous fluid motion, laminar flow and the basic characteristics of turbulence, the principle of similarity and dimensional analysis, etc.

Learning points:

(1) Familiar with the basic characteristics of viscous fluid motion, the stress state of viscous fluid motion, and the generalized Newton's internal friction theorem (constitutive relationship);
(2) Familiar with the derivation process of the viscous flow differential equations (Navier–Stokes equations) and the physical meaning of each item;
(3) Familiar with the basic properties of viscous fluid motion and the physical meaning of laminar flow, turbulence, and its energy loss;
(4) Master the similarity principle and dimensional analysis method.

5.1 The Viscosity of Fluid and Its Influence on Flow

Viscous fluid mechanics is a branch of fluid mechanics, which is the study of the law of motion of viscous fluid and its interaction with solids. Viscosity is one of the basic physical properties of fluids, reflecting the ability of fluids to resist shear deformation. In the early stages of the development of theoretical fluid mechanics, ideal fluid models were used for fluids with very small viscosities. The ideal fluid has no viscosity and thermal conductivity, which brings great convenience to the mathematical solution, so that a series of analytical solutions to flow problems can be obtained. The ideal fluid model is an approximation to the real fluid. Experiments have found that when solving the lift problem that is not affected by the viscosity,

© Science Press 2022
P. Liu, *Aerodynamics*, https://doi.org/10.1007/978-981-19-4586-1_5

the results are still accurate; but when solving the drag problem, the results obtained by the ideal fluid model are very different from the experimental results, and even the opposite conclusion. For example, the flow around a two-dimensional cylinder without resistance is a typical example. This chapter mainly discusses the stress state of viscous fluid, the constitutive relationship with fluid motion, the basic differential equations of viscous flow—Navier–Stokes equations derivation, the basic properties of viscous fluid motion, and the basic characteristics of laminar and turbulent flow.

5.1.1 Viscosity of Fluid

In a static state, the fluid cannot withstand shear forces. But in the state of motion, the fluid can bear the shear force, and the magnitude of the shear force is different for different fluids. The viscosity of the fluid reflects the ability of the fluid to resist shear deformation in motion. The shear deformation of fluid refers to the relative movement between fluid particles. Therefore, the viscosity of a fluid refers to the ability of the fluid to resist relative movement between the mass points. The ability of a fluid to resist shear deformation can be manifested by the shear force between fluid layers. (This shear force is called internal friction). In the process of fluid flow, it must overcome internal friction to do work, so fluid viscosity is the root cause of fluid loss of mechanical energy.

As early as 1686, Newton proposed the law of internal friction for fluid motion in his book The Mathematical Principles of Natural Philosophy, which is

$$\tau = \mu \frac{\mathrm{d}u}{\mathrm{d}y} \tag{5.1}$$

5.1.2 Characteristics of Viscous Fluid Movement

Fluids in nature are viscous, so the effect of viscosity on fluid movement is universal. But for specific flow problems, viscosity does not necessarily play the same role. Especially for small viscous fluids like water and air, the effect of viscosity is not always negligible. The following examples illustrate the effect of viscosity on flow.

(1) Flow around the plate

When the ideal fluid bypasses the plate (no thickness), the plate does not have any effect on the flow. The fluid particles on the surface of the plate are allowed to slide through the plate, but not to penetrate the plate (usually called the non-penetration condition). The plate has no blocking effect on the flow, and the resistance of the plate is zero (Fig. 5.1).

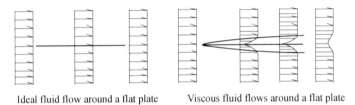

Ideal fluid flow around a flat plate   Viscous fluid flows around a flat plate

Fig. 5.1 Flow around a flat plate of different fluids

But if it is a viscous fluid, the situation is different. Due to the viscosity, the fluid particles close to the surface of the plate adhere to the plate, and there is no relative movement with the surface of the plate (neither penetration nor sliding is allowed). That is to say, the fluid particles on the boundary surface must meet the requirements. Slip condition. As the distance from the plate increases, the velocity of the fluid increases rapidly from the zero value at the wall to the velocity of the incoming flow. In this way, there is a flow with a large velocity gradient in the vicinity of the flat plate, so the viscous shear stress between the flow layers cannot be ignored, which hinders the flow. This area is called the boundary layer area. The plate has a blocking effect on the flow, and the resistance of the plate is not zero, and the resistance is equal to the integral of the frictional shear stress on the plate surface, which is

$$D_f = 2 \int_0^L \tau_0 dx \tag{5.2}$$

(2) Flow around the cylinder

The ideal fluid flows around a cylinder, and there are front stagnant point A and back stagnant point D as well as maximum velocity points B and C on the cylinder. The central streamline diverges at the front stagnation point and merges at the rear stagnation point, as shown in Fig. 5.2. According to Bernoulli's equation, the process by that the fluid particle goes through the cylinder is in the $A - B$ (C) zone. The fluid particle goes from the stagnation point A to a point where the flow velocity is zero and the pressure is the largest. Later, the pressure of the particle decreases along the way, and the flow velocity increases along the way. When reaching point B, the flow velocity is the largest and the pressure is the smallest. This zone belongs to the increasing speed and decreasing pressure zone, following the pressure gradient zone; in the $B(C) - D$ zone, the pressure of the fluid particle increases along the way, and the flow velocity decreases along the way. Point D has the highest pressure and zero-flow rate. This zone belongs to the decreasing speed and increasing pressure zone, which is the adverse pressure gradient zone. In the process of fluid particles circumventing the cylinder, there is only mutual conversion of kinetic energy and pressure energy, without loss of mechanical energy. The pressure distribution on the cylindrical surface is symmetrical, and there is no resistance (D'Alembert's paradox),

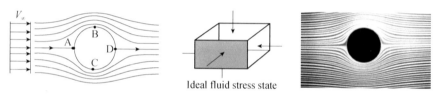

Ideal fluid stress state

Fig. 5.2 Flow around an ideal fluid cylinder

which is

$$D = \oint_{2\pi R} (-p_s \cos\theta)\mathrm{d}s = 0 \tag{5.3}$$

The flow around viscous fluid is very different from the flow around ideal fluid. Due to the adhesion conditions between the fluid and the solid wall surface, a boundary layer will be formed in the near area of the object surface. Due to the blocking effect of the fluid viscosity, the fluid particles will consume part of the kinetic energy to overcome the friction in the flow from point A to point B. The resistance does work so that it cannot meet the pressure increase requirements from point B to point D. As a result, the fluid particles in the BD process flow over a certain distance, and all the kinetic energy will be consumed (part of it is converted into pressure energy, and part of the friction resistance is overcome. Do work), so the velocity at a certain point on the wall becomes zero (point S), and the particles of the fluid flowing from here will leave the object surface and enter the main flow field. This point is called the separation point. This phenomenon is called boundary layer separation, as shown in Fig. 5.3. The fluid particles in the cavity between the separation points flow backwards, and flow from the downstream high-pressure area to the low-pressure area, thus forming a vortex area behind the cylinder. The appearance of this vortex zone changes the pressure distribution on the cylindrical wall, which is asymmetric (for example, the pressure at the front stagnation point is significantly greater than the pressure at the rear stagnation point, as shown in Fig. 5.4), so resistance D appears.

$$D = \oint_{2\pi R} \tau_0 \sin\theta\,\mathrm{d}s + \oint_{2\pi R} -p_s \cos\theta\,\mathrm{d}s = D_f + D_p \neq 0 \tag{5.4}$$

The overall conclusion is as follows:

(1) The adhesion condition between the viscous friction shear stress and the object surface (no-slip condition) is the main sign that the motion of the viscous fluid is different from the ideal fluid motion.
(2) The existence of viscosity is the main cause of resistance.
(3) The necessary conditions for the separation of the boundary layer are the viscosity of the fluid and the reverse pressure gradient.

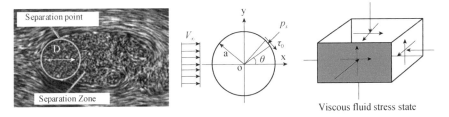

Viscous fluid stress state

Fig. 5.3 Flow and separation of viscous fluid around a cylinder

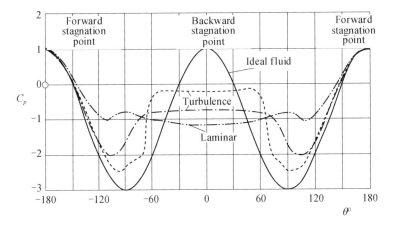

Fig. 5.4 The pressure distribution on the surface of the viscous fluid flowing around a cylinder

(4) Viscosity plays a leading role in the study of resistance, boundary layer and its separation, and vortex diffusion, and cannot be ignored.

5.2 Deformation Matrix of a Differential Fluid Element

During the movement of fluid clusters, not only rigid body motion (translation and rotation), but also deformation motion (line deformation and angular deformation motion) occur. The appearance of such compound motions, especially rotation and deformation motions, is determined by the gradient matrix of the velocity field, which is

$$[I] = \begin{bmatrix} \dfrac{\partial u}{\partial x} & \dfrac{\partial u}{\partial y} & \dfrac{\partial u}{\partial z} \\[2mm] \dfrac{\partial v}{\partial x} & \dfrac{\partial v}{\partial y} & \dfrac{\partial v}{\partial z} \\[2mm] \dfrac{\partial w}{\partial x} & \dfrac{\partial w}{\partial y} & \dfrac{\partial w}{\partial z} \end{bmatrix} \tag{5.5}$$

According to the velocity decomposition theorem of Helmholtz (1821–1894), the German physicist, the deformation rate matrix of the velocity field is

$$[\varepsilon] = \begin{bmatrix} \varepsilon_x & \varepsilon_{yx} & \varepsilon_{zx} \\ \varepsilon_{xy} & \varepsilon_y & \varepsilon_{zy} \\ \varepsilon_{xz} & \varepsilon_{yz} & \varepsilon_z \end{bmatrix} = \begin{bmatrix} \dfrac{\partial u}{\partial x} & \dfrac{1}{2}\left(\dfrac{\partial u}{\partial y} + \dfrac{\partial v}{\partial x}\right) & \dfrac{1}{2}\left(\dfrac{\partial u}{\partial z} + \dfrac{\partial w}{\partial x}\right) \\ \dfrac{1}{2}\left(\dfrac{\partial v}{\partial x} + \dfrac{\partial u}{\partial y}\right) & \dfrac{\partial v}{\partial y} & \dfrac{1}{2}\left(\dfrac{\partial v}{\partial z} + \dfrac{\partial w}{\partial y}\right) \\ \dfrac{1}{2}\left(\dfrac{\partial w}{\partial x} + \dfrac{\partial u}{\partial z}\right) & \dfrac{1}{2}\left(\dfrac{\partial w}{\partial y} + \dfrac{\partial v}{\partial z}\right) & \dfrac{\partial w}{\partial z} \end{bmatrix} \tag{5.6}$$

The velocity field rotation matrix is

$$[\omega] = \begin{bmatrix} 0 & -\dfrac{1}{2}\left(\dfrac{\partial v}{\partial x} - \dfrac{\partial u}{\partial y}\right) & \dfrac{1}{2}\left(\dfrac{\partial u}{\partial z} - \dfrac{\partial w}{\partial x}\right) \\ \dfrac{1}{2}\left(\dfrac{\partial v}{\partial x} - \dfrac{\partial u}{\partial y}\right) & 0 & -\dfrac{1}{2}\left(\dfrac{\partial w}{\partial y} - \dfrac{\partial v}{\partial z}\right) \\ -\dfrac{1}{2}\left(\dfrac{\partial u}{\partial z} - \dfrac{\partial w}{\partial x}\right) & \dfrac{1}{2}\left(\dfrac{\partial w}{\partial y} - \dfrac{\partial v}{\partial z}\right) & 0 \end{bmatrix} \tag{5.7}$$

Obviously, Eqs. (5.6) and (5.7) are decompositions of Eqs. (5.5) and (5.6) are symmetric matrices, and Eq. (5.7) are antisymmetric matrices. Since the deformation rate matrix characterizes the degree of relative deformation of the fluid micelles per unit time, the viscous stress of the fluid micelles has a direct relationship with this deformation rate matrix. In this deformation rate matrix, the size of each component is related to the choice of the coordinate system. If you find the eigenvalues of this deformation rate matrix, it is found that three quantities are invariants that have nothing to do with the choice of the coordinate axis. The characteristic equation of the deformation rate matrix is

$$\begin{vmatrix} \varepsilon_x - \lambda & \varepsilon_{xy} & \varepsilon_{xz} \\ \varepsilon_{yx} & \varepsilon_y - \lambda & \varepsilon_{yz} \\ \varepsilon_{zx} & \varepsilon_{zy} & \varepsilon_z - \lambda \end{vmatrix} = 0 \tag{5.8}$$

Expand the cubic equation about the characteristic quantity λ as

$$D = -\lambda^3 + I_1\lambda^2 - I_2\lambda + I_3 = 0 \tag{5.9}$$

Among them, the expression of the three invariants is

$$I_1 = \varepsilon_x + \varepsilon_y + \varepsilon_z \tag{5.10}$$

$$I_2 = \begin{vmatrix} \varepsilon_x & \varepsilon_{yx} \\ \varepsilon_{xy} & \varepsilon_y \end{vmatrix} + \begin{vmatrix} \varepsilon_y & \varepsilon_{zy} \\ \varepsilon_{yz} & \varepsilon_z \end{vmatrix} + \begin{vmatrix} \varepsilon_z & \varepsilon_{xz} \\ \varepsilon_{zx} & \varepsilon_x \end{vmatrix}$$

$$= \varepsilon_x \varepsilon_y + \varepsilon_y \varepsilon_z + \varepsilon_x \varepsilon_z - \varepsilon_{xy}^2 - \varepsilon_{yz}^2 - \varepsilon_{zx}^2 \tag{5.11}$$

$$I_3 = \begin{vmatrix} \varepsilon_x & \varepsilon_{xy} & \varepsilon_{xz} \\ \varepsilon_{yx} & \varepsilon_y & \varepsilon_{yz} \\ \varepsilon_{zx} & \varepsilon_{zy} & \varepsilon_z \end{vmatrix} \tag{5.12}$$

Among them, the first invariant I_1 has a clear physical meaning. Represents the divergence of the velocity field, or the relative volume expansion rate of fluid clusters, which is

$$I_1 = \varepsilon_x + \varepsilon_y + \varepsilon_z = \frac{\partial u}{\partial x} + \frac{\partial v}{\partial y} + \frac{\partial w}{\partial z} = \nabla \cdot \vec{V} \tag{5.13}$$

If the selected coordinate axes are the three main axes of the deformation rate matrix, then the off-diagonal component of the deformation rate matrix is zero at this time, and the corresponding deformation rate matrix and invariant are

$$[\varepsilon] = \begin{bmatrix} \varepsilon_1 & 0 & 0 \\ 0 & \varepsilon_2 & 0 \\ 0 & 0 & \varepsilon_3 \end{bmatrix} \tag{5.14}$$

$$\begin{aligned} I_1 &= \varepsilon_1 + \varepsilon_2 + \varepsilon_3 \\ I_2 &= \varepsilon_1 \varepsilon_2 + \varepsilon_2 \varepsilon_3 + \varepsilon_1 \varepsilon_3 \\ I_3 &= \varepsilon_1 \varepsilon_2 \varepsilon_3 \end{aligned} \tag{5.15}$$

5.3 Stress State of Viscous Fluid

The fluid is in a static state, can only withstand pressure, can hardly withstand tension and shear, and does not have the ability to resist shear deformation. In an ideal fluid state, there can be relative motion between fluid particles, but it does not have the ability to resist shear deformation. Therefore, the force acting on any surface inside the fluid is only normal force, not tangential force.

When a viscous fluid is in motion, there can be relative motion between fluid particles, and the fluid has the ability to resist shear deformation. Therefore, the force acting on any surface inside the fluid has both normal force and tangential force. At this time, in the flow field, the surface force per unit area passing through any point is not perpendicular to the acting surface (the normal force and the surface force exist like a solid element), and the resultant force in each direction is not necessarily

Fig. 5.5 Stress state at a
point in the movement of
viscous fluid

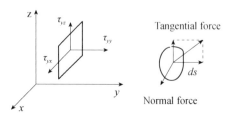

equal. Therefore, the resultant stress acting on the micro-element area in any direction
can be decomposed into normal stress and tangential stress. If the normal direction
of the action surface coincides with the coordinate axis, the resultant stress can be
decomposed into three components, including the normal stress perpendicular to the
action surface and two shear stresses parallel to the action surface. The two tangential
stresses are parallel to the action surface. The other two coordinate axes are the
projected components of the shear stress on the coordinate axis. As shown in Fig. 5.5,
take the action surface of the normal y-axis as an example to illustrate the value of each
quantity. The normal direction of the action surface and the projection direction of the
force are respectively indicated at the subscript position of the stress. For example,
the first subscript y of τ_{yx} represents the normal direction of the action surface, the
second subscript x represents the projection direction of the stress component, and
other directions can be deduced by analogy. For τ_{yy} represents the stress component
acting on the y-plane along the y-direction, that is, the normal force (normal stress) on
the y-plane, and the tension is positive. In this way, the stress on the three coordinate
planes crossing a point can be written as 9 stress components, which is
The resultant stress on the x-plane is

$$\vec{\tau}_x = \tau_{xx}\vec{i} + \tau_{xy}\vec{j} + \tau_{xz}\vec{k} \tag{5.16}$$

Resultant stress on y-plane is

$$\vec{\tau}_y = \tau_{yx}\vec{i} + \tau_{yy}\vec{j} + \tau_{yz}\vec{k} \tag{5.17}$$

Resultant stress on z-plane is

$$\vec{\tau}_z = \tau_{zx}\vec{i} + \tau_{zy}\vec{j} + \tau_{zz}\vec{k} \tag{5.18}$$

If the stress on three mutually perpendicular coordinate planes is given at the
same point, the stress on the acting surface in any direction passing through the
point can be uniquely determined by coordinate transformation. Therefore, the nine
stress components on the three coordinate planes are often called the stress state of
the point, and the matrix composed of these nine stress components is called the
stress matrix. According to the shear reciprocity theorem, only six of the nine stress
components are independent of each other, including three normal stresses and three
tangential stresses. This stress matrix is a symmetric matrix like the deformation rate

matrix.

$$[\tau] = \begin{bmatrix} \tau_{xx} & \tau_{xy} & \tau_{xz} \\ \tau_{yx} & \tau_{yy} & \tau_{yz} \\ \tau_{zx} & \tau_{zy} & \tau_{zz} \end{bmatrix}, \quad \tau_{xy} = \tau_{yx}, \ \tau_{xz} = \tau_{zx}, \ \tau_{yz} = \tau_{zy} \tag{5.19}$$

Similarly, from the eigenvalue equation of the stress matrix, the invariant can be obtained as.

The characteristic equation of the stress matrix is

$$\begin{vmatrix} \tau_{xx} - \lambda & \tau_{xy} & \tau_{xz} \\ \tau_{yx} & \tau_{yy} - \lambda & \tau_{yz} \\ \tau_{zx} & \tau_{zy} & \tau_{zz} - \lambda \end{vmatrix} = 0 \tag{5.20}$$

After expansion, from the eigenvalue λ equation of the stress matrix, $D = -\lambda^3 + I_1\lambda^2 - I_2\lambda + I_3 = 0$, the expressions of the three invariants are obtained as

$$I_1 = \tau_{xx} + \tau_{yy} + \tau_{zz} \tag{5.21}$$

$$I_2 = \begin{vmatrix} \tau_{xx} & \tau_{yx} \\ \tau_{xy} & \tau_{yy} \end{vmatrix} + \begin{vmatrix} \tau_{yy} & \tau_{zy} \\ \tau_{yz} & \tau_{zz} \end{vmatrix} + \begin{vmatrix} \tau_{zz} & \tau_{xz} \\ \tau_{zx} & \tau_{xx} \end{vmatrix}$$
$$= \tau_{xx}\tau_{yy} + \tau_{yy}\tau_{zz} + \tau_{xx}\tau_{zz} - \tau_{xy}^2 - \tau_{yz}^2 - \tau_{zx}^2 \tag{5.22}$$

$$I_3 = \begin{vmatrix} \tau_{xx} & \tau_{xy} & \tau_{xz} \\ \tau_{yx} & \tau_{yy} & \tau_{yz} \\ \tau_{zx} & \tau_{zy} & \tau_{zz} \end{vmatrix} \tag{5.23}$$

For the first invariant I_1, its clear physical meaning is that in the movement of a viscous fluid, the sum of the normal stresses in the directions of the three vertical axes does not change with the choice of the coordinate system. If the pressure at a point is defined as the average of this invariant, then there is

$$p = -\frac{I_1}{3} = -\frac{\tau_{xx} + \tau_{yy} + \tau_{zz}}{3} \tag{5.24}$$

where the negative sign indicates that p points perpendicular to the action surface, and the pressure is positive. The pressure defined in this way is not the real pressure in the movement of the viscous fluid, so it is called the nominal pressure, and its value does not change with the orientation of the action surface and is isotropic. The characteristics is the same as the nature of real pressure in ideal fluid motion.

$$[\tau] = \begin{bmatrix} -p & 0 & 0 \\ 0 & -p & 0 \\ 0 & 0 & -p \end{bmatrix} + \begin{bmatrix} \tau_{xx} + p & \tau_{xy} & \tau_{xz} \\ \tau_{yx} & \tau_{yy} + p & \tau_{yz} \\ \tau_{zx} & \tau_{zy} & \tau_{zz} + p \end{bmatrix} \tag{5.25}$$

On the right side of formula (5.25), the first part is the stress part that has nothing to do with viscosity; the second part is the stress part related to viscosity, which is called the deviatoric stress matrix. Can also be expressed as

$$[\tau] = -p[E] + [D] \tag{5.26}$$

where $[E]$ is the identity matrix and $[D]$ is the deviatoric stress matrix. which is

$$[E] = \begin{bmatrix} 1 & 0 & 0 \\ 0 & 1 & 0 \\ 0 & 0 & 1 \end{bmatrix}, [D] = \begin{bmatrix} \tau_{xx} + p & \tau_{xy} & \tau_{xz} \\ \tau_{yx} & \tau_{yy} + p & \tau_{yz} \\ \tau_{zx} & \tau_{zy} & \tau_{zz} + p \end{bmatrix} \tag{5.27}$$

In summary, the following conclusions can be drawn:

(1) In an ideal fluid, there is no shear stress, and the three normal stresses are equal and equal to the negative value of the pressure at that point. Its stress matrix is

$$\tau_{xx} = \tau_{yy} = \tau_{zz} = -p \,[\tau] = -p \begin{bmatrix} 1 & 0 & 0 \\ 0 & 1 & 0 \\ 0 & 0 & 1 \end{bmatrix} \tag{5.28}$$

(2) In a viscous fluid, the normal stress components on the three mutually perpendicular planes passing through any point are not necessarily equal, but the sum is an invariant and has nothing to do with the choice of the coordinate system. Define the negative of the average value of this invariant to be the average pressure at that point. The expression is

$$p = -\frac{\tau_{xx} + \tau_{yy} + \tau_{zz}}{3}$$

(3) In viscous fluids, the stress matrix related to viscosity is called the deviatoric stress matrix. The deformation rate on any surface inside the fluid is generally not zero, so the shear stress is not zero. That is, in the movement of viscous fluid, under normal circumstances.

5.4 Generalized Newton's Internal Friction Theorem (Constitutive Relationship)

According to Newton's internal friction law, when a viscous fluid moves in a linear laminar flow, the shear stress between the flow layers is proportional to the velocity gradient. For a Newtonian fluid, in plane flow, the relationship between shear stress and velocity gradient is shown in Eq. (5.1). As shown in Fig. 5.6, for the plane motion of the viscous fluid, the velocity component is

$$u(x, y) = u(y), v(x, y) = 0, w = 0 \tag{5.29}$$

From Eq. (5.6), it can be seen that the deformation rate component is

$$\varepsilon_{xy} = \varepsilon_{yx} = \frac{1}{2}\left(\frac{\partial v}{\partial x} + \frac{\partial u}{\partial y}\right) = \frac{1}{2}\frac{du}{dy}, \varepsilon_x = \frac{\partial u}{\partial x} = 0, \varepsilon_y = \frac{\partial v}{\partial y} = 0 \tag{5.30}$$

Substituting Eq. (5.30) into Eq. (5.1), Newton's internal friction law can be written as a more general expression, which is

$$\tau_{yx} = \tau_{xy} = 2\mu\varepsilon_{yx} = \mu\left(\frac{\partial u}{\partial y} + \frac{\partial v}{\partial x}\right) = \mu\frac{du}{dy} \tag{5.31}$$

It shows that the shear stress of fluid clusters is proportional to the corresponding component of shear deformation rate.

For the general three-dimensional viscous flow, the British mechanic and mathematician Stokes (1819–1903) adopted three assumptions in 1845 to extend Newton's law of internal friction to the more general case of viscous fluid motion, and proposed the famous generalized Newton's law of internal friction. The three hypotheses suggested by Stokes are:

(1) The fluid is continuous, and its deviator stress matrix has a linear relationship with the deformation rate matrix and has nothing to do with the translation and rotation of the fluid.
(2) The fluid is isotropic, and the relationship between the deviator stress matrix and the deformation rate matrix has nothing to do with the choice of the coordinate system.

Fig. 5.6 Simple plane flow velocity distribution

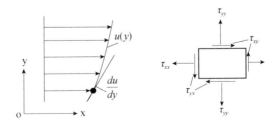

(3) The relationship established is not only suitable for sports situations. When the fluid is at rest, the deformation rate is zero, and the stress in the fluid is the static pressure.

From the assumption of the third condition, it can be known that in the static state, the fluid stress is only normal stress and no shear stress. That is, the stress at rest is

$$\tau_{xx} = \tau_{yy} = \tau_{zz} = -p_0 \tag{5.32}$$

In the formula, p_0 is the hydrostatic pressure. In the static state, the stress matrix of the fluid is

$$[\tau] = \begin{bmatrix} \tau_{xx} & \tau_{xy} & \tau_{xz} \\ \tau_{yx} & \tau_{yy} & \tau_{yz} \\ \tau_{zx} & \tau_{zy} & \tau_{zz} \end{bmatrix} = \begin{bmatrix} -p_0 & 0 & 0 \\ 0 & -p_0 & 0 \\ 0 & 0 & -p_0 \end{bmatrix}$$

$$= -p_0 \begin{bmatrix} 1 & 0 & 0 \\ 0 & 1 & 0 \\ 0 & 0 & 1 \end{bmatrix} = -p_0[E] \tag{5.33}$$

According to Stokes' first assumption and inspired by the third assumption, the deviator stress matrix and the deformation rate matrix can be written as a linear relationship (constitutive relationship). Using formula (5.26) and formula (5.27), there are

$$[D] = \begin{bmatrix} \tau_{xx} + p & \tau_{xy} & \tau_{xz} \\ \tau_{yx} & \tau_{yy} + p & \tau_{yz} \\ \tau_{zx} & \tau_{zy} & \tau_{zz} + p \end{bmatrix} = a \begin{bmatrix} \varepsilon_x & \varepsilon_{yx} & \varepsilon_{zx} \\ \varepsilon_{xy} & \varepsilon_y & \varepsilon_{zy} \\ \varepsilon_{xz} & \varepsilon_{yz} & \varepsilon_z \end{bmatrix} + b \begin{bmatrix} 1 & 0 & 0 \\ 0 & 1 & 0 \\ 0 & 0 & 1 \end{bmatrix} \tag{5.34}$$

The total stress matrix is

$$[\tau] = -p[E] + [D] = -p[E] + a[\varepsilon] + b[E] \tag{5.35}$$

In the formula, the coefficients a and b are both scalar quantities that have nothing to do with the choice of the coordinate system. With reference to Newton's internal friction law (5.31), the coefficient a is only related to the physical properties of the fluid, which is desirable

$$a = 2\mu \tag{5.36}$$

Because the coefficient b has nothing to do with the choice of coordinate system. Therefore, it can be inferred that in order to maintain a linear relationship between the deviator stress and the deformation rate, the coefficient b can only be composed of those linear invariants (such as the first invariant) in the stress matrix and the deformation rate matrix, that is, assuming

$$b = b_1(\tau_{xx} + \tau_{yy} + \tau_{zz}) + b_2(\varepsilon_x + \varepsilon_y + \varepsilon_z) + b_3 \tag{5.37}$$

And using formula (5.13) and formula (5.24), we can get

$$b = -3b_1 p + b_2 \nabla \cdot \vec{V} + b_3 \tag{5.38}$$

In the formula, b_1, b_2, b_3 are undetermined coefficients. Substituting a and b into Eq. (5.35), we can get

$$[\tau] = -p[E] + 2\mu[\varepsilon] + \left\{-3pb_1 + b_2 \nabla \cdot \vec{V} + b_3\right\}[E] \tag{5.39}$$

Taking the sum of the three components on the main diagonal of the two sides of the matrix in the above formula, and using formula (5.24), we can get

$$-3p = -3p + 2\mu \nabla \cdot \vec{V} + 3(-3pb_1 + b_2 \nabla \cdot \vec{V} + b_3) \tag{5.40}$$

After finishing, get

$$-3pb_1 + \left(b_2 + \frac{2\mu}{3}\right)\nabla \cdot \vec{V} + b_3 = 0 \tag{5.41}$$

In the static state, the divergence of the velocity is zero, substituting the above formula to get

$$\nabla \cdot \vec{V} = 0, \quad -3pb_1 + b_3 = 0, \quad b_3 = 3pb_1 \tag{5.42}$$

Substituting b_3 in Eq. (5.42) into Eq. (5.41), we get

$$\left(b_2 + \frac{2\mu}{3}\right)\nabla \cdot \vec{V} = 0 \tag{5.43}$$

We can get

$$b_2 = -\frac{2}{3}\mu \tag{5.44}$$

The coefficient b_2 characterizes the influence of the volume expansion rate of the fluid micelles on the normal stress. Substituting the b_3 coefficient Eq. (5.42) and b_2 coefficient Eq. (5.44) into Eq. (5.39), we get

$$[\tau] = -p[E] + 2\mu[\varepsilon] - \frac{2\mu}{3}\nabla \cdot \vec{V}[E] = 2\mu[\varepsilon] - \left(p + \frac{2\mu}{3}\nabla \cdot \vec{V}\right)[E] \tag{5.45}$$

Equation (5.45) is called the generalized Newtonian internal friction law, which establishes the constitutive relationship of Newtonian fluid motion, that is, the relationship between the relative motion stress and deformation rate of viscous fluid. Written in matrix form as

$$
\begin{bmatrix} \tau_{xx} & \tau_{xy} & \tau_{xz} \\ \tau_{yx} & \tau_{yy} & \tau_{yz} \\ \tau_{zx} & \tau_{zy} & \tau_{zz} \end{bmatrix} = 2\mu \begin{bmatrix} \varepsilon_x & \varepsilon_{yx} & \varepsilon_{zx} \\ \varepsilon_{xy} & \varepsilon_y & \varepsilon_{zy} \\ \varepsilon_{xz} & \varepsilon_{yz} & \varepsilon_z \end{bmatrix} - \left(p + \frac{2\mu}{3} \nabla \cdot \vec{V} \right) \begin{bmatrix} 1 & 0 & 0 \\ 0 & 1 & 0 \\ 0 & 0 & 1 \end{bmatrix} \tag{5.46}
$$

For incompressible fluids, there are $\nabla \cdot \vec{V} = 0$, the formula (5.45) is simplified as

$$
[\tau] = -p[E] + 2\mu[\varepsilon] \tag{5.47}
$$

Using the deformation rate matrix formula (5.6), write the formula (5.47) in a quantitative form, which are.

The tangential stress is

$$
\tau_{xy} = 2\mu\varepsilon_{xy} = \mu \left(\frac{\partial v}{\partial x} + \frac{\partial u}{\partial y} \right) \tag{5.48}
$$

$$
\tau_{yz} = 2\mu\varepsilon_{yz} = \mu \left(\frac{\partial w}{\partial y} + \frac{\partial v}{\partial z} \right) \tag{5.49}
$$

$$
\tau_{zx} = 2\mu\varepsilon_{zx} = \mu \left(\frac{\partial u}{\partial z} + \frac{\partial w}{\partial x} \right) \tag{5.50}
$$

The normal stress is

$$
\tau_{xx} = -p + 2\mu \frac{\partial u}{\partial x} = -p + 2\mu\varepsilon_x \tag{5.51}
$$

$$
\tau_{yy} = -p + 2\mu \frac{\partial v}{\partial y} = -p + 2\mu\varepsilon_y \tag{5.52}
$$

$$
\tau_{zz} = -p + 2\mu \frac{\partial w}{\partial z} = -p + 2\mu\varepsilon_z \tag{5.53}
$$

5.5 Differential Equations of Viscous Fluid Motion—Navier–Stokes Equations

5.5.1 The Basic Differential Equations of Fluid Motion

Use Newton's second theorem to derive the differential equation of viscous fluid motion expressed in the form of stress. As shown in Fig. 5.7, take a differential hexahedral fluid micro-cluster in the flow field. Its side lengths are respectively $\Delta x, \Delta y, \Delta z$, and the mass of the micro-cluster is $\Delta m = \rho \Delta x \Delta y \Delta z$. Take the x-direction as an example to establish the differential equation of motion, which is

$$\sum F_x = m \frac{du}{dt} \tag{5.54}$$

Try to sum the surface of the fluid cluster along the x-direction, there is

$$\sum F_{sx} = \left[\left(\tau_{xx} + \frac{\partial \tau_{xx}}{\partial x} \Delta x \right) - \tau_{xx} \right] \Delta y \Delta z$$
$$+ \left[\left(\tau_{yx} + \frac{\partial \tau_{yx}}{\partial y} \Delta y \right) - \tau_{yx} \right] \Delta z \Delta x$$
$$+ \left[\left(\tau_{zx} + \frac{\partial \tau_{zx}}{\partial z} \Delta z \right) - \tau_{zx} \right] \Delta x \Delta y \tag{5.55}$$

Acting on the micelles, the mass force in the x-direction is

$$F_{mx} = \rho f_x \Delta x \Delta y \Delta z \tag{5.56}$$

Substituting formula (5.55) and formula (5.56) into formula (5.54), we can get

Fig. 5.7 Force of viscous fluid clusters along the x-direction

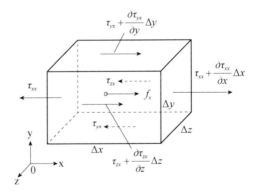

$$\sum F_x = \left(\frac{\partial \tau_{xx}}{\partial x} + \frac{\partial \tau_{yx}}{\partial y} + \frac{\partial \tau_{zx}}{\partial z} \right) \Delta x \Delta y \Delta z + \rho f_x \Delta x \Delta y \Delta z \qquad (5.57)$$

Finally from the formula (5.54), we have

$$\left(\frac{\partial \tau_{xx}}{\partial x} + \frac{\partial \tau_{yx}}{\partial y} + \frac{\partial \tau_{zx}}{\partial z} \right) \Delta x \Delta y \Delta z + \rho f_x \Delta x \Delta y \Delta z = \rho \frac{du}{dt} \Delta x \Delta y \Delta z \qquad (5.58)$$

Divide both sides of the above formula by $\Delta x \Delta y \Delta z$, and take the limit $\Delta x \to 0$, $\Delta y \to 0$, $\Delta z \to 0$, the differential equation of motion in the x-direction is

$$\frac{du}{dt} = f_x + \frac{1}{\rho} \left(\frac{\partial \tau_{xx}}{\partial x} + \frac{\partial \tau_{yx}}{\partial y} + \frac{\partial \tau_{zx}}{\partial z} \right) \qquad (5.59)$$

Doing a similar derivation in the same way, the differential equations of motion in the y- and z-directions can be obtained, which is

$$\begin{cases} \dfrac{du}{dt} = f_x + \dfrac{1}{\rho} \left(\dfrac{\partial \tau_{xx}}{\partial x} + \dfrac{\partial \tau_{yx}}{\partial y} + \dfrac{\partial \tau_{zx}}{\partial z} \right) \\[2mm] \dfrac{dv}{dt} = f_y + \dfrac{1}{\rho} \left(\dfrac{\partial \tau_{xy}}{\partial x} + \dfrac{\partial \tau_{yy}}{\partial y} + \dfrac{\partial \tau_{zy}}{\partial z} \right) \\[2mm] \dfrac{dw}{dt} = f_z + \dfrac{1}{\rho} \left(\dfrac{\partial \tau_{xz}}{\partial x} + \dfrac{\partial \tau_{yz}}{\partial y} + \dfrac{\partial \tau_{zz}}{\partial z} \right) \end{cases} \qquad (5.60)$$

This is a system of differential equations of fluid motion expressed in the form of stress, which has a universal meaning and is suitable for both ideal fluids and viscous fluids. But this is a set of unclosed equations. When the mass force is known, there are six more stress components in the equations. In order to obtain a closed form, a constitutive relationship must be introduced, for example, the generalized law of Newtonian friction for a viscous fluid.

5.5.2 Navier–Stokes Equations (Differential Equations of Viscous Fluid Motion)

The study of fluid motion was given a qualitative description by Italian all-round scientist Da Vinci (1452–1519) before 1500. In 1755, Swiss scientist Euler (1707–1783) quantitatively deduced the differential equations of ideal fluid motion. In 1822, French engineer Navier (1785–1836) first considered the influence of fluid viscosity. After the efforts of many scientists, the British scientist Stokes (1819–1903) completed the final derivation in 1845 and obtained the current form of the differential equations for the motion of viscous fluids, called Navier–Stokes equations, or N–S equations for short. From 1755 to 1845, it lasted 90 years.

Taking the equation in the x-direction as an example, the derivation is given. For the differential equation in the x-direction (5.60)

$$f_x + \frac{1}{\rho}\left(\frac{\partial \tau_{xx}}{\partial x} + \frac{\partial \tau_{yx}}{\partial y} + \frac{\partial \tau_{zx}}{\partial z}\right) = \frac{du}{dt}$$

Introduce the generalized Newton's internal friction law (5.45), namely,

$$\tau_{xx} = -p + 2\mu\frac{\partial u}{\partial x} - \frac{2}{3}\mu\nabla\cdot\vec{V},\ \tau_{yx} = \mu\left(\frac{\partial v}{\partial x} + \frac{\partial u}{\partial y}\right),\ \tau_{zx} = \mu\left(\frac{\partial w}{\partial x} + \frac{\partial u}{\partial z}\right) \tag{5.61}$$

Get

$$\frac{du}{dt} = f_x - \frac{1}{\rho}\frac{\partial p}{\partial x} + \frac{1}{\rho}\frac{\partial}{\partial x}\left(2\mu\frac{\partial u}{\partial x} - \frac{2}{3}\mu\nabla\cdot\vec{V}\right)$$
$$+ \frac{1}{\rho}\frac{\partial}{\partial y}\left[\mu\left(\frac{\partial v}{\partial x} + \frac{\partial u}{\partial y}\right)\right] + \frac{1}{\rho}\frac{\partial}{\partial z}\left[\mu\left(\frac{\partial w}{\partial x} + \frac{\partial u}{\partial z}\right)\right] \tag{5.62}$$

In the same way, the equations in the y- and z-directions are

$$\frac{dv}{dt} = f_y - \frac{1}{\rho}\frac{\partial p}{\partial y} + \frac{1}{\rho}\frac{\partial}{\partial x}\left[\mu\left(\frac{\partial v}{\partial x} + \frac{\partial u}{\partial y}\right)\right]$$
$$+ \frac{1}{\rho}\frac{\partial}{\partial y}\left(2\mu\frac{\partial v}{\partial y} - \frac{2}{3}\mu\nabla\cdot\vec{V}\right) + \frac{1}{\rho}\frac{\partial}{\partial z}\left[\mu\left(\frac{\partial w}{\partial y} + \frac{\partial v}{\partial z}\right)\right] \tag{5.63}$$

$$\frac{dw}{dt} = f_z - \frac{1}{\rho}\frac{\partial p}{\partial z} + \frac{1}{\rho}\frac{\partial}{\partial x}\left[\mu\left(\frac{\partial w}{\partial x} + \frac{\partial u}{\partial z}\right)\right]$$
$$+ \frac{1}{\rho}\frac{\partial}{\partial y}\left[\mu\left(\frac{\partial w}{\partial y} + \frac{\partial v}{\partial z}\right)\right] + \frac{1}{\rho}\frac{\partial}{\partial z}\left(2\mu\frac{\partial w}{\partial z} - \frac{2}{3}\mu\nabla\cdot\vec{V}\right) \tag{5.64}$$

These are the N–S equations describing the movement of viscous fluids, which are suitable for compressible and incompressible fluids.

For non-shrinkable fluids, and the viscosity coefficient and density are regarded as constants, the equations can be simplified. Take the x-direction as an example. Substituting the continuity equation of incompressible fluid into Eq. (5.62), we get

$$\frac{1}{\rho}\frac{\partial}{\partial x}\left(2\mu\frac{\partial u}{\partial x} - \frac{2}{3}\mu\nabla\cdot\vec{V}\right) + \frac{1}{\rho}\frac{\partial}{\partial y}\left[\mu\left(\frac{\partial v}{\partial x} + \frac{\partial u}{\partial y}\right)\right] + \frac{1}{\rho}\frac{\partial}{\partial z}\left[\mu\left(\frac{\partial w}{\partial x} + \frac{\partial u}{\partial z}\right)\right]$$
$$= \frac{2\mu}{\rho}\frac{\partial^2 u}{\partial x^2} + \frac{\mu}{\rho}\left(\frac{\partial^2 v}{\partial x\partial y} + \frac{\partial^2 u}{\partial y^2}\right) + \frac{\mu}{\rho}\left(\frac{\partial^2 w}{\partial x\partial z} + \frac{\partial^2 u}{\partial z^2}\right)$$
$$= \frac{\mu}{\rho}\left(\frac{\partial^2 u}{\partial x^2} + \frac{\partial^2 u}{\partial y^2} + \frac{\partial^2 u}{\partial z^2}\right) + \frac{\mu}{\rho}\frac{\partial}{\partial x}\left(\frac{\partial u}{\partial x} + \frac{\partial v}{\partial y} + \frac{\partial w}{\partial z}\right)$$

$$= \nu \left(\frac{\partial^2 u}{\partial x^2} + \frac{\partial^2 u}{\partial y^2} + \frac{\partial^2 u}{\partial z^2} \right) \tag{5.65}$$

Thus, the differential equations of motion of the incompressible fluid clusters are obtained, namely,

$$\begin{aligned}
\frac{du}{dt} &= f_x - \frac{1}{\rho}\frac{\partial p}{\partial x} + \nu \left(\frac{\partial^2 u}{\partial x^2} + \frac{\partial^2 u}{\partial y^2} + \frac{\partial^2 u}{\partial z^2} \right) \\
\frac{dv}{dt} &= f_y - \frac{1}{\rho}\frac{\partial p}{\partial y} + \nu \left(\frac{\partial^2 v}{\partial x^2} + \frac{\partial^2 v}{\partial y^2} + \frac{\partial^2 v}{\partial z^2} \right) \\
\frac{dw}{dt} &= f_z - \frac{1}{\rho}\frac{\partial p}{\partial z} + \nu \left(\frac{\partial^2 w}{\partial x^2} + \frac{\partial^2 w}{\partial y^2} + \frac{\partial^2 w}{\partial z^2} \right)
\end{aligned} \tag{5.66}$$

Together with the incompressible continuity equations, this set of equations forms a closed differential equation set describing the motion of incompressible fluids. The continuity differential equation of incompressible fluid is

$$\frac{\partial u}{\partial x} + \frac{\partial v}{\partial y} + \frac{\partial w}{\partial z} = 0 \tag{5.67}$$

That is, four unknowns (u, v, w, p) and four differential equations. Write Eqs. (5.66) and (5.67) in vector form as

$$\frac{d\vec{V}}{dt} = \vec{f} - \frac{1}{\rho}\nabla p + \nu \Delta \vec{V} \tag{5.68}$$

$$\nabla \cdot \vec{V} = 0 \tag{5.69}$$

Compared with Euler's equations, the N–S equations are a set of second-order quasi-linear partial differential equations. In order to study the vortex motion of the fluid, formula (5.68) is written as a Lamb (British mechanic, 1849–1934) type equation.

$$\frac{\partial \vec{V}}{\partial t} + \nabla \left(\frac{V^2}{2} \right) + 2\vec{\omega} \times \vec{V} = \vec{f} - \frac{1}{\rho}\nabla p + \nu \Delta \vec{V} \tag{5.70}$$

5.5.3 Bernoulli Integral

Similar to the Bernoulli integral of the differential equation of ideal fluid motion, the N-S equations are now integrated along the streamline. Assumptions: (1) The fluid is an incompressible viscous fluid. (2) The flow is a steady flow. (3) The mass force is potential (gravity). (4) The integral is along the streamline, as shown in Fig. 5.8.

Fig. 5.8 Integrating the N–S equation along the streamline

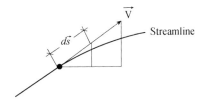

Streamline

Integrate the N–S equation along the streamline to obtain the energy equation of the viscous fluid. Different from the ideal fluid energy, there is an additional loss term due to viscosity in the equation, which represents the energy consumed by fluid particles to overcome viscous stress. As shown in Fig. 5.8, in the viscous incompressible steady flow, any streamline can be selected, and a micro-segment ds is taken somewhere on the streamline $d\vec{s} = dx\vec{i} + dy\vec{j} + dz\vec{k}$, where the corresponding flow velocity is $\vec{u} = u\vec{i} + v\vec{j} + w\vec{k}$, and the N–S equation is integrated along the streamline to obtain

$$\frac{du}{dt}dx = \left(f_x - \frac{1}{\rho}\frac{\partial p}{\partial x} + v\Delta u \right)dx$$

$$\frac{dv}{dt}dy = \left(f_y - \frac{1}{\rho}\frac{\partial p}{\partial y} + v\Delta v \right)dy$$

$$\frac{dw}{dt}dz = \left(f_z - \frac{1}{\rho}\frac{\partial p}{\partial z} + v\Delta w \right)dz \tag{5.71}$$

Written in vector form as

$$\frac{d\vec{V}}{dt}\cdot d\vec{s} = \left(\vec{f} - \frac{1}{\rho}\nabla p + v\Delta\vec{V} \right)\cdot d\vec{s} \tag{5.72}$$

In the case of steady flow, the trace and the streamline coincide. The relationship between streamline micro-segment and velocity is

$$dx = udt,\ dy = vdt,\ dz = wdt \tag{5.73}$$

So get

$$\frac{d\vec{V}}{dt}\cdot d\vec{s} = \frac{du}{dt}dx + \frac{dv}{dt}dy + \frac{dw}{dt}dz$$

$$= \frac{dx}{dt}du + \frac{dy}{dt}dv + \frac{dz}{dt}dw$$

$$= udu + vdv + wdw$$

$$= d\left(\frac{u^2}{2}\right) + d\left(\frac{v^2}{2}\right) + d\left(\frac{w^2}{2}\right)$$

$$= d\left(\frac{V^2}{2}\right) \tag{5.74}$$

Under the condition of strong mass force, we can get

$$f_x dx + f_y dy + f_z dz = -\left(\frac{\partial \Pi}{\partial x}dx + \frac{\partial \Pi}{\partial y}dy + \frac{\partial \Pi}{\partial z}dz\right) = -d\Pi \tag{5.75}$$

For incompressible steady flow, there are

$$-\left(\frac{1}{\rho}\frac{\partial p}{\partial x}dx + \frac{1}{\rho}\frac{\partial p}{\partial y}dy + \frac{1}{\rho}\frac{\partial p}{\partial z}dz\right) = -d\left(\frac{p}{\rho}\right) \tag{5.76}$$

The sum of the sticky terms is expressed as

$$dP_w = -\mu(\Delta u dx + \Delta v dy + \Delta w dz)$$
$$= -\mu(u\Delta u + v\Delta v + w\Delta w)dt \tag{5.77}$$

On the streamline micro-segment, the differential form is

$$d\left(\Pi + \frac{p}{\rho} + \frac{V^2}{2}\right) + \frac{dP_w}{\rho} = 0 \tag{5.78}$$

Compared with the ideal fluid energy differential equation, there is an additional term related to viscosity in the above formula $\frac{dP_w}{\rho}$. Physically, it represents the mechanical energy consumed by the fluid particle per unit mass to overcome the viscous stress. This part of the energy loss can no longer be used by the mechanical movement of the fluid particle, so it is called the mechanical energy loss per unit mass fluid, or energy loss.

For the case where the mass force is gravity, Eq. (5.78) can be written as

$$d\left(gz + \frac{p}{\rho} + \frac{V^2}{2}\right) + \frac{dP_w}{\rho} = 0 \tag{5.79}$$

Divide both sides of the equation by g to get

$$d\left(z + \frac{p}{\gamma} + \frac{V^2}{2g}\right) + \frac{dP_w}{\gamma} = 0 \tag{5.80}$$

where $\frac{dP_w}{\gamma}$ represents the mechanical energy consumed by the unit weight of the fluid on the micro-segment, which can also make

$$dh_w = \frac{dP_w}{\gamma} \tag{5.81}$$

Fig. 5.9 Energy equation of viscous fluid

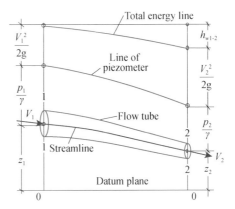

The formula (5.80) is written as

$$d\left(z + \frac{p}{\gamma} + \frac{V^2}{2g}\right) + dh_w = 0 \tag{5.82}$$

Integrate along the streamline (5.82), we get

$$z_1 + \frac{p_1}{\gamma} + \frac{V_1^2}{2g} = z_2 + \frac{p_2}{\gamma} + \frac{V_2^2}{2g} + h_{w1-2} \tag{5.83}$$

The above formula shows that in a viscous fluid, the mechanical energy (the sum of potential energy, pressure energy, and kinetic energy) of the fluid per unit time and unit weight along the same streamline decreases along the way and cannot be maintained (in an ideal fluid, the total mechanical energy is kept conserved and there is no loss of mechanical energy). The reduced part represents the mechanical energy consumed by the fluid particles to overcome the viscous stress and do work, as shown in Fig. 5.9. The Bernoulli Equation (5.83) of viscous fluid shows that no matter how the potential energy, pressure energy, and kinetic energy are transformed in the flow of viscous fluid, the total mechanical energy decreases along the way, always flowing from the place where the mechanical energy is high to the place where the mechanical energy is low. Generally speaking, the flow of water from high to low, high pressure to low pressure, is incomplete.

5.6 Exact Solutions of Navier–Stokes Equations

The Navier–Stokes equations plus appropriate initial and boundary conditions can theoretically solve any viscous fluid motion law. However, because the Navier–Stokes equations are a second-order nonlinear partial differential equations, it is generally

very difficult to solve them accurately. This section only discusses the case where
the Navier–Stokes equations have exact solutions under certain conditions.

5.6.1 Couette Flow (Shear Flow)

Now discuss a kind of flow between parallel plates with infinite length at both ends.
One plate is at rest and one moves horizontally. There is fluid between the plates.
After the moving plate is stable, find the velocity distribution between the plates as
shown in Fig. 5.10. This question was raised by the French physicist Couette in 1890.

Assuming that the distance between the plates is h, the constant velocity of the
upper plate movement is U, the two sections are infinitely long, and the end effects
are negligible. The flow is a two-dimensional parallel flow, the flow field is steady,
and the velocity field is

$$\begin{cases} u = u(x, y) \\ v \approx 0 \end{cases} \tag{5.84}$$

The boundary conditions are

$$u = u(x, 0) = 0, \ u(x, h) = U \tag{5.85}$$

From the continuous equation, we know

$$\frac{\partial u}{\partial x} + \frac{\partial v}{\partial y} = 0, \ \frac{\partial u}{\partial x} = 0 \tag{5.86}$$

Show that $u(x, y) = u(y)$. From the Navier–Stokes equation in the x-direction,
without considering the mass force, the flow is steady, and the pressure gradient
along the x-direction is zero, we can get

$$\nu \frac{d^2 u}{dy^2} = 0$$

Fig. 5.10 The flow of
Couette

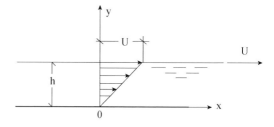

Integrate along the y-direction to get

$$u(y) = C_1 y + C_2 \tag{5.87}$$

From $y = 0$, $u = 0$, we get $C_2 = 0$. When $y = h$, $u = U$, we get $C_1 = U/h$. Then the exact solution is

$$u(y) = U \frac{y}{h} \tag{5.88}$$

The wall shear stress is

$$\tau_0 = \mu \frac{du}{dy} \bigg|_{y=h} = \mu \frac{U}{h} \tag{5.89}$$

This formula is Newton's internal friction law. If the board field is b, the resistance of the board is

$$F_D = \tau_0 b = \mu \frac{U}{h} b \tag{5.90}$$

The power required for tablet movement is

$$P_w = F_D U = \mu \frac{U^2}{h} b \tag{5.91}$$

5.6.2 Poiseuille Flow (Pressure Gradient Flow)

Now discuss a flow between parallel plates with infinite length at both ends. The upper and lower plates are static, but the pressure gradient between the plates along the x-direction is constant. When the flow reaches a stable level, find the velocity distribution between the plates. As shown in Fig. 5.11, this question was raised by the French physiologist Poiseuille in 1840 who studied the flow of pipes.

Assuming that the distance between the plates is h, the upper plate is stationary, the two sections are infinitely long, and the end effects are negligible, the flow is a two-dimensional parallel flow, the flow field is steady, and the velocity field

$$\begin{cases} u = u(x, y) \\ v \approx 0 \end{cases}$$

The boundary conditions are

$$u = u(x, 0) = 0, \ u(x, h) = 0 \tag{5.92}$$

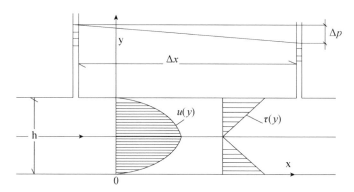

Fig. 5.11 Poiseuille flow

From the continuous equation, we know

$$\frac{\partial u}{\partial x} + \frac{\partial v}{\partial y} = 0, \ \frac{\partial u}{\partial x} = 0$$

Show that $u(x, y) = u(y)$. From the Navier–Stokes equation in the x-direction, without considering the mass force, the flow is constant, and the pressure gradient along the x-direction is a constant, we can get

$$-\frac{1}{\rho}\frac{\partial p}{\partial x} + \nu\frac{d^2 u}{dy^2} = 0 \tag{5.93}$$

Let, the pressure gradient along the x-direction is

$$\frac{\partial p}{\partial x} = \frac{\Delta p}{\Delta x} = k \tag{5.94}$$

Integral formula (5.93), we can get

$$\frac{du}{dy} = \frac{k}{\mu}y + C_1, \ u = \frac{1}{2}\frac{k}{\mu}y^2 + C_1 y + C_2 \tag{5.95}$$

From $y = 0$, $u = 0$, we get $C_2 = 0$. When $y = h$, $u = 0$, we have

$$C_1 = -\frac{1}{2}\frac{k}{\mu}h \tag{5.96}$$

The exact solution is

$$u(y) = \frac{1}{2}\frac{k}{\mu}y^2 - \frac{1}{2}\frac{k}{\mu}hy = -\frac{1}{2\mu}\frac{\Delta p}{\Delta x}h^2\left(1 - \frac{y}{h}\right)\frac{y}{h} \tag{5.97}$$

Explain that the velocity distribution along the y-direction is a parabolic distribution. The maximum speed of the center is

$$u_{max} = u\left(\frac{h}{2}\right) = -\frac{1}{2\mu}\frac{\Delta p}{\Delta x}h^2\left(1 - \frac{1}{2}\right)\frac{1}{2} = -\frac{1}{8\mu}\frac{\Delta p}{\Delta x}h^2 \qquad (5.98)$$

The formula (9.97) can also be expressed as

$$\frac{u(y)}{u_{max}} = \left(1 - \frac{y}{h}\right)\frac{y}{h} \qquad (5.99)$$

The wall shear stress is

$$\tau_0 = \mu\frac{du}{dy}\bigg|_{y=0} = -\frac{1}{2\mu}\frac{\Delta p}{\Delta x}h \qquad (5.100)$$

The volume flow through is

$$Q = \int_0^h u(y)dy = -\frac{1}{2\mu}\frac{\Delta p}{\Delta x}\frac{h^3}{6} \qquad (5.101)$$

The power consumed by the flow is

$$P_w = Q(-\Delta p) = \frac{1}{2\mu}\frac{\Delta p^2}{\Delta x}\frac{h^3}{6} \qquad (5.102)$$

5.6.3 Couette Flow and Poiseuille Flow Combination

This flow is equivalent to the combined flow of shear and pressure gradient. Because the pressure gradient is different between the forward pressure gradient and the reverse pressure gradient, there are positive and negative combinations in combination with the shear flow.

(1) Flow along the pressure gradient (forward combination)

Now discuss a flow between parallel plates with infinite length at both ends. The upper plate moves at a speed U, the lower plate is stationary, and the pressure gradient along the x-direction between the plates (the pressure gradient is decreasing along the flow direction) is constant, when the flow is stable, find the velocity distribution between the plates, as shown in Fig. 5.12. Assuming that the distance between the plates is h, the upper plate moves with U, the two sections are infinitely long, and the end effects are negligible, the flow is a two-dimensional parallel flow, and the flow field is

Fig. 5.12 Combined flow of
Poiseuille flow and Couette
flow under the condition of
favorable pressure gradient

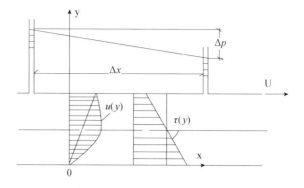

steady. The velocity field formula of the Navier–Stokes accurate solution is obtained
(5.95) is

$$u(y) = \frac{1}{2\mu}\frac{\Delta p}{\Delta x}y^2 + C_1 y + C_2$$

From $y = 0$, $u = 0$, we get $C_2 = 0$. When $y = h$, $u = U$, we have

$$C_1 = \frac{1}{h}\left[U - \frac{1}{2\mu}\frac{\Delta p}{\Delta x}h^2\right]$$

(5.103)

The exact solution obtained is

$$u(y) = U\frac{y}{h} - \frac{1}{2\mu}\frac{\Delta p}{\Delta x}h^2\left(1 - \frac{y}{h}\right)\frac{y}{h}$$

(5.104)

Explain that the velocity distribution along the y-direction is also a parabolic
distribution, and it is the superposition of formula (5.88) and formula (5.97), the
combined velocity distribution is shown in Fig. 5.12.

The shear stress of the lower wall is

$$\tau_{0d} = \mu\frac{du}{dy}\bigg|_{y=0} = \frac{U}{h} - \frac{1}{2\mu}\frac{\Delta p}{\Delta x}h$$

(5.105)

The shear stress of the upper wall is

$$\tau_{0u} = \mu\frac{du}{dy}\bigg|_{y=h} = \frac{U}{h} + \frac{1}{2\mu}\frac{\Delta p}{\Delta x}h$$

(5.106)

The shear stress is

$$\tau(y) = \mu\frac{du}{dy} = \mu\frac{U}{h} - \frac{1}{2}\frac{\Delta p}{\Delta x}h\left(1 - 2\frac{y}{h}\right)$$

(5.107)

Since $\Delta p < 0$, the above two formulas show that the upper wall shear stress is smaller than the lower wall shear stress, as shown in Fig. 5.12.

The volume flow through is

$$Q = \int_0^h u(y)\mathrm{d}y = \frac{Uh}{2} - \frac{1}{2\mu}\frac{\Delta p}{\Delta x}\frac{h^3}{6} \tag{5.108}$$

When the shear stress of the upper wall is zero, it can be obtained by formula (5.106)

$$\tau_{0u} = \mu\left.\frac{\mathrm{d}u}{\mathrm{d}y}\right|_{y=h} = \frac{U}{h} + \frac{1}{2\mu}\frac{\Delta p}{\Delta x}h = 0, \quad \frac{\Delta p}{\Delta x} = -\frac{2\mu U}{h^2} \tag{5.109}$$

Substituting into Eq. (5.104), the velocity distribution is obtained as

$$u(y) = 2U\frac{y}{h} - U\frac{y^2}{h^2} \tag{5.110}$$

The shear stress of the lower wall is

$$\tau_{0d} = \mu\left.\frac{\mathrm{d}u}{\mathrm{d}y}\right|_{y=0} = \frac{U}{h} - \frac{1}{2\mu}\frac{\Delta p}{\Delta x}h = 2\frac{U}{h} \tag{5.111}$$

(2) Flow against pressure gradient (negative combination)

Now discuss a flow between parallel plates with infinite length at both ends. The upper plate moves at a speed U, the lower plate is stationary, and the inverse pressure gradient (pressure gradient increases along the flow direction) between the plates along the x-direction is constant. When the flow is stable, find the velocity distribution between the plates, as shown in Fig. 5.13.

Fig. 5.13 Combined flow of Poiseuille flow and Couette flow under the condition of adverse pressure gradient

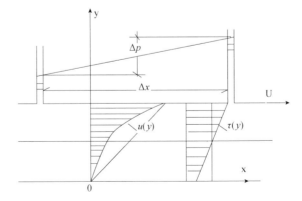

Assuming that the distance between the plates is h, the upper plate moves in U, the two sections are infinitely long, and the end effects are negligible, the flow is a two-dimensional parallel flow, and the flow field is steady. The velocity field obtained by the Navier–Stokes accurate solution is Formula (5.104). Because of $\Delta p > 0$, it shows that the effect of the reverse pressure gradient reduces the flow rate. Among them, from Eqs. (5.105) and (5.106), it can be seen that the shear stress of the lower wall decreases, and the shear stress of the upper wall increases. Let P be the dimensionless coefficient of the pressure gradient as

$$P = -\frac{h^2}{2\mu U}\frac{\Delta p}{\Delta x} \qquad (5.112)$$

Substituting P into the velocity distribution formula (5.104), there is

$$\frac{u(y)}{U} = \frac{y}{h} + P\left(1 - \frac{y}{h}\right)\frac{y}{h} \qquad (5.113)$$

When the shear stress of the lower wall is zero, it can be obtained by formula (5.105)

$$\tau_{0d} = \mu\frac{du}{dy}\bigg|_{y=0} = \frac{U}{h} - \frac{1}{2\mu}\frac{\Delta p}{\Delta x}h = 0, \quad \frac{\Delta p}{\Delta x} = \frac{2\mu U}{h^2}, \quad P = -1 \qquad (5.114)$$

Under the action of different P, the velocity distribution curve is shown in Fig. 5.14. The volume flow through is

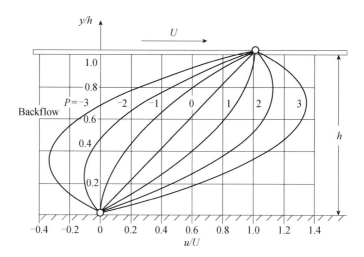

Fig. 5.14 The velocity distribution of different pressure gradients

$$Q = \int_0^h u(y)\mathrm{d}y = \frac{Uh}{2} + \frac{Uh}{6}P \tag{5.115}$$

Analytical formula (5.113) and Fig. 5.14 can be obtained:

(1) $P = 0$, $\Delta p = 0$, is a flat flow with zero pressure gradient;
(2) $P > 0$, $\Delta p < 0$, the pressure decreases along the flow direction and flows along the favorable pressure gradient;
(3) $P < 0$, $\Delta p > 0$, the pressure increases along the flow direction and flows against the adverse pressure gradient;
(4) $P = -1$, the shear stress of the bottom wall is zero;
(5) $P < -1$, $\Delta p > 0$, back flow near the bottom plate;
(6) $P = -3$, the flow through the pipe is zero, indicating that the reverse flow is equal to the positive flow.

5.6.4 Vortex Column and Its Induced Flow Field

Assuming that in an infinite flow field, there is an infinitely long vortex core that rotates steadily, the radius of the vortex core is R, and the rotation angle of the vortex core is.

Since the flow is an incompressible steady two-dimensional flow, the mass force is not considered, and the flow field is assumed to be

$$u_r = 0, \ u_\theta = u_\theta(r, \theta), \ u_z = 0, \ p = p(r) \tag{5.116}$$

Substitute this formula into the continuous equation of the cylindrical coordinate system

$$\frac{\partial u_r}{\partial r} + \frac{1}{r}\frac{\partial u_\theta}{\partial \theta} + \frac{\partial u_z}{\partial z} + \frac{u_r}{r} = 0 \tag{5.117}$$

Get

$$\frac{\partial u_\theta}{\partial \theta} = 0, \ u_\theta(r, \theta) = u_\theta(r) \tag{5.118}$$

The momentum equation in the r-direction is

$$u_r\frac{\partial u_r}{\partial r} + \frac{u_\theta}{r}\frac{\partial u_\theta}{\partial \theta} + u_z\frac{\partial u_z}{\partial z} - \frac{u_\theta^2}{r}$$
$$= -\frac{1}{\rho}\frac{\partial p}{\partial r} + v\left(\frac{\partial^2 u_r}{\partial r^2} + \frac{1}{r}\frac{\partial u_r}{\partial \theta} + \frac{1}{r^2}\frac{\partial^2 u_r}{\partial \theta^2} + \frac{\partial^2 u_r}{\partial z^2} - \frac{2}{r^2}\frac{\partial u_\theta}{\partial \theta} - \frac{u_r}{r^2}\right)$$

Simplified to

$$\frac{\partial p}{\partial r} = \rho \frac{u_\theta^2}{r}$$

(5.119)

The momentum equation in the θ direction is

$$u_r \frac{\partial u_\theta}{\partial r} + \frac{u_\theta}{r} \frac{\partial u_\theta}{\partial \theta} + u_z \frac{\partial u_\theta}{\partial z} + \frac{u_r u_\theta}{r}$$
$$= -\frac{1}{\rho} \frac{1}{r} \frac{\partial p}{\partial \theta} + v \left(\frac{\partial^2 u_\theta}{\partial r^2} + \frac{1}{r} \frac{\partial u_\theta}{\partial r} + \frac{1}{r^2} \frac{\partial^2 u_\theta}{\partial \theta^2} + \frac{\partial^2 u_\theta}{\partial z^2} + \frac{2}{r^2} \frac{\partial u_r}{\partial \theta} - \frac{u_\theta}{r^2} \right)$$

(5.120)

Simplified to get

$$\frac{\partial^2 u_\theta}{\partial r^2} + \frac{1}{r} \frac{\partial u_\theta}{\partial r} - \frac{u_\theta}{r^2} = 0$$

(5.121)

Since $u_\theta = u_\theta(r)$, the above formula can be changed to

$$\frac{d^2 u_\theta}{dr^2} + \frac{1}{r} \frac{du_\theta}{dr} - \frac{u_\theta}{r^2} = 0$$

(5.122)

Organize the above formula

$$\frac{d^2 u_\theta}{dr^2} + \frac{d}{dr} \left(\frac{u_\theta}{r} \right) = 0$$

Integrate, we get

$$\frac{du_\theta}{dr} + \frac{u_\theta}{r} = C_1, \quad \frac{1}{r} \frac{d}{dr} (r u_\theta) = C_1$$

Integrate again, there is

$$u_\theta = \frac{C_1}{2} r + \frac{C_2}{r}$$

(5.123)

where C_1 and C_2 are integral constants. Determined by boundary layer conditions.

(1) In the vortex core area

When $r = R$, $u_\theta = u_R$; when $r = 0$, $u_\theta = u_R$. We get $C_2 = 0$, $C_1 = \frac{2u_R}{R}$. There is

$$u_\theta = u_R \frac{r}{R}$$

(5.124)

Since $u_R = R\omega$, the equation above is

$$u_\theta = \omega r \qquad (5.125)$$

(2) In the outer zone of the vortex core

When $r = R$, $u_\theta = u_R$; when $r \to \infty$, $u_\theta \to 0$. We get $C_1 = 0$, $C_2 = u_R R$. There is

$$u_\theta = u_R \frac{R}{r} = \omega \frac{R^2}{r} \qquad (5.126)$$

Find the velocity circle around r ($>R$) circle, then we have

$$\Gamma = \oint_r^{2\pi} u_\theta r \mathrm{d}\theta = \int_0^{2\pi} \omega \frac{R^2}{r} r \mathrm{d}\theta = 2\pi \omega R^2, \, u_\theta = \frac{\Gamma}{2\pi r} \qquad (5.127)$$

In summary, it can be seen that the theoretical solution for the strong vortex core and its induced velocity field is

$$\begin{cases} u_\theta = \dfrac{\Gamma}{2\pi R} \dfrac{r}{R}, \, r \le R \\ u_\theta = \dfrac{\Gamma}{2\pi r}, \, r > R \end{cases} \qquad (5.128)$$

This theoretical solution is consistent with the Rankine composite vortex, and it can be said to be an approximate model of a tornado. As shown in Fig. 5.15, inside the vortex core is an equal vorticity cylindrical rotating flow field, and outside the vortex core is a free vortex induced flow field.

Discuss:

Fig. 5.15 Vortex core and its induced velocity field

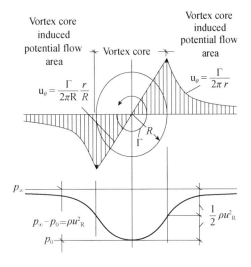

(1) Vortex core area

According to the vortex core velocity field

$$u_\theta = \frac{\Gamma}{2\pi R} \frac{r}{R}$$

The rotational angular velocity of the fluid cluster in the vortex core region is

$$\omega_z = \frac{1}{2}\left[\frac{1}{r}\frac{\partial(ru_\theta)}{\partial r} - \frac{1}{r}\frac{\partial u_r}{\partial \theta}\right] = \frac{\Gamma}{\pi R^2} \neq 0 \qquad (5.129)$$

The vortex core area has vortex flow, and the vorticity is constant, which is similar to the velocity field of a rigid rotating cylinder.

Fluid micelles, the linear deformation rate is zero, and the angular deformation rate is

$$\varepsilon_{r\theta} = \frac{1}{2}\left[\frac{\partial u_\theta}{\partial r} + \frac{1}{r}\frac{\partial u_r}{\partial \theta} - \frac{u_\theta}{r}\right] = 0$$

It shows that the fluid cluster in the vortex core is a movement without deformation, so it is also non-viscous. The vortex area is vortex non-viscous flow.

Substituting the velocity of the vortex core area into Eq. (5.119), we have

$$\frac{\partial p}{\partial r} = \rho\frac{u_\theta^2}{r} = \rho\frac{1}{r}\left(\frac{\Gamma}{2\pi R}\frac{r}{R}\right)^2 = \rho\left(\frac{\Gamma}{2\pi R^2}\right)^2 r \qquad (5.130)$$

Integrate, we get

$$p = p_0 + \frac{1}{2}\rho\left(\frac{\Gamma}{2\pi R^2}r\right)^2 = p_0 + \frac{1}{2}\rho\left(\frac{\Gamma}{2\pi R}\frac{r}{R}\right)^2 = p_0 + \frac{1}{2}\rho u_\theta^2 \qquad (5.131)$$

where p_0 is the pressure at the center of the vortex.

(2) The potential flow area induced outside the vortex core

According to the potential flow area velocity field

$$u_\theta = \frac{\Gamma}{2\pi r}$$

The rotational angular velocity of the fluid cluster outside the vortex core is

$$\omega_z = \frac{1}{2}\left[\frac{1}{r}\frac{\partial(ru_\theta)}{\partial r} - \frac{1}{r}\frac{\partial u_r}{\partial \theta}\right] = 0 \qquad (5.132)$$

Outside the vortex core, the rotational angular velocity of the fluid cluster is zero, indicating that the area outside the vortex core is a potential flow zone.

Fluid micelles, the linear deformation rate is zero, and the angular deformation rate is

$$\varepsilon_{r\theta} = \frac{1}{2}\left[\frac{\partial u_\theta}{\partial r} + \frac{1}{r}\frac{\partial u_r}{\partial \theta} - \frac{u_\theta}{r}\right] = -\frac{\Gamma}{2\pi r^2} \neq 0 \tag{5.133}$$

It shows that outside the vortex nucleus is the velocity field induced by the vortex nucleus, and the fluid cluster is a non-rotational and viscous flow. The shear stress in this zone is

$$\tau_{r\theta} = 2\mu\varepsilon_{r\theta} = \mu\left[\frac{\partial u_\theta}{\partial r} + \frac{1}{r}\frac{\partial u_r}{\partial \theta} - \frac{u_\theta}{r}\right] = -\frac{\mu\Gamma}{\pi r^2} \tag{5.134}$$

The torque produced is

$$M_z = \int_0^{2\pi} \tau_{r\theta} R^2 d\theta = \int_0^{2\pi} R^2\left(-\frac{\mu\Gamma}{\pi R^2}\right)d\theta = -2\mu\Gamma \tag{5.135}$$

Power is

$$P_w = M_z\omega = 2\mu\Gamma\frac{\Gamma}{2\pi R^2} = \frac{\mu\Gamma^2}{\pi R^2} \tag{5.136}$$

Substituting the velocity field induced outside the vortex core into Eq. (5.119), we get

$$\frac{\partial p}{\partial r} = \rho\frac{u_\theta^2}{r} = \rho\left(\frac{\Gamma}{2\pi r}\right)^2\frac{1}{r} \tag{5.137}$$

Integrate, we get

$$p = p_\infty - \frac{1}{2}\rho\left(\frac{\Gamma}{2\pi r}\right)^2 = p_\infty - \frac{1}{2}\rho u_\theta^2 \tag{5.138}$$

The above formula can also be obtained from the Bernoulli equation of potential flow, namely, $p_\infty = p + \frac{1}{2}\rho u_\theta^2$. At the boundary of the vortex core, the pressure should satisfy

$$p_0 + \frac{1}{2}\rho u_R^2 = p_\infty - \frac{1}{2}\rho u_R^2$$

$$p_\infty - p_0 = \rho u_R^2 = \rho\left(\frac{\Gamma}{2\pi R}\right)^2 \tag{5.139}$$

It shows that there is a pressure funnel in the vortex core, and the height of the pressure funnel is proportional to the square of the vortex intensity.

5.6.5 Parallel Flow Along an Infinitely Long Slope Under Gravity

As shown in Fig. 5.16, on an infinitely long inclined (inclination angle θ) flat plate, a liquid layer with a thickness of h flows in parallel at a steady constant speed under the action of gravity, and the velocity distribution of the liquid layer is obtained. Taking the coordinate system shown in Fig. 5.16, the x-axis is along the downward direction of the inclined plane, and the y-axis is perpendicular to the x-axis. Due to the infinite length, ignoring the influence of the two ends, assuming that the flow is a two-dimensional parallel flow, the flow field is steady, and the velocity field is

$$u = u(x, y), v \approx 0, \frac{\partial u}{\partial x} = 0$$

Since it is a free surface flow, the pressure p has nothing to do with x, which is

$$\frac{\partial p}{\partial x} \approx 0 \qquad (5.140)$$

The force per unit mass is only gravity, and its component is

$$f_x = g \sin \theta, \ f_y = -g \cos \theta \qquad (5.141)$$

From the two-dimensional Navier–Stokes equations, simplified to

$$0 = g \sin \theta + v \frac{\partial^2 u}{\partial y^2} \qquad (5.142)$$

Fig. 5.16 Parallel flow of infinite slope

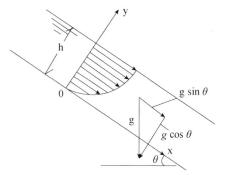

$$0 = -g \cos \theta - \frac{1}{\rho} \frac{\partial p}{\partial y} \tag{5.143}$$

Integrate the formula (5.143) along the y-direction, and get

$$p = -\gamma y \cos \theta + c \tag{5.144}$$

On the free surface, $y = h$, $p = p_a$, substituting the above formula to get c, and finally the formula (5.144) becomes

$$p = p_a + \gamma (h - y) \cos \theta \tag{5.145}$$

Integrate the formula (5.142) along the y-direction, and get

$$u = -\frac{\rho g}{2\mu} y^2 \sin \theta + C_1 y + C_2 \tag{5.146}$$

Substitute boundary conditions

$$y = 0, u = 0, \ y = h, \ \frac{du}{dy} = 0(\tau = 0) \tag{5.147}$$

Get

$$C_2 = 0, C_1 = \frac{\rho g}{\mu} h \sin \theta \tag{5.148}$$

Substituting into the formula (5.146), we get

$$u = \frac{\rho g h^2 \sin \theta}{2\mu} \frac{y}{h} \left(2 - \frac{y}{h} \right) \tag{5.149}$$

The flow rate discharged along the inclined plate is

$$q = \int_0^h u \, dy = \frac{\rho g h^2 \sin \theta}{2\mu} \int_0^h \frac{y}{h} \left(2 - \frac{y}{h} \right) dy = \frac{\rho g h^3 \sin \theta}{3\mu} \tag{5.150}$$

As shown in Fig. 5.17, there is a belt collection device. The belt inclination angle is θ. The thickness of the oil layer on the belt is h. The oil layer flows in parallel at a constant speed under the action of gravity. The upward speed of the belt is U. Oil flow as shown in Fig. 5.17, the component of gravity per unit mass is

$$f_x = -g \sin \theta, \ f_y = -g \cos \theta \tag{5.151}$$

From the two-dimensional Navier–Stokes equations, simplified to

Fig. 5.17 The device of the
belt oil collector

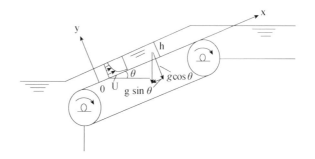

$$0 = -g\sin\theta + v\frac{\partial^2 u}{\partial y^2} \tag{5.152}$$

Integrate the formula (5.152) along the y-direction, and get

$$u = \frac{\rho g}{2\mu}y^2\sin\theta + C_1 y + C_2 \tag{5.153}$$

The boundary conditions of this problem are

$$y = 0, u = U; \; y = h, \frac{du}{dy} = 0(\tau = 0) \tag{5.154}$$

Using formula (5.153), the integral constant is obtained as

$$C_1 = -\frac{\rho g h}{\mu}\sin\theta, \; C_2 = U \tag{5.155}$$

Substituting into the formula (5.153), we get

$$u = U - \frac{\rho g h^2\sin\theta}{2\mu}\frac{y}{h}\left(2 - \frac{y}{h}\right) \tag{5.156}$$

Integrate along the y-direction, we get

$$q = \int_0^h u\,dy = Uh - \frac{\rho g h^2\sin\theta}{2\mu}\int_0^h \frac{y}{h}\left(2 - \frac{y}{h}\right)dy = Uh - \frac{\rho g h^3\sin\theta}{3\mu} \tag{5.157}$$

5.7 Basic Properties of Viscous Fluid Motion

Due to the shear stress in the motion of viscous fluid, compared with the motion of ideal fluid, the main characteristics of the motion during the motion are rotation of motion, diffusibility of vortex, and dissipation of energy.

5.7.1 Vorticity Transport Equation of Viscous Fluid Motion

According to the three theorems of German fluid mechanics Helmholtz (1821–1894), for an ideal positive pressure fluid, under the condition of the mass force, the vorticity remains conserved, that is, once there is no rotation, there will be no rotation forever. But in the case of viscous fluid, oblique pressure, or mass force being potentialless, the vorticity cannot be kept constant, and the vorticity can be generated or extinguished during the flow. Especially in the flow of viscous fluids, the vorticity is always changing during the generation and disappearance of the flow. Therefore, with very few exceptions, almost all have rotational motion.

In order to discuss the nature and laws of vortices in the flow of viscous fluids, it is first necessary to derive the vorticity transport equation. For incompressible fluids, the Gromic-Lamb type equation

$$\frac{\partial \vec{V}}{\partial t} + \nabla\left(\frac{V^2}{2}\right) + 2\vec{\omega} \times \vec{V} = \vec{f} - \frac{1}{\rho}\nabla p + \nu\Delta\vec{V}$$

Introducing vorticity $\vec{\Omega} = 2\vec{\omega}$, the above formula can be written as

$$\frac{\partial \vec{V}}{\partial t} + \nabla\left(\frac{V^2}{2}\right) + \vec{\Omega} \times \vec{V} = \vec{f} - \frac{1}{\rho}\nabla p + \nu\Delta\vec{V} \tag{5.158}$$

Take the curl on both sides of the above formula to get

$$\nabla \times \left\{\frac{\partial \vec{V}}{\partial t} + \nabla\left(\frac{V^2}{2}\right) + \vec{\Omega} \times \vec{V}\right\} = \nabla \times \left\{\vec{f} - \frac{1}{\rho}\nabla p - \nu\Delta\vec{V}\right\} \tag{5.159}$$

Calculate each item to get

$$\nabla \times \frac{\partial \vec{V}}{\partial t} = \frac{\partial(\nabla \times \vec{V})}{\partial t} = \frac{\partial\vec{\Omega}}{\partial t} \tag{5.160}$$

$$\nabla \times \nabla\left(\frac{V^2}{2}\right) = 0 \tag{5.161}$$

$$\nabla \times \vec{\Omega} \times \vec{V} = -(\vec{\Omega} \cdot \nabla)\vec{V} + (\vec{V} \cdot \nabla)\vec{\Omega} \ (u \sin g \nabla \cdot \vec{V} = 0) \tag{5.162}$$

$$\nabla \times \nu \Delta \vec{V} = \nu \Delta (\nabla \times \vec{V}) = \nu \Delta \vec{\Omega} \tag{5.163}$$

$$\nabla \times \left(\frac{1}{\rho}\nabla p\right) = \frac{1}{\rho}\nabla \times \nabla p = 0 \ (\text{Foroblique pressure fluid, it is not zero}) \tag{5.164}$$

Substituting formula (5.160)–(5.164) into formula (5.159), the vorticity transport equation of the incompressible viscous fluid motion is obtained after finishing

$$\frac{\partial \vec{\Omega}}{\partial t} + (\overleftarrow{V} \cdot \nabla)\vec{\Omega} = (\vec{\Omega} \cdot \nabla)\vec{V} + \nabla \times \vec{f} + \nu \Delta \vec{\Omega} \tag{5.165}$$

This is the Helmholtz vorticity transport equation for incompressible viscous fluid motion. The left side of the equation is the satellite derivative of the vorticity. The first term on the right is the vorticity change caused by the stretching and shear deformation of the fluid clusters. The second term is caused by mass force, if it is influential, this term is zero. The third term is the diffusion of vorticity caused by viscosity. This equation shows that the viscosity of fluid, mass forcelessness, and baroclinic fluid are the three major sources of vortex conservation. Among the three, the most common is the sticky effect. For example, (1) The flow of viscous fluid around the near wall of the object is the vortex layer. (2) For non-positive pressure fluids, the density is a function of pressure and temperature, and stratified atmospheric flow can form vortices, and in some extreme cases, strong cyclones can be formed. (3) Mass force has no potential, the air current on the earth can form a vortex under the action of Coriolis force (mass force has no potential). Discuss separately as follows.

(1) If the mass force is potent, incompressible fluid, and inviscid, the vorticity transport equation is simplified to

$$\nabla \times \vec{f} = \nabla \times (-\nabla \Pi) = 0 \tag{5.166}$$

$$\frac{d\vec{\Omega}}{dt} - (\vec{\Omega} \cdot \nabla)\vec{V} = 0 \tag{5.167}$$

This equation is the Helmholtz vorticity conservation equation of an ideal fluid.

(2) If the mass force has a potential and an incompressible viscous fluid, the vorticity transport equation becomes

$$\frac{d\vec{\Omega}}{dt} = (\vec{\Omega} \cdot \nabla)\vec{V} + \nu \Delta \vec{\Omega} \tag{5.168}$$

(3) For the two-dimensional flow of incompressible viscous fluid ($\vec{\Omega} = \Omega_z \vec{k}$), formula (5.168) is simplified as

$$(\vec{\Omega} \cdot \nabla)\vec{V} = \Omega_x \frac{\partial \vec{V}}{\partial x} + \Omega_y \frac{\partial \vec{V}}{\partial y} + \Omega_z \frac{\partial \vec{V}}{\partial z} = \Omega_z \frac{\partial \vec{V}}{\partial z} = 0, \vec{V} = \vec{V}(x, y, t)$$

(5.169)

$$\frac{d\Omega_z}{dt} = \nu \Delta \Omega_z$$

(5.170)

5.7.2 Rotation of Viscous Fluid Motion

The ideal fluid motion can be non-rotational or rotational, but the vorticity remains unchanged. The movement of viscous fluids is generally rotational. Use contradiction to illustrate this point. For incompressible viscous fluids, the equations of motion (N–S equations) are

$$\nabla \cdot \vec{V} = 0$$
$$\frac{d\vec{V}}{dt} = \vec{f} - \frac{1}{\rho}\nabla p + \nu \Delta \vec{V}$$

From field theory knowledge

$$\Delta \vec{V} = \nabla(\nabla \cdot \vec{V}) - \nabla \times (\nabla \times \vec{V}) = -\nabla \times \vec{\Omega}$$

(5.171)

Substituting formula (5.171) into the N–S equations, we get

$$\nabla \cdot \vec{V} = 0$$
$$\frac{d\vec{V}}{dt} = \vec{f} - \frac{1}{\rho}\nabla p - \nu \nabla \times \vec{\Omega}$$

(5.172)

If the flow is non-rotating, the above equation is simplified to

$$\nabla \cdot \vec{V} = 0$$
$$\frac{d\vec{V}}{dt} = \vec{f} - \frac{1}{\rho}\nabla p$$

(5.173)

This set of equations is exactly the same as the Euler equations for the motion of an ideal incompressible fluid. The disappearance of the effect of viscous force indicates that the flow of viscous fluid is exactly the same as that of ideal fluid, and the mathematical properties of the original equation have also changed. From the original second-order quasi-linear partial differential equation system to the first-order quasi-linear partial differential equation system. But the problem lies in the boundary of the solid wall. In the movement of viscous fluid, the boundary conditions of the solid wall surface are non-penetration condition and non-slip condition. That is, on the solid wall surface, it needs to meet

$$V_n = 0 \text{ (Non-penetration)}$$
$$V_\tau = 0 \text{ (Non-slip)} \tag{5.174}$$

It is generally impossible to require the reduced-order Eq. (5.173) to satisfy these two boundary conditions at the same time. This shows that under normal circumstances, the flow of viscous fluid is always swirling.

But there are special cases. If the tangential velocity of the solid wall is exactly equal to the tangential velocity of the fluid to be processed by the solid wall, that is, there is no relative slip between the solid wall and the ideal fluid particle, and the non-slip condition is naturally satisfied. In this way, the ideal fluid equation automatically satisfies the boundary conditions of the solid wall. In this case, the viscous fluid flow can be vortex-free. The velocity field induced by a point vortex is a typical viscous vortex-free current field. But under normal circumstances, there is always a relative slip between the solid wall surface and the ideal fluid particle, and a vortex is inevitably generated due to the viscosity of the fluid. It can be concluded that the vortex of the viscous fluid is generated by the viscous interaction between the relative moving solid wall surface and the fluid, or when two layers of viscous fluids with different speeds are mixed, the vortex will also be generated due to the effect of viscosity.

5.7.3 Diffusion of Viscous Fluid Vortex

Homogeneous, incompressible, and inviscid fluid under the influence of force, the strength of the vortex remains unchanged, and the vortex in the flow field is like "frozen" on the vortex line. However, in a viscous fluid, the size of the vortex can not only be generated, developed, decayed, and disappeared over time, but also diffused. The vorticity spreads from the strong place to the weaker place until the vortex intensity is evenly distributed. Take the diffusion law of a spatial isolated vortex line as an example to illustrate. Since the flow field is axisymmetric, in the case of polar coordinates

$$u_\theta = u_\theta(r, t), \ v_r = 0 \tag{5.175}$$

The expression of vorticity Ω_z is

$$\Omega_z = \frac{1}{r}\frac{\partial(ru_\theta)}{\partial r} - \frac{1}{r}\frac{\partial v_r}{\partial \theta} = \frac{1}{r}\frac{\partial(ru_\theta)}{\partial r} = \Omega_z(r, t) \tag{5.176}$$

Formula (5.170) in polar coordinates, there are

$$\frac{\partial \Omega_z}{\partial t} + v_r \frac{\partial \Omega_z}{\partial r} + \frac{u_\theta}{r} \frac{\partial \Omega_z}{\partial \theta} = \frac{v}{r} \frac{\partial}{\partial r} \left(r \frac{\partial \Omega_z}{\partial r} \right) + \frac{1}{r^2} \frac{\partial^2 \Omega_z}{\partial \theta^2} \tag{5.177}$$

Under the condition of axisymmetric, simplified to

$$\frac{\partial \Omega_z}{\partial t} = \frac{v}{r} \frac{\partial}{\partial r} \left(r \frac{\partial \Omega_z}{\partial r} \right) \tag{5.178}$$

In this way, the definite solution problem of the vortex line intensity diffusion process is

$$\begin{cases} \dfrac{\partial \Omega_z}{\partial t} = \dfrac{v}{r} \dfrac{\partial}{\partial r} \left(r \dfrac{\partial \Omega_z}{\partial r} \right) \\ t = 0, \ r > 0, \ \Omega_z = 0 \\ t \geq 0, \ r \rightarrow \infty, \ \Omega_z = 0 \end{cases} \tag{5.179}$$

This problem characterizes a cylinder with a radius of $r0$, and its axis z is perpendicular to the flow surface. If the rotation angular velocity of the cylinder around the z-axis is $\Omega_0/2$, then in the case of steady flow, the velocity field induced outside the rotating cylinder is

$$u_{\theta 0} = \frac{\Gamma_0}{2\pi r} = \frac{\Omega_0 r_0}{2} \frac{r_0}{r}, \ \Gamma_0 = \Omega_0 \pi r_0^2 \tag{5.180}$$

Solving formula (5.179) can be solved by the similar transformation method in mathematical and physical equations. Namely: Introduce non-dimensional similar variables composed of independent variables, and transform partial differential equations into ordinary differential equations for solution. Introduce the dimensionless variable η, namely,

$$\eta = \frac{r^2}{vt} \tag{5.181}$$

Let the vorticity at any moment be

$$\Omega_z = \frac{\Gamma_0}{vt} F(\eta) \tag{5.182}$$

Among them, $F(\eta)$ is an unknown function of the dimensionless vorticity function. Substituting the above formula into formula (5.179), we get

$$F(\eta) + \eta F'(\eta) + 4 \left[F'(\eta) + \eta F''(\eta) \right] = 0 \tag{5.183}$$

$$\frac{d\left[F(\eta) + 4F'(\eta)\right]}{F(\eta) + 4F'(\eta)} + \frac{d\eta}{\eta} = 0 \tag{5.184}$$

Integrate, and get

$$\eta\left[F(\eta) + 4F'(\eta)\right] = C_1 \tag{5.185}$$

Considering that $F(\eta)$ and $F'(\eta)$ are both finite values at the vortex center, the integral constant C_1 should be taken as zero. From this

$$F(\eta) + 4F'(\eta) = 0, \ F = Ce^{-\frac{r^2}{4\nu t}} \tag{5.186}$$

Solve from this

$$\Omega_z = \frac{\Gamma_0}{\nu t}Ce^{-\frac{r^2}{4\nu t}} \tag{5.187}$$

In order to determine the integral constant C, using Stokes integral in the radius r area, we get

$$\int_0^{2\pi} u_\theta r d\theta = \iint_r \Omega_z d\sigma, \ u_\theta = \frac{1}{2\pi r}\int_0^r \frac{\Gamma_0}{\nu t}Ce^{-\frac{r^2}{4\nu t}}2\pi r dr \tag{5.188}$$

Integrate, and get

$$u_\theta = \frac{2\Gamma_0 C}{r}\left(1 - e^{-\frac{r^2}{4\nu t}}\right) \tag{5.189}$$

Using the initial conditional Eq. (5.180), we get, substituting into Eqs. (5.187) and (5.189), by

$$\Omega_z = \frac{\Gamma_0}{4\pi \nu t}e^{-\frac{r^2}{4\nu t}} \tag{5.190}$$

$$u_\theta = \frac{\Gamma}{2\pi r} = \frac{\Gamma_0}{2\pi r}\left(1 - e^{-\frac{r^2}{4\nu t}}\right) \tag{5.191}$$

Integrate Ω_z to get the vortex intensity at any time as

$$\Gamma = \int_0^r \Omega_z 2\pi r dr = \Gamma_0\left(1 - e^{-\frac{r^2}{4\nu t}}\right) \tag{5.192}$$

It can be seen that in viscous flow, the vorticity diffuses over time, and the range of diffusion $\delta \approx \sqrt{4\nu t}$ is shown in Fig. 5.18.

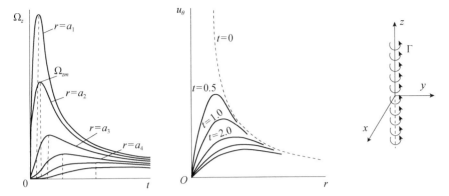

Fig. 5.18 The spread of vorticity and circumferential velocity over time

5.7.4 Dissipation of Viscous Fluid Energy

In viscous fluids, the movement of fluid clusters must overcome viscous stress and consume mechanical energy. This energy generates heat through friction and cannot be recovered from mechanical energy to do useful work. It is an irreversible process, so it is called energy loss. Obviously, this energy is mainly determined by the viscosity of the fluid and the deformation movement of the micelles. For example, the greater the deformation rate during the re-movement of the fluid cluster, the greater the dissipation of mechanical work.

5.8 Laminar Flow, Turbulent Flow and Its Energy Loss

5.8.1 Force of Viscous Fluid Clusters and Its Influence on Flow

The main difference between viscous fluid motion and ideal fluid motion is that in addition to inertial force, there is also viscous force on the micro-clusters, which is reflected in the force behavior of fluid micro-clusters in addition to normal stress (pressure), as well as tangential stress. (Viscous shear stress). Therefore, the motion behavior of the viscous fluid cluster is essentially the result of the interaction between the inertial force and the viscous force. By definition, the role of viscous force is to prevent the relative movement of fluid clusters, while the role of inertial force is the opposite of the role of viscous force, which aggravates the movement of fluid clusters. In fluid mechanics, two extreme conditions of force have been given high attention. One is that the effect of viscous force is much greater than that of inertial force, and the other is that the effect of inertial force is much greater than that of viscous force. The movement characteristics of the fluid clusters are very different in these two cases. This leads to the concept of laminar flow and turbulent flow.

5.8.2 Reynolds Transition Test

There are a lot of flows in nature that are not laminar, but turbulent. This type of flow is extremely complex, and practical applications are even more urgent. Therefore, the formation and development mechanism of turbulence has continuously attracted people's attention. This involves the transition problem of layer loss stability and the problem of fully developed turbulence. Regarding the problem of transition, as early as 1839, German scholar Hagen discovered that the flow characteristics in a circular tube were related to the speed. In 1869, it was discovered that the flow characteristics of two different flow regimes were different. In 1880, the British fluid mechanics Osborne Reynolds (1842–1912, as shown in Fig. 5.19) conducted the famous round tube flow transition experiment (Fig. 5.20). In 1883, he proposed the concepts of laminar flow and turbulent flow. It is recommended to use a dimensionless number (hereinafter referred to as Reynolds number) as the criterion, and the given round tube transition Reynolds number is 2000 (now 2320). The observation that the boundary layer is a turbulent flow state has long been recorded. In 1872, the British scholar William Froude (1810–1879, as shown in Fig. 5.21) observed that the resistance of the plate is proportional to the 1.85 power of the velocity, rather than laminar flow. To the first power. In 1914, Prandtl (1875–1953), the world's master of fluid mechanics, proposed the concept of turbulent boundary layer when studying the resistance of the sphere.

Layer loss stably forms turbulence, and one of the most obvious characteristics is the randomness of turbulence. It has been found that the randomness of turbulence does not only come from the various disturbances and excitations of external

Fig. 5.19 British fluid mechanics Osborne Reynolds (1842–1912) and transitional experimental device

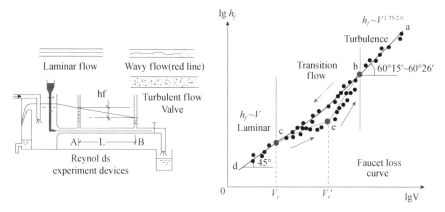

Fig. 5.20 Reynolds transition experiment results

Fig. 5.21 William Froude, British fluid mechanics (1810–1879)

boundary conditions, but more importantly, it comes from the internal nonlinear mechanism. The discovery of chaos has greatly impacted the "determinism". The definite equation system is not like the famous scientist Laplace said. As long as the definite solution conditions are given, everything in the future can be determined, but a definite system can produce uncertainty. the result of. Chaos organically connects determinism and randomism, making us more convinced that the definite Navier–Stokes equations can be used to describe turbulence (that is, a dissipative system is affected by nonlinear inertial forces, which may occur under certain conditions. Multiple nonlinear bifurcations (Bifurcation) eventually become a chaotic structure).

5.8.3 The Criterion of Flow Pattern—Critical Reynolds Number

The experiment found that the critical velocity of laminar transition is related to pipe diameter, fluid density, and dynamic viscosity coefficient. The critical flow rate is directly proportional to the dynamic viscosity coefficient and inversely proportional to the pipe diameter and fluid density. It can be obtained by dimensional analysis that

$$V_c = f(\rho, d, \mu) \tag{5.193}$$

According to the experimental results, the critical speed can be written as

$$V_c = C \frac{\mu}{\rho d} \tag{5.194}$$

Among them, the coefficient C is a dimensionless coefficient. Its value can be expressed as

$$C = \frac{V_c \rho d}{\mu} \tag{5.195}$$

This dimensionless coefficient C is called the critical Reynolds number. Experiments have found that this critical Reynolds number is affected by incoming flow disturbances and boundary conditions, which is

$$\mathrm{Re}_c = \frac{\rho V_c d}{\mu} = \frac{V_c d}{\nu} = f(\text{incoming disturbances, boundary conditions}) \tag{5.196}$$

For flows with the same boundary feature, the transition from laminar flow to turbulent flow is called the upper critical Reynolds number, and the transition from turbulent flow to laminar flow is called the lower critical Reynolds number, and the former is generally greater than the latter. The lower critical Reynolds number is given to Reynolds number for circular pipe flow as $\mathrm{Re}_c = 2000$, and some scholars later gave a result of 2320 (currently considered to be more accurate, 2300 is generally used). The upper critical Reynolds number is a variable that is directly related to the disturbance of the incoming flow.

In addition to the critical Reynolds number, it can also be calculated with any pipe speed and pipe diameter characteristic scale, and the general meaning of the Reynolds number can be obtained as

$$\mathrm{Re} = \frac{\rho V d}{\mu} = \frac{V d}{\nu} \tag{5.197}$$

Reynolds number is the ratio of inertial force and viscous force of fluid flow. If the characteristic scale of the fluid mass is L, the mass is $m = \rho L3$, the characteristic velocity of the fluid mass movement is V, the characteristic time is T, then the inertial force of the fluid mass is

$$F_I = m\frac{dV}{dt} = \rho L^3 \frac{V}{T} = \rho L^2 V^2$$

Viscosity is

$$F_\mu = \mu \frac{dV}{dy} L^2 = \mu L V$$

Therefore

$$Re = \frac{F_I}{F_\mu} = \frac{\rho V^2 L^2}{\mu V L} = \frac{V L}{\nu}$$

For fluid clusters, the effect of inertial force is to promote the instability of the particle and enlarge the disturbance. The role of viscous force is to restrain the particle and restrain the disturbance. A large Reynolds number means that the inertial force of the particle is greater than the viscous force, the flow loses stability, and the flow is turbulent. A small Reynolds number means that the viscous force of the fluid particle is greater than the inertial force, the flow is stable, the layers are distinct, and the flow is laminar.

5.8.4 Resistance Loss Classification

In the flow of viscous fluids, mechanical energy loss is inevitable. The loss in the pipeline can be:

(1) Loss along the way, h_f, refers to the mechanical energy loss caused by the work done by the fluid overcoming the frictional resistance of the solid wall and the internal frictional resistance between the fluid layers along the way.
(2) The local loss h_j refers to the mechanical energy lost by the internal frictional resistance caused by the fluid bypassing the sudden change of the pipe wall and causing the flow to change sharply.

A large number of pipeline experiments found that the loss h_f along the pipeline is a function of the fluid density ρ, the pipeline velocity V, the pipeline diameter d, the pipeline length L, the acceleration of gravity g, the fluid viscosity coefficient μ and the pipeline roughness Δ, which is

$$h_f = f(\rho, V, d, L, g, \mu, \Delta) \tag{5.198}$$

Experiments have found that the resistance loss is proportional to the following variables, which is

$$h_f \propto \frac{V^2}{2g}, \; h_f \propto L, \; h_f \propto \frac{1}{d} \tag{5.199}$$

Taken together, get

$$h_f \propto \frac{L}{d}\frac{V^2}{2g}, \; h_f = \lambda\frac{L}{d}\frac{V^2}{2g} \tag{5.200}$$

where λ is a dimensionless coefficient, which is related to the flow Reynolds number and the relative roughness of the pipe wall. This formula was first proposed by the German scientist Julius Weisbach (1806–1871) in 1850, and the French scientist Henry Darcy (1803–1858, as shown in Fig. 5.22) used the experiment in 1858. The method has been verified, so it is called the Darcy–Weisbach formula, also called the general formula for head loss along the way, or Darcy's formula for short. The Darcy–Weisbach formula is suitable for fully developed laminar and turbulent flows in smooth and rough pipes with any cross-sectional shape and has important engineering application value.

The general expression of local loss is proportional to the local kinetic energy. The proportional coefficient ζ is called the local drag coefficient, and its value is related to the shape of the flow area and flow separation, which is

$$h_j = \zeta\frac{V^2}{2g} \tag{5.201}$$

Fig. 5.22 French engineer
Henry Darcy (1803–1858)

5.8.5 Definition of Turbulence

The earliest research on the phenomenon of turbulence can be traced back to Da Vinci (AD 1452–1519), a scientific and artistic all-rounder in the Italian Renaissance. As shown in Fig. 5.23, he made a careful observation of the turbulent flow. In a famous painting about turbulence, he wrote that: dark clouds were torn apart by violent winds, sand was raised from the beach, and trees bent down. These phenomena constitute the main characteristics of modern wall turbulence, that is, the splitting and breaking of vortices, the entrainment of vortices, and the shearing effect in the near-wall area.

In 1883, the British fluid mechanics Reynolds defined turbulence as tortuous motion (waves). In 1937, British mechanic Taylor (1886–1975, as shown in Fig. 5.24) and American scientist von Kármán (1881–1963, as shown in Fig. 5.25) defined turbulence as turbulence is a kind of non-Regular motion, when the fluid flows over the solid surface, or when the adjacent fluids of the same kind flow or bypass, this kind of irregular motion will generally appear in the fluid (this definition highlights the irregularity of turbulence). In 1959, Hinze (Dutch scientist) defined: Turbulence is an irregular flow state, but its various physical quantities change with time and space coordinates to show randomness, so different statistical averages can be distinguished. The famous Chinese scientist Zhou Peiyuan (1902–1993, as shown in Fig. 5.26) has always maintained that turbulence is an irregular vortex motion. General textbooks define turbulence as a chaotic, intermixed, irregular random motion, which characterizes the trajectory of fluid particles in turbulence. The generally accepted view is that turbulence is composed of vortex structures of different sizes and different frequencies, so that their physical quantities exhibit irregular randomness to changes in time and space. With the deepening of understanding in recent years, people believe that turbulence contains both ordered large-scale vortex structures and disordered and random small-scale vortex structures. The random pulsation of turbulent physical quantities is the result of the combined action of these vortices of different sizes.

Fig. 5.23 Da Vinci's famous paintings of turbulence

Fig. 5.24 G. I. Taylor,
British mechanics (1886
– 1975)

Fig. 5.25 Theodore von
Kármán, American scientist
(1881–1963)

Fig. 5.26 Zhou Peiyuan, the
famous Chinese fluid
mechanics (1902–1993)

5.8.6   Basic Characteristics of Turbulence

(1)   The vortex of turbulence and the theory of vortex cascade

Turbulence is accompanied by vortex motions of different sizes and frequencies, and vortices are the main source of random pulsation of turbulent physical quantities. It is generally believed that the random change process of turbulent physical quantities is produced by these vortices of different sizes. Obviously, in the process of a physical quantity change, the large vortex produces large fluctuations, and the small vortex produces small fluctuations. If there are small vortices in the large vortex, there will be small fluctuations in the large fluctuation, as shown in Fig. 5.27. From a formal point of view, the speed directions around these vortices are relative (opposite). This indicates that there is a considerable velocity gradient in the fluid layer between the vortex bodies. Large vortices obtain energy from basic flow (also called time-average flow or average flow) and are the main energetic vortex of turbulent energy. Then through the process of viscosity and dispersion (instability) cascade splitting into small vortices of different scales, and the energy is gradually transferred to the small-scale vortices in the process of splitting and breaking of these vortices, until viscous dissipation is reached. This process is the vortex cascade theory proposed by British meteorologist Richardson in 1922, as shown in Fig. 5.28.

(2)   Irregularity and randomness of turbulence

The movement of fluid particles in turbulent flow is a random, disorderly and irregular movement. However, because the turbulence field contains vortices of different scales, there is no characteristic scale in theory. Therefore, this random motion must be accompanied by transitions of various scales. The physical quantities of fluid particles in the turbulent field are random variables in time and space, and their statistical average values obey certain regularities. In recent years, with the advent of fractal and chaotic science and the rapid development of nonlinear mechanics, people have gained a new understanding of this randomness. That is, the randomness of turbulence does not only come from the various disturbances and excitations

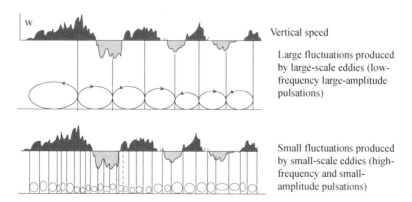

Fig. 5.27   Pulsation and vortex structure of turbulent physical quantities

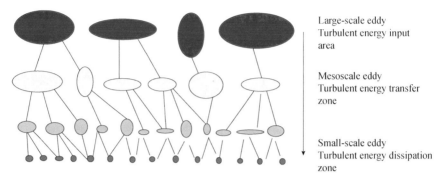

Fig. 5.28 Cascade view of turbulent vortex

of external boundary conditions, but more importantly, it comes from the internal nonlinear mechanism. The discovery of chaos greatly impacted "determinism." In other words, the definite system of equations is not like the famous scientist Laplace said, as long as the definite solution conditions are given, everything in the future can be determined, but the definite system can produce uncertain results. Chaos establishes an organic connection between determinism and stochastic theory, which convinces people that the definite Navier–Stokes equations can be used to describe turbulence (that is, a dissipative system is affected by nonlinear inertial forces, and under certain conditions, more may occur). Sub-nonlinear bifurcation eventually becomes a chaotic structure.

(3) The diffusivity of turbulence and the dissipation of energy

Due to the pulsation and mixing of fluid particles in turbulent flow, the diffusion of momentum, energy, heat, mass, concentration, and other physical quantities in turbulent flow is greatly increased, much greater than in laminar flow.

Small-scale vortices in turbulent flow will produce large instantaneous velocity gradients, which will cause greater viscous dissipation. This is caused by turbulent vortices, which are much larger than laminar flow.

(4) Coherent structure of turbulence

Classical turbulence theory believes that the pulsation in turbulence is a completely disorderly irregular motion. But since Brown and Roshko used the shading instrument to discover the coherent structure in free shear turbulence in the 1970s (as shown in Fig. 5.29), people have realized that the pulsation in turbulence is not entirely irregular motion, but on the surface. It seems that there are still detectable and orderly movements in irregular movements. This coherent structure plays a leading role in the generation and development of shear turbulence pulsation. For example, the discovery of coherent structures in free shear turbulence (turbulent mixing layer, far-field turbulent jets, turbulent wakes, etc.) clearly describes the mixing of coherent large-scale vortices in turbulence and the effects of turbulent jets. Entrainment. The discovery of the band structure in wall shear turbulence reveals the mechanism of turbulence generation near the wall.

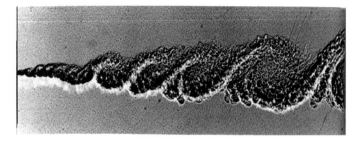

Fig. 5.29 Large-scale coherent structure in turbulence

Fig. 5.30 Small-scale
turbulent vortex structure

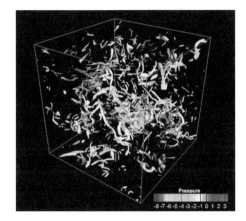

(5) Intermittent turbulence

The earliest discovery of the intermittent characteristics of turbulence is in the areas
where turbulence and non-turbulent flow intersect, such as the outer area of the
turbulent boundary layer, the entrainment area of turbulent jets, etc., where turbulent
and non-turbulent flows alternately appear. However, recent studies have shown that
it is intermittent, even in the interior of turbulence. This is because the energy of the
large vortex will eventually cascade to the small vortices with its leading role in the
turbulent vortex. The vortex occupies only a small area in the space field. Therefore,
the intermittent characteristics of turbulence is common and strange, as shown in
Fig. 5.30.

5.8.7 The Concept of Reynolds Time Mean

Taking into account the randomness of turbulence, in 1895, the British fluid
mechanics Reynolds first regarded instantaneous turbulence as a combination of
time-averaged motion (describes the average trend of the flow) and pulsating motion

Fig. 5.31 The stationary
random process of turbulent
flow

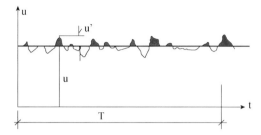

(the degree of deviation from the time-averaged motion). Later, people proposed
space Methods such as decomposition and statistical decomposition.

(1) Time decomposition method (Reynolds' time mean concept)

If the turbulent motion is a stable random process (Fig. 5.31), the instantaneous
velocity u at any point in the turbulent field can be decomposed into the sum of the
time-averaged velocity and the pulsating velocity, which is

$$u = \bar{u} + u' \tag{5.202}$$

where the time-average speed $\bar{u}$ is defined as

$$\bar{u} = \frac{1}{T} \int_0^T u\,\mathrm{d}t = \int_{T\to\infty} \frac{1}{T} \int_0^T u\,\mathrm{d}t \tag{5.203}$$

In the formula, the time T for taking the time average is required to be much
larger than the integral time scale of the pulsating motion. For non-stationary random
processes (as shown in Fig. 5.32), strictly speaking, the time-average decomposition
method cannot be used. But if the characteristic time of time-averaged motion is much
greater than the characteristic time of pulsating motion, and when the mean time T is
much smaller than the characteristic time of time-averaged motion and much greater
than the characteristic time of pulsating motion, the time-average decomposition is
still approximately true.

(2) Spatial decomposition method (spatial average method)

Fig. 5.32 Non-stationary
random process of turbulent
flow

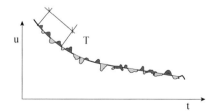

If the turbulence field is a random field with spatial uniformity, the spatial averaging method can be used to spatially decompose the instantaneous quantity of turbulence, which is

$$\bar{u} = \frac{1}{L} \int_0^L u \, dx \qquad (5.204)$$

(3) Ensemble average method (decomposition in the sense of probability)

If the turbulent motion is neither time-stable nor spatially uniform, then the instantaneous motion of turbulence can also be decomposed in the sense of probability, which is

$$\bar{u} = \frac{1}{N} \sum_{i=1}^{\infty} u_i = \int_{-\infty}^{\infty} pu \, du \qquad (5.205)$$

Although the above three decomposition methods are proposed for turbulent fields with different properties, they are statistically equivalent under certain conditions. According to the ergodic theorem of probability theory, all possible values of a random variable in repeated experiments will also appear many times in one experiment in a long time (or within a large space). And the probability of occurrence is the same. Therefore, for a turbulent field that is stable in time and uniform in space, the average values of the physical quantities obtained by the above three decomposition methods are equal.

5.8.8 Reynolds Time-Averaged Motion Equations

In the Cartesian coordinate system, the Navier–Stokes equations describing the instantaneous motion of an incompressible viscous fluid are

$$\frac{\partial u}{\partial t} + \frac{\partial u^2}{\partial x} + \frac{\partial uv}{\partial y} + \frac{\partial uw}{\partial z} = f_x - \frac{1}{\rho}\frac{\partial p}{\partial x} + \nu\left(\frac{\partial^2 u}{\partial x^2} + \frac{\partial^2 u}{\partial y^2} + \frac{\partial^2 u}{\partial z^2}\right)$$

$$\frac{\partial v}{\partial t} + \frac{\partial uv}{\partial x} + \frac{\partial v^2}{\partial y} + \frac{\partial vw}{\partial z} = f_y - \frac{1}{\rho}\frac{\partial p}{\partial y} + \nu\left(\frac{\partial^2 v}{\partial x^2} + \frac{\partial^2 v}{\partial y^2} + \frac{\partial^2 v}{\partial z^2}\right)$$

$$\frac{\partial w}{\partial t} + \frac{\partial uw}{\partial x} + \frac{\partial vw}{\partial y} + \frac{\partial w^2}{\partial z} = f_z - \frac{1}{\rho}\frac{\partial p}{\partial z} + \nu\left(\frac{\partial^2 w}{\partial x^2} + \frac{\partial^2 w}{\partial y^2} + \frac{\partial^2 w}{\partial z^2}\right) \qquad (5.206)$$

The continuity differential equation of incompressible fluid is

$$\frac{\partial u}{\partial x} + \frac{\partial v}{\partial y} + \frac{\partial w}{\partial z} = 0 \qquad (5.207)$$

Reynolds is based on the concept of time average, assuming that the instantaneous motion of turbulent flow satisfies the N–S equations of incompressible viscous fluid motion, and then calculates the time average of the instantaneous motion equations to obtain the Reynolds equations describing the turbulent time-averaged motion, which is

$$u = \bar{u} + u', v = \bar{v} + v', w = \bar{w} + w', p = \bar{p} + p' \tag{5.208}$$

and

$$\overline{u^2} = \overline{(\bar{u} + u')^2} = \bar{u}^2 + \overline{u'^2}, \overline{uv} = \overline{(\bar{u} + u')(\bar{v} + v')} = \bar{u}\bar{v} + \overline{u'v'} \tag{5.209}$$

Taking the time-averaged operation of formula (5.206), we get

$$\frac{\partial \bar{u}}{\partial t} + \bar{u}\frac{\partial \bar{u}}{\partial x} + \bar{v}\frac{\partial \bar{u}}{\partial y} + \bar{w}\frac{\partial \bar{u}}{\partial z} = \bar{f}_x - \frac{1}{\rho}\frac{\partial \bar{p}}{\partial x} + \frac{1}{\rho}\frac{\partial}{\partial x}\left(\mu\frac{\partial \bar{u}}{\partial x} - \rho\overline{u'^2}\right)$$
$$+ \frac{1}{\rho}\frac{\partial}{\partial y}\left(\mu\frac{\partial \bar{u}}{\partial y} - \rho\overline{u'v'}\right) + \frac{1}{\rho}\frac{\partial}{\partial z}\left(\mu\frac{\partial \bar{u}}{\partial z} - \rho\overline{u'w'}\right) \tag{5.210}$$

$$\frac{\partial \bar{v}}{\partial t} + \bar{u}\frac{\partial \bar{v}}{\partial x} + \bar{v}\frac{\partial \bar{v}}{\partial y} + \bar{w}\frac{\partial \bar{v}}{\partial z} = \bar{f}_y - \frac{1}{\rho}\frac{\partial \bar{p}}{\partial y} + \frac{1}{\rho}\frac{\partial}{\partial x}\left(\mu\frac{\partial \bar{v}}{\partial x} - \rho\overline{u'v'}\right)$$
$$+ \frac{1}{\rho}\frac{\partial}{\partial y}\left(\mu\frac{\partial \bar{v}}{\partial y} - \rho\overline{v'^2}\right) + \frac{1}{\rho}\frac{\partial}{\partial z}\left(\mu\frac{\partial \bar{v}}{\partial z} - \rho\overline{v'w'}\right) \tag{5.211}$$

$$\frac{\partial \bar{w}}{\partial t} + \bar{u}\frac{\partial \bar{w}}{\partial x} + \bar{v}\frac{\partial \bar{w}}{\partial y} + \bar{w}\frac{\partial \bar{w}}{\partial z} = \bar{f}_z - \frac{1}{\rho}\frac{\partial \bar{p}}{\partial z} + \frac{1}{\rho}\frac{\partial}{\partial x}\left(\mu\frac{\partial \bar{w}}{\partial x} - \rho\overline{u'w'}\right)$$
$$+ \frac{1}{\rho}\frac{\partial}{\partial y}\left(\mu\frac{\partial \bar{w}}{\partial y} - \rho\overline{v'w'}\right) + \frac{1}{\rho}\frac{\partial}{\partial z}\left(\mu\frac{\partial \bar{w}}{\partial z} - \rho\overline{w'^2}\right) \tag{5.212}$$

Time-averaged motion continuity equation

$$\frac{\partial \bar{u}}{\partial x} + \frac{\partial \bar{v}}{\partial y} + \frac{\partial \bar{w}}{\partial z} = 0 \tag{5.213}$$

where $\bar{u}$ is the time-averaged velocity component in the x-direction; u' is the pulsating velocity component in the x-direction; $\bar{p}$ is the time-averaged pressure; $\bar{f}_x$ is the time-average value of the volume force per unit mass in the x-direction; $-\rho\overline{u'v'}$ and other terms are the second-order related terms of the pulsating velocity, called the Reynolds stress term or the turbulent stress term, which is physically explained as the momentum exchange term caused by the pulsating motion of fluid particles.

$$\left[-\rho\overline{u_i'u_j'}\right] = \begin{bmatrix} -\rho\overline{u'^2} - \rho\overline{u'v'} - \rho\overline{u'w'} \\ -\rho\overline{u'v'} - \rho\overline{v'^2} - \rho\overline{v'w'} \\ -\rho\overline{u'w'} - \rho\overline{v'w'} - \rho\overline{w'^2} \end{bmatrix} \qquad (5.214)$$

The matrix composed of these 6 turbulent stresses is called the Reynolds stress matrix.

5.9 Turbulent Eddy Viscosity and Prandtl Mixing Length Theory

Based on the principle of phenomenology, in 1877, the French mechanic Joseph Valentin Boussinesq (1842–1929, as shown in Fig. 5.33) first compared the additional shear stress (later called Reynolds stress) generated by turbulent pulsation with viscous stress and proposed the famous. The hypothesis of eddy viscosity establishes the analogy relationship between the Reynolds stress and the time-average velocity gradient. Although the concept of vortex viscosity predates the emergence of the Reynolds equations, it laid the foundation for later engineering turbulence. For a simple time-averaged two-dimensional flow near the wall (Fig. 5.34), the turbulent stress (Reynolds stress) can be expressed as

$$\tau_t = -\rho\overline{u'v'} = \rho v_t \frac{\partial \overline{u}}{\partial y} \qquad (5.215)$$

Fig. 5.33 French scientist Joseph Valentin Boussinesq (1842–1929)

Fig. 5.34 Shear turbulence
near the wall

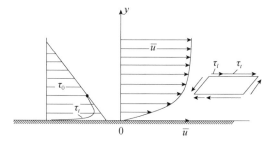

In the formula, ν_t is the eddy viscosity coefficient. In contrast, the viscous shear stress generated by the time-average flow is

$$\tau_l = \rho \nu \frac{\partial \overline{u}}{\partial y} \tag{5.216}$$

The total shear stress acting between the fluid layers is

$$\tau_0 = \tau_t + \tau_l = \rho(\nu + \nu_t)\frac{\partial \overline{u}}{\partial y} \tag{5.217}$$

where ν_t is not a physical property of the fluid, but a function of the turbulent motion state.

How to determine the size and distribution of ν_t? At first Boussinesq thought that ν_t is a constant. Later, people discovered that ν_t not only has different values for different flow problems, but also has different values for the same flow problem at different times and in different regions. According to the turbulent motion characteristics, ν_t can change significantly in the flow field. According to the results of dimensional analysis and turbulence research, the vortex viscosity ν_t is proportional to the product of the characteristic length scale of the energy-carrying vortex and the characteristic velocity scale, which is

$$\nu_t \propto l_t V_t \tag{5.218}$$

The ratio of turbulent stress to viscous stress is

$$\frac{\tau_t}{\tau_l} = \frac{-\rho \overline{u'v'}}{\rho \nu \frac{\partial u}{\partial y}} = \frac{\rho \nu_t \frac{\partial \overline{u}}{\partial y}}{\rho \nu \frac{\partial \overline{u}}{\partial y}} = \frac{\nu_t}{\nu} = \frac{l_t V_t}{\nu} = \mathrm{Re}_t \tag{5.219}$$

In the formula, Re_t represents the Reynolds number of large-scale turbulent motion characteristics, generally $\mathrm{Re}_t = 103$–105. In 1925, Prandtl proposed the theory of mixing length based on the analogy of molecular motion theory. In 1932, the German scholar Nicholas (Nikuradse) Based on the results of the sand pipe resistance experiment, the problems of the time-averaged velocity distribution and the resistance loss of the pipe turbulence are solved, and the well-known logarithmic velocity distribution formula is derived. According to Prandtl's mixing length theory, for shear turbulence, Prandtl believes that the characteristic velocity V_t of the turbulent vortex is proportional to the product of the time-average velocity gradient and the mixing length, which is

$$V_t \propto l_m \left| \frac{\partial \overline{u}}{\partial y} \right| \qquad (5.220)$$

Using the above formula and absorbing the proportional coefficient in the mixing length, you can get

$$\tau_t = -\overline{\rho u' v'} = \rho l_m^2 \frac{\partial \overline{u}}{\partial y} \left| \frac{\partial \overline{u}}{\partial y} \right|, \; v_t = l_m^2 \left| \frac{\partial \overline{u}}{\partial y} \right| \qquad (5.221)$$

In the near-wall turbulent flow (as shown in Fig. 5.35), the pulsation velocity is very small and the turbulent shear stress is small due to the influence of the wall, but the flow velocity gradient is large, the viscous shear stress plays a leading role, and the velocity distribution is linear. The zone is called the viscous bottom zone. The outer zone of the viscous bottom layer is the core zone of turbulence. At this time, the turbulent shear stress plays a leading role, and the velocity distribution conforms to the logarithmic or power distribution. There is a transition zone between the turbulent core zone and the viscous bottom zone. The viscous bottom layer is neither laminar nor turbulent. There are turbulent spots in this layer. The thickness of the viscous bottom layer and the roughness of the wall directly affect the energy loss along the way. In the turbulent region near the wall, assuming that the turbulent shear stress is approximately equal to the wall shear stress τ_w, and assuming that the mixing length is proportional to the distance y from the particle to the wall, that is $l_m = ky$ (k is the Karman constant ≈ 0.4),

$$\frac{\tau_w}{\rho} = k^2 y^2 \left(\frac{d\overline{u}}{dy} \right)^2 \qquad (5.222)$$

Fig. 5.35 Turbulent flow structure near the wall

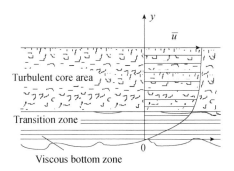

Turbulent core area

Transition zone

Viscous bottom zone

Integrate the above formula to get the well-known logarithmic distribution curve of time-averaged velocity near the wall, which is

$$\frac{\overline{u}}{u^*} = \frac{1}{k} \ln \frac{u^* y}{\nu} + C \tag{5.223}$$

where C is a constant, $u^* = \sqrt{\frac{\tau_w}{\rho}}$ is the friction speed. For a smooth wall, $C = 5.5$, then the velocity distribution of the smooth wall is

$$\frac{\overline{u}}{u^*} = \frac{1}{k} \ln \frac{u^* y}{\nu} + 5.5 \tag{5.224}$$

If it is a rough wall, the time-average velocity distribution is

$$\frac{\overline{u}}{u^*} = \frac{1}{k} \ln \frac{y}{\Delta} + 8.5 \tag{5.225}$$

where Δ is the rough height of the wall. The mixing length theory establishes the relationship between the vortex viscosity coefficient and the local time-average velocity gradient, and l_m can be determined experimentally for different flows. For some common shear layer flows, there are corresponding empirical relationships of mixing lengths, so it will not be repeated here.

5.10 Similarity Principle and Dimensionless Differential Equations

5.10.1 Principles of Dimensional Analysis-π Theorem

Fluid mechanics tests are generally divided into two categories: physical tests and model tests. Physical tests (various prototype observation tests) will not cause distortion of models and environment simulations. It has always been the final means of identifying physical flow dynamics and observing flow fields (such as airplane flight test tests), but the cost of the test is expensive and the test conditions are difficult to control. The model test uses a model that is geometrically similar to the real object and is carried out under artificial control. In order for the model test results to be applied to the actual situation, it is necessary to make the flow around the model and around the object similar. In this way, their dimensionless hydrodynamic properties can be the same. This requires that the similar forces acting on the volume element have the same ratio at all similar corresponding points. The ratio of these dimensionless numbers is called similarity parameter or similarity criterion. There are many similar parameters, such as Mach number, Reynolds number, Freud number, and so on.

The method of deriving the similarity criterion from the physical equation is called the similarity transformation method. The specific steps of the similarity transformation method to derive the similarity criterion are:

(1) List the physical equations;
(2) List the similar transformation formulas of various physical quantities and substitute them into the physical equations;
(3) Get the similarity index and similarity criterion composed of similar numbers.

Take length dimension L, time dimension T, and mass dimension M as basic dimensions, and other physical dimensions as derived dimensions, and the corresponding dimension expression is

$$[q] = M^x L^y T^z \tag{5.226}$$

where x, y, and z are dimensional indices, which can be determined by physical theorems or definitions. If in the dimensional expression of a physical quantity, the dimensional index of all the quantities is zero, then the physical quantity is a dimensionless quantity, otherwise it is a dimensional quantity. Dimensionless quantities are different from pure numbers and have specific physical meanings and quantitative properties. The value of a dimensional quantity changes with different units, and the value of a non-dimensional quantity does not change with different units.

According to the principle of dimensional harmony, for any physical process, there are n-dimensional physical quantities. If m of them is selected as the basic physical quantities, then the physical process can be described by n-independent non-dimensional quantities. Because these dimensionless quantities are expressed by π, they are called π theorem. The π theorem was developed in 1914 by the American physicist E. Buckingham (1867–1940, as shown in Fig. 5.36). It is the basis of dimensional analysis and is also the Buckingham theorem.

Fig. 5.36 E. Buckingham, American physicist (1867–1940)

If A_1, A_2, A_3, ...A_n is used to represent n physical quantities of a physical process (such as speed, pressure, temperature, etc.), the functional relationship is

$$f(A_1, A_2, A_3, ..., A_n) = 0 \tag{5.227}$$

If you choose m basic physical quantities, you can use dimensionless quantities to give the following expressions, which is

$$f(\pi_1, \pi_2, \pi_3, ..., \pi_{n-m}) = 0 \tag{5.228}$$

When determining dimensionless variable π_i, first select m independent variables from n physical quantities (called basic physical quantities). In the International System of Units, generally up to three independent variables are selected, because it contains three basic variables: length, quality, and time. Dimension), and then repeatedly use the basic physical quantities and the remaining physical quantities one by one to establish a dimensionless quantity π_i using formula (5.217), which is

$$\pi_i = \frac{A_i}{A_{01}^x A_{02}^y A_{03}^z} \tag{5.229}$$

where A_{01}, A_{02}, A_{03} is the selected three basic physical quantities, and it is best to choose the main physical quantities that affect the physical process. The following example illustrates the application of the π theorem.

If an airplane flies in a straight line at a uniform speed, the test found that the main physical quantities that affect the airplane drag D include air density ρ, flight speed V_∞, wing length b, air viscosity coefficient μ, air wave speed a, gravitational acceleration g, the thickness of the wing c, the angle of attack α of the aircraft, ... If written as a function, then the relationship is

$$D = f(\rho, V_\infty, b, \mu, a, g, c, \alpha, ...) \tag{5.230}$$

If air density ρ, flight speed V_∞, and wing span b are selected as the basic physical quantities (including the three basic dimensions of L, M, and T), these three basic physical quantities are independent and cannot be mutually exclusive. Because the dimensional index ranks of these three basic physical quantities are not zero, they are independent of each other and cannot be expressed with each other, which is

$$[\rho] = M^1 L^{-3} T^0, \ [V_\infty] = M^0 L^1 T^{-1}, \ [b] = M^0 L^1 T^0 \tag{5.231}$$

And then

$$\begin{vmatrix} 1 & -3 & 0 \\ 0 & 1 & -1 \\ 0 & 1 & 0 \end{vmatrix} = 1 \neq 0 \tag{5.232}$$

In this way, the dimensionless quantity expression is

$$\pi_D = f(\pi_c, \pi_\mu, \pi_a, \pi_g, \pi_\alpha) \tag{5.233}$$

Each dimensionless quantity is determined as follows.

$$\pi_D = \frac{D}{\rho^x V_\infty^y b^z} \tag{5.234}$$

According to dimensional analysis, determine the exponents x, y, and z, which is

$$[\pi_D] = \frac{[D]}{[\rho]^x [V_\infty]^y [b]^z} \tag{5.235}$$

Because π_D is a dimensionless quantity, and the dimension of resistance D is

$$[\pi_D] = M^0 L^0 T^0, \quad [D] = M^1 L^1 T^{-2} \tag{5.236}$$

Substituting formula (5.231) and formula (5.236) into formula (5.235), we can get

$$M^0 L^0 T^0 = \frac{M^1 L^1 T^{-2}}{[ML^{-3}]^x [LT^{-1}]^y [L]^z} = M^{1-x} L^{1+3x-y-z} T^{-2+y} \tag{5.237}$$

Comparing the exponents of M, L, and T on both sides of the equation, we get

$$\begin{aligned} 1 - x &= 0 \\ 1 - 3x - y - z &= 0 \\ -2 + y &= 0 \end{aligned} \tag{5.238}$$

Solve it and get: $x = 1$, $y = 2$, $z = 2$. Then π_D is

$$\pi_D = \frac{D}{\rho^x V_\infty^y b^z} = \frac{D}{\rho V_\infty^2 b^2} \tag{5.239}$$

It can also be solved by another simpler method. Determine the basic dimension from Eq. (5.231), then substitute into Eq. (5.236) to directly derive π_D. From the formula (5.231) we can see

$$\begin{aligned} \rho &= M^1 L^{-3} T^0 \\ V_\infty &= M^0 L^1 T^{-1} \\ b &= M^0 L^1 T^0 \end{aligned} \tag{5.240}$$

Solve for M, L, and T as

$$M = \rho b^3$$
$$T = b/V_\infty \tag{5.241}$$
$$L = b$$

Substituting Eq. (5.237) to get the same result as Eq. (5.240).

$$[D] = \rho b^3 b(V_\infty/b)^2 = \rho V_\infty^2 b^2, \ \pi_D = \frac{D}{\rho V_\infty^2 b^2} \tag{5.242}$$

Dimensionless quantities of other physical quantities can be obtained in the same way, which is

$$[\mu] = M^1 L^{-1} T^1, \ \pi_\mu = \frac{\mu}{\rho b^3 b^{-1}(b/V_\infty)^{-1}} = \frac{\mu}{\rho b V_\infty} \tag{5.243}$$

$$[a] = LT^{-1}, \ \pi_a = \frac{a}{b(b/V_\infty)^{-1}} = \frac{a}{V_\infty} \tag{5.244}$$

$$[g] = LT^{-2}, \ \pi_a = \frac{g}{b(b/V_\infty)^{-2}} = \frac{gb}{V_\infty^2} \tag{5.245}$$

$$[c] = L, \ \pi_c = \frac{c}{b} \tag{5.246}$$

The angle of attack α of the aircraft itself is a dimensionless quantity, without calculation, $\pi_\alpha = \alpha$. Substituting formula (5.242) to formula (5.246) into formula (5.233), we get

$$\pi_D = \frac{D}{\rho V_\infty^2 b^2} = f\left(\frac{\mu}{\rho b V_\infty}, \frac{a}{V_\infty}, \frac{gb}{V_\infty^2}, \frac{c}{b}, \alpha\right) \tag{5.247}$$

Or it can be written as

$$D = \rho \frac{V_\infty^2}{2} Sf(\mathrm{Re}, \mathrm{Ma}, \mathrm{Fr}^2, \frac{c}{b}, \alpha)$$
$$\mathrm{Re} = \frac{\rho b V_\infty}{\mu}, \ \mathrm{Ma} = \frac{V_\infty}{a}, \ \mathrm{Fr} = \frac{V_\infty}{\sqrt{gb}}, \ S = b^2 (\text{represent as the area of wing}) \tag{5.248}$$

Among them, Re represents the Reynolds number, Ma represents the Mach number, and Fr represents the Freud number.

5.10.2 Dimensionless N–S Equations

Based on two similar flows, they must be described by the same physical equation, and the N–S equations that characterize the incompressible flow can be expressed as dimensionless equations. For dimensional incompressible fluid N–S group (mass force only gravity) is

$$\frac{\partial \vec{V}}{\partial t} + (\vec{V} \cdot \nabla)\vec{V} = \vec{g} - \frac{1}{\rho}\nabla p + v\Delta\vec{V}$$

$$\nabla \cdot \vec{V} = 0$$

If you want to become a dimensionless form, then carry out a dimensionless transformation on each quantity in the equation system. Take L, T, V_0, p_0 as the characteristic length, time, speed, and pressure, and do the following dimensionless transformation

$$t^* = \frac{t}{T}, x^* = \frac{x}{L}, u^* = \frac{u}{V_0}, p^* = \frac{p}{p_0}\ldots \tag{5.249}$$

Substituting into the continuity equation, the dimensionless continuity equation is obtained, which is

$$\left(\frac{\partial u^*}{\partial x^*} + \frac{\partial v^*}{\partial y^*} + \frac{\partial w^*}{\partial z^*}\right) = 0 \tag{5.250}$$

Substituting into the N–S equation in the x-direction, we get

$$\frac{V_0}{T}\frac{\partial u^*}{\partial t^*} + \frac{V_0^2}{L}\left(u^*\frac{\partial u^*}{\partial x^*} + v^*\frac{\partial u^*}{\partial y^*} + w^*\frac{\partial u^*}{\partial z^*}\right)$$

$$= g - \frac{p_0}{\rho L}\frac{\partial p^*}{\partial x^*} + v\frac{V_0}{L^2}\left(\frac{\partial^2 u^*}{\partial x^{*2}} + \frac{\partial^2 u^*}{\partial y^{*2}} + \frac{\partial^2 u^*}{\partial z^{*2}}\right) \tag{5.251}$$

Organized into an infinite form

$$\mathrm{Sh}\frac{\partial u^*}{\partial t^*} + u^*\frac{\partial u^*}{\partial x^*} + v^*\frac{\partial u^*}{\partial y^*} + w^*\frac{\partial u^*}{\partial z^*}$$

$$= \frac{1}{Fr^2} - \mathrm{Eu}\frac{\partial p^*}{\partial x^*} + \frac{1}{\mathrm{Re}}\left(\frac{\partial^2 u^*}{\partial x^{*2}} + \frac{\partial^2 u^*}{\partial y^{*2}} + \frac{\partial^2 u^*}{\partial z^{*2}}\right) \tag{5.252}$$

Among them, Sh is the Strouhal number, which is

$$\mathrm{Sh} = \frac{L}{V_0 T} \tag{5.253}$$

Fr is the Freud number, namely,

$$Fr = \frac{V_0}{\sqrt{gL}} \tag{5.254}$$

Eu is Euler's number, namely,

$$Eu = \frac{p_0}{\rho V_0^2} \tag{5.255}$$

Re is the Reynolds number, namely,

$$Re = \frac{V_0 L}{\nu} \tag{5.256}$$

Obviously, if two flows are similar, they must be described by the same dimensionless physical equation. For the similar flow of prototypes and models. If the similar scales of the basic physical quantities are

$$\lambda_L = \frac{L_p}{L_m}, \lambda_T = \frac{T_p}{T_m}, \lambda_M = \frac{M_p}{M_m} \tag{5.257}$$

The subscript p in the formula represents the prototype, and the subscript m represents the model. Among them, λ_L is the length scale, λ_T is the time scale, and λ_M is the mass scale. According to the dimensionless equation, the similarity criterion of the prototype and the model can be obtained:

(1) The similarity criterion of Sh Strohal number is

$$Sh_p = Sh_m, \frac{L_p}{V_p T_p} = \frac{L_m}{V_m T_m}, \frac{\lambda_L}{\lambda_V \lambda_T} = 1 \tag{5.258}$$

(2) The similarity criterion of Fr Freud number is

$$Fr_p = Fr_m, \frac{V_p}{\sqrt{g_p L_p}} = \frac{V_m}{\sqrt{g_m L_m}}, \frac{\lambda_V}{\sqrt{\lambda_g \lambda_L}} = 1 \tag{5.259}$$

(3) The Euler number similarity criterion is

$$Eu_p = Eu_m, \frac{p_p}{\rho_p V_p^2} = \frac{p_m}{\rho_m V_m^2}, \frac{\lambda_p}{\lambda_\rho \lambda_V^2} = 1 \tag{5.260}$$

(4) The similarity criterion of Re Reynolds number is

$$\mathrm{Re}_p = \mathrm{Re}_m, \ \frac{V_p L_p}{\nu_p} = \frac{V_m L_m}{\nu_m}, \ \frac{\lambda_V \lambda_L}{\lambda_\nu} = 1 \qquad (5.261)$$

For compressible flow, in addition to the above-mentioned similarity criterion, the Mach Ma similarity criterion that characterizes compressibility will be derived, namely,

$$\mathrm{Ma}_p = \mathrm{Ma}_m, \ \frac{V_p}{a_p} = \frac{V_m}{a_m}, \ \frac{\lambda_V}{\lambda_a} = 1 \qquad (5.262)$$

where a is the speed of sound waves.

Exercises

A. Thinking questions

1. Under what circumstances can the viscosity of the fluid be reflected? From the perspective of the transport properties of fluid movement, what kind of transport characteristics does fluid viscosity characterize?
2. Please explain the physical meaning of Newton's internal friction law? What is the physical mechanism of the dynamic viscosity coefficient?
3. Write down the deformation rate matrix of fluid clusters and its three invariants? Explain the physical meaning of divergence?
4. Write down the stress matrix of the fluid cluster and its three invariants? Explain the physical meaning of the nominal pressure acting on the viscous fluid clusters? In the movement of viscous fluid, what is the physical factor that causes the normal stress of each coordinate axis of the micelle to be different?
5. What are the three assumptions of Stokes regarding the generalized Newton's internal friction law? And explain the physical meaning?
6. Write down the constitutive relationship of the incompressible viscous fluid micro-cluster motion. In the steady flow of a straight pipe with constant cross section, what is the constitutive relationship of the three normal stresses?
7. Write down the constitutive relationship of the three normal stresses in the steady flow of a shrinking pipe?
8. Write down the component expressions of the differential equations of motion of incompressible viscous fluids (Navier–Stokes equations)? Vector form? And the physical meaning of each item?

9. Why is the motion of viscous fluid generally vortex motion? For the motion of incompressible viscous fluid, where does the vortex mainly occur under the condition of mass force advantage? why?

10. Please explain that the flow field induced by an infinitely long rotating cylinder is viscous fluid motion, but non-rotating motion. That is, viscous potential flow?

11. Write the vorticity transport equation for the motion of incompressible viscous fluid, and point out the physical meaning of each item?

12. Write down the Gromic-Lamb type differential equation of motion for the incompressible viscous fluid, and explain the physical meaning of each item?

13. Under the action of gravity, for the Bernoulli equation of the N–S equations of incompressible fluid motion along streamlines, explain the physical meaning of the Bernoulli integral, and explain the physical mechanism of energy loss?

14. Explain whether the statement "Water always flows from a high place to a low place" is correct and why? If it is not correct, please give the correct statement?

15. Compared with ideal fluid motion, what are the main characteristics of viscous fluid motion?

16. Why do we say that viscosity and inverse pressure gradient are necessary conditions for boundary layer separation?

17. The adhesion condition of the object surface is the difference between viscous fluid turbulence and ideal fluid turbulence? Please explain the physical reason for the resistance caused by the spoiler? And pointed out why there are frictional resistance and differential pressure resistance in the turbulence of low-speed objects?

18. Please indicate the main measures to reduce the resistance of low-speed objects moving around the flow?

B. **Calculation questions**

1. The gauge pressure of the oil in the cylinder $p = 29.418 \times 10^4$ Pa, the viscosity of the oil $\mu = 0.1$ Pa s, the diameter of the plunger $d = 50$ mm, the radial gap between the plunger and the sleeve $\delta = 50$ mm, the length of the sleeve $l = 300$ mm. Suppose that the plunger is pushed by the force F to keep it still, and the oil leakage flow q_v and the force F is calculated.

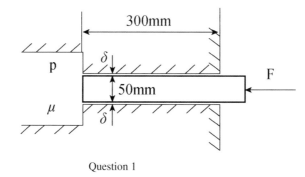

300mm

p

δ

50mm

F

μ

δ

Question 1

2. Water at 20 °C will be sucked out through a pipe with a length of 1 m and a diameter of 2 mm, as shown in the figure. Is there a height H that the flow may not be laminar? If $H = 50$ cm, what is the flow velocity? Ignore the tube curvature

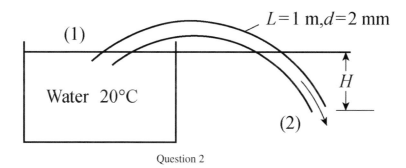

(1)

$L = 1$ m, $d = 2$ mm

Water 20°C

H

(2)

Question 2

3. The relative density of the oil $d = 0.85$, the viscosity $\mu = 3 \times 103$ Pa s, and it flows through the annular pipe with $r_1 = 15$ cm and $r_2 = 7.5$ cm. If the pressure drop per meter of pipe length is 196 Pa when the pipeline is placed horizontally, try to find (1) the flow of oil; (2) the tangential stress on the outer pipe wall; (3) the axial force acting on the inner pipe per meter.

4. The figure shows that the two plates move in the opposite direction $v_1 = 2v_2 = 2$ m/s. If $P_1 = P_2 = 9.806 \times 10^4$ Pa, $a = 1.5$ mm $\mu = 0.49$ Pa s find the effect on each Tangential stress of a flat plate.

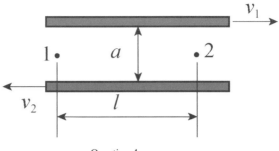

Question 4

5. Consider the incompressible viscous flow of air between two infinitely long parallel plates at a distance h. The bottom plate is stationary, and the top plate moves in the direction of the plate at a constant speed u_e. It is assumed that there is no pressure gradient in the flow direction. (a) Expression of speed change between boards. (b) If $T = $ constant $ = 320$ k, $u_e = 30$ m/s, $h = 0.01$ m, calculate the shear stress of the upper and lower plates.

6. The incompressible viscous flow of air between two infinitely long parallel plates with a distance h. Both parallel plates are stationary, but there is a constant pressure gradient in the flow direction (i.e., $dp/dx = $ constant).

   a. Obtain the expression of the velocity change between the two plates.
   b. Obtain the dp/dx expression of the shear stress on the plate.

7. Two concentric circular tubes with an outer diameter of $2r_1$ and an inner diameter of $2r_2$ rotate in the same direction at angular velocities ω_1 and ω_2. Try to prove The velocity distribution between the two circular tubes is

$$v_\theta = \frac{1}{r_1^2 - r_2^2}\left[r\left(r_1^2 w_1 - r_2^2 w_2\right) - \frac{r_1^2 r_2^2}{r_1^2 - r_2^2}(w_1 - w_2)\right]$$

8. In the case of the same Reynolds number Re_l, try to find the ratio of frictional resistance when water at 20 °C and air at 30 °C flow parallel through a long plate.

9. A flat plate with a length of 6 m and a width of 2 m is placed in a parallel and static air flow at a speed of 60 m/s at 40 °C, and the critical Reynolds number of the transition from laminar flow to turbulent flow in the boundary layer of the plate is $\text{Re} = 10^6$. Try to calculate the frictional resistance of the plate.

10. As shown in the figure, the oil flows upward through the inclined pipe at a distance of $\rho = 900$ kg/m³ and $v = 0.0002$ m²/s. The distance between Section 1 and Section 2 is 10 m. Assuming that the laminar flow is stable, (a) calculate the h_f between 1 and 2, calculate (b) Q, (c) V, and (d) Re_d. Is the flow laminar?

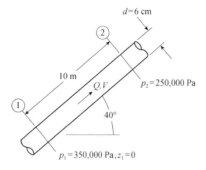

11. Petroleum ($\rho = 890$ kg/m^3 and $\mu = 0.07$ kg/m s) flows through a 15 m long horizontal pipeline. The power delivered to the flow is 1 hp.
    (a) If the flow is at the transition point of laminar flow, what is the appropriate pipe diameter? In this case, the unit of m^3/h (b) What is Q; and (c) τ_w, in kilopascals?

12. The shaft power N of the water pump is related to the torque M and the angular velocity ω of the pump shaft. Use the Rayleigh method to derive the expression of the shaft power.

13. The sound velocity a in water is related to the volume modulus K and density ρ. Use the Rayleigh method to derive the sound velocity expression.

14. It is known that the venturi flow rate y is related to the flow rate difference Δp, the main pipe diameter d_1, the throat diameter d_2, and the fluid density ρ and kinematic viscosity v. Try the π theorem to prove that the flow velocity relationship is

$$v = \sqrt{\frac{\Delta p}{\rho}} \varphi\left(Re, \frac{d_2}{d_1}\right)$$

15. The wing chord length of a certain aircraft is $b = 150$ mm, and it is flying at a speed of $v = 180$ km/h in the atmosphere with air pressure $P_a = 10^5$ Pa and temperature $t = 10\,°C$. It is planned to be used in a wind tunnel for model tests. To determine the airfoil resistance, use the length scale $k_l = 1/3$.
    (a) If an open wind tunnel is used, it is known that the air pressure in the test section is $P'_a = 101{,}325$ Pa, and the temperature $t' = 25\,°C$. What is the wind speed in the test section? What is the problem with such a test? (b) If pressure is used In the wind tunnel, the air pressure in the test section is $P''_a = 1$ MPa, the temperature is $t'' = 30\,°C$, $\mu'' = 1.854 \times 10^5$ Pa s, what should the wind speed in the test section be equal to?

Chapter 6
Boundary Layer Theory and Its Approximation

6.1 Boundary Layer Approximation and Its Characteristics

6.1.1 The Influence of the Viscosity of the Flow Around a Large Reynolds Number Object

Since the N-S equation was derived in 1845, people have been searching for its exact solution, but because the equation system is a set of nonlinear second-order partial differential equations. Exact solutions in the general sense are extremely difficult in mathematics. It is said that only a few examples of accurate solutions of NS have been found so far. Famous examples include the Couette flow caused by the drag of an uncompressed flat plate (French physicist, Couette) and Fully developed laminar pipe flow (Poiseuille flow, French physiologist Poiseuille, 1799~1869), Stokes (Stokes, 1819–1903) solution for small Reynolds number spheres, etc. A large number of problems in practice can only be solved by approximate methods.

Since the French physicist, D'Alembert put forward the D'Alembert's paradox that an ideal fluid flows around an arbitrary three-dimensional object without resistance in 1752, people have raised doubts about the classical theory based on the ideal fluid model. In the first half of the nineteenth century, with the perfection of the ideal potential flow theory, classical fluid mechanics was in a low state, especially the conclusion that there is no resistance to the flow around a cylinder with this model. This prompted people to consider the solution of the N-S equation for the motion of viscous fluids. A thorny problem encountered is how to solve the effect of the viscous effect of the flow around the object under a large Reynolds number? According to the recognized facts at that time, if the incoming Reynolds number calculated from the current flow velocity and cylinder diameter is greater than 10^4, the effect of viscosity can be ignored, that is to say, the effect of viscosity can be ignored. This brings us back to the old proposition of ideal fluid flow around. If the effect of viscosity is not ignored, then how to understand the concept of large Reynolds number, and at that time, it is impossible to solve the whole N-S equation system more accurately.

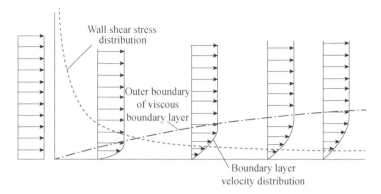

Fig. 6.1 Laminar boundary layer under zero pressure gradient

This problem did not get a convincing solution until Ludwig Prandtl (1875~1953) put forward the famous boundary layer theory in 1904. Counting from the question of D'Alembert in 1752, 152 years have passed. Counting from the derivation of the N-S equations in 1845, it has also lingered for 59 years. Now it seems to be a simple problem, that is, the relationship between the overall flow and the local flow, which belongs to the size of the area affected by the viscosity of the near-wall viscous flow around the object with a large Reynolds number. In 1904, Prandtl published a paper on the motion of small viscous fluids at the Third International Annual Conference of Mathematics in Heidelberg, Germany, and proposed the famous boundary layer concept (as shown in Fig. 6.1). It profoundly elaborates the boundary layer flow characteristics and control equations of the surface of the circumfluence object affected by the viscosity under large Reynolds numbers, and cleverly solves the problems of overall flow and local flow. In other words, the incoming Reynolds number calculated from the flow velocity and the cylinder diameter can only characterize the overall flow characteristics, and cannot characterize the local flow near the wall of the flowing object (flow in the boundary layer). The incoming Reynolds number can only control the viscous effect on the boundary layer. The influence of external flow and the influence of viscosity in the boundary layer can only be determined by the characteristics of the flow in the boundary layer. And on this basis, the boundary layer separation and control is proposed (as shown in Fig. 6.2), and the matching relationship between the viscous flow near the wall and the inviscid outflow far away from the wall is found, thus finding a new solution for the problem of viscous flow. Ways play the role of milestones.

6.1.2   The Concept of Boundary Layer

Prandtl found through a lot of experiments; for small viscous fluids such as air and water, when the incoming Reynolds number is large enough, the effect of viscosity

Fig. 6.2 Separation of the
boundary layer of the flow
around a cylinder

is only limited to the thin layer of fluid close to the surface area of the circumfluence
object. The ideal flow is far apart, there is a large velocity gradient along the normal
direction, and the viscous force cannot be ignored. Prandtl calls the thin layer in which
the viscous force plays an important role near the object surface as the boundary
layer. The introduction of the concept of boundary layer opens up ideas for how to
account for the effect of viscosity under the condition of diurnal Reynolds number
circumfluence, and puts forward the idea of partitioning the overall flow, which is

(1) Under the large flow Reynolds number, the overall flow area can be divided
    into an ideal fluid flow area (potential flow area) and a viscous fluid flow area
    (viscous flow area).
(2) In the ideal fluid flow area far away from the object, the influence of viscosity
    can be ignored, and it is treated according to the theory of potential flow.
(3) In the viscous flow area, it is limited to the thin layer near the object surface,
    which is called the boundary layer area. In this area, an important feature is that
    the effect of viscous stress cannot be ignored, which is of the same magnitude as
    the inertial force, and the fluid particles move in a rotational motion. Based on
    the assumption that the viscous force and inertial force in the boundary layer are
    of the same magnitude, the thickness of the boundary layer can be estimated.
    Take the flow around a flat plate as an example. Suppose the incoming flow
    velocity is, the length in the x-direction is L, and the boundary layer thickness
    is δ. In the boundary layer, the inertial force of the fluid cluster is

$$F_J = m\frac{\mathrm{d}u}{\mathrm{d}t} \propto \rho L^2 \delta \frac{V_\infty}{T} = \rho L^2 \delta \frac{V_\infty}{L/V_\infty} = \rho L V_\infty^2 \delta \qquad (6.1)$$

The viscous force of fluid micelles is

$$F_\mu = \rho v A \frac{\mathrm{d}u}{\mathrm{d}y} \propto \rho L^2 v \frac{V_\infty}{\delta} = \rho L^2 v \frac{V_\infty}{\delta} \qquad (6.2)$$

According to the assumption that the inertial force and the viscous force are of the same magnitude, we get

$$F_J \approx F_\mu, \ \rho L \delta V_\infty^2 \approx \rho L^2 v \frac{V_\infty}{\delta}$$

$$\frac{\delta}{L} \approx \frac{1}{\sqrt{\mathrm{Re}_L}}, \ \mathrm{Re}_L = \frac{V_\infty L}{v} \tag{6.3}$$

Equation (6.3) shows that the ratio of the boundary layer thickness to the plate length L is inversely proportional to the root of the overall Re_L number calculated from the flow velocity and the plate length. If the incoming air velocity $= 14.6$ m/s, the plate length $L = 1.0$ m, the air movement viscosity coefficient is $v = 1.46 \times 10^{-5}$ m²/s, $\mathrm{Re}_L = 106$ is calculated, and the boundary layer thickness is on the order of millimeters, which is equivalent to 1/1000 of the board length. The theoretical approximate solution of the flat plate laminar boundary layer is $\delta \approx 5.0$ mm.

6.1.3 Various Thicknesses and Characteristics of the Boundary Layer

(1) Definition of boundary layer

It is known that near the object surface, due to the adhesion conditions between the fluid and the object surface, the velocity of the fluid in direct contact with the object surface drops to zero, and the velocity of the adjacent fluid will also be affected by this layer of fluid. Decelerate, but as the distance from the object surface increases, the fluid velocity increases rapidly. When reaching a certain distance from the object surface, the fluid velocity reaches the velocity value of the external flow field, and the effect of viscosity is negligible. This point is the boundary point between the boundary layer and the external flow zone, and the corresponding thickness is called the boundary layer thickness. Strictly speaking, there is no obvious boundary between the boundary layer area and the outflow area. Prandtl stipulates that the outer boundary of the boundary layer should be 0.99 times the velocity of the outflow area V_∞, correspondingly from the outer boundary of the boundary layer to the vertical of the object surface. The distance is called the nominal thickness of the boundary layer, and is expressed by as shown in Fig. 6.3.

Fig. 6.3 Boundary layer thickness and development

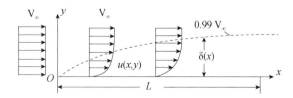

(2) Vorticity of the boundary layer

The movement of viscous fluid is always accompanied by the generation, diffusion, and attenuation of vorticity. The boundary layer is the vortex layer. When the fluid bypasses the object surface, the non-slip boundary condition is equivalent to making the object surface a continuously distributed vortex source with a certain strength. Take two-dimensional flow as an example. At this time, according to Newton's law of internal friction, the intensity of the vortex source on the object surface is

$$\Omega_z = \left(\frac{\partial v}{\partial x} - \frac{\partial u}{\partial y}\right) = -\frac{\partial u}{\partial y} = -\frac{\tau_0}{\mu} \tag{6.4}$$

For incompressible fluids, the vorticity transport equation for two-dimensional flow is

$$\frac{d\Omega_z}{dt} = v\Delta\Omega_z = v\left(\frac{\partial^2 \Omega_z}{\partial x^2} + \frac{\partial^2 \Omega_z}{\partial y^2}\right) \tag{6.5}$$

The above formula shows that under the influence of viscosity, the vorticity on the object surface spreads along the vertical streamline direction on the one hand, and on the other hand, the vorticity migrates along the main flow direction and gradually attenuates accordingly. The diffusion speed of the vorticity is related to the viscosity, and the migration speed of the vorticity depends on the flow speed.

(3) Definition and magnitude estimation of the thickness of each boundary layer

1. Boundary layer thickness

The thickness of the boundary layer can be estimated from the viscous force in the boundary layer and the inertial force of the same order. For the flow around the plate, it can be seen from Eq. (6.3) that the relationship between the thickness of the boundary layer and the plate length is

$$\frac{\delta}{L} \approx \frac{1}{\sqrt{Re_L}}, \quad Re_L = \frac{V_\infty L}{v}$$

It can be seen that at high Re numbers, the thickness of the boundary layer is much smaller than the characteristic length of the object being circumvented.

2. Layer displacement thickness of the boundary

Affected by the viscosity, the fluid velocity in the boundary layer is smaller than the incoming flow velocity, so the mass flow of the fluid passing through the boundary layer is smaller than the incoming mass flow of the same thickness. This part of the mass flow difference is equivalent to the effect of the boundary layer to make the incoming flow outward. The displacement distance is called the displacement thickness of the boundary layer. The specific calculation is as follows:

For the boundary layer of the flow around a flat plate, suppose the incoming flow velocity is V_∞, the fluid density is ρ, the thickness of the local boundary layer is δ, and the mass flow rate of the ideal fluid passing through the boundary layer is

$$m_i = \int_0^\delta \rho V_\infty dy \tag{6.6}$$

Due to the retarding effect of viscosity, the velocity in the boundary layer is less than the incoming flow velocity, and the mass flow rate of the viscous fluid is

$$m_e = \int_0^\delta \rho u dy \tag{6.7}$$

The difference between the above two indicates the loss of flow in the boundary layer due to the viscous effect. This part of the flow is pushed out of the boundary layer, which is equivalent to a certain distance thickening of the ideal fluid around the surface of the object, as shown in Fig. 6.4. Suppose the increased thickness is δ_1, then the flow rate in this thickness is equal to the flow rate lost in the boundary layer, which is

$$\rho V_\infty \delta_1 = \int_0^\delta \rho(V_\infty - u)dy \tag{6.8}$$

Get

$$\delta_1 = \int_0^\delta \left(1 - \frac{u}{V_\infty}\right)dy \tag{6.9}$$

3.   Momentum loss thickness of the boundary layer

Similarly, due to the retardation of viscosity, the momentum of the viscous fluid in the boundary layer is smaller than the momentum of the ideal fluid through the

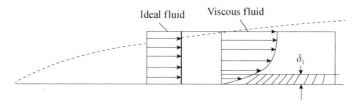

Fig. 6.4   The displacement thickness of the boundary layer

boundary layer, and the difference between the two is the momentum loss in the boundary layer. If the ideal fluid velocity is used to calculate the converted thickness of this loss, it is called the momentum loss thickness.

The momentum of the ideal fluid passing through the boundary layer is

$$K_i = \int_0^\delta \rho V_\infty^2 dy \tag{6.10}$$

The momentum passed by the viscous fluid is

$$K_e = \int_0^\delta \rho u^2 dy \tag{6.11}$$

The difference between the above two indicates the momentum lost by the viscous effect. If this part of the momentum loss is used to calculate the flow V_∞, and the converted momentum loss thickness is δ_{20}, then

$$\rho V_\infty^2 \delta_{20} = \int_0^\delta \rho(V_\infty^2 - u^2) dy \tag{6.12}$$

Expand the above formula, get

$$\rho V_\infty^2 \delta_{20} = \int_0^\delta \rho(V_\infty + u)(V_\infty - u) dy$$

$$= \int_0^\delta \rho V_\infty(V_\infty - u) dy + \int_0^\delta \rho u(V_\infty - u) dy \tag{6.13}$$

Substituting into formula (6.8), we get

$$\rho V_\infty^2 \delta_{20} = \rho V_\infty^2 \int_0^\delta \left(1 - \frac{u}{V_\infty}\right) dy + \rho V_\infty^2 \int_0^\delta \frac{u}{V_\infty}\left(1 - \frac{u}{V_\infty}\right) dy \tag{6.14}$$

$$\delta_{20} = \delta_1 + \delta_2$$

where δ_{20} is the thickness of absolute momentum loss, and δ_2 is called the thickness of relative momentum loss. The above formula shows that the thickness of absolute momentum loss includes two parts, one is caused by mass loss, and the other is loss caused by speed difference in addition to mass. If the mass loss is deducted, the momentum loss thickness is

$$\delta_2 = \int_0^\delta \frac{u}{V_\infty}\left(1 - \frac{u}{V_\infty}\right)dy \tag{6.15}$$

Formula (6.15) can also be derived in this way. The mass flow rate of the boundary layer is the momentum passing through the ideal fluid velocity as

$$K_{i1} = V_\infty \int_0^\delta \rho u\, dy \tag{6.16}$$

The only difference between Eqs. (6.11) and (6.16) is obtained by converting with δ_2

$$K_{i1} - K_e = V_\infty \int_0^\delta \rho u\, dy - \int_0^\delta \rho u^2 dy = \rho V_\infty^2 \delta_2 \tag{6.17}$$

Reorganizing the above formula, we can get formula (6.15).

4. **The thickness of the kinetic energy loss of the boundary layer**

Similarly, the kinetic energy of the viscous fluid passing through the boundary layer is less than the kinetic energy of the ideal fluid passing through the boundary layer, and the difference between the two is the kinetic energy loss in the boundary layer. If the ideal fluid velocity is used to calculate the converted thickness of this loss, it is called the kinetic energy loss thickness.

The kinetic energy of an ideal fluid passing through the boundary layer is

$$E_i = \int_0^\delta \rho \frac{V_\infty^2}{2} V_\infty dy \tag{6.18}$$

The kinetic energy of the viscous fluid is

$$E_e = \int_0^\delta \rho \frac{u^2}{2} u\, dy \tag{6.19}$$

The difference between the above two formulas represents the kinetic energy lost due to the viscous effect. If this part of the kinetic energy loss is used to calculate the flow V_∞, the converted kinetic energy loss thickness is δ_{30}, then

$$\rho \frac{1}{2} V_\infty^3 \delta_{30} = \int_0^\delta \rho \frac{1}{2}\left(V_\infty^3 - u^3\right)dy \tag{6.20}$$

Expand the above formula

$$\rho V_\infty^3 \delta_{30} = \int_0^\delta \rho(V_\infty - u)(V_\infty^2 + V_\infty u + u^2)dy$$

$$= \int_0^\delta \rho V_\infty^2(V_\infty - u)dy + \int_0^\delta \rho u(V_\infty - u)(V_\infty + u)dy$$

$$= \rho V_\infty^3 \int_0^\delta \left(1 - \frac{u}{V_\infty}\right)dy + \rho V_\infty^3 \int_0^\delta \frac{u}{V_\infty}\left(1 - \frac{u^2}{V_\infty^2}\right)dy \qquad (6.21)$$

Substituting into formula (6.9), we get

$$\delta_{30} = \delta_1 + \delta_3 \qquad (6.22)$$

where δ_3 is expressed as

$$\delta_3 = \int_0^\delta \frac{u}{V_\infty}\left(1 - \frac{u^2}{V_\infty^2}\right)dy \qquad (6.23)$$

Among them, δ_{30} is the absolute kinetic energy loss thickness, and δ_3 is called the relative kinetic energy loss thickness. The above formula shows that the thickness of absolute kinetic energy loss includes two parts, one is caused by mass loss, and the other is loss caused by speed difference in addition to mass. If the mass loss is deducted, the thickness of the kinetic energy loss obtained is represented by δ_3. If the quality loss part is not considered, the derivation of Eq. (6.23) is as follows.

The kinetic energy of the mass flow in the boundary layer when passing at the ideal fluid velocity is

$$E_{i1} = \frac{1}{2}V_\infty^2 \int_0^\delta \rho u \, dy \qquad (6.24)$$

Using the difference between formulas (6.19) and (6.24), δ_3 can be obtained, which is

$$E_{i1} - E_e = \frac{1}{2}V_\infty^2 \int_0^\delta \rho u \, dy - \int_0^\delta \rho \frac{1}{2} u^3 dy = \frac{1}{2}\rho V_\infty^3 \delta_3 \qquad (6.25)$$

Reorganizing the above formula, we can get formula (6.23).

If the object surface is not a flat plate, but a curved surface (the air flow is not separated), and the fluid density is not constant, the thickness of the boundary layer can be obtained by similar derivation

$$\delta_1 = \int_0^\delta \left(1 - \frac{\rho u}{\rho_e u_e}\right) \mathrm{d}y \tag{6.26}$$

$$\delta_2 = \int_0^\delta \frac{\rho u}{\rho_e u_e}\left(1 - \frac{u}{u_e}\right) \mathrm{d}y \tag{6.27}$$

$$\delta_3 = \int_0^\delta \frac{\rho u}{\rho_e u_e}\left(1 - \frac{u^2}{u_e^2}\right) \mathrm{d}y \tag{6.28}$$

where ρ_e is the density on the outer edge of the boundary layer; u_e is the velocity on the outer boundary of the boundary layer.

5. **A few explanations**

(1) In actual flow, the airflow in the boundary layer and the ideal airflow are asymptotically transitioned. The outer boundary line of the boundary layer does not actually exist. Therefore, the outer boundary line of the boundary layer is not a streamline, but is passed by the fluid, allowing the fluid flows across the boundary line, but at infinity, the outer boundary line of the boundary layer is close to the streamline. In the boundary layer, the streamlines are skewed outward, as shown in Fig. 6.5.
(2) The definitions of various thicknesses of the boundary layer are applicable to both laminar flow and turbulent flow. As long as the boundary layer is not separated, given the velocity distribution, it can be calculated.
(3) The various thicknesses of the boundary layer are related to the velocity distribution in the boundary layer. However, the order of the thickness is the thickness of the boundary layer is greater than the displacement thickness of the boundary layer and the momentum loss thickness of the boundary layer, as shown in

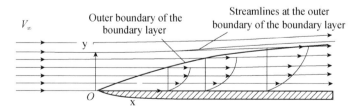

Fig. 6.5 The boundary line of the boundary layer

Fig. 6.6 Area corresponding to the thickness of the boundary layer displacement and the thickness of the momentum loss

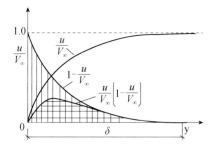

Fig. 6.6. The velocity distribution in the boundary layer is convex, and the relationship between the thickness of the boundary layer is approximately.

$$\frac{\delta_1}{\delta} \approx \begin{cases} \dfrac{1}{3} & \text{(Laminar boundary layer)} \\ \dfrac{1}{7} & \text{(Turbulent boundary layer)} \end{cases} \tag{6.29}$$

$$\frac{\delta_2}{\delta} \approx \begin{cases} \dfrac{1}{8} & \text{(Laminar boundary layer)} \\ \dfrac{1}{10} & \text{(Turbulent boundary layer)} \end{cases} \tag{6.30}$$

The rate of change of the boundary layer with the flow direction coordinate x is

$$\frac{d\delta}{dx} \approx \begin{cases} 0.0152 & \left(\text{Laminar boundary layer, Re}\, x = 3 \times 10^4\right) \\ 0.015 & \text{(Turbulent boundary layer), Outer edge angle of outer} \\ & \text{boundary line is } 1° \end{cases} \tag{6.31}$$

6.2 Laminar Boundary Layer Equations of Incompressible Fluids

For fluids with small viscosity, when the velocity gradient in the boundary layer is large, the thickness of the boundary layer is much smaller than the characteristic length of the surface of the flowing object. The NS equations can be simplified by the comparison of the magnitudes to obtain a relative comparison that is suitable for the boundary layer flow. Easy-to-solve governing equations. For simplicity, the following derivation assumes that the fluid is incompressible, the incoming flow is uniform, and the viscosity coefficient is constant.

6.2.1 Boundary Layer Equation on the Wall of a Flat Plate

First, derive a simpler boundary layer equation on the surface of the plate. According to the Prandtl boundary layer hypothesis, through the comparison of magnitudes, the N-S equations are simplified, and the approximate equations of the boundary layer can be obtained. For the motion of a two-dimensional incompressible fluid, the N-S equations are

$$\begin{cases} \dfrac{\partial u}{\partial x} + \dfrac{\partial v}{\partial y} = 0 \\[2mm] \dfrac{\partial u}{\partial t} + u\dfrac{\partial u}{\partial x} + v\dfrac{\partial u}{\partial y} = f_x - \dfrac{1}{\rho}\dfrac{\partial p}{\partial x} + v\left(\dfrac{\partial^2 u}{\partial x^2} + \dfrac{\partial^2 u}{\partial y^2}\right) \\[2mm] \dfrac{\partial v}{\partial t} + u\dfrac{\partial v}{\partial x} + v\dfrac{\partial v}{\partial y} = f_y - \dfrac{1}{\rho}\dfrac{\partial p}{\partial y} + v\left(\dfrac{\partial^2 v}{\partial x^2} + \dfrac{\partial^2 v}{\partial y^2}\right) \end{cases} \qquad (6.32)$$

Choose the object surface length L as the characteristic length, the mainstream velocity outside the boundary layer V_∞ as the characteristic velocity, and take the time characteristic scale as $T = \frac{L}{V_\infty}$, and now do the following magnitude comparison and dimensional analysis.

(1) According to the boundary layer definition, the longitudinal partial derivative is much smaller than the lateral partial derivative. In other words, in the flow of the two-dimensional boundary layer, the change of the physical quantity of the fluid along the flow direction is much slower than the change along the normal direction, because the velocity gradient in the normal direction is very large, which is

$$\frac{\delta}{L} \propto \frac{1}{\sqrt{\text{Re}}}, \ \delta << L, \ \text{Re} = \frac{V_\infty L}{v}$$

$$\frac{\partial}{\partial x} \propto \frac{1}{L}, \ \frac{\partial}{\partial y} \propto \frac{1}{\delta}, \ \frac{\partial}{\partial x} << \frac{\partial}{\partial y} \qquad (6.33)$$

(2) Due to the impermeability of the object surface, the normal velocity is much smaller than the longitudinal velocity.

$$v \propto \frac{\delta}{T} = \frac{\delta}{L/V_\infty} = \frac{\delta}{L} V_\infty, \ \frac{v}{V_\infty} \propto \frac{1}{\sqrt{\text{Re}}}, \ \frac{v}{\delta} \propto \frac{V_\infty}{L}$$

$$u \propto \frac{L}{T} = V_\infty, \ v << u \qquad (6.34)$$

(3) The pressure in the boundary layer is proportional to the square of the outflow velocity.

$$p \propto \rho V_\infty^2 \qquad (6.35)$$

Substitute these magnitude relations into the N-S equations (6.32), and compare the magnitudes of the equations. The magnitudes of f_x and f_y are temporarily ignored, and the magnitudes of each item in the equation are as follows.

Continuous equation and comparison of various magnitudes

$$\frac{\partial u}{\partial x} + \frac{\partial v}{\partial y} = 0$$

$$\frac{V_\infty}{L} \qquad \frac{V_\infty}{L} \tag{6.36}$$

$$O(1) + O(1)$$

Comparison of momentum equation in x-direction with various magnitudes

$$\frac{\partial u}{\partial t} + u\frac{\partial u}{\partial x} + v\frac{\partial u}{\partial y} = f_x - \frac{1}{\rho}\frac{\partial p}{\partial x} + v\left(\frac{\partial^2 u}{\partial x^2} + \frac{\partial^2 u}{\partial y^2}\right)$$

$$\frac{V_\infty^2}{L} \quad \frac{V_\infty^2}{L} \quad \frac{V_\infty^2}{L} \qquad \frac{V_\infty^2}{L} \quad v\frac{V_\infty}{L^2} \quad v\frac{V_\infty}{\delta^2} \tag{6.37}$$

$$O(1) \quad O(1) \quad O(1) \qquad O(1) \quad O\left(\frac{1}{Re}\right) \quad O(1)$$

Comparison of momentum equation in y-direction and various magnitudes

$$\frac{\partial v}{\partial t} + u\frac{\partial v}{\partial x} + v\frac{\partial v}{\partial y} = f_y - \frac{1}{\rho}\frac{\partial p}{\partial y} + v\left(\frac{\partial^2 v}{\partial x^2} + \frac{\partial^2 v}{\partial y^2}\right)$$

$$\frac{V_\infty^2}{L\sqrt{Re}} \quad \frac{V_\infty^2}{L\sqrt{Re}} \quad \frac{V_\infty^2}{L\sqrt{Re}} \quad \frac{\sqrt{Re}u_e^2}{L} \quad \frac{V_\infty^2}{LRe\sqrt{Re}} \quad \frac{V_\infty^2}{L\sqrt{Re}}$$

$$O\left(\frac{1}{\sqrt{Re}}\right) \quad O\left(\frac{1}{\sqrt{Re}}\right) \quad O\left(\frac{1}{\sqrt{Re}}\right) \quad O\left(\sqrt{Re}\right) \quad O\left(\frac{1}{Re\sqrt{Re}}\right) \quad O\left(\frac{1}{\sqrt{Re}}\right)$$

$$\tag{6.38}$$

Through the comparison of various magnitudes, small quantities are ignored in the case of high Re number, and the following simplified equations are obtained. Get

$$\begin{cases} \dfrac{\partial u}{\partial x} + \dfrac{\partial v}{\partial y} = 0 \\[2mm] \dfrac{\partial u}{\partial t} + u\dfrac{\partial u}{\partial x} + v\dfrac{\partial u}{\partial y} = f_x - \dfrac{1}{\rho}\dfrac{\partial p}{\partial x} + v\dfrac{\partial^2 u}{\partial y^2} \\[2mm] 0 = f_y - \dfrac{1}{\rho}\dfrac{\partial p}{\partial y} \end{cases} \tag{6.39}$$

This simplified equation set is the boundary layer equation set by Prandtl in 1904. Compared with the N-S equations, it is greatly simplified, and the type of the equations has also changed. The original N-S equations are elliptic equations, and the simplified Prandtl boundary layer equations are parabolic equations.

If the mass force is neglected, it is obtained from the third equation

$$0 = -\frac{1}{\rho}\frac{\partial p}{\partial y} \tag{6.40}$$

This shows that in the case of a high Re number, the pressure in the boundary layer does not change along the normal direction. In other words, the pressure in the boundary layer is equal to the pressure on the outer boundary of the boundary layer on the normal line. That is to say, p has nothing to do with y, it is only a function of x and t, which is

$$p = p_e(x, t) \tag{6.41}$$

This shows that in the boundary layer equation, the pressure is not an unknown quantity, but is determined by the conditions of the potential flow outside the boundary layer. Since the area outside the boundary layer is a potential flow field, the velocity field can be determined by solving the potential flow. For steady flow, the Bernoulli equation can be used to determine the pressure in the boundary layer, which is

$$p_e(x, t) + \frac{1}{2}\rho u_e^2 = p_\infty + \frac{1}{2}\rho V_\infty^2 \tag{6.42}$$

For unsteady flow, it can be determined by Euler's equation of external potential flow, which is

$$\frac{\partial u_e}{\partial t} + u_e\frac{\partial u_e}{\partial x} = -\frac{1}{\rho}\frac{\partial p_e}{\partial x} \tag{6.43}$$

In the formula, u_e is the velocity on the outer boundary of the boundary layer. If the mass force is neglected, Prandtl's boundary layer equations become

$$\begin{cases} \dfrac{\partial u}{\partial x} + \dfrac{\partial v}{\partial y} = 0 \\[2mm] \dfrac{\partial u}{\partial t} + u\dfrac{\partial u}{\partial x} + v\dfrac{\partial u}{\partial y} = -\dfrac{1}{\rho}\dfrac{\partial p}{\partial x} + \nu\dfrac{\partial^2 u}{\partial y^2} \\[2mm] -\dfrac{1}{\rho}\dfrac{\partial p}{\partial y} = 0 \end{cases} \tag{6.44}$$

Boundary conditions

$$\begin{cases} y = 0, & u = 0, \quad v = 0 \\ y = \infty, & u = V_\infty \end{cases} \tag{6.45}$$

The pressure in the boundary layer is determined according to the ideal fluid potential flow equation. Substituting Eq. (6.43) into Eq. (6.44), we get

$$\begin{cases} \dfrac{\partial u}{\partial x} + \dfrac{\partial v}{\partial y} = 0 \\[2mm] \dfrac{\partial u}{\partial t} + u\dfrac{\partial u}{\partial x} + v\dfrac{\partial u}{\partial y} = \dfrac{\partial u_e}{\partial t} + u_e\dfrac{\partial u_e}{\partial x} + v\dfrac{\partial^2 u}{\partial y^2} \end{cases} \tag{6.46}$$

In the case of steady flow, there are

$$\begin{cases} \dfrac{\partial u}{\partial x} + \dfrac{\partial v}{\partial y} = 0 \\[2mm] u\dfrac{\partial u}{\partial x} + v\dfrac{\partial u}{\partial y} = u_e\dfrac{\partial u_e}{\partial x} + v\dfrac{\partial^2 u}{\partial y^2} \end{cases} \tag{6.47}$$

In summary, the basic characteristics of the boundary layer can be summarized as follows:

(1) The thickness of the boundary layer is much smaller than the length scale of the object surface, namely $\frac{\delta}{L} \propto \frac{1}{\sqrt{Re}}$;
(2) The normal velocity in the boundary layer is much smaller than the flow velocity, namely $\frac{v}{V_\infty} \propto \frac{1}{\sqrt{Re}}$;
(3) The velocity gradient along the flow direction in the boundary layer is much smaller than the velocity gradient along the normal direction, namely $\frac{\partial}{\partial x} << \frac{\partial}{\partial y}$;
(4) The pressure gradient along the normal direction in the boundary layer is zero, that is $\frac{\partial p}{\partial y} = 0$;
(5) The pressure in the boundary layer is only a function of flow direction and time, namely $p = p_e(x, t)$.

6.2.2 Boundary Layer Equation on Curved Wall

The two-dimensional boundary layer equations on the wall of the flat plate were pushed earlier. However, the object surface encountered in actual flow is often curved, so it is more universal to derive the boundary layer equation on the curved wall. In the derivation process, the boundary layer coordinate system on the curved wall is used. Among them, the x-axis is attached to the wall, and the y-axis is perpendicular to the wall. Take any point M in the boundary layer, and its coordinates are $x = ON$, $y = NM$, as shown in Fig. 6.7. M' is the neighbor of M, and the arc length ds of MM' is

$$ds = \sqrt{(MM'')^2 + (M''M')^2} \tag{6.48}$$

At x, let the radius of curvature of the wall be $R(x)$, there is

Fig. 6.7 Curved coordinate
system

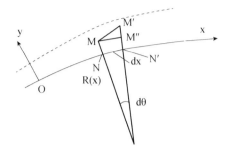

$$MM'' = (R + y)\mathrm{d}\theta = \frac{R + y}{R}\mathrm{d}x, \quad M''M' = M'N' - MN = \mathrm{d}y \qquad (6.49)$$

Substituting into Eq. (6.48), we get

$$\mathrm{d}s^2 = \left(MM''\right)^2 + \left(M''M'\right)^2 = \left(\frac{R + y}{R}\right)^2 \mathrm{d}x^2 + \mathrm{d}y^2 = (H_1 \mathrm{d}x)^2 + (H_2 \mathrm{d}y)^2$$

$$H_1 = \frac{R + y}{R}, \quad H_2 = 1$$

$$(6.50)$$

Still using u and v to represent the velocity components in the x and y directions in the boundary layer coordinate system, the N-S equations in the orthogonal curvilinear coordinate system are

The continuous fraction equation is

$$\frac{R}{R + y}\frac{\partial u}{\partial x} + \frac{\partial v}{\partial y} + \frac{v}{R + y} = 0 \qquad (6.51)$$

The equation of motion is

$$\frac{\partial u}{\partial t} + \frac{R}{R + y}u\frac{\partial u}{\partial x} + v\frac{\partial u}{\partial y} + \frac{uv}{R + y} = -\frac{R}{R + y}\frac{1}{\rho}\frac{\partial p}{\partial x}$$

$$+ v\left[\begin{array}{l}\left(\dfrac{R}{R + y}\right)^2\dfrac{\partial^2 u}{\partial x^2} + \dfrac{\partial^2 u}{\partial y^2} + \dfrac{1}{R + y}\dfrac{\partial u}{\partial y} - \dfrac{u}{(R + y)^2} \\[2mm] + \dfrac{2R}{(R + y)^2}\dfrac{\partial v}{\partial x} - \dfrac{Rv}{(R + y)^3}\dfrac{\mathrm{d}R}{\mathrm{d}x} + \dfrac{Ry}{(R + y)^3}\dfrac{\mathrm{d}R}{\mathrm{d}x}\dfrac{\partial u}{\partial x}\end{array}\right] \qquad (6.52)$$

$$\frac{\partial v}{\partial t} + \frac{R}{R+y}u\frac{\partial v}{\partial x} + v\frac{\partial v}{\partial y} - \frac{u^2}{R+y} = -\frac{1}{\rho}\frac{\partial p}{\partial y}$$

$$+v\left[\begin{array}{l}\left(\frac{R}{R+y}\right)^2\frac{\partial^2 v}{\partial x^2} + \frac{\partial^2 v}{\partial y^2} + \frac{1}{R+y}\frac{\partial v}{\partial y} - \frac{v}{(R+y)^2} \\[2mm] -\frac{2R}{(R+y)^2}\frac{\partial u}{\partial x} + \frac{Ru}{(R+y)^3}\frac{dR}{dx} + \frac{Ry}{(R+y)^3}\frac{dR}{dx}\frac{\partial v}{\partial x}\end{array}\right] \qquad (6.53)$$

Assuming that the radius of curvature $R(x)$ of the object surface is of the same magnitude as the characteristic length L in the x-direction, and the magnitude of y is the same magnitude as the thickness of the boundary layer, so

$$\delta \ll L, \ \delta \ll R, \ \frac{dR}{dx} \approx 1, \ \frac{d\delta}{dx} \ll 1, \ H_1 = 1 + \frac{y}{R} \approx 1, \ R + y \approx R$$

$$\frac{\partial}{\partial x} \propto \frac{1}{L}, \ \frac{\partial}{\partial y} \propto \frac{1}{\delta}, \ \frac{\partial}{\partial x} \ll \frac{\partial}{\partial y}, \ \frac{y}{R} \propto \frac{1}{\sqrt{Re}} \qquad (6.54)$$

$$v \propto \frac{\delta}{t} \propto \frac{\delta}{L/u_e} = \frac{\delta}{L}u_e, \ \frac{v}{u_e} \propto \frac{1}{\sqrt{Re}}, \ u \propto \frac{L}{t} = u_e, \ v \ll u \qquad (6.55)$$

In order of magnitude comparison, the simplified boundary layer equations are

$$\begin{cases} \dfrac{\partial u}{\partial x} + \dfrac{\partial v}{\partial y} = 0 \\[3mm] \dfrac{\partial u}{\partial t} + u\dfrac{\partial u}{\partial x} + v\dfrac{\partial u}{\partial y} = -\dfrac{1}{\rho}\dfrac{\partial p}{\partial x} + v\dfrac{\partial^2 u}{\partial y^2} \\[3mm] \rho\dfrac{u^2}{R} = \dfrac{\partial p}{\partial y} \end{cases} \qquad (6.56)$$

This is the boundary layer equation on the curved wall. Compared with the equation on the flat wall, only the equation in the y-direction is different. In order to balance the centrifugal force generated by the bending of the flow, there must be a pressure gradient in the y-direction. The following estimates the magnitude of this pressure gradient, and initially assumes that the velocity distribution in the boundary layer is linear.

$$u = u_e\frac{y}{\delta}, \ \frac{\partial p}{\partial y} = \frac{\rho}{R}\left(u_e\frac{y}{\delta}\right)^2 = \frac{\rho u_e^2}{R\delta^2}y^2 \qquad (6.57)$$

Integrate from $y = 0$ to $y = \delta$, therefore

$$\Delta p = p(\delta) - p(0) = \frac{1}{3}\rho u_e^2\frac{\delta}{R} \qquad (6.58)$$

$$\frac{\Delta p}{\rho u_e^2} = \frac{1}{3}\frac{\delta}{R} \qquad (6.59)$$

In the case of $R \gg \delta$, this pressure difference is a small amount and can be ignored. It can still be concluded that the normal pressure is also constant in the boundary layer on the curved wall. This shows that when the radius of curvature is large enough, the boundary layer equation on the curved wall is exactly the same as the boundary layer equation on the flat wall.

6.3 Similar Solutions to the Laminar Boundary Layer on a Flat Plate

Compared with the N-S equations, the boundary layer equations have only limited simplifications, neither linearizing the original equations nor reducing the order, only changing the type of the original equations, so it is still difficult to solve them. In 1908, the German fluid mechanics Blasius (a student of Prandtl) first gave a solution to the boundary layer of an uncompressed gradient slab. Blasius introduced the assumption of the similarity of velocity distribution in the boundary layer, solved the laminar boundary layer equation of the plate, obtained the approximate solution of the plate boundary layer, and obtained the law that the plate resistance is proportional to the 1.5th power of the incoming flow velocity. This is the first time in the history of fluid mechanics that the frictional resistance of the flow around a large Reynolds number can be solved by theoretical methods. Therefore, the boundary layer solution of a flat plate has a special significance in boundary layer theory. For zero pressure gradient, steady flow, laminar flow of incompressible fluid around a flat plate, the boundary layer equation is

$$\begin{cases} \dfrac{\partial u}{\partial x} + \dfrac{\partial v}{\partial y} = 0 \\[3mm] u\dfrac{\partial u}{\partial x} + v\dfrac{\partial u}{\partial y} = \nu\dfrac{\partial^2 u}{\partial y^2} \end{cases} \qquad (6.60)$$

The boundary conditions are

$$\begin{cases} y = 0,\ u = 0,\ v = 0 \\ y = \infty,\ u = V_\infty \end{cases} \qquad (6.61)$$

This is a second-order quasi-linear partial differential equation system. The unknowns in the equation system are u and v. The two equations are solved simultaneously. Since the type of equations has changed to parabolic equations, the principle can be solved from the known value of $y = 0$, but the difficulty is that the first derivative of u at $y = 0$ is unknown, but the velocity on the outer edge of the boundary

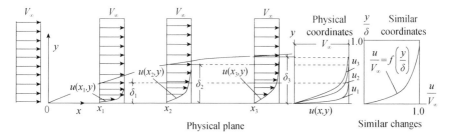

Fig. 6.8 Similarity transformation of plate boundary layer

layer is given V_∞. Therefore, when solving, it is necessary to first assume the first derivative of u at $y = 0$, then proceed with the solution, and finally check whether the velocity value on the outer boundary of the boundary layer is satisfied.

In order to convert the partial differential equations into ordinary differential equations for further solution, Blasius assumed: the velocity distribution in the boundary layer on the flat plate is similar, that is, the dimensionless velocity $\frac{u}{V_\infty}$ is only the dimensionless coordinate $\frac{y}{\delta}$ function, as shown in Fig. 6.8.

In the physical coordinate system, speed is a function of x and y, which is

$$u = f(x, y) \tag{6.62}$$

If the velocity V_∞, the flow direction length L, and the vertical length δ are selected as the characteristic quantities of similar transformation, the dimensionless expression of Eq. (6.72) is

$$\frac{u}{V_\infty} = f\left(\frac{x}{L}, \frac{y}{\delta}\right) \tag{6.63}$$

Using Eq. (6.62) to arrange the velocity distribution of different x stations in the physical coordinate system, it is found that this infinite velocity distribution curve has nothing to do with the station coordinates, according to Eq. (6.63). It is only a function of y/δ. This is the assumption of similarity in the velocity distribution of the boundary layer, namely

$$\frac{u}{V_\infty} = f\left(\frac{y}{\delta}\right) \tag{6.64}$$

According to the magnitude comparison, the magnitude of the boundary layer thickness is

$$\delta \propto \frac{x}{\sqrt{Re_x}} = \frac{x}{\sqrt{\frac{V_\infty x}{\nu}}} = \sqrt{\frac{x\nu}{V_\infty}} \tag{6.65}$$

According to hypothesis of similarity

$$\frac{u}{V_\infty} = f(\eta), \quad \eta = \frac{y}{\delta} = y\sqrt{\frac{V_\infty}{\nu x}} \tag{6.66}$$

Introducing the flow function (plane incompressible flow, there is a flow function. But because the flow in the boundary layer is vortex, there is no velocity potential function), the continuity equation can be eliminated. According to the definition, through the integral formula (6.66), we can get

$$\psi = \int u\,\mathrm{d}y = \int V_\infty f(\eta)\sqrt{\frac{\nu x}{V_\infty}}\,\mathrm{d}\eta = \sqrt{\nu x\, V_\infty}\, F(\eta) \tag{6.67}$$

Defined by the stream function, we get

$$u = \frac{\partial \psi}{\partial y} = \frac{\partial \psi}{\partial \eta}\frac{\partial \eta}{\partial y} = V_\infty \frac{\mathrm{d}F(\eta)}{\mathrm{d}\eta} = V_\infty F'(\eta) = V_\infty f(\eta) \tag{6.68}$$

$$v = -\frac{\partial \psi}{\partial x} = -\frac{\partial}{\partial x}\left[\sqrt{\nu x\, V_\infty}\, F(\eta)\right] = -\left[\sqrt{\nu x\, V_\infty}\frac{\partial F}{\partial \eta}\frac{\partial \eta}{\partial x} + F(\eta)\frac{\partial}{\partial x}\sqrt{\nu x\, V_\infty}\right]$$

$$= \frac{V_\infty}{2}\frac{1}{\sqrt{Rex}}\left(\eta F' - F\right) \tag{6.69}$$

where $Rex = \frac{V_\infty x}{\nu}$. Through calculation, we can get

$$u\frac{\partial u}{\partial x} = -\frac{V_\infty^2}{2x}\eta F' F'' \tag{6.70}$$

$$v\frac{\partial u}{\partial y} = \frac{1}{2}\frac{V_\infty^2}{x}\left[\eta F' - F\right]F'' \tag{6.71}$$

$$v\frac{\partial^2 u}{\partial y^2} = v\frac{V_\infty^2}{x\nu}F''' = \frac{V_\infty^2}{x}F''' \tag{6.72}$$

Substituting formula (6.70) to formula (6.72) into the second formula of the system of equations (6.60), we get

$$-\frac{1}{2}\frac{V_\infty^2}{x}\eta F' F'' + \frac{1}{2}\frac{V_\infty^2}{x}\left[\eta F' - F\right]F'' = \frac{V_\infty^2}{x}F''' \tag{6.73}$$

After simplification becomes

$$F F'' + 2F''' = 0 \tag{6.74}$$

The boundary conditions are

$$\begin{cases} \eta = 0, \ F' = 0, \ F = 0 \\ \eta = \infty, \ F' = 1.0 \end{cases} \tag{6.75}$$

Blasius solved Eq. (6.74) with an infinite series. Suppose

$$F(\eta) = A_0 + A_1\eta + \frac{A_2}{2!}\eta^2 + \frac{A_3}{3!}\eta^3 + \cdots + \frac{A_n}{n!}\eta^n + \cdots \tag{6.76}$$

where $A_0, A_1, A_2, \ldots A_n \ldots$ are the undetermined coefficients. Find the first and second derivatives of F, and we get

$$\frac{\mathrm{d}F}{\mathrm{d}\eta} = A_1 + \frac{A_2}{1}\eta + \frac{A_3}{2!}\eta^2 + \cdots + \frac{A_n}{(n-1)!}\eta^{n-1} + \cdots \tag{6.77}$$

$$\frac{\mathrm{d}^2 F}{\mathrm{d}\eta^2} = A_2 + \frac{A_3}{1!}\eta^1 + \cdots + \frac{A_n}{(n-2)!}\eta^{n-2} + \cdots \tag{6.78}$$

$$\frac{\mathrm{d}^3 F}{\mathrm{d}\eta^3} = A_3 + \frac{A_4}{1}\eta \cdots + \frac{A_n}{(n-3)!}\eta^{n-3} + \cdots \tag{6.79}$$

Substituting Eq. (6.75) into Eqs. (6.76) and (6.77), we get $A_0 = 0$, $A_1 = 0$, and F becomes

$$F(\eta) = \frac{A_2}{2!}\eta^2 + \frac{A_3}{3!}\eta^3 + \cdots + \frac{A_n}{n!}\eta^n + \cdots \tag{6.80}$$

Substituting the above formulas into formula (6.74), we get

$$\left(\frac{A_2}{2!}\eta^2 + \frac{A_3}{3!}\eta^3 + \cdots + \frac{A_n}{n!}\eta^n + \cdots \right)\left(A_2 + A_3\eta + \cdots + \frac{A_n}{(n-2)!}\eta^{n-2} + \cdots \right)$$
$$+ 2\left(A_3 + A_4\eta \cdots + \frac{A_n}{(n-3)!}\eta^{n-3} + \cdots \right) = 0 \tag{6.81}$$

On expanding, there is

$$2A_3 + 2A_4\eta + \frac{\eta^2}{2!}\left(A_2^2 + 2A_5 \right) + \frac{\eta^3}{3!}(4A_2A_3 + 2A_6)$$
$$+ \frac{\eta^4}{4!}\left(7A_2A_4 + 4A_3^2 + 2A_7 \right) + \frac{\eta^5}{5!}(11A_2A_5 + 15A_3A_4 + 2A_8)\ldots = 0 \tag{6.82}$$

Since any power of η needs to satisfy formula (6.82), all coefficients are required to be zero, so

$$A_3 = 0, \; A_4 = 0, \; A_5 = -\frac{A_2^2}{2}, \; A_6 = 0, \; A_7 = 0, \; A_8 = \frac{11}{4}A_2^3, \ldots \quad (6.83)$$

Substituting the above formula into formula (6.76), we have

$$F(\eta) = A_2^{1/3}\left[\frac{1}{2!}(A_2^{1/3}\eta)^2 - \frac{1}{2}\frac{1}{5!}(A_2^{1/3}\eta)^5 + \frac{1}{2^2}\frac{11}{8!}(A_2^{1/3}\eta)^8 - \frac{1}{2^3}\frac{375}{11!}(A_2^{1/3}\eta)^{11}\cdots\right]$$

$$= A_2^{1/3}\sum_{n=0}^{\infty}\left(-\frac{1}{2}\right)^n\frac{C_n}{(3n+2)!}(A_2^{1/3}\eta)^{(3n+2)} \quad (6.84)$$

The coefficients are

$$C_0 = 1, \; C_1 = 1, \; C_2 = 11, \; C_3 = 375, \; C_4 = 27{,}897, \; C_5 = 3{,}817{,}137\ldots \quad (6.85)$$

Calculating the first derivative of F on Eq. (6.84), we get

$$F'(\eta) = A_2\eta - \frac{1}{2\times 4!}A_2^2\eta^4 + \frac{11}{2^2\times 7!}A_2^3\eta^7 - \frac{375}{2^3\times 10!}A_2^4\eta^{10}\cdots \quad (6.86)$$

According to boundary condition, $\lim\limits_{\eta\to\infty}F'(\eta) = 1$, the result given by Blasius is

$$F''(0) = A_2 = 0.3321 \quad (6.87)$$

Substituting the value of A_2 into Eqs. (6.69), (6.84), and (6.86), we get

$$F(\eta) = 0.1661\eta^2 - 0.00046\eta^5 + 0.0000025\eta^8 - \cdots \quad (6.88)$$

$$\frac{u}{V_\infty} = F'(\eta) = 0.3321\eta - 0.0023\eta^4 + 0.00002\eta^7 - \cdots \quad (6.89)$$

$$\frac{v}{V_\infty} = \frac{1}{\sqrt{Rex}}\frac{(\eta F' - F)}{2}$$

$$= \frac{1}{\sqrt{Rex}}\left(0.08305\eta^2 - 0.00092\eta^5 + 0.00002\eta^8 - \cdots\right) \quad (6.90)$$

$$F''(\eta) = 0.3321 - 0.0092\eta^3 + 0.00014\eta^6 - \cdots \quad (6.91)$$

Using the series solution of Eqs. (6.88) to (6.91), the values under different η are shown in Table 6.1. Using this table, the velocity distribution in the boundary layer can be determined and the velocity at a certain point can be obtained. At the same time, the relationship between the dimensionless velocity $\frac{u}{u_e}$ and the dimensionless coordinates $\frac{y}{\delta}$, namely $\frac{u}{u_e} = f\left(\frac{y}{\delta}\right) = F'(\eta)$, can be obtained, as shown in Fig. 6.9.

(1) Boundary layer thickness

According to the definition of the boundary layer thickness, take u/V_∞, look up Table 6.1 to know that $\eta = 5.0$, from the following formula

$$\eta = \frac{y}{\delta} = y\sqrt{\frac{V_\infty}{\nu x}}$$

Get

$$\delta = \frac{5x}{\sqrt{Re_x}} \tag{6.92}$$

(2) Boundary layer displacement thickness

Using Eq. (6.9), the displacement thickness δ_1 can be obtained, which is

$$\delta_1 = \int_0^\delta \left(1 - \frac{u}{V_\infty}\right) dy = 1.7208 \frac{x}{\sqrt{Re_x}} \tag{6.93}$$

(3) Momentum loss thickness of the boundary layer

Similarly, using Eq. (6.15), the momentum thickness δ_2 can be obtained, which is

$$\delta_2 = \int_0^\delta \frac{u}{V_\infty}\left(1 - \frac{u}{V_\infty}\right) dy = 0.664 \frac{x}{\sqrt{Re_x}} \tag{6.94}$$

(4) Wall shear stress

$$\tau_0 = \mu \frac{\partial u}{\partial y}\bigg|_{y=0} = 0.332 \rho V_\infty^2 \frac{1}{\sqrt{Re_x}}, \quad \tau_0 \rightarrow \frac{1}{\sqrt{x}} \tag{6.95}$$

(5) Wall friction stress coefficient

$$C_f = \frac{\tau_0}{0.5\rho V_\infty^2} = 0.664 \frac{1}{\sqrt{Re_x}} \tag{6.96}$$

Table 6.1 $F(\eta)$ and its derivative of the boundary layer of the flow around the plate wall

$\eta = y\sqrt{\frac{u_e}{\nu x}}$	$F(\eta)$	$F'(\eta)$	$F''(\eta)$
0.00E+00	0.00E+00	0.00E+00	0.332060
0.25	1.04E−02	8.30E−02	0.331917
0.50	4.15E−02	0.165887	0.330914
0.75	9.33E−02	0.248321	0.328208
1.00	0.165575	0.329783	0.32301
1.25	0.258037	0.409560	0.314636
1.50	0.370144	0.486793	0.302583
1.75	0.501142	0.560523	0.286602
2.00	0.650032	0.629770	0.266754
2.25	0.815576	0.693610	0.243445
2.50	0.996322	0.751264	0.217413
2.75	1.190646	0.802172	0.189663
3.00	1.396821	0.846049	0.161362
3.25	1.613085	0.882906	0.133704
3.50	1.837715	0.913044	0.107774
3.75	2.069094	0.937008	8.44E−02
4.00	2.305766	0.955522	6.42E−02
4.25	2.546470	0.969408	4.74E−02
4.50	2.790157	0.979517	3.40E−02
4.75	3.035983	0.986656	2.36E−02
5.00	3.283299	0.991544	1.59E−02
5.25	3.531620	0.994791	1.04E−02
5.50	3.780600	0.996882	6.58E−03
5.75	4.029997	0.998186	4.04E−03
6.00	4.279651	0.998976	2.41E−03
6.25	4.529459	0.999439	1.39E−03
6.50	4.779355	0.999703	7.76E−04
6.75	5.02930	0.999848	4.21E−04
7.00	5.279273	0.999926	2.21E−04
7.25	5.529260	0.999966	1.13E−04
7.50	5.779254	0.999987	5.58E−05
7.75	6.029253	0.999996	2.68E−05
8.00	6.279252	1.000001	1.25E−05

Fig. 6.9 Velocity distribution of the laminar boundary layer of a flat plate at zero angle of attack

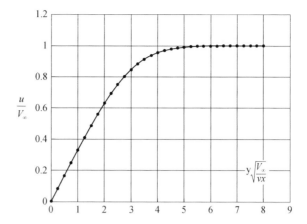

(6) Average wall friction total resistance coefficient

$$C_{\mathrm{Df}} = \frac{1}{L} \int_0^L C_{\mathrm{f}} \mathrm{d}x = 2C_{\mathrm{f}}(L) = 1.328 \frac{1}{\sqrt{\mathrm{Re}_L}} \qquad (6.97)$$

Guo Yonghuai (1953) gave a correction to the front edge point of the plate, 得

$$C_{\mathrm{Df}} = \frac{1.328}{\sqrt{\mathrm{Re}_L}} + \frac{4.10}{\mathrm{Re}_L} \qquad (6.98)$$

The applicable range of this formula is $3 \times 10^5 < \mathrm{Re}_L < 3 \times 10^6$.

6.4 Boundary Layer Momentum Integral Equation

6.4.1 Derivation of Karman Momentum Integral Equation

Although Prandtl's boundary layer differential equations have been greatly simplified, it is still difficult to obtain accurate solutions due to the nonlinear characteristics of the equations. In order to facilitate engineering applications, in 1921, American Aerodynamics Von. Kármán (1881~1963) derived the well-known boundary layer momentum integral relationship based on the momentum integral equation. This integral equation can quickly and easily give the approximate characteristics of the boundary layer. Momentum integral relation is suitable for laminar boundary layer and turbulent boundary layer, and as long as the velocity distribution is selected appropriately, the solution result has certain accuracy. The specific derivation is given below.

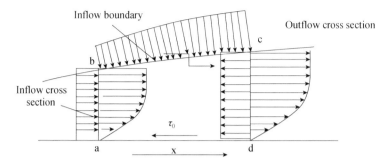

Fig. 6.10   Momentum integral relationship of the boundary layer

As shown in Fig. 6.10, for the two-dimensional boundary layer problem, assuming that the flow is constant and the fluid density is constant, any control body *abcd* in the boundary layer, the length of the control body is dx, and the control surfaces are A_{ab}, A_{bc}, A_{cd}, A_{da}. The momentum integral equation is now applied to the control body.

$$\frac{\partial}{\partial t} \iiint_\tau \rho \vec{V} d\tau + \oiint_S \rho \vec{V}\left(\vec{V} \cdot \vec{n}\right) dS = \iiint_\tau \rho \vec{f} d\tau + \oiint_S \vec{p}_n dS$$

Because the flow is constant, the mass force is not considered, and the above formula is projected along the *x*-direction to obtain

$$\oiint_S \rho \vec{V}_x\left(\vec{V} \cdot \vec{n}\right) dS = \oiint_S \vec{p}_{nx} dS \qquad (6.99)$$

The mass flowing into the control body from the A_{ab} surface is

$$m_{ab} = \int_0^{\delta(x)} \rho u \, dy \qquad (6.100)$$

The mass flowing into the control body from the A_{cd} surface is

$$m_{cd} = m_{ab} + \frac{\partial}{\partial x}\left(\int_0^{\delta(x)} \rho u \, dy\right) dx \qquad (6.101)$$

According to the law of conservation of mass, the mass flowing into the control body through A_{bc} is

$$m_{bc} = m_{cd} - m_{ab} = \frac{\partial}{\partial x}\left(\int_0^{\delta(x)} \rho u \, dy\right) dx \qquad (6.102)$$

The momentum flowing into the control body from the A_{ab} surface is

$$K_{ab} = \int_0^{\delta(x)} \rho u^2 dy \qquad (6.103)$$

The momentum flowing out of the control body from the A_{cd} surface is

$$K_{cd} = K_{ab} + \frac{\partial}{\partial x}\left(\int_0^{\delta(x)} \rho u^2 dy\right) dx \qquad (6.104)$$

The x-direction component of the momentum flowing into the control body through A_{bc} is

$$K_{bc} = u_e \frac{\partial}{\partial x}\left(\int_0^{\delta(x)} \rho u \, dy\right) dx \qquad (6.105)$$

The force along the x-direction on the A_{ab} plane is

$$F_{ab} = p_e \delta(x) \qquad (6.106)$$

The acting force along the x-direction on the A_{cd} surface is

$$F_{cd} = -\left(p_e + \frac{dp_e}{dx} dx\right)(\delta(x) + d\delta) \qquad (6.107)$$

The force along the x-direction on the A_{bc} plane is

$$F_{bc} = \left(p_e + \frac{dp_e}{dx}\frac{dx}{2}\right) d\delta \qquad (6.108)$$

The shear force on the wall of A_{ad} is

$$F_{ad} = -\tau_0 dx \qquad (6.109)$$

Establish the momentum integral equation in the x-direction for the control body as

$$p_e \delta(x) - \left(p_e + \frac{\mathrm{d}p_e}{\mathrm{d}x} \mathrm{d}x \right) (\delta(x) + \mathrm{d}\delta) + \left(p_e + \frac{\mathrm{d}p_e}{\mathrm{d}x} \frac{\mathrm{d}x}{2} \right) \mathrm{d}\delta - \tau_0 \mathrm{d}x$$

$$= K_{ab} + \frac{\partial}{\partial x} \left(\int_0^{\delta(x)} \rho u^2 \mathrm{d}y \right) \mathrm{d}x - K_{ab} - u_e \frac{\partial}{\partial x} \left(\int_0^{\delta(x)} \rho u \mathrm{d}y \right) \mathrm{d}x \qquad (6.110)$$

After sorting, we get

$$\frac{\mathrm{d}p_e}{\mathrm{d}x} \delta(x) + \tau_0 = u_e \frac{\partial}{\partial x} \left(\int_0^{\delta(x)} \rho u \mathrm{d}y \right) - \frac{\partial}{\partial x} \left(\int_0^{\delta(x)} \rho u^2 \mathrm{d}y \right) \qquad (6.111)$$

Considering that $\int_0^{\delta(x)} \rho u \mathrm{d}y$ and $\int_0^{\delta(x)} \rho u^2 \mathrm{d}y$ in the differential sign on the right side of the equal sign of the above equation are both x functions, the partial differential can be replaced by total differential, then the above equation becomes

$$\frac{\mathrm{d}p_e}{\mathrm{d}x} \delta(x) + \tau_0 = u_e \frac{\mathrm{d}}{\mathrm{d}x} \left(\int_0^{\delta(x)} \rho u \mathrm{d}y \right) - \frac{\mathrm{d}}{\mathrm{d}x} \left(\int_0^{\delta(x)} \rho u^2 \mathrm{d}y \right) \qquad (6.112)$$

Using Bernoulli's equation on the outer boundary of the boundary layer, $\frac{1}{2} \rho u_e^2 + p_e = $ Constant, the two ends are differentiated to get

$$u_e \frac{\mathrm{d}u_e}{\mathrm{d}x} = -\frac{1}{\rho} \frac{\mathrm{d}p_e}{\mathrm{d}x} \qquad (6.113)$$

Using the above formula, the first term on the left side of formula (6.112) can be written as

$$\frac{\mathrm{d}p_e}{\mathrm{d}x} \delta(x) = -\rho u_e \frac{\mathrm{d}u_e}{\mathrm{d}x} \int_0^{\delta(x)} \mathrm{d}y \qquad (6.114)$$

The first term on the right side of formula (6.112) can be written as

$$u_e \frac{\mathrm{d}}{\mathrm{d}x} \left(\int_0^{\delta(x)} \rho u \mathrm{d}y \right) = \frac{\mathrm{d}}{\mathrm{d}x} \left(\int_0^{\delta(x)} \rho u u_e \mathrm{d}y \right) - \frac{\mathrm{d}u_e}{\mathrm{d}x} \int_0^{\delta(x)} \rho u \mathrm{d}y \qquad (6.115)$$

Substituting formulas (6.114) and (6.115) into formula (6.111), and after sorting, we get

$$\frac{\tau_0}{\rho} = \frac{d}{dx}\left(u_e^2 \int_0^{\delta(x)} \frac{u}{u_e}\left(1 - \frac{u}{u_e}\right)dy\right) + u_e\frac{du_e}{dx}\int_0^{\delta(x)}\left(1 - \frac{u}{u_e}\right)dy \qquad (6.116)$$

Using the definitions of boundary layer displacement thickness and boundary layer momentum loss thickness, it can be simplified as substituting formulas (6.114) and (6.115) into formula (6.111), and after sorting out

$$\frac{\tau_0}{\rho} = \frac{d}{dx}\left(u_e^2\delta_2\right) + u_e\delta_1\frac{du_e}{dx} \qquad (6.117)$$

This is the momentum integral equation of the boundary layer. This equation is a first-order ordinary differential equation, which is suitable for laminar and turbulent boundary layers. Expand and recombine the differential term on the right side of the above formula to get

$$\frac{\tau_0}{\rho} = u_e^2\frac{d\delta_2}{dx} + u_e(2\delta_2 + \delta_1)\frac{du_e}{dx} \qquad (6.118)$$

If it is written in a dimensionless form, there are

$$\frac{C_f}{2} = \frac{d\delta_2}{dx} + (2 + H)\frac{\delta_2}{u_e}\frac{du_e}{dx}, \quad H = \frac{\delta_1}{\delta_2}, \quad C_f = \frac{\tau_0}{\frac{1}{2}\rho u_e^2} \qquad (6.119)$$

In the formula, H is called the shape factor of the boundary layer velocity distribution function, which is determined by the boundary layer longitudinal velocity distribution function.

6.4.2 Derivation of Boundary Layer Momentum Integral Equation from Differential Equation

For two-dimensional incompressible fluid boundary layer equation

$$\begin{cases} \dfrac{\partial u}{\partial x} + \dfrac{\partial v}{\partial y} = 0 \\[2mm] \dfrac{\partial u}{\partial t} + u\dfrac{\partial u}{\partial x} + v\dfrac{\partial u}{\partial y} = \dfrac{\partial u_e}{\partial t} + u_e\dfrac{\partial u_e}{\partial x} + v\dfrac{\partial^2 u}{\partial y^2} \end{cases}$$

Multiply the continuous equation by u_e, and use the continuous equation to rewrite the momentum equation as

$$\begin{cases} \dfrac{\partial u_e u}{\partial x} + \dfrac{\partial u_e v}{\partial y} = u \dfrac{\partial u_e}{\partial x} \\[3mm] \dfrac{\partial u}{\partial t} + \dfrac{\partial uu}{\partial x} + \dfrac{\partial uv}{\partial y} = \dfrac{\partial u_e}{\partial t} + u_e \dfrac{\partial u_e}{\partial x} + \dfrac{1}{\rho}\dfrac{\partial \tau}{\partial y}, \quad \left(\tau = \rho v \dfrac{\partial u}{\partial y}\right) \end{cases} \tag{6.120}$$

Subtracting between the two formulas, we get

$$\frac{\partial}{\partial t}(u_e - u) + \frac{\partial}{\partial x}(u_e u - uu) + \frac{\partial}{\partial y}(u_e v - uv) + (u_e - u)\frac{\partial u_e}{\partial x} = -\frac{1}{\rho}\frac{\partial \tau}{\partial y} \tag{6.121}$$

Integrating the above formula, there are

$$\frac{\partial}{\partial t}\int_0^\infty (u_e - u)\mathrm{d}y + \frac{\partial}{\partial x}\int_0^\infty (u_e u - uu)\mathrm{d}y$$

$$+ \int_0^\infty \frac{\partial}{\partial y}(u_e v - uv)\mathrm{d}y + \frac{\partial u_e}{\partial x}\int_0^\infty (u_e - u)\mathrm{d}y = -\frac{1}{\rho}\int_0^\infty \frac{\partial \tau}{\partial y}\mathrm{d}y \tag{6.122}$$

Get after sorting

$$\frac{\partial u_e \delta_1}{\partial t} + u_e^2 \frac{\partial \delta_2}{\partial x} + u_e (2\delta_2 + \delta_1)\frac{\partial u_e}{\partial x} = \frac{\tau_0}{\rho} \tag{6.123}$$

For a steady flow, $\frac{\partial u_e \delta_1}{\partial t} = 0$, there are

$$u_e^2 \frac{\partial \delta_2}{\partial x} + u_e (2\delta_2 + \delta_1)\frac{\partial u_e}{\partial x} = \frac{\tau_0}{\rho} \tag{6.124}$$

6.5 The Solution of the Momentum Integral Equation of Laminar Boundary Layer on a Flat Plate

In the momentum integral equation, there are three unknowns, displacement thickness, momentum loss thickness, and wall shear stress. Therefore, two complementary relations must be sought to obtain the integral solution. The first supplementary relationship is the velocity distribution in the boundary layer which is directly related to the three unknowns. Obviously, the accuracy of the integral equation solution depends on the rationality of the velocity distribution in the boundary layer. The second supplementary relationship is the relationship between wall shear stress and velocity distribution. For the laminar boundary layer, Newton's law of internal friction is used. Assuming that the flow in the boundary layer is a steady incompressible flow, the velocity distribution in the boundary layer is

$$\frac{u}{u_e} = f(\eta) = a_0 + a_1\eta + a_2\eta^2 + a_3\eta^3 + a_4\eta^4 \tag{6.125}$$

Differential equation of motion in plane boundary layer

$$\begin{cases} \dfrac{\partial u}{\partial x} + \dfrac{\partial v}{\partial y} = 0 \\[2mm] u\dfrac{\partial u}{\partial x} + v\dfrac{\partial u}{\partial y} = u_e\dfrac{\partial u_e}{\partial x} + v\dfrac{\partial^2 u}{\partial y^2} \end{cases}$$

At the wall of the plate, $y = 0$, $u = v = 0$, we get

$$\frac{\partial^2 u}{\partial y^2} = -\frac{u_e}{v}\frac{\partial u_e}{\partial x} \tag{6.126}$$

$$\tau_0 = \mu\frac{\partial u}{\partial y} \tag{6.127}$$

Calculate the partial derivative of Eq. (6.126) along the y-direction to obtain

$$\frac{\partial^3 u}{\partial y^3} = 0 \tag{6.128}$$

On the outer boundary of the boundary layer

$$y = \infty, \ u = u_e, \ \frac{\partial^n u_e}{\partial y^n} = 0, \ n = 1, 2, 3, \ldots \tag{6.129}$$

Taken together, the conditions for determining the coefficients in formula (6.125) are

$$y = 0, \ u = v = 0, \ \frac{\partial u}{\partial y} = \frac{\tau_0}{\mu}, \ \frac{\partial^2 u}{\partial y^2} = -\frac{u_e u_e'}{v}, \ \frac{\partial^3 u}{\partial y^3} = 0$$

$$y = \infty, \ u = u_e, \ \frac{\partial^n u}{\partial y^n} = 0, \ n = 1, 2, 3, \ldots \tag{6.130}$$

For a steady flow in the boundary layer of a flat plate with zero pressure gradient, the mainstream velocity outside the boundary layer is a constant value, namely

$$u_e = V_\infty = \text{const}, \ \frac{du_e^2}{dx} = 0, \ \frac{dp_e}{dx} = 0 \tag{6.131}$$

The integral momentum Eq. (6.124) is simplified to

$$\frac{\tau_0}{\rho} = V_\infty^2 \frac{d\delta_2}{dx} \tag{6.132}$$

Assuming that the velocity distribution in the boundary layer is a quadratic curve, that is

$$\frac{u}{V_\infty} = f(\eta) = \eta(2 - \eta) \tag{6.133}$$

Obtain the momentum loss thickness of the boundary layer as

$$\delta_2 = \int_0^{\delta(x)} \frac{u}{V_\infty}\left(1 - \frac{u}{V_\infty}\right)dy$$

$$= \delta \int_0^1 f(1 - f)d\eta = \delta \int_0^1 (2\eta - \eta^2)(1 - 2\eta + \eta^2)d\eta = \frac{2}{15}\delta \tag{6.134}$$

In addition, the wall shear stress is defined as

$$\tau_0 = \mu \left.\frac{\partial u}{\partial y}\right|_{y=0} = \mu \frac{V_\infty}{\delta} \left.\frac{\partial f}{\partial \eta}\right|_{\eta=0} = 2\mu \frac{V_\infty}{\delta} \tag{6.135}$$

Substituting formulas (6.134) and (6.135) into formula (6.132), we get

$$V_\infty^2 \frac{2}{15} \frac{d\delta}{dx} = 2\frac{\mu}{\rho} \frac{V_\infty}{\delta}$$

Integrate the above formula to get

$$\frac{\delta}{x} = \frac{\sqrt{30}}{\sqrt{Re_x}} = \frac{5.477}{\sqrt{Re_x}}, \quad Re_x = \frac{V_\infty x}{\nu} \tag{6.136}$$

The wall shear stress is

$$\tau_0 = \mu \left.\frac{\partial u}{\partial y}\right|_{y=0} = \frac{2}{\sqrt{30}} \rho V_\infty^2 \frac{1}{\sqrt{Re_x}} = 0.3651\rho V_\infty^2 \frac{1}{\sqrt{Re_x}} \tag{6.137}$$

The other calculation results of different velocity distributions are shown in Table 6.2.

The momentum integral equation is used to solve the frictional resistance on the flat surface. Although the calculation is simple, as long as the velocity distribution is selected appropriately, better results can be obtained.

Table 6.2 Calculation results of different velocity distributions

Speed distribution	$\frac{u}{u_e} = f(\eta)$	$\frac{\delta}{x}$	$\frac{\tau_0}{\rho V_\infty^2}$
One-time	$f(\eta) = \eta$	$\frac{3.464}{\sqrt{Re_x}}$	$\frac{0.289}{\sqrt{Re_x}}$
Quadratic	$f(\eta) = 2\eta - \eta^2$	$\frac{5.477}{\sqrt{Re_x}}$	$\frac{0.3651}{\sqrt{Re_x}}$
Cubic	$f(\eta) = \frac{3}{2}\eta - \frac{1}{2}\eta^3$	$\frac{4.641}{\sqrt{Re_x}}$	$\frac{0.323}{\sqrt{Re_x}}$
Quartic	$f(\eta) = 2\eta - 2\eta^3 + \eta^4$	$\frac{5.835}{\sqrt{Re_x}}$	$\frac{0.343}{\sqrt{Re_x}}$
Sinusoidal	$f(\eta) = \sin\left(\frac{\pi}{2}\eta\right)$	$\frac{4.795}{\sqrt{Re_x}}$	$\frac{0.328}{\sqrt{Re_x}}$
Blasius	Series solution	$\frac{5.0}{\sqrt{Re_x}}$	$\frac{0.3321}{\sqrt{Re_x}}$

6.6 Solution of the Momentum Integral Equation of the Turbulent Boundary Layer on a Flat Plate

For the turbulent boundary layer flow, the momentum integral equation is used to solve the wall resistance and the development of the boundary layer. The experimental results (as shown in Fig. 6.11) show that the time-average velocity distribution in the wall turbulence zone is the same as the 1932 German scholar Nikuradse. The turbulent time-averaged velocity distribution in the pipeline is very similar. As long as the maximum velocity of the centerline of the pipe is taken as the external flow velocity u_e, and the pipe radius is taken as the boundary layer thickness δ, the time-average velocity distribution in the turbulent zone above the plate wall satisfies the exponential law, which is

$$\frac{\overline{u}}{u_e} = \left(\frac{y}{\delta}\right)^{1/n} \tag{6.138}$$

Fig. 6.11 Velocity distribution in the turbulent boundary layer

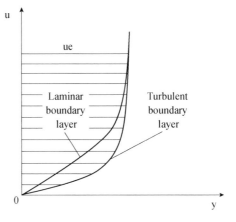

The index n is between 6 and 10, often taken as 7, and the Reynolds number $(\mathrm{Re} = \frac{u_e \delta}{\nu})$ is related. If the wall friction velocity u_τ is introduced as a dimensionless parameter, the exponential law of velocity distribution can be expressed as

$$\frac{\overline{u}}{u_\tau} = K \left(\frac{u_\tau y}{\nu} \right)^{1/n} \tag{6.139}$$

In the formula, K is a dimensionless coefficient with a value of 8.74–11.5. For $n = 7$, $K = 8.74$. The friction speed is

$$u_\tau = \sqrt{\frac{\tau_w}{\rho}} \tag{6.140}$$

When $y = \delta$, formula (6.139) becomes

$$\frac{u_e}{u_\tau} = K \left(\frac{u_\tau \delta}{\nu} \right)^{1/n} \tag{6.141}$$

The ratio of formulas (6.139) and (6.141) can be obtained by formula (6.138). If Eq. (6.141) is taken to the power of n, multiplied by u_e, and then raised to the power of $(n + 1)$, we get

$$\frac{u_e}{u_\tau} = K^{\frac{n}{n+1}} \left(\frac{u_e \delta}{\nu} \right)^{\frac{1}{n+1}} \tag{6.142}$$

For flat-plate boundary layer flow, the velocity of the outer boundary of the boundary layer is $u_e = V_\infty$, and the integral equation of the momentum of the boundary layer is

$$\frac{\tau_0}{\rho} = V_\infty^2 \frac{d\delta_2}{dx}, \quad \frac{d\delta_2}{dx} = \frac{u_\tau^2}{V_\infty^2} \tag{6.143}$$

Substituting formula (6.142) into formula (6.143), we get

$$\frac{d\delta_2}{dx} = K^{-\frac{2n}{n+1}} \left(\frac{V_\infty \delta}{\nu} \right)^{-\frac{2}{n+1}} \tag{6.144}$$

Using formula (6.138), the momentum loss thickness of the boundary layer is

$$\delta_2 = \delta \int_0^1 \frac{\overline{u}}{V_\infty} \left(1 - \frac{\overline{u}}{V_\infty} \right) d\eta = \delta \int_0^1 \eta^{1/n} \left(1 - \eta^{1/n} \right) d\eta = \frac{n}{(n+1)(n+2)} \delta \tag{6.145}$$

Substituting into formula (6.44), we get

$$\frac{n}{(n+1)(n+2)}\frac{\mathrm{d}\delta}{\mathrm{d}x} = K^{-\frac{2n}{n+1}}\left(\frac{V_\infty\delta}{\nu}\right)^{-\frac{2}{n+1}} \tag{6.146}$$

Integrate the above formula (assuming that $x = 0$, and there is full turbulence on the plate), we get

$$\frac{\delta}{x} = \left[\frac{(n+3)(n+2)}{n}\right]^{\frac{n+1}{n+3}} K^{-\frac{2n}{n+3}}\left(\frac{V_\infty x}{\nu}\right)^{-\frac{2}{n+3}} \tag{6.147}$$

The wall friction resistance coefficient is

$$C_f = \frac{2\tau_w}{\rho V_\infty^2} = 2\frac{u_\tau^2}{V_\infty^2} = 2K^{-\frac{2n}{n+3}}\left[\frac{(n+2)(n+3)}{n}\right]^{-\frac{2}{n+3}}\left(\frac{V_\infty x}{\nu}\right)^{-\frac{2}{n+3}} \tag{6.148}$$

Let $n = 7$, $K = 8.74$, $\mathrm{Re}x = \frac{V_\infty x}{\nu}$, we get

$$\frac{\delta}{x} = 0.371\left(\frac{V_\infty x}{\nu}\right)^{-\frac{1}{5}} = 0.371\mathrm{Re}x^{-\frac{1}{5}} \tag{6.149}$$

The momentum loss thickness of the boundary layer is

$$\frac{\delta_2}{x} = \frac{n}{(n+1)(n+2)}\frac{\delta}{x} = \frac{7}{72} \times 0.371\left(\frac{V_\infty x}{\nu}\right)^{-\frac{1}{5}} = 0.0361\mathrm{Re}x^{-\frac{1}{5}} \tag{6.150}$$

The displacement thickness of the boundary layer is

$$\frac{\delta_1}{x} = \frac{1}{(n+1)}\frac{\delta}{x} = \frac{1}{8} \times 0.371\left(\frac{V_\infty x}{\nu}\right)^{-\frac{1}{5}} = 0.0464\mathrm{Re}x^{-\frac{1}{5}} \tag{6.151}$$

The wall friction resistance coefficient is

$$C_f = \frac{2\tau_w}{\rho V_\infty^2} = 2\frac{u_\tau^2}{V_\infty^2} = 0.0288\left(\frac{V_\infty x}{\nu}\right)^{-\frac{1}{5}} = 0.0577\mathrm{Re}x^{-\frac{1}{5}} \tag{6.152}$$

$$\tau_w = 0.0289\rho V_\infty^2\mathrm{Re}x^{-\frac{1}{5}} \tag{6.153}$$

The total resistance of the plate is

$$D_f = \int_0^L \tau_w\mathrm{d}x = \int_0^L 0.0289\rho V_\infty^2\mathrm{Re}x^{-\frac{1}{5}}\mathrm{d}x = 0.361\rho V_\infty^2 L\mathrm{Re}_L^{-\frac{1}{5}} \tag{6.154}$$

The plate resistance coefficient is

$$C_{\mathrm{Df}} = \frac{D_{\mathrm{f}}}{\frac{1}{2}\rho V_{\infty}^2 L} = 0.0722\mathrm{Re}_L^{-\frac{1}{5}} \qquad (6.155)$$

Compared with the experimental data, the coefficient of 0.074 is in better agreement.

$$C_{\mathrm{Df}} = \frac{D_{\mathrm{f}}}{\frac{1}{2}\rho V_{\infty}^2 L} = 0.074\mathrm{Re}_L^{-\frac{1}{5}} \qquad (6.156)$$

The formula is Prandtl (1927) formula of the resistance coefficient of the plate. The above derivation borrowed the 1/7 exponential law of the pipeline turbulence velocity distribution and the wall shear stress relationship. The above results are adapted to $\mathrm{Re}_L = 5 \times 10^5 - 10^7$. For the case where Re_L is greater than 107, the velocity distribution of the wall turbulent boundary layer is adapted to the law of logarithm, namely

$$\frac{\overline{u}}{u_\tau} = 5.756\lg\frac{u_\tau y}{v} + 5.5 \qquad (6.157)$$

Prandtl (1927) and Schlichting (1932, as shown in Fig. 6.12, German fluid mechanics, 1907~1982, under the tutelage of Professor Prandtl), through appropriate corrections and derivations, obtained the empirical formula for the resistance coefficient of the plate is

$$C_{\mathrm{Df}} = \frac{0.455}{(\lg\mathrm{Re}_L)^{2.58}} \qquad (6.158)$$

The empirical formula given by the German scholar Schultz-Gnunow (1940) is

$$C_{\mathrm{Df}} = \frac{0.427}{(\lg\mathrm{Re}_L - 0.407)^{2.64}} \qquad (6.159)$$

The relationship curve between drag coefficient and Reynolds number is given in Fig. 6.13. Among them, curve (1) is the laminar flow resistance coefficient formula, (2) is the turbulent resistance coefficient Prandtl formula, curve (3) is the Prandtl-Schlichting formula, and curve (4) is Schultz-Gnunow's formula.

6.7 Boundary Layer Separation

The boundary layer approximation theory discussed above is obviously only suitable for the case where the boundary layer is not separated, because once the flow

Fig. 6.12 Hermann
Schlichting, German fluid
mechanics (1907~1982)

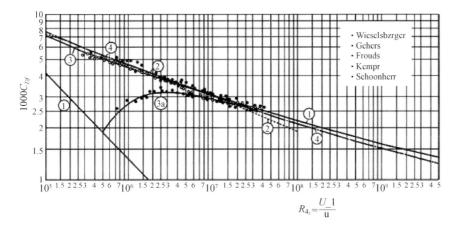

Fig. 6.13 Flat resistance coefficient curve

is separated, the boundary layer approximation assumption becomes invalid. This
section mainly discusses the separation of the boundary layer, the reasons for the
separation, and the influence of pressure gradient on the flow pattern.

6.7.1 Boundary Layer Separation Phenomenon of Flow Around Cylinder

In the boundary layer with steady flow around a cylinder, the fluid clusters will be affected by inertial force, viscous force, and differential pressure force. First of all, in the flow process, the fluid micelles have to overcome the viscous force to do work and consume mechanical energy. The effect of the differential pressure force depends on the different areas of the circular flow around the cylinder. It helps the fluid micelles to accelerate in the pressure gradient area, while in the reverse pressure gradient area it hinders the fluid movement (deceleration movement). The separation of the boundary layer on the cylindrical surface means that the fluid micelles in the boundary layer consume kinetic energy due to the effect of viscosity. In the area where the pressure increases along the flow direction (i.e., the reverse pressure gradient zone), it cannot continue to flow along the object surface. As a result, backflow occurred, causing the fluid to leave the object surface. As shown in Fig. 6.14, for the flow of ideal fluid clusters around the cylinder, acceleration and decompression appear on the windward side of the cylinder, and deceleration and pressure increase on the leeward side. If the influence of viscosity is taken into account, the boundary layer on the cylindrical surface will not always be attached to the object surface, but the boundary layer separation occurs at a certain point on the cylindrical surface.

For the viscous fluid to flow around, in the above-mentioned energy conversion process, due to the effect of viscosity, the fluid particles in the boundary layer will have to overcome the viscous force to perform work and consume mechanical energy. In the counterpressure gradient zone, the fluid clusters cannot reach the stagnation point, but the velocity drops to zero at a point on the leeward surface, and the later particles will reroute into the main flow, separating the incoming boundary layer from the wall. In the area downstream of the separation point, reverse flow occurs due to the reverse pressure gradient. The separation point is defined as the boundary point between the downstream and reverse flow areas next to the wall. In the vicinity of the separation point and in the separation zone, the boundary layer hypothesis is no longer valid due to the greatly increased thickness of the boundary layer. Moreover, the separated boundary layer and the wakes formed by it will also interfere with the

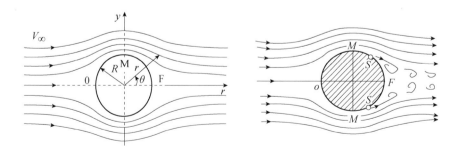

Fig. 6.14 The flow around a cylinder is separated from the boundary layer

main flow. The necessary conditions for the separation of the boundary layer are the blocking effect of the reverse pressure gradient and the viscosity of the object surface. There is only a viscous retarding effect without an inverse pressure gradient, and no boundary layer separation occurs because there is no back thrust to make the boundary layer fluid enter the outflow zone. This shows that the boundary layer separation is unlikely to occur in the flow along the pressure gradient. Similarly, there is only an inverse pressure gradient without viscous blocking effect, and no separation phenomenon will occur, because without blocking effect, it is impossible for the moving fluid to consume kinetic energy and stagnate. Experiments have found that in the flow around a cylinder, the separation angle of the laminar boundary layer flow is about 82° (the angle between the direction of the incoming flow), and the separation angle of the turbulent boundary layer is about 120°. The flow separation flow pattern of the actual flow around a cylinder is directly related to the flow around the Reynolds number (Re = $\frac{V_\infty d}{\nu}$), as shown in Fig. 6.15.

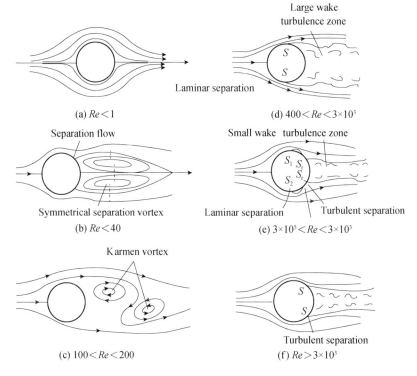

Fig. 6.15 Separation patterns of flow around cylinders with different Reynolds numbers

6.7.2 Airfoil Separation Phenomenon

The airfoil circumfluence is usually separated by trailing edge and leading edge of bubble separation. As shown in Fig. 6.16, once the airfoil flow is separated, the flow structure is complicated, and an obvious vortex zone appears after the separation zone. At the same time, the resistance of the flow object is greatly increased. The trailing edge separation is often seen in the flow around a low-speed airfoil with a high angle of attack or above, and the separation occurs in a large backpressure gradient area near the trailing edge. Experiments have found that the trailing edge separation is often turbulent separation, and as the angle of attack increases, the separation point gradually moves upstream, the separation zone continues to increase, and a wake zone is formed downstream of the separation zone. The stall state that appears from the separation of the trailing edge is called trailing edge stall, which is characterized by a relatively slow development of the stall. The semi-empirical condition for trailing edge separation given by Crabtree is $Re_\delta > 2700$, where Re_δ is the Reynolds number of the separation point based on the thickness of the boundary layer. For the steady flow around a thin airfoil, there is a large backpressure gradient near the leading edge, when Re_δ exceeds a certain value, separation bubbles will form near the leading edge of the airfoil, that is, a closed separation zone. The outer boundary of the bubble is a streamline, called the separation streamline, and the intersection point with the object surface is the separation point and the reattachment point respectively. The flow in the bubble is a vortex flow. Experiments have found that, according to the size of the separation bubbles, it can be divided into short bubbles and long bubbles. Short bubbles occur under the condition of $Re_\delta > 400$–500, and their length is about 1% of the chord length (or about 100 times the thickness of the displacement of the separation point). As the angle of attack increases, the short bubbles will suddenly burst, and the airflow will not be attached, leading to a leading edge stall. Long bubbles occur when $Re_\delta < 400$–500, and are common in thin airfoil flows, and their length is about 2–3% chord length (or 10^4 times or longer than the displacement thickness of the boundary layer at the separation point). As the angle of attack increases, the long bubbles will continue to increase, and the reattached point will move downstream as the angle of attack increases, until it reaches the trailing edge, and the separation appears to be completely separated. It should be pointed out that the appearance of separation bubbles plays a triggering role in the transition from laminar flow to turbulent flow.

6.7.3 Velocity Distribution Characteristics of the Boundary Layer in Different Pressure Gradient Areas

In order to more clearly reveal the formation and development process of the boundary layer separation phenomenon, it is necessary to analyze the characteristics of the boundary layer velocity distribution under the pressure gradient. This section focuses

Fig. 6.16 Flow around a separated airfoil at the trailing edge of a high angle of attack

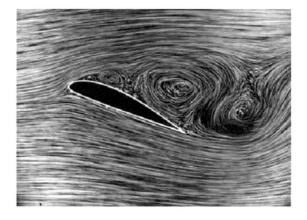

on the analysis of the velocity profile and its rate of change along the normal direction and the characteristics of the change of curvature in the near-wall area. According to the boundary layer differential Eq. (6.70), we can get

$$\frac{\partial^2 u}{\partial y^2} = \frac{1}{\mu}\frac{\partial p}{\partial x} = -\frac{u_e}{\nu}\frac{\partial u_e}{\partial x} \qquad (6.160)$$

Therefore, the pressure gradient has an impact on the flow velocity distribution in the boundary layer. As shown in Fig. 6.17, for the case of the forward pressure gradient, there are

$$\frac{\partial p}{\partial x} < 0, \ \frac{\partial^2 u}{\partial y^2} < 0, \ \frac{\partial u_e}{\partial x} > 0 \qquad (6.161)$$

For the case of backpressure gradient, there are

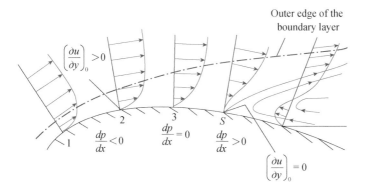

Fig. 6.17 Boundary layer on a curved surface

$$\frac{\partial p}{\partial x} > 0, \quad \frac{\partial^2 u}{\partial y^2} > 0, \quad \frac{\partial u_e}{\partial x} < 0 \tag{6.162}$$

For the case of zero pressure gradient, there are

$$\frac{\partial p}{\partial x} = 0, \quad \frac{\partial^2 u}{\partial y^2} = 0, \quad \frac{\partial u_e}{\partial x} = 0 \tag{6.163}$$

It can be seen that as the pressure gradient changes sign, the curvature of the boundary layer velocity distribution will change sign.

(1) In the favorable pressure gradient zone, the pressure decreases along the way and the speed increases along the way. At the wall, there are

$$\left.\frac{\partial u}{\partial y}\right|_{y=0} > 0, \quad \left.\frac{\partial^2 u}{\partial y^2}\right|_{y=0} < 0 \tag{6.164}$$

On the other hand, on the outer boundary of the boundary layer, there are

$$\left.\frac{\partial u}{\partial y}\right|_{y=\delta} = 0, \quad \left.\frac{\partial^2 u}{\partial y^2}\right|_{y=0-\delta} < 0 \tag{6.165}$$

This shows that in the favorable pressure gradient zone, the velocity in the boundary layer increases monotonously along the y-direction, the distribution curve has no inflection point, it is a smooth curve convex outwards, and the flow is stable, as shown in Fig. 6.18.

(2) In the adverse pressure gradient zone (before the separation point), the pressure increases along the way, and the speed decreases along the way. On the wall, there are

$$\left.\frac{\partial u}{\partial y}\right|_{y=0} > 0, \quad \left.\frac{\partial^2 u}{\partial y^2}\right|_{y=0} > 0 \tag{6.166}$$

On the other hand, on the outer boundary of the boundary layer, as shown in Fig. 6.19, there are

Fig. 6.18 Flow along the pressure gradient (steady velocity distribution)

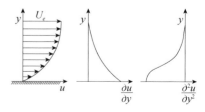

Fig. 6.19 Velocity distribution in the reverse pressure gradient zone

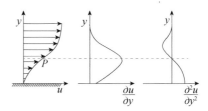

$$\frac{\partial u}{\partial y}\bigg|_{y=\delta} = 0, \quad \frac{\partial^2 u}{\partial y^2}\bigg|_{y=\delta} < 0 \tag{6.167}$$

Therefore, in the boundary layer $0 < y < \delta$, the curvature of the velocity distribution changes from positive to negative, and there must be

$$\frac{\partial^2 u}{\partial y^2}\bigg|_{y=y_0} = 0 \tag{6.168}$$

This point is the inflection point of the velocity distribution. The appearance of the inflection point changes the shape of the velocity distribution. Above the inflection point it is convex, and below the inflection point it is concave. The velocity distribution type with inflection points is unstable.

(3) At the minimum pressure point, there are

$$\frac{dp}{dx} = 0, \quad \frac{\partial^2 u}{\partial y^2}\bigg|_{y=0} = 0 \tag{6.169}$$

It shows that the inflection point is on the object surface. As the fluid particle flows downstream, the inflection point moves to the outer boundary, and the velocity distribution in the near area of the object surface becomes thinner and thinner, but when the inflection point moves to a certain point, the object surface appears

$$\frac{\partial u}{\partial y}\bigg|_{y=0} = 0, \quad \frac{\partial^2 u}{\partial y^2}\bigg|_{y=0} > 0 \tag{6.170}$$

This point is called as a separation point. In the downstream area of the separation point, there are

$$\frac{dp}{dx} > 0, \quad \frac{\partial^2 u}{\partial y^2}\bigg|_{y=0} > 0, \quad \frac{\partial u}{\partial y}\bigg|_{y=0} < 0 \tag{6.171}$$

A backflow occurs, which pushes the main flow away from the wall, and the boundary layer hypothesis fails. From the above analysis, it can be seen that the larger the reverse pressure gradient, the more advanced the boundary layer separation. After the boundary layer is separated, the flow characteristics have changed.

In summary, the general two-dimensional separation flow characteristics can be summarized as follows:

(a) The separation point generally appears in the reverse pressure gradient zone, that is, the deceleration and pressure increase zone.

(b) At the separation point, the velocity along the wall is $\left(\frac{\partial u}{\partial y}\right)_s = 0$. S is the separation point, and y is the coordinate perpendicular to the object plane.

(c) Near the separation point, the boundary layer thickness increases rapidly, and the boundary layer approximation assumption fails.

(d) Reverse flow occurs downstream of the separation point, forming an unstable vortex zone, which makes the main flow zone from the original non-vortex zone to a vortex zone.

(e) The pressure distribution on the object surface changes from the original almost symmetrical distribution to an asymmetrical distribution, and a low-pressure zone (or negative pressure zone) appears after the separation point, which greatly increases the pressure difference resistance of the circling object.

6.8   Separated Flow and Characteristics of Two-Dimensional Steady Viscous Fluid

6.8.1   Separation Mode-Prandtl Image

The so-called separation mode refers to the physical image of the separation flow. The correct separation mode is established on the basis of experimental observations. For the steady two-dimensional flow near the solid wall (as shown in Fig. 6.20), Prandtl's separation mode refers to the fluid moving forward from the original attachment surface, if at a certain point 0 suddenly leaves the object surface, and thereafter a backflow zone appears. It is said that the flow has been separated, and the 0 point is the separation point. The flow line 0A that separates the incoming flow and the return flow is called the separation flow line, and this separation image is called the Prandtl separation mode.

Fig. 6.20 Prandtl separation mode

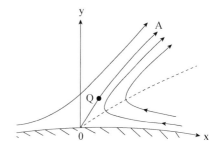

6.8.2 Necessary Conditions for Flow Separation

It is not general, you can choose the separation point 0 as the origin of the coordinates, the object plane is the x-axis, the normal direction is the y-axis, the direction of the incoming flow is the x-axis as the positive direction, and the right-hand coordinate system is taken (as shown in Fig. 6.20), the streamline equation is

$$\frac{\mathrm{d}y}{h_x \mathrm{d}x} = \frac{v}{u} \tag{6.172}$$

In the formula, h_x is the Lame coefficient in the x-direction. Because the fluid satisfies the non-slip condition on the object surface, that is, when $y = 0$, there is

$$u = v = 0 \tag{6.173}$$

In order to determine the slope of the separation line 0A, it is necessary to use the L'Hospital law to determine the limit value of the 0/0 type to formula (6.172). That is, because the streamline 0A suddenly lifts and leaves the object surface at 0 point, there is

$$tg\theta_0 = \frac{1}{h_x} \frac{\mathrm{d}y}{\mathrm{d}x} \neq 0 \tag{6.174}$$

where θ_0 is the inclination angle of the 0A streamline at 0 point. According to the L'Hospital law

$$tg\theta_0 = \left(\frac{\mathrm{d}y}{h_x \mathrm{d}x}\right)_0 = \frac{\left(\frac{\mathrm{d}v}{\mathrm{d}s}\right)_0}{\left(\frac{\mathrm{d}u}{\mathrm{d}s}\right)_0} \tag{6.175}$$

where $\mathrm{d}/\mathrm{d}s$ is the derivative along the direction of the separation streamline 0A. Since the directional derivative in the curvilinear coordinate system is

$$\left(\frac{\mathrm{d}}{\mathrm{d}s}\right)_0 = \left(\frac{h_x \mathrm{d}x}{\mathrm{d}s}\right)_0 \left(\frac{\partial}{h_x \partial x} + \frac{\mathrm{d}y}{h_x \mathrm{d}x} \frac{\partial}{\partial y}\right)_0 = \left(\frac{h_x \mathrm{d}x}{\mathrm{d}s}\right)_0 \left(\frac{\partial}{h_x \partial x} + tg\theta_0 \frac{\partial}{\partial y}\right)_0 \tag{6.176}$$

Substituting formula (6.176) into formula (6.175), we have

$$tg\theta_0 = \left(\frac{\mathrm{d}y}{h_x \mathrm{d}x}\right)_0 = \frac{\left(\frac{\mathrm{d}v}{\mathrm{d}s}\right)_0}{\left(\frac{\mathrm{d}u}{\mathrm{d}s}\right)_0} = \frac{\left(\frac{\partial v}{h_x \partial x}\right)_0 + tg\theta_0 \left(\frac{\partial v}{\partial y}\right)_0}{\left(\frac{\partial u}{h_x \partial x}\right)_0 + tg\theta_0 \left(\frac{\partial u}{\partial y}\right)_0} \tag{6.177}$$

According to the object surface conditional formula (6.173), we can see

$$\left(\frac{\partial u}{h_x \partial x}\right)_0 = \left(\frac{\partial v}{h_x \partial x}\right)_0 = 0 \tag{6.178}$$

Substituting formula (6.177), we get

$$tg\theta_0 = \left(\frac{dy}{h_x dx}\right)_0 = \frac{\left(\frac{\partial v}{\partial y}\right)_0}{\left(\frac{\partial u}{\partial y}\right)_0} \tag{6.179}$$

According to the continuity equation, there are

$$\frac{\partial(\rho u)}{\partial x} + \frac{\partial(\rho v h_x)}{\partial y} = 0 \tag{6.180}$$

可得

$$\left(\frac{\partial v}{\partial y}\right)_0 = -\left(\frac{1}{\rho h_x}\right)_0 \left[\frac{\partial(\rho u)}{\partial x} + \frac{\partial(\rho h_x)}{\partial y}\right]_0 = 0 \tag{6.181}$$

According to formulas (6.179) and (6.181), it can be seen that to obtain a non-zero solution $tg\theta_0$, only the following formula holds, which is

$$\left(\frac{\partial u}{\partial y}\right)_0 = 0 \tag{6.182}$$

This condition is the flow separation criterion and the first necessary condition for the existence of the separation streamline, but it does not specify the second necessary condition for the separation streamline to leave the object surface. In other words, at the separation point 0, the separation streamline is required to leave the object surface, that is, $v > 0$. Since, at the separation point 0, $v_0 = 0$, $\left(\frac{\partial v}{\partial y}\right)_0 = 0$, the condition for $v > 0$ is

$$\left(\frac{\partial^2 v}{\partial y^2}\right)_0 > 0 \tag{6.183}$$

Calculate the partial derivative of Eq. (6.181) in the y-direction to obtain

$$
\begin{aligned}
\left(\frac{\partial^2 v}{\partial y^2}\right)_0 &= -\frac{\partial}{\partial y}\left(\frac{1}{\rho h_x}\right)_0 \left[\frac{\partial(\rho u)}{\partial x} + v\frac{\partial(\rho h_x)}{\partial y}\right]_0 \\
&\quad -\left(\frac{1}{\rho h_x}\right)_0 \left[\frac{\partial}{\partial y}\left(u\frac{\partial \rho}{\partial x}\right) + \frac{\partial \rho}{\partial y}\frac{\partial u}{\partial x} + \rho\frac{\partial^2 u}{\partial y \partial x} + v\frac{\partial^2(\rho h_x)}{\partial y^2} + \frac{\partial v}{\partial y}\frac{\partial(\rho h_x)}{\partial y}\right]_0 \\
&= -\left(\frac{1}{h_x}\right)_0 \left[\frac{\partial^2 u}{\partial y \partial x}\right]_0 \tag{6.184}
\end{aligned}
$$

From Eq. (6.183), the second necessary condition for boundary layer separation is obtained, namely

$$\left[\frac{\partial}{\partial x}\left(\frac{\partial u}{\partial y}\right)\right]_0 < 0 \tag{6.185}$$

In summary, the necessary conditions for separation of flow are

$$\begin{cases} \left(\dfrac{\partial u}{\partial y}\right)_0 = 0 \\ \left[\dfrac{\partial}{\partial x}\left(\dfrac{\partial u}{\partial y}\right)\right]_0 < 0 \end{cases} \tag{6.186}$$

According to the wall friction stress, $\tau_w = \mu \frac{\partial u}{\partial y}$, formula (6.186) can also be written as

$$\begin{cases} (\tau_w)_0 = 0 \\ \left(\dfrac{\partial \tau_w}{\partial x}\right)_0 < 0 \end{cases} \tag{6.187}$$

According to the same analysis, the flow on the surface of the reattachment after separation can be obtained. The necessary conditions for flow at reattachment

$$\begin{cases} \left(\dfrac{\partial u}{\partial y}\right)_0 = 0 \\ \left[\dfrac{\partial}{\partial x}\left(\dfrac{\partial u}{\partial y}\right)\right]_0 > 0 \end{cases} \tag{6.188}$$

The wall friction stress is expressed as

$$\begin{cases} (\tau_w)_0 = 0 \\ \left(\dfrac{\partial \tau_w}{\partial x}\right)_0 > 0 \end{cases} \tag{6.189}$$

6.8.3 Sufficient Conditions for Flow Separation

If the N-S equations are satisfied in the field of separation point 0, the flow velocity components u and v are required to be continuously differentiable near the 0 point, and there will be no singularities. However, it can be seen from the following analysis that for the flow described by the boundary layer equation, u and v are not analytic near the separation point. Therefore, when establishing the necessary conditions for the separation of flows, the L'Hospital law was adopted, and the requirements for physical quantities were relaxed, making them applicable to flows described by the N-S equations as well as those described by the boundary layer equations. In the following, we will discuss the problem under the condition that u and v are analytic functions. Using formula (6.186), u and v near the separation point 0 can be expressed as

$$
\begin{cases}
u = \left(\dfrac{\partial^2 u}{\partial x \partial y}\right)_0 xy + \dfrac{1}{2}\left(\dfrac{\partial^2 u}{\partial y^2}\right)_0 y^2 + \cdots \\[3mm]
v = \dfrac{1}{2}\left(\dfrac{\partial^2 v}{\partial y^2}\right)_0 y^2 + \cdots
\end{cases}
\tag{6.190}
$$

Substituting formula (6.190) into formula (6.172), we get

$$
\frac{1}{h_x}\frac{dy}{dx} = \frac{v}{u} = \frac{\left(\frac{\partial^2 v}{\partial y^2}\right)_0 y}{\left(\frac{2}{h_x}\frac{\partial^2 u}{\partial x \partial y}\right)_0 (h_x)_0 x + \left(\frac{\partial^2 u}{\partial y^2}\right)_0 y}
\tag{6.191}
$$

Introduce an independent variable ζ, let

$$
\begin{cases}
\dfrac{dy}{d\varsigma} = \left(\dfrac{\partial^2 v}{\partial y^2}\right)_0 y = dy \\[3mm]
\dfrac{h_x dx}{d\varsigma} = \left(\dfrac{2}{h_x}\dfrac{\partial^2 u}{\partial x \partial y}\right)_0 (h_x)_0 x + \left(\dfrac{\partial^2 u}{\partial y^2}\right)_0 y = a(h_x)_0 x + by
\end{cases}
\tag{6.192}
$$

Using the singularity theory of ordinary differential equations, if let

$$
\begin{aligned}
q &= \left(\frac{\partial^2 v}{\partial y^2}\right)_0 \cdot \left(\frac{2}{h_x}\frac{\partial^2 u}{\partial x \partial y}\right)_0 \\[3mm]
p &= -\left[\left(\frac{2}{h_x}\frac{\partial^2 u}{\partial x \partial y}\right)_0 + \left(\frac{\partial^2 v}{\partial y^2}\right)_0\right]
\end{aligned}
\tag{6.193}
$$

Then when $q > 0$, 0 point is a node or spiral point singularity, and when $p < 0$, the node or spiral point is unstable, and when $p > 0$, the node or spiral point is stable; when $q < 0$, 0 point is a saddle-point singularity. From the continuity equation

$$
\frac{\partial}{\partial y}\left[\frac{1}{h_x}\frac{\partial u}{\partial x} + \frac{\partial v}{\partial y}\right]_0 = 0, \quad \left(\frac{\partial^2 v}{\partial y^2}\right)_0 = -\left(\frac{1}{h_x}\frac{\partial^2 u}{\partial x \partial y}\right)_0
\tag{6.194}
$$

Substituting into Eq. (6.193), we can get

$$q = -2\left(\frac{1}{h_x}\frac{\partial^2 u}{\partial x \partial y}\right)_0^2 < 0 \qquad (6.195)$$

It can be seen that the 0 point is a saddle-point singularity. Since 0 is on the wall, the flow near the 0 point is a half-saddle point image. And because of $\left(\frac{1}{h_x}\frac{\partial^2 u}{\partial x \partial y}\right)_0 < 0$, the flow points to the 0A direction (as shown in Fig. 6.20), $\left(\frac{\partial^2 v}{\partial y^2}\right)_0 > 0$, the 0 point flow must point outward, and the zero point must be the separation point. When $\left(\frac{1}{h_x}\frac{\partial^2 u}{\partial x \partial y}\right)_0 > 0$, the flow points to 0 point, $\left(\frac{\partial^2 v}{\partial y^2}\right)_0 < 0$, the flow of 0 point points inward, and the 0 point is the reattachment point.

6.8.4 Flow Characteristics Near the Separation Point

1. **A streamline passing the separation point OA**

If formula (6.191) represents a streamline passing the separation point 0, then z and x in the formula should be on the streamline in question, and when (x, y) is very close to point 0, $\frac{1}{h_x}\frac{dy}{dx} = \left(\frac{1}{h_x}\frac{y}{x}\right)_0 = tg\theta_0$, where θ_0 represents the angle between the streamline at the separation point 0 and the x-axis. From Eq. (6.191), we get

$$tg\theta_0 = \frac{\left(\frac{\partial^2 v}{\partial y^2}\right)_0 tg\theta_0}{\left(\frac{2}{h_x}\frac{\partial^2 u}{\partial x \partial y}\right)_0 + \left(\frac{\partial^2 u}{\partial y^2}\right)_0 tg\theta_0} \qquad (6.196)$$

Therefore, we get

$$tg\theta_0\left[\left(\frac{\partial^2 u}{\partial y^2}\right)_0 tg\theta_0 + \left(\frac{2}{h_x}\frac{\partial^2 u}{\partial x \partial y}\right)_0 - \left(\frac{\partial^2 v}{\partial y^2}\right)_0\right] = 0 \qquad (6.197)$$

Let

$$\begin{cases} tg\theta_0 = 0 \\ tg\theta_0 = \dfrac{\left(\frac{\partial^2 v}{\partial y^2}\right)_0 - \left(\frac{2}{h_x}\frac{\partial^2 u}{\partial x \partial y}\right)_0}{\left(\frac{\partial^2 u}{\partial y^2}\right)_0} \end{cases} \qquad (6.198)$$

It can be seen that the streamline at the 0 point has two directions: one is along the object surface; the other has an inclination angle with the object surface, and the latter is the separation line 0A. Using Eq. (6.194) obtained from the continuous equation, the slope of the separation line 0A is

$$tg\theta_0 = \frac{-3\left(\frac{1}{h_x}\frac{\partial^2 u}{\partial x \partial y}\right)_0}{\left(\frac{\partial^2 u}{\partial y^2}\right)_0}, \quad tg\theta_0 = \frac{3\left(\frac{\partial^2 v}{\partial y^2}\right)_0}{\left(\frac{\partial^2 u}{\partial y^2}\right)_0} \tag{6.199}$$

Using the surface conditions, we can know from the N-S equations

$$\begin{cases} \left(\dfrac{\partial^2 u}{\partial y^2}\right)_0 = \dfrac{1}{\mu}\dfrac{1}{h_x}\dfrac{\partial p}{\partial x} \\[3mm] \left(\dfrac{\partial^2 v}{\partial y^2}\right)_0 = \dfrac{1}{\mu}\dfrac{\partial p}{\partial y} \end{cases} \tag{6.200}$$

Substituting formula (6.200) and the wall shear stress expression into formula (6.199), we obtain

$$tg\theta_0 = -3\frac{\left(\frac{\partial \tau_w}{\partial x}\right)_0}{\left(\frac{\partial p}{\partial x}\right)_0}, \quad tg\theta_0 = \frac{3\left(\frac{\partial p}{\partial y}\right)_0}{\left(\frac{1}{h_x}\frac{\partial p}{\partial x}\right)_0} \tag{6.201}$$

In the case of separate streams, due to $\left(\frac{\partial \tau_w}{\partial x}\right)_0 < 0$ and $tg\theta_0 > 0$, $\left(\frac{\partial p}{\partial x}\right)_0 > 0$ is required at the separation point, indicating that the separation point is in the reverse pressure gradient zone. In addition, in the separation point area, $\left(\frac{\partial p}{\partial y}\right)_0 > 0$.

2. The zero u line passing the separation point

Because there is a backflow zone downstream of the separation point, there must be a curve with velocity component $u = 0$, which is called the zero u line (the dotted line shown in Fig. 6.20). Now determine the slope of the zero u line, from Eq. (6.198) we can see

$$y\left[\left(\frac{\partial^2 u}{\partial x \partial y}\right)_0 x + \frac{1}{2}\left(\frac{\partial^2 u}{\partial y^2}\right)_0 y + \cdots\right] = 0 \tag{6.202}$$

Let the slope of the zero u line at 0 point be $tg\theta_u$, there is $tg\theta_u = \frac{1}{h_x}\frac{dy}{dx} = \frac{1}{h_x}\frac{y}{x}$, so the above formula (6.202) can be obtained

$$tg\theta_u\left[\left(\frac{1}{h_x}\frac{\partial^2 u}{\partial x \partial y}\right)_0 + \frac{1}{2}\left(\frac{\partial^2 u}{\partial y^2}\right)_0 tg\theta_u\right] = 0 \tag{6.203}$$

Therefore, we get

$$\begin{cases} tg\theta_u = 0 \\[3mm] tg\theta_u = \frac{-\left(\frac{2}{h_x}\frac{\partial^2 u}{\partial x \partial y}\right)_0}{\left(\frac{\partial^2 u}{\partial y^2}\right)_0} = \frac{2}{3}tg\theta_u \end{cases} \tag{6.204}$$

Explanation: A zero u line coincides with the object surface, which is naturally true, because the object surface adhesion condition gives $u = 0$. The other is inclined to the object surface, and its slope is 2/3 of the slope of the separation streamline 0A.

3. **The zero vorticity line passing the separation point**

In a two-dimensional flow, the vorticity is defined as

$$\Omega_z = \frac{1}{h_x}\frac{\partial v}{\partial x} - \frac{\partial u}{\partial y} \tag{6.205}$$

It can be seen that at the separation point

$$\Omega_{z0} = \left(\frac{1}{h_x}\frac{\partial v}{\partial x}\right)_0 - \left(\frac{\partial u}{\partial y}\right)_0 = 0 \tag{6.206}$$

And it indicates that the separation point is the zero vorticity point. Some literatures also use the zero vorticity criterion as the criterion for the separation point. The so-called zero vorticity line refers to the curve where the vorticity is zero. Now substituting Eq. (6.190) into Eq. (6.205), we get

$$\Omega_z = \frac{1}{h_x}\frac{\partial v}{\partial x} - \frac{\partial u}{\partial y} = -\left[\left(\frac{\partial^2 u}{\partial x \partial y}\right)_0 x + \left(\frac{\partial^2 u}{\partial y^2}\right)_0 y + \cdots\right] \tag{6.207}$$

Since $\Omega_z = 0$, the slope of the zero vorticity line at the separation point 0 can be obtained as

$$tg\theta_\Omega = \frac{1}{h_x}\frac{dy}{dx} = \frac{1}{h_x}\frac{y}{x} = -\frac{\left(\frac{1}{h_x}\frac{\partial^2 u}{\partial x \partial y}\right)_0}{\left(\frac{\partial^2 u}{\partial y^2}\right)_0} = \frac{1}{3}tg\theta_0 \tag{6.208}$$

Explanation: The slope of the zero vortex line passing through 0 is equal to 1/3 of the slope of the separation streamline 0A (shown in Fig. 6.21).

Fig. 6.21 A streamline passing through the separation point, zero u line, and zero vorticity line

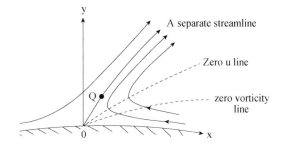

y

A separate streamline

Zero u line

zero vorticity line

Q

0

x

6.8.5 Singularity of Boundary Layer Equation (Goldstein Singularity)

The separation point is actually the singularity of the solution of the boundary layer equation (Goldstein singularity). The following proves this point. In the boundary layer object surface coordinates, the boundary layer equation of the incompressible fluid is

$$\begin{cases} \dfrac{\partial u}{\partial x} + \dfrac{\partial v}{\partial y} = 0 \\[2mm] u\dfrac{\partial u}{\partial x} + v\dfrac{\partial u}{\partial y} = -\dfrac{1}{\rho}\dfrac{\partial p}{\partial x} + v\dfrac{\partial^2 u}{\partial y^2} \\[2mm] -\dfrac{1}{\rho}\dfrac{\partial p}{\partial y} = 0 \end{cases} \tag{6.209}$$

Among them, assuming that the radius of curvature of the object surface is large, approximately $h_x = 1.0$, and the object surface condition is

$$y = 0, \quad u = v = 0 \tag{6.210}$$

From the derivation of the second formula of formula (6.209) along the y-direction, we get

$$u\frac{\partial^2 u}{\partial x \partial y} + v\frac{\partial^2 u}{\partial y^2} + \frac{\partial u}{\partial y}\left(\frac{\partial u}{\partial x} + \frac{\partial v}{\partial y}\right) = v\frac{\partial^3 u}{\partial y^3} \tag{6.211}$$

Using the continuous equation, we get

$$u\frac{\partial^2 u}{\partial x \partial y} + v\frac{\partial^2 u}{\partial y^2} = v\frac{\partial^3 u}{\partial y^3} \tag{6.212}$$

In seeking the partial derivative of y, we have

$$\frac{\partial u}{\partial y}\frac{\partial^2 u}{\partial x \partial y} + u\frac{\partial}{\partial x}\left(\frac{\partial^2 u}{\partial y^2}\right) + v\frac{\partial^3 u}{\partial y^3} + \frac{\partial v}{\partial y}\frac{\partial^2 u}{\partial y^2} = v\frac{\partial^4 u}{\partial y^4} \tag{6.213}$$

Using the surface conditions and the continuity equation, it can be known that on the surface, $u = v = 0$, $\frac{\partial v}{\partial y} = 0$. From the above formula, there is

$$\frac{\partial u}{\partial y}\frac{\partial^2 u}{\partial x \partial y} = v\frac{\partial^4 u}{\partial y^4} \tag{6.214}$$

Since, at the separation point, there is $\left(\dfrac{\partial u}{\partial y}\right)_0 = 0$, and

$$\mu\left(\frac{\partial^4 u}{\partial y^4}\right)_0 = \left(\frac{\partial^3 \tau_w}{\partial y^3}\right)_0 \neq 0 \tag{6.215}$$

Therefore,

$$\frac{\partial}{\partial x}\left(\frac{\partial u}{\partial y}\right)_0 = \infty \tag{6.216}$$

This shows that when solving the traditional boundary layer equations, once you enter the field of separation points, $x = 0$, $\frac{\partial}{\partial x}\left(\frac{\partial u}{\partial y}\right)_0$ appears singularity. In this case, the approximation of the boundary layer equation does not hold and the advance calculation is interrupted, indicating that the boundary layer equation cannot describe the separation point and its downstream viscous flow. From the further study of the flow characteristics at the separation point, from Eq. (6.214), we can see

$$\left[\frac{\partial}{\partial x}\left(\frac{\partial u}{\partial y}\right)^2\right]_0 = 2v\left(\frac{\partial^4 u}{\partial y^4}\right)_0 \tag{6.217}$$

Because, at the separation point, $\left(\frac{\partial u}{\partial y}\right)_0^2 = 0$, if let $\left(\frac{\partial u}{\partial y}\right)^2$ to perform Taylor series expansion at separation point 0, then

$$\left(\frac{\partial u}{\partial y}\right)^2 = \left(\frac{\partial u}{\partial y}\right)_0^2 + \left\{\frac{\partial}{\partial x}\left[\left(\frac{\partial u}{\partial y}\right)^2\right]\right\}_0 x + \cdots = 2v\left(\frac{\partial^4 u}{\partial y^4}\right)_0 x + \cdots \tag{6.218}$$

In the upstream of the separation point $x \leq 0$, it can be seen from the above formula that there must be

$$\left(\frac{\partial^4 u}{\partial y^4}\right)_0 < 0 \tag{6.219}$$

Let

$$2v\left(\frac{\partial^4 u}{\partial y^4}\right)_0 = -k^2 \tag{6.220}$$

In the formula, $k > 0$. This style (6.218) becomes

$$\left(\frac{\partial u}{\partial y}\right)^2 = -2v\left(\frac{\partial^4 u}{\partial y^4}\right)_0 |x| = k^2|x|$$

$$\left(\frac{\partial u}{\partial y}\right) = \sqrt{-2v\left(\frac{\partial^4 u}{\partial y^4}\right)_0} |x|^{\frac{1}{2}} = k|x|^{\frac{1}{2}} \tag{6.221}$$

Taking the derivative of the above formula along the x-direction, we get

$$\left(\frac{\partial^2 u}{\partial x \partial y}\right) = -\frac{1}{2}\sqrt{-2v\left(\frac{\partial^4 u}{\partial y^4}\right)_0} |x|^{-\frac{1}{2}} = -\frac{1}{2}k|x|^{-\frac{1}{2}} \qquad (6.222)$$

Indicates: when the incoming flow approaches the separation point from upstream, $\left(\frac{\partial^2 u}{\partial x \partial y}\right)$ tends to infinity according to $|x|^{-\frac{1}{2}}$, This singularity is called Goldstein singularity, and the two-dimensional separation point is called Goldstein singularity.

Near the upstream of the separation point $(x < 0)$

$$u = \left(\frac{\partial u}{\partial y}\right) y = \sqrt{-2v\left(\frac{\partial^4 u}{\partial y^4}\right)_0} |x|^{\frac{1}{2}} y = k|x|^{\frac{1}{2}} y + O(y^2) \qquad (6.223)$$

From the continuous equation, we can get

$$v = -\int \frac{\partial u}{\partial x} dy = -\frac{1}{4}\sqrt{-2v\left(\frac{\partial^4 u}{\partial y^4}\right)_0} |x|^{-\frac{1}{2}} y^2 = -\frac{1}{4}k|x|^{-\frac{1}{2}} y^2 + O(y^3) \quad (6.224)$$

Therefore, the streamline near the separation point (upstream, $x < 0$) is

$$\frac{dy}{dx} = \frac{v}{u} = -\frac{1}{4}\frac{y}{x} \qquad (6.225)$$

Integrate the above formula to get

$$y^4 x = C \qquad (6.226)$$

It indicates the streamline near $x < 0$ (as shown in Fig. 6.22), which has been verified by numerical calculations.

Summarizing the discussion above, we get.

Fig. 6.22 The streamline shape in the upstream area of the separation point

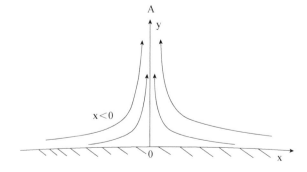

(1) For the actual viscous flow described by the N-S equations, physical quantities such as u and v are analytic at the separation point. However, the boundary layer approximately fails near the separation point, which makes the flow described by the boundary layer equation have Goldstein singularity.

(2) Starting from the Prandtl separation image, it is necessary and sufficient to establish the separation criterion based on the L'Hospital law. It is applicable to the flow described by the N-S equation and the boundary layer equation.

(3) For the flow described by the boundary layer equation, the separation line is perpendicular to the object surface, and the streamline near the separation point takes the y-axis as the asymptote. When $y > 0$, v at $x = 0$ appears singularity. But the flow described by the N-S equation is non-singular.

6.8.6 Critical Point Analysis of Two-Dimensional Steady Separated Flow

1. Autonomous system, phase plane, critical point

In the theory of differential equations, for first-order ordinary differential equations

$$\begin{cases} \dfrac{dx}{dt} = f(x, y) \\ \dfrac{dy}{dt} = g(x, y) \end{cases} \tag{6.227}$$

If the independent variable only appears in the form of a derivative, and does not appear in f and g, such a system of equations is called as an autologous system, and the physical system controlled by it is called an autologous system.

The plane composed of rectangular coordinates x and y is called the phase plane. According to the basic theorem of uniqueness of simultaneous differential equations, when formula (6.235) $f(x, y)$ and $g(x, y)$ are analytic functions, under given initial conditions, the solution of formula (6.235) is unique. However, at some points, $f(x, y) = 0$ and $g(x, y) = 0$, these points are called critical points, also called zero points or singularities in the phase plane. The other points are called as ordinary points or constant points. If there is no other critical point near a given critical point, the critical point is called as an isolated singularity.

2. Linear approximation of the area near the critical point

In the Cartesian coordinate system, the streamline equation is

$$\frac{dy}{dx} = \frac{v}{u} = \frac{v(x, y)}{u(x, y)} \tag{6.228}$$

In the formula, u and v are the velocity components in the x and y directions. The so-called critical point refers to the point where $u(x, y) = v(x, y) = 0$. Suppose 0 is the critical point, and the origin of the coordinates is 0, so u and v near 0 can be expanded into

$$\begin{cases} u = \left(\dfrac{\partial u}{\partial x}\right)_0 x + \left(\dfrac{\partial u}{\partial y}\right)_0 y + \cdots \\ v = \left(\dfrac{\partial v}{\partial x}\right)_0 x + \left(\dfrac{\partial v}{\partial y}\right)_0 y + \cdots \end{cases} \tag{6.229}$$

At 0 point, using the continuous equation $\left(\frac{\partial v}{\partial y}\right)_0 = -\left(\frac{\partial u}{\partial x}\right)_0$, the above formula can be expressed as

$$\begin{cases} u = \left(\dfrac{\partial u}{\partial x}\right)_0 x + \left(\dfrac{\partial u}{\partial y}\right)_0 y + \cdots \\ v = \left(\dfrac{\partial v}{\partial x}\right)_0 x - \left(\dfrac{\partial u}{\partial x}\right)_0 y + \cdots \end{cases} \tag{6.230}$$

Substituting formula (6.230) into formula (6.228), we get

$$\frac{dy}{dx} = \frac{v}{u} = \frac{cx + dy}{ax + by} \tag{6.231}$$

$$\begin{cases} \dfrac{dx}{dt} = ax + by + \cdots \\ \dfrac{dy}{dt} = cx + dy + \cdots \end{cases} \tag{6.232}$$

where the coefficients are

$$a = \left(\frac{\partial u}{\partial x}\right)_0, \ b = \left(\frac{\partial u}{\partial y}\right)_0, \ c = \left(\frac{\partial v}{\partial x}\right)_0, \ d = \left(\frac{\partial v}{\partial y}\right)_0 = -a \tag{6.233}$$

Assume that the unusual solution of Eq. (6.232) is

$$\begin{cases} x = Re^{\lambda t} \\ y = Se^{\lambda t} \end{cases} \tag{6.234}$$

In the formula, R and S are coefficients, and λ is the characteristic value. Substituting formula (6.234) into formula (6.232), we get

$$\begin{cases} (a - \lambda)R + bS = 0 \\ cx + (d - \lambda)S = 0 \end{cases} \tag{6.235}$$

Then its characteristic equation is

$$\begin{vmatrix} a - \lambda & b \\ c & d - \lambda \end{vmatrix} = 0 \tag{6.236}$$

That is,

$$\lambda^2 - (a + d)\lambda + (ad - cb) = 0 \tag{6.237}$$

Let

$$p = -(a + d) = -\left[\left(\frac{\partial u}{\partial x}\right)_0 + \left(\frac{\partial v}{\partial y}\right)_0\right]$$

$$q = ad - cb = \left(\frac{\partial u}{\partial x}\right)_0 \left(\frac{\partial v}{\partial y}\right)_0 - \left(\frac{\partial u}{\partial y}\right)_0 \left(\frac{\partial v}{\partial x}\right)_0 \tag{6.238}$$

When $\Delta = p^2 - 4q \neq 0$, characteristic values are

$$\lambda_{1,2} = \frac{-p \pm \sqrt{\Delta}}{2} = \frac{-p \pm \sqrt{p^2 - 4q}}{2} \tag{6.239}$$

The general solution of formula (6.232) is

$$\begin{cases} x(t) = R_1 e^{\lambda_1 t} + R_2 e^{\lambda_2 t} \\ y(t) = S_1 e^{\lambda_1 t} + S_2 e^{\lambda_2 t} \end{cases} \tag{6.240}$$

The corresponding streamline equation is

$$\frac{dy}{dx} = \frac{\lambda_1 S_1 e^{\lambda_1 t} + \lambda_2 S_2 e^{\lambda_2 t}}{\lambda_1 R_1 e^{\lambda_1 t} + \lambda_2 R_2 e^{\lambda_2 t}} \tag{6.241}$$

3. The nature of the critical point

(1) Node

If $\Delta > 0, q > 0$, this critical point is a node. At this time, λ_1 and λ_2 are two unequal real roots with the same sign.

(2) Saddle point

If $\Delta > 0, q < 0$, this critical point is a saddle point. At this time, $\lambda_1 \lambda_2 = q$ is the real root of two different signs. If $\Delta = 0$, this critical point is a degenerate node. In this case, $\lambda_1 = \lambda_2 = $ Real.

(3) Focus point

If $\Delta < 0$, this critical point is the focus. In this case, λ_1 and λ_2 are conjugate complex numbers. If $a < 0$, $p > 0$ it is a stable spiral point (contracted spiral point). If $a > 0$, $p < 0$, it is an unstable spiral point (extended spiral point).

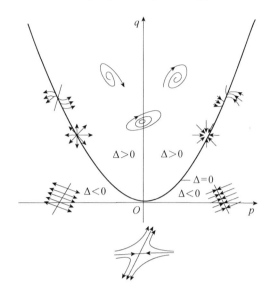

(4) **Center point**

If $\Delta < 0, p = 0$, this critical point is the center point. The solution in the phase plane
is an ellipse centered on the critical point (Fig. 6.23).

6.9 Introduction to the Steady Three-Dimensional Separated Flow Over any Object

6.9.1 Overview

Compared with the two-dimensional separation flow, the three-dimensional steady
separation flow of viscous fluid is much more complicated. At present, there is a lot
of work to be done in both theory, experiment and numerical calculation. In terms of
experimental research, obtaining the flow spectrum of the surface limit streamline and
the motion of the vortex in the separation zone through the flow field visualization
technology is still the most important method for the study of three-dimensional
separated flow. Especially in terms of theoretical research, the various concepts of
three-dimensional separation, the topological law of the circumfluence map and the
critical point theory, to a large extent, are established with the help of experimental
photos of wind tunnels and water tunnels. In terms of numerical calculations, in
the early 1970s, Wang Guozhang first used the boundary layer equation to calculate
the separation line of the three-dimensional ellipsoid. Since then, work in this area
has made great progress. For smaller separation areas, such as bubbles appearing
on the leeward side of the aircraft and some small corner flows, the N-S method

or simplified N-S equation can be effectively used to numerically simulate the flow characteristics of the separation area. For turbulent flow separation, the concept of time averaging can be used to study time-averaged flow separation. However, the problem of unsteady separation and flow is currently in the exploratory stage due to its complexity.

6.9.2 Limit Streamlines and Singularities

In the object surface coordinate system (as shown in Fig. 6.24, x and y are along the object surface, and the z-axis is perpendicular to the object surface), expand the streamline close to the object surface according to the velocity component near the object surface, and we get

$$\frac{h_y dy}{h_x dx} = \frac{v_{z=0} + \left(\frac{\partial v}{\partial z}\right)_{z=0} z + \frac{1}{2}\left(\frac{\partial^2 v}{\partial z^2}\right)_{z=0} z^2 + \cdots}{u_{z=0} + \left(\frac{\partial u}{\partial z}\right)_{z=0} z + \frac{1}{2}\left(\frac{\partial^2 u}{\partial z^2}\right)_{z=0} z^2 + \cdots} \quad (6.242)$$

where h_x and h_z are the Lame coefficient of the object plane coordinate system along the coordinate axis. When $z \to 0$, for a stationary wall, $u = v = 0$, then

$$\left(\frac{h_y dy}{h_x dx}\right)_{z=0} = \left(\frac{\partial v/\partial z}{\partial u/\partial z}\right)_{z=0} = \left(\frac{\tau_y}{\tau_x}\right)_{z=0} \quad (6.243)$$

This is the equation of the limit streamline, which shows that in the Newtonian fluid, the limit streamline is the wall friction line.

If the parameter t is introduced, formula (6.243) can be written as

$$\frac{h_y dy}{dt}\bigg|_{z=0} = \tau_y(x, y)\big|_{z=0} \quad (6.244)$$

$$\frac{h_x dx}{dt}\bigg|_{z=0} = \tau_x(x, y)\big|_{z=0} \quad (6.245)$$

At those places where $\tau_x(x_s, y_s) = \tau_y(x_s, y_s) = 0$, the point (x_s, y_s) is called the critical point (or zero point, also called the singularity). The above formula can be

Fig. 6.24 Object coordinate system

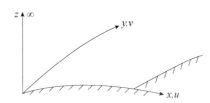

expanded by Taylor series at the critical point as

$$\frac{dy}{dt} = a(x - x_s) + b(y - y_s) \tag{6.246}$$

$$\frac{dx}{dt} = c(x - x_s) + d(y - y_s) \tag{6.247}$$

Among them, a, b, c, and d are coefficients. If $p = -(a + d)$ and $q = ad - bc$, the nature of the critical point in the limit streamline can be determined as

(1) When $q < 0$, it is a saddle point;
(2) When $q > 0$, but $\Delta = p2 - 4q > 0$, it is a node. Among them, $p > 0$ is a stable node, and $p < 0$ is not.

Stable node

(3) When $q > 0$, but $\Delta < 0$, it is the focus. Where $p > 0$ is a stable focus, and $p < 0$ is an unstable focus;
(4) When $\Delta = 0$, it degenerates into a node, where $p > 0$ is a stable node, and $p < 0$ is an unstable node;
(5) When $\Delta < 0$, but $p = 0$, it is the center.

Figure 6.25 shows the critical point distribution of the limit streamline on the leeward side of the missile head at a certain angle of attack.

Fig. 6.25 Flow spectrum of the leeward surface of the missile head. F-focus; N-node; S-saddle point

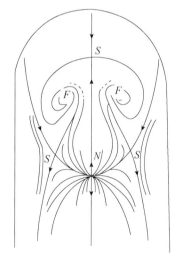

6.9.3 The Concept of Three-Dimensional Separation

As shown in Fig. 6.26, take a rectangular cross-section flow tube near the wall. A_1A_2 and B_1B_2 are friction lines (limit flow lines), and $A_1'A_2'$ and $B_1'B_2'$ are flow lines in the boundary layer. The initial width of the flow tube is n_1 and the height is h_1. From the continuity equation, the volume flow through any section of the flow tube is

$$\dot{Q} \approx hnV_{\mathrm{a}} \tag{6.248}$$

where V_{a} is the average velocity through the section of the flow tube, which can be written as

$$V_{\mathrm{a}} = \frac{1}{2}h\sqrt{\left(\frac{\partial u}{\partial z}\right)^2 + \left(\frac{\partial v}{\partial z}\right)^2}\Bigg|_{z=0} \tag{6.249}$$

Then, we get

$$\dot{Q} = \frac{1}{2v}\sqrt{\tau_x^2 + \tau_y^2}\Big|_{z=0} h^2 n \tag{6.250}$$

According to the conservation of mass, in the same flow tube Eq. (6.248) is a constant, so when $\tau_x = \tau_y = 0$, h will tend to infinity, that is, the thickness of the boundary layer will tend to infinity, and the flow will separate at this time, which happens to be the starting point of the critical point. According to formula (6.250), when $\sqrt{\tau_x^2 + \tau_y^2}\Big|_{z=0}$ is not equal to 0, since the flow lines A_1A_2 and B_1B_2 are infinitely

Fig. 6.26 The concept of three-dimensional separation

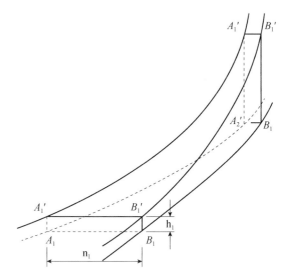

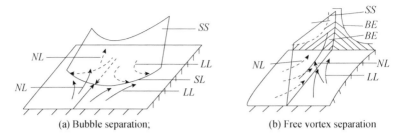

(a) Bubble separation;   (b) Free vortex separation

Fig. 6.27 Maskell's three-dimensional separation concept LL-limit streamline; SL-separation line; SS-separation surface; NL-non-viscous streamline; 1-viscous zone; BE-outer edge of viscous layer. **a** Bubble separation; **b** Free vortex separation

close, that is, when $n \to 0$, $h \to$ infinity, the separation starting point of the flow at this time is not the critical point, but the normal point.

Obviously, when the boundary layer leaves the object surface, a separation surface will be produced. The intersection line between the separation surface and the object surface is called the separation line.

Lighthill derived formula (6.250), and he defined the separation line as a section of friction line from the beginning of the critical point and except for the critical point, the friction lines on both sides of the separation line will be the separation line As the asymptote.

Maskell divides the three-dimensional separation into two types: bubble separation and free vortex separation, and defines the separation line as the envelope of the limit streamline. The bubble separation starts from the critical point, while the free vortex separation starts from the normal point, as shown in Fig. 6.27.

Inspired by the concept of Maskell's three-dimensional separation, Wang Guozhang put forward the concepts of closed separation and open separation through a large number of experiments and numerical simulations. Among them, the closed separation is consistent with the usual two-dimensional separation concept. The separation lines on both sides of the object surface are connected in the front of the leeward, and pass through the singularity of the limit streamlines, and the limit streamlines on both sides of the separation line are converged on the separation line., But the limit streamline starting from the previous stagnation point can only be gathered at the separation line from one side, as shown in Fig. 6.28a. Open separation is different. The separation lines on both sides of the object surface are not closed on the leeward front. They start from the normal point, that is, the wall friction vector at the separation starting point is not zero. And the limit streamline emitted by the front stagnation point will be gathered at the separation line on both sides of the separation line, that is, part of the limit streamline can extend to the downstream of the separation line without separation, as shown in Fig. 6.28b. This is widely recognized.

Peake and Tobak called open separation as Local Separation, but advocated starting from the previous stagnation point in order to maintain Lighthill's definition of separation line. They call the separation with the critical point as the starting point as global separation. The reason is that according to the topological point of

(a) Closed separation   (b) Open separation

Fig. 6.28 Wang Guozhang's three-dimensional separation concept (SL-separation line)

view, a pattern of unseparated flow can be transformed into a partially separated flow through continuous deformation, but it cannot become a pattern of overall separated flow. Their definition is essentially the same as that of Wang Guozhang's open and closed separation. They are mainly defined from a different perspective. They also pointed out a type of focus separation, which also belongs to the overall separation, as shown in Fig. 6.29.

The Chinese scholar Zhang Hanxin has studied the characteristics of various separated streamlines. He concluded that starting from the N-S equation, the separation line itself is a part of the surface friction line. In other words, this is the asymptote of the adjacent friction line. Starting from the boundary layer equation, the separation line is the envelope of the friction line of the object surface. There are four basic forms of separation line, namely the saddle point start (as shown in Fig. 6.30a), the normal point start (as shown in Fig. 6.30b), and the node or focus is the end point (as shown in Fig. 6.30c). The normal point is the end (as shown in Fig. 6.30d). The above-mentioned basic forms can form separate flows in various poses. Among them, the separation line extends from the saddle point to both sides (one side ends with a normal point, and the other side ends with a focal point or node), similar to the focal separation of Pique and Toback. The separation line starts from the saddle point and ends at the node or focal point, which conforms to the definition of Lighthill

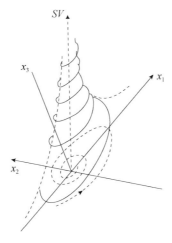

Fig. 6.29 Three-dimensional separation of focus types (SV-eigenvector)

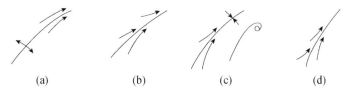

Fig. 6.30 Four basic forms of separation lines

separation line. The separation line starts from the normal point and is similar to the open separation proposed by Wang Guozhang.

6.9.4 Topological Law of Three-Dimensional Separation

Topology is a branch of mathematics. The object of discussion is the property that geometric figures undergo single-valued and continuous transformations in both the forward and inverse directions without changing. This transformation is called topological transformation, and this property is called topological property. Topology is more esoteric; and here, only the topological properties of graphs in two-dimensional real space are studied. In fact, it still belongs to the critical point theory in the qualitative theory of differential equations. We only give topological laws without proofs.

(1) In a three-dimensional flow field, any cross-section can be taken, then the velocity at any point on the cross-section can be decomposed into two components, the vertical cross-section and the tangent to the cross-section, the latter is called the cross-sectional velocity. A vector curve family can be constructed in the section, so that the curves are tangent to the section velocity. These curves are called section streamlines, and the vector field composed of this section velocity is called the section flow field. Obviously, only when the cross-section is orthogonal to the object surface, the cross-sectional contour line (that is, the line of intersection with the object surface) is the cross-sectional streamline. The point where the section velocity is zero is called the singularity (i.e. critical point), and the zero velocity point on the profile of the section is called the semi-singularity. The reason is that half of the flow attached to the point is in the surface of the object, and the semi-singularity is impossible. It is the focus and the center. Generally speaking, in the cross-sectional flow field, except for individual cases, there are central points, but they are all nodes, focal points and saddle points, and satisfy

$$\left(\sum N + \sum N' \right) - \left(\sum S + \frac{1}{2} \sum S' \right) = 1 - n \qquad (6.251)$$

Fig. 6.31 Flow around a
plane-cylinder combination
(A-reattach point;
S-separation point; N-node)

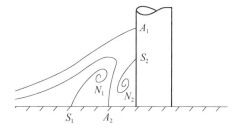

In the formula, ΣN and ΣS respectively represent the total number of nodes (including focal points) and the total number of saddle points in the cross-sectional flow field. $\Sigma N'$ and $\Sigma S'$ represent the total number of half-nodes and the total number of half-saddle points, respectively, $n = 1$ is a single connected domain, $n = 2$ is a complex connected domain. This law was first discovered and proved by Hunt. Figure 6.31 shows the flow spectrum of the plane-cylindrical combination flowing around the symmetry plane, and the double vortex separation structure appears due to the height of the cylinder. Among them, there are two vortices in front of the cylinder in the double vortex structure, one is the separation main vortex N_1, the other is the secondary separation vortex N_2, S_1 is the main separation point, A_1 is the reattachment point, S_2 and A_2 are the secondary separation points, respectively. And then attach a point. In this example, $n = 1$, $\Sigma N = 2$, $\Sigma N' = 0$, $\Sigma S = 0$, $\Sigma S' = 4$, and the result satisfies formula (6.259).

Figure 6.32 is the flow pattern of the AA section in the flow around the cone at the medium angle of attack. In the figure, S_1, A_1 are the main separation point and reattachment point, S_2, A_2 are the secondary separation point and reattachment point, N_1, N_3 are the main separation vortex, N_2 and N_4 are the second separation vortex, and A is the leading edge stagnation point (Half-saddle point), S is a saddle point, $n = 2$, then $\Sigma N = 4$, $\Sigma N' = 0$, $\Sigma S = 1$, $\Sigma S' = 8$, so formula (6.251) is satisfied.

(2) On the surface of the object, the critical point of the surface friction force line equation (or the limit streamline equation) satisfies

$$\sum N - \sum S = 2(2 - n) \tag{6.252}$$

For a single connected surface, $n = 1$. This law was first applied to separated flows by Lighthill.

(3) On the separation line, nodes and saddle points are alternately distributed, and the total number of nodes (including focal points) differs from the total number of saddle points by at most 1.

Fig. 6.32 Flow around the
AA section of a cone at a
large angle of attack
(*A*-reattachment point;
S-separation point; *N*-node)

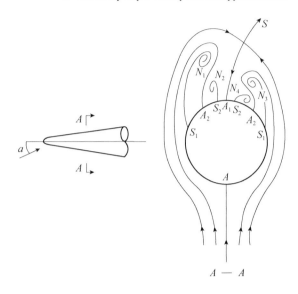

6.10   Resistance Over Objects

6.10.1   *The Resistance Over Any Object*

For a two-dimensional flow around an arbitrary object, the forces acting on any micro-segment on the surface of the object include pressure force and shear force. The projection of these acting on the flow direction is the resistance on the micro-segment, as shown in Fig. 6.33, which is

$$dD = (p \sin \theta + \tau_0 \cos \theta)ds \tag{6.253}$$

$$D = \oint (p \sin \theta + \tau_0 \cos \theta)ds = D_f + D_p$$

$$D_p = \oint (p \sin \theta)ds, \quad D_f = \oint (\tau_0 \cos \theta)ds \tag{6.254}$$

Fig. 6.33 The resistance of
any object to flow around

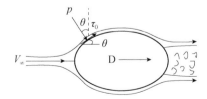

It shows that the fluid resistance acting on any object can be divided into the resistance produced by D_p due to the difference in surface pressure, which is called differential pressure resistance. Especially after the separation of the flow around the object, since the pressure in the separation zone is smaller than the pressure on the windward side, there is differential pressure resistance after integration, which is also called shape resistance because it has a direct relationship with the shape of the object surface. The resistance produced by D_f due to surface shear stress is called surface friction resistance.

Generally speaking, for objects placed downstream such as slender bodies, when there is no separation (minor separation) in the boundary layer of the circumfluence, the friction resistance is the main resistance in the circumfluence of the object. For the flow around the blunt body, the boundary layer of the flow is separated, and the pressure difference resistance will become the main one. If it is used to characterize the resistance of the flow energy and the windward area A of the circumfluence object, then there is

$$D = \frac{1}{2}\rho V_\infty^2 A C_D \tag{6.255}$$

where C_D is the drag coefficient of the circumfluence object, and A is the projected area (windward area) of the object perpendicular to the direction of the incoming flow. Similarly, for differential pressure resistance and friction resistance, they can be written as

$$D_p = \frac{1}{2}\rho V_\infty^2 A_p C_p, \quad D_f = \frac{1}{2}\rho V_\infty^2 A_f C_f \tag{6.256}$$

where C_f is the frictional resistance coefficient, C_p is the differential pressure resistance coefficient; A_f is the projected area along the flow direction, and A_p is the projected area perpendicular to the flow direction. Experiments show that for viscous fluids, the resistance of the object is related to the viscosity of the fluid, the shape of the object, the speed of the incoming flow, etc., especially when the object is flowing at low speed, the main parameter that affects the resistance coefficient is the flow Reynolds number, namely

$$C_D = f(\text{Re}) = f\left(\frac{\rho V_\infty d}{\mu}\right) \tag{6.257}$$

Figure 6.34 shows the distribution of the pressure coefficient on the wall of the flow around the cylinder under different incoming Reynolds numbers. The drag coefficients and corresponding incoming Reynolds numbers of objects of different shapes such as plates and cylinders are shown in Fig. 6.35.

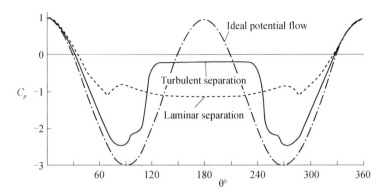

Fig. 6.34 Pressure distribution of flow around a cylinder under different Reynolds numbers

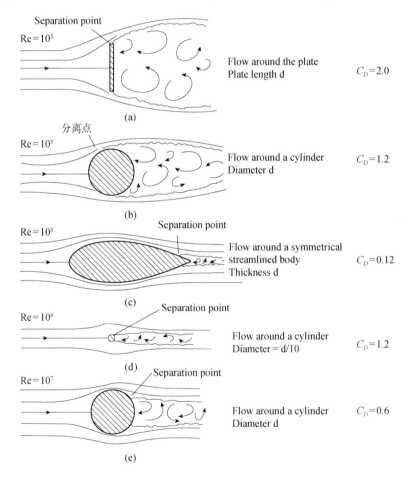

Fig. 6.35 The drag coefficient of circumfluence objects with different shapes and incoming Reynolds numbers

6.10.2 Two-Dimensional Flow Resistance Around a Cylinder

As we all know, the flow around a cylinder is a complicated separated flow problem. Under different numbers of circumfluence Re, circumfluence flow pattern, separation angle, vortex shedding frequency, and drag coefficient are all different. When Re < 150–300, it is laminar flow separation. At this time, the boundary layer is laminar flow. After separation, the shear layer and vortex are both laminar flow. After the cylinder is laminar flow vortex street, the separation frequency of separation vortex is

$$\text{St} = \frac{fd}{V_\infty} = 0.212(1 - 21.2/\text{Re}) \qquad (6.258)$$

In the formula, St is the Strouhal number (Czech physicist Vincenc Strouhal, 1850–1922). In 1878 when studying the sound of the string wind, it represents the infinite shedding of the separation vortex of the flow around the cylinder Gang frequency. The relationship between the shedding frequency of the separation vortex of the flow around a cylinder and the Reynolds number is shown in Fig. 6.36.

When $300 < \text{Re} < 10^5$ is a subcritical zone, the boundary layer on the surface of the cylinder is laminar flow, but the shear layer and vortex are turbulent after separation, and the turbulent vortex street behind the cylinder, drag coefficient $C_d \approx 1 - 1.2$, St ≈ 0.2 separation angle $\theta_s \approx \pm 80°$ (calculated from the front stagnation point). When $10^5 < \text{Re} < 3.5 \times 10^5$ is a subcritical zone, the drag coefficient is reduced from 1.2 to 0.4. The laminar boundary layer on the surface of the cylinder is separated and then transformed into turbulent flow, and then attached to form bubbles. The boundary layer transformed into turbulent flow improves the ability to resist reverse pressure gradient. The position of the separation point is increased from 80° in the laminar flow state to (110° ~ 120°), St ≈ 0.2 – 0.7. When $3.5 \times 10^5 < \text{Re} < 3.5 \times 10^6$ is a subcritical zone, the drag coefficient is increased from 0.4 to 0.7–0.8. The boundary layer around the cylinder is turbulent except for the small area of the front stagnation point. The boundary layer separation is turbulent separation, and the

Fig. 6.36 The relationship between the Reynolds number of the flow around a cylinder and St

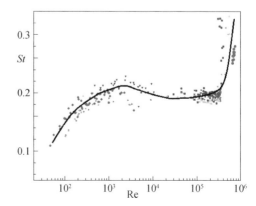

Fig. 6.37 The relationship
between the resistance
coefficient of the flow around
a cylinder and the Reynolds
number

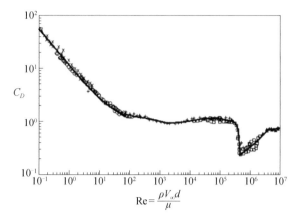

separation angle $\theta_s \approx \pm 110°$, St $\approx 0.2 - 0.7$. When Re $> 3.5 \times 10^6$ is a subcritical zone, the drag coefficient becomes constant again, about 0.7–0.8, the separation angle $\theta_s \approx \pm 110°$, St $\approx 0.3 - 0.4$. The relationship between the Reynolds number of the flow around a cylinder and the drag coefficient is shown in Fig. 6.37.

6.11 Aircraft Drag and Drag Reduction Technology

6.11.1 Composition of Aircraft Drag

As we all know, when the airflow bypasses the aircraft, the drag experienced by the aircraft is defined as the integral of the component force of the resultant force of the airflow acting on the surface of the aircraft and the frictional shear stress in the direction of the incoming flow (as shown in Fig. 6.38). The integral of the force component perpendicular to the direction of the incoming flow is called lift. When the aircraft is cruising, the gravity and lift of the aircraft are balanced, and the thrust and drag of the aircraft engine are balanced. Because the pressure and friction shear stress on the aircraft surface are related to the aircraft's flight speed, attitude angle, aircraft size, surface shape, and roughness, etc., the drag of the aircraft must be affected by these factors. Aircraft drag can be divided into two categories from a major perspective. One is the drag generated by the integral projection of the pressure in the incoming flow direction; the other is the drag generated by the integral surface friction stress, which is called frictional resistance. Specifically, according to the main reason for the resistance, the resistance obtained from the surface pressure integration is divided into the induced drag caused by the trailing edge of the wing dragging the free wake vortex to induce the downwash, and the differential pressure drag caused by the different shape of the aircraft (including wing body interference resistance, bottom resistance, resistance around exposed parts, etc.), and resistance

Fig. 6.38 Schematic diagram of the force on the aircraft during cruise

caused by shock waves generated by aircraft flying at speeds above high subsonic speeds.

In the direction parallel to the incoming flow, the thrust generated by the engine is balanced with the drag of the aircraft; in the direction perpendicular to the incoming flow, the lift generated by the wing is balanced with the aircraft's own gravity.

If the flow velocity and the characteristic area of the wing are used to express the resistance, it can be written as

$$D = \frac{1}{2}\rho_\infty V_\infty^2 S C_D \tag{6.259}$$

where D is the total drag of the aircraft, ρ_∞ is the air density, V_∞ is the incoming flow velocity, S is the characteristic area of the wing, and C_D is the total drag coefficient. According to the specific causes of aircraft drag, the total drag coefficient can be subdivided into

$$C_D = C_f + C_{dp} + C_i + C_{sw} \tag{6.260}$$

where C_f is the friction resistance coefficient, C_{dp} is the differential pressure resistance coefficient caused by different viscous boundary layers, C_i is the induced resistance coefficient, and C_{sw} is the shock wave resistance coefficient. In aircraft design, the sum of frictional resistance and viscous pressure difference resistance is also called parasitic resistance, or waste resistance. The range of the aircraft can be estimated using the Breguet relation, which is

$$R = \frac{C_L}{C_D} \frac{V_\infty}{\text{SFC}} \ln\left(\frac{W_L + W_F}{W_L}\right) \tag{6.261}$$

where R is the range, C_L/C_D is the lift-to-drag ratio, SFC is the fuel consumption ratio (the fuel required to generate a unit thrust per unit time), W_F is the total fuel weight, and W_L is the basic weight. It can be seen from the above expression that the larger the lift-to-drag ratio, the farther the aircraft's range will be. Therefore, reducing the aircraft drag will directly help improve flight performance. Studies have shown that reducing aircraft drag can increase range, reduce take-off weight, increase cruise lift-to-drag ratio, save fuel, increase payload, and reduce aircraft direct operating costs. In addition, by reducing resistance and reducing fuel exhaust emissions, thereby reducing air pollution.

According to relevant data, for a large aircraft with high subsonic speed during cruise, the surface frictional resistance of the aircraft accounts for 50% of the total resistance, the induced resistance accounts for 30%, the shock wave resistance accounts for 5%, and the differential pressure resistance accounts for 15%. Figures 6.39 and 6.40 respectively show the drag coefficients of a typical car and a large aircraft, but it should be noted that the characteristic area for the resistance coefficient of a car is the maximum windward area, while the characteristic area of an aircraft is the exposed area of the wing. The related technologies and flow mechanisms for reducing frictional resistance, induced resistance and shock wave resistance will be introduced below.

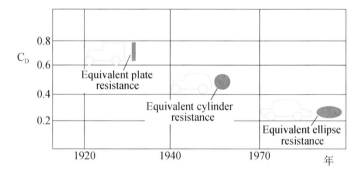

Fig. 6.39 Drag coefficients of cars with different shapes

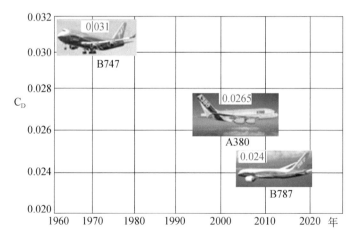

Fig. 6.40 Drag coefficients of different aircraft

6.11.2   *Technology to Reduce Laminar Flow Resistance*

First consider the surface frictional resistance of the aircraft, which accounts for the largest proportion of the total resistance. The frictional resistance is directly related to the boundary layer of the aircraft surface. The boundary layer is caused by the viscosity of the air and the relative motion of the aircraft and the air. Moreover, depending on the Reynolds number, a transition from laminar flow to turbulent flow also occurs in the boundary layer. If the Reynolds number Re_x is calculated with the flow direction length x of the object surface as the characteristic length and the boundary layer outflow velocity V_∞ as the characteristic velocity, the experiment finds that the Reynolds number for the transition of the laminar boundary layer lies between 3.5×10^5 and 3.5×10^6 for the flow around a flat plate. The frictional shear stress of the aircraft surface is related to the flow regime in the boundary layer. Generally, the frictional shear stress of the laminar boundary layer is 1/7–1/8 of the frictional shear stress of the turbulent boundary layer, as shown in Figs. 6.41 and 6.42. Therefore, the best way to reduce drag is to delay the transition of the boundary layer and keep laminar flow on the surface of the wing and fuselage as much as possible. Thus, a technology to control drag reduction through laminar flow is proposed.

The most effective way to reduce aircraft drag is to reduce the frictional drag on the surface of the aircraft. Since the frictional resistance of the turbulent boundary layer is much greater than that of the laminar boundary layer, the basic idea of reducing the frictional resistance includes two aspects: one is to delay the occurrence of transition as much as possible and expand the laminar flow area of the surface. The second is to reduce the frictional resistance in the turbulent boundary layer flow area. In the past few decades, domestic and foreign scholars have proposed many control technologies to reduce frictional resistance and conducted a lot of research on this. However, these technologies are still in the research stage, and almost no control technology is used in actual aircraft. Among the many control technologies, laminar flow control is one

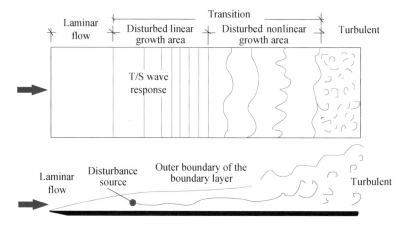

Fig. 6.41 Flat transition

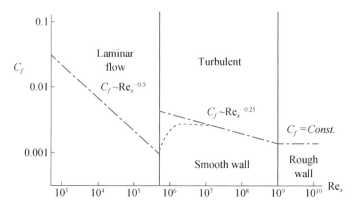

Fig. 6.42 Frictional resistance coefficient of laminar and turbulent boundary layer

of the effective methods to reduce frictional resistance. This technology is to delay the transition of the boundary layer by adopting control measures and expand the laminar flow area of the object surface to achieve the purpose of drag reduction. For an aircraft, the main areas where the wing, engine pod, nose, horizontal tail, and vertical tail get laminar flow, as shown in Fig. 6.43.

The research of laminar flow control technology has a history of more than 70 years. Existing studies have shown that in the design of laminar airfoil and wing, the purpose of control in the boundary layer is to delay the boundary layer transition under the condition of not affecting aerodynamic performance and structures as much as possible. According to the different control methods, there are three different control technologies (as shown in Fig. 6.44). One is passive control or natural laminar flow control (NLF), that is, by adjusting the shape to increase the range of the surface pressure gradient, thereby delaying the transition. This method has poor aerodynamic performance under non-design conditions. The second is active control or laminar flow control (LFC), that is, manipulation at a specific location in the

Fig. 6.43 The main laminar flow control area on the aircraft surface (including the outer surface of the aircraft main wing, vertical tail, flat tail, and engine nacelle)

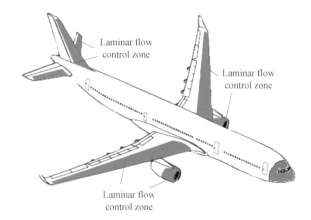

boundary layer (such as aspiration technology) to delay the transition. The third is the Hybrid Laminar Flow Control Technology (HLFC), which combines the advantages of natural laminar flow control (passive control) and laminar flow control (active control, such as suction technology), which can effectively reduce the complexity of the suction volume and control system. The characteristics of mixed laminar flow control are (1) Only the leading edge needs to be sucked. (2) It is only necessary to modify the surface geometry near the leading edge to achieve a favorable pressure gradient. (3) The mixed laminar flow control wing design has good turbulence performance. As shown in Fig. 6.45, the active control technology mainly includes suction, wall cooling, and active compliant wall technology. The passive control technology mainly includes wall surface modification, surface roughness distribution, passive compliant wall, and porous wall technology. The current development trend is the hybrid laminar flow control technology (HLFC), and the most widely used is the combination of wall modification (to maintain a better pressure gradient) and suction technology.

(1) Natural laminar flow control

For example, the early designed NACA-6 series airfoil is a representative of natural laminar flow. The results of early laminar airfoil designs are usually not satisfactory. For example, the NACA 632-215 airfoil can achieve low drag, but the usable lift range is much smaller than that of the turbulent airfoil NACA 23015. Of course, with the advancement of airfoil design technology, the performance that the natural laminar airfoil can achieve is getting better and better. As shown in Fig. 6.46, the engine nacelle of Boeing's B787 aircraft is designed with laminar flow control technology.

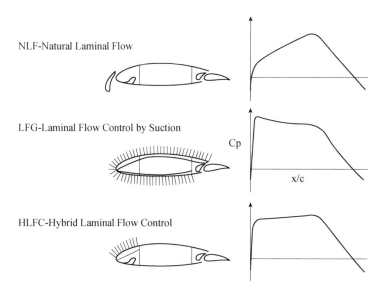

Fig. 6.44 Conceptual design of NLF, LFC, and HLFC

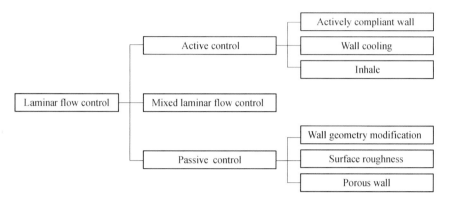

Fig. 6.45 Laminar flow control technology

But only relying on the laminar flow airfoil design cannot solve the laminar flow control problem of large transport aircraft.

(2) Active laminar flow control

Active laminar flow control methods include suction control, temperature control, active flexible wall control, plasma control, etc. The suction control technology is relatively mature, and it has undergone a large number of flight tests. And the drag reduction effect is relatively obvious. The principle of inhalation control can be simply understood as changing the average velocity profile of the local boundary layer, thereby suppressing the growth of related unstable disturbances. There are usually two ways to inhale, one is channel inhalation, and the other is small hole inhalation. In order to test the actual effect of the air intake control, NASA carried

Fig. 6.46 Laminar flow control of B787 aircraft engine nacelle

Fig. 6.47 NASA C-140
Jetstar aircraft and suction
control devices installed on
the left and right sides

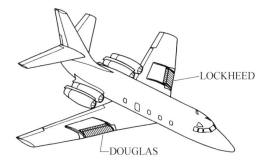

Fig. 6.47 NASA C-140
Jetstar aircraft and suction
control devices installed on
the left and right sides

out a research on the leading edge flight test project. Two air intake control devices
were installed on the leading edges of the wings on both sides of the C-140 Jetstar
aircraft (shown as Fig. 6.47), and conducted a large number of flight tests. The results
show that laminar fluidization can be achieved in most flight conditions, covering
nearly 65–75% of the chord length.

If active control is added on the basis of natural laminar flow, this control method
is called mixed laminar flow control. For the hybrid laminar flow control technology,
NASA conducted a test on the B757 (shown in Fig. 6.48) aircraft. It can be clearly
seen from Fig. 6.49 that the flow at the edge of the laminar flow control area has
become turbulent, while the laminar flow is maintained in the active control area.
The results of the flight test under the condition of Mach number $M = 0.8$ show
that only 1/3 of the designed inspiratory volume is needed to achieve a laminar flow
coverage of 65% of the chord length. The result was better than expected. And after
calculation, the mixed laminar flow control reduces the drag of the aircraft wing by
29% and the entire aircraft by about 6%.

In addition to the main wings of the aircraft, laminar flow control tests were also
carried out on the outer surface of the aircraft engine nacelle and the vertical tail. For
example, on the vertical tail of the A320 aircraft, Airbus Europe applies laminar flow
control to the basic vertical tail to obtain 40% chord-length laminar flow coverage.

Fig. 6.48 B757 aircraft (the
black area on the left wing is
the laminar flow control test
area)

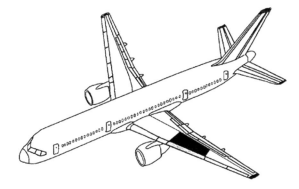

Fig. 6.49 B757 aircraft
wing mixed laminar flow
control results (the open
circle indicates the laminar
flow state and the solid circle
indicates the turbulent state,
and the number on the left
indicates the percentage of
position relative to the chord
length)

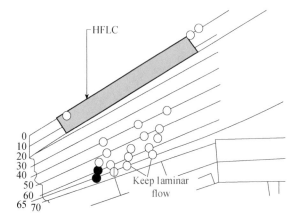

The HLFC vertical tail can obtain 50% on-site coverage, which can reduce the drag of the entire aircraft by 1–1.5%.

(3)   Wall roughness control

Research on improving wall structure control and delaying the transition of the boundary layer has been developed late, and most of the technologies are still in the experimental research stage. Studies have shown that this laminar flow control is based on the principle of cross-flow transition of the swept-back wing. The development of the cross-current vortex is directly related to the spanwise wavelength. The spanwise wavelength is determined by the tiny rough structure on the wall. If no control is applied, then the most unstable wavelength will become dominant and transition earlier. If the distributed rough array is added, the development of the cross-current vortex is determined by the rough array. Therefore, we can find a rough array that can delay the transition by adjusting the parameters.

6.11.3   Technology to Reduce Turbulence Resistance

A large number of studies have shown that improving the turbulent structure near the wall is an effective method to reduce the frictional resistance of the wall in the turbulent boundary layer. Turbulent drag reduction is the control of the turbulent eddy structure in the near-wall area, specifically, the control of the large-scale vortex structure in the turbulent boundary layer. The main characteristics of the coherent structure in the near-wall turbulent flow are (1) Low-speed bands in the viscous bottom layer. (2) The jetting behavior of low-velocity flow in the wall area causes the low-velocity strip to rise. (3) The high-speed fluid at the outer edge of the boundary layer sweeps toward the wall, causing the outer area to flow in. (4) Various forms of turbulent structures appear. (5) The inclination of the near-wall shear structure is manifested by the concentration of spanwise vorticity. (6) A near-wall "vortex envelope" structure

Fig. 6.50 Shark skin groove structure

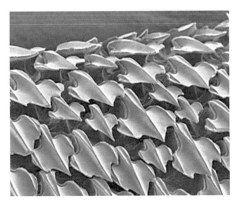

appears. (7) The "backward movement" of the shear layer caused by the movement of the large-scale turbulent structure in the outer zone of the boundary layer results in the discontinuity of the flow velocity. These complex near-wall turbulent structures make the physical quantities in the boundary layer appear as spatial and temporal uncertainties, so the control of the turbulent boundary layer is much more difficult than the control of the laminar boundary layer.

Experiments have found that effectively controlling the spanwise pulsation in the turbulent boundary layer can reduce the frictional resistance of the turbulent boundary layer. The current wall groove control is a passive control technology. In the study of turbulent boundary layer control, the drag reduction mechanism of shark skin grooves has received extensive attention. As shown in Fig. 6.50, the microstructure of shark skin is actually a complex groove.

Someone's measurement of the turbulent structure on the surface of the groove found that under certain groove parameters, a drag reduction effect of about 8% can be achieved, but it was also noted that the groove does not have a significant effect on the frequency of turbulent burst events. The existence of grooves can obviously inhibit the exchange of spanwise pulsation energy in the turbulent boundary layer. Figure 6.51 shows a typical wall groove. This control technology does not require energy input, so it is a passive boundary layer control technology. It is found that the dimensionless number of groove spacing with good drag reduction effect is $S^+ = 10 \sim 20$. Among them,

$$S^+ = \frac{u^* s}{\nu}, \quad u^* = \sqrt{\tau_0 / \rho} \tag{6.262}$$

where u^* is the wall friction speed. Under flying conditions, the actual distance is generally $25 \sim 75$ μm (human hair is about 70 μm). The groove drag reduction is about $5 \sim 15\%$.

Fig. 6.51 Schematic
diagram of the cross-section
of the groove surface

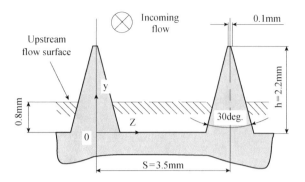

6.11.4 Technology to Reduce Induced Resistance

In the cruising state of a large aircraft, the proportion of induced drag of wing to the total drag is second only to frictional drag, which is induced by the downwashing of the aircraft itself by the shedding vortex of the trailing edge of the wing. Reducing induced drag can be achieved by expanding the wingspan, but the expansion of the span will be limited by the wing structure. For this reason, Richard T. Whitcomb (1921~2009), director of the NASA Langley Wind Tunnel Laboratory in the United States in the 1970s, imitated the shape of the swan's wing tip feathers when they were flying (as shown in Fig. 6.52), invented a wing tip upturning device called a winglet. The winglet is mainly proposed to reduce the induced drag when the aircraft is cruising. Soon after the winglet was invented, the U.S. Air Force tested the effect of the winglet on the KC-135 tanker. The test results show that the addition of winglets can reduce the resistance of the cruise state by about 7%, and it is estimated that this improvement can save the KC-135 fleet of billions of dollars in the next two decades. Now the large aircraft produced by Boeing and European Space Company are equipped with winglets, and my country's ARJ21 passenger aircraft and C919 aircraft are also equipped with winglets. Due to the pressure difference on the wing tip surface, the air tends to flow outward along the lower surface of the wing tip and inward along the upper surface. After the winglet is installed, the wing tip vortex will have an end plate effect on the spanwise flow of the wing, and the winglet vortex will diffuse the wing tip vortex, so that the wing tail vortex will be reduced. The effect of the downwash is weakened, the angle of the downwash is reduced, and the induced resistance is reduced. The main features of the winglet are (1) The end plate effect blocks the flow from the lower surface to the upper surface of the wing, weakens the strength of the wingtip vortex, and increases the effective aspect ratio of the wing. (2) Dissipate the effect of the wingtip vortex of the main wing. Because the winglet itself is also a small wing, it can also produce wingtip vortices, although the direction is the same as that of the main wing wingtip vortex. But because the distance is very close, a strong shearing effect is formed at the intersection of the two vortices, which causes large viscous dissipation, prevents the winding of the main vortex, plays the role of spreading the main vortex, and also achieves the purpose of reducing induced drag (such as Fig. 6.53 shown). (3) Increase the wing lift and forward thrust. The upper winglet can use the three-dimensional distortion flow field to generate the wing lift and thrust components (as shown in Fig. 6.54). (4) Delay the premature separation of the wing tip airflow and increase the stall angle of attack. Generally speaking, the three-dimensional effect of the wing tip of the swept-back wing is more obvious. The flow tube shrinks. When the air flow passes through, it first accelerates sharply, the pressure decreases, and then the pressure recovers sharply. The sharp boundary layer separates, causing a stall. However, the wingtip installed at the wingtip can use the favorable pressure gradient generated by it to offset part of the wingtip backpressure field, so that the pressure distribution becomes relaxed and the backpressure gradient is reduced. If properly designed, the airflow separation at the wingtips of the wings can be delayed, and the stall angle of attack and buffeting lift coefficient of the aircraft can be improved.

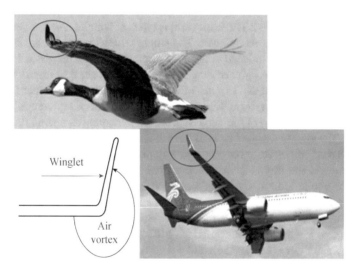

Fig. 6.52 Winglet

Fig. 6.53 Shear dissipation of wingtip vortex and winglet vortex

Fig. 6.54 Winglet increase and thrust effect

The effect of the winglet: There are various types of winglets such as single upper winglet and upper and lower winglet. The single winglet is used more because of its simple structure. The induced drag of the aircraft accounts for about 15–30% of the cruise drag. Reducing induced drag is of great significance to improving cruise economy. The greater the aspect ratio of the wing, the lower the induced drag. An excessively large aspect ratio will make the wing too heavy, so there is a limit to increasing the aspect ratio of the wing. In addition to serving as a wing tip end plate to increase the effective aspect ratio of the wing, a winglet can also reduce induced drag due to the "pull effect" produced by the deflection of the airflow at the wing tip. The wind tunnel and flight test results show that the winglet can reduce the induced drag of the whole aircraft by 20–30%, which is equivalent to an increase of the lift-to-drag ratio by 5%. Winglets, as an advanced aerodynamic design measure to improve flight economy and save fuel, have been adopted on many aircraft. There are also types of wingtips: wingtip vortex diffusers (as shown in Fig. 6.55), shark fin winglets (as shown in Fig. 6.56), and wingtip sails (5 in my country, as shown in Fig. 6.57).

Fig. 6.55 Wingtip vortex diffuser (A320)

Fig. 6.56 Shark fin winglet (B787)

Fig. 6.57 Wingtip sail piece (Yun 5, China)

6.11.5 Technology to Reduce Shock Wave Resistance

For a long time, reducing the flight resistance of transonic aircraft and increasing the resistance divergence Mach number have always been technical difficulties in aircraft design. As early as the 1950s, NASA's Whitcomb et al. found through wind tunnel tests that when the flying speed was near the speed of sound, the zero-lift wave resistance of the aircraft would be greatly affected by the longitudinal distribution of its cross-sectional area. And the zero-lift wave resistance of the rotating body with the same cross-sectional area distribution is the same. That is to say, the shape of the cross-sectional area of the aircraft in the longitudinal position has no effect on the wave resistance. But the size of the cross-sectional area in the longitudinal distribution is influential. When the traditional straight fuselage passes the wing, it will cause a significant increase in wave resistance. If the bee waist structure is adopted, the wave resistance can be greatly reduced. Therefore, an effective method for reducing the zero-lift wave resistance by modifying the airframe with the transonic area law is proposed. Experiments have found that the application of area law can reduce the transonic zero-lift wave resistance by 25–30%. However, as the Mach number

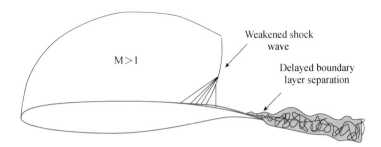

Fig. 6.58 Schematic diagram of shock wave bulge control shock wave intensity (red line is shock wave bulge)

increases, the drag reduction effect of the area law gradually weakens. When the Mach number is between 1.8 and 2.0, the area law effect is almost zero.

In addition to the overall design of the aircraft to reduce shock resistance, in recent years, a new type of shock wave has been developed to control the strength of the shock wave on the wing by adding Contour bumps to the airfoil, thereby reducing the shock resistance technology. In the design of supercritical wing, a swept-back wing that generates a weak shock wave in the cruise design state is given, but when the flight state deviates from the design state, the shock wave resistance will rise sharply. Someone proposed the aerodynamic performance of the airfoil with a bulge in the transonic state, and proposed to improve the transonic drag characteristics of the airfoil by adding a bulge (as shown in Fig. 6.58). Subsequent studies have shown that the range of shock drum kits can be extended from 20% chord length to 40% chord length, and the drum kit shape can be dynamically adjusted as needed. Both Europe (Euroshock Project) and the United States (NASA Aircraft Deformation Project) have carried out systematic research on bulge control shock wave intensity. Numerical simulation and experimental tests have found that a properly designed drum kit can effectively improve the lift-to-drag ratio of the aircraft when the flight speed is close to the speed of sound.

Exercises

A. Thinking questions

1. For fluids with small viscosity coefficients, please explain why the viscous effect in the flow near the wall cannot be ignored.
2. Please explain the basic assumptions of the boundary layer approximation.
3. Write down the displacement thickness of the boundary layer, the thickness of the momentum loss, and the thickness of the energy loss. Explain the physical meaning.

4. If the velocity distribution functions of the boundary layer are $\frac{u}{V_\infty} = \frac{y}{\delta}$ and
    $\frac{u}{V_\infty} = 2\frac{y}{\delta} - \left(\frac{y}{\delta}\right)^2$, please calculate the displacement, momentum loss, and
    energy loss thickness of the boundary layer.
5. The motion of the viscous fluid in the boundary layer is vortex motion.
    Please write down the relationship between wall vorticity and shear stress.
6. Please write down the vorticity diffusion equation in the boundary layer.
7. Please write down Prandtl's boundary layer equation and explain how to
    calculate the pressure in the boundary layer.
8. Point out the physical meaning of the boundary layer similarity hypothesis.
9. Write the Karman momentum integral equation, and point out the condi-
    tions that need to be supplemented to solve the momentum integral
    equation.
10. Point out why the inverse pressure gradient and surface adhesion conditions
    are necessary conditions for boundary layer separation.
11. Why is the velocity distribution in the boundary layer near the object surface
    stable under the pressure gradient, and there is no velocity inflection point?
12. Why is the velocity distribution in the boundary layer near the object
    surface unstable under the reverse pressure gradient, and there is a velocity
    inflection point in the boundary layer?
13. Please indicate the conditions of the separation point in the boundary layer.
    Why does the resistance of the object flow around increase significantly
    after the boundary layer is separated?
14. Explain the basic measures for the proposed separation of the boundary
    layer.

B. **Calculation questions**

1. If the velocity distribution functions of the boundary layer are $\frac{u}{V_\infty} = \frac{y}{\delta}$ and
    $\frac{u}{V_\infty} = 2\frac{y}{\delta} - \left(\frac{y}{\delta}\right)^2$. Try to calculate the displacement, momentum loss and
    energy loss thickness of the boundary layer.
2. Air with a temperature of 25 °C will pass a thin plate longitudinally at
    a speed of 30 m/s. The pressure is atmospheric pressure. Calculate the
    thickness of the boundary layer at a distance of 200 mm from the front
    edge of the plate?
3. A flat plate with a chord length L of 3.5 m, $Re_L = 10^5$. Try to estimate
    the thickness of the boundary layer at the rear edge of the slab (all laminar
    flow).
4. An oil flow with a temperature of 20 °C and $\rho = 925$ kg/m^3 flows longi-
    tudinally around a thin flat plate with a width of 15 cm and a length of
    50 cm at a speed of 0.6 m/s. Try to find the total frictional resistance and
    the thickness of the boundary layer. The oil $= 7.9 \times 10^{-5}$ m^2/s at 20 °C.
5. The wings of a general aviation aircraft are rectangular (extended length
    9.75 m, chord length 1.6 m). The aircraft flies at a constant speed at sea
    level at a cruising speed of 227 km/h. Assuming that the surface frictional
    resistance on the wing can be approximated to the resistance on a flat plate

of the same size. Try to calculate the surface frictional resistance (the flow is completely laminar).

6. In the case of the same Reynolds number Re_l, try to find the ratio of the frictional resistance generated when water at 20 °C and air at 30 °C flow in parallel through a plate of length L.

7. A flat plate with a length of 6 m and a width of 2 m is placed parallel and static in an air flow at a speed of 60 m/s at 40 °C. The critical Reynolds number of the transition from laminar flow to turbulent flow in the boundary layer of the flat plate is $Re_x = 10^6$. Try to calculate the frictional resistance of the plate.

8. If the velocity at the outer boundary of the boundary layer is $u_\delta = V_0 x^m$, V_0 is a constant. Try to prove that the corresponding pressure change is

$$\frac{\partial p}{\partial x} = -m\rho V_0^2 x^{2m-1}$$

Therefore, $m > 0$ represents the forward pressure gradient, and $m < 0$ represents the reverse pressure gradient.

9. For the laminar boundary layer on a two-dimensional curved surface with a radius of curvature of R, set the velocity distribution in the boundary layer as

$$\frac{u}{u_\delta} = 2\left(\frac{y}{\delta}\right) - \left(\frac{y}{\delta}\right)^2, \quad 0 \le y \le \delta$$

The streamlines in the boundary layer have the same curvature as the curved surface. Try to establish the equilibrium condition between pressure and centrifugal force, and integrate laterally along the boundary layer to prove that the pressure change is

$$\Delta p = \frac{8}{15}\frac{\delta}{R}\rho u_\delta^2$$

If $\delta = 0.01$ m, $R = 0.3$ m, and at the outer of the boundary layer, $u_\delta = 100$ m/s, the pressure is the standard atmospheric pressure at sea level. Try to prove that the pressure change along the transverse direction of the boundary layer (the normal direction of the object surface) is 218 N/m² (much less than the pressure at the outer boundary of the boundary layer).

10. The velocity distribution in the laminar boundary layer of a flat plate is as follows: $\frac{u}{u_\delta} = 2\frac{y}{\delta} - \left(\frac{y}{\delta}\right)^2$. Try to find the relationship between the thickness of the boundary layer and the frictional resistance coefficient and the Reynolds number Re.

11. If the velocity distribution in the laminar boundary layer of a flat plate is a sinusoidal curve, $u = u_\delta \sin\left(\frac{\pi y}{2\delta}\right)$. Try to find the relationship between δ and C_f and Re.

12. According to the exponential law of turbulence velocity distribution in the boundary layer, $\frac{u}{u_\delta} = \left(\frac{y}{\delta}\right)^{\frac{1}{9}}$ and $\lambda = 0.185\mathrm{Re}^{-\frac{1}{5}}$. Try to find the thickness of the turbulent boundary layer δ.

13. Assume that the velocity distribution in the boundary layer of the flat plate (length L) is

$$\frac{u}{u_\delta} = \frac{3}{2}\left(\frac{y}{\delta}\right) - \frac{1}{2}\left(\frac{y}{\delta}\right)^3$$

where δ is the thickness of the boundary layer. Try the momentum integral relational method to solve and calculate

(a) $(\delta_2/x)\sqrt{\mathrm{Re}_x}$, (b) $(\delta_1/x)\sqrt{\mathrm{Re}_x}$, (c) $(\delta/x)\sqrt{\mathrm{Re}_x}$, (d) $C_f\sqrt{\mathrm{Re}_x}$; (e) $C_f\sqrt{\mathrm{Re}_x}$

14. Try to find the power to overcome air resistance when a car is driving at a speed of 60 km/h. It is known that the projected area of the car perpendicular to the direction of motion is 2 m², the drag coefficient is 0.3, and the temperature of still air is assumed to be 0 °C.

15. Suppose a low-speed aircraft is flying at an altitude of 3000 m at an altitude of 360 km/h. If the wing area is 40 m², the average chord length is 2.5 m. Try to estimate the frictional resistance of the wing (calculated by complete turbulence) using the calculation formula of the two-dimensional flat plate boundary layer.

16. In the wind tunnel, blow vertically to a disc with a diameter of 50 cm at a wind speed of 10 m/s, and try to find the resistance of the disc. The air temperature is 20 °C.

17. For a flat plate placed downstream in a two-dimensional incompressible flow, try the momentum integral method to find the wall friction stress and the friction resistance on one side of the plate F, the plate width b, length L, it is recommended to assume that the velocity distribution in the boundary layer is $0 \sim 90°$ sine curve. Try to compare the result with the Blasius solution.

18. If the 1/7 exponential law is used for the velocity profile

$$\frac{u}{u_\delta} = \left(\frac{y}{\delta}\right)^{\frac{1}{7}}$$

Prove that the displacement thickness and momentum loss thickness of the boundary layer are $\delta_1 = \frac{\delta}{8}$ and $\delta_2 = \frac{7}{72}\delta$, respectively. Therefore, for the flat, $H = \frac{\delta_1}{\delta_2} \approx 1.3$.

Chapter 7
Fundamentals of Compressible Aerodynamics

This chapter introduces the fundamentals of compressible aerodynamics, including the first law of thermodynamics, the differential form of the energy equation for the motion of gas clusters, sound velocity and Mach number, one-dimensional compressible steady flow theory, the propagation characteristics of small disturbances (Mach cone and Mach wave), expansion wave and supersonic wall external corner flow, compression wave and shock wave, compressible fluid boundary layer flow, the interaction between shock wave and boundary layer, compressible one-dimensional friction pipe flow, and the working performance of contraction nozzle and Laval nozzle.

Learning points:

(1) Learn the first law of thermodynamics of compressible air movement, the derivation process of energy equation of gas micro mass movement and various physical meanings.
(2) Learn the definition of acoustic velocity, Mach number, and the propagation characteristics of small disturbances (basic concepts and physical significance such as Mach cone and Mach wave).
(3) Learn the derivation process and application of one-dimensional compressible steady flow theory, expansion wave and shock wave, and other basic equations.
(4) Understand the boundary layer flow of compressible fluid, the interference between shock wave and boundary layer, the flow of compressible one-dimensional friction pipe, and the working performance of contraction nozzle and Laval nozzle.

7.1 Thermodynamic System and the First Law

Thermodynamics is a subject that studies the conversion between thermal energy and mechanical energy, including the characteristics of various working media used to realize the conversion. This section mainly introduces the thermodynamics

© Science Press 2022
P. Liu, *Aerodynamics*, https://doi.org/10.1007/978-981-19-4586-1_7

foundation related to high-speed flow, focusing on the first and second laws of thermodynamics and the energy equation of gas motion.

The gas system in thermodynamics refers to a gas system of arbitrary shape isolated from the ambient gas by a certain mass of gas within a specified boundary. The size of the gas system must be macroscopic, and its relationship with the outside world is as follows:

1. The system without mass exchange and energy exchange is called an isolation system.
2. The system without mass exchange but with energy exchange is called a closed system.
3. The system with mass exchange and energy exchange is called an open system.

In high-speed aerodynamics, most of the flow problems are the flow problems of air system in isolated system or closed system. Classical thermodynamics deals with the system in equilibrium.

7.1.1   Equation of State and Perfect Gas Hypothesis

The pressure, density, and absolute temperature of any gas are not independent, and there is a certain functional relationship among them, which is called equation of state. After two of them are confirmed, the third one is confirmed. Namely,

$$f(\rho, p, T) = 0 \tag{7.1}$$

This function is called the equation of state. The specific form will be given according to the type of medium.

Ideal gas is a kind of gas which neglects the molecular volume and intermolecular force, and its equation of state satisfies Clapeyron's equation (Benoit Pierre Emile Clapeyron, 1799–1864).

$$\frac{p}{\rho} = RT \tag{7.2}$$

where R is the gas constant, $R = 287.053$ N m / (kg K). At room temperature and atmospheric pressure, the volume of actual gas molecules and the interaction between molecules can also be ignored, and the state parameters meet the equation of state of ideal gas, so the actual gas is often simplified as ideal gas in aerodynamics, and the ideal gas with constant specific heat ratio is called perfect gas. Perfect gas is suitable for low velocity flow. But in the high-speed flow, if the temperature of the gas flow is high, the rotational energy and vibration energy of the gas molecules are excited with the increase of the temperature, and the specific heat ratio is no longer constant. In the temperature range of 1500–2000 K, air can be regarded as a perfect gas with variable specific heat ratio.

7.1.2 *Internal Energy and Enthalpy*

The internal energy of gas refers to the sum of kinetic energy contained in molecular micro-thermal motion (related to temperature) and internal potential energy due to the interaction between molecules. For a perfect gas, there is no force between molecules, and the internal energy e per unit mass of gas is only the result of molecular thermal motion and a function of temperature.

In thermodynamics, another parameter h (enthalpy), which represents the heat content, is often introduced.

$$h = e + \frac{p}{\rho} \tag{7.3}$$

Because $\frac{p}{\rho}$ represents the pressure energy per unit mass of gas, enthalpy h represents the sum of internal energy and pressure energy per unit mass of gas.

7.1.3 *The First Law of Thermodynamics*

The first law of thermodynamics is the specific application of the law of conservation of energy in thermodynamics. Its physical meaning is that the heat transferred from the outside to a closed material system is equal to the sum of the increment of the internal energy of the closed system and the mechanical work done by the system to the outside. For a small change process,

$$dQ = dE + pd\forall \tag{7.4}$$

This is the first law of thermodynamics for stationary systems. Here, $d\forall$ is the volume increment of the system and p is the pressure of the system. If the above equation is removed by the mass of the system, it becomes the energy equation per unit mass. Namely,

$$dq = de + pd\left(\frac{1}{\rho}\right) \tag{7.5}$$

where the reciprocal of density is the volume per unit mass. The physical meaning of the above formula: the heat transferred from the outside to the unit mass gas dq is equal to the sum of the internal energy increment of the unit mass gas and the expansion work done by the pressure to the unit mass gas.

Because the pressure, density, and temperature of a gas are functions of space points, there is a certain functional relationship between them, but it has nothing to do with the change process, which represents a thermodynamic state. p, T, ρ, e, h represent thermodynamic state parameters. Two thermodynamic parameters can determine one thermodynamic state. If the independent variables are t and R, the relationship between other state variables is

$$
\begin{cases}
p = p(\rho, T) \\
e = e(\rho, T) \\
h = e + p/\rho
\end{cases}
\tag{7.6}
$$

If we do the differential of enthalpy, we get

$$
\mathrm{d}h = \mathrm{d}e + pd\left(\frac{1}{\rho}\right) + \frac{1}{\rho}\mathrm{d}p
\tag{7.7}
$$

The formula shows that in the differential section, the increment of gas enthalpy is equal to the sum of internal energy increment, gas expansion work, and work done by pressure difference.

For a flowing gas element, the energy equation becomes

$$
\mathrm{d}q = \mathrm{d}e + pd\left(\frac{1}{\rho}\right) + \frac{\mathrm{d}p}{\rho} + d\left(\frac{V^2}{2}\right)
\tag{7.8}
$$

Compared with the energy equation of static gas (7.5), there are two more terms in the energy equation of moving gas. The term $\frac{\mathrm{d}p}{\rho}$ represents the unique work of gas element in the process of flow and represents the work of gas element caused by pressure change under the condition of constant volume (the work of fluid element overcoming pressure difference). The other term $d\left(\frac{V^2}{2}\right)$ is the kinetic energy increment of the macro-motion of the unit mass gas. Namely,

$$
d\left(\frac{V^2}{2}\right) = d\left(\frac{u^2 + v^2 + w^2}{2}\right)
\tag{7.9}
$$

If Eq. (7.7) is substituted by Eq. (7.8), then

$$
\mathrm{d}q = \mathrm{d}h + d\left(\frac{V^2}{2}\right)
\tag{7.10}
$$

where, in the differential period, $\mathrm{d}q$ is the heat transferred from the outside to the gas, which comes from conduction and thermal radiation, and can also generate heat through chemical changes such as combustion.

7.2 Thermodynamic Process

7.2.1 *Reversible and Irreversible Processes*

In thermodynamics, if the change process is reversed step by step, all the thermodynamic parameters of the gas will return to the initial state, and the external state
will also return to the old state. Such a process is called a reversible process, otherwise it is an irreversible process (for example, heat transfer from high temperature
to low temperature, mechanical work generating heat through friction, etc. are all
irreversible processes). In mathematics, reversible process is also called quasi-static
process or continuous equilibrium process.

7.2.2 *Isovolumetric Process*

In Eq. (7.5), the internal energy e is a function of the state, but q is not a function of
the state, because the work of pressure expansion is not only determined by the start
and end of the process, but also related to the change process. As shown in Fig. 7.1,
on the $p - 1/\rho$ diagram, the work done during the whole change process can be
expressed as

$$W_p = \int_1^2 pd\left(\frac{1}{\rho}\right) \tag{7.11}$$

Different curves from 1 to 2 point represent different thermodynamic processes,
and the expansion work of these different processes is different.

If the volume of gas per unit mass remains unchanged during the change, such
a process is called isovolumetric process. The expansion work of the gas is zero.
All the heat added by the outside is used to increase the internal energy of the gas.
Namely,

Fig. 7.1 Diagram of
expansion work

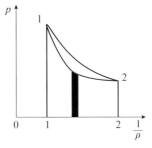

Fig. 7.2 Isovolumetric
process

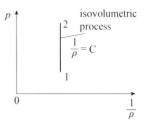

$$dq = du = C_V dT \tag{7.12}$$

where C_V is called constant volume specific heat. Its physical meaning is: in the case of constant volume, the heat required for each degree rise of unit mass gas temperature, as shown in Fig. 7.2.

From Eq. (7.12), it can be concluded that

$$e = \int_0^T C_V dT \tag{7.13}$$

$$C_V = \left(\frac{dq}{dT} \right)_\rho \tag{7.14}$$

7.2.3   Constant Pressure Process

If the pressure of the gas remains constant in the process of change, such a process is called constant pressure process. The expansion work of the gas is not equal to zero. Part of the heat added by the outside is used to increase the internal energy of the gas, and the other part is used for the expansion work of the gas. In the constant pressure process, the heat required for each degree rise of unit mass gas temperature is called constant pressure specific heat. As shown in Fig. 7.3, when the gas changes at constant pressure, $p =$ constant, $dp = 0$, which can be obtained from Eqs. (7.5) and (7.7)

$$C_p = \frac{dq}{dT} = \frac{dh}{dT} \tag{7.15}$$

$$C_p = \left(\frac{\partial h}{\partial T} \right)_p \tag{7.16}$$

The ratio of specific heat at constant pressure to specific heat at constant volume is called specific heat ratio of gas. Namely,

Fig. 7.3 Constant pressure
process

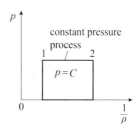

$$\gamma = \frac{C_p}{C_V} \tag{7.17}$$

In aerodynamics, when the temperature is less than 300 °C and the pressure is not high, generally, C_p, C_V, γ are constants. For the differential at both ends of Eq. (7.3), we get

$$dh = de + d\left(\frac{p}{\rho}\right) \tag{7.18}$$

Using the equation of state $p/\rho = RT$, the above equation can be transformed into

$$dh = de + RdT \tag{7.19}$$

According to the definition Eqs. (7.13) and (7.15), it can be obtained from the above equation

$$C_p = \frac{\gamma}{\gamma - 1} R \tag{7.20}$$

$$C_v = \frac{1}{\gamma - 1} R \tag{7.21}$$

The work per unit mass of gas is

$$W_P = \int_1^2 pd(\frac{1}{\rho}) = \int_1^2 d\left(\frac{p}{\rho}\right) \int_1^2 RdT = R(T_2 - T_1)$$

Fig. 7.4 Isothermal process

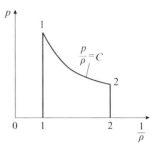

7.2.4 Isothermal Process

If the temperature of the gas remains constant in the process of change, such a process is called isothermal process. As shown in Fig. 7.4, in the isothermal process, the internal energy remains unchanged, and the heat and expansion work are equal. The work per unit mass of gas is

$$q = W_P = \int_1^2 pd\left(\frac{1}{\rho}\right) = \int_1^2 RT\rho d\left(\frac{1}{\rho}\right) = RT \ln \frac{\rho_1}{\rho_2} \qquad (7.22)$$

Because the temperature is invariable, the internal energy is invariable, $de = 0$.

7.2.5 Adiabatic Process

In the process of thermodynamic change, there is no heat exchange with the outside world, which is called adiabatic process. As shown in Fig. 7.5, it is obtained from the energy Eq. (7.5)

$$de + pd\left(\frac{1}{\rho}\right) = 0 \qquad (7.23)$$

$$C_V dT + pd\left(\frac{1}{\rho}\right) = 0 \qquad (7.24)$$

By using the equation of state of ideal gas, we have the following results:

$$pd\left(\frac{1}{\rho}\right) + \frac{1}{\rho}dp = RdT \qquad (7.25)$$

Substituting Eq. (7.24) into the above equation, we get

Fig. 7.5 Adiabatic process

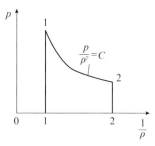

$$(C_V + R)pd\left(\frac{1}{\rho}\right) + C_V \frac{1}{\rho}dp = 0 \tag{7.26}$$

or

$$C_p pd\left(\frac{1}{\rho}\right) + C_V \frac{1}{\rho}dp = 0 \tag{7.27}$$

Divide the two sides by C_V at the same time to get

$$\gamma \frac{d\left(\frac{1}{\rho}\right)}{\frac{1}{\rho}} + \frac{dp}{p} = 0 \tag{7.28}$$

After integral, we get

$$p = C\rho^\gamma \tag{7.29}$$

Equation (7.29) is the relationship between pressure and density in the adiabatic process. In the adiabatic process, the work per unit mass of gas is

$$W_P = \int_1^2 pd\left(\frac{1}{\rho}\right) = \int_1^2 C\rho^\gamma d\left(\frac{1}{\rho}\right) = \frac{-1}{\gamma - 1}\left(\frac{p_2}{\rho_2} - \frac{p_1}{\rho_1}\right) \tag{7.30}$$

The change of internal energy is

$$\Delta e = C_V(T_2 - T_1) \tag{7.31}$$

7.3 The Second Law of Thermodynamic and Entropy

In nature, there are many spontaneous physical processes, which are unidirectional and never go back spontaneously. For example, water flows from high to low, high-pressure gas expands to low-pressure area, and heat transfers from high-temperature object to low-temperature object. These irreversible phenomena are spontaneous physical processes.

The second law of thermodynamics is based on the spontaneous conduction direction of heat, which is a necessary limiting law to characterize the success of heat conversion. There are many versions of the second law of thermodynamics. Although the description is different, the essence is the same thing, that is, to have a heat engine circulation system to convert heat energy into mechanical work, there must be a cold source as well as a heat source. Only part of the heat energy from the heat source can be converted into mechanical work, and the other part is discharged into the cold source with lower temperature. In other words, when the heat energy is converted into mechanical work in the heat engine of cyclic operation, it will be divided into the available part (that is, the part converted into mechanical work) and the unusable part (that is, the part discharged into the cold source). For example, for an adiabatic process, the positive work is equal to the reduction of the internal energy.

As far as the first law of thermodynamics is concerned, all kinds of energy are equally effective, so the energy balance relationship is established. But in fact, different kinds of energy are different from each other. Work is force multiplied by distance, which indicates the interaction between systems. The internal energy is a function of state, and the change of state is determined by work. Heat is determined by work and internal energy. In the first law of thermodynamics, the inequivalence of thermal work is not mentioned. In fact, the transformation of different kinds of energy is directional and irreversible. For example:

1. The heat always transfers from the high-temperature object to the low-temperature object, the reverse direction does not hold.
2. The two gases will not separate spontaneously after mixing.
3. Friction mechanical work can be transformed into heat, but heat cannot be transformed 100%.
4. It is impossible to make a continuous running machine, which only absorbs heat from a single heat source and converts it into equal work.

In order to point out the inequivalence of energy conversion, the second law of thermodynamics stipulates the direction of energy conversion. That is, if the change process in a certain direction can be realized, but the change process in the opposite direction cannot be realized or can only be realized under specific conditions. There are many ways to express the second law of thermodynamics. For example:

1. Clausius said: It is impossible to create a kind of circulating working heat engine to transfer heat from the object with lower temperature to the object with higher temperature.

2. Kelvin Planck said: It is impossible to build a cycle working heat engine, which takes heat from a single heat source and turns it into useful work without any other effect.

In the following, the concept of entropy is introduced to characterize the limitation of the second law of thermodynamics in irreversible process. Entropy is an index of the available part of heat energy. The entropy increase per unit mass of gas is defined as

$$ds = \frac{dq}{T} \tag{7.32}$$

Among them, dq and dq/T are two different quantities. dq is related to the integral path and dq/T is a quantity independent of the integral path, which can be expressed as the total differential of a function. Namely,

$$ds = \frac{dq}{T} = \frac{1}{T}\left(de + pd\left(\frac{1}{\rho}\right)\right) = C_V\frac{dT}{T} + R\rho d\left(\frac{1}{\rho}\right) \tag{7.33}$$

$$ds = \frac{dq}{T} = d\left(C_V \ln T + R \ln \frac{1}{\rho}\right) \tag{7.34}$$

In the study of thermodynamics, the most significant is the entropy increment, that is, the entropy increment from state 1 to state 2.

$$\Delta s = s_2 - s_1 = \int_1^2 ds = \int_1^2 \frac{dq}{T} = C_V \ln \frac{T_2}{T_1} + R \ln \frac{\rho_2}{\rho_1} \tag{7.35}$$

If you use $R = C_p - C_V$ and $p = R\rho T$, you get

$$\Delta s = \int_1^2 ds = \int_1^2 \frac{dq}{T} = C_V \ln\left[\frac{p_2}{p_1}\left(\frac{\rho_1}{\rho_2}\right)^\gamma\right] \tag{7.36}$$

The second law of thermodynamics points out: for an isolated system, if the process is reversible in the adiabatic change process, the entropy value remains unchanged, $ds = 0$, which is called isentropic process. It can be seen from Eq. (7.36)

$$\frac{p_2}{p_1}\left(\frac{\rho_1}{\rho_2}\right)^\gamma = 1, \quad \frac{p_2}{\rho_2^\gamma} = \frac{p_1}{\rho_1^\gamma} = c, \quad p = c\rho^\gamma \tag{7.37}$$

It can be seen that this equation is consistent with Eq. (7.29) for the adiabatic process of perfect gas.

If the process is irreversible, the entropy will increase, $ds > 0$. Therefore, the second law of thermodynamics is also called entropy law. By introducing the concept of entropy, we can judge whether the process is reversible or not and measure the degree of irreversibility. In the process of high-speed gas flow, irreversibility is caused by viscous friction, shock wave, and temperature gradient. Generally, in most areas of the flow field, the velocity gradient and temperature gradient are not large. The flow field can be approximately regarded as adiabatic reversible, with constant entropy, which is called isentropic flow. If the entropy of a streamline is constant, it is called isentropic flow along the streamline. If the entropy of the whole flow field is constant, it is called isentropic flow. In the boundary layer and its wake region, the flow region of shock wave, the region of gas viscosity, and heat conduction can't be ignored. The flow is an adiabatic irreversible process (entropy increase), and the isentropic relation can't be used.

Example 7.1 There is 1.5 kg air, from the starting point of 1 atmospheric pressure and 21 °C, after adiabatic compression, the pressure reaches 4.08 atmospheric pressure. Calculate the volume of starting gas, volume of terminal gas, terminal temperature, work done by the outside world on the medium, heat added, and change of internal energy.

Solution:

Volume of starting point:

Using equation of state $\frac{p}{\rho} = RT$,

$$\rho_1 = \frac{p_1}{RT_1} = \frac{101325}{287 \times (273.15 + 21)} = 1.2 \, \text{kg/m}^3$$

$$\forall_1 = \frac{m_1}{\rho_1} = \frac{1.5}{1.2} = 1.25 \, \text{m}^3$$

Volume of terminal gas:

Since it is an adiabatic compression process, the relationship between pressure and density is

$$p = C\rho^\gamma \quad p_1 = C\rho_1^\gamma \quad p_2 = C\rho_2^\gamma$$

Therefore

$$\frac{\rho_2}{\rho_1} = \left(\frac{p_1}{p_2}\right)^{\frac{1}{\gamma}}$$

Then

$$\rho_2 = \rho_1 \left(\frac{p_2}{p_1}\right)^{\frac{1}{\gamma}} = 1.2 \times \left(\frac{4.08}{1.0}\right)^{\frac{1}{1.4}} = 3.276 \, \text{kg/m}^3$$

$$\forall_2 = \frac{m}{\rho_2} = \frac{1.5}{3.276} = 0.458 \, \text{m}^3$$

Terminal temperature:

$$T_2 = \frac{p_2}{R\rho_2} = \frac{4.08 \times 101325}{287 \times 3.276} = 439.69\,\text{K}$$

Work done by the outside world on the medium:

$$W_P = -\int_1^2 p d\left(\frac{1}{\rho}\right) = -\int_1^2 C\rho^\gamma d\left(\frac{1}{\rho}\right) = \frac{1}{\gamma - 1}\left(\frac{p_2}{\rho_2} - \frac{p_1}{\rho_1}\right)$$

$$W_P = 1.5 \times \frac{1}{1.4 - 1} \times \left(\frac{4.08 \times 101325}{3.276} - \frac{101325}{1.2}\right) = 156580\,\text{Nm}$$

Heat added:

$$Q = 0$$

The change of internal energy,

$$
\begin{aligned}
\Delta U &= mC_V(T_2 - T_1) \\
&= 1.5 \times 716 \times (439.69 - 294.15) \\
&= 156310\,\text{Nm}
\end{aligned}
$$

In the process of adiabatic compression, all the work done by the outside world on the gas becomes the internal energy of the gas.

7.4 Energy Equation of Viscous Gas Motion

7.4.1 Physical Meaning of Energy Equation

Energy equation is the expression of the first theorem of thermodynamics in moving fluid. The first theorem of thermodynamics shows that the sum of the work done by all forces acting on the system in unit time and the heat input into the system in unit time is equal to the change rate of the total energy of the system. Namely,

$$\frac{dE}{dt} = Q + W \tag{7.38}$$

where Q is the total heat input into the system per unit time, including heat radiation and heat conduction. W is the work of all forces acting on the system per unit time, including surface force and volume force.

7.4.2 Derivation Process of Energy Equation

In the viscous fluid region, if any differential parallelepiped fluid element is taken as the system and the hexahedron itself is taken as the control body, the change rate of the total energy per unit time of the fluid system should be equal to the sum of the work of all forces acting on the system and the heat transferred to the system from the outside per unit time. The internal energy per unit mass of fluid is expressed by e, the kinetic energy per unit mass of fluid is $\frac{V^2}{2}$, and the total energy per unit mass of fluid (internal energy + kinetic energy) is

$$e_0 = e + \frac{V^2}{2} \tag{7.39}$$

The change rate of the total energy per unit time is

$$\frac{dE}{dt} = \rho \frac{d}{dt}\left(e + \frac{V^2}{2}\right) dx dy dz \tag{7.40}$$

As shown in Fig. 7.6, the forces acting on the system include the surface force acting on the system through the control surface and the mass force acting on the system. The work of all forces on the system per unit time is as follows:

The power of the mass force is

$$W_1 = (f_x u + f_y v + f_z w)\rho dx dy dz = \rho \vec{f} \cdot \vec{V} dx dy dz \tag{7.41}$$

The power of surface force in X-direction is

Fig. 7.6 Surface stress diagram

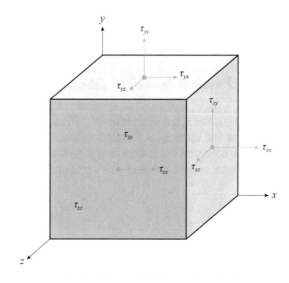

$$W_{2x} = \left(\tau_{xx} u + \frac{\partial(\tau_{xx} u)}{\partial x} dx - \tau_{xx} u \right) dy dz$$

$$+ \left(\tau_{yx} u + \frac{\partial(\tau_{yx} u)}{\partial y} dy - \tau_{yx} u \right) dx dz$$

$$+ \left(\tau_{zx} u + \frac{\partial(\tau_{zx} u)}{\partial z} dz - \tau_{zx} u \right) dx dy \tag{7.42}$$

$$W_{2x} = \left[\frac{\partial(\tau_{xx} u)}{\partial x} + \frac{\partial(\tau_{yx} u)}{\partial y} + \frac{\partial(\tau_{zx} u)}{\partial z} \right] dx dy dz \tag{7.43}$$

In the same way, the power of surface force in Y- and Z-directions is

$$W_{2y} = \left[\frac{\partial(\tau_{xy} v)}{\partial x} + \frac{\partial(\tau_{yy} v)}{\partial y} + \frac{\partial(\tau_{zy} v)}{\partial z} \right] dx dy dz \tag{7.44}$$

$$W_{2z} = \left[\frac{\partial(\tau_{xz} w)}{\partial x} + \frac{\partial(\tau_{yz} w)}{\partial y} + \frac{\partial(\tau_{zz} w)}{\partial z} \right] dx dy dz \tag{7.45}$$

The power of the total surface force is

$$W_2 = W_{2x} + W_{2y} + W_{2z}$$

$$W_2 = \left[\frac{\partial(\tau_{xx} u)}{\partial x} + \frac{\partial(\tau_{yx} u)}{\partial y} + \frac{\partial(\tau_{zx} u)}{\partial z} \right] dx dy dz$$

$$+ \left[\frac{\partial(\tau_{xy} v)}{\partial x} + \frac{\partial(\tau_{yy} v)}{\partial y} + \frac{\partial(\tau_{zy} v)}{\partial z} \right] dx dy dz$$

$$+ \left[\frac{\partial(\tau_{xz} w)}{\partial x} + \frac{\partial(\tau_{yz} w)}{\partial y} + \frac{\partial(\tau_{zz} w)}{\partial z} \right] dx dy dz \tag{7.46}$$

In unit time, the total heat Q transferred from outside to the system includes heat radiation and heat conduction. Let q denote the heat transferred to the unit mass fluid by thermal radiation per unit time, and the total radiation heat is

$$Q_R = \rho q \, dx dy dz \tag{7.47}$$

According to the Fourier theorem, the heat transferred to the system through the control surface can be obtained. As shown in Fig. 7.7, for the X-direction, the heat transferred into the system per unit time through the control surface heat conduction is

$$Q_{kx} = \left[q_{kx} - \left(q_{kx} + \frac{\partial q_{kx}}{\partial x} dx \right) \right] dy dz = - \frac{\partial q_{kx}}{\partial x} dx dy dz$$

Fig. 7.7 Heat conduction diagram

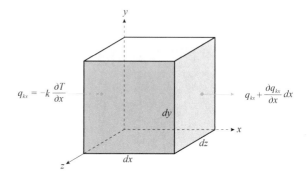

$$= -\frac{\partial}{\partial x}\left(-k\frac{\partial T}{\partial x}\right)\mathrm{d}x\mathrm{d}y\mathrm{d}z = \frac{\partial}{\partial x}\left(k\frac{\partial T}{\partial x}\right)\mathrm{d}x\mathrm{d}y\mathrm{d}z \qquad (7.48)$$

In the same way, the heat conduction in Y- and Z-directions is

$$Q_{ky} = \frac{\partial}{\partial y}\left(k\frac{\partial T}{\partial y}\right)\mathrm{d}x\mathrm{d}y\mathrm{d}z \qquad (7.49)$$

$$Q_{kz} = \frac{\partial}{\partial z}\left(k\frac{\partial T}{\partial z}\right)\mathrm{d}x\mathrm{d}y\mathrm{d}z \qquad (7.50)$$

The results show that the total heat conduction per unit time is

$$
\begin{aligned}
Q_k &= Q_{kx} + Q_{ky} + Q_{kz} \\
&= \left[\frac{\partial}{\partial x}\left(k\frac{\partial T}{\partial x}\right) + \frac{\partial}{\partial y}\left(k\frac{\partial T}{\partial y}\right) + \frac{\partial}{\partial z}\left(k\frac{\partial T}{\partial z}\right)\right]\mathrm{d}x\mathrm{d}y\mathrm{d}z \\
&= \nabla \cdot (k\nabla T)\mathrm{d}x\mathrm{d}y\mathrm{d}z \qquad (7.51)
\end{aligned}
$$

The energy equation based on the differential form of the system is obtained by substituting the above formulas into Eq. (7.38),

$$\rho\frac{d}{dt}\left(e + \frac{V^2}{2}\right) = \rho\vec{f}\cdot\vec{V} + \nabla\cdot([\tau]\cdot\vec{V}) + \rho q + \nabla\cdot(k\nabla T) \qquad (7.52)$$

In which,

$$
\begin{aligned}
\nabla\cdot([\tau]\cdot\vec{V}) &= \frac{\partial(\tau_{xx}u + \tau_{xy}v + \tau_{xz}w)}{\partial x} \\
&\quad + \frac{\partial(\tau_{yx}u + \tau_{yy}v + \tau_{yz}w)}{\partial y} + \frac{\partial(\tau_{zx}u + \tau_{zy}v + \tau_{zz}w)}{\partial z} \qquad (7.53)
\end{aligned}
$$

In addition, u, v, and w are, respectively, multiplied by the corresponding equations of the differential equations of motion (5.60) expressed in the form of stress, and the sum is obtained

$$\begin{cases} \rho u \frac{du}{dt} = \rho u f_x + u\left(\frac{\partial \tau_{xx}}{\partial x} + \frac{\partial \tau_{yx}}{\partial y} + \frac{\partial \tau_{zx}}{\partial z}\right) \\ \rho v \frac{dv}{dt} = \rho v f_y + v\left(\frac{\partial \tau_{xy}}{\partial x} + \frac{\partial \tau_{yy}}{\partial y} + \frac{\partial \tau_{zy}}{\partial z}\right) \\ \rho w \frac{dw}{dt} = \rho w f_z + w\left(\frac{\partial \tau_{xz}}{\partial x} + \frac{\partial \tau_{yz}}{\partial y} + \frac{\partial \tau_{zz}}{\partial z}\right) \end{cases} \tag{7.54}$$

$$\rho u \frac{du}{dt} + \rho v \frac{dv}{dt} + \rho w \frac{dw}{dt} = \rho \frac{d}{dt}\left(\frac{V^2}{2}\right)$$

$$\rho u f_x + \rho v f_y + \rho w f_z = \rho \vec{f} \cdot \vec{V} \tag{7.55}$$

$$u\left(\frac{\partial \tau_{xx}}{\partial x} + \frac{\partial \tau_{yx}}{\partial y} + \frac{\partial \tau_{zx}}{\partial z}\right) + v\left(\frac{\partial \tau_{xy}}{\partial x} + \frac{\partial \tau_{yy}}{\partial y} + \frac{\partial \tau_{zy}}{\partial z}\right) + w\left(\frac{\partial \tau_{xz}}{\partial x} + \frac{\partial \tau_{yz}}{\partial y} + \frac{\partial \tau_{zz}}{\partial z}\right)$$

$$= \frac{\partial}{\partial x}\left(\tau_{xx} u + \tau_{yx} v + \tau_{zx} w\right) + \frac{\partial}{\partial y}\left(\tau_{xy} u + \tau_{yy} v + \tau_{zy} w\right) + \frac{\partial}{\partial z}\left(\tau_{xz} u + \tau_{yz} v + \tau_{zz} w\right)$$

$$- \left(\tau_{xx} \frac{\partial u}{\partial x} + \tau_{yx} \frac{\partial u}{\partial y} + \tau_{zx} \frac{\partial w}{\partial z}\right) - \left(\tau_{yx} \frac{\partial v}{\partial x} + \tau_{yy} \frac{\partial v}{\partial y} + \tau_{zy} \frac{\partial v}{\partial z}\right)$$

$$- \left(\tau_{xz} \frac{\partial w}{\partial x} + \tau_{yz} \frac{\partial w}{\partial y} + \tau_{zz} \frac{\partial w}{\partial z}\right)$$

$$= \nabla \cdot ([\tau] \cdot \vec{V}) - \Phi \tag{7.56}$$

$$\Phi = \tau_{xx} \frac{\partial u}{\partial x} + \tau_{yy} \frac{\partial v}{\partial y} + \tau_{zz} \frac{\partial w}{\partial z}$$

$$+ \tau_{xy}\left(\frac{\partial u}{\partial y} + \frac{\partial v}{\partial x}\right) + \tau_{zx}\left(\frac{\partial u}{\partial z} + \frac{\partial w}{\partial x}\right) + \tau_{yz}\left(\frac{\partial v}{\partial z} + \frac{\partial w}{\partial y}\right)$$

$$= [\tau] \cdot [\varepsilon] \tag{7.57}$$

Substituting Eqs. (7.55), (7.56), and (7.57) into the summation expression of Eq. (7.54), we get

$$\rho \frac{d}{dt}\left(\frac{V^2}{2}\right) = \rho \vec{f} \cdot \vec{V} + \nabla \cdot ([\tau] \cdot \vec{V}) - \Phi \tag{7.58}$$

where Φ is the work function of surface force.

Substituting Eq. (7.58) into Eq. (7.52), we get

$$\rho \frac{de}{dt} = \rho q + \nabla \cdot (k \nabla T) + \Phi \tag{7.59}$$

The above formulas are more concise in tensor form. The tensor form of Eq. (7.52) is

$$\rho \frac{d}{dt}\left(e + \frac{u_i u_i}{2}\right) = \rho f_i u_i + \frac{\partial(\tau_{ij} u_j)}{\partial x_i} + \rho q + \frac{\partial}{\partial x_i}\left(k \frac{\partial T}{\partial x_i}\right) \tag{7.60}$$

Subscripts i and j are operation indexes, $i = 1, 2, 3, j = 1, 2, 3$. In one formula, the same two indexes represent the sum. If we multiply u_i by the system of differential equations of motion, we have

$$\rho u_i \frac{du_i}{dt} = \rho f_i u_i + u_i \frac{\partial \tau_{ji}}{\partial x_j}$$

$$\rho \frac{d}{dt}\left(\frac{u_i u_i}{2}\right) = \rho f_i u_i + \frac{\partial(\tau_{ji} u_i)}{\partial x_j} - \tau_{ji} \frac{\partial u_i}{\partial x_j}$$

$$\rho \frac{d}{dt}\left(\frac{u_i u_i}{2}\right) = \rho f_i u_i + \frac{\partial(\tau_{ji} u_i)}{\partial x_j} - \tau_{ji} \frac{1}{2}\left(\frac{\partial u_i}{\partial x_j} + \frac{\partial u_j}{\partial x_i}\right)$$

$$\rho \frac{d}{dt}\left(\frac{u_i u_i}{2}\right) = \rho f_i u_i + \frac{\partial(\tau_{ji} u_i)}{\partial x_j} - \tau_{ji}\varepsilon_{ji} \tag{7.61}$$

Substituting the energy Eq. (7.60), the tensor form of the energy equation is obtained

$$\rho \frac{de}{dt} = \tau_{ij}\varepsilon_{ij} + \rho q + \frac{\partial}{\partial x_i}\left(k \frac{\partial T}{\partial x_i}\right) \tag{7.62}$$

The work function of the surface force is

$$\Phi = \tau_{ij}\varepsilon_{ij} \tag{7.63}$$

The physical meaning of Eq. (7.59) or (7.62) is: in unit time, the change rate of the internal energy of the unit volume fluid element with time is equal to the sum of the deformation power of the surface force of the unit volume fluid element and the heat from the outside. Among them, the deformation power of surface force includes normal force power and shear force power. The normal force power represents the expansion power of normal force when the fluid element deforms. The shear force power represents the power consumed to overcome the friction when the fluid element shears. This part is caused by the viscosity of the fluid, which irreversibly converts the mechanical energy of the fluid element into heat energy.

By using the generalized Newton's internal friction theorem, Formulas (5.45) and (7.57), it can be concluded that

$$\Phi = -p\nabla \cdot \vec{V} + \phi$$

$$\phi = -\frac{2}{3}\mu\left(\frac{\partial u}{\partial x} + \frac{\partial v}{\partial y} + \frac{\partial w}{\partial z}\right)^2 + 2\mu\left[\left(\frac{\partial u}{\partial x}\right)^2 + \left(\frac{\partial v}{\partial y}\right)^2 + \left(\frac{\partial w}{\partial z}\right)^2\right]$$

$$+ \mu \left(\frac{\partial v}{\partial x} + \frac{\partial u}{\partial y} \right)^2 + \mu \left(\frac{\partial u}{\partial z} + \frac{\partial w}{\partial x} \right)^2 + \mu \left(\frac{\partial w}{\partial y} + \frac{\partial v}{\partial z} \right)^2$$

$$= -\frac{2}{3} \mu \left(\nabla \cdot \vec{V} \right)^2 + 2\mu[\varepsilon] \cdot [\varepsilon] \tag{7.64}$$

where φ is the power due to surface viscous stress which is called dissipation function. It can be expressed as tensor

$$\Phi = \tau_{ij} \varepsilon_{ij} = -p \frac{\partial u_i}{\partial x_i} + \phi$$

$$\phi = 2\mu \varepsilon_{ij} \varepsilon_{ij} - \frac{2}{3} \mu \left(\frac{\partial u_i}{\partial x_i} \right)^2 \tag{7.65}$$

By substituting Eq. (7.64) into Eq. (7.59), the energy equation can be written as

$$\rho \frac{de}{dt} = -p \nabla \cdot \vec{V} + \rho q + \nabla \cdot (k \nabla T) + \phi \tag{7.66}$$

The tensor form is

$$\rho \frac{de}{dt} = -p \frac{\partial u_i}{\partial x_i} + \rho q + \frac{\partial}{\partial x_i} \left(k \frac{\partial T}{\partial x_i} \right) + \phi \tag{7.67}$$

It is shown that the change rate of internal energy per unit volume of fluid is equal to the sum of expansion power, external heat, and mechanical power consumed due to viscosity. From the continuous equation, we have

$$\nabla \cdot \vec{V} = -\frac{1}{\rho} \frac{d\rho}{dt}, \quad -p \nabla \cdot \vec{V} = \frac{p}{\rho} \frac{d\rho}{dt} \tag{7.68}$$

The differential of enthalpy is obtained as follows:

$$\frac{dh}{dt} = \frac{d}{dt} \left(e + \frac{p}{\rho} \right) = \frac{de}{dt} + p \frac{d}{dt} \left(\frac{1}{\rho} \right) + \frac{1}{\rho} \frac{dp}{dt} \tag{7.69}$$

$$\rho \frac{dh}{dt} = \rho \frac{de}{dt} - \frac{p}{\rho} \frac{d\rho}{dt} + \frac{dp}{dt} \tag{7.70}$$

Substituting Eqs. (7.68) and (7.70) into the energy Eq. (7.66), we get

$$\rho \frac{dh}{dt} = \frac{dp}{dt} + \rho q + \nabla \cdot (k \nabla T) + \phi \tag{7.71}$$

Then $dh = C_p dT$, $de = C_V dT$, the energy equation of the following form is obtained:

$$\rho C_V \frac{dT}{dt} = -p \nabla \cdot \overrightarrow{V} + \rho q + \nabla \cdot (k \nabla T) + \phi \qquad (7.72)$$

$$\rho C_p \frac{dT}{dt} = \frac{dp}{dt} + \rho q + \nabla \cdot (k \nabla T) + \phi \qquad (7.73)$$

For the ideal compressible fluid, $\phi = 0$, and we get

$$\rho C_V \frac{dT}{dt} = -p \nabla \cdot \overrightarrow{V} + \rho q + \nabla \cdot (k \nabla T) \qquad (7.74)$$

$$\rho C_p \frac{dT}{dt} = \frac{dp}{dt} + \rho q + \nabla \cdot (k \nabla T) \qquad (7.75)$$

The dot product of Euler equations for the ideal compressible fluid and velocity vector is

$$\rho \frac{d}{dt} \left(\frac{V^2}{2} \right) = \rho \overrightarrow{f} \cdot \overrightarrow{V} - \overrightarrow{V} \cdot \nabla p \qquad (7.76)$$

The result of adding Eqs. (7.75) and (7.76) is

$$\rho \frac{d}{dt} \left(C_p T + \frac{V^2}{2} \right) = \rho \overrightarrow{f} \cdot \overrightarrow{V} + \frac{\partial p}{\partial t} + \rho q + \nabla \cdot (k \Delta T) \qquad (7.77)$$

Or, Eq. (7.77) is written as

$$\rho \frac{d}{dt} \left(h + \frac{V^2}{2} \right) = \rho \overrightarrow{f} \cdot \overrightarrow{V} + \frac{\partial p}{\partial t} + \rho q + \nabla \cdot (k \Delta T) \qquad (7.78)$$

For the adiabatic motion of an ideal compressible fluid, we can obtain

$$\phi = 0, \ \rho q = 0, \ k = 0$$

$$\rho C_V \frac{dT}{dt} = -p \nabla \cdot \overrightarrow{V} \qquad (7.79)$$

$$\rho C_p \frac{dT}{dt} = \frac{dp}{dt} \qquad (7.80)$$

$$\rho \frac{d}{dt} \left(h + \frac{V^2}{2} \right) = \rho \overrightarrow{f} \cdot \overrightarrow{V} + \frac{\partial p}{\partial t} \qquad (7.81)$$

For the viscous incompressible fluid, we can obtain

$$\rho C_V \frac{dT}{dt} = \rho q + k \Delta T + \phi \qquad (7.82)$$

$$\rho C_p \frac{\mathrm{d}T}{\mathrm{d}t} = \frac{\mathrm{d}p}{\mathrm{d}t} + \rho q + k\Delta T + \phi \tag{7.83}$$

7.5   Speed of Sound and Mach Number

7.5.1   *Propagation Velocity of Disturbance Wave in Elastic Medium*

It has been found that for any elastic medium given any disturbance, the disturbance will automatically propagate around, and as long as the disturbance is not too strong, its propagation speed is constant and does not change because of the specific form of disturbance. Sound velocity is essentially the propagation velocity of small distur-bance in elastic medium. It is illustrated by a mass ball and spring system. There is a system connected by a mass ball and a spring. The ball is rigid and the spring is massless. Someone tapped the ball on the left with a small hammer. Now look at the motion of the system (Fig. 7.8).

After the first ball is hit, it moves slightly to the right, thus compressing the first spring. After the first spring is compressed, it produces an elastic force. This force pushes the second mass ball, causing the second ball to move slightly to the right, thus compressing the second spring and so on. The disturbance of the small hammer is passed down step by step from left to right until the last ball. In this process, it is necessary to distinguish between the motion of the ball and the transfer of disturbance. The motion of each ball is small, but the disturbance is transmitted through the spring ball by ball, and its propagation speed is totally different from the ball's motion speed.

The analysis of the behavior of the system shows that the propagation velocity of the disturbance is different from that of the medium itself. In order of magnitude, the propagation velocity of disturbance is much higher than that of the medium itself caused by disturbance. In the case of small disturbance, the velocity of the medium

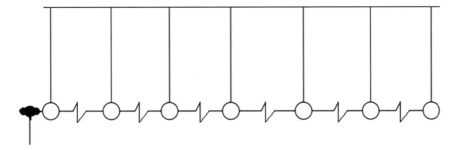

Fig. 7.8 Propagation velocity of disturbance wave

is also small, but the propagation velocity of small disturbance is not small, but a constant, which is much larger than the velocity of medium particles. It is found that the value is related to the elasticity and mass of the medium (a function of physical properties), but not to the amplitude of the disturbance. Air is a kind of elastic medium, in which any small disturbance will propagate in all directions. Of course, the propagation speed depends on the physical properties and state of the medium. Sound is a kind of propagation of audible disturbance, which is consistent with the propagation speed of inaudible disturbance.

7.5.2 Micro-Disturbance Propagation Velocity—Speed of Sound

The propagation of micro-disturbance in elastic medium is in the form of pressure wave. Its propagation velocity (also known as sound velocity) is closely related to the elasticity of medium, and it is also an important physical quantity to study compressible flow field. It is found that the propagation velocity is smaller when the medium is elastic (compressible). On the contrary, if the elasticity is small (compressibility is small), the propagation speed will be large. For a rigid body (or incompressible fluid), the propagation velocity is infinite, and the transfer of disturbance is instantaneous. For the sake of simplicity, the wave velocity in a long thin tube is derived. Suppose there is a long and thin pipe with a cross-sectional area of A and a piston at the left end of the pipe. Now push the piston to the right with a small speed dv to make the air in the pipe produce a small compression disturbance, as shown in Fig. 7.9. Let the disturbance propagate to the right at a constant wave velocity a and advance to the right with wave front A-A in the pipe. The gas on the right side of the wave front is undisturbed, and its pressure, density, temperature, and velocity are p, ρ, T, and $v = 0$, respectively. When the gas on the left side of the wave front is disturbed, its pressure, density, temperature, and velocity become $p + \mathrm{d}p$, $+ \rho\mathrm{d}\rho$, $T + \mathrm{d}T$, and dv, respectively. As shown in Fig. 7.9, because the movement of the piston is small, the disturbance generated in the pipeline is small, namely,

$$\frac{\mathrm{d}p}{p} \ll 1, \quad \frac{\mathrm{d}\rho}{\rho} \ll 1, \quad \frac{\mathrm{d}T}{T} \ll 1, \quad \frac{\mathrm{d}v}{a} \ll 1 \qquad (7.84)$$

For the convenience of analysis, it is assumed that the observer moves to the right with the wave front, and then the whole flow problem changes from the original unsteady flow to a steady flow. As shown in Fig. 7.10, at this time, the wave front does not move, the undisturbed gas moves to the left at the wave speed a, and the airflow continuously crosses the A–A plane into the disturbed area, while the disturbed airflow leaves the A–A plane to the left at the a-dv speed. Taking a control volume 1234 around the A–A plane, we can get the mass conservation equation

$$\rho A a = (\rho + \mathrm{d}\rho)(a - \mathrm{d}V)A \qquad (7.85)$$

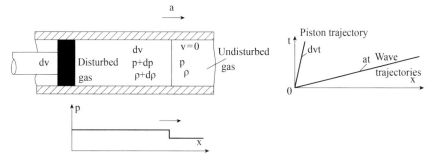

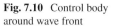

Fig. 7.9 Propagation velocity of disturbance in long thin pipe

Fig. 7.10 Control body
around wave front

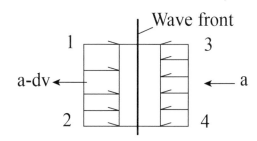

If we ignore the second-order small quantity, we have

$$a\mathrm{d}\rho = \rho\mathrm{d}V \tag{7.86}$$

From the theorem of momentum, it can be concluded that

$$pA - (p + \mathrm{d}p)A = \rho a A(a - \mathrm{d}V - a) \tag{7.87}$$

After simplification, we get

$$\mathrm{d}p = \rho a \mathrm{d}V \tag{7.88}$$

It can be obtained by solving Eqs. (7.86) and (7.88)

$$a^2 = \frac{\mathrm{d}p}{\mathrm{d}\rho} \tag{7.89}$$

Equation (7.89) is the propagation velocity of small disturbance in fluid medium, which is not only applicable to gas, but also applicable to liquid. It shows that the propagation velocity of disturbance wave depends on the ratio of dp and dρ in fluid medium. By definition, the bulk modulus of elasticity E of a fluid is defined as

$$E = -\frac{dp}{d\left(\frac{1}{\rho}\right)/\left(\frac{1}{\rho}\right)} = \rho\frac{dp}{d\rho} \tag{7.90}$$

In the sound wave Formula (7.89), we get

$$a = \sqrt{\frac{E}{\rho}} \tag{7.91}$$

In physics, because of the fast propagation speed and small disturbance, in the process of disturbance propagation, the gas can be considered as an adiabatic isentropic process without heat exchange and friction. At this time, the pressure–density relationship is as follows:

$$p = C\rho^\gamma, \quad p = \rho RT, \quad a = \sqrt{\gamma RT} \tag{7.92}$$

For the sea level standard atmosphere, $R = 287.053$ N m/(kg K), $T = 288.15$ K, $\gamma = 1.4$, the results are as follows:

$$a = \sqrt{1.4 \times 287.035 \times 288.15} = 340.3\,\text{m/s}$$

For water at normal temperature and pressure, substituting Eq. (7.91) to get

$$a_\text{w} = \sqrt{\frac{E}{\rho}} = \sqrt{\frac{E_\text{w} = 2.1 \times 10^9\,\text{N/m}^2}{1000}} = 1449.1\,\text{m/s}$$

The water wave velocity is 1482.0 m/ s at 20 °C.

7.5.3 Mach Number

Mach number is the ratio of air velocity V to local sound velocity a, $Ma = V/a$. It represents a basic physical parameter of compressible fluid. However, in the flow field, the same *Mach* number does not necessarily mean the same velocity because of the different sound velocity a at each point. For example, when a fighter flies at an altitude of 10 km, $Ma = 2$ means $V = 600$ m/s. If the same aircraft flies at sea level with Ma = 2, the flight speed is 682 m/s. Generally speaking, the velocity and sound velocity of each point in the flow field are different, so *Ma* refers to the local value, which is called the local Mach number. In aerodynamics, the incoming Mach number $Ma_\infty = \frac{V_\infty}{a_\infty}$ is the ratio of the incoming velocity V_∞ to the sound velocity a_∞ corresponding to the incoming velocity.

Mach number is also a dimensionless parameter to characterize the compressibility of the flow field. It is an important basic physical parameter in high-speed aerodynamics and a similarity criterion to reflect the compressibility of the flow field. The compressibility of gas can be expressed by the change of relative density, which is closely related to Ma number. Namely,

$$a^2 = \frac{dp}{d\rho} \propto \frac{\Delta p}{\Delta \rho} \propto \frac{\rho V^2}{\Delta \rho} \quad \frac{\Delta \rho}{\rho} \propto \frac{V^2}{a^2} = Ma^2 \quad (7.93)$$

It shows that the greater the Ma number, the greater the compressibility of the gas. When $Ma < 0.3$, the gas density changes little, so it can be treated as incompressible fluid. In addition, Ma number also represents the ratio of kinetic energy to internal energy per unit mass of gas. Namely,

$$\frac{V^2/2}{C_v T} = \frac{V^2/2}{\frac{\gamma RT}{\gamma(\gamma-1)}} = \frac{\gamma(\gamma-1)}{2} \frac{V^2}{a^2} = \frac{\gamma(\gamma-1)}{2} Ma^2 \quad (7.94)$$

7.5.4 Assumption of Incompressible Flow

Strictly speaking, any gas is compressible. According to Bernoulli's equation, the velocity of gas flow will affect the pressure, the pressure will affect the density, and the density will affect the flow. Therefore, compressibility is universal. But when the incoming Mach number is small enough, the influence of gas compressibility on the flow is small and can be ignored. At this time, the gas flow can be regarded as incompressible flow. By definition, the flow dynamic pressure can be written as

$$\rho \frac{V_\infty^2}{2} = \frac{1}{2} \frac{\rho_\infty}{\gamma p_\infty} \gamma p_\infty V_\infty^2 = \frac{1}{2} \frac{1}{\gamma RT_\infty} \gamma p_\infty V_\infty^2 = \frac{1}{2} \frac{V_\infty^2}{a_\infty^2} \gamma p_\infty = \frac{1}{2} Ma_\infty^2 \gamma p_\infty \quad (7.95)$$

In compressible flow, the pressure at any point (see Sect. 7.6) is

$$\frac{p_0}{p_\infty} = \left(1 + \frac{\gamma-1}{2} Ma_\infty^2\right)^{\frac{\gamma}{\gamma-1}} \quad (7.96)$$

where p_0 is the stagnation point pressure and $\gamma = 1.4$. And Eq. (7.96) is expanded by Taylor series,

$$\frac{p_0}{p_\infty} = \left(1 + \frac{\gamma-1}{2} Ma_\infty^2\right)^{\frac{\gamma}{\gamma-1}} = \left(1 + 0.2 Ma_\infty^2\right)^{3.5}$$

$$= 1 + \frac{7}{11} Ma_\infty^2 + \frac{7}{40} Ma_\infty^4 + \frac{7}{400} Ma_\infty^6 + \frac{7}{16000} Ma_\infty^8 + \cdots \quad (7.97)$$

The pressure coefficient of stagnation point is

$$C_{p_0} = \frac{p_0 - p_\infty}{\frac{1}{2}\rho_\infty V_\infty^2} = \frac{p_\infty}{\frac{1}{2}\rho_\infty V_\infty^2}\left(\frac{p_0}{p_\infty} - 1\right) = \frac{2}{\gamma Ma_\infty^2}\left[\left(1 + \frac{\gamma - 1}{2}Ma_\infty^2\right)^{\frac{\gamma}{\gamma-1}} - 1\right]$$

(7.98)

Substituting Eq. (7.97) into Eq. (7.98), we get

$$\begin{aligned} C_{p_0} &= \frac{2}{\gamma Ma_\infty^2}\left[\left(1 + \frac{\gamma - 1}{2}Ma_\infty^2\right)^{\frac{\gamma}{\gamma-1}} - 1\right] = \frac{1}{0.7Ma_\infty^2}\left[\left(1 + 0.2Ma_\infty^2\right)^{3.5} - 1\right] \\ &= \frac{10}{7Ma_\infty^2}\left(1 + \frac{7}{10}Ma_\infty^2 + \frac{7}{40}Ma_\infty^4 + \frac{7}{400}Ma_\infty^6 + \frac{7}{16000}Ma_\infty^8 + \cdots - 1\right) \\ &= 1 + \frac{1}{4}Ma_\infty^2 + \frac{1}{40}Ma_\infty^4 + \frac{1}{1600}Ma_\infty^6 + \cdots \end{aligned}$$

(7.99)

It can be seen from the above formula that the flow is incompressible ($C_{p_0} = 1.0$) in the case of $Ma_\infty = 0$. In this case $Ma_\infty = 0.3$, Formula (7.99) shows $C_{p_0} = 1.0227$ that compared with the incompressible result, the error is about 2%, which can be ignored in engineering application. Therefore, in aerodynamics, the flow with Mach number less than 0.3 is regarded as incompressible flow or low-speed flow.

7.6   One-Dimensional Compressible Steady Flow Theory

Compared with the incompressible flow, the parameters of one-dimensional compressible steady flow are p, ρ, T, and V, we need four basic equations to solve. In addition to the equation of state, the continuity equation and the momentum equation of ideal flow (Euler equation), the energy equation needs to be added.

7.6.1   Energy Equation of One-Dimensional Compressible Steady Adiabatic Flow

For one-dimensional compressible steady flow, the energy Eq. (7.8) is

$$dq = d\left(e + \frac{p}{\rho} + \frac{V^2}{2}\right)$$

(7.100)

In the condition of adiabatic flow, the energy equation is integrated along the streamline

$$h + \frac{V^2}{2} = C$$

(7.101)

Using $h = C_p T$, we can get

$$C_p T + \frac{V^2}{2} = C \qquad (7.102)$$

$$\frac{\gamma R T}{\gamma - 1} + \frac{V^2}{2} = C \qquad (7.103)$$

$$\frac{a^2}{\gamma - 1} + \frac{V^2}{2} = C \qquad (7.104)$$

$$\frac{\gamma}{\gamma - 1} \frac{p}{\rho} + \frac{V^2}{2} = C \qquad (7.105)$$

For the adiabatic steady flow of ideal fluid, which is also isentropic flow, the above energy equation can also be obtained by integrating Euler equation along the streamline. Using the isentropic relation, the Euler equation is integrated along the streamline without considering the mass force

$$\frac{V^2}{2} + \int \frac{dp}{\rho} = C \qquad (7.106)$$

By using the isentropic relation $p = C\rho^\gamma$ and substituting the above formula, we can get the following result:

$$\frac{V^2}{2} + \frac{\gamma}{\gamma - 1} \frac{p}{\rho} = C \qquad (7.107)$$

In thermodynamics, adiabatic process and isentropic process are two different processes. The adiabatic flow of ideal fluid must be isentropic. In the case of viscous fluid, when there is friction between the flow layers, although it is adiabatic, the friction converts the mechanical energy into heat energy and increases the entropy of the gas flow, so the adiabatic is not equal to the entropy. In adiabatic flow, the effect of viscous friction can't change the sum of kinetic energy and enthalpy of gas, but part of kinetic energy is converted into enthalpy. (The above energy equation is suitable for adiabatic flow and isentropic flow.)

7.6.2 Basic Relations Between Parameters of One-Dimensional Compressible Adiabatic Steady Flow

For one-dimensional compressible steady adiabatic flow, the relationship of flow parameters along streamline can be determined. The parameter values of reference points are often needed, and the reference points can be stagnation points or critical points.

1. **Parameter relation using stagnation point as reference value**

Stagnation point refers to the point where the flow velocity or kinetic energy is zero, which can exist in the flow field or be a virtual reference point. According to the energy equation of one-dimensional compressible adiabatic steady flow, the maximum enthalpy of fluid at stagnation point is called total enthalpy h_0, the corresponding temperature is called total temperature T_0, and the pressure is called total pressure P_0. Using the stationary point condition, the energy equation can be written as

$$C_p T + \frac{V^2}{2} = h_0 \tag{7.108}$$

$$T + \frac{V^2}{2C_p} = T_0 \tag{7.109}$$

Among them, $h_0 = C_p T_0$ represents the total energy of one-dimensional adiabatic flow, T_0 is the total temperature, and the temperature T at $V \neq 0$ point is called static temperature, and the relationship between total temperature and static temperature is

$$\frac{T_0}{T} = 1 + \frac{\gamma - 1}{2} Ma^2 \tag{7.110}$$

In one-dimensional adiabatic viscous flow, the total pressure at any point on the streamline is defined as P_0, which is the pressure at which the velocity isentropic drops to zero. Namely,

$$\frac{p_0}{p} = \frac{\rho_0^\gamma}{\rho^\gamma} \tag{7.111}$$

$$\frac{p_0}{p} = \frac{\rho_0}{\rho} \frac{T_0}{T} \tag{7.112}$$

$$\frac{p_0}{p} = \left(\frac{T_0}{T}\right)^{\frac{\gamma}{\gamma-1}} = \left(1 + \frac{\gamma - 1}{2} Ma^2\right)^{\frac{\gamma}{\gamma-1}} \tag{7.113}$$

The entropy increase between points 1 and 2 on the streamline is

$$\Delta s = \int_1^2 ds = \int_1^2 \frac{dq}{T} = C_v \ln\left[\frac{p_2}{p_1}\left(\frac{\rho_1}{\rho_2}\right)^\gamma\right] \tag{7.114}$$

$$\frac{p_2}{p_1}\left(\frac{\rho_1}{\rho_2}\right)^{\gamma} = \left(\frac{p_{02}}{p_{01}}\right)^{-(\gamma-1)} \tag{7.115}$$

$$\Delta s = -C_v(\gamma - 1) \ln\left(\frac{p_{02}}{p_{01}}\right) \tag{7.116}$$

According to the second law of thermodynamics, if the entropy increases along the flow direction and $ds > 0$, there will be $p_{02} < p_{01}$, indicating that although the total temperature T_0 remains unchanged along the flow direction, the total pressure decreases (representing the total mechanical energy in the airflow).

For one-dimensional isentropic flow, the total temperature and total pressure at any point on the streamline are equal, so

$$\begin{cases} \frac{T_0}{T} = 1 + \frac{\gamma-1}{2} Ma^2 \\ \frac{p_0}{p} = \left(\frac{T_0}{T}\right)^{\frac{\gamma}{\gamma-1}} = (1 + \frac{\gamma-1}{2} Ma^2)^{\frac{\gamma}{\gamma-1}} \\ \frac{\rho_0}{\rho} = \left(\frac{p_0}{p}\right)^{\frac{1}{\gamma}} = (1 + \frac{\gamma-1}{2} Ma^2)^{\frac{1}{\gamma-1}} \end{cases} \tag{7.117}$$

2. The critical point as reference quantity

In one-dimensional adiabatic flow, the velocity along the streamline is exactly equal to the local sound velocity ($Ma = 1$), which is called critical point or critical section. The physical quantity at this point is called the critical parameter, which is expressed by the subscript "*". From the energy equation of one-dimensional adiabatic isentropic flow, we can get

$$\frac{T_*}{T_0} = \frac{2}{\gamma + 1} = 0.833 \tag{7.118}$$

$$\frac{p_*}{p_0} = (\frac{2}{\gamma + 1})^{\frac{\gamma}{\gamma-1}} = 0.528 \tag{7.119}$$

$$\frac{\rho_*}{\rho_0} = (\frac{2}{\gamma + 1})^{\frac{1}{\gamma-1}} = 0.634 \tag{7.120}$$

$$\frac{a_*^2}{a_0^2} = \frac{2}{\gamma + 1} = 0.833 \tag{7.121}$$

From the one-dimensional adiabatic flow energy equation, we can get

$$\frac{V^2}{2} + \frac{a^2}{\gamma - 1} = \frac{\gamma + 1}{\gamma - 1} \frac{a_*^2}{2} \tag{7.122}$$

Define velocity coefficient

$$\lambda = \frac{V}{a_*} \tag{7.123}$$

Since the sound velocity at the critical point is only a function of the total temperature, the greatest advantage of the introduction of velocity coefficient is that its denominator is constant at a given total temperature, so all kinds of operations on velocity coefficient are only for molecules.

The relationship between Ma number and velocity coefficient is as follows:

$$Ma^2 = \frac{V^2}{a^2} = \frac{V^2}{a_*^2} \frac{a_*^2}{a_0^2} \frac{a_0^2}{a^2}$$

$$Ma^2 = \lambda^2 \frac{2}{\gamma + 1} \left(1 + \frac{\gamma - 1}{2} Ma^2 \right) \tag{7.124}$$

So

$$\lambda^2 = \frac{\frac{\gamma+1}{2} Ma^2}{1 + \frac{\gamma-1}{2} Ma^2} \tag{7.125}$$

$$Ma^2 = \frac{\lambda^2 \frac{2}{\gamma+1}}{1 - \lambda^2 \frac{\gamma-1}{\gamma+1}} \tag{7.126}$$

$$\begin{cases} \frac{T}{T_0} = 1 - \frac{\gamma-1}{\gamma+1}\lambda^2 = \tau(\lambda) \\[2mm] \frac{p}{p_0} = \left(1 - \frac{\gamma-1}{\gamma+1}\lambda^2\right)^{\frac{\gamma}{\gamma-1}} = \pi(\lambda) \\[2mm] \frac{\rho}{\rho_0} = \left(1 - \frac{\gamma-1}{\gamma+1}\lambda^2\right)^{\frac{1}{\gamma-1}} = \varepsilon(\lambda) \end{cases} \tag{7.127}$$

As shown in Fig. 7.11, the relationship curve between $\tau(\lambda)$, $\pi(\lambda)$, $\varepsilon(\lambda)$, and λ is given.

Fig. 7.11 The relationship curve

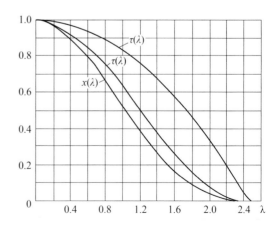

Fig. 7.12 The relation
between velocity coefficient
λ and Mach number

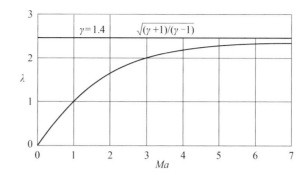

It should be pointed out that in the case of $Ma \to \infty$, the speed coefficient λ takes the maximum value. Namely,

$$\lambda_{max} = \sqrt{\frac{\gamma + 1}{\gamma - 1}} = \sqrt{6} \qquad (7.128)$$

When Ma is less than 1, the velocity coefficient is greater than Ma number; when Ma number is greater than 1, the velocity coefficient is less than Ma number, as shown in Fig. 7.12.

For one-dimensional steady isentropic pipe flow, the flow rate through each section is constant, which is expressed as mass flow rate

$$G = \rho A V \qquad (7.129)$$

Here, take $\rho = \rho_0 \varepsilon(\lambda)$, $\rho_0 = \frac{p_0}{RT_0}$, the velocity is

$$V = \lambda a_* = \lambda \sqrt{\frac{2\gamma}{1 + \gamma} RT_0} \qquad (7.130)$$

$$G = \frac{p_0 A}{\sqrt{T_0}} \sqrt{\frac{2\gamma}{1 + \gamma} \frac{1}{R}} \lambda \varepsilon(\lambda) = C \frac{p_0 A}{\sqrt{T_0}} q(\lambda) \qquad (7.131)$$

In which

$$q(\lambda) = \left(\frac{\gamma + 1}{2}\right)^{\frac{1}{\gamma - 1}} \lambda \varepsilon(\lambda), \quad C = \sqrt{\frac{\gamma}{R} \left(\frac{2}{1 + \gamma}\right)^{\frac{\gamma + 1}{\gamma - 1}}} \qquad (7.132)$$

For the air with low temperature and pressure, $\gamma = 1.4$, $R = 287$, $C = 0.04042$. When $\lambda = 1.1$, $q(\lambda) = 1$. When $\lambda = 0$ and $\lambda = \lambda_{max}$, $q(\lambda) = 0$. In Eq. (7.131), the static parameters are used to express the mass flow rate

$$G = C \frac{p_0 A}{\sqrt{T_0}} q(\lambda) = C \frac{pA}{\sqrt{T_0}} \frac{q(\lambda)}{\pi(\lambda)} = C \frac{pA}{\sqrt{T}} \frac{q(\lambda)\sqrt{\tau(\lambda)}}{\pi(\lambda)} \qquad (7.133)$$

Using the Ma number expression (7.132), we get

$$G = C \frac{p_0 A}{\sqrt{T_0}} q(\lambda) = C \frac{pA}{\sqrt{T_0}} \frac{q(\lambda)}{\pi(\lambda)} = C \frac{pA}{\sqrt{T}} \frac{q(\lambda)\sqrt{\tau(\lambda)}}{\pi(\lambda)} \qquad (7.134)$$

Because of the conservation of mass through the pipe, we get

$$G = \rho V A = \rho^* V^* A^*, \quad \frac{A^*}{A} = \frac{\rho}{\rho^*} \frac{V}{V^*}$$

$$\frac{A^*}{A} = q(\lambda) \qquad (7.135)$$

$$\frac{A^*}{A} = q(Ma) = Ma \left[\frac{2}{\gamma + 1} \left(1 + \frac{\gamma - 1}{2} Ma^2 \right) \right]^{-\frac{\gamma+1}{2(\gamma-1)}} \qquad (7.136)$$

As shown in Fig. 7.13, the curve of the relationship between $q(\lambda)$ and λ is given.

Example 7.2 There is an aircraft flying at $H = 5000$ m and $Ma_\infty = 0.8$. The inlet cross section of the intake $A_1 = 0.5$ m^2, $Ma_1 = 0.4$, and the outlet cross section $Ma_2 = 0.2$. Try to calculate the total parameters of the incoming flow and the p_1, ρ_1, T_1, and mass flow m_1 at the inlet section.

Solution

From the standard atmosphere table, according to $h = 5000$ m, it is found that

$$p_H = 54020 \, \text{N/m}^2, \ \rho_H = 0.73612 \, \text{kg/m}^3, \ T_H = 255.65 \, \text{K}$$

Then, the results are calculated by $Ma_\infty = 0.8$

Fig. 7.13 The curve of the relationship between $q(\lambda)$ and λ

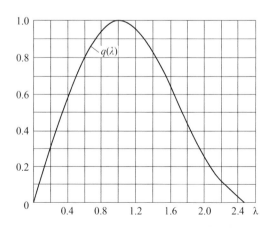

$$\frac{p_\infty}{p_0} = \frac{p_H}{p_0} = 0.6560, \quad \frac{\rho_\infty}{\rho_0} = \frac{\rho_H}{\rho_0} = 0.7400, \quad \frac{T_\infty}{T_0} = \frac{T_H}{T_0} = 0.8865$$

So

$$p_0 = 82347.6 \, \text{N/m}^2, \, \rho_0 = 0.99476 \, \text{kg/m}^3, \, T_0 = 288.36 \, \text{K}$$

Due to $Ma_1 = 0.4$,

$$\frac{p_1}{p_0} = 0.8956, \quad \frac{\rho_1}{\rho_0} = 0.9243, \quad \frac{T_1}{T_0} = 0.9690$$

So

$$p_1 = 73750.5 \, \text{N/m}^2, \, \rho_1 = 0.91946 \, \text{kg/m}^3$$
$$T_1 = 279.44K, \, a_1 = 335.1 \, \text{m/s}$$
$$V_1 = M_1 a_1 = 0.4 \times 335.1 = 134.033 \, \text{m/s}$$
$$m_1 = p_1 V_1 A_1 = 0.91946 \times 134.033 \times 0.5 = 61.62 \, \text{kg/s}$$

7.6.3 Relationship Between Velocity and Cross Section of One-Dimensional Steady Isentropic Pipe Flow

1. Relationship between the change of pipe section and velocity

In the previous chapter, only the relationship between the flow parameters along the streamline is given, and the cross-sectional area of the pipe flow is not taken into account. Now we study the changing law of velocity and other parameters when the cross-sectional area of pipeline changes. In order to highlight the change of cross-sectional area, one-dimensional steady isentropic flow in pipeline is analyzed. From the equation of continuity, it can be concluded that

$$\frac{d\rho}{\rho} + \frac{dV}{V} + \frac{dA}{A} = 0 \tag{7.137}$$

By substituting the relation of sound velocity $a^2 = \frac{dp}{d\rho}$ into Euler equation $dp = -\rho V dV$, we can get the following result:

$$\frac{d\rho}{\rho} = -Ma^2 \frac{dV}{V} \tag{7.138}$$

Substituting into the continuous equation, we get

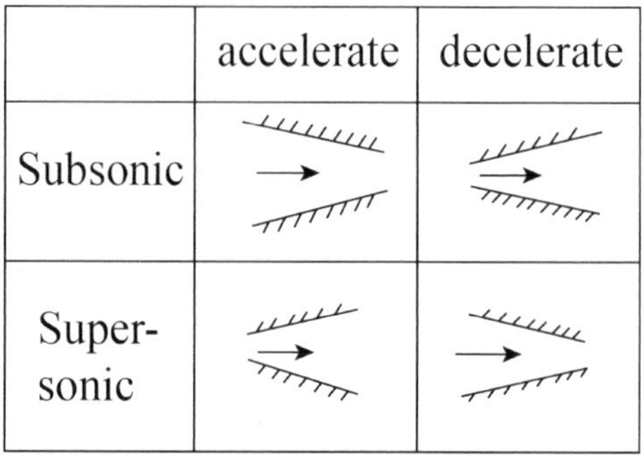

Fig. 7.14 The relationship between pipe section and velocity

	accelerate	decelerate
Subsonic		
Super-sonic		

$$(Ma^2 - 1)\frac{\mathrm{d}V}{V} = \frac{\mathrm{d}A}{A} \tag{7.139}$$

From Eq. (7.139), it can be concluded that

$$Ma < 1, \ \mathrm{d}A > 0, \ \mathrm{d}V < 0; \ \mathrm{d}A < 0, \ \mathrm{d}V > 0_{\circ}$$
$$Ma > 1, \ \mathrm{d}A > 0, \ \mathrm{d}V > 0; \ \mathrm{d}A < 0, \ \mathrm{d}V < 0_{\circ}$$
$$Ma = 1, \ \mathrm{d}A/A = 0, \ A \text{ has extremum.}$$

As shown in Fig. 7.14, it can be concluded that

(1) For subsonic (including low speed) flow, if the cross section of the pipe shrinks, the flow velocity increases, the area expands, and the flow velocity decreases.
(2) For supersonic flow, if the cross section of the pipe shrinks, the velocity decreases, the area expands, and the velocity increases.
(3) The reason is that the variation of velocity and area in supersonic flow is opposite to that in subsonic flow, and the contribution of density to the continuity equation is different. At subsonic speed, the change of density is slower than that of velocity; at supersonic speed, the change of density is faster than that of velocity. Therefore, in order to increase the velocity, the cross-sectional area should be reduced at subsonic speed and enlarged at supersonic speed.

2. **Laval nozzle**

For one-dimensional isentropic steady pipe flow, in order to continuously accelerate the flow from subsonic flow to supersonic flow along the pipe axis, that is, to keep $\mathrm{d}V > 0$, the pipe should first shrink and then expand. There is a minimum section in the middle, which is called throat. The pipe with such shape is called Laval nozzle, as shown in Fig. 7.15. Laval (1845–1913, as shown in Fig. 7.16), a Swedish engineer, successfully obtained supersonic airflow through the first contraction and then expansion pipes in 1889 and manufactured the impact steam turbine, so Laval was the inventor of the single-stage impact steam turbine. Laval nozzle is the main device used to produce high-speed airflow in various industrial technology fields. It is an important part of aerospace vehicle power plant and related experimental equipment (calibration wind tunnel). The cross-sectional area of the flow passage of the convergent nozzle is gradually reduced. Under the effect of the pressure difference between the inlet and outlet of the nozzle, the internal energy of the high-temperature gas is transformed into kinetic energy, which produces a lot of thrust, but the air velocity accelerates to the speed of sound at most. With Laval nozzle, the flow can continue to accelerate from sonic section to supersonic flow after passing through the throat.

Laval found that the conditions for a nozzle to produce a supersonic flow with $Ma > 1$ at the exit section are as follows: (1) the shape of the pipe should be Laval pipe shape with first contraction and then expansion; (2) the pressure ratio should be large enough in the upstream and downstream of the nozzle. When the nozzle outlet reaches the design Ma number and the outlet pressure is equal to the external atmospheric pressure, the nozzle is in the design state, and the corresponding upstream and downstream pressure ratio (i.e., the ratio of upstream total pressure to outlet

Fig. 7.15 Working principle of Laval nozzle

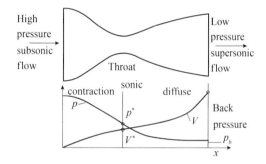

Fig. 7.16 Laval, Swedish Engineer (1845–1913)

atmospheric back pressure) is called the design pressure ratio. If the downstream pressure is too high or too low, shock wave or expansion wave will appear inside and outside the nozzle outlet.

7.7 Small Disturbance Propagation Region, Mach Cone, Mach Wave

When an object moves in still air, different moving speeds have different influence ranges and influence methods on the air. The so-called disturbance refers to the phenomenon that causes changes in the speed, density, and pressure of the airflow. It is known that the propagation of disturbance and its propagation range are different for subsonic flow field and supersonic flow field. In a uniform flow field, it is assumed that the gas is stationary and the disturbance source is moving. The amount of disturbance from the disturbance source will propagate around at the speed of sound. The area of influence will be determined according to the relative size of the speed of sound and the speed of the object's movement. There are the following four situations.

(1) Static disturbance source ($Ma = 0$), $V = 0$

From a certain moment, the disturbance wave front emitted in the first i second is a concentric sphere with the disturbance source O as the center and the radius ia. As long as the time is long enough, any point in space will be affected by the disturbance source, that is, the influence area of the disturbance source is the full flow field, as shown in Fig. 7.17a.

(2) Subsonic disturbance source ($Ma < 1$), $V < a$

When an object moves, the propagation of the disturbance is affected by the speed of the object's movement. The spherical wave with radius ia emitted by the disturbance source in the first i seconds will move downstream with the moving speed of the disturbance source, and the disturbance source will move downstream from O to the point Oi, $OOi = iV$. Since $iV < ia$, the disturbance can still spread throughout the flow field, as shown in Fig. 7.17b. Small disturbances in the subsonic flow field can cover the entire flow field, and the airflow has been felt its existence before the disturbance source arrives. Therefore, when the airflow flows to the disturbance source, the flow direction and airflow parameters will be gradually changed to meet the requirements of the disturbance source.

(3) Sound velocity disturbance source ($Ma = 1$), $V = a$

When the disturbance source moves in the flow field at the speed of sound, the small disturbance wave will not be transmitted upstream of the disturbance source, which means that the airflow does not feel any disturbance before the disturbance source arrives, so the existence of the disturbance source cannot be detected in advance, as shown in Fig. 7.17c.

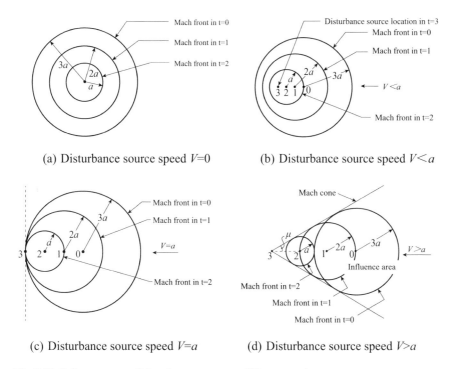

(a) Disturbance source speed $V=0$

(b) Disturbance source speed $V<a$

(c) Disturbance source speed $V=a$

(d) Disturbance source speed $V>a$

Fig. 7.17 Influence range of disturbance sources at different speeds

(4) Supersonic disturbance source ($Ma > 1$), $V > a$

When the disturbance source moves at supersonic speed in the flow field, the disturbance wave not only cannot propagate upstream, but also the area propagating downstream is concentrated in a certain range. When the supersonic airflow is slightly disturbed, it will propagate out at the speed of sound, but because the speed of the disturbance source is greater than the speed of sound, the disturbing spherical wave will form an envelope surface downstream. In aerodynamics, this envelope surface is called a Mach wave array, referred to as Mach wave for short. (Strictly speaking, Mach waves are not wave fronts, but envelope surfaces of different disturbance wave fronts, which belong to the boundary line of disturbance waves.) For point disturbance sources, this front is tapered, so it is also called Mach cone (different shapes, the shape of the Mach wave surface is different, for example, the influence zone of thin wedge-shaped objects is wedge-shaped; for slender pointed cones, Mach cone is of course conical). The airflow outside the Mach wave line is not affected, and the airflow inside the Mach wave line is affected by disturbance.

The half apex angle of the Mach cone is called the Mach angle and is represented by μ. As shown in Fig. 7.18, according to the geometric relationship, since the normal velocity of the Mach wave line is the speed of sound a, then

Fig. 7.18 Mach wave line
and Mach angle

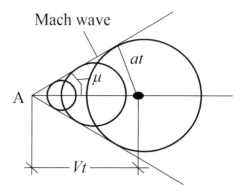

$$\sin(\mu) = \frac{V_n}{V} = \frac{a}{V} \tag{7.140}$$

$$\mu = \arcsin\left(\frac{a}{V}\right) = \arcsin\left(\frac{1}{Ma}\right) \tag{7.141}$$

It can be seen that the larger the incoming Mach number, the smaller the range of the Mach cone.

7.8   Expansion Wave and Supersonic Flow Around the Wall at an Outer Angle

7.8.1   Mach Wave (Expansion Wave)

For high-speed airflow, the influence of density changes on the flow cannot be ignored. In aerodynamics, the process of increasing pressure and density is called the compression process; the process of decreasing pressure and density is called the expansion process. In the high subsonic flow, although there are compression and expansion processes, because there is no perturbation boundary, the perturbation will affect the entire flow field. However, in supersonic flow, both compression and expansion processes have perturbation boundaries, and this perturbation boundary surface is related to Mach waves.

As shown in Fig. 7.19, in supersonic parallel flow, if the wall at point O deflects outward by a slight angle dδ, the flow area will be enlarged, which is called the expansion angle. Obviously, there is a slight disturbance caused by the outward folding of the wall at point O. The propagation range of the disturbance is the downstream area of the Mach wave OL emitted from point O. The result of the disturbance is as follows: the airflow is also deflected outward by the angle of dδ and increases speed, reduces the pressure, and the airflow expands.

Fig. 7.19 Supersonic flow at outer corners

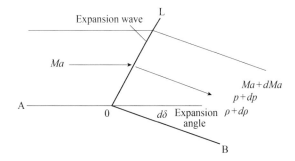

Because the wall surface is folded outward, it is equivalent to the expansion of the airflow channel. For supersonic airflow, expanding the cross-sectional area of the channel will increase the airflow speed; reduce the pressure, density, and temperature; and expand the airflow. Hence, the function of the Mach wave line OL is to increase and reduce the speed of the supersonic airflow, and the airflow undergoes adiabatic acceleration and expansion process, so the Mach wave OL is called an expansion wave.

After the expansion wave, the changing trends of the airflow parameters are as follows:

(1) The flow velocity V keeps increasing, $dV > 0$, so we get

$$d\left(\frac{V^2}{2}\right) > 0 \qquad (7.142)$$

(2) The pressure p decreases, $dp < 0$. From the energy equation of adiabatic flow (7.106), we know that

$$d\left(\frac{V^2}{2}\right) + \frac{dp}{\rho} = 0, \ dp = -\rho d\left(\frac{V^2}{2}\right) < 0 \qquad (7.143)$$

(3) The temperature T decreases, $dT < 0$. From the energy equation of adiabatic flow (7.108), we know that

$$d\left(\frac{V^2}{2}\right) + C_p dT = 0, \quad dT = -\frac{1}{C_p}d\left(\frac{V^2}{2}\right) < 0 \qquad (7.144)$$

(4) Density ρ decreases, $d\rho < 0$. Because, substituting in (7.143), we get

$$d\rho = -\frac{a^2}{\rho}d\left(\frac{V^2}{2}\right) < 0 \qquad (7.145)$$

7.8.2 The Relationship Between the Physical Parameters of the Mach Wave

As shown in Fig. 7.20, choose a micro-element control body abcd along the Mach wave line *OL* and across it, where ab is parallel to *OL*, ac is perpendicular to the *OL* line, the upstream of the control body is the wave front airflow parameter, and the downstream of the control body is a post-wave airflow parameter. For a straight supersonic airflow of *Ma*, there is a small deflection angle at point *O* on the *AOB* wall, and a Mach wave *OL* is emitted at point *O*. The angle between the Mach wave line and the incoming flow *AO* line is the Mach angle, and its magnitude is determined by Eq. (7.141). After the airflow passes through the Mach wave line *OL*, the deflection angle is parallel to the *OB* wall. The airflow parameters of the Mach wave front are *Ma*, *p*, *T*, *ρ*, etc., and the airflow parameters after the wave are equal. For the control body taken, it can be obtained from the continuity equation, and the mass per unit time and unit area of the control body is

$$m = \rho V_n = (\rho + \mathrm{d}\rho)(V_n + \mathrm{d}V_n) \tag{7.146}$$

Omitting the second order and second-order small quantities, we get

$$\mathrm{d}V_n = -V_n \frac{\mathrm{d}\rho}{\rho} \tag{7.147}$$

In the formula, V_n is the velocity component of the wave front perpendicular to the Mach wave line *OL*. At the same time, since there is no pressure change in the *OL* direction parallel to the Mach wave line (that is, the pressure on the ac and bd surfaces are equal), the tangential momentum is conserved, and we get

$$m V_t' - m V_t = 0, \ \ V_t' = V_t \tag{7.148}$$

where V_t is the velocity component of the wave front parallel to the Mach wave line OL and V_t' is the velocity component of the wave back parallel to *OL*. From the momentum equation perpendicular to the *OL* line, we get

Fig. 7.20 Mach wave flow analysis

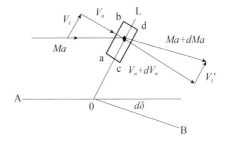

$$dp = -m \, dV_n \tag{7.149}$$

Substituting $m = \rho V_n$ and Eq. (7.147) into Eq. (7.149), we get

$$a^2 = \frac{dp}{d\rho} = V_n^2, \; V_n = a \tag{7.150}$$

It shows that the normal velocity of the Mach wave front airflow is equal to the local speed of sound. From the geometric relationship of the velocity triangle in Fig. 7.20, we get

$$V_n = V \sin \mu, \; V_t = V \cos \mu \tag{7.151}$$

$$V_t' = (V + dV) \cos(\mu + d\delta) \tag{7.152}$$

Substituting the above two equations into Eq. (7.148), we get

$$V_t' = (V + dV) \cos(\mu + d\delta) = V_t = V \cos \mu \tag{7.153}$$

Expand and simplify to

$$V \cos \mu = (V + dV) \cos(\mu + d\delta) = (V + dV)(\cos \mu \cos d\delta - \sin \mu \sin d\delta)$$

$$\frac{dV}{V} = (tg \, \mu) d\delta = \frac{d\delta}{\sqrt{Ma^2 - 1}} \tag{7.154}$$

Using the differential form of Euler's equation and the definition of wave speed, we get

$$dp = -\rho V dV, \; p = c\rho^\gamma, \; a^2 = \frac{dp}{d\rho} = \gamma \frac{p}{\rho} \tag{7.155}$$

We get

$$\frac{dV}{V} = -\frac{1}{\gamma Ma^2} \frac{dp}{p} \tag{7.156}$$

Substituting Eq. (7.156) into Eq. (7.154), we have

$$\frac{dp}{p} = -\frac{\gamma Ma^2}{\sqrt{Ma^2 - 1}} d\delta \tag{7.157}$$

If substituting $\frac{dp}{d\rho} = a^2$, $p = \rho RT$ into equation above, we have

$$\frac{d\rho}{\rho} = -\frac{\gamma Ma^2}{\sqrt{Ma^2 - 1}} d\delta$$

$$\frac{dT}{T} = -(\gamma - 1)\frac{Ma^2}{\sqrt{Ma^2 - 1}} d\delta \qquad (7.158)$$

It can be seen from Eqs. (7.154), (7.156)–(7.158) that if the wall is folded outward $d\delta(>0)$ with the increase of airflow speed, the pressure, density, and temperature decrease, and the airflow expands, so this Mach wave is an expansion wave. If the wall is folded inwardly $d\delta$ (<0), through contraction, the flow rate decreases; the pressure, density, and temperature increase; and the airflow is compressed. At this time, the Mach wave is a compression wave.

The pressure coefficient on the wall after passing through the Mach wave is

$$C_p = \frac{p + dp - p}{\frac{1}{2}\rho V^2} = \frac{dp}{\frac{1}{2}\frac{pa^2}{RT}Ma^2} = \frac{2dp}{\gamma pMa^2} = -\frac{2d\delta}{\sqrt{Ma^2 - 1}} \qquad (7.159)$$

The derivation results of the above one-dimensional flow theory are only suitable for small deflection angles and cannot be directly applied to the results of large deflection angles.

7.8.3 Flow Around the Outer Corner of the Supersonic Wall (Prandtl–Meyer Flow)

If the outer corner of the wall is not a small amount, but a finite value. In order to obtain the relationship of the aerodynamic parameters, although the previous micro-quantity formula cannot be used directly, it can be obtained by summing a series of small deflection angles for multiple small external deflection angles, as shown in Fig. 7.21.

As shown in Fig. 7.20, assuming that at point O_1, the wall is offset by $d\delta_1$, through the expansion wave OL_1, the airflow parameter becomes

$$Ma_2 = Ma_1 + dMa_1 \qquad (7.160)$$

Fig. 7.21 Different outward deflection of airflow

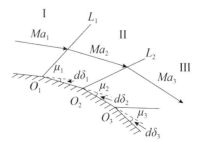

At the O_2 point, the wall is offset by $d\delta_2$, and the airflow expands further through the expansion wave OL_2, and the airflow parameter becomes

$$Ma_3 = Ma_2 + dMa_2 \tag{7.161}$$

The angle between the first expansion wave and the direction of the incoming flow is

$$\mu_1 = \arcsin \frac{1}{Ma_1} \tag{7.162}$$

The angle between the second expansion wave and the direction of the incoming flow is

$$\mu_2 = \arcsin \frac{1}{Ma_2} \tag{7.163}$$

As the airflow expands, $Ma_2 > Ma_1$, then $\mu_2 < \mu_1$, that is to say, the angle between the second expansion wave and the wave front airflow direction is smaller than the inclination angle of the first expansion wave. In this way, the inclination angle of each subsequent expansion wave relative to the original airflow is smaller than the previous one, so each expansion wave cannot intersect with each other, but forms a continuous scattering expansion area.

As shown in Fig. 7.21, when the deflection angle of limited size is actually processed, it is divided into countless tiny angles, and then the fixed points of O_1, O_2… are approached wirelessly, and the outer corners are summed, as shown in Fig. 7.22. The parameter relation of the flow around the angle was first derived by Prandtl, the world master of fluid mechanics, and his student Meyer in 1908. Later generations called this flow around the angle Prandtl–Meyer flow, which is

$$\delta = \sum_{n=1}^{\infty} d\delta_n \tag{7.164}$$

Fig. 7.22 Prandtl–Meyer flow

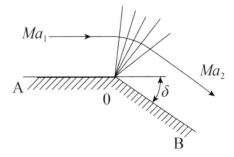

7.8.4 The Calculation Formula for the Flow Around the Outer Corner of the Supersonic Wall

According to Eq. (7.154), the relationship between the small external deflection angle $d\delta$ and the velocity increment is

$$d\delta = \sqrt{Ma^2 - 1}\frac{dV}{V} \tag{7.165}$$

Integrating the above equation, the angle is from zero to δ, the Mach number increases from Ma_1 to Ma_2, and then we get

$$\int_0^\delta d\delta = \int_{Ma_1}^{Ma_2} \sqrt{Ma^2 - 1}\frac{dV}{V} \tag{7.166}$$

Using Eqs. (7.123) and (7.126) and introducing the speed coefficient, Eq. (7.166) can be written as

$$\int_0^\delta d\delta = \int_{Ma_1}^{Ma_2} \sqrt{Ma^2 - 1}\frac{d\lambda}{\lambda} \tag{7.167}$$

Using,

$$Ma^2 = \frac{\frac{2}{\gamma+1}\lambda^2}{1 - \frac{\gamma-1}{\gamma+1}\lambda^2} \tag{7.168}$$

Equation (7.167) becomes

$$\delta = \int \sqrt{\frac{\lambda^2 - 1}{1 - \frac{\gamma-1}{\gamma+1}\lambda^2}}\frac{d\lambda}{\lambda} + C \tag{7.169}$$

Introducing variables

$$t^2 = \frac{\lambda^2 - 1}{1 - \frac{\gamma-1}{\gamma+1}\lambda^2}, \quad K^2 = \frac{\gamma+1}{\gamma-1}, \quad t^2 = \frac{\lambda^2 - 1}{1 - \lambda^2/K^2} \tag{7.170}$$

we get

$$\lambda^2 = \frac{K^2(1 + t^2)}{K^2 + t^2}, \quad \frac{d\lambda}{\lambda} = \left(\frac{t}{1 + t^2} - \frac{t}{K^2 + t^2}\right)dt \tag{7.171}$$

Substituting Eqs. (7.170) and (7.171) into Eq. (7.169), we get

$$\delta = \int t \left(\frac{t}{1+t^2} - \frac{t}{K^2 + t^2} \right) dt = \int \left(\frac{K^2}{K^2 + t^2} - \frac{1}{1+t^2} \right) dt$$

$$\delta = K \arctan \frac{t}{K} - \arctan t + C \tag{7.172}$$

Changing the variable t back to λ, we get

$$\delta = \sqrt{\frac{\gamma + 1}{\gamma - 1}} \arctan \sqrt{\frac{\gamma - 1}{\gamma + 1} \frac{\lambda^2 - 1}{1 - \frac{\gamma - 1}{\gamma + 1} \lambda^2}} - \arctan \sqrt{\frac{\lambda^2 - 1}{1 - \frac{\gamma - 1}{\gamma + 1} \lambda^2}} + C \tag{7.173}$$

The integral constant can be determined by the initial conditions. It is stipulated that when $\lambda = 1$, the direction angle of the airflow is zero, and $C = 0$.

$$\delta = \sqrt{\frac{\gamma + 1}{\gamma - 1}} \arctan \sqrt{\frac{\gamma - 1}{\gamma + 1} \frac{\lambda^2 - 1}{1 - \frac{\gamma - 1}{\gamma + 1} \lambda^2}} - \arctan \sqrt{\frac{\lambda^2 - 1}{1 - \frac{\gamma - 1}{\gamma + 1} \lambda^2}} \tag{7.174}$$

Substituting Eq. (7.168) into the equation above, we get

$$\delta = \sqrt{\frac{\gamma + 1}{\gamma - 1}} \arctan \sqrt{\frac{\gamma - 1}{\gamma + 1} (Ma^2 - 1)} - \arctan \sqrt{Ma^2 - 1} \tag{7.175}$$

For the case where the original air velocity is the speed of sound ($\lambda = 1$), the above equation gives the functional relationship between the local velocity coefficient anywhere in the expansion wave and the local airflow angle δ (calculated from $\lambda = 1$). As long as the local airflow angle δ is known, the local velocity coefficient λ can be uniquely determined, and vice versa.

According to the energy equation, the total energy of the airflow is equal to the kinetic energy plus the enthalpy. The two can be converted to each other, the flow rate increases, and the enthalpy value decreases. When all energy is converted into kinetic energy, the flow rate reaches $V_{\max}$, which is the maximum corresponding velocity coefficient.

$$V_{\max} = \sqrt{2C_p T_0}, \quad \frac{V^2}{2} + \frac{a^2}{\gamma - 1} = \frac{V_{\max}^2}{2}, \quad a_*^2 = \frac{\gamma - 1}{\gamma + 1} V_{\max}^2$$

$$\lambda_{\max} = \frac{V_{\max}}{a_*} = \sqrt{\frac{\gamma + 1}{\gamma - 1}} = \sqrt{6} \tag{7.176}$$

At this time, the Ma number of the supersonic airflow reaches infinity, and the Mach angle approaches zero. The corresponding maximum possible turning angle is

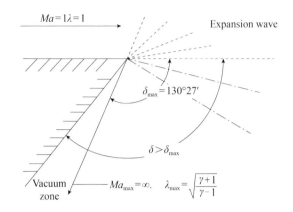

Fig. 7.23 Flow around the maximum angle outside corner

$$\delta_{max} = \left(\sqrt{\frac{\gamma + 1}{\gamma - 1}} - 1 \right) \frac{\pi}{2} \tag{7.177}$$

When γ is 1.4, we can get

$$\delta_{max} = 130.27' \tag{7.178}$$

λ increases with the increase of δ. However, when λ reaches λ_{max}, the airflow expands to an absolute vacuum state where the pressure, temperature, and density are all reduced to zero, and the corresponding airflow angle δ_{max} is called the maximum angle. If the actual angle is greater than δ_{max}, after the airflow is turned to δ_{max}, the airflow cannot continue to expand and accelerate, and it will no longer flow against the surface of the object, and an absolute vacuum zone appears between the airflow and the wall, as shown in Fig. 7.23.

When using Eq. (7.175) to calculate, the solution is more cumbersome. For simplification, when the external declination angle is not large (for example, the declination angle is less than 10°), you can use the Taiwanese series expansion to find the second-order approximate solution. It is known that the incoming flow Mach number and the outer deflection angle are M_∞ and δ, respectively, and the flow Mach number after the outer inflection angle δ is Ma and the pressure is p. Since,

$$\frac{p}{p_\infty} = \frac{p}{p_0} \frac{p_0}{p_\infty} = \left[\frac{1 + \frac{\gamma-1}{2} Ma_\infty^2}{1 + \frac{\gamma-1}{2} Ma^2} \right]^{\frac{\gamma}{\gamma-1}} \tag{7.179}$$

According to Taiwanese labor series of δ, we have

$$\frac{p}{p_\infty} = 1 - \frac{Ma_\infty^2}{\sqrt{Ma_\infty^2 - 1}} \delta + \frac{\gamma Ma_\infty^2}{2} \left[\frac{(\gamma + 1) Ma_\infty^4 - 4(Ma_\infty^2 - 1)}{2(Ma_\infty^2 - 1)^2} \right] \delta^2 + \cdots \tag{7.180}$$

Fig. 7.24 Critical flow around 10° of outer bend angle

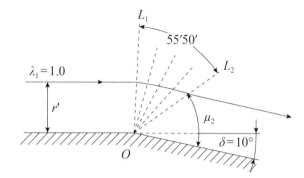

The pressure coefficient is

$$
\begin{aligned}
C_p &= \frac{2}{\gamma M a_\infty^2}\left(\frac{p}{p_\infty} - 1\right) \\
&= -\frac{2\delta}{\sqrt{M a_\infty^2 - 1}} + \left[\frac{(\gamma + 1)M a_\infty^4 - 4(M a_\infty^2 - 1)}{2(M a_\infty^2 - 1)^2}\right]\delta^2 + \cdots
\end{aligned}
\tag{7.181}
$$

Solved by Eq. (7.179), the Ma number is

$$
Ma = \sqrt{\frac{2}{\gamma - 1}\left[\frac{1 + \frac{\gamma - 1}{2}M a_\infty^2}{\left(\frac{p}{p_\infty}\right)^{\frac{\gamma - 1}{\gamma}}} - 1\right]}
\tag{7.182}
$$

Example The flow with a critical angle of 10° around the outer bend is shown in Fig. 7.24. Knowing the airflow of $\lambda = 1.0$ ($\gamma = 1.4$), around the outer bending angle of 10°, $p_1 = 1.0$ pa (1 atmosphere pressure), try to find the λ and ρ of the airflow after the expansion.

Solution: When $\delta = 10°$, substituting into Eq. (7.175), we get

$$
\lambda_2 = 1.323; \ \frac{p_2}{p_0} = 0.299, \ \frac{p_1}{p_0} = 0.528, \ p_2 = \left(\frac{p_2}{p_0}\right) \cdot \left(\frac{p_0}{p_1}\right) \cdot p_1 = \frac{0.299}{0.528} = 0.565.
$$

As shown in Fig. 7.25, it is a flow of supersonic velocity around an outward turning angle of 10°. Knowing that $\lambda_1 = 1.323$, the outward turning angle is 10°, and Ma_2 is calculated.

Solution: When the $\lambda_1 \neq 1$ of the original airflow, calculate according to the following steps:

(1) According to the given value of λ_1, calculate the imaginary fold angle δ' corresponding to $\lambda_1 = 1$, and then superimpose it with the external fold angle to obtain the total fold angle $\Sigma\delta$.

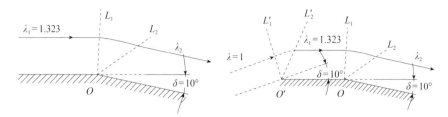

Fig. 7.25 Supersonic flow around 10° of outward turning angle

(2) According to $\Sigma\delta$, calculate the value of λ_2.

According to $\lambda_1 = 1.323$, we calculate the turning angle is 10°, when $\lambda_1 = 1$, so that λ_2 is equivalent to the total turning angle of the airflow from $\Sigma\delta = 20°$, when $\lambda_1 = 1$, again calculating $\lambda_2 = 1.523$, $Ma_2 = 1.775$.

7.9 Compression Wave and Shock Wave

7.9.1 Compression Wave

As shown in Fig. 7.26, in supersonic parallel flow, if the wall at point O deflects inward by a slight angle $d\delta$, the flow area will shrink, which is called the compression angle. Obviously, there is a slight disturbance at point O due to the inflection of the wall surface. The propagation range of the disturbance is the downstream area of the Mach wave OL emitted by point O. The result of the disturbance is: the airflow is deflected inward by $d\delta$ angle, and the speed is reduced, the pressure increases, and the airflow is compressed.

The wall surface is folded inward, which is equivalent to the contraction of the airflow channel. For supersonic airflow, the contraction of the cross-sectional area of the channel will decelerate the airflow; increase the pressure, density, and temperature; and compress the airflow. At this time, the function of the Mach wave line OL is to pressurize and decelerate the supersonic airflow, and the airflow undergoes an adiabatic compression process, so the Mach wave OL is called a compression wave.

After the compression wave, the changing trends of the parameters of the airflow are as follows:

Fig. 7.26 Supersonic flow at inner corners

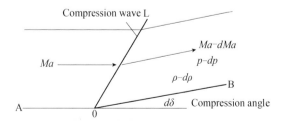

(1) The flow velocity V decreases, $dV < 0$. So we get

$$d\left(\frac{V^2}{2}\right) < 0 \tag{7.183}$$

(2) The pressure p increases, $dp > 0$, which is

$$d\left(\frac{V^2}{2}\right) + \frac{dp}{\rho} = 0, \ dp = -\rho d\left(\frac{V^2}{2}\right) < 0 \tag{7.184}$$

(3) The temperature T increases, $dT > 0$, which is

$$d\left(\frac{V^2}{2}\right) + C_p dT = 0, \ dT = -\frac{1}{C_p} d\left(\frac{V^2}{2}\right) < 0 \tag{7.185}$$

(4) The density ρ increases $d\rho > 0$, which is

$$d\rho = -\frac{a^2}{\rho} d\left(\frac{V^2}{2}\right) < 0 \tag{7.186}$$

7.9.2 The Formation Process of Shock Waves

In supersonic airflow, two kinds of wave phenomena are the most important, one is expansion wave, the other is compression wave, and shock wave is an extreme form of compression wave. Expansion wave is to increase the speed of airflow and decompress, compression wave is to decelerate and pressurize airflow, and shock wave is a wave that causes sudden compression of airflow. This section starts with the simplest normal shock wave, analyzes the shock wave phenomenon and parameter calculation, and understands the physical laws that the shock wave satisfies. Then, on this basis, it is applied to oblique shock waves. Finally, the internal structure of shock waves and the interference between shock waves and boundary layers will be briefly introduced. The shock wave phenomenon is a common phenomenon in the aerospace field. As shown in Fig. 7.27, the shock wave formed by the jet nozzle of a fighter jet is shown.

Suppose there is a very long tube, the left end of the tube is sealed with a piston, the tube is filled with static gas, the pressure is p_1, the density ρ_1, and the temperature is T_1. The piston starts from a standstill and makes a sharp acceleration movement to the right, compressing the gas in the tube. From $t = 0$ to $t = t_1$, the piston accelerates sharply to the right and moves forward at a constant speed after t_1. During the acceleration from $t = 0$ to $t = t_1$, the gas on the right side of the piston is compressed more and more strongly, and the gas pressure on the contact surface of the piston and the pressurized gas continues to increase. When $t = t_1$, it is assumed that the pressure of the gas on the piston surface rises from p_1 to p_2. As shown in Fig. 7.28, the AA

Fig. 7.27 Shock wave
formed by fighter jet nozzle

interface is the place reached by the first disturbance, the right is the undisturbed gas, the left is the compressed gas, and the more the piston is compressed, the more the gas pressure from p_1 at AA continuously rises to p_2 at the piston. After a certain period of time, all subsequent waves have caught up with the first wave, resulting in the length of the entire compression wave zone A–B being almost reduced to zero. Numerous tiny compression waves are superimposed, and each wave increases the pressure by a Δp, each small step of the compression wave propagates to the right at the local speed of sound. When the piston first moves, the first wavelet advances to the right at a speed of $a_1 = \sqrt{\gamma R T_1}$, and the pressure and temperature of the gas swept by this wave are slightly increased. The speed of the second compression wave advancing to the right is $a_1 + \Delta a = \sqrt{\gamma R(T_1 + \Delta T)}$ faster than the first wave. The third wave is after the second, and each subsequent wave is chasing the wave before it. The length from AA to BB must become shorter and shorter as time.

The weak compression waves are stacked together to form a sudden compression surface S–S with a certain strength. When S–S is not there, the gas is not compressed at all, and as soon as S–S arrives, the gas is suddenly compressed, and the pressure suddenly increases from p_1 to p_2. Such a sudden compression surface S–S is called a shock wave. Because the S–S surface is perpendicular to the direction of the airflow, this shock wave is called a normal shock wave. The above discussion did not consider the moving speed of the gas clusters. The gas was originally at rest. After the first wave is compressed, the gas clusters have a little rightward moving speed, so the speed of the second wave should also be superimposed on the gas moving speed.

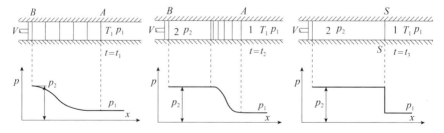

Fig. 7.28 Formation of the normal shock wave in the tube

Both factors make the second wave faster than the first wave. Shock wave formation is inevitable.

7.9.3 Propulsion Speed of Shock Wave

When countless tiny compression waves are superimposed together to form a shock wave, its wave front advances to the right at a certain speed. Now, using the continuous equation and momentum equation in integral form, the shock wave propulsion velocity V_s is derived. As shown in Fig. 7.28, when the piston is advancing to the right at a constant speed V_g, when the airflow is stabilized, the speed of the piston moves together with the air compressed by the shock wave, and the resulting shock wave propulsion speed is advancing to the right at V_s. The air before the shock wave is static, the flow parameters are speed $V_1 = 0$, the pressure is p_1, the density is ρ_1, etc., the compressed air speed after the shock is V_g, the pressure is p_2, the density is ρ_2, etc., if the observer moves with the shock wave. It can be seen that the airflow flows steadily relative to the shock wave, surrounding the shock wave front. In the control body ABCD shown in Fig. 7.29, the airflow is constant relative to the flow of the control body. The inflow parameters of the airflow through the CD section are velocity over V_s, the pressure is p_1, the density is ρ_1, etc.; the outflow parameters of the airflow through the AB section are velocity over $(V_s - V_g)$, the pressure is p_2, and the density is ρ_2. From the continuous equation in integral form, we can get

$$\rho_2(V_s - V_g) - \rho_1 V_s = 0 \tag{7.187}$$

From the momentum equation in integral form, we get

$$\rho_2(V_s - V_g)(V_s - V_g - V_s) = p_1 - p_2$$
$$\rho_2(V_s - V_g)V_g = p_2 - p_1 \tag{7.188}$$

Fig. 7.29 Shock wave propulsion speed

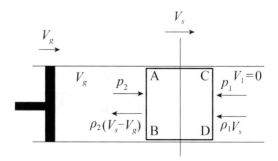

Solve the continuum equation and the momentum integral equation simultaneously to get

$$\begin{cases} V_g = \frac{\rho_2 - \rho_1}{\rho_2} V_s \\ V_s = \sqrt{\frac{\rho_2}{\rho_1} \frac{p_2 - p_1}{p_2 - p_1}} \\ V_g = \sqrt{\frac{(p_2 - p_1)(\rho_2 - \rho_1)}{\rho_2 \rho_1}} \end{cases} \tag{7.189}$$

In order to obtain the density ratio from the pressure ratio, an integral energy equation needs to be applied to the control body, which is

$$\frac{\partial}{\partial t} \iiint_\tau \rho(e + \frac{V^2}{2}) d\tau + \oiint_S \rho(e + \frac{V^2}{2})(\vec{V} \cdot \vec{n}) dS$$

$$= \oiint_S q_\lambda dS + \iiint_\tau \rho q_R d\tau + \iiint_\tau \rho \vec{f} \cdot \vec{V} d\tau + \oiint_S \vec{p_n} \cdot \vec{V} dS \tag{7.190}$$

Ignoring the heat conduction and heat radiation, neglecting the mass force, the flow is constant, we can get

$$\oiint_S \rho(e + \frac{V^2}{2})(\vec{V} \cdot \vec{n}) dS = \oiint_S \vec{p_n} \cdot \vec{V} dS \tag{7.191}$$

With reference to Fig. (7.29), Eq. (7.191) can be expanded to obtain

$$\oiint_S \rho\left(e + \frac{V^2}{2}\right)(\vec{V} \cdot \vec{n}) dS = \rho_2 A\left(e_2 + \frac{(V_s - V_g)^2}{2}\right)(V_s - V_g) - \rho_1 A\left(e_1 + \frac{V_s^2}{2}\right)V_s \tag{7.192}$$

$$\oiint_S \vec{p_n} \cdot \vec{V} dS = p_1 V_s A - p_2(V_s - V_g)A \tag{7.193}$$

Using continuous Eq. (7.187), $\rho_1 V_s = \rho_2(V_s - V_g)$, substituting Eqs. (7.193) and (7.192) into Eq. (7.191), we can get

$$\rho_2\left(e_2 + \frac{(V_s - V_g)^2}{2}\right) - \rho_2\left(e_1 + \frac{V_s^2}{2}\right) = \frac{p_1}{\rho_1}\rho_2 - p_2 \tag{7.194}$$

Using Eq. (7.189) to organize and simplify to get

$$\rho_1\left(e_2 - e_1 + \frac{V_g^2}{2}\right)V_s = p_2 V_g \tag{7.195}$$

The internal energy can be expressed as

$$e = C_v T = \frac{p}{\rho(\gamma - 1)} \tag{7.196}$$

Substituting Eq. (7.195), and using Eq. (7.189) to replace V_s and V_g, the relationship between pressure and density is obtained as

$$\frac{\rho_2}{\rho_1} = \frac{\frac{\gamma+1}{\gamma-1}\frac{p_2}{p_1} + 1}{\frac{p_2}{p_1} + \frac{\gamma+1}{\gamma-1}} \tag{7.197}$$

The advancing speed of the shock wave and the moving speed of the gas behind the wave are also expressed by the ratio of pressure to density as

$$V_s = \frac{a_1}{\sqrt{\gamma}}\sqrt{\frac{\rho_2}{\rho_1}\frac{p_2/p_1 - 1}{\rho_2/\rho_1 - 1}} \tag{7.198}$$

$$V_g = \frac{a_1}{\sqrt{\gamma}}\sqrt{\frac{(p_2/p_1 - 1)(\rho_2/\rho_1 - 1)}{\rho_2/\rho_1}} \tag{7.199}$$

If the intensity of the shock wave p_2/p_1 is specified, the shock wave advancing speed can be obtained. If order

$$\frac{p_2}{p_1} = \frac{p_1 + \Delta p}{p_1} = 1 + \frac{\Delta p}{p_1}, \quad \Delta p = p_2 - p_1 \tag{7.200}$$

Substituting into Eq. (7.197), we have

$$\frac{p_2}{p_1} = \frac{\frac{\gamma+1}{\gamma-1}\frac{p_2}{p_1} + 1}{\frac{p_2}{p_1} + \frac{\gamma+1}{\gamma-1}} = \frac{\frac{\gamma+1}{\gamma-1}(1 + \frac{\Delta p}{p_1}) + 1}{1 + \frac{\Delta p}{p_1} + \frac{\gamma+1}{\gamma-1}} = 1 + \frac{2\frac{\Delta p}{p_1}}{2\gamma + \frac{\Delta p}{p_1}(\gamma - 1)} > 1 \tag{7.201}$$

The shock wave velocity is

$$V_s = \frac{a_1}{\sqrt{\gamma}}\sqrt{\frac{p_2}{p_1}\left(\gamma + \frac{\gamma - 1}{2}\frac{\Delta p}{p_1}\right)} = a_1\sqrt{1 + \frac{\gamma + 1}{2\gamma}\frac{\Delta p}{p_1}} > a_1 \tag{7.202}$$

This means that the advancing speed of the shock wave is always greater than the sound speed of the wave front. If the relative increase in pressure is a small amount ε_p, then we get

$$\frac{p_2}{p_1} = \frac{p_1 + \Delta p}{p_1} = 1 + \frac{\Delta p}{p_1} = 1 + \varepsilon_p \tag{7.203}$$

Substituting into Eq. (7.202), we get

$$V_s = a_1 \sqrt{1 + \frac{\gamma + 1}{2\gamma} \varepsilon_p} \approx a_1 \tag{7.204}$$

This shows that the advancing speed of the shock wave is always greater than the sound speed of the wave front, that is, relative to the shock wave, the wave front gas must be supersonic. Since the gas after the shock wave already has a velocity V_g, relative to the gas velocity $V_s - V_g$ after the shock wave, it must be subsonic, that is, $(V_s - V_g) < a_2$. The proof is as follows.

From Eq. (7.189), we can get

$$V_s - V_g = V_s - \frac{\rho_2 - \rho_1}{\rho_2} V_s = \frac{\rho_1}{\rho_2} V_s = \frac{\rho_1}{\rho_2} \sqrt{\frac{\rho_2}{\rho_1} \frac{p_2 - p_1}{\rho_2 - \rho_1}} = \sqrt{\frac{\rho_1}{\rho_2} \frac{p_2 - p_1}{\rho_2 - \rho_1}} \tag{7.205}$$

Using $a_2 = \sqrt{\gamma R T_2} = \sqrt{\gamma \frac{p_2}{\rho_2}}$ and Eq. (7.201), we get

$$V_s - V_g = \sqrt{\frac{p_2}{\rho_2} \frac{\rho_1}{\rho_2} \frac{p_2/p_1 - 1}{\rho_2/\rho_1 - 1}} = a_2 \sqrt{1 - \frac{\gamma + 1}{2\gamma} \frac{p_2/p_1 - 1}{p_2/p_1}} \tag{7.206}$$

This formula shows that the velocity of the airflow after the shock wave is less than the sound velocity of the compressed air after the shock wave, indicating that the relative velocity of the airflow after the shock wave is subsonic.

Example Let the pressure p_1 of the still air in the long pipe be 1 atmosphere, $\rho_1 = 1.225$ kg/m^3, $T_1 = 288$ K. A shock wave is generated by compressing air with a piston. The compressed air pressure p_2 is two atmospheres. Find the advancing speed V_s of the shock wave, the advancing speed V_g of the piston, and the sound speed a_2 of the compressed air, relative to the incoming Mach number Ma_1 of the shock wave and the shock wave relative to the Mach number after the wave Ma_2. Let $= \gamma 1.4$, $R = 287.053$ m^2/s^2/K.

Solution: $a_1 = \sqrt{\gamma R T_1} = 340.1$ m/s

$$\frac{\rho_2}{\rho_1} = \frac{\frac{\gamma+1}{\gamma-1} \frac{p_2}{p_1} + 1}{\frac{p_2}{p_1} + \frac{\gamma+1}{\gamma-1}} = \frac{\frac{1.4+1}{1.4-1} \times 2 + 1}{2 + \frac{1.4+1}{1.4-1}} = 13/8 = 1.625$$

$$V_s = \frac{a_1}{\sqrt{\gamma}} \sqrt{\frac{\rho_2}{\rho_1} \frac{(p_2/p_1 - 1)}{(\rho_2/\rho_1 - 1)}} = a_1 \sqrt{\frac{1.625}{1.4 \times 0.625}} = 1.363 a_1 = 463.6 \, \text{m/s}$$

$$V_g = \frac{a_1}{\sqrt{\gamma}} \sqrt{\frac{(p_2/p_1 - 1)(\rho_2/\rho_1 - 1)}{\rho_2/\rho_1}} = a_1 \sqrt{\frac{1 \times 0.625}{1.4 \times 1.625}} = 0.524 a_1 = 178.3 \, \text{m/s}$$

$$T_2 = \frac{p_2 \rho_1}{p_1 \rho_2} T_1 = 288 \times 2/1.625 = 354.5 \, \text{K}$$

$a_2 = \sqrt{\gamma R T_2} = \sqrt{1.4 \times 287 \times 354.5} = 377.4 \, \text{m/s}$

$V_s - V_g = 463.6 - 178.3 = 285.3 \, \text{m/s}$

The airflow Mach number relative to the shock wave is given below:
The incoming Mach number is

$$Ma_1 = \frac{V_s}{a_1} = \frac{463.6}{340.1} = 1.353$$

The relative Mach number of the airflow after the shock is

$$Ma_2 = \frac{V_s - V_g}{a_2} = \frac{285.3}{377.4} = 0.756$$

7.9.4 Normal Shock Wave

The wave front of the Normal Shock Wave is perpendicular to the direction of the airflow. In order to facilitate the analysis, we will now examine the coordinate system that follows the shock wave front and establish the relationship between the airflow parameters before and after the shock wave in the coordinate system relative to the shock wave front. The advantage of using relative coordinates is that the airflow is steady relative to the wave front, and the basic equations of steady flow can be directly applied. As shown in Fig. 7.30, taking the control body 1122 around the shock wave front, the airflow parameter of the shock wave front is $Ma_1(V_1)$, p_1, ρ_1, and the shock wave airflow parameter is $Ma_2(V_2)$, p_2, ρ_2. Applying the continuous equation, we get

$$\rho_1 V_1 = \rho_2 V_2 \tag{7.207}$$

Using the momentum equation, we get

$$-\rho_1 V_1^2 + \rho_2 V_2^2 = p_1 - p_2 \tag{7.208}$$

Fig. 7.30 Control volume relative to shock wave front

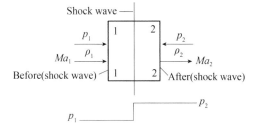

Applying the energy equation of adiabatic flow to this control surface, we get

$$V_1^2 + \frac{2}{\gamma - 1}a_1^2 = V_2^2 + \frac{2}{\gamma - 1}a_2^2 = \frac{\gamma + 1}{\gamma - 1}a_*^2 \tag{7.209}$$

Using the definition of sound speed, $a^2 = \gamma p / \rho$, the above equation is

$$\frac{p}{\rho} = \frac{\gamma + 1}{2\gamma}a_*^2 - \frac{\gamma - 1}{2\gamma}V^2 \tag{7.210}$$

From the continuity equation and the momentum equation, we get

$$V_1 - V_2 = \frac{p_2}{\rho_2 V_2} - \frac{p_1}{\rho_1 V_1} \tag{7.211}$$

Substituting Eq. (7.210) into Eq. (7.211),

$$\begin{aligned} V_1 - V_2 &= \frac{1}{V_2}\left[\frac{\gamma + 1}{2\gamma}a_*^2 - \frac{\gamma - 1}{2\gamma}V_2^2\right] - \frac{1}{V_1}\left[\frac{\gamma + 1}{2\gamma}a_*^2 - \frac{\gamma - 1}{2\gamma}V_1^2\right] \\ &= (V_1 - V_2)\left(\frac{\gamma + 1}{2\gamma}\frac{a_*^2}{V_1 V_2} + \frac{\gamma - 1}{2\gamma}\right) \end{aligned} \tag{7.212}$$

Solving Eq. (7.212), one solution is $V_1 = V_2$, which represents the flow of no change, ordinary solution. Another situation is

$$\frac{\gamma + 1}{2\gamma}\frac{a_*^2}{V_1 V_2} + \frac{\gamma - 1}{2\gamma} = 1 \tag{7.213}$$

Solution is

$$\frac{a_*^2}{V_1 V_2} = 1 \quad \text{or} \quad \lambda_1 \lambda_2 = 1 \tag{7.214}$$

This formula is the famous Prandtl shock wave formula. It represents the relationship between the wave front and back-wave velocity coefficients. It shows that the airflow velocity coefficient λ_2 after the normal shock is exactly the reciprocal of the wave front airflow velocity coefficient λ_1. Since the wave front is a supersonic flow, $\lambda_1 > 1$, there must be a velocity coefficient $\lambda_2 < 1$ behind the wave, that is to say, the supersonic flow must be a subsonic flow after passing through the normal shock wave. Using the relationship between the velocity coefficient and the Mach number, the relationship between the airflow parameters before and after the shock can be established.

Since,

$$\lambda^2 = \frac{\frac{\gamma+1}{2} Ma^2}{1 + \frac{\gamma-1}{2} Ma^2}$$

Mach number relationship before and after shock is

$$Ma_2^2 = \frac{1 + \frac{\gamma-1}{2} Ma_1^2}{\gamma Ma_1^2 - \frac{\gamma-1}{2}} \tag{7.215}$$

Density ratio relationship is

$$\frac{\rho_2}{\rho_1} = \frac{V_1}{V_2} = \frac{\lambda_1}{\lambda_2} = \lambda_1^2 = \frac{\frac{\gamma+1}{2} Ma_1^2}{1 + \frac{\gamma-1}{2} Ma_1^2} \tag{7.216}$$

Pressure ratio relationship is

$$\frac{p_2}{p_1} = 1 + \gamma Ma_1^2 \left(1 - \frac{\rho_1}{\rho_2}\right) = \frac{2\gamma}{\gamma+1} Ma_1^2 - \frac{\gamma-1}{\gamma+1} \tag{7.217}$$

The relationship between static temperature and temperature can be obtained as

$$\frac{T_2}{T_1} = \frac{p_2}{p_1} \frac{\rho_1}{\rho_2} = \frac{2 + (\gamma-1)Ma_1^2}{(\gamma+1)Ma_1^2} \left(\frac{2\gamma}{\gamma+1} Ma_1^2 - \frac{\gamma-1}{\gamma+1}\right) \tag{7.218}$$

The relationship of total pressure ratio, using Eq. (7.127) $p_2 = p_{02}\pi(\lambda_2)$, $p_1 = p_{01}\pi(\lambda_1)$, we get

$$\sigma = \frac{p_{02}}{p_{01}} = \frac{p_2}{\pi(\lambda_2)} \frac{\pi(\lambda_1)}{p_1} = \lambda_1^2 \left[\frac{1 - \frac{\gamma-1}{\gamma+1}\lambda_1^2}{1 - \frac{\gamma-1}{\gamma+1}\frac{1}{\lambda_1^2}}\right]^{\frac{1}{\gamma-1}} \tag{7.219}$$

or

$$\sigma = \left(\frac{2\gamma}{\gamma+1} Ma_1^2 - \frac{\gamma-1}{\gamma+1}\right)^{-\frac{1}{\gamma-1}} \left[\frac{(\gamma+1)Ma_1^2}{(\gamma-1)Ma_1^2 + 2}\right]^{\frac{\gamma}{\gamma-1}} \tag{7.220}$$

The increase in entropy through the shock is

$$\frac{\Delta S}{C_v} = -(\gamma-1)\ln\frac{p_{02}}{p_{01}} = \ln\left(\frac{2\gamma}{\gamma+1} Ma_1^2 - \frac{\gamma-1}{\gamma+1}\right)\left[\frac{(\gamma-1)Ma_1^2 + 2}{(\gamma+1)Ma_1^2}\right]^{-\gamma} \tag{7.221}$$

 **Fig. 7.31** Entropy increase in shock wave

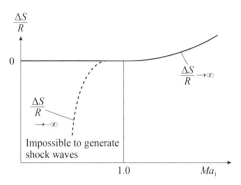

The above relation equations are known as equations of Rankine and Hugoniot, which were deduced by the British scientist Rankine in 1870 and by the French scientist Hugoniot in 1887, respectively. After the shock wave, the total temperature does not change, the total pressure decreases, and the entropy increases, as shown in Fig. 7.31. This is consistent with the conclusion that the entropy in the isolated system can only increase according to the second law of thermodynamics.

When $Ma_1 > 1$, entropy increment is always positive; when $Ma_1 < 1$, the entropy increment is always negative. This shows that for complete gas, shock waves can only be generated in supersonic flow, and shock waves are impossible to generate in subsonic flow. Because the increase in entropy of an isolated system cannot be negative, generating shock waves in a subsonic flow violates the second law of thermodynamics. Of course, in practical problems, weak shock waves can be regarded as isentropic. Let's discuss this issue below. The shock wave intensity P is defined as the ratio of the pressure increment of the passing shock wave to the wave front pressure, which is

$$P = \frac{p_2 - p_1}{p_1} = \frac{p_2}{p_1} - 1 = \frac{2\gamma}{\gamma + 1}(Ma_1^2 - 1) \qquad (7.222)$$

The so-called weak shock wave refers to the shock wave whose intensity P approaches zero. It can be seen from the above formula that the Ma_1 of the weak shock wave must approach 1. The weak shock wave can be regarded as an isentropic wave, that is, an isentropic compression wave. The proof is as follows:

$$\frac{\Delta S}{R} = \frac{\gamma + 1}{12\gamma^2}P^3 - \frac{\gamma^2 - 1}{8\gamma^2}P^4 + \cdots$$

$$\frac{\Delta S}{R} = \frac{2}{3}\frac{\gamma}{(\gamma + 1)^2}(Ma_1^2 - 1)^3 - \frac{2\gamma^2}{(\gamma + 1)^2}(Ma_1^2 - 1)^4 + \cdots \qquad (7.223)$$

When the shock wave intensity is very weak, the entropy increase caused by the shock wave is the third power of the shock wave intensity. Therefore, in the first-order approximate calculation, the entropy increment caused by the weak shock wave can be completely ignored, and the shock wave can be treated as an isentropic wave.

How large is Ma_1 as a weak shock? If the total pressure loss is specified to not exceed 1%, the wave front Mach number is allowed to reach 1.2.

7.9.5 Oblique Shock Wave

Experiments have found that the shapes of shock waves generated by objects with different head shapes are different when flying at supersonic speed. For example, for an airplane with a diamond-shaped wing shape, when flying at supersonic speed, it is actually observed that under a certain $Ma_1 > 1$. If the top angle 2δ of the leading edge of the wing is not too large, the upper and lower two formed a simple oblique shock wave, its wave front is at a certain oblique angle to the direction of movement, and the shock wave is attached to the tip of the object (also called an attached shock wave), as shown in Fig. 7.32. This kind of shock wave is different from the normal shock wave in form. This kind of shock wave whose wave front obliquely crosses the direction of the incoming flow is called an oblique shock wave. In an oblique shock wave, the angle β between the shock wave front and the direction of the incoming flow is called the shock angle. Similarly, the airflow direction after the oblique shock wave is not perpendicular to the wave front, nor parallel to the wave front airflow direction, but parallel to the wedge (the angle δ), which is called the airflow angle, which means that the airflow passes through the oblique shock wave. The angle after turning is shown in Fig. 7.32.

As shown in Fig. 7.33, the control body 1234 is taken around the oblique shock wave front. The surfaces 12 and 34 are parallel to the wave front and they are very close. According to the direction of the wave front, the velocity is decomposed into components perpendicular and parallel to the wave front. For the surface 12 as the inflow surface, the incoming flow velocity is V_1 and the components are V_{1t} and V_{1n}; for the 34 surface as the outflow surface, the combined velocity is V_2, and the components are V_{2t} and V_{2n}. From the geometric relationship,

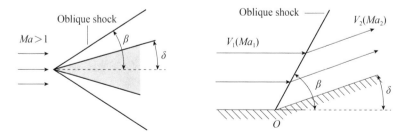

Fig. 7.32 Oblique shock wave concept

Fig. 7.33 Oblique shock
wave relationship

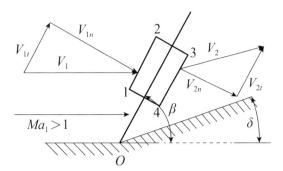

$$V_{1n} = V_1 \sin \beta, \;\; V_{1t} = V_1 \cos \beta$$
$$V_{2n} = V_2 \sin(\beta - \delta), \;\; V_{2t} = V_2 \cos(\beta - \delta) \tag{7.224}$$

Using the mass equation in integral form, we get

$$\rho_1 V_{1n} = \rho_2 V_{2n} \tag{7.225}$$

From the tangential momentum integral equation (the pressures are equal on the surfaces 14 and 23), we get

$$-\rho_1 V_{1n} V_{1t} + \rho_2 V_{2n} V_{2t} = 0 \tag{7.226}$$

From this, we get

$$V_{1t} = V_{2t} = V_t \tag{7.227}$$

This shows that when the airflow crosses the oblique shock wave, the tangential velocity is constant. The oblique shock wave can be regarded as the superposition of the flow field of the normal shock wave and a straight uniform flow field parallel to the shock wave surface. Through the oblique shock wave front, the normal partial velocity of the airflow has a sudden change, the tangential partial velocity is unchanged, and the airflow behind the wave is parallel to the object surface. Therefore, for a given incoming flow Ma_1, how much abrupt change occurs in its normal velocity depends on the oblique angle δ of the object surface. We get,

$$\beta = f(Ma_1, \delta) \tag{7.228}$$

If the wedge half apex angle δ is small, the disturbance caused by this thin wedge to the supersonic airflow must also be a small disturbance; obviously, the disturbance wave must be the Mach wave and the disturbance angle is the Mach angle. As the wedge angle δ increases, β also gradually increases. The larger the δ, the larger the β. From this point of view, for shock waves, the increments of the various parameters are fixed, as long as Ma_1 is determined, while for oblique shock waves, it is necessary to determine the shock angle β from the two parameters Ma_1 and δ. And then, according to the intensity of the shock wave and other physical quantities determined by β.

From the normal momentum integral equation, we get

$$\rho_2 V_{2n}^2 - \rho_1 V_{1n}^2 = p_1 - p_2 \tag{7.229}$$

or

$$p_1 + \rho_1 V_{1n}^2 = p_2 + \rho_2 V_{2n}^2 \tag{7.230}$$

The energy equation of adiabatic motion is

$$T_0 = T_1 + \frac{V_1^2}{2C_p} = T_2 + \frac{V_2^2}{2C_p} \tag{7.231}$$

Because of $V^2 = V_n^2 + V_t^2$, reducing V_t, we get

$$T_1 + \frac{V_{1n}^2}{2C_p} = T_2 + \frac{V_{2n}^2}{2C_p} = T_{0n} \tag{7.232}$$

Here, T_{0n} is the total temperature excluding V_t. It indicates that only the part of the total temperature that takes into account the normal velocity. The corresponding critical sound velocity is

$$a_{*n}^2 = \frac{2\gamma}{\gamma + 1} R T_{0n} \quad \text{or} \quad a_*^2 = a_{*n}^2 + \frac{\gamma - 1}{\gamma + 1} V_t^2 \tag{7.233}$$

From Eq. (7.229), we can get

$$V_{1n} - V_{2n} = \frac{p_2}{\rho_2 V_{2n}} - \frac{p_1}{\rho_1 V_{1n}} \tag{7.234}$$

Finally, we get

$$V_{1n} - V_{2n} = (V_{1n} - V_{2n}) \left(\frac{\gamma + 1}{2\gamma} \frac{a_{*n}^2}{V_{1n} V_{2n}} + \frac{\gamma - 1}{2\gamma} \right) \tag{7.235}$$

We get,

$$\frac{a_{*n}^2}{V_{1n} V_{2n}} = 1, \quad V_{1n} V_{2n} = a_{*n}^2 = a_*^2 - \frac{\gamma - 1}{\gamma + 1} V_t^2 \tag{7.236}$$

From the momentum integral equation, we get

$$\frac{p_2}{p_1} = 1 + \frac{\rho_1 V_{1n}^2}{p_1}(1 - \frac{V_{2n}}{V_{1n}}) = 1 + \gamma \frac{V_{1n}^2}{a_1^2}(1 - \frac{\rho_1}{\rho_2}) \tag{7.237}$$

Using the geometric relationship $V_{1n} = V_1 \sin \beta$, we get

$$\frac{p_2}{p_1} = 1 + \frac{\rho_1 V_{1n}^2}{p_1}(1 - \frac{V_{2n}}{V_{1n}}) = 1 + \gamma Ma_1^2 \sin^2 \beta (1 - \frac{\rho_1}{\rho_2}) \tag{7.238}$$

The relationship between pressure and density ratio remains unchanged, and we get

$$\frac{\rho_2}{\rho_1} = \frac{\frac{\gamma+1}{\gamma-1}\frac{p_2}{p_1} + 1}{\frac{p_2}{p_1} + \frac{\gamma+1}{\gamma-1}} \tag{7.239}$$

Substituting the above equation into Eq. (7.238), we get

$$\frac{p_2}{p_1} = \frac{2\gamma}{\gamma + 1}Ma_1^2 \sin^2 \beta - \frac{\gamma - 1}{\gamma + 1} \tag{7.240}$$

Using the same derivation method as the parameter changes before and after the normal shock wave, we can have obtained a series of parameter changes before and after the oblique shock wave.

The density ratio is

$$\frac{\rho_2}{\rho_1} = \frac{\frac{\gamma+1}{\gamma-1}}{1 + \frac{2}{\gamma-1}\frac{1}{Ma_1^2 \sin^2 \beta}} = \frac{(\gamma + 1)Ma_1^2 \sin^2 \beta}{(\gamma - 1)Ma_1^2 \sin^2 \beta + 2} \tag{7.241}$$

The temperature ratio is

$$\frac{T_2}{T_1} = \frac{p_2}{p_1}\frac{\rho_1}{\rho_2} = \left(\frac{\gamma + 1}{\gamma - 1}\right)^2 \left(\frac{2\gamma}{\gamma - 1}Ma_1^2 \sin^2 \beta - 1\right)\left(\frac{2}{\gamma - 1}\frac{1}{Ma_1^2 \sin^2 \beta} + 1\right) \tag{7.242}$$

The total pressure ratio is

$$\sigma = \frac{p_{02}}{p_{01}} = \left(\frac{2\gamma}{\gamma + 1}Ma_1^2 \sin^2 \beta - \frac{\gamma - 1}{\gamma + 1}\right)^{-\frac{1}{\gamma-1}}\left[\frac{(\gamma + 1)Ma_1^2 \sin^2 \beta}{(\gamma - 1)Ma_1^2 \sin^2 \beta + 2}\right]^{\frac{\gamma}{\gamma-1}} \tag{7.243}$$

The entropy increment is

$$\frac{\Delta S}{C_v} = -(\gamma - 1)\ln \sigma = \ln\left(\frac{2\gamma}{\gamma + 1}Ma_1^2 \sin^2 \beta - \frac{\gamma - 1}{\gamma + 1}\right)\left[\frac{(\gamma - 1)Ma_1^2 \sin^2 \beta + 2}{(\gamma + 1)Ma_1^2 \sin^2 \beta}\right]^{-\gamma}$$

$$(7.244)$$

Ma number after the wave is

$$Ma_2^2 = \frac{Ma_1^2 + \frac{2}{\gamma - 1}}{\frac{2\gamma}{\gamma - 1}Ma_1^2 \sin^2 \beta - 1} + \frac{\frac{2}{\gamma - 1}Ma_1^2 \cos^2 \beta}{Ma_1^2 \sin^2 \beta + \frac{2}{\gamma - 1}}$$

$$(7.245)$$

After the oblique shock wave, the tangential velocity remains unchanged and the total temperature remains unchanged. It can be seen from the formula of pressure ratio that for a given number of incoming Ma, the shock intensity is related to the shock angle. The greater the shock angle, the greater the shock intensity. When the shock wave angle is equal to 90°, the shock wave intensity is the largest. This shows that the normal shock wave is the strongest shock wave. The other extreme is a weak shock wave, the shock wave intensity approaches zero, and we get

$$\frac{p_2}{p_1} = \frac{2\gamma}{\gamma + 1}Ma_1^2 \sin^2 \beta - \frac{\gamma - 1}{\gamma + 1} \approx 1$$

$$(7.246)$$

$$\sin \beta = \frac{1}{Ma_1}$$

$$(7.247)$$

Comparing the above two equations with the relationship between the parameters of the Mach wave before and after the wave, it is not difficult to find that the weakest shock wave is the Mach wave at a given Ma number. Now we use the geometric relationship to determine the relationship between the shock angle and the airflow angle, which is

$$\begin{aligned} V_{1n}V_{2n} &= a_*^2 - \frac{\gamma - 1}{\gamma + 1}V_t^2 \\ &= \frac{\gamma - 1}{\gamma + 1}V_1^2 + \frac{2a_1^2}{\gamma + 1} - \frac{\gamma - 1}{\gamma + 1}V_t^2 \\ &= \frac{2a_1^2}{\gamma + 1} + \frac{\gamma - 1}{\gamma + 1}V_{1n}^2 \end{aligned}$$

$$(7.248)$$

After simplification, we get

$$\frac{V_{2n}}{V_{1n}} = \frac{2}{\gamma + 1}\frac{1}{Ma_1^2 \sin^2 \beta} + \frac{\gamma - 1}{\gamma + 1}$$

$$(7.249)$$

According to the geometric relationship, we get

$$\frac{V_{2n}}{V_{1n}} = \frac{V_{2n}}{V_t}\frac{V_t}{V_{1n}} = \frac{tg(\beta - \delta)}{tg\beta} \tag{7.250}$$

Substituting the above formula to solve

$$tg\delta = \frac{Ma_1^2 \sin^2 \beta - 1}{\left[Ma_1^2\left(\frac{\gamma+1}{2} - \sin^2 \beta\right) + 1\right]tg\beta} \tag{7.251}$$

(Fig. 7.34)
Under the condition of given Ma_1, by use of Eq. (7.251), from

$$\frac{d\delta}{d\beta} = 0 \tag{7.252}$$

we can get β_{max} at δ_{max} That is,

$$\sin^2 \beta_{max} = \frac{1}{\gamma Ma_1^2}\left[\frac{\gamma+1}{4}Ma_1^2 - 1 + \sqrt{(\gamma+1)\left(1 + \frac{\gamma-1}{2}Ma_1^2 + \frac{\gamma+1}{16}Ma_1^4\right)}\right] \tag{7.253}$$

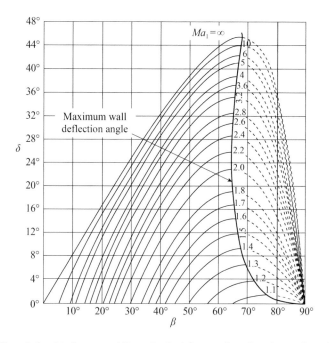

Fig. 7.34 The relationship between oblique shock airflow angle and wedge angle under different Mach numbers

(1) For a given Ma_1 and δ, there are two different values of β and Ma_2. The reason is: for a certain Ma_1, when the airflow passes through the normal shock wave, the direction remains unchanged, that is, $\delta = 0°$, while when the airflow passes through the Mach wave (infinitely weak compression wave), $\delta = 0°$. Therefore, when the shock angle β increases from the Mach angle μ to $90°$, there must be a certain maximum bending angle δ_{max} in the middle. When the shock angle β gradually increases from μ, δ gradually increases from $0°$ accordingly. To δ_{max}, while β continues to increase to $90°$, the airflow angle δ gradually decreases from δ_{max}.

(2) Under the same Ma_1, one δ value corresponds to two β. Larger β represents a stronger shock, which is called a strong shock; a small β represents a weaker shock, which is called a weak shock. The maximum wall deflection line in the figure represents the connection line corresponding to each point of δ_{max}. This line divides each figure into two parts. The part with the larger β is the strong wave, and the part with the smaller β is the weak wave. Whether the actual problem is a strong wave or a weak wave is determined by the specific boundary conditions that produce the shock wave. According to experimental observations, the oblique shock wave determined by the direction will always appear only with weak waves and no strong waves.

In the ultrasonic airflow, there are three situations in which shock waves are generated.

(1) Shock wave determined by airflow reversal

In the supersonic airflow, a wedge is placed, the inclined surface of the wedge squeezes the airflow channel smaller, the airflow is compressed, and a shock wave occurs. This shock wave is determined by the angle of the inclined plane, as shown in Fig. 7.35.

(2) Shock wave determined by pressure conditions

A shock wave on the free boundary determined by the pressure condition. For example, when the pressure at the outlet of the supersonic nozzle is lower than the outside atmospheric pressure, then the airflow will generate a shock wave to increase

Fig. 7.35 Shock wave generated by airflow turning

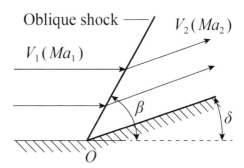

Fig. 7.36 Shock wave of
supersonic jet

the pressure. Figure 7.36 is a Schlieren diagram of the shock wave phenomenon of
a supersonic jet. It can be clearly seen from the figure that there will be shock waves
generated and reflected on the jet exit and the jet boundary, which are all determined
by the pressure conditions.

(3) Congestion shock wave

In pipelines (such as in supersonic wind tunnels and jet engine pipelines), a kind
of congestion may occur. That is, a certain section of the pipeline restricts the flow,
making the upstream part unable to circulate. This will force the upstream supersonic
airflow to generate shock waves and adjust the airflow. This shock wave is neither
dictated by the direction nor by the negative pressure.

7.9.6 Isolated Shock Wave

In the oblique shock wave graph 7.34, for a given Ma_1, after an oblique shock wave,
the turning angle of the airflow has a maximum value. Even when Ma_1 approaches
infinity, the maximum airflow angle is only 45.58°. When Ma_1 approaches infinity,
we get

$$\lim_{Ma_1 \to \infty} tg\delta = \frac{\sin^2 \beta}{\left(\frac{\gamma+1}{2} - \sin^2 \beta\right) tg\beta} \tag{7.254}$$

From extremum condition $\frac{d\delta}{d\beta} = 0$, we get

$$\sin^2 \beta = \frac{\gamma + 1}{2\gamma} \tag{7.255}$$

Therefore,

$$tg\delta_{\max} = \frac{1}{\sqrt{(\gamma - 1)(\gamma + 1)}} \tag{7.256}$$

When $\gamma = 1.4$, the extreme values of δ and β can be obtained

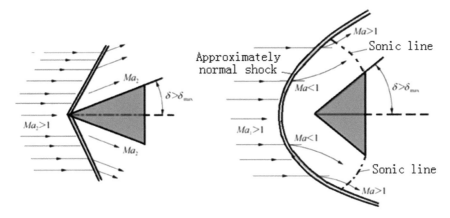

Fig. 7.37 The attached shock wave and the out-of-body shock wave

$$\delta_{\max} = 45.58^0, \beta = 67.79^0 \tag{7.257}$$

If under a certain *Ma* number, the required airflow angle of the object surface is greater than the maximum angle under this *Ma* number, then there is no solution. This shows that it is impossible for an oblique shock wave generated by the airflow to bypass the object in this case. The experiment found that an in-body shock wave appeared on the head of the object at this time, and the wave front was curved in a bow-shaped state, called a bow-shaped shock wave, with a normal shock wave in the middle and curved sides on both sides. There is a certain distance between the shock wave position and the head of the object. The larger the incoming flow *Ma*, the closer is the wave to the object. As shown in Fig. 7.37, when the wedge angle is small, it is an attached oblique shock wave. When the wedge angle is greater than the maximum wedge angle, the shock wave is out of the body, and the shock wave in front is approximately a normal shock wave. Figure 7.38 shows the Schlieren diagrams of the attached shock wave and the isolated shock wave.

7.9.7 The Internal Structure of Shock Waves

Treating the shock wave as a jump surface (discontinuous surface) with no thickness is possible to deal with general flow problems without causing large errors. However, in a viscous gas, there is no shock wave without thickness. Since the velocity has a certain change after the shock wave, the thickness of the shock wave is zero and the velocity gradient is infinite. At this time, the viscosity has a great influence. Under the action of the viscosity, the velocity cannot be controlled in a plane without thickness. v_1 suddenly accelerates to v_2, that is to say, there must be a transition zone, and the thickness of this transition zone is the thickness of the shock wave. However, this thickness is a very small amount, the same order of magnitude as the molecular mean

Fig. 7.38 The shock wave image of the attached body and in vitro

Fig. 7.39 The law of velocity change in the shock wave obtained using the continuum assumption

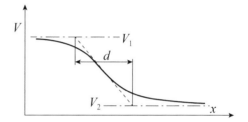

free path. In the sea atmosphere, the molecular free path is 70×10^{-6} mm. When Ma = 3, the shock wave thickness calculated by the continuum theory is 66×10^{-6} mm. Someone used the equation of the continuum medium considering viscosity and heat transfer to analyze the change process of airflow parameters in the shock wave, and found that the parameters such as speed change continuously in the shock wave, as shown in Fig. 7.39. The velocity change from v_1 at the front of the wave to v_2 at the back of the wave is a gradual process, so the shock thickness cannot be clearly defined. The tangent of the curve is usually made at the inflection point of the v-x curve, and the distance between it and the intersection of the horizontal line where v_1 and v_2 are constant is the shock thickness.

7.10 Boundary Layer Approximation of a Compressible Flow

When $Ma < 5$, the concept of the boundary layer is still available, but it needs to be considered because of the high velocity in the boundary layer, viscous internal friction causes kinetic energy loss and generates a lot of frictional heat, which increases the temperature in the layer. In other words, not only the compressibility, but also the temperature boundary layer must be considered. As the temperature in the boundary layer increases, the density and viscosity coefficient of the gas change. At the same time, heat conduction problems between the air layers and between the air and the surface of the object appear due to the normal temperature gradient. Because density and temperature become variables, not only the solution of the compressible boundary layer problem is more complicated than the incompressible boundary layer problem, but also the high-temperature gradient formed by viscous friction introduces a large amount of heat to the aircraft wall, causing aerodynamic heating problems. The structural design of the aircraft brings difficulties, coupled with the interference of shock waves and boundary layers in the airflow, resulting in significant changes in the pressure distribution on the object surface.

7.10.1 Temperature Boundary Layer

The frictional heat generated by the airflow increases the temperature in the boundary layer, which mainly limits the boundary layer area affected by viscosity. The airflow near the object surface is severely blocked, the speed and temperature gradient are very large, a lot of heat is generated, and the temperature is quite high. As the height from the object surface increases, the blocking effect of the airflow becomes smaller, and the temperature gradually drops. In the boundary layer of the object surface, the heat conduction problem caused by the temperature gradient must be considered. Like the concept of boundary layer, there is a thin layer of temperature change near the object surface, which is called temperature boundary layer. Usually the temperature in the layer reaches the height of 99% of the surface temperature T_e corresponding to the ideal gas, which is called the outer boundary of the temperature boundary layer which is

$$y|_{\frac{T}{T_e}=0.99} = \delta_T \qquad (7.258)$$

Generally, the temperature boundary layer is smaller than the velocity boundary layer. The velocity distribution in the temperature boundary layer depends to a large extent on the heat transfer of the wall. If heat is transferred to the air through the wall, it is called a hot wall, as shown in Fig. 7.40. Due to,

$$q_w = -\lambda \frac{\partial T}{\partial y}\bigg|_{y=0} > 0 \qquad (7.259)$$

Let,

$$\frac{\partial T}{\partial y}\bigg|_{y=0} < 0 \qquad (7.260)$$

Fig. 7.40 Temperature
distribution and heat transfer
in the wall boundary layer

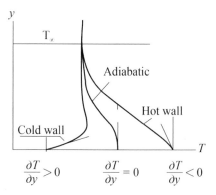

If the heat is transferred to the wall through the air, it is called a cold wall, then we get

$$\left.\frac{\partial T}{\partial y}\right|_{y=0} > 0 \tag{7.261}$$

In view of the fact that the velocity boundary layer is closely related to the temperature boundary layer, an important dimensionless parameter is now introduced. That is, the Prandtl number is defined as

$$\Pr = \frac{\mu C_p}{\lambda} \tag{7.262}$$

In the equation, μ is the viscosity coefficient of the gas, λ is the thermal conductivity coefficient, and C_p is the constant pressure specific heat of the gas. It characterizes the ratio of the momentum diffusion of gas clusters to the thermal diffusion. In the boundary layer, it represents the ratio of the heat generated by friction to the heat dissipated by heat conduction. Theoretically, it can be proved that the ratio of laminar boundary layer thickness δ to temperature boundary layer thickness δ_T is of the same magnitude with $\sqrt{\Pr}$. If $\Pr = 1$, δ and δ_T are in the same magnitude; $\Pr > 1.0$, δ is larger than δ_T. For air, its Prandtl number $\Pr = 0.72$, δ is less than δ_T, as shown in Fig. 7.41.

7.10.2 Recovery Temperature and Recovery Factor of Adiabatic Wall

When insulating the wall, the air temperature close to the surface of the object (also called the surface temperature) is actually always lower than the total temperature of the mainstream. Although the air velocity close to the object surface is zero, heat will not be transferred to the object surface when the insulation wall is used. But its

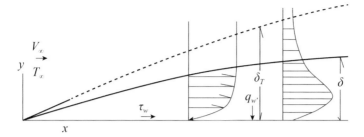

Fig. 7.41 Comparison of velocity boundary layer and temperature boundary layer of airflow

Fig. 7.42 Temperature distribution on the adiabatic wall

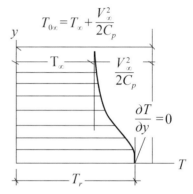

temperature is higher than that of the air slightly far away from the object surface, and the air near the object surface will transfer heat to the upper air, so that the wall temperature is lower than the total temperature. The temperature of the adiabatic wall excluding the effect of thermal radiation is called the recovery temperature, which is represented by T_r, as shown in Fig. 7.42.

In order to express the size of the recovery temperature, the rewarming factor R is introduced, which is defined as

$$R = \frac{T_r - T_\infty}{T_{0\infty} - T_\infty} \tag{7.263}$$

In the equation, $T_{0\infty} - T_\infty$ represents the temperature increase when the airflow stagnates adiabatically, and $T_r - T_\infty$ represents the increase of the stagnation temperature of the airflow on the surface due to wall friction and heat conduction. Experiments show that R does not change much with Mach number and Reynolds number, and is only related to the Pr number of the airflow. For Pr $= 1.0$, $R = 1$, $T_r = T_{0\infty}$, Pr < 1, $R < 1$, $T_r < T_{0\infty}$. At this time, the recovery temperature is

$$T_r = T_\infty \left(1 + R \frac{\gamma - 1}{2} Ma_\infty^2 \right) \tag{7.264}$$

7.10.3 Boundary Layer Equation of Adiabatic Wall

For the steady and compressible two-dimensional boundary layer flow, without considering the mass force, using the boundary layer approximation (Chap. 6), the continuity equation, momentum equation, and energy equation are simplified as

$$\frac{\partial(\rho u)}{\partial x} + \frac{\partial(\rho v)}{\partial y} = 0 \tag{7.265}$$

$$\rho\left(u\frac{\partial u}{\partial x} + v\frac{\partial u}{\partial y}\right) = -\frac{1}{\rho}\frac{\partial p}{\partial x} + \frac{\partial}{\partial}\left(\mu\frac{\partial u}{\partial y}\right) \tag{7.266}$$

Using Eqs. (7.83) and (7.64), the energy equation can be simplified to

$$\rho C_p\left(u\frac{\partial T}{\partial x} + v\frac{\partial T}{\partial y}\right) = u\frac{\partial p}{\partial x} + \frac{\partial}{\partial y}\left(\lambda\frac{\partial T}{\partial y}\right) + \mu\left(\frac{\partial u}{\partial y}\right)^2 \tag{7.267}$$

If the pressure in the boundary layer uses the value outside the boundary layer, then the unknown airflow parameters in the boundary are u, v, ρ, T, and a supplementary equation (equation of state) is needed, $p = \rho R T$.

The boundary conditions are

$$\text{exist } y = 0, \quad u = v = 0, T = T_r; \quad \text{exist } y = \infty, \quad u = V_\infty, T = T_\infty \tag{7.268}$$

7.11 Shock Wave and Boundary Layer Interference

The study of the interference between shock waves and compressible boundary layer flows was first conducted by American fluid mechanics Liepmann in 1946 and Swiss aerodynamics Ackeret (1898–1981). The research began in 1947, and the development was slow for a period of time after that. But with the advent of supersonic vehicles, people's attention to this issue has begun to increase.

Especially in the past 10 years, under the impetus of the development of supersonic vehicles, transonic transport aircraft, and reusable aerospace hypersonic vehicles, the rapid development of computational and experimental fluid mechanics, these push the study of supersonic compressible boundary layer flow and shock wave interference to a new climax. According to the different Reynolds number of the flow around, there are also laminar and turbulent flows in the boundary layer. And their effects on the frictional resistance and heat conduction performance of the wall are completely different. If there is interference between shock waves and the compressible boundary layer, the flow in the boundary layer is more complicated, and complicated flow problems such as laminar flow, transition, turbulence, separation,

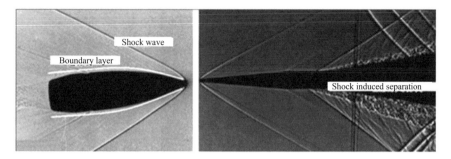

Fig. 7.43 The interaction between shock wave and boundary layer

and reattachment may occur (as shown in Fig. 7.43) which seriously affect the lift, drag, and surface thermal protection of the aircraft.

7.11.1 Interference Between Normal Shock Wave and Laminar Boundary Layer

It is known that the occurrence of a positive shock in a supersonic airflow will reduce the mainstream Mach number to a subsonic value. This deceleration process is accompanied by a rapid increase in pressure, density, and temperature along the flow direction. If you encounter a wall boundary layer, the effect of the shock wave is similar to applying a sudden reverse pressure gradient to the boundary layer, which will seriously change the flow characteristics of the boundary layer. If the shock wave is strong, the interference between the shock wave and the boundary layer will cause the boundary layer behind the wave to separate, and it will also cause strong unsteady flow and strong heat conduction in local areas, which will seriously affect the performance of the aircraft. Therefore, studying the mechanism of shock wave and boundary layer interference must be of great significance to predict the boundary layer transition position, separation position, control shock wave oscillation, control separation, and other complex issues.

For supersonic flow around objects, the flow in the area outside the boundary layer of the object is supersonic flow. However, due to the viscosity of the fluid in the inner zone of the boundary layer, the fluid velocity rapidly decreases to zero velocity on the wall. Therefore, there are subsonic and supersonic regions in the layer, and the shock boost outside the boundary layer will propagate upstream through the subsonic region in the layer to the shock front area. This reverse pressure gradient acting from downstream to upstream obviously changes the state of the boundary layer upstream of the shock wave, causing the boundary layer in the disturbance zone to become thicker, and the velocity, temperature, pressure, and density distribution within the layer change, and the frictional resistance decreases. It also changes the local shock wave structure in the near-wall area. In the flow around a transonic airfoil,

Fig. 7.44 Interaction between shock wave and boundary layer of transonic airfoil

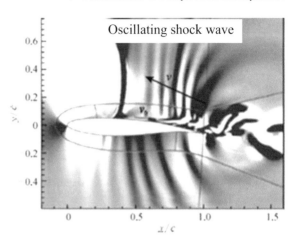

when the incoming Mach number is greater than the critical Mach number, a supersonic flow zone will appear when the airflow bypasses the upper airfoil. Obviously, this ultrasonic zone must be connected to the downstream subsonic flow through almost normal shock wave form. It is impossible for this shock wave pressurization and deceleration behavior to travel backward in the supersonic region outside the boundary layer. However, when encountering a near-wall subsonic boundary layer zone, the shock wave pressurization will travel upstream along the boundary layer countercurrent, resulting in thickening of the boundary layer. In severe cases, the boundary layer will separate (separation bubbles or complete separation after the wave appears, and the shock intensity is related to the characteristics of the boundary layer) shock wave oscillation and other complex phenomena, as shown in Fig. 7.44.

The interference characteristics of the normal shock wave and the boundary layer are closely related to the intensity of the shock wave and the characteristics of the boundary layer. For example, in a strong interference area, the definition and magnitude of the boundary layer are usually no longer applicable, because the flow velocity gradient and the normal velocity gradient have the same magnitude. The interfering flow in the laminar or turbulent flow zone is also significantly different. For example, when the flow conditions at the outer edge of the boundary layer are the same as the intensity of the shock wave, then the distance that the reverse pressure propagates upstream in the laminar boundary layer is farther than the turbulent boundary layer, and the laminar boundary layer has a weaker ability to resist flow separation. The interference characteristics of the normal shock wave and the laminar boundary layer are related to the Mach number, Reynolds number, and shock wave intensity outside the boundary layer. According to the shock wave intensity, three different interference situations may occur.

(1) **Weak interference**

In the flow around a transonic airfoil, the shock wave appears relatively weak in the upper airfoil area of the airfoil, and interference between the shock wave and the laminar boundary layer will occur. The weak shock wave pressurization will slowly thicken the wave front boundary layer. No transition and separation will happen. This thickened wave front boundary layer deflects the airflow inward, reflecting a series of weak compression waves on the wave front boundary layer, and converging with the main shock wave to form a so-called λ wave, leading to the main wave front and the boundary. The angle between the layers is less than 90°, as shown in Fig. 7.45. At the reattachment surface after the main shock, the thickness reduces, and the outer boundary deviates to the wall surface. When the airflow turns outward, the subsonic region becomes the supersonic region again, resulting in a fan-shaped supersonic expansion wave system, and the pressure drops. Then a second shock wave appeared again, which caused the shock wave and the laminar boundary layer to interfere again. The interference characteristics are basically similar, but the intensity decreased significantly. If the conditions are right, this situation may repeat several times to form a series of λ wave systems.

(2) **Moderate interference**

If the incoming flow velocity increases, the wave front Mach number and Reynolds number will increase. Although the wave front is a laminar boundary layer, under the action of a strong shock wave (increasing the inverse pressure gradient), this causes the boundary layer behind the wave to separate, and it quickly turns into turbulent flow, and then attaches to the wall. The interference results are (1) separation bubbles appear on the wall; (2) a series of reflected compression waves appear in front of the main shock, forming a λ-shaped wave system with the main shock; (3) after the main shock, the boundary layer is attached to the wall again, the outer boundary

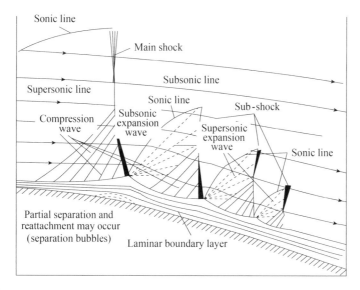

Fig. 7.45 Interference between weak shock waves on the curved surface and laminar boundary layer (laminar boundary layer)

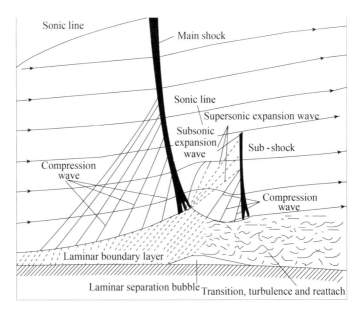

Fig. 7.46 Interference between a medium shock wave and a laminar boundary layer on a curved surface (transition)

deviates to the wall, the airflow turns outward and a fan-shaped supersonic expansion wave system appears, the pressure drops, and a secondary shock wave appears; (4) when the downstream boundary layer flattens, a series of compression wave systems appear again; (5) when the incoming flow has a larger Reynolds number, the shock wave is stronger, and the back wave generally transitions to a turbulent boundary layer, as shown in Fig. 7.46.

(3) **Strong interference**

As the incoming flow Mach number increases, the intensity of the shock wave also increases, and its interference with the boundary layer is enough to cause the separation of the laminar boundary layer. As a result, the direction of the main flow outside the boundary layer changes significantly, and a stable oblique shock wave appears in front of the main shock wave, thus forming an obvious λ shock wave above the boundary layer. Because the boundary layer can no longer be attached, the secondary shock wave no longer exists, as shown in Fig. 7.47. This kind of strong interference will cause the airfoil to separate suddenly, the lift will drop, the drag will increase suddenly, and the shock wave will induce the stall.

Fig. 7.47 The interference (separation) between the strong shock wave on the curved surface and the laminar boundary layer

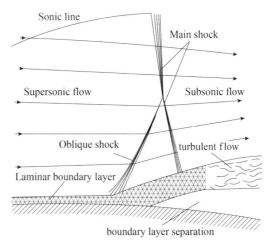

7.11.2 Interference Between Oblique Shock Wave and Boundary Layer

One of the main differences between supersonic flow and subsonic flow is that as the cross-sectional area of the flow tube changes along the flow direction, the change in pressure gradient along the flow direction is inverse. In the supersonic flow in the shrinking pipe, the airflow is compressed, the pressure increases, and the speed decreases. In the expansion pipe, the supersonic airflow is expanded, the airflow is accelerated, and the pressure is reduced. For subsonic flow, the flow characteristics in contracting and expanding pipes are exactly the opposite of supersonic flow behavior. Therefore, for supersonic flow around the expansion angle, a series of expansion waves appear outside the boundary layer, and the pressure gradient is forward pressure, and the boundary layer around the angle will not separate. In this way, the interference between the supersonic expansion wave and the boundary layer is weak, and the mutual influence is not large, as shown in Fig. 7.48. If it flows around at a compression angle, then the airflow is compressed outside the boundary layer and an oblique shock wave appears. The interference between this oblique shock wave and the boundary layer will cause the boundary layer to thicken at the wave front, and may cause a partial separation of the boundary layer at the corners behind the wave, forming a series of compression waves outside the boundary layer, as shown in Fig. 7.49. If the oblique shock wave is strong and the thickness of the laminar boundary layer is thin, the shock wave penetrates deeply, and the forward propagation area of the high-pressure backflow after the wave is small, and a dense reflection wave system is formed in front of the incident point, which quickly converges into the first reflection wave. Affected by the reverse pressure gradient, the cambium boundary layer separates bubbles. An expansion wave appears downstream of the incident wave, and then a second reflected wave appears. After the wave is reflected,

Fig. 7.48 Interference between the expansion wave system and the boundary layer of the flow around the supersonic expansion angle

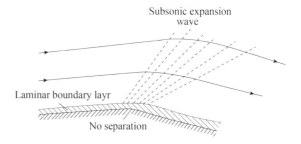

Fig. 7.49 Oblique shock wave around supersonic compression angle and boundary layer interference (transition turbulence)

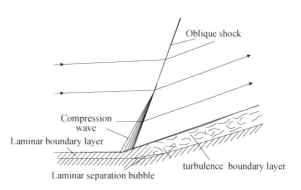

the laminar boundary layer transforms into a turbulent boundary layer, as shown in Fig. 7.50.

In addition, the interference of oblique shock wave and turbulent boundary layer is different from that of laminar boundary layer. Under the same incoming flow conditions, the time-average velocity distribution in the turbulent boundary layer is full, and the subsonic velocity zone in the boundary layer is thinner than that in the laminar boundary layer, which results in the shock wave penetrating deeply and the

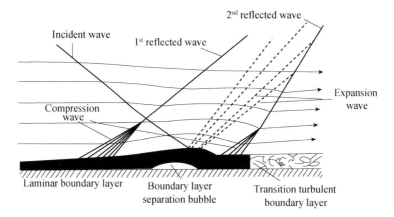

Fig. 7.50 Supersonic strong oblique shock wave and laminar boundary layer interference (separation of bubbles, transition turbulence)

Fig. 7.51 Interference between oblique shock wave and turbulent boundary layer (turbulent boundary layer without separation)

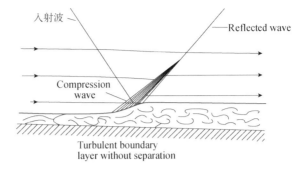

Fig. 7.52 Interference between oblique shock wave and turbulent boundary layer (turbulent boundary layer separation bubble)

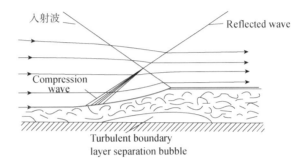

back-wave high-pressure backflow forward passage is small. Therefore, the distance that the pressure flows backward is smaller than that of the laminar boundary layer. The momentum exchange caused by the pulsation of wall turbulence can suppress the effect of shock inverse pressure gradient and make the boundary layer difficult to separate. If the boundary layer does not separate after the interaction (as shown in Fig. 7.51), a λ-shaped wave system will appear, forming a dense reflected wave system in front of the incident point and quickly converge into a reflected wave close to the ideal flow. If the incident wave is strong, the turbulent boundary layer will separate and reattach locally, forming separation bubbles, as shown in Fig. 7.52. At this time, the boundary layer bulges greatly, and a large range of compression wave system appears before the incident point, which merges to form a reflected wave and passes through the incident wave to form a λ-shaped shock wave.

7.11.3 Head Shock and Boundary Layer Interference

The interference between the bow shock wave and the boundary layer of the wedge-shaped body supersonic flow around the head is shown in Fig. 7.53. Affected by the rapid thickening of the boundary layer of the flow around the head, the head of the flow around the boundary layer becomes blunt, and an isolated shock wave (bow shock wave) appears, and a small area of subsonic velocity zone is formed after the

Fig. 7.53 Interference between wedge head shock wave and boundary layer

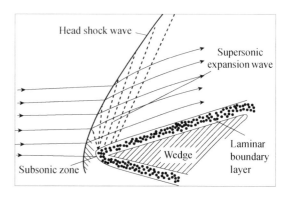

Fig. 7.54 Interference between the flat head shock wave and the boundary layer

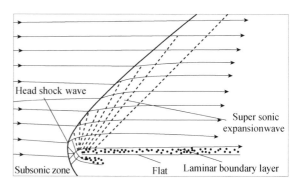

bow shock wave. The expansion wave from the boundary layer of the head intersects the shock wave, weakening and bending the shock wave. Similarly, the action of the bow shock wave will also change the boundary layer. If the boundary layer is thin, the mutual influence mainly occurs on the head. For the supersonic flow around the flat plate, a weak separated shock wave will also appear at the head, but it will quickly weaken to a Mach wave, as shown in Fig. 7.54.

7.12   Compressible One-Dimensional Friction Pipe Flow

7.12.1   The Effect of Friction in Straight Pipes on Airflow

The effect of friction on the compressible pipe flow will reduce the total pressure of the airflow. The following analyzes the size of the loss caused by friction in the pipe flow of equal diameter. Assuming that the tube wall is adiabatic, there is no energy exchange between the gas flow and the outside, and heat exchange between gas clusters is not considered. Still using a one-dimensional assumption, as shown in Fig. 7.55, takes a control body with a length of dx along the pipe axis, the diameter of the pipe is D, the area is A, and the flow parameters of the upstream section of the control body are V, ρ, p, T. The flow parameters of the downstream section are $V + dV, \rho + d\rho, p + dp, T + dT$, and the frictional stress on the pipe wall is τ_0. Assuming that the flow through the pipe is a steady flow, the mass conservation equation is (Fig. 7.56)

$$d(\rho V A) = 0, \quad \frac{d\rho}{\rho} + \frac{dV}{V} + \frac{dA}{A} = 0 \tag{7.269}$$

In the flow, $dA = 0$, we get

$$\frac{d\rho}{\rho} + \frac{dV}{V} = 0 \tag{7.270}$$

The momentum conservation equation (pipe level or without mass force) is

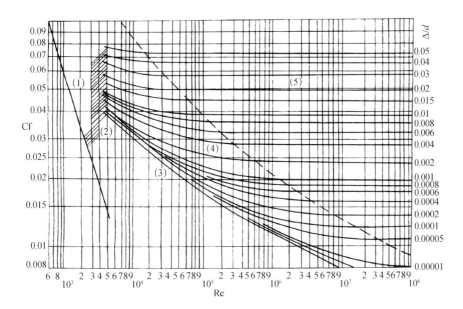

Fig. 7.55 Resistance coefficient along the pipeline

Fig. 7.56 Experimental device for shrinking nozzle

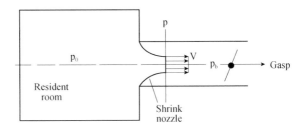

$$pA - (p + dp)A - \tau_0 P dx = \rho V A (V + dV - V) \tag{7.271}$$

After simplification, we get

$$\rho V dV + dp + \tau_0 \frac{P}{A} dx = 0 \tag{7.272}$$

Assuming that the wall shear stress can be expressed as

$$\tau_0 = \frac{1}{8} C_f \rho V^2 \tag{7.273}$$

Among them, C_f is the resistance coefficient along the pipeline (check the table, as shown in Fig. 7.55) and R is the hydraulic radius ($=A/P = D/4$). Substituting Eq. (7.270), we get

$$\rho V dV + dp + C_f \frac{1}{2} \rho V^2 \frac{dx}{D} = 0 \tag{7.274}$$

From the steady adiabatic flow, the energy equation is

$$dh + d(V^2/2) = 0 \tag{7.275}$$

Among them is the enthalpy of the gas, $dh = C_p \, dT$, then the above equation has

$$C_p dT + V dV = 0 \tag{7.276}$$

The equation of state for an ideal gas is

$$\frac{p}{\rho} = RT \tag{7.277}$$

Solving Eqs. (7.270), (7.274), (7.276), and (7.277) simultaneously, we can get

$$dp = -\rho \left(\frac{\gamma - 1}{\gamma} + \frac{1}{\gamma Ma^2} \right) V dV \tag{7.278}$$

$$(1 - Ma^2) \frac{dV}{V} = \frac{1}{2} \gamma Ma^2 C_f \frac{dx}{D} \tag{7.279}$$

The above equation shows that in the subsonic flow zone $dV > 0$, friction accelerates the subsonic airflow; when the Mach number is greater than 1, $dV < 0$ can be obtained, indicating that friction decelerates the supersonic airflow. Using the continuity Eq. (7.270), the above equation can be changed to

$$(Ma^2 - 1) \frac{d\rho}{\rho} = \frac{1}{2} \gamma Ma^2 C_f \frac{dx}{D} \tag{7.280}$$

Using the energy Eq. (7.276), the above equation becomes

$$\frac{Ma^2 - 1}{(\gamma - 1)Ma^2} \frac{dT}{T} = \frac{1}{2}\gamma Ma^2 C_f \frac{dx}{D} \qquad (7.281)$$

Using Eq. (7.278), the above equation can be changed to

$$\frac{Ma^2 - 1}{1 + (\gamma - 1)Ma^2} \frac{dp}{p} = \frac{1}{2}\gamma Ma^2 C_f \frac{dx}{D} \qquad (7.282)$$

It can be seen from these relations that with subsonic airflow ($Ma < 1$) friction causes $d\rho < 0$, $dT < 0$, and $dp < 0$, which is a kind of expansion change. On the contrary, supersonic flow ($Ma > 1$), friction makes $d\rho > 0$, $dT > 0$, $dp > 0$, which is a kind of compression change.

7.12.2 Distribution of Flow Velocity Along the Length of the Pipe

If we take the drag coefficient along the way as a constant, we can integrate (7.279). Introducing the velocity coefficient and integrating from λ_1 where $x = 0$ to λ, we get

$$\int_{\lambda_1}^{\lambda} \left(\frac{1}{\lambda^2} - 1\right) \frac{d\lambda}{\lambda} = \frac{\gamma}{\gamma + 1} \frac{C_f}{D} dx \qquad (7.283)$$

After integration, we get

$$\left(\frac{1}{\lambda_1^2} - \frac{1}{\lambda^2}\right) - \ln\frac{\lambda^2}{\lambda_1^2} = \frac{\gamma}{\gamma + 1} \frac{2C_f}{D} x \qquad (7.284)$$

If let,

$$\eta = 2C_f \frac{\gamma}{\gamma + 1} \frac{x}{D} \qquad (7.285)$$

It is called the reduced tube length, which is

$$\left(\frac{1}{\lambda_1^2} - \frac{1}{\lambda^2}\right) - \ln\frac{\lambda^2}{\lambda_1^2} = \eta \qquad (7.286)$$

For a given λ_1, the change of λ with η is related to whether the λ_1 is greater than or less than 1. For the case when λ_1 is less than 1, the value of λ increases with the increase of η; for the case when λ_1 is greater than 1, the value of λ decreases with the increase of η. This shows that friction causes the speed coefficient to tend to 1.0. For a given λ_1, the limit tube length η_{max} ($\lambda = 1$) is

$$\eta_{max} = \left(\frac{1}{\lambda_1^2} - 1\right) + \ln \lambda_1^2 \tag{7.287}$$

7.13 Working Performance of Shrinking Nozzle, Laval Nozzle, and Supersonic Wind Tunnel

7.13.1 Working Performance of Shrink Nozzle

According to the theory of one-dimensional isentropic pipe flow, when the subsonic airflow (or low-speed airflow) passes through the shrinking nozzle, the flow velocity in the pipe increases along the way, the pressure and density decrease, and the airflow expands, while the supersonic airflow passes through the shrinking nozzle, the airflow speed decreases along the way, the pressure and density increase, and the airflow compresses. In the following, the working performance of the shrinking nozzle is analyzed based on the one-dimensional isentropic flow relationship. For the experimental device of the shrinking nozzle shown in Fig. 7.56, there is a large container in front of the nozzle, and the velocity of the gas in it is very small. It can be regarded as the kinetic energy tends to zero. The total pressure of the gas in the container is p_0, the total temperature is T_0, the total density is ρ_0, the nozzle outlet area is A, the outlet pressure is p, and the pressure (or back pressure) p_b of the back pressure chamber can be adjusted. Assuming that the nozzle outlet pressure ratio β is

$$\beta = \frac{p}{p_0} \tag{7.288}$$

The working performance of the shrink nozzle includes the following:

(1) When $p_* \leq p(= p_b) \leq p_0$, the flow in the shrinking nozzle is all subsonic flow, where p* is the outlet pressure corresponding to the nozzle outlet velocity reaching the speed of sound ($V = a$). The outlet flow rate of the nozzle can be determined according to the one-dimensional isentropic flow relationship. From Eq. (7.113), the relationship between the nozzle outlet Mach number and the pressure ratio is

$$Ma = \frac{V}{a} = \sqrt{\frac{2}{\gamma - 1}\left(\beta^{-\frac{\gamma-1}{\gamma}} - 1\right)} \tag{7.289}$$

The outlet speed is

$$V = \sqrt{\frac{2\gamma}{\gamma - 1} \frac{p}{\rho} \left(\beta^{-\frac{\gamma-1}{\gamma}} - 1 \right)} \tag{7.290}$$

The relationship between the outlet flow rate and the pressure ratio is

$$G = \rho V A = A \sqrt{\frac{2\gamma}{\gamma - 1} p_0 \rho_0 \left(\beta^{\frac{2}{\gamma}} - \beta^{\frac{\gamma+1}{\gamma}} \right)} \tag{7.291}$$

Or the flow rate expressed by the outlet Mach number is

$$G = \sqrt{\frac{\gamma}{R} \left(\frac{2}{1+\gamma} \right)^{\frac{\gamma+1}{\gamma-1}}} \frac{p_0 A}{\sqrt{T_0}} \mathrm{Ma} \left[\frac{2}{\gamma + 1} \left(1 + \frac{\gamma - 1}{2} Ma^2 \right) \right]^{-\frac{\gamma+1}{2(\gamma-1)}} \tag{7.292}$$

It can be seen that for a given shrinking nozzle, when the outlet area of the nozzle and the total pressure are given, the outlet flow is only a function of the pressure ratio β. If the outlet pressure of the nozzle is p_*, the outlet velocity is the speed of sound, the Mach number $Ma = 1$, and the nozzle flow reaches the maximum G_{max}, at this time, it is obtained by Formula (7.292)

$$G_{max} = \sqrt{\frac{\gamma}{R} (\frac{2}{1+\gamma})^{\frac{\gamma+1}{\gamma-1}}} \frac{p_0 A}{\sqrt{T_0}} \tag{7.293}$$

The critical pressure at the nozzle outlet is

$$\frac{p_*}{p_0} = \left(\frac{2}{\gamma + 1} \right)^{\frac{\gamma}{\gamma-1}} \tag{7.294}$$

As shown in Fig. 7.57, at $p_* \le p(= p_b) \le p_0$, the pressure curve is between 1 and 3, the mass flow is between 0 and Gmax, and the nozzle Mach number is between 0 and 1.

(2) When $p_b < p \le p_*$, the airflow at the outlet of the nozzle is in a critical state, and after leaving the nozzle, the airflow will continue to expand until the pressure reaches the back pressure. Because the nozzle outlet velocity reaches the speed of sound, the downstream disturbance cannot be transmitted upstream, so no matter how the back pressure changes, the nozzle flow rate is constant, and the pressure change curve along the way is shown in Fig. 7.57, curve 4.

From the above analysis, it can be seen that: (1) Given the incoming flow parameters and nozzle exit area before shrinking the nozzle,
The outlet pressure of the maximum flow rate G_{max} through the nozzle is the critical pressure p*. If the back pressure $p_b > p_*$, the flow rate will decrease; if the back pressure $p_b < p_*$, the flow rate will not change; (2) the flow rate of the airflow through the shrinking nozzle is proportional to the outlet area and the total temperature p_0

Fig. 7.57 Working
performance of shrink nozzle

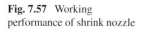

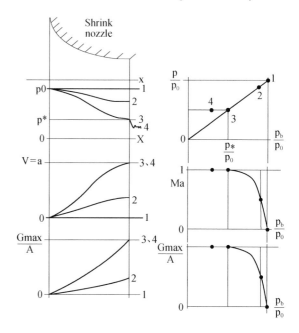

and is the square root of the total temperature T_0 inversely proportional; (3) in order
to increase the maximum flow rate of the nozzle under a given outlet cross-sectional
area, it can only be achieved by increasing the total airflow pressure in front of the
nozzle.

The pipe section whose cross-sectional area gradually expands is called a diffuser
pipe. When the subsonic airflow passes through the diffuser tube, because the cross-
sectional area increases, then the speed decreases along the way, and the pressure
and density increase along the way; when the supersonic airflow passes through the
diffuser tube, because the cross-sectional area increases, then the speed increases
along the way. The pressure and density decrease, and the airflow expands.

7.13.2 Working Performance of Laval Nozzle

It is known that the use of a Laval nozzle can make the airflow exceed the speed of
sound and continue to accelerate to the supersonic airflow. In terms of structure, the
Laval nozzle is a pipe that shrinks first and then expands. Its working performance
is more complicated than that of a shrinking nozzle, as shown in Fig. 7.58. The
experimental device is shown in Fig. 7.58.

The working performance of Laval nozzle is as follows (as shown in Fig. 7.59):

(1) When $p_2 \le p(= p_b) \le p_0$, then the flow in the Laval nozzle is subsonic flow,
      where p_2 is the outlet pressure corresponding to the velocity of the Laval nozzle

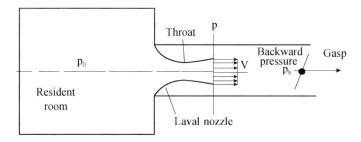

Fig. 7.58 Experimental setup of Laval nozzle

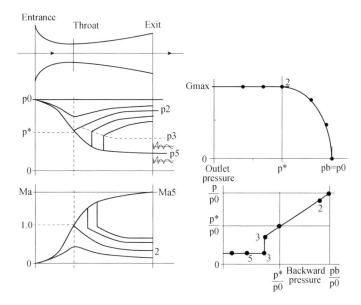

Fig. 7.59 Working performance of the Rafar nozzle

throat reaching the speed of sound ($V = a$). The airflow through the Laval nozzle is similar to the flow of a Venturi flowmeter. The airflow accelerates and decompresses before the throat, and the airflow decelerates and pressurizes behind the throat. The flow velocity is the largest at the throat, but does not exceed the speed of sound. The outflow can be determined according to the one-dimensional isentropic flow relationship. p_2 is the corresponding outlet pressure when the speed of sound appears in the throat.

(2)   When $p_3 \leq p(= p_b) \leq p_2$, then supersonic flow appears in the Laval nozzle, and shock waves appear in the expansion section, and the nozzle exit is subsonic flow. Among them, p_3 is the shock wave at the exit of the Laval nozzle, and the pressure at the nozzle exit after the shock wave.

(3) When $p_5 \leq p \leq p_3$, and the nozzle outlet pressure $p < p_b$ (backward pressure), then all supersonic flow appears in the expansion section of the Laval nozzle, and there is no shock wave in the expansion section. The outlet of the nozzle is supersonic flow, but because the outlet pressure is less than the back pressure, a compression wave will be generated after the nozzle flows out. (Oblique shock wave) transition to back pressure, so the outflow from the nozzle is over-expansion. Among them, p_5 is that the Laval nozzle outlet is supersonic flow, and the outlet pressure is equal to the back pressure, that is, $p_5 = p_b$. This situation is called the nozzle design condition.

(4) When $p_b \leq p \leq p_5$, and the nozzle outlet pressure $p > p_b$ (backward pressure), then supersonic flow appears in the expansion section of the Laval nozzle, and there is no shock wave in the expansion section. The nozzle outlet is supersonic flow, but because the outlet pressure is stronger than the back pressure, the jet, after the pipe flows out, will continue to expand (expansion wave appears) and transitions to back pressure, so the nozzle flow is under-expanded.

7.13.3   Working Performance of Supersonic Wind Tunnel

The supersonic wind tunnel is a piping system that simulates supersonic airflow. The test section can achieve a uniform airflow field with a given Mach number. As shown in Fig. 7.60, it is a schematic diagram of an ideal continuous supersonic wind tunnel. Suppose the total pressure of the high-pressure gas source is p_0 and the total temperature is T_0. The supersonic airflow is accelerated to a given Mach number in the test section through the contraction and expansion nozzle airflow, and then leaves the test section and enters the wind tunnel diffusion section. The supersonic airflow first decelerates through the contraction section. It reaches the speed of sound, and then further slows down to a low-pressure gas source through the diffusion section. From the low-pressure air source to the high-pressure air source, the airflow is completed by the cooler and the compressor to overcome the friction loss of the airflow and keep the total temperature of the airflow unchanged.

In a supersonic wind tunnel, a critical flow with Mach number $Ma = 1$ appears in the nozzle and the throat of the diffuser section. Considering the total pressure loss after the shock wave, according to Formula (7.283), the product of the total pressure at the throat and the area of the throat is a constant, which is

$$p_0 A_c = C \tag{7.295}$$

Suppose the area of the first throat is A_{c1}, total pressure is p_{01}, the area of the second throat is A_{c2}, and total pressure is p_{02}. Obtaining by Formula (7.293)

$$\frac{p_{01}}{p_{02}} = \frac{A_{c2}}{A_{c1}} > 1 \tag{7.296}$$

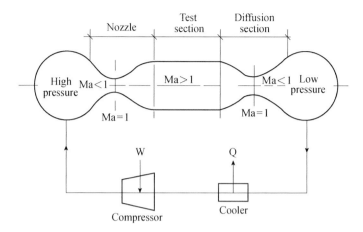

Fig. 7.60 Working principle of supersonic wind tunnel

Due to shock loss, $p_{01} > p_{02}$ and $A_{c1} < A_{c2}$. It shows that to maintain the supersonic flow in the test section, the area of the second throat is required to be larger than the area of the first throat, and the supersonic flow in the test section can be stabilized by adjusting the area of the second throat.

If the area of the first throat is equal to the area of the second throat, the airflow appears in the first throat with a sonic flow with a Mach number $Ma = 1$, and the airflow entering the expansion section of the nozzle decelerates and maintains a subsonic flow in the test section. Entering the diffusion section, the airflow in the second throat accelerates to the speed of sound $Ma = 1$, and the airflow is blocked. If the back pressure is reduced enough, supersonic flow appears downstream of the second throat of the diffusion section, and a positive shock wave appears in the expansion section of the diffusion section. If the back pressure is further reduced, the shock wave continues to move downstream, and at the nozzle exit and the test section is always flowing at subsonic speed. However, if the area of the second throat is larger than the area of the first throat, the blockage occurs downstream of the first throat instead of the second throat, so that the airflow appears at the first throat with a speed of sound $Ma = 1$, but still in the second throat, it is a subsonic flow. As the back pressure decreases, supersonic airflow appears in the expansion section downstream of the first throat, and then transitions to subsonic airflow through the normal shock wave. If the back pressure continues to decrease, the shock wave moves downstream to the downstream of the second throat. At this time, the entire test section forms a stable supersonic flow.

Exercises

A. Thinking questions

1. Please explain the physical meaning of the complete gas state equation.
2. What is the internal energy, pressure energy, kinetic energy, and enthalpy of a system?
3. Please write down the mathematical expression and physical meaning of the first law of thermodynamics in a static state.
4. Please explain the mathematical expression and physical meaning of the first law of thermodynamics in motion. Point out the difference with the steady flow energy equation of incompressible fluid (without mass force).
5. Write down the work done by the unit mass system to the outside world in the process of isovolume, pressure, and temperature.
6. Explain the physical meaning of adiabatic, reversible, and irreversible processes.
7. Write down the relationship between pressure and density in the adiabatic process.
8. What is the concept of entropy and why is dq/T a function of state?
9. For steady adiabatic flow (non-isentropic), please write down the relationship between the ratio of static pressure at any two points along the streamline and the Mach number.
10. For steady adiabatic flow (isentropic), please write down the relationship between the ratio of static pressure at any two points along the streamline and the Mach number.
11. What is the stagnation point parameter? Why does the total enthalpy remain unchanged and the total pressure decreases in the steady adiabatic friction pipeline flow?
12. Explain the physical phenomenon that the total temperature of a system remains unchanged and the total pressure decreases. Explain a physical phenomenon in which the total temperature and total pressure of a system remain unchanged.
13. For an adiabatic steady flow in a sudden expansion pipe, please explain the relationship between the speeds at the sudden expansion section of unequal entropic flow, isentropic flow, and incompressible flow. Why?
14. Please derive the relationship between the relative increment of steady flow velocity and the relative increment of area in a variable cross-sectional pipeline?
15. Please derive the relationship between the ratio of the minimum cross-sectional area of steady flow to the cross-sectional area of any cross section and the Mach number in a variable cross-sectional pipeline?
16. What is Mach wave? Explain the characteristics of Mach wave. Write down the pressure coefficient through Mach wave.
17. Write the continuous equation, tangential momentum equation, and normal momentum equation that pass through the Mach wave line?

18. Why is the Mach wave line emitted from the disturbance source when passing through the expansion angle? What wave is Mach wave?

19. What is the Mach wave when passing through the compression angle? Why does the Mach wave line converge?

20. Please explain the physical mechanism of shock wave formation. Why is the velocity of the shock wave greater than the velocity of the gas before the shock wave?

21. Why is the shock wave advancing speed less than the wave speed of the gas after the shock wave?

22. Write the governing equations (continuity equation, momentum equation, energy equation) through the shock wave.

23. Please write the expression of entropy increase through normal shock wave. Why is there an increase in entropy?

24. What is an oblique shock wave? Please point out the difference between oblique shock and normal shock.

25. Please explain how shock waves are generated in supersonic airflow.

B. **Calculation questions**

1. Consider a room with a rectangular floor measuring 5 m by 7 m and a ceiling height of 3.3 m. The indoor air pressure and temperature are 101,325 Pa and 25 °C, respectively. Calculate the internal energy and enthalpy of indoor air.

2. Assuming that a Boeing 747 is flying at a standard altitude of 10,972 m, the pressure at a certain point on the wing is 19152 Pa. Assuming that there is an isentropic flow on the wing, calculate the temperature at this time.

3. The pressure, temperature, and velocity at a certain point in the airflow are 101,325 Pa, 46.85 °C, and 1000 m/s. Calculate the total temperature and total pressure at that point.

4. An airplane flies at a standard altitude of 3048 m. The Pitot tube installed in the nose measures 105,336 Pa pressure. The plane is flying at subsonic speeds higher than 483 km per hour. The flow should be considered compressible. Calculate the speed of the airplane.

5. Considering the normal shock wave in the air, its upstream flow characteristics are $u_1 = 680\,\text{m/s}$, $T_1 = 288\,\text{K}$, and $p_1 = 101325\,\text{Pa}$. Calculate the velocity, temperature, and pressure downstream of the shock wave.

6. Calculate the ratio of kinetic energy to internal energy at a certain point in the airflow, where the Mach number is (a) $Ma = 2$ and (b) $Ma = 20$.

7. An airplane flies at a standard altitude of 3048 m. The Pitot tube installed in the nose measures 106,294 Pa pressure. The plane is flying at subsonic speeds higher than 482 km per hour. Calculate the speed of the airplane.

8. The temperature and pressure at the stagnation point of the high-speed missile are 245 °C and 790,335 Pa, respectively. Calculate the density at this location.

9. At the upstream of the shock wave, the air temperature and pressure are 14.85 °C and 101,325 Pa, respectively; just downstream of the shock wave, the air temperature and pressure are 417 °C and 877,069.2 Pa, respectively. Calculate the changes in the enthalpy, internal energy, and entropy of the wave.

10. Consider the isentropic flow on the airfoil. The free-flow condition is $T\infty = -28.15\,°C$, $P\infty = 4.35 \times 10^4\,N/m^2$. At a certain point on the airfoil, the pressure is $3.6 \times 10^4\,N/m^2$. Calculate the density at this point.

11. Consider the isentropic flow through the nozzle of a supersonic wind tunnel. $T_0 = 227\,°C$, $P_0 = 1013250\,Pa$. If $P = 1013250\,Pa$ at the nozzle exit, calculate the exit temperature and density.

12. Consider the air at a pressure of 20,265 Pa. Calculate the values of τ_T and τ_s.

13. Consider the water pour point where the flow velocity and temperature are 396 m/s and −6 °C, respectively. Calculate the total enthalpy at this time.

14. In the supersonic wind tunnel, the speed is negligible and the temperature is 727 °C. The nozzle outlet temperature is 327 °C. Assuming an adiabatic flow through the nozzle, calculate the exit velocity.

15. The airfoil is in free flow, where $p_\infty = 61,808.25\,Pa$, $\rho_\infty = 0.819\,kg/m^3$, and $V_\infty = 300\,m/s$. At a certain point on the airfoil surface, the pressure is 50662.5 Pa. Assuming an isentropic flow, calculate the velocity at that point.

16.
   (a) Suppose that a 300-m-long pipe is filled with air at a temperature of 47 °C, and one end of the pipe generates sound waves. How long does it take for the wave to reach the other end?
   (b) If a tube is filled with helium at a temperature of 320 K and a sound wave is generated at one end of the tube, how long does it take for the sound wave to reach the other end? For monoatomic gases, such as helium, $\gamma = 1.67$. Note: Helium $R = 2078.5\,J/(Kg\,K)$.

17. Suppose an airplane is flying at a speed of 250 m per second. If it is flying at a standard altitude of (a) sea level, (b) 5 km, and (c) 10 km, calculate its Mach number.

18. Assuming a point in the airflow, the pressure and density are 70927.5 Pa and 0.98 kg/m³, respectively. (A) Calculate the corresponding value of the isentropic compression coefficient. (B) Calculate the speed of sound at a certain point in the flow based on the value of the isentropic compression coefficient.

19. As shown in the figure, consider a wedge with a half angle of 15° in a water flow with a Mach number of 5. Calculate the drag coefficient of this wedge. (Assume that the pressure on the substrate is equal to the free-flow static pressure, as shown in the figure.)

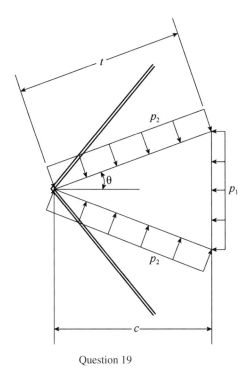

Question 19

20. Consider the separated curved bow shock wave in front of the two-dimensional parabolic bluff body shown in the figure. Free flow is Mach 8. Consider the two streamlines that pass through the impact at points a and b shown in the figure. The wave angle at point a is 90° and the wave angle at point b is 60°. Calculate and compare the entropy values of streamlines a and b in the airflow after impact (relative to free flow).

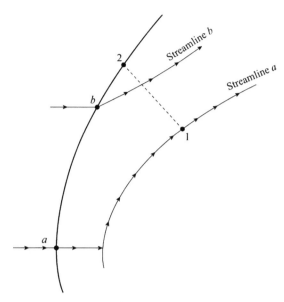

Question 20

21. The supersonic airflow with $M_1 = 1.5$, $p_1 = 101325\,\text{Pa}$, $T_1 = 14.8\,^\circ\text{C}$ expands at a deflection angle of 15° at a sharp corner (see picture). Calculate M_2, p_2, T_2, $p_{0,2}$, $T_{0,2}$, and the angle of the front and rear Mach lines relative to the upstream flow direction.

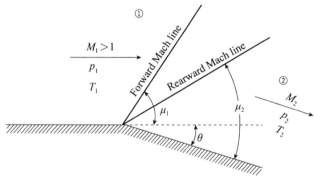

Question 21

22. Consider the isentropic compression surface drawn in the figure. The Mach number and upstream pressure of the wave are $M_1 = 10$ and $p_1 = 1\,\text{atm}$ (1 atm $= 101325\,\text{Pa}$), respectively. The water flows through an angle of 15°. Calculate the Mach number and pressure in the 2 zone after the compression wave.

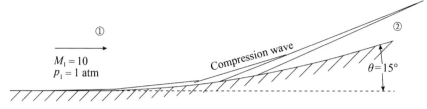

Question 22

Part II
Applied Aerodynamics

Chapter 8
Aerodynamic Characteristics of Flow Over Low-Speed Airfoils

This chapter introduces the aerodynamic characteristics of two-dimensional low-speed airfoil, including airfoil geometric parameters, airfoil development history, airfoil aerodynamic coefficient, airfoil flow, and aerodynamic characteristics, determination of Kutta–Joukowski trailing-edge condition and circulation, generation mechanism of airfoil lift, development of airfoil near boundary layer, solution of steady incompressible potential flow around airfoil. Theory of thin airfoil and aerodynamic characteristics of practical low-speed airfoil.

Learning points:

(1) Familiar with the characteristics of low-speed airfoil flow, airfoil geometry and aerodynamic parameters, airfoil flow characteristics, etc.
(2) Master the determination of Kutta–Joukowski trailing-edge condition and circulation, the generation mechanism of airfoil lift, the development of airfoil near boundary layer, the solution of steady incompressible potential flow around airfoil and thin airfoil theory.
(3) Understand the aerodynamic characteristics of low-speed airfoil and practical airfoil.

8.1 Geometric Parameters of Airfoil and Its Development

8.1.1 Development of Airfoil

In all kinds of flight conditions, the wing is the main part of the aircraft to bear the lift, and the vertical tail and horizontal tail are the aerodynamic parts to maintain the stability and maneuverability of the aircraft. Generally, an aircraft has a plane of symmetry. If a blade is cut parallel to the plane of symmetry at any position in the spanwise direction of the wing, the geometrically cut wing section is called

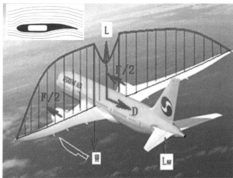

Fig. 8.1 Definition of airfoil

wing section or airfoil, and only the two-dimensional flow on the wing section is considered in aerodynamics. Airfoil is an important part of wing and tail forming, which directly affects the aerodynamic performance and flight quality of aircraft. As shown in Fig. 8.1, the same is true for bird wing profiles.

Generally, aircraft design requires that the lift of wing and tail should be as large as possible, the drag should be small, and the zero-lift moment should be small. Therefore, for different flight speeds, the airfoil shape of the wing is different. For low subsonic aircraft, in order to improve the lift coefficient, the shape of airfoil is round nose and pointed tail; for high subsonic aircraft, in order to improve the Ma number of drag divergence, the supercritical airfoil is adopted, which is characterized by full leading edge, flat upper airfoil and concave trailing edge. For supersonic aircraft, in order to reduce the shock drag, a sharp nose and a sharp tail airfoil are used. For the first time, the earliest wings imitated kites, with a piece of cloth on the skeleton, which was basically flat. In practice, it is found that the bending plate is better than the flat plate and can be used in a larger range of angles of attack. In 1903, the Wright brothers developed a thin airfoil with positive camber. In 1909, Joukowski first studied the steady flow around an ideal fluid airfoil by using the conformal transformation method of complex function, and proposed the famous Joukowski airfoil. It was clear that the low-speed airfoil should be round nose and sharp tail airfoil with better aerodynamic effect, and the round nose airfoil can adapt to a larger range of angles of attack. As shown in Fig. 8.2, the lift-to-drag ratio of airfoil increases with the improvement of airfoil shape.

During the first World War (from July 1914 to November 1918), all the warring countries explored some airfoils with better performance in practice. For example, the Joukowski airfoil, the Gottingen airfoil of Germany, and the RAF airfoil of Britain. After that, it was changed to Rae airfoil Royal Aircraft Institute, Clark-y, etc. After the 1930s, NACA airfoil (National Advisory Committee for Aeronautics) was established in March 1915. In October 1958, it was changed to NASA (National Aeronautics and Space Administration) and the former Soviet Union central air fluid research institute ЦАГИ Airfoil as shown in Fig. 8.3. In the late 1930s, the National

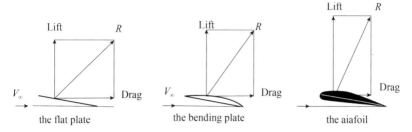

Fig. 8.2 Lift-to-drag ratio of airfoils with different shapes

Aviation Advisory Committee of the United States conducted a systematic study on the airfoil performance of various countries and proposed NACA four-digit airfoil family and five-digit airfoil family. They found that: (1) if the airfoil is not too thick, the effect of thickness and camber can be considered separately. (2) If the camber is straightened, that is to say, it is changed into a symmetrical airfoil, and converted into the same relative thickness, the thickness distribution of the airfoil is almost coincident. Therefore, it is suggested that the thickness distribution of NACA airfoil family is the best one at that time. After the 1950s, with the increase of aircraft speed, airfoils adapted to high subsonic, transonic, supersonic, supercritical airfoils, and laminar airfoils developed after the 1960s began to appear.

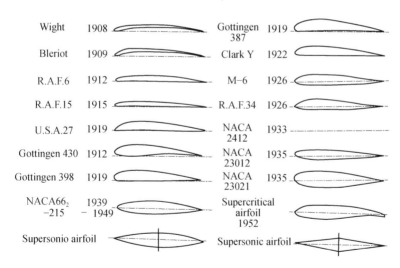

Fig. 8.3 Development of different airfoils

8.1.2 Definition and Geometric Parameters of Airfoil

As shown in Fig. 8.4, the geometric parameters of airfoil include chord length, camber, thickness, middle arc, etc. The leading-edge point is the most forward point of the airfoil, and the trailing-edge point is the last end point. The leading-edge point can also be defined as: the trailing-edge point is the center of the circle, draw an arc, and the tangent point between the arc and the airfoil is the leading-edge point. The line between the leading-edge and trailing-edge points is called the geometric chord of the airfoil. For some airfoils whose bottom surfaces are mostly straight lines, this line is also defined as a geometric chord. The distance between the leading-edge and trailing-edge points of the airfoil is called the chord length of the airfoil, which is expressed by b. For an airfoil with a flat bottom, the chord length is defined as the distance between the projections of the leading and trailing edges on the chord line.

The curves of upper and lower surfaces (upper and lower edges) of general airfoils are expressed as functions of relative coordinates of chord length. Taking the leading-edge point as the coordinate origin, the chord length direction as the x axis and the vertical direction as the y axis, the upper and lower airfoil curves are defined as the following functions.

$$x = \frac{\overline{x}}{b}, \quad y_u(x) = \frac{\overline{y}_u(x)}{b}, \quad y_d(x) = \frac{\overline{y}_d(x)}{b} \tag{8.1}$$

where x and y are relative values based on chord length b. $y_u(x)$ is the relative coordinates of the upper wing curve and $y_d(x)$ is that of the lower wing curve. Half of the distance between the upper and lower airfoils is called the airfoil thickness distribution function.

$$y_c(x) = \frac{1}{2}(y_u(x) - y_d(x)) \tag{8.2}$$

For the dimensionless maximum y-direction height between the upper and lower wings, it is called the relative thickness of the airfoil, and the x corresponding to the maximum thickness is called the position of the maximum thickness of the airfoil, which is usually expressed by x.

$$C = \max(y_u(x) - y_d(x)), \quad x = x_p \tag{8.3}$$

Fig. 8.4 Definition of geometric parameters of airfoil

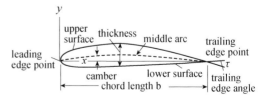

For example, $c = 9\%$, which means that the thickness of airfoil is 9% of chord length. The line between the middle points of the upper and lower airfoil curves is called the middle arc of the airfoil. The middle arc curve is

$$y_f(x) = \frac{1}{2}(y_u(x) + y_d(x)) \tag{8.4}$$

$y_f(x)$ is called the curvature function. The maximum value of $y_f(x)$ is called the camber of airfoil, which is represented by f, and the corresponding value of x is represented by x_f.

$$f = \max(y_f(x)), \quad x = x_f \tag{8.5}$$

If the middle arc is a straight line (in line with the chord), the airfoil is called a symmetric airfoil. If the middle arc is a curve, the airfoil must be curved. The curvature is represented by the y coordinate of the highest point on the middle arc. This value is also usually expressed relative to chord length. In addition, the leading edge of the airfoil is circular. To draw the airfoil curve near the leading edge accurately, the leading-edge radius is usually given. The center of the circle tangent to the leading edge is on the tangent line of the leading edge point of the middle arc. The angle between the tangent lines of the upper and lower surfaces of the airfoil at the trailing edge is called the trailing edge angle (Fig. 8.4).

For the general cambered airfoil, the upper and lower edge curvilinear coordinates are expressed as

$$x_u = x - y_c \sin\theta$$
$$y_u = y_f + y_c \cos\theta$$
$$x_d = x + y_c \sin\theta$$
$$y_d = y_f - y_c \cos\theta \tag{8.6}$$

where θ is the inclination of the middle arc at position x. Considering small of θ, the following simplified coordinate curve can be used.

Fig. 8.5 Airfoil coordinate curve construction

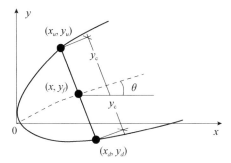

$$x_u = x$$
$$y_u = y_f + y_c$$
$$x_d = x$$
$$y_d = y_f - y_c \tag{8.7}$$

For a round nose airfoil, the leading-edge bluntness is expressed by the radius of a circle tangent to the leading edge, and the center of the circle is on the tangent line of the leading edge point of the middle arc, which is called the relative leading edge radius r_L. The relative leading edge radius is defined as $r_L = \frac{\bar{r}_L}{b}$. The angle between the tangent lines of the upper and lower surfaces of the airfoil at the trailing edge is called the trailing edge angle τ. The value of τ represents the sharp angle of the trailing edge.

8.1.3 NACA Airfoil Number and Structure

(1) NACA four-digit airfoil

In the late 1930s, the National Aeronautical Advisory Council (NACA, now NASA) conducted a systematic study on the performance of airfoils and proposed NACA four-digit wing family and five-digit wing family. The thickness distribution of NACA airfoil family with the best airfoil thickness distribution is given.

$$y_c(x) = \pm \frac{C}{0.2}(0.2969\sqrt{x} - 0.1260x - 0.3516x^2 + 0.2843x^3 - 0.1015x^4) \tag{8.8}$$

where C is the airfoil thickness. The leading edge radius is $r = 1.1019c^2$. The middle arc takes two sections of parabola, which are tangent to each other at the highest point of the middle arc.

$$y_f = \frac{f}{x_f^2}(2x_f x - x^2), \, x < x_f$$

$$y_f = \frac{f}{(1 - x_f)^2}\left[(1 - 2x_f) + (2x_f x - x^2)\right], \, x > x_f \tag{8.9}$$

where, f is the ordinate of the highest point of the middle arc, and x_f is the chordal position of the highest point of the arc. The height f (i.e., the curvature) of the highest point of the middle arc and the chordal position of the point are artificially determined. Given a series of values of f, x_f and thickness C, the airfoil family is obtained.

The form of NACA four-digit wing family is shown in Fig. 8.6.

The first digit represents the camber f, which is the percentage of chord length. The second digit represents x_f, which is the tenth fraction of chord length. The last

Fig. 8.6 NACA four-digit
airfoil nomenclature

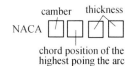

Fig. 8.7 NACA2415 airfoil

two digits represent the thickness as a percentage of chord length. For example, NACA 0012 is a symmetrical airfoil with no camber and 12% thickness. NACA 2415 (as shown in Fig. 8.7) indicates that the camber $f = 0.02$, the chord position of camber is 0.4, and the thickness is 0.15. There are 6, 8, 9, 10, 12, 15, 18, 21 and 24% airfoils of NACA four-digit wing family with available experimental data.

(2) NACA five-digit airfoil

The thickness distribution of five-digit airfoil is the same as that of four-digit airfoil, and the thickness distribution function is shown in Eq. (8.8), but the difference is the middle arc. It is found that adjusting the chordal position of the highest point of the middle arc can improve the maximum lift coefficient of the airfoil, but moving backward or forward will affect the torque coefficient. If the backward displacement is too large, a large pitching moment will be produced. If the forward displacement is large, the middle arc of the original four-digit airfoil cannot be used, and other middle arc equations must be used instead. The characteristic of this kind of middle arc is that its curvature gradually decreases from leading edge to trailing edge, and then it becomes zero after skipping the highest point, and then it becomes a straight line. The calculation formula of middle arc is

$$y_f = \frac{1}{6}k_1\left[x^3 - 3mx^2 + m^2(3 - m)x\right], \quad 0 < x < m$$
$$y_f = \frac{1}{6}k_1m^3(1 - x), \quad m < x < 1.0 \tag{8.10}$$

where m and k_1 are the values varying with the chord position x_f of the highest point of the middle arc. For the case of lift coefficient 0.3, the values are shown in Table 8.1.

The numerical meaning of the five-digit airfoil is as follows (as shown in Fig. 8.8).

The first digit represents the camber, but it is not a direct geometric parameter. It is expressed by the design lift coefficient. The number multiplied by 3/2 is equal to ten times of the design lift coefficient. The second and third digits are $2x_f$, expressed as a percentage of chord length; the last two digits are still percent thickness. For example, the design lift coefficient of NACA 23012 airfoil is $(2) \times 3/20 = 0.30$, $X_f = 30/2$, that is, the chordal position of the highest point of the middle arc is at 15% of the chord length, and the thickness is still 12%.

Table 8.1 Values of m and k_1 at different chordal positions

Serial number	x_f	m	k_1
1	0.05	0.0580	361.4
2	0.10	0.1260	51.64
3	0.15	0.2025	15.957
4	0.20	0.2900	6.643
5	0.25	0.3910	3.230

Fig. 8.8 Meaning of NACA five-digit airfoil code

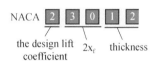

NACA 2 3 0 1 2

the design lift coefficient $2x_f$ thickness

The five-digit wing families with ready-made experimental data are 230 series, the design lift coefficient is 0.30, the chord position x_f of the highest point of the middle arc is at 15% chord length, and the thickness is 12, 15, 18, 21, and 24. Other modified five-digit airfoils will not be introduced here.

(3) Laminar airfoil

Laminar flow airfoil is designed to reduce the turbulent friction drag. It makes the positive pressure gradient region of the upper airfoil increase as much as possible, reduces the pressure gradient region and reduces the turbulent range. In general, in the region of positive pressure gradient $\frac{dp}{ds} < 0$, the thin boundary layer flow is not easy to transition into turbulence, while in the region of negative pressure gradient $\frac{dp}{ds} > 0$, the thick boundary layer flow is easy to transition into turbulence. Because the normal airfoil has a small positive pressure gradient region, NACA proposes a laminar airfoil family. The thickness distribution of laminar airfoil is designed separately from the middle arc, and the chord positions of the maximum thickness point are 0.35, 0.4, 0.45, and 0.5. The shape of the middle arc is designed according to the load distribution. The load distribution takes a broken line. There is a constant load a from the leading-edge point to a certain position in the chord direction, and then to the trailing edge, the load coefficient decreases linearly to zero. The NACA6 series laminar airfoils commonly used at present (as shown in Fig. 8.9) are represented by a six-digit number with a description of the most central arc. For example, NACA65₃-218, $a = 0.5$, in which the first 6 represents 6 series, the second 5 represents that in the case of zero degree of attack and symmetrical airfoil, the lowest pressure point is at the chord length of 0.5 (10 fractions of chord length), and the third after the comma represents that there is still a favorable pressure distribution on the airfoil in the range of 3/10 above and below the design lift coefficient. The first number after the follow bar is ten times the design lift coefficient, $C_L = 0.2$. The range of favorable pressure distribution is $0.2 \pm 3/10$, that is, $C_L = -0.1$ to 0.5, and the last two digits are the percentage of thickness (0.18). If the thickness distribution curve

Fig. 8.9 Naming of NACA6 series airfoils

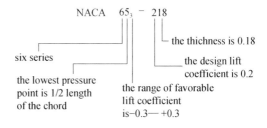

Fig. 8.10 Laminar flow range on different airfoil surfaces

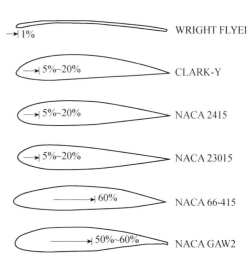

of the airfoil uses different curves, it can be expressed as NACA65$_3$-218, $a = 0.5$. The laminar flow region of different airfoils is different. The laminar flow region of conventional airfoils is less than 20%, and that of laminar airfoils can reach 50–60%, as shown in Fig. 8.10.

8.1.4 Supercritical Airfoil

Supercritical airfoil is a special airfoil proposed by Whitcomb (1921–2009), director of wind tunnel laboratory of NASA Langley Research Center, in 1967 to improve the Ma number of drag divergence of subsonic transport aircraft, as shown in Fig. 8.11. Compared with the conventional airfoil, the drag divergence Mach number can be increased by 0.05–0.12. Compared with the conventional laminar airfoil (peaked airfoil), the leading edge of supercritical airfoil is blunt, the upper surface is flat, and the trailing edge of the lower surface is thin and downward curved. Because of its flat upper surface, the acceleration process of air flow is slowed down, so the lift is also reduced. Therefore, the lack of lift can be made up by increasing the downward bend of the trailing edge of the lower airfoil, as shown in Fig. 8.11. Supercritical airfoil

Fig. 8.11 Common and
supercritical airfoils

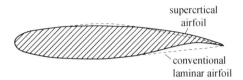

supercrtical
airfoil

conventional
laminar airfoil

was successfully applied to A320 in the 1980s, which is the core technology of large
aircraft wing design (supercritical wing). It is widely used in the new generation of
transonic civil and military transport aircraft because of its advantages of large Mach
number, high aerodynamic efficiency and relative thickness of wings.

8.1.5   Typical Airfoil Data

(1)   RAF-6E airfoil

This is a flat bottom airfoil with excellent aerodynamic performance proposed by
RAF in 1930s. The relative coordinates are given in Table 8.2, and the profile curve
is given in Fig. 8.12.

Table 8.2 Relative
coordinates of RAF-6E airfoil

$x(\%)$	$y_u(\%)$	$y_d(\%)$
0.0	1.15	0.0
1.25	3.19	0.0
2.5	4.42	0.0
5.00	6.10	0.0
7.50	7.24	0.0
10.00	8.09	0.0
15.00	9.28	0.0
20.00	9.90	0.0
30.00	10.30	0.0
40.00	10.22	0.0
50.00	9.80	0.0
60.00	8.98	0.0
70.00	7.70	0.0
80.00	5.91	0.0
90.00	3.79	0.0
95.00	2.58	0.0
100.00	0.76	0.0

Fig. 8.12 RAF-6E airfoil

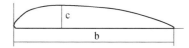

(2) Clark-Y airfoil

This is another flat bottom airfoil with excellent aerodynamic performance proposed by NACA in 1930s. The relative coordinates are given in Table 8.3, and the profile curve is given in Fig. 8.13.

(3) NACA0012 airfoil

This is a four-digit standard airfoil proposed by NACA in 1930s. The relative coordinates are given in Table 8.4 and the profile curve is given in Fig. 8.14.

Table 8.3 Relative coordinates of Clark-Y airfoil

$x(\%)$	$y_u(\%)$	$y_d(\%)$
0.0	2.99	0.0
1.25	4.66	0.0
2.5	5.56	0.0
5.00	6.75	0.0
7.50	7.56	0.0
10.00	8.22	0.0
15.00	9.14	0.0
20.00	9.72	0.0
30.00	10.00	0.0
40.00	9.75	0.0
50.00	9.00	0.0
60.00	7.82	0.0
70.00	6.28	0.0
80.00	4.46	0.0
90.00	2.39	0.0
95.00	1.27	0.0
100.00	0.10	0.0

Fig. 8.13 Clark-y airfoil

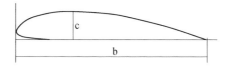

Table 8.4 Relative coordinates of NACA0012 airfoil

$x_u(\%)$	$Y_u(\%)$	$Y_d(\%)$
0.00	0.00	0.00
5.00	3.55	−3.55
10.00	4.68	−4.68
15.00	5.35	−5.35
20.00	5.74	−5.74
25.00	5.94	−5.94
30.00	6.00	−6.00
35.00	5.95	−5.95
40.00	5.80	−5.80
45.00	5.58	−5.58
50.00	5.29	−5.29
55.00	4.95	−4.95
60.00	4.56	−4.56
65.00	4.13	−4.13
70.00	3.66	−3.66
75.00	3.16	−3.16
80.00	2.62	−2.62
85.00	2.05	−2.05
90.00	1.45	−1.45
95.00	0.81	−0.81
100.00	0.13	−0.13

Fig. 8.14 NACA0012 airfoil

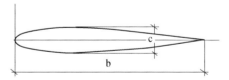

8.2 Aerodynamics and Aerodynamic Coefficients on Airfoils

8.2.1 Relationship Between Airfoil Aerodynamics and Angle of Attack

In the plane of the airfoil, the angle between the incoming flow and the chord is defined as the geometric angle of attack of the airfoil. For chords, the upward deviation of incoming flow is positive and the downward deviation is negative, so the angle of attack can be divided into positive and negative. Positive angle of attack means

Fig. 8.15 Force analysis of any point on airfoil

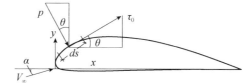

upward deviation of incoming flow, and negative angle of attack means downward deviation of incoming flow. The flow over the airfoil is a plane flow, and the aerodynamic force on the airfoil can be regarded as the aerodynamic force on the wing with infinite span in spanwise direction. When the air flows over the airfoil, there are forces at every point on the airfoil surface, including pressure P (perpendicular to the airfoil) and friction shear stress τ (tangent to the airfoil). The resultant force generated by this distributed force is R, and the point of action of the resultant force is called the pressure center. The component of the resultant force in the incoming direction is resistance D, and the component perpendicular to the incoming direction is lift L.

As shown in Fig. 8.15, take the fixed point of the airfoil as the origin of the curve coordinates, and take any micro segment ds at a certain point on the airfoil surface. The forces acting on the segment include the force caused by pressure and the force caused by wall shear stress, which are projected in the x direction and y direction respectively.

$$dF_x = (p \sin \theta + \tau_0 \cos \theta)ds \tag{8.11}$$

$$dF_y = (-p \cos \theta + \tau_0 \sin \theta)ds \tag{8.12}$$

According to the above two equations, the resultant force of the distributed force system is obtained.

$$F_x = \oint (p \sin \theta + \tau_0 \cos \theta)ds \tag{8.13}$$

$$F_y = \oint (-p \cos \theta + \tau_0 \sin \theta)ds \tag{8.14}$$

Because the coordinate system xy is taken on the airfoil, it is called volume coordinate system, and the corresponding component force is called component force in body axis coordinate system. The resultant force R is

$$R = \sqrt{F_x^2 + F_y^2} \tag{8.15}$$

Take the moment to the origin of the coordinate (make the airfoil head up to be positive), and the resultant moment is

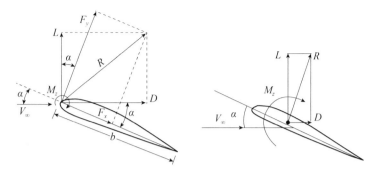

Fig. 8.16 Resultant force and resultant moment (different moment taking points)

$$M_z = \oint (p \sin \theta + \tau_0 \cos \theta) y \mathrm{d}s - \oint (-p \cos \theta + \tau_0 \sin \theta) x \mathrm{d}s \qquad (8.16)$$

where the subscript z is the axis of the moment. In this way, a distributed force system is combined into a concentrated force and a moment to the origin. M_z is the origin, and the magnitude of the moment is taken, and the head up is positive. As shown in Fig. 8.16. By projecting F_x and F_y of the body axis to the direction of the flow and perpendicular to the direction of the incoming flow (called the wind axis coordinate system), the lift and drag of the airfoil can be obtained as follows.

$$L = F_y \cos \alpha - F_x \sin \alpha$$
$$D = F_y \sin \alpha + F_x \cos \alpha \qquad (8.17)$$

The magnitude of aerodynamic moment M_z depends on the position of moment point. If the moment taking point is located at the leading edge of the airfoil, it is called the leading-edge moment. The moment value is expressed by M_z, and the distance from the leading-edge point is $x_0 = 0$. If the moment point is located at the centroid of the distributed force system, it is called the pressure center, and the distance X_p from the leading-edge point, the moment value M_{zp} is zero. If the moment taking point is located at the point where the moment does not change with the angle of attack, the distance x_a from the leading-edge point, and the moment value is expressed by M_{za}, it is called the aerodynamic center of the airfoil. At this time, the moment of the resultant force to the aerodynamic center does not change with the angle of attack within a certain range of angle of attack. The experimental results show that the aerodynamic center of thin airfoil is $0.25b$ (b is the chord length of airfoil), most airfoils are between $0.23b$ and $0.24b$, and laminar airfoils are between $0.26b$ and $0.27b$.

As can be seen from Fig. 8.17, the resultant force and resultant moment for the above three moment points can be expressed as follows.

(1) The leading-edge point is taken as the moment taking point $x = x_0 = 0$, and the resultant force and resultant moment are L, D, M_z.

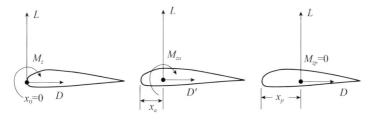

Fig. 8.17 Moment of resultant force and translation of resultant force

(2) Taking the aerodynamic center as the moment taking point $x = x_a$, the resultant force and resultant moment are L, D, M_{za}.

(3) Taking the pressure center as the moment taking point $x = x_p$, the resultant force and resultant moment are L, D, $M_{zp} = 0$.

If the upper and lower surfaces of the airfoil are integrated separately in the process of calculating the resultant force, the contribution of the upper and lower surfaces to the lift and drag can be investigated, especially the contribution of the upper and lower surfaces to the lift can be analyzed.

$$dx = ds \cos\theta, \, dy = ds \sin\theta = tg\theta dx$$

Among the above, there are

$$L = L_u + L_d$$

$$L_u = \int_0^b (-p_u + \tau_{u0}tg\theta)dx \cos\alpha - \int_0^b (p_utg\theta + \tau_{0u})dx \sin\alpha$$

$$L_d = \int_0^b (p_d - \tau_{d0}tg\theta)dx \cos\alpha + \int_0^b (p_dtg\theta + \tau_{0d})dx \sin\alpha \qquad (8.18)$$

$$D = D_u + D_d$$

$$D_u = \int_0^b (-p_u + \tau_{u0}tg\theta)dx \sin\alpha + \int_0^b (p_utg\theta + \tau_{0u})dx \cos\alpha$$

$$D_d = \int_0^b (p_d - \tau_{d0}tg\theta)dx \sin\alpha + \int_0^b (p_dtg\theta + \tau_{0d})dx \cos\alpha \qquad (8.19)$$

For small angles of attack and thin airfoils, θ and α both of them are small (angle of attack is less than 8–100, relative thickness of airfoil is less than 10–12%), so the above equations can be simplified as

Fig. 8.18 Aerodynamic
integration of upper and
lower wings

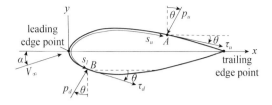

$$L_u = \int_0^b (-p_u)dx \cos\alpha, \quad L_d = \int_0^b (p_d)dx \cos\alpha$$

$$L = L_u + L_d = \int_0^b (-p_u + p_d)dx \cos\alpha = \int_0^b (-p_u + p_d)dx \tag{8.20}$$

$$D_u = \int_0^b (\tau_{0u})dx, \quad D_d = \int_0^b (\tau_{0d})dx$$

$$D = \int_0^b (\tau_{0u} + \tau_{0d})dx \tag{8.21}$$

$$M_z = -\int_0^b x(-p_u + p_d)dx \cos\alpha = -\int_0^b x(-p_u + p_d)dx \tag{8.22}$$

It can be seen that the lift of the airfoil is mainly determined by the pressure distribution on the upper and lower surfaces, and the drag of the airfoil is mainly determined by the friction shear stress on the upper and lower surfaces of the airfoil. Therefore, it is very important to study the flow field around the airfoil, because the surface pressure can be determined according to Bernoulli equation (Fig. 8.18).

8.2.2 Aerodynamic Coefficient

For the convenience of analysis and application, infinity coefficients are often used to express the aerodynamic coefficients in aerodynamics. The lift L (perpendicular to V_∞), drag D (parallel to V_∞) and moment M_z (pitching moment around a reference point) acting on the airfoil are expressed by dimensionless coefficients with infinite inflow V_∞ and airfoil chord length b as characteristic quantities. For example, the lift coefficient is

$$C_L = \frac{L}{\frac{1}{2}\rho V_\infty^2 b} = \frac{L}{q_\infty b} \tag{8.23}$$

where $q_\infty = \frac{1}{2}\rho V_\infty^2$ is the flowing pressure.

$$C_D = \frac{D}{\frac{1}{2}\rho V_\infty^2 b} = \frac{D}{q_\infty b} \tag{8.24}$$

The moment coefficient is

$$C_m = \frac{M_z}{\frac{1}{2}\rho V_\infty^2 b^2} = \frac{M_z}{q_\infty b^2}, \quad C_{ma} = \frac{M_{za}}{q_\infty b^2} \tag{8.25}$$

where C_m and C_{ma} represent the moment coefficients around the leading edge and aerodynamic center of the airfoil, respectively.

The pressure coefficient is

$$C_p = \frac{p - p_\infty}{\frac{1}{2}\rho V_\infty^2} = \frac{p - p_\infty}{q_\infty} \tag{8.26}$$

The friction coefficient of the wing surface is

$$C_f = \frac{\tau_0}{\frac{1}{2}\rho V_\infty^2} = \frac{\tau_0}{q_\infty} \tag{8.27}$$

By using the above definition of dimensionless coefficient and substituting into Eqs. (8.20) and (8.21), we get

$$C_L = \int_0^1 (-C_{pu} + C_{pd})dx \cos\alpha = \int_0^1 (-C_{pu} + C_{pd})dx \tag{8.28}$$

$$C_D = \int_0^1 (C_{fu} + C_{fd})dx \tag{8.29}$$

where C_{pu} and C_{pd} are the pressure coefficients of the upper and lower wings. C_{fu} and C_{fd} are the friction coefficients of the upper and lower surfaces. According to Bernoulli equation, for incompressible fluid the pressure coefficient can be determined by the velocity distribution near the surface. That is, from Bernoulli equation we can get

$$p + \frac{1}{2}\rho V^2 = p_\infty + \frac{1}{2}\rho V_\infty^2$$

Fig. 8.19 Velocity values at
the outer boundary of the
boundary layer

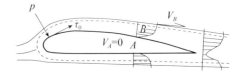

According to the boundary layer theory, when the actual air bypasses the airfoil,
there is a boundary layer on the airfoil. The flow velocity near the object surface
refers to the velocity V_B (potential flow velocity of the external flow field) on the
external boundary of the local boundary layer. Because the pressure in the boundary
layer is constant along the normal direction, the pressure on the airfoil is equal to the
pressure on the external boundary of the boundary layer, as shown in Fig. 8.19. The
pressure coefficient can be written as

$$C_p = \frac{p - p_\infty}{\frac{1}{2}\rho V_\infty^2} = \frac{p - p_\infty}{q_\infty} = 1 - \left(\frac{V}{V_\infty}\right)^2 \tag{8.30}$$

Experimental and theoretical studies show that different airfoil shapes directly
affect the velocity distribution around the airfoil, and thus affect the pressure distribution on the airfoil. As can be seen from Fig. 8.20, for NACA2412 airfoil flow, the
ratio of the velocity near the airfoil surface to the incoming flow velocity at the airfoil
head is 1.8, which is substituted into Eq. (8.30) to obtain $C_{pu} = -2.24$. The ratio
of velocity to inflow velocity is 0.75, and the pressure coefficient is $C_{pd} = 0.4375$.
The pressure coefficient distribution on NACA2412 airfoil is shown in Fig. 8.21. It
can be seen that the pressure distribution on the airfoil can be changed by changing
the shape of the airfoil, thus changing the lift. It is concluded that when the velocity
near the airfoil is greater than the velocity of the incoming flow, the pressure on the
airfoil is less than the atmospheric pressure around, and the pressure coefficient is
negative, which plays a role of suction in the contribution of lift (the surrounding air
adsorbs the airfoil). When the velocity near the airfoil is less than the velocity of the
incoming flow, the pressure on the airfoil is greater than the atmospheric pressure
around it, and the pressure coefficient is positive, which plays a supporting role in the
contribution of lift. Wind tunnel experiments show that for low-speed airfoil flow, the
suction of the upper airfoil accounts for about 70% of the total lift, while the jacking
force of the lower airfoil accounts for about 30% (as shown in Fig. 8.22). It shows
that the establishment of Bernoulli equation provides an important basis for people
to correctly understand the contribution of upper wing suction to lift. Figure 8.23
shows the flow over the airfoil and the variation characteristics of lift and drag at
different angles of attack.

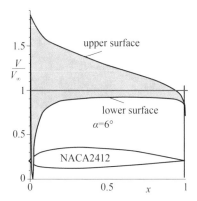

Fig. 8.20 The velocity distribution of the surface of NACA2412 airfoil

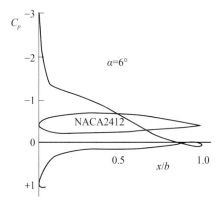

Fig. 8.21 Pressure coefficient distribution on NACA2412 airfoil

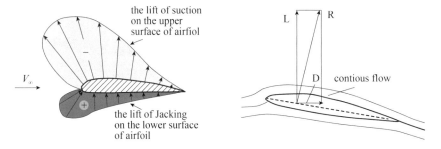

Fig. 8.22 Contribution of pressure distribution on upper and lower wings to lift

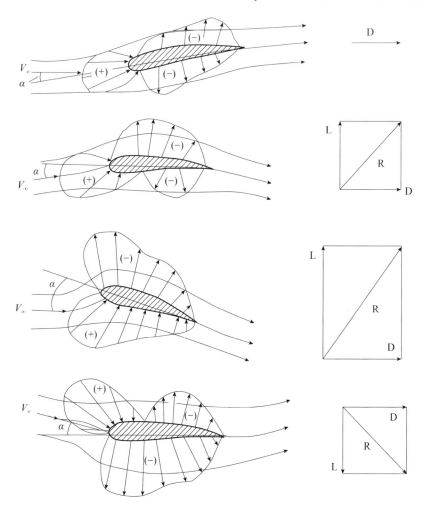

Fig. 8.23 Aerodynamic characteristics of airfoil at different angles of attack

8.2.3   *Dimensional Analysis of Lift Coefficient*

A large number of wind tunnel experiments show that the main physical parameters affecting airfoil lift L include air density ρ, Flight speed V_∞, airfoil chord length b, air viscosity coefficient μ, sound velocity a, airfoil thickness c, airfoil camber f, and angle of attack α. If it is written as a function, the relation is

$$L = f(\rho, V_\infty, b, \mu, a, c, f, \alpha) \tag{8.31}$$

Select air density ρ, flight speed V_∞, and airfoil chord length b are taken as the basic physical quantities (selected from the main physical quantities affecting lift, in principle, the dimensional expressions of the three physical quantities cannot be derived from each other). According to the π theorem, the dimensionless expression of Eq. (8.31) is

$$\pi_L = f(\pi_\mu, \pi_a, \pi_c, \pi_f, \pi_\alpha) \tag{8.32}$$

The equation can be gotten by dimensional analysis

$$\pi_L = \frac{L}{\rho V_\infty^2 b^2} = f\left(\frac{\mu}{\rho b V_\infty}, \frac{a}{V_\infty}, \frac{c}{b}, \frac{f}{b}, \alpha\right) \tag{8.33}$$

Expressed by dimensionless coefficient.

$$C_L = f\left(\mathrm{Re}, \mathrm{Ma}, \frac{c}{b}, \frac{f}{b}, \alpha\right) \tag{8.34}$$

where Re is the Reynolds number and Ma is the Mach number.

$$\mathrm{Re} = \frac{\rho b V_\infty}{\mu}, \quad \mathrm{Ma} = \frac{V_\infty}{a} \tag{8.35}$$

Similarly, the drag coefficient and moment coefficient can also be expressed as Eq. (8.36)

$$C_D = f_D\left(\mathrm{Re}, \mathrm{Ma}, \frac{c}{b}, \frac{f}{b}, \alpha\right)$$

$$C_m = f_m\left(\mathrm{Re}, \mathrm{Ma}, \frac{x_p}{b}, \frac{c}{b}, \frac{f}{b}, \alpha\right) \tag{8.36}$$

where x_p is the position of the moment. For low-speed airfoil flow, the compressibility of air is neglected, but the viscosity of air must be considered. Therefore, the aerodynamic coefficient is actually a function of the angle of attack and Re number. The specific form of the function can be given by experiment or theoretical analysis. For high-speed flow, the effect of compressibility must be taken into account, so Ma is the main parameter.

8.3   Overview of Flow and Aerodynamic Characteristics of Low-Speed Airfoil

8.3.1   Phenomenon of Flow Over a Low-Speed Airfoil

A large number of wind tunnel experiments show that for the compressible steady flow around the airfoil, as shown in Fig. 8.24, the phenomenon can be summarized as follows:

(1) The flow around the airfoil always adheres to the airfoil surface, and there is no separation on the whole airfoil surface. This kind of flow close to the airfoil surface is called attached flow. There is a boundary layer on the upper and lower airfoil surface (the larger the Reynolds number of the incoming flow, the thinner the boundary layer), and the wake area at the trailing edge of the airfoil is also very thin.

(2) The front stagnation point (the point with zero flow velocity) is not far from the leading edge of the lower wing. The streamline flowing through the stagnation point is divided into two parts. One part flows from the stagnation point to the upper wing through the leading edge, and the other part flows from the stagnation point to the lower wing through the wall. The flow at the trailing edge converges smoothly and flows downward.

(3) On the outer boundary line of the boundary layer, the velocity of the fluid particle on the upper wing accelerates rapidly to the maximum from the zero value of the front stagnation point and then decelerates gradually until it leaves the wing. According to Bernoulli equation, the pressure distribution is maximum at stagnation point and minimum at maximum velocity point, and then the pressure increases gradually (after the minimum pressure point is the negative pressure gradient area). The fluid particle decelerates continuously in the negative pressure environment. The flow in the boundary layer belongs to the negative pressure gradient boundary layer flow on curved wall, and the boundary layer is relatively thick. On the contrary, in the lower wing of the airfoil, the velocity of the fluid particles accelerates from the stagnation point to the trailing edge, and the fluid particles accelerate to the trailing edge in the environment of the

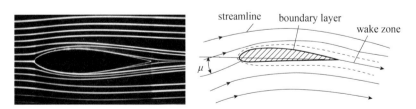

Fig. 8.24   Smoke flow experiment of steady airfoil flow in wind tunnel

positive pressure gradient. The flow in the boundary layer belongs to the positive pressure gradient boundary layer flow on the curved surface, and the boundary layer is relatively thin. The upper and lower airfoil flow, after converging near the trailing-edge point, smoothly leaves the airfoil and flows downstream, as shown in Fig. 8.24.

(4) With the increase of the angle of attack, the stagnation point gradually moves backward, and the maximum velocity point is close to the leading-edge point. The larger the maximum velocity is, the larger the area surrounded by the pressure difference between the upper and lower wings, so the greater the lift is, as shown in Fig. 8.25.

(5) The upper and lower wing flow converges at the trailing-edge point and leaves smoothly. There is no large velocity in the downstream area near the trailing-edge point, which leads to obvious wake vortex region. In addition, the trailing-edge points are not necessarily all post-stop points, as long as the smooth can be guaranteed, as shown in Fig. 8.26.

(6) When the angle of attack increases to a certain value, the flow first separates from the upper wing region of the trailing edge (because the boundary layer of the upper wing is in the negative pressure gradient region, so the boundary layer is easy to separate). Compared with the case without separation, the negative

Fig. 8.25 Pressure distribution of NACA0012 airfoil at 9 degrees of attack

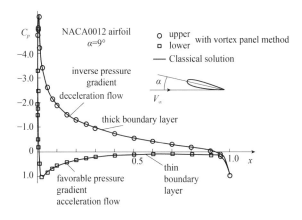

Fig. 8.26 Attached flow around airfoil at different angles of attack

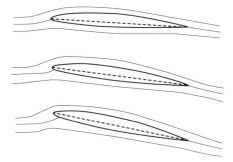

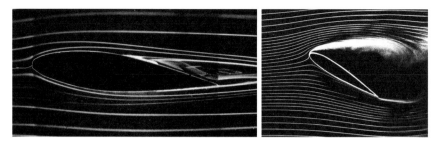

Fig. 8.27 Flow separation around airfoil at large angle of attack (partial separation and complete separation)

pressure of potential flow on the wing in the separation region disappears, so the lift is weakened. At the same time, the wake region of the flow around the airfoil increases significantly, and a large number of separated vortices appear, as shown in Fig. 8.27.

8.3.2   Curve of Aerodynamic Coefficient of Airfoil Flow

The aerodynamic forces on airfoil are usually expressed by dimensionless aerodynamic coefficients. For incompressible airfoil flow, according to Eqs. (8.33)–(8.35), the angle of attack α is usually used as an independent variable, the Reynolds number Re of the incoming flow is plotted as a parameter to draw the curve of lift coefficient C_L versus α, curve of drag coefficient C_D versus α and curve of moment coefficient C_m versus α. In addition, to facilitate the use of aircraft aerodynamic design, there are two curves with C_L as the independent variable, namely the curve of drag coefficient C_D versus C_L (called polar curve) and the curve of moment coefficient C_m versus C_L.

At small angle of attack, according to Eq. (8.27), the lift coefficient of thin airfoil is

$$C_L = \int_0^1 (-C_{pu} + C_{pd}) dx \cos \alpha$$

where $C_{pu} = \frac{p_u - p_\infty}{\frac{1}{2}\rho V_\infty^2}$, $C_{pl} = \frac{p_l - p_\infty}{\frac{1}{2}\rho V_\infty^2}$ represent the pressure coefficients of the upper and lower wings, respectively, and the lift coefficient can be obtained by integrating the area of the pressure coefficient distribution curve of the upper and lower wings. In the wind tunnel experiment, the aerodynamic force and moment can also be directly measured by the six component balance. Figures 8.28 and 8.29 show the aerodynamic coefficient curve of NACA23012 airfoil and NACA63$_1$-212 airfoil from wind tunnel test with the Reynolds number is Re $= 6 \times 10^6$.

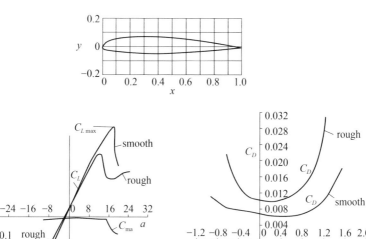

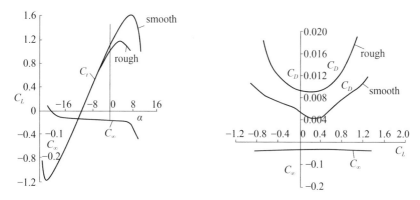

Fig. 8.28   Aerodynamic characteristic curve of NACA23012 (Re $= 6 \times 10^6$)

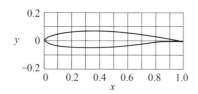

Fig. 8.29   Aerodynamic characteristic curve of NACA63$_1$-212 (Re $= 6 \times 10^6$)

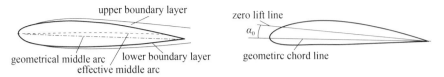

Fig. 8.30 Effect of upper and lower wing boundary layers on camber

According to the experimental data (Figs. 8.28 and 8.29), the following characteristics can be obtained:

(1) In the curve of lift coefficient and angle of attack, C_L is a straight line in a certain range of angle of attack, and the slope of this straight line is denoted as

$$C_L^\alpha = \frac{dC_L}{d\alpha} \tag{8.37}$$

The theoretical value of thin wing is 2π/radian, and the experimental value is slightly smaller equaling to 0.10965/degree. The slope of NACA23012 is 0.105/degree, and that of NACA 63_1-212 is 0.106/degree. The reason why the experimental value is slightly smaller is due to the viscous effect of the actual air flow. At the positive angle of attack, the displacement thickness of the upper and lower wings is different. Under the action of the negative pressure gradient, the displacement thickness of the upper wing is larger than that of the lower wing. The effect is equivalent to reducing the camber of the airfoil and changing the position of the trailing edge, thus reducing the effective angle of attack, as shown in Fig. 8.30. The slope of lift line is a very important parameter. When calculating the performance of aircraft, the lift coefficient should be calculated according to the angle of attack.

(2) For airfoils with different cambers, the lift coefficient curve does not pass through the origin. Usually, the angle of attack with zero-lift coefficient is called zero-lift angle of attack α_0 means. A straight line through the trailing-edge point with an angle of α_0 to the geometric chord is called the zero-lift line. Generally, the larger the curvature is, the larger the α_0 is. When the camber is positive (downward), the zero-lift angle of attack is negative. When the camber is negative (upward), the zero-lift angle of attack is positive, as shown in Fig. 8.31.

(3) When the angle of attack is larger than a certain value, the lift line begins to bend. If it is larger, it will reach its maximum value. This value is recorded as the maximum lift coefficient C_{Lmax}, which is the maximum lift coefficient that can be obtained by increasing the angle of attack. The corresponding angle of attack is called the critical angle of attack α_J. If the angle of attack is increased, the lift coefficient begins to decrease, which is called stall. This critical angle of attack is also called stall angle of attack. To sum up, the shape of airfoil lift coefficient curve is shown in Fig. 8.32. The influence of camber on lift coefficient and drag coefficient is shown in Fig. 8.33. The camber makes the lift coefficient curve shift upward to the left and reduces the zero-lift angle of attack. The curvature makes the drag coefficient curve move to the right.

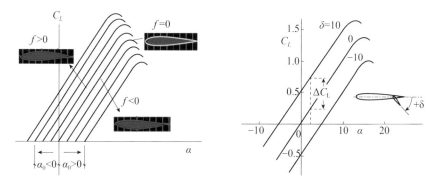

Fig. 8.31 Relationship between lift curve and camber

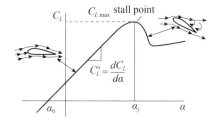

Fig. 8.32 Typical curve of lift coefficient for low-speed airfoil

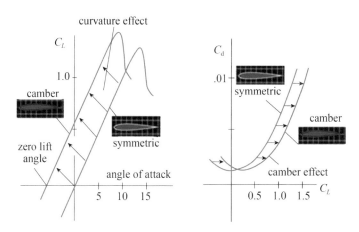

Fig. 8.33 Effect of airfoil camber on aerodynamic lift and drag coefficients

(4) There is a minimum drag coefficient in the drag coefficient curve, and then the drag coefficient increases with the change of the angle of attack, which is approximately a quadratic curve with the angle of attack. For symmetric airfoils, the lift coefficient corresponding to the minimum drag coefficient is zero, and

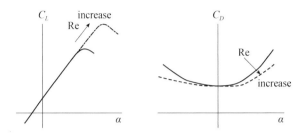

Fig. 8.34 Trend of aerodynamic coefficient of airfoil with Reynolds number

the main contribution is the frictional drag. For the airfoil with camber, the lift coefficient corresponding to the minimum drag coefficient is a small positive value, and it also has the contribution of pressure difference. But it should be pointed out that both frictional drag and pressure drag are related to viscosity. Therefore, the drag coefficient is closely related to Re number. With the increase of Reynolds number, the boundary layer on the airfoil becomes thinner, which delays the separation, increases the maximum lift coefficient and decreases the drag coefficient. As shown in Fig. 8.34.

(5) C_{ma} is the moment coefficient of aerodynamic center (for thin wing, it is the moment coefficient of quarter chord). It is basically a straight line and a horizontal line below the stall angle of attack. However, when the angle of attack exceeds the stall angle of attack, the flow around the airfoil enters into a large area of separated flow region, the bow moment increases greatly, and the moment curve decreases. The reason why the moment coefficient does not change is that with the increase of the angle of attack, the lift increases, the pressure center moves forward, and the distance from the pressure center to the aerodynamic center shortens. As a result, the product of the force and the force arm, that is, the pitching moment, remains unchanged. As shown in Fig. 8.35, the moment coefficient characteristic curve of NACA23012 airfoil is obtained.

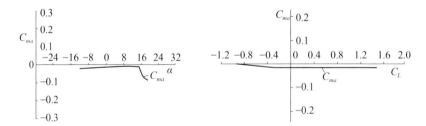

Fig. 8.35 The moment coefficient curve of NACA23012 (Re $= 6 \times 10^6$)

8.3.3 Separation Phenomenon of Flow Around Airfoil

With the increase of angle of attack, the lift coefficient of airfoil will be the largest and then decrease. This is the result of separation of the airflow around the airfoil. Stall characteristics of airfoil refer to the aerodynamic performance near the maximum lift coefficient. The separation of airfoil is closely related to the flow and pressure distribution on the airfoil. At a certain angle of attack, when the low-speed air flow bypasses the airfoil, it can be seen from the pressure distribution and velocity change of the upper wing surface that the air begins to accelerate rapidly and depressurize to the maximum speed point after passing the stagnation point (flow in the positive pressure gradient) after passing the front stop point, and then the deceleration is started to pressurize to the trailing-edge point of the airfoil (negative pressure gradient flow), as shown in Fig. 8.24. With the increase of the angle of attack, the front stagnation moves backward, and the suction peak around the near leading-edge increases, which makes it difficult for the airflow to flow backward against the negative pressure gradient after the peak point, and the more serious the deceleration of the airflow is. This not only makes the boundary layer thicken and becomes turbulent, but also when the angle of attack is large to a certain extent, when the negative pressure gradient reaches a certain value, the airflow cannot be able to slow down against the negative pressure, and the separation occurs. At this time, the airflow is divided into the separation zone flow and the attachment flow zone, as shown in Fig. 8.27.

Because on the outer boundary of the separation zone (called free boundary), the static pressure is equal everywhere. After separation, the main stream no longer decelerates and pressurizes. Due to the viscous effect of the main flow on the free boundary, the mass of the airflow in the separation zone is continuously carried away, and the air flow in the central part is constantly filled from the back, forming the reverse flow in the central part. Therefore, the pressure on the wall of the separation zone is smaller than that in the free flow zone. According to a large number of experiments, airfoil separation at large Reynolds number can be divided into (1) trailing-edge separation (turbulent separation), (2) leading-edge separation (leading-edge short bubble separation), and (3) thin wing separation (leading-edge long bubble separation). Different separation forms have different effects on lift coefficient, as shown in Fig. 8.36.

(1) Trailing-edge separation (turbulent separation)

The thickness of the airfoil corresponding to this separation is greater than 12%, and the negative pressure at the airfoil head is not particularly large (according to the potential flow theory, the centrifugal force of the air flow around the leading edge is balanced with the pressure, so the pressure is inversely proportional to the radius of the airfoil head). The separation starts from the trailing edge of the upper airfoil surface. With the increase of the angle of attack, the separation point gradually develops to

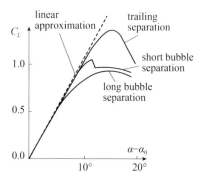

Fig. 8.36 Effect of different separation forms on lift coefficient

the leading edge. When the angle of attack reaches a certain value and the separation point develops to a certain position on the upper wing (about half of the wing), the lift coefficient reaches the maximum and then decreases. The development of trailing-edge separation is relatively slow, generally turbulent separation, the change of flow spectrum is continuous, the lift curve of stall area also changes slowly, and the stall characteristics are good, as shown in Fig. 8.37.

(2) Leading-edge separation (leading-edge short bubble separation)

For the airfoil with medium thickness (6–9%), the leading-edge radius is small, and the negative pressure is very large when the air flows around the leading edge, which results in a large negative pressure gradient in the local region. Even if the angle of attack is small, the flow separation occurs near the leading edge, and the separated boundary layer turns into turbulence to obtain energy from the outflow, and then attaches to the airfoil to form separated bubbles. At first, this kind of short bubble is very short, only 0.5–1% of the chord length. When the angle of attack reaches the stall angle, the short bubble suddenly opens and the air flow can no longer attach

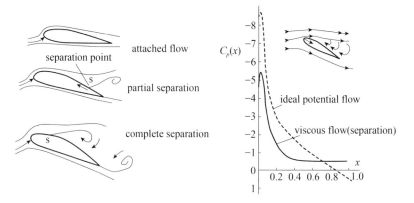

Fig. 8.37 Trailing-edge separation

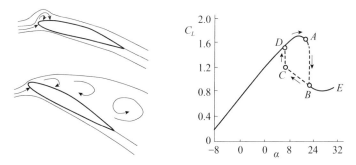

Fig. 8.38 Leading-edge short bubble separation

to it, resulting in the sudden complete separation of the upper wing surface and the sudden change of lift and torque, as shown in Fig. 8.38.

(3) Thin wing separation (leading-edge long bubble separation)

For thin airfoils (4–6% in thickness), due to the smaller leading-edge radius, the negative pressure of airflow around the leading edge is greater, resulting in a large negative pressure gradient. Even at a small angle of attack, flow separation occurs near the leading edge, and the separated boundary layer turns into turbulence to obtain energy from the outflow. After flowing for a long distance, it attaches to the airfoil and forms separated bubbles, as shown in Fig. 8.39. At first, the bubble is not long, only 2–3% of the chord length. However, as the angle of attack increases, the reattachment point moves downstream. When the stall angle of attack is reached, the bubble reaches the maximum and the lift also reaches the maximum. As the angle of attack continues to increase, the bubble opens, the upper wing enters the complete separation zone, and the lift decreases gradually.

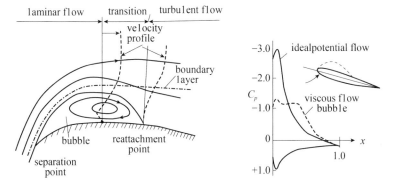

Fig. 8.39 Leading edge long bubble separation

(4) In addition to the above three kinds of separation, there may be mixed separation, and the flow around the airfoil may have both leading-edge and trailing-edge separation.

8.3.4 Stall Characteristics of Airfoil Flow

The phenomenon of stall characteristics for airfoil refers to the region where the lift coefficient of airfoil reaches the maximum. Stall is caused by separation, but separation does not necessarily for stall. Stall occurs only when separation reaches a certain degree. The reason for the maximum lift coefficient is closely related to the separation and development of the upper wing boundary layer, as shown in Fig. 8.40. When the angle of attack increases to a certain value, if there is separation in the trailing-edge region of the upper wing, the separated wing will lose potential flow lift, making the lift increment in this region negative, and the larger the separation region is, the greater the lift reduction is. With the increase of angle of attack, the increment of lift coefficient can be written as

$$C_L(\alpha + \Delta\alpha) = C_L(\alpha) + \Delta C_L = C_L(\alpha) + \frac{dC_L}{d\alpha}\Delta\alpha \qquad (8.38)$$

At the point of maximum lift coefficient, there is

$$\frac{dC_L}{d\alpha} = 0 \qquad (8.39)$$

This requires that at the maximum lift coefficient, the increased lift coefficient is equal to the decreased lift coefficient. Obviously, this kind of situation can be realized only when the wing separation zone reaches a certain degree. This is because, in the region of non-separation, the lift of potential flow always increases with the increase of angle of attack.

If it is assumed that the increment of lift coefficient of the potential flow without separation ΔC_{L1} is greater than 0, and the lift coefficient of the loss in the separation zone ΔC_{L2} is less than 0, the total lift coefficient increment is

$$\Delta C_L = \Delta C_{L_1} + \Delta C_{L2} \qquad (8.40)$$

Fig. 8.40 Lift increment after separation of upper airfoil surface

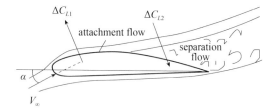

With the increase of angle of attack, if there is no separation on the upper wing (ΔC_{L1} is greater than zero and ΔC_{L2} equal to zero), the lift coefficient increases linearly. If there is a separation zone on the upper wing, $\Delta C_{L1}(>0)$ is the contribution of the potential flow in the non-separation zone, and $\Delta C_{L2}(<0)$ is the lift lost in the separation zone. Therefore, after separation, the total lift increment deviates from the linear value and becomes a curve, which makes the lift coefficient lower than the linear value. With the increase of the angle of attack, the separation zone will develop to the leading edge, resulting in the potential flow lift increment of the front non-separation zone becoming smaller and smaller, and the lift loss of the back separation zone becoming larger and larger. Finally, when the angle of attack reaches the critical angle of attack, when the increased lift and the lost lift are equal, the total increment is zero and the lift reaches the maximum. Beyond the critical angle of attack, the increased lift force is less than the lost lift force, resulting in the decrease of the total lift coefficient.

8.4   Kutta–Joukowski Trailing-Edge Condition and Determination of Circulation

8.4.1   Kutta–Joukowski Trailing-Edge Condition

How to use potential flow theory to build lift model? The best way to solve this problem is to compare the flow around an airfoil with that around a circular cylinder. It has been known that for the potential flow around a cylinder, if there is a velocity circulation around the cylinder, there will be a lift on the cylinder, and the lift is proportional to the velocity circulation. Because the boundary condition of potential flow solution is that the object surface is a streamline, any circulation on the cylinder satisfies the boundary condition, so the flow around the cylinder with any circulation is a solution of potential flow, and the lift on the cylinder satisfies Kutta–Joukowski trailing-edge condition. That is for the steady flow of ideal incompressible fluid, the circulation flow of straight uniform flow around the object with arbitrary cross-section shape will be subject to the lift (lateral force) perpendicular to the incoming flow direction, and the value of the lift is

$$L = \rho V_\infty \Gamma \tag{8.41}$$

The lift direction is the direction indicated by counter circulation rotation $90°$, as shown in Fig. 8.41. Where L is the lift acting on the flow around the body, ρ is the incoming air density, V_∞ is the incoming velocity, Γ is the velocity circulation around the flow object.

This rule holds for the flow around anybody and is also applicable to the flow around airfoil. Obviously, similar to the flow around a cylinder, any circulation added to an airfoil satisfies the condition that the airfoil is a streamline. No matter what the

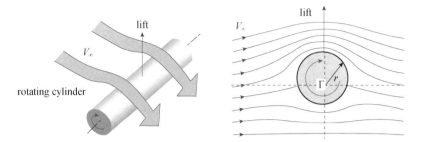

Fig. 8.41 Kutta–Joukowski lift circulation rule

shape of the airfoil is, as long as the circulation is not zero, the lift of the flow around the airfoil is not zero. Carefully observe the flow around the airfoil, for different circulation values, in addition to the lift, the stagnation points of the flow around the airfoil are also different. That is to say, for a given airfoil, under a certain angle of attack, it is impossible to determine the circulation value around the airfoil according to this rule because any value satisfies the boundary condition that the airfoil surface is a streamline. But in fact, for a given airfoil, the lift is uniquely determined at a certain angle of attack. That is to say, for the actual airfoil flow, there is only one definite circulation value, and other circulation values are not correct. So, how to determine the circulation value? It seems that we have to find another way. Kutta (1867–1944), a German mathematician, first gave the answer to this question. In 1902, he proposed the conditions for determining the circulation of airfoil. In 1906, Russian physicist Joukowski (1847–1921) independently gave the same conditions.

When different circulation values are added to the flow around the airfoil, according to the potential flow superposition principle, the front and rear stagnation points move. With the increase of circulation value, the front stagnation point moves from the lower wing to the leading-edge point, and the rear stagnation point moves from the upper wing to the trailing-edge point, as shown in Fig. 8.42. According to the potential flow theory, the stagnation point may be located at the upper wing, lower wing and trailing-edge points. As far as the stagnation point is located on the upper and lower wings, the airflow must flow around the trailing edge of the tip. According to the potential flow theory, there will be infinite velocity and negative pressure when the flow around the sharp corner, which is physically impossible. Therefore, restricted by the viscosity of the air, it is impossible for the air flow on the upper airfoil to leave the lower airfoil around the trailing-edge point without separation under the steady flow, and it is also impossible for the airflow on the lower airfoil to leave the upper airfoil around the trailing-edge point without separation. That is to say, the viscous effect makes the stagnation point around the airfoil can only be located at the trailing-edge point without separation. Therefore, Kutta–Joukowski proposed that the actual air flow on the upper and lower surface around the trailing edge leaves smoothly at the trailing edge (as shown in Fig. 8.42d), and the trailing-edge velocity remains limited. This is the famous Kutta–Joukowski unique trailing-edge condition for determining the value of circulation.

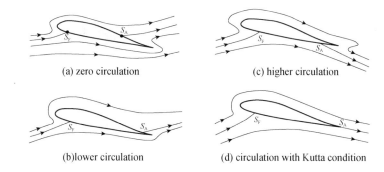

(a) zero circulation   (c) higher circulation

(b)lower circulation   (d) circulation with Kutta condition

Fig. 8.42 Flow pattern of airfoil with different circulation values

Once the circulation value is determined, the flow field characteristics of the cylinder with circulation and the airfoil with angle of attack can be further compared. As shown in Fig. 8.43, the flow field around a circular cylinder with circulation is an up-down asymmetric flow field, which can be expressed as the superposition of a symmetric flow field around a circular cylinder without circulation and an anti-symmetric flow field induced by a rotating cylinder. The superposition results in an asymmetric flow field with small velocity below the cylinder and large velocity above the cylinder. Thus, from Bernoulli's equation, it can be obtained that the pressure below the cylinder is high and the velocity above the cylinder is high which leads to an upward lift. In contrast, the flow around the airfoil does not rotate, but by changing the angle of attack and the shape asymmetry, it is equivalent to superimposing an antisymmetric flow field on the flow around the airfoil, which leads to the up-down asymmetry of the flow around the airfoil, resulting in lift.

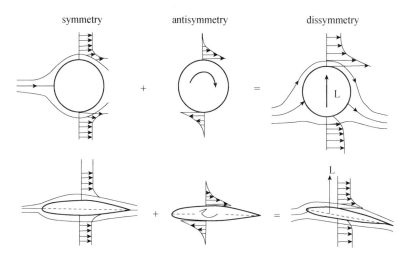

Fig. 8.43 Flow pattern around the cylinder and the airfoil with circulation values

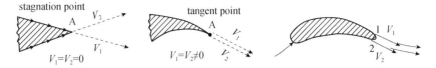

Fig. 8.44 Kutta–Joukowski trailing-edge condition

To sum up, as shown in Fig. 8.44, the Kutta–Joukowski trailing-edge condition is expressed as follows.

(1) For a given airfoil and angle of attack, the circulation around the airfoil should be smooth away from the trailing edge.
(2) If the trailing-edge angle of the airfoil is greater than 0, the trailing-edge point is the stagnation point. That is $V_1 = V_2 = 0$.
(3) If the trailing-edge angle of the airfoil is 0, the velocity at the trailing edge is finite, and the magnitude and direction are the same $V_1 = V_2 \neq 0$.
(4) The trailing edge of a real airfoil is not a sharp angle, but a small arc. The actual flow is separated at two points near the back of the upper and lower wings, and the separation zone is very small. The proposed conditions are $P_1 = P_2$, $V_1 = V_2$.

8.4.2 Incipient Vortex and the Generation of Circulation Value

According to Helmholtz's rule of conservation of vortices, for an ideal incompressible fluid, the velocity circulation on the closed circumference composed of the same fluid particles does not change with time under the conservative force, $d\Gamma/dt = 0$. According to the rule of conservation of vortices, the velocity circulation caused by airfoil should be as zero as at rest. However, according to Kutta–Joukowski rule of circulation of lift, the lift is produced by circulation, which is almost contradictory. Therefore, the problem of circulation is put forward.

In order to solve this problem, a large fluid contour around the airfoil is taken as a closed curve when the airfoil is stationary. According to Helmholtz's rule of conservation of vortices, it can be concluded as follows.

(1) In the static state, the velocity loop around the girth is zero. As shown in Fig. 8.45.

(2) When the airfoil starts up, because the viscous boundary layer has not formed on the airfoil, the velocity circulation around the airfoil is zero, and the stagnation point is not at the trailing edge, but at a certain point on the upper airfoil. The airflow will flow around the trailing edge to the upper airfoil. With the development of time, the upper boundary layer is formed on the wing, and the lower wing will form a large velocity and low pressure when the airflow around the trailing edge. As a result, there is a large negative pressure gradient from the

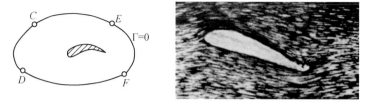

Fig. 8.45 Velocity circulation around the shroud is zero (static flow field)

trailing edge to the stagnation point, resulting in the separation of the boundary layer and the generation of a counterclockwise circulation, which is called the start vortex, as shown in Fig. 8.46.

(3) The starting vortices leave the trailing edge and flow downstream with the flow, and the closed fluid line also moves with the flow, but always surrounds the airfoil and the starting vortices. According to the vorticity retention rule, a clockwise equal circulation vortex (because it is attached to the airfoil, it is called the attached vortex) must be generated around the airfoil, so that the total circulation around the closed contour is zero, as shown in Fig. 8.47. In this way, the stagnation point of the airfoil moves backward under the guidance of the attached vortex. As long as the trailing stagnation point has not moved to the trailing-edge point, there will be a continuous counterclockwise vortex shedding from the trailing edge of the airfoil, so the amount of attached circulation around the airfoil will continue to increase until the air flow smoothly leaves from the trailing edge point (the trailing stagnation point moves to the trailing edge point), and the attached vortex and start vortex will reach the maximum, as shown in Fig. 8.48.

From the above discussion, it can be concluded that.

(1) The viscosity of the fluid and the sharp trailing edge of the airfoil are the physical causes of the start vortex. The value of the attached vortex circulation around the airfoil is always equal to the value of the start vortex circulation, but the direction is opposite.

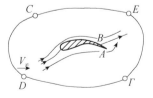

Fig. 8.46 Appearance of start vortex

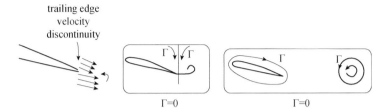

Fig. 8.47 Generation of start vortex

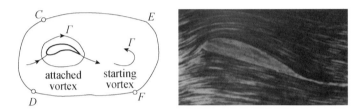

Fig. 8.48 The maximum value of the attached vortex (the air flows smoothly away from the trailing edge)

(2) For an airfoil with a certain shape, as long as the velocity and angle of attack are given, there is a fixed velocity circulation corresponding to it, and the determined condition is Kutta condition.

(3) If the velocity and angle of attack change, the velocity circulation will be readjusted to ensure that the flow always leaves the trailing edge smoothly when it bypasses the airfoil.

(4) The vortices representing the circulation around the airfoil are always attached to the airfoil. According to the rule of lift circulation, the lift produced by the straight uniform flow with a certain strength of attached vortex is exactly the same as that produced by an airfoil with circulation in the straight uniform flow.

8.5   Lift Generation Mechanism of Airfoil

When the flow around the airfoil, it will act on the airfoil lift and drag. In addition to the velocity, geometry, and size of the airfoil, the lift is also related to the angle of attack between the airfoil and the flow direction. At present, the mechanism of airfoil lift is as follows.

(1) Skipping Stone Theory (principle of force and reaction)

In 1686, British physicist Newton applied the mechanics principle and deduction method and suggested that the force exerted by moving objects in the fluid is proportional to the square of the velocity of the object, the characteristic area of the object and the product of the fluid density. Newton proposed the so-called Skipping Stone

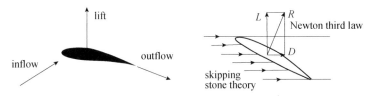

Fig. 8.49 Newton's skipping stone theory

Theory for the lift of airfoil according to the principle of force and reaction. This theory holds that the lift force is derived from the air reaction to the lower wing surface of the airfoil. Like water drift, when the stone slides over the water rapidly, the water body will be drained to obtain the reverse force to leave the water body. The aircraft pushes the air down continuously during flight, thus obtaining the lift by the reaction force, as shown in Fig. 8.49. It is concluded that the lift force is mainly from the lower surface of the airfoil, and the contribution of the upper surface of the airfoil can be ignored. It is thus extended that the change in the shape of the upper surface does not result in a lift change if the lower surface of the airfoil is unchanged which is obviously not true. A typical example is the spoiler of an aircraft. When the spoiler on the upper surface of the wing is opened, the lower surface is unchanged, and the upper surface can be said to have little change in shape, but it has a great influence on aerodynamic force.

(2) Bernoulli principle

In 1738, Swiss mathematician and fluid physicist Daniel Bernoulli gave the energy equation of ideal fluid, established the quantitative relationship between pressure and velocity, and provided a theoretical basis for understanding the mechanism of lift generation. In particular, according to Bernoulli equation, the lift on the airfoil is not only related to the air jacking force acting on the lower airfoil but also related to the suction of the upper airfoil (as shown in Fig. 8.23). Later wind tunnel tests confirmed that the suction of the upper airfoil accounts for about 60–70% of the total lift of the airfoil. According to Bernoulli equation, the suction of the upper wing is due to the flow velocity of the upper wing is greater than that of the incoming flow, while the jacking force of the lower wing is due to the flow velocity of the lower wing is less than that of the incoming flow. Therefore, in order to make clear the generation of lift, it can be attributed to the change of velocity on the upper and lower surfaces of the airfoil. There are two ways to say this.

(a) Long distance theory or isochronous theory. According to this theory, the flow is divided into two parts at the leading edge of the airfoil and finally converges at the trailing edge of the airfoil. Because the shape of the upper and lower surfaces of the airfoil is asymmetric, the distance of the air flow along the upper surface of the airfoil is long, and the natural velocity is fast. According to Bernoulli theorem, the pressure of the fast air pressure is small, so that the pressure of the lower airfoil is greater than that of the upper airfoil, and the lift force is generated.

Since this theory mainly depends on Bernoulli principle, later generations call it Bernoulli school. The key of this theory is that the asymmetry of the upper and lower surfaces of the airfoil is the source of lift. However, for the purpose of reducing shock wave intensity, the length of the lower surface of supercritical airfoil, which is widely used in modern aircraft, is actually longer than that of the upper surface, so this explanation is also suspected. At the same time, this theory cannot explain the reason why the plane flies backwards.

(b) Flow tube change theory. When the air flows through the upper and lower surfaces, due to the convex of the upper surface, the spacing of the upper streamline becomes narrower, while the lower surface is flat, and the spacing of the streamline becomes wider. According to the continuity equation of fluid, when the fluid continuously flows through a pipe of different thickness, the fluid in any part of the channel cannot be interrupted or accumulated, so at the same time, The mass of fluid flowing into any section is equal to the mass of fluid flowing out of another section, which leads to the flow velocity on the upper surface is greater than that on the lower surface, so that the pressure on the upper wing is less than that on the lower wing, and the lift force is generated. The doubtful point is that this theory can only be established in two-dimensional environment, the real wing is three-dimensional flow, and the flow tube shrinkage deformation is not obvious.

(3) Momentum theory

There is a downward trend in the flow direction of the flow around the airfoil. According to the momentum theory, there is a lift that produces reaction force. This part of the lift does exist, known as impact lift, but it accounts for relatively small proportion of the whole airfoil lift. For the supercritical airfoil used in large passenger aircraft, the after-load effect is to generate downwash flow by bending the rear edge of the wing to provide lift.

All the above views are cognition under the condition of non-viscous flow, and there are limitations. There is no point of view that can explain the cause of lift of airfoil satisfactorily. In fact, the theories are adapted to different regions of airfoil flow, as shown in Fig. 8.50.

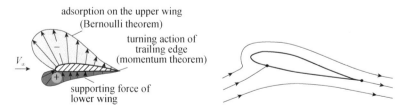

Fig. 8.50 Theoretical adaptability of airfoil flow based on ideal fluid

8.6 Development of Boundary Layer Near Airfoil Surface and Determination of Circulation Value

In the unbounded flow field, when the infinite cylinder rotating at a constant speed reaches equilibrium, a resultant force perpendicular to the flow direction acts on the cylinder, which is called lift. If the rotating cylinder is regarded as the vortex core, the flow field inside the vortex core is a vortex field with equal vorticity, and the flow field outside the vortex core is a cylinder induced flow field without vortex field. Such a flow model is a typical Rankin vortex model. At this time, the circulation of the rotating cylinder acts on the fluid through the cylinder edge interface in contact with the fluid, so as to induce the flow field outside the cylinder. The flow is intuitive and easy to understand, which has also been verified by experiments. However, for the ideal fluid flow around the low-speed airfoil, the airfoil does not rotate, so how does the circulation occur? This involves the physical mechanism of the formation process of the flow circulation (attached vortex) around the airfoil. At the beginning of the twentieth century, the physical explanation of the concepts of starting vortex and attached vortex based on ideal fluid motion belongs to the classical theory recognized by aerodynamics, but how does the attached vortex exist near the airfoil surface? How is it applied to the flow around the airfoil? What is the relationship with the development of wing boundary layer? Will the attitude and shape of the airfoil change the characteristics of the boundary layer? Will it change the size of the attached vortex? These problems can only be explained by Prandtl's theory.

8.6.1 Characteristics of Boundary Layer and Velocity Circulation Around Airfoil in a Viscous Steady Flow Field

As shown in Fig. 8.51, for the steady viscous flow over the airfoil, a fully developed boundary layer flow will be formed in the near wall area of the upper and lower surfaces of the airfoil, which is in a stable equilibrium state. At this time, the airfoil flow can be divided into the boundary layer viscous flow (vortex flow) in the near wall area and the potential flow outside the boundary layer. Now, take any closed contour around the airfoil in a clockwise direction outside the airfoil flow boundary layer, and the velocity circulation value around the airfoil in this contour can be obtained from the Stokes integral formula, i.e.,

$$\Gamma = \oint_C \vec{V} \cdot d\vec{s} = \iint_A 2(\omega_u - \omega_d) d\sigma \qquad (8.42)$$

The velocity circulation around the airfoil is positive clockwise, and the vorticity in the boundary layer near the wall of the upper airfoil is $2\omega_u$. Rotating clockwise (ω_u) is

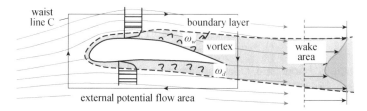

Fig. 8.51 Characteristics of boundary layer and velocity circulation around steady airfoil

the rotational angular velocity) is a positive contribution to the velocity loop. Vorticity $2\omega_d$ in the boundary layer near the wall of the lower wing. Rotating counterclockwise makes a negative contribution. This can be written as

$$\Gamma = \oint_C \vec{V} \cdot d\vec{s} = \int_0^b \left[\int_0^{\delta_u(x)} 2\omega_u dy - \int_0^{\delta_d(x)} 2\omega_d dy \right] dx = \int_0^b \gamma(x)dx$$

$$\gamma(x) = \int_0^{\delta_u(x)} 2\omega_u dy - \int_0^{\delta_d(x)} 2\omega_d dy = \gamma_u(x) - \gamma_d(x)$$

$$\gamma_u(x) = \int_0^{\delta_u(x)} 2\omega_u dy, \quad \gamma_d(x) = \int_0^{\delta_d(x)} 2\omega_d dy \qquad (8.43)$$

where $\delta_u(x)$, $\delta_d(x)$, respectively, represent the local thickness of the boundary layer of the upper and lower wing surfaces. $\gamma(x)$ is the surface vortex intensity distributed along the chord. γ_u is the value of the upper wing surface (positive contribution), γ_d is the value of the lower wing surface (negative contribution).

At the trailing edge of the airfoil, according to the Kutta–Joukowski trailing-edge conditions, it is necessary to keep the air flow smoothly away from the trailing edge, that is, when the air flow on the upper and lower airfoils leaves the trailing edge, there is no velocity shear and no vortex shedding, which requires that the surface vortex intensity of the airfoil is zero at $x = b$.

$$\gamma(b) = 0.0 \qquad (8.44)$$

Available from Eq. (8.43)

$$\gamma_u = \gamma_d, \int_0^{\delta_u(b)} 2\omega_u dy = \int_0^{\delta_d(b)} 2\omega_d dy \qquad (8.45)$$

As an approximation, as shown in Fig. 8.52, Eq. (8.46) can be obtained by definition

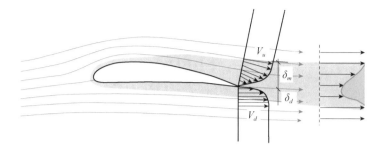

Fig. 8.52 Kutta–Joukowski trailing-edge condition for steady airfoil flow

$$2\omega_u \approx \frac{V_u}{\delta_u(b)}, \quad 2\omega_d \approx \frac{V_d}{\delta_d(b)} \tag{8.46}$$

Thus,

$$\int_0^{\delta_u(b)} \frac{V_u}{\delta_u(b)} dy \approx \int_0^{\delta_d(b)} \frac{V_d}{\delta_d(b)} dy, V_u \approx V_d \tag{8.47}$$

This is the Kutta–Joukowski trailing-edge condition. That is, leaving the trailing edge smoothly requires that the velocity of leaving the trailing edge on the outer boundary line of the upper and lower wing boundary layer is approximately equal.

At the trailing edge of the airfoil, if it is assumed that the surface vortex intensity in the upper wing boundary layer is

$$\gamma_0 = \int_0^{\delta_u(b)} 2\omega_u dy \tag{8.48}$$

From Eq. (8.43), it can be gotten

$$\Gamma = \oint_C \vec{V} \cdot d\vec{s} = \int_0^b \left[\gamma_u - \gamma_d\right] dx = \int_0^b \left[\gamma_u - \gamma_0\right] dx + \int_0^b \left[\gamma_0 - \gamma_d\right] dx = \Gamma_u + \Gamma_d$$

$$L = \rho V_\infty \Gamma_u + \rho V_\infty \Gamma_d \tag{8.49}$$

The contribution of the upper and lower wings to the lift can be obtained. As shown in Fig. 8.53, the calculation equation of lift coefficient acting on an airfoil based on pressure coefficient distribution at small angle of attack is

$$C_L = \int_0^1 (C_{pd} - C_{pu}) d\xi = \int_0^1 C_{pd} dx + \int_0^1 (-C_{pu}) d\xi$$

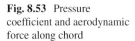

Fig. 8.53 Pressure
coefficient and aerodynamic
force along chord

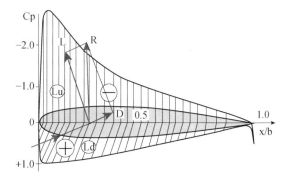

$$L = \rho V_\infty \Gamma_u + \rho V_\infty \Gamma_d = \frac{1}{2}\rho V_\infty^2 b C_L$$

$$\Gamma_d = \frac{1}{2}V_\infty b \int_0^1 C_{pd}d\xi, \quad \Gamma_u = \frac{1}{2}V_\infty b \int_0^1 (-C_{pu})d\xi \qquad (8.50)$$

Among them, C_{Pu} and C_{Pd} are pressure coefficients acting on the upper and lower wings, respectively. ξ is x/b.

Obviously, anywhere from the leading edge, the magnitude of $\gamma(x)$ depends on the difference between the vorticity integral values in the local upper and lower wing boundary layer. According to the velocity distribution characteristics in the upper and lower wing boundary layer, the distribution of $\gamma(x)$ along the chord line shall be a variation curve gradually decreasing from the leading edge to the trailing edge, as shown in Fig. 8.54. It can be seen that the attached vortex based on the concept of ideal fluid flow actually refers to the difference between the vorticity integral values in the boundary layer of the upper and lower airfoils. If the ideal fluid flow model is used to replace the boundary layer flow, it should be seen as the shape of the ideal fluid bypassing the boundary curve of the airfoil surface and superimposing the displacement thickness of the boundary layer. At the same time, the velocity circulation value generated by the viscous boundary layer is added to this boundary, which is the attached vortex. This shows that the attached vortex is added to the external potential flow through the wing boundary layer flow.

8.6.2   Vorticity Characteristics in Boundary Layer of Upper and Lower Wing Surfaces

For the flow in the boundary layer near the wall, the motion of viscous fluid is always accompanied by the generation, diffusion and dissipation of vorticity. In the case of large Reynolds number of incoming flow, the boundary layer near the wall is thin, which meets Prandtl's boundary layer approximation. Under the condition of no-slip

Fig. 8.54 Surface vortex intensity distribution along chord

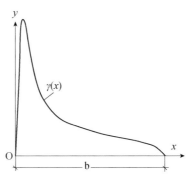

boundary, it is equivalent to forming a vortex surface source with certain intensity distribution in the near area of the wing surface, and its surface vortex intensity is Ω_b (positive clockwise) and the wall shear stress τ_b relationship is

$$\Omega_b = 2\omega_b = \left(\frac{\partial u}{\partial y} - \frac{\partial v}{\partial x}\right)_b = \left(\frac{\partial u}{\partial y}\right)_b = \frac{\tau_b}{\mu} \tag{8.51}$$

where u and v are the flow velocity components in the boundary layer. It can be seen that the vorticity on the wing surface is related to the wall shear stress, indicating that the vorticity in the boundary layer is the largest on the wing surface and decreases in the area away from the object surface, which is caused by the viscous diffusion and dissipation of vorticity. For incompressible viscous fluid flow around airfoil, the diffusion equation of vorticity Ω is

$$\frac{\partial \Omega}{\partial t} + u\frac{\partial \Omega}{\partial x} + v\frac{\partial \Omega}{\partial y} = v\left(\frac{\partial^2 \Omega}{\partial x^2} + \frac{\partial^2 \Omega}{\partial y^2}\right) \tag{8.52}$$

In the steady boundary layer flow, Eq. (8.50) is simplified as

$$u\frac{\partial \Omega}{\partial x} + v\frac{\partial \Omega}{\partial y} = v\frac{\partial^2 \Omega}{\partial y^2} \tag{8.53}$$

In the boundary layer, on the one hand, the vorticity migrates along the mainstream direction and gradually attenuates. On the other hand, vorticity diffuses along the vertical direction, and its vertical diffusion speed and attenuation speed depend on the fluid viscosity coefficient. The migration velocity of vorticity depends on the flow velocity in the horizontal direction, so the vorticity generated on the material surface will not diffuse to the whole field and can only be limited to the boundary layer. The distance order of vorticity diffusion in the normal direction of the wall is $\sqrt{vt}$, and the distance of vorticity migration along the flow direction is $v_{\infty}t$. For an airfoil with chord length b, the time required for vorticity migration from the leading edge to the trailing edge is b/v_{∞}, so the thickness of the boundary layer is

$$\delta \propto \sqrt{v\frac{b}{V_\infty}} = \sqrt{v\frac{b^2}{V_\infty b}} = b\sqrt{\frac{v}{V_\infty b}}$$

$$\frac{\delta}{b} \propto \frac{1}{\sqrt{Re}} \quad \left(Re = \frac{V_\infty b}{v}\right) \tag{8.54}$$

8.6.3 Evolution Mechanism of Boundary Layer During Airfoil Starting

(1) Governing equations of boundary layer formation

The unsteady flow problem during airfoil starting belongs to the formation and development process of viscous boundary layer on the airfoil. The physical mechanism is complex, involving the transformation of inviscid flow and viscous flow, momentum diffusion caused by viscosity, the movement of separation point in the trailing edge of the upper airfoil, and the evolution and development process of separation zone and separation vortex. Obviously, in the process of airfoil starting, the increase of velocity circulation is a process of the formation and development of unsteady boundary layer and finally achieves the fully developed boundary layer flow of airfoil steady flow. Obviously, the development controlling this process is the incompressible two-dimensional unsteady laminar boundary layer differential equation.

$$\frac{\partial u}{\partial x} + \frac{\partial v}{\partial y} = 0$$

$$\frac{\partial u}{\partial t} + u\frac{\partial u}{\partial x} + v\frac{\partial u}{\partial y} = \frac{\partial V_e}{\partial t} + V_e\frac{\partial V_e}{\partial x} + v\frac{\partial^2 u}{\partial y^2} \tag{8.55}$$

where, V_e is the outflow velocity of the boundary layer.

At the initial stage when the airfoil starts from rest (as shown in Fig. 8.55), the boundary layer has not been formed, the viscous shear force is large, the migration inertial force is small, and the unsteady inertial force of the outflow field is the main. The above equation can be simplified as

$$\frac{\partial u}{\partial x} + \frac{\partial}{\partial y} = 0$$

$$\frac{\partial u}{\partial t} - v\frac{\partial^2 u}{\partial y^2} = \frac{\partial V_e}{\partial t} \tag{8.56}$$

For the later stage of airfoil starting process, the boundary layer is basically formed and close to the stable state. At this time, the unsteady inertial force is in the secondary

Fig. 8.55 Initial stage of
starting process

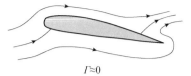

$\Gamma \approx 0$

position, and the boundary layer equation can be simplified as

$$\frac{\partial u}{\partial x} + \frac{\partial v}{\partial y} = 0$$

$$\frac{\partial u}{\partial t} - v\frac{\partial^2 u}{\partial y^2} = V_e\frac{\partial V_e}{\partial x} - u\frac{\partial u}{\partial x} - v\frac{\partial u}{\partial y} \tag{8.57}$$

Now, according to the formation process of boundary layer of unsteady airfoil flow, the evolution of separation and vortex shedding, and combined with the physical mechanism of viscous flow, the starting process of airfoil flow can be divided into the following stages.

(2) Analysis of the formation process of fully developed boundary layer

    (1) Initial potential flow stage

        When the airfoil starts (as shown in Fig. 8.55), the flow around the airfoil almost does not form a boundary layer flow, which can be seen mainly ideal fluid. The rear stagnation point is located in the trailing edge area of the upper airfoil, and the air flow on the lower airfoil reaches the rear stagnation point around the trailing-edge point. The rear edge point does not coincide with the rear stagnation point, and there is no separation in the rear edge area of the airfoil. At this time, the upper boundary layer on the airfoil is almost not formed, and the attached vorticity, lift are almost zero. In this case, the velocity near the object surface, the lower wing surface is slightly larger than the upper wing surface.

    (2) Separation bubble stage

        Due to the influence of centrifugal inertial force around the trailing-edge point, the negative pressure gradient of the flow around the trailing-edge point to the rear stagnation point continues to increase, resulting in the separation of the trailing-edge area and the formation of separation bubbles (as shown in Fig. 8.56). At the same time, the rear stagnation point of the upper wing surface moves to the end of the separation bubbles. A separation point appears in the trailing-edge area of the upper wing surface and the separation point does not coincide with the trailing-edge point (after separation, the upper wing has only separation points, but no stagnation points). At this time, the viscous flow near the trailing edge of the airfoil begins to form, but the overall flow is still dominated by the flow around the ideal fluid, the attached vorticity is almost zero (the two counter rotating vortices in the separation bubble counteract each other), and the lift also approaches zero.

(3) Periodic shedding stage of trailing-edge separation vortex

With the increase of airfoil speed, the energy of vortex motion in the separation bubble increases continuously, so that the flow kinetic energy accumulation in the bubble cannot be consumed by itself, so that the separation bubble opens and vortex shedding is formed (as shown in Fig. 8.56). At the same time, the centrifugal inertial force around the trailing-edge point increases and the negative pressure near the trailing-edge point increases due to the increase of air flow speed. As a result, the separation point in the trailing-edge area of the upper wing moves to the trailing-edge point (flow from high pressure to low pressure). At this time, the boundary layer around the upper wing begins to form. Relatively speaking, the lower wing is slightly slower, which increases the vorticity in the boundary layer, and the boundary layer flow around the near area of the airfoil surface begins to play a role. In this way, the attached vorticity is not equal to zero, and the lift begins to appear. With the increasing speed of the airfoil, the vortices generated from the trailing edge continue to fall off and throw downstream with the flow. At the same time, the flow velocity around the trailing edge point continues to increase, and the increase of the centrifugal inertial force at the trailing edge point continuously enhances the negative pressure near the trailing-edge point, resulting in the further movement of the separation point in the trailing-edge area of the upper wing surface to the trailing-edge point. At this time, the flow boundary layer around the upper wing surface continues to develop, and the boundary layer around the lower wing also began to form and develop. With the acceleration of the airfoil, the velocity in the outflow region of the upper and lower airfoil boundary layer increases, the attached vorticity around the airfoil continues to increase, and the lift also continues to increase.

(4) Boundary layer stability stage

The periodic shedding of trailing-edge separation vortices is repeated until the airfoil reaches a uniform speed and no acceleration. At this time, the

Fig. 8.56 Periodic shedding stage of trailing-edge separated vortex (acceleration)

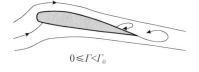

$$0 \leqslant \Gamma < \Gamma_0$$

Fig. 8.57 Boundary layer stability stage (uniform velocity)

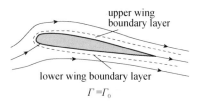

upper wing boundary layer

lower wing boundary layer

$$\Gamma = \Gamma_0$$

Fig. 8.58 Periodic shedding stage (deceleration) of trailing-edge separation vortex

separation point of the upper wing surface moves to the trailing-edge point, the vortex shedding stops, the air flow around the trailing-edge point smoothly leaves, the boundary layer of the upper and lower wings forms a stable state, the velocity difference in the outflow area of the boundary layer reaches the maximum, the attached vorticity around the airfoil reaches the maximum, and the lift reaches the maximum. The airfoil flow around the airfoil completes the starting process, as shown in Fig. 8.57.

It should be noted that if the airfoil accelerates or decelerates again and changes from one stable boundary layer to another, the shedding of vortex at the trailing edge will continue until a new stable boundary layer is formed (as shown in Fig. 8.58). After reaching the new equilibrium state, the flow around the trailing edge will return to smoothly leaving the trailing edge. It is just that the direction of shedding vortex is different between accelerating airfoil and decelerating airfoil.

8.7   General Solution of the Steady Incompressible Potential Flow Around Airfoil

For the attached flow around the airfoil with small angle of attack, it can be seen from Eqs. (8.20) and (8.22) that the lift and torque characteristic curves are obtained by pressure integration, which can be solved by potential flow theory according to the boundary layer theory. This shows that the effect of viscosity on the lift and moment coefficients is small and can be ignored. However, it has great influence on the drag coefficient, maximum lift coefficient and aerodynamic characteristic curve of separated airfoil, which can not be ignored.

8.7.1   Conformal Transformation Method

In 1909, the Russian physicist Joukowski considered that for the two-dimensional steady incompressible potential flow around the airfoil, there were velocity potential function and flow function, both of which satisfied the Laplace equation. Firstly, the complex variable function theory was introduced to solve the airfoil flow problem,

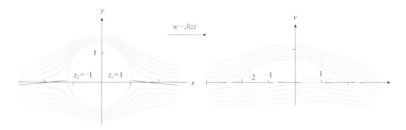

Fig. 8.59 Joukowski transform and airfoil

and the famous Joukowski airfoil theory was proposed, as shown in Fig. 8.59. The main idea of conformal transformation method is to form complex velocity by using stream function and velocity potential function. Through complex variable function transformation, the airfoil in the physical plane is transformed into a circle in the calculation plane (as shown in Fig. 8.59), then the complex potential function around the circle is obtained, and then the transformation formula is returned to the complex potential function in the physical plane. On this basis, the first variable density wind tunnel was proposed and built in 1922 by German aerodynamics scientist Max M. Munk (1890–1986), Prandtl student. Max M. Munk and Hermann Glauert (1892–1934, as shown in Fig. 8.60), a British aerodynamic scientist, developed and improved the thin airfoil theory in 1924. Using the singularity superposition principle of velocity potential function and small disturbance hypothesis, a potential flow method for solving the flow around arbitrary thin airfoil was proposed. It is concluded that for the steady flow around an ideal incompressible thin airfoil at a small angle of attack, the disturbed velocity potential, surface boundary conditions and pressure coefficient can be linearly superimposed, and the lift and torque acting on the thin airfoil can be regarded as the sum of curvature, thickness and angle of attack. Therefore, the flow around the thin airfoil can be superimposed by three simple flows.

8.7.2 Numerical Calculation of Airfoil—Panel Method

(1) Basic idea of potential flow superposition method around airfoil

In plane potential flow, according to the superposition principle of potential flow and isolated singular point flow, the flow around some regular objects can be obtained. For example, through the superposition of direct uniform flow with point source and point sink, the circular flow around a cylinder can be obtained; through the superposition of straight uniform flow, point source, point sink and point vortex, the flow around a cylinder with circulation can be obtained, and then the lift of the body around the flow can be calculated. For the flow around an object of arbitrary shape, it is certainly impossible to be so simple. However, such a solution idea is desirable.

Fig. 8.60 Hermann Glauert,
British aerodynamics
scientist, 1892–1934

For the flow around an airfoil of any shape and thickness at a certain angle of attack, the basic idea of using the potential flow superposition method is:

1) Continuously distributed point sources $q(s)$ are arranged along the airfoil surface and superimposed with the straight uniform flow to meet the condition that the airfoil surface is a streamline, so as to simulate the effect of airfoil thickness without lift, as shown in Fig. 8.61.
2) The continuously distributed surface vortex function $\gamma(s)$ is arranged along the airfoil surface and superimposed with the straight uniform flow to meet the conditions that the airfoil surface is a streamline and Joukowski conditions, so as to simulate the lift effect caused by the angle of attack, airfoil camber and determine the lift of the airfoil, as shown in Fig. 8.62.

Fig. 8.61 Source function
of airfoil surface layout

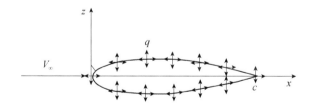

Fig. 8.62 Vorticity function
of airfoil surface layout

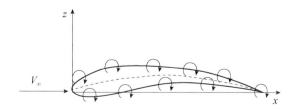

3) The key to the potential flow solution of the superposition of surface sources and surface vortices arranged on the wing and direct uniform flow is to find the distribution function $q(s)$ and $\gamma(s)$ satisfying the boundary conditions and Kutta–Joukowski trailing-edge conditions. The surface source of the arrangement is determined by the fact that the wing surface is a streamline. The arrangement surface vortex not only meets the condition that the wing surface is a streamline but also needs to meet the Kutta–Joukowski trailing-edge conditions. It is not easy to give the exact distribution of surface source and surface vortex functions for arbitrary airfoils. However, it is usually possible to solve the numerical value by numerical calculation method. The wing surface is divided into several differential segments (panel elements), and the undetermined surface source and surface vortex distribution functions are arranged on each panel element. The non-penetration conditions are satisfied at the selected control points and the trailing-edge conditions are satisfied at the trailing edge, so as to determine the distribution function. Finally, the distribution function is used to calculate the surface pressure distribution, lift and torque characteristics.

(2) The basic characteristics of the surface source function

As shown in Fig. 8.63, if the unit length of the surface source strength is q, the source strength above the ds micro segment is $q\,ds$, and the velocity induced at point P of the flow field (distance r from point P) is

$$dV_r = \frac{q\,ds}{2\pi r} \quad d\varphi = \int \vec{V}\cdot d\vec{r} = \int \frac{q\,ds}{2\pi r}\,dr = \frac{q\,ds}{2\pi}\ln r \qquad (8.58)$$

The velocity potential function generated by the whole area source and the area source intensity are

$$\varphi = \int_a^b d\varphi = \int_a^b \frac{q\,ds}{2\pi}\ln r \quad Q = \int_a^b q\,ds \qquad (8.59)$$

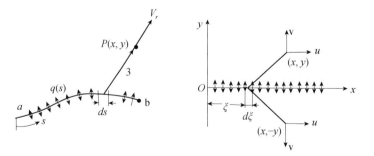

Fig. 8.63 Velocity field induced by area source

Fig. 8.64 Control volume of area source micro segment

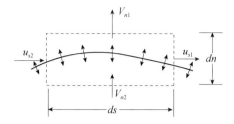

Except for the surface source line, any point in the flow field satisfies the continuity equation. However, the normal velocity of fluid particles on the surface source line is discontinuous due to the influence of surface source intensity. As shown in Fig. 8.63, the induced velocity generated by the area source intensity of the horizontal line is

$$u(x, y) = u(x, -y)$$
$$v(x, y) = -v(x, -y) \tag{8.60}$$

$$y \rightarrow \pm 0, \ u(x, 0) = u(x, -0), \ v(x, 0) = -v(x, -0) \tag{8.61}$$

It is concluded that the normal velocity of area source is discontinuous and the tangential velocity is continuous. The same is true for surface area source layout. As shown in Fig. 8.64, take the micro segment control volume and establish the continuity equation.

$$q ds = (v_{n1} - v_{n2}) ds + (u_{s1} - u_{s2}) dn \tag{8.62}$$

Considering the continuity of $u_s(s)$, there are

$$u_{s2} = u_s - \frac{\partial u_s}{\partial s} \frac{ds}{2}, \ u_{s1} = u_s + \frac{\partial u_s}{\partial s} \frac{ds}{2} \tag{8.63}$$

Take the limit and get

$$q = v_{n1} - v_{n2} \tag{8.64}$$

This shows that the area source line is the normal velocity discontinuity, and the sudden jump value of the local normal velocity passing through the area source is equal to the local area source strength. For linear surface source functions, there are

$$q = v(x, 0) - v(x, -0)$$
$$v(x, 0) = -v(x, -0)$$
$$v(x, 0) = -v(x, -0) = q/2 \tag{8.65}$$

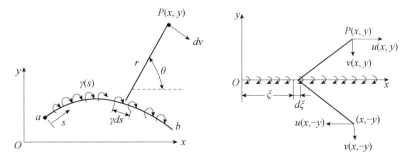

Fig. 8.65 Velocity field induced surface vortex

(3) Basic characteristics of surface vortex

As shown in Fig. 8.65, assuming that the surface vortex intensity per unit length is γ, the vortex intensity above the ds micro segment is γds, and its induced velocity at point P of the flow field is (distance r from point P)

$$dv = \frac{\gamma ds}{2\pi r}, \quad d\varphi = \int \vec{V} \cdot d\vec{r} = -\int \frac{\gamma ds}{2\pi r} r d\theta = \frac{\gamma ds}{2\pi} \theta \qquad (8.66)$$

The velocity potential function and surface vortex intensity produced by the whole surface vortex are

$$\varphi = \int d\varphi = -\int \frac{\gamma ds}{2\pi} \theta, \quad \Gamma = \int_a^b \gamma ds \qquad (8.67)$$

Except for the surface vortex line, any point in the flow field satisfies the continuity equation. However, the tangential velocity of fluid particles on the surface vortex line is discontinuous due to the strength of the surface vortex. As shown in Fig. 8.65, the induced velocity generated by the surface vortex intensity of the horizontal line is

$$u(x, y) = -u(x, -y)$$
$$v(x, y) = v(x, -y) \qquad (8.68)$$

$$y \to \pm 0, \quad u(x, 0) = -u(x, -0), \quad v(x, 0) = v(x, -0) \qquad (8.69)$$

It is concluded that the upper and lower tangential velocity of the surface vortex is discontinuous, but the normal velocity is continuous. The same is true for the arrangement of surface vortices on surfaces. As shown in Fig. 8.66, take a closed curve around the micro segment, and it is positive clockwise around the circumference. According to the Stokes integral, it is obtained

Fig. 8.66 Closed contour of surface vortex micro segment

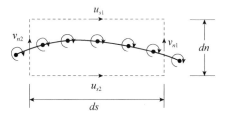

$$\gamma ds = (u_{s1} - u_{s2})ds + (v_{n2} - v_{n1})dn \tag{8.70}$$

Considering the continuity of $v_n(n)$, there is

$$v_{n2} = v_n - \frac{\partial v_n}{\partial s} \frac{ds}{2}, \quad v_{n1} = v_n + \frac{\partial v_n}{\partial s} \frac{ds}{2} \tag{8.71}$$

Take the limit and get

$$\gamma(s) = u_{s1} - u_{s2} \tag{8.72}$$

This shows that the surface vortex line is the discontinuity of tangential velocity, and the sudden jump of local tangential velocity passing through the surface vortex is equal to the local surface vortex strength. For the vortex distribution function on a straight plane, there are

$$\gamma = u(x, 0) - u(x, -0)$$
$$u(x, 0) = -u(x, -0)$$
$$u(x, 0) = -u(x, -0) = \gamma/2 \tag{8.73}$$

(4) Surface source method and surface vortex method

1) When solving the flow around a body without lift, including the flow around a non-lift airfoil considering the influence of thickness, the surface source method can be used.
2) If the flow around an airfoil with lift (simulating the influence of camber and angle of attack), the surface vortex method can be used. In addition to satisfying that the airfoil surface is a streamline, the trailing edge of the airfoil is required to meet the Kutta–Joukowski trailing-edge conditions, $\gamma = 0$.

8.8 Theory of Thin Airfoil

For the airfoil flow of ideal steady incompressible fluid, if the angle of attack, airfoil thickness and airfoil curvature are very small, the flow field around the airfoil is a small disturbed potential flow field. At this time, the boundary conditions and pressure coefficient on the wing surface can be linearized, and the effects of thickness, camber and angle of attack can be considered separately. This method is called thin wing theory.

8.8.1 Decomposition of Flow Around Thin Airfoils

(1) Disturbance velocity potential function

Take the xoy coordinate system as shown in Fig. 8.67. Φ represents the velocity potential function around the airfoil, satisfies the two-dimensional Laplace equation, and the superposition principle of potential function is established. Let Φ be composed of a straight uniform flow potential function and a perturbation velocity potential function caused by the presence of an airfoil.

$$\Phi = \varphi_\infty + \varphi \tag{8.74}$$

The perturbation velocity potential function also satisfies the Laplace equation.

$$\frac{\partial^2 \Phi}{\partial x^2} + \frac{\partial^2 \Phi}{\partial y^2} = 0 \quad \frac{\partial^2 (\varphi_\infty + \varphi)}{\partial x^2} + \frac{\partial^2 (\varphi_\infty + \varphi)}{\partial y^2} = 0$$

$$\frac{\partial^2 \varphi_\infty}{\partial x^2} + \frac{\partial^2 \varphi_\infty}{\partial y^2} = 0 \quad \frac{\partial^2 \varphi}{\partial x^2} + \frac{\partial^2 \varphi}{\partial y^2} = 0 \tag{8.75}$$

The perturbation velocity potential function also satisfies the superposition principle.

(2) Linear expression of boundary conditions of wing

Assuming that the disturbance velocities on the airfoil are, respectively, u'_w, v'_w thus the velocity component is shown in Eq. (8.76) at a small angle of attack.

$$u_w = V_\infty \cos\alpha + u'_w \approx V_\infty + u'_w$$
$$v_w = V_\infty \sin\alpha + v'_w \approx V_\infty\alpha + v'_w \tag{8.76}$$

The boundary condition of airfoil streamline is

Fig. 8.67 Flow around thin airfoil

$$\frac{dy_w}{dx} = \frac{v_w}{u_w} = \frac{V_\infty \alpha + v'_w}{V_\infty + u'_w}, \quad v'_w = V_\infty \frac{dy_w}{dx} + u'_w \frac{dy_w}{dx} - V_\infty \alpha \tag{8.77}$$

For thin airfoil, the thickness and camber of airfoil are very small, and the first-order small quantity is retained

$$v'_w = V_\infty \frac{dy_w}{dx} - V_\infty \alpha \tag{8.78}$$

Because the structure of the airfoil is

$$y_w|_{ul} = y_f \pm y_c \tag{8.79}$$

where y_f is the radian of the airfoil and y_c is the thickness of the airfoil. Then there is

$$v'_w|_{ul} = V_\infty \frac{dy_f}{dx} \pm V_\infty \frac{dy_c}{dx} - V_\infty \alpha \tag{8.80}$$

The above Equation shows that under small disturbance, the velocity in y direction on the airfoil can be approximately expressed as the linear sum of the contributions of camber, thickness and angle of attack.

(3) Perturbation velocity potential function decomposition

According to the perturbation velocity potential function equation and the linearization of wing velocity in y direction, the perturbation velocity potential function can be decomposed into the sum of velocity potential functions of camber, thickness and angle of attack.

$$\varphi = \varphi_f + \varphi_c + \varphi_\alpha \tag{8.81}$$

where, φ_f represents the velocity potential function caused by camber, φ_c represents the velocity potential function caused by thickness, and φ_α represents the velocity potential function caused by angle of attack. The partial derivative of y direction is obtained

$$v'_w = \frac{\partial \varphi_w}{\partial y} = \left(\frac{\partial \varphi_f}{\partial y}\right)_w + \left(\frac{\partial \varphi_c}{\partial y}\right)_w + \left(\frac{\partial \varphi_\alpha}{\partial y}\right)_w$$

$$v'_w = v'_{wf} + v'_{wc} + v'_{w\alpha}$$

$$= V_\infty \frac{dy_f}{dx} \pm V_\infty \frac{dy_c}{dx} - V_\infty \alpha \tag{8.82}$$

It can be seen that the disturbance velocity potential function and boundary conditions can be decomposed into the sum of disturbance velocity potential function when curvature, thickness and angle of attack exist alone.

(4) Linear expression of pressure coefficient C_p

For ideal incompressible potential flow, according to Bernoulli equation, the pressure coefficient is

$$C_p = \frac{p - p_\infty}{\frac{1}{2}\rho V_\infty^2} = 1 - \left(\frac{V}{V_\infty}\right)^2 \tag{8.83}$$

By substituting the disturbed velocity field, it can be gotten

$$C_p = 1 - \frac{(V_\infty \cos\alpha + u\prime)^2 + (V_\infty \sin\alpha + v\prime)^2}{V_\infty^2} \tag{8.84}$$

Under the assumption that the camber, thickness and angle of attack are small, if only the first-order small quantity is retained,

$$C_p = -\frac{2u\prime}{V_\infty}, \quad \begin{aligned} u\prime &= \frac{\partial\varphi}{\partial x} = \frac{\partial\varphi_f}{\partial x} + \frac{\partial\varphi_c}{\partial x} + \frac{\partial\varphi_\alpha}{\partial x} \\ &= u\prime_f + u\prime_c + u\prime_\alpha \\ C_p &= -2\frac{u\prime_f + u\prime_c + u\prime_\alpha}{V_\infty} = C_{pf} + C_{pc} + C_{p\alpha} \\ C_{pw} &= C_{pfw} + C_{pcw} + C_{p\alpha w} \end{aligned} \tag{8.85}$$

It can be seen that under small disturbance, the disturbance velocity potential equation, surface boundary conditions and wing pressure coefficient can be linearly superimposed.

8.8.2 Potential Flow Decomposition of Thin Airfoil at Small Angle of Attack

At a small angle of attack, for the incompressible flow around a thin airfoil, the disturbed velocity potential, surface boundary conditions and pressure coefficient can be linearly superimposed. The lift and torque acting on the thin airfoil can be regarded as the sum of the results of curvature, thickness and angle of attack. Therefore, the flow around a thin airfoil can be superimposed by three simple flows. That is, the flow around thin airfoil = camber problem (the flow around the middle arc bending plate at zero angle of attack) + thickness problem (the flow around the thickness distribution y_t symmetrical airfoil at zero angle of attack) + angle of attack problem (the flow around the flat plate with non-zero angle of attack), as shown in Fig. 8.68.

For the thickness problem, because the airfoil is symmetrical and the pressure distribution on the airfoil is symmetrical up and down, there is no lift and torque. The flow caused by camber and angle of attack is asymmetric up and down, and the resulting pressure difference obtains lift and torque. The combination of camber and

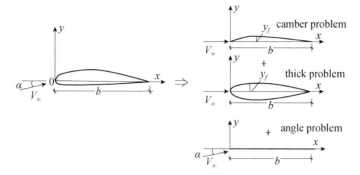

Fig. 8.68 Breakdown of flow around thin airfoil

angle of attack is called the problem of angle of attack and camber. Therefore, for the flow around thin airfoil at small angle of attack, the lift and torque can be determined by the flow around arc bend at small angle of attack.

8.8.3 Problem of Angle of Attack and Camber

The problem of angle of attack and camber is equivalent to solving the potential flow problem of straight uniform flow bypassing the middle arc bending plate with angle of attack. It needs to be solved by the surface vortex method. As shown in Fig. 8.69, the key point is to determine the distribution function $\gamma(s)$ of the surface vortex intensity arranged on the middle arcs. The streamline condition and Kutta–Joukowski trailing-edge condition are required to be satisfied.

$$v'_w = V_\infty \left(\frac{dy_f}{dx} - \alpha \right) \tag{8.86}$$

(1) Integral equation of surface vortex intensity function

Because the curvature of airfoil is very small and the gap between the middle arc and chord line is not large, the vortex intensity function $\gamma(\xi)$ of the plane is arranged approximately on the string line instead of $\gamma(s)$. According to Taylor series expansion,

$$v'_w = v'_w(x, y_f) = v'(x, 0) + \frac{\partial v'}{\partial y} y_f + \cdots \tag{8.87}$$

Omit the small amount and get

$$v'(x, y_f) = v'(x, 0) \tag{8.88}$$

Fig. 8.69 Problem simplification of angle of attack and camber

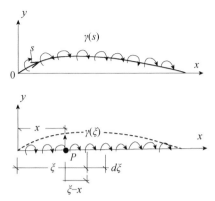

Under the first-order approximation, the problem of solving the lift and moment of thin airfoil can be summarized as the distribution function of surface vortex strength along the chord satisfying the following conditions $\gamma(\xi)$.

1) Infinite boundary (perturbation velocity potential) condition

$$u'_\infty = 0, v'_\infty = 0 \tag{8.89}$$

2) Surface boundary condition

$$v'_w(x, 0) = V_\infty \left(\frac{\mathrm{d}y_f}{\mathrm{d}x} - \alpha \right) \tag{8.90}$$

3) Kutta–Joukowski trailing-edge condition

$$\gamma(b) = 0 \tag{8.91}$$

On the chord, the surface vortex intensity at a point is $\gamma(\xi)$ (clockwise is positive), the vortex intensity on the segment $\mathrm{d}\xi$ is $\gamma(\xi)\mathrm{d}\xi$, and the induced velocity at point x on the chord is

$$\mathrm{d}v'(x, 0) = \frac{\gamma(\xi)\mathrm{d}\xi}{2\pi(\xi - x)} \tag{8.92}$$

The induced velocity of the whole vortex surface is

$$v'(x, 0) = \int_0^b \frac{\gamma(\xi)\mathrm{d}\xi}{2\pi(\xi - x)} \tag{8.93}$$

Substitute into Eq. (8.90) to obtain

$$\int\limits_0^b \frac{\gamma(\xi)d\xi}{2\pi(\xi - x)} = V_\infty\left(\frac{dy_f}{dx} - \alpha\right) \tag{8.94}$$

This is an integral equation of the unknown function of the surface vortex intensity arranged on the chord.

(2) Trigonometric series solution of surface vortex intensity function
Do variable replacement and make

$$\xi = \frac{b}{2}(1 - \cos\theta), x = \frac{b}{2}(1 - \cos\theta_1)$$

$$d\xi = \frac{b}{2}\sin\theta d\theta, \xi = 0, \theta = 0, \xi = b, \theta = \pi \tag{8.95}$$

$$-\int\limits_0^\pi \frac{\gamma(\theta)\sin\theta d\theta}{2\pi(\cos\theta - \cos\theta_1)} = V_\infty\left(\frac{dy_f}{dx} - \alpha\right) \tag{8.96}$$

Then make

$$\gamma(\theta) = 2V_\infty\left(A_0 ctg\frac{\theta}{2} + \sum_{n=1}^\infty A_n \sin n\theta\right) \tag{8.97}$$

In this series, there are two points to be explained: (1) the first term on the right is necessary for the ideal fluid to bypass the infinite negative pressure (i.e., infinite velocity) at the leading edge of the thin airfoil; (2) at the trailing edge, this series is equal to zero, so the series of sinusoidal function is used. The reduction of the surface vortex intensity to zero at the trailing edge is determined by the Kutta–Joukowski trailing-edge condition. Substitute Eq. (8.97) into Eq. (8.96) to obtain

$$-\frac{1}{2\pi}\int\limits_0^\pi \frac{\gamma(\theta)\sin\theta d\theta}{\cos\theta - \cos\theta_1}$$

$$= -\frac{1}{2\pi}\int\limits_0^\pi \frac{2V_\infty\left[A_0 ctg\frac{\theta}{2} + \sum_{n=1}^\infty A_n \sin n\theta\right]\sin\theta d\theta}{\cos\theta - \cos\theta_1}$$

$$= -\frac{V_\infty}{\pi}\int\limits_0^\pi \frac{\left[A_0(1 + \cos\theta) + \sum_{n=1}^\infty A_n \sin\theta \sin n\theta\right]d\theta}{\cos\theta - \cos\theta_1}$$

$$= -\frac{V_\infty}{\pi} \int_0^\pi \frac{\left\{ A_0(1 + \cos\theta) + \frac{1}{2}\sum_{n=1}^{\infty} A_n[\cos(n-1)\theta - \cos(n+1)\theta] \right\} d\theta}{\cos\theta - \cos\theta_1}$$

(8.98)

Using generalized integral formula

$$I_n = \int_0^\pi \frac{\cos n\theta\, d\theta}{\cos\theta - \cos\theta_1} = \pi \frac{\sin n\theta_1}{\sin\theta_1} (n = 0, 1, 2, \ldots)$$

(8.99)

Substitute Eq. (8.98) into Eq. (8.99) to obtain

$$\alpha - A_0 + \sum_{n=1}^{\infty} A_n \cos n\theta_1 = \frac{\mathrm{d}y_f}{\mathrm{d}x}$$

(8.100)

Integrate both sides of Eq. (8.100) along the direction of θ_1, and the integral limit is from 0 to π.

$$\int_0^\pi \left(\alpha - A_0 + \sum_{n=1}^{\infty} A_n \cos n\theta_1 \right) d\theta_1 = \int_0^\pi \frac{\mathrm{d}y_f}{\mathrm{d}x} d\theta_1$$

$$A_0 = \alpha - \frac{1}{\pi} \int_0^\pi \frac{\mathrm{d}y_f}{\mathrm{d}x} d\theta_1$$

(8.101)

Then multiply both sides of Eq. (8.100) by $\cos(n\theta_1)$ $(n = 1, 2, \ldots)$, the integral is obtained

$$\int_0^\pi \left(\alpha - A_0 + \sum_{n=1}^{\infty} A_n \cos n\theta_1 \right) \cos m\theta_1 d\theta_1 = \int_0^\pi \frac{\mathrm{d}y_f}{\mathrm{d}x} \cos m\theta_1 d\theta_1$$

$$\int_0^\pi A_m \cos^2(m\theta_1) d\theta_1 = \int_0^\pi \frac{\mathrm{d}y_f}{\mathrm{d}x} \cos m\theta_1 d\theta_1$$

$$A_m = \frac{2}{\pi} \int_0^\pi \frac{\mathrm{d}y_f}{\mathrm{d}x} \cos m\theta_1 d\theta_1 (m = 1, 2, 3, \ldots)$$

(8.102)

For the flow around a thin airfoil with a given camber and angle of attack, $A_0, A_1, A_2, \ldots$ can be determined by Eqs. (8.101) and (8.102).

(3) Aerodynamic characteristics of flow around thin airfoils

Using the known surface vortex intensity function, the forces and moments acting on thin airfoils can be easily obtained. The pressure coefficient on the airfoil surface is

$$C_p = -2\frac{u'(x, \pm 0)}{V_\infty} = \mp\frac{\gamma(x)}{V_\infty} \tag{8.103}$$

$$C_p = -2\frac{u'(x, \pm 0)}{V_\infty} = \mp 2\left(A_0 ctg\frac{\theta}{2} + \sum_{n=1}^{\infty} A_n \sin n\theta\right) \tag{8.104}$$

The velocity circulation around the airfoil is

$$\Gamma = \int_0^b \gamma(x)dx = bV_\infty \int_0^\pi \left(A_0 ctg\frac{\theta}{2} + \sum_{n=1}^{\infty} A_n \sin n\theta\right) \sin\theta d\theta$$

$$= \pi bV_\infty\left(A_0 + \frac{A_1}{2}\right) \tag{8.105}$$

The lift of the airfoil is

$$L = \rho\pi bV_\infty^2\left(A_0 + \frac{A_1}{2}\right) \tag{8.106}$$

1) The lift coefficient is

$$C_L = \frac{L}{\frac{1}{2}\rho V_\infty^2 b} = \frac{\rho\pi bV_\infty^2\left(A_0 + \frac{A_1}{2}\right)}{\frac{1}{2}\rho V_\infty^2 b} = 2\pi\left(A_0 + \frac{A_1}{2}\right) \tag{8.107}$$

Substitute A_0 and A_1 into the above Equation to obtain

$$C_L = 2\pi\left(\alpha + \frac{1}{\pi}\right)\int_0^\pi \frac{dy_f}{dx}(\cos\theta_1 - 1)d\theta_1 \tag{8.108}$$

2) The slope of the lift line is

$$\frac{dC_L}{d\alpha} = 2\pi \tag{8.109}$$

The above equation shows that for thin airfoils, the slope of the lift line is independent of the shape of the airfoil. If the general expression is

$$C_L = \frac{dC_L}{d\alpha}(\alpha - \alpha_0) = 2\pi(\alpha - \alpha_0) \tag{8.110}$$

where α_0 is the zero-lift angle of attack of the airfoil, which is determined by the mid arc shape of the airfoil (camber effect). For symmetrical airfoils, $\alpha_0 = 0$; Asymmetric airfoil $\alpha_0 <$ or > 0.

$$\alpha_0 = \frac{1}{\pi} \int_0^\pi \frac{dy_f}{dx}(1 - \cos\theta_1)d\theta_1 \tag{8.111}$$

Take the leading edge as reference point, and the pitch moment is

$$M_z = -\int_0^b x\,dL = -\int_0^b \rho V_\infty \gamma x\,dx$$

$$= -\frac{\rho}{2}V_\infty^2 b^2 \int_0^\pi \left[A_0\left(1 - \cos^2\theta_1\right) \right.$$

$$\left. + \sum_1^\infty A_n \sin n\theta_1 \sin\theta_1(1 - \cos\theta_1) \right]d\theta_1$$

$$= -\frac{\pi}{4}\rho V_\infty^2 b^2\left(A_0 + A_1 - \frac{A_2}{2}\right) \tag{8.112}$$

3) The leading-edge moment coefficient is

$$C_m = \frac{M_z}{\frac{1}{2}\rho V_\infty^2 b^2} = -\frac{\pi}{2}\left(A_0 + A_1 - \frac{A_2}{2}\right)$$

$$= -\frac{\pi}{2}\left[\left(A_0 + \frac{A_1}{2}\right) + \frac{1}{2}(A_1 - A_2)\right]$$

$$= \frac{\pi}{4}(A_2 - A_1) - \frac{C_L}{4} = C_{m0} - \frac{C_L}{4} \tag{8.113}$$

where c_{m0} is the zero-lift moment coefficient. The expression is

$$C_{m0} = \frac{\pi}{4}(A_2 - A_1) = \frac{1}{2}\int_0^\pi \frac{dy_f}{dx}(\cos 2\theta_1 - \cos\theta_1)d\theta_1 \tag{8.114}$$

Take the point of $b/4$ as the reference point to obtain

$$M_{\frac{1}{4}} = M_z + \frac{b}{4}L = -\frac{\pi}{4}\rho V_\infty^2 b^2\left(A_0 + A_1 - \frac{A_2}{2}\right)$$

$$+ \pi\rho V_\infty^2 b\left(A_0 + \frac{1}{2}A_1\right)\frac{b}{4}$$

$$= \frac{\pi}{4}\rho V_\infty^2 b^2 \left(-A_0 - A_1 + \frac{A_2}{2} + A_0 + \frac{1}{2}A_1 \right)$$

$$= \frac{\pi}{4}\rho V_\infty^2 b^2 \frac{A_2 - A_1}{2} \tag{8.115}$$

$$C_{m\frac{1}{4}} = \frac{M_{1/4}}{\frac{1}{2}\rho V_\infty^2 b^2} = \frac{\pi}{4}(A_2 - A_1) = C_m + \frac{C_L}{4} = C_{m0} = C_{ma} \tag{8.116}$$

where, if there is no angle of attack, it means that this moment is constant (does not change with the angle of attack). Even if the lift is zero, there is still this moment, which can be called the residual moment. For the flow around a thin airfoil, as long as the reference point is taken as the quarter chord point, the moment is equal to the zero-lift moment, which shows that the quarter chord point is the position of the aerodynamic center. Because the zero-lift moment does not change with the angle of attack, the aerodynamic center also represents the action point of the lift increment. In addition, there is a special point, called the pressure center, which represents the point where the aerodynamic force acts, and the torque at this point is zero. The distance between the aerodynamic center and the pressure center is assumed Δ_{xp}, as shown in Fig. 8.70.

$$0 = x_p L + M_z$$
$$M_{za} = Lx_a + M_z$$
$$0 = M_{za} + L\Delta x_p \tag{8.117}$$

$$\frac{x_p}{b} = -\frac{M_z}{L} = \frac{1}{4} - \frac{C_{m0}}{C_L}$$
$$\frac{x_a}{b} = \frac{M_{za}}{L} - \frac{M_z}{L} = \frac{C_{m0}}{C_L} + \frac{1}{4} - \frac{C_{m0}}{C_L} = \frac{1}{4}$$
$$\Delta x_p = -\frac{M_{za}}{L} = -\frac{C_{m0}}{C_L} \tag{8.118}$$

It can be seen from Eq. (8.118) that the distance between the pressure center and the aerodynamic center depends on the ratio of zero-lift moment coefficient to lift coefficient, and the distance decreases with the increase of lift coefficient, as shown in Fig. 8.70.

When $C_{m0} < 0$ and $\Delta x_p > 0$, the pressure center is behind the aerodynamic center for positive curved airfoils;

When $C_{m0} > 0$ and $\Delta x_p < 0$, the pressure center is located before the aerodynamic center for negative curved airfoils (upward bending);

When $C_{m0} = 0$ and $\Delta x_p = 0$, the aerodynamic pressure coincides with the pressure center for symmetrical airfoils;

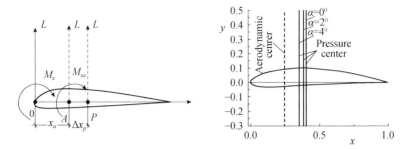

Fig. 8.70 Distance between pressure center and aerodynamic center

(3) Leading-edge suction and suction coefficient

    1) Leading-edge suction and suction coefficient of flow around a flat plate

        Firstly, taking the flow around a flat plate as an example, the physical concepts of leading-edge suction and leading-edge suction coefficient of ideal fluid bypassing a thin airfoil are explained. As shown in Fig. 8.71, for incoming flow V_∞, angle of attack α, the aerodynamic force on the plate is calculated. According to the thin wing theory, since the mid arc of the flat plate is zero, i.e., $\frac{dy_f}{dx} = 0$. Substituting into Eqs. (8.101) and (8.102), the surface vortex intensity distribution function of the flow around the flat plate is obtained by substituting into Eq. (8.97)

$$\gamma(\theta) = 2V_\infty \alpha ctg\frac{\theta}{2} = 2V_\infty \alpha \sqrt{\frac{b-x}{x}} \tag{8.119}$$

        The distribution curve of the surface vortex intensity distribution function along the x direction is shown in Fig. 8.71. At the leading-edge point $x = 0$, $\gamma(0)$ tends to infinity large; At the trailing edge of $x = b$, $\gamma(b) = 0$, the trailing-edge condition is met. From Eq. (8.108), the lift coefficient is

$$C_L = 2\pi\alpha \tag{8.120}$$

Fig. 8.71 Flow around an
ideal fluid plate at small
angle of attack

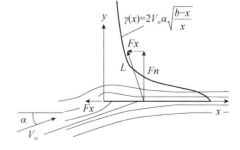

From Eq. (8.113), the moment coefficient is

$$C_m = -\frac{C_L}{4} = -\frac{\pi}{2}\alpha, \quad C_{ma} = 0 \tag{8.121}$$

Since the pressure acting on the plate is perpendicular to the plate surface, the resultant force obtained from the pressure integration on the plate surface must be perpendicular to the plate surface as F_n, and the projection of the force in the flow direction is $F_n \sin \alpha$, indicating that there is drag, which is not accord with the conclusion that there is no drag around the ideal fluid. The problem is that the above analysis does not include the adsorption effect of flow around the leading edge, because for a plate with small thickness, the leading-edge radius tends to zero. When the air flow bypasses the leading-edge head from below the leading edge of the plate, the leading-edge velocity tends to infinity due to the action of centrifugal force. According to Bernoulli equation, the pressure at the leading-edge tends to negative infinity and the action surface tends to zero. The product of infinitely large force and infinitely small area is a finite value, which is the suction F_x at the leading edge, and the direction is along the plate surface. The resultant force of suction F_x and normal force F_n is the lift L acting on the plate, and the drag $D = 0$.

$$L = \sqrt{F_x^2 + F_n^2} \tag{8.122}$$

As shown in Fig. 8.71, it can be seen that

$$F_n = L \cos \alpha \approx L$$
$$F_x = F_n tg\alpha = L\alpha = \frac{1}{2}\rho V_\infty^2 b(2\pi\alpha^2) \tag{8.123}$$

The leading-edge suction coefficient of the plate is

$$C_{Fx} = \frac{F_x}{\frac{1}{2}\rho V_\infty^2 b} = 2\pi\alpha^2 \tag{8.124}$$

2) Leading-edge suction and suction coefficient of flow around curved airfoil
    As shown in Fig. 8.72, take a micro segment ds at any bending line, and the pressure difference force on the upper and lower parts is

$$dF_n = (p_d - p_u)ds \tag{8.125}$$

According to the geometric relationship given in Fig. 8.72, and considering the small amount of α, θ so it is approximate that

Fig. 8.72 Flow of ideal fluid
around bending plate

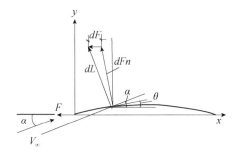

$$ds = \sqrt{dx^2 + dy^2} \approx dx$$

$$dF_n = dL \cos(\alpha - \theta) \approx dL$$

$$dF = dL \sin \alpha - dF_n \sin \theta \approx dF_n \left(\alpha - \frac{dy_f}{dx} \right) \tag{8.126}$$

Substituting to Eq. (8.125), it can be gotten that

$$F = \int_0^b dF = \int_0^b (p_d - p_u)\left(\alpha - \frac{dy_f}{dx} \right) dx \tag{8.127}$$

According to Kutta–Joukowski rule, there is

$$dL = \rho V_\infty \gamma(x) dx = (p_d - p_u) dx \tag{8.128}$$

From the boundary streamline Eq. (8.90), there is

$$\alpha - \frac{dy_f}{dx} = -\frac{v'_w(x, 0)}{V_\infty} \tag{8.129}$$

Substituting Eqs. (8.128) and (8.129) into Eq. (8.127), it can be gotten
that

$$F = \int_0^b dF = -\int_0^b \rho \gamma(x) v'_w(x, 0) dx \tag{8.130}$$

The leading-edge suction coefficient is

$$C_F = \frac{F}{\frac{1}{2}\rho V_\infty^2 b} = -\frac{2}{V_\infty^2 b} \int_0^b \gamma(x) v'_w(x, 0) dx$$

$$= \frac{2}{V_\infty b} \int_0^b \gamma(x) \left(\alpha - \frac{dy_f}{dx} \right) dx \qquad (8.131)$$

A curved plate airfoil is set, and the plate surface function is parabola, i.e.,

$$y_f = 4f \frac{x}{b} \left(1 - \frac{x}{b} \right) \qquad (8.132)$$

Due to

$$\frac{dy_f}{dx} = 4\frac{f}{b} \left(1 - 2\frac{x}{b} \right) \qquad (8.133)$$

$$A_0 = \alpha - \frac{1}{\pi} \int_0^\pi 4\frac{f}{b} [1 - (1 - \cos\theta_1)] d\theta_1 = \alpha \qquad (8.134)$$

$$A_1 = \frac{2}{\pi} \int_0^\pi 4\frac{f}{b} \left(1 - 2\frac{x}{b} \right) \cos\theta_1 d\theta_1$$

$$= \frac{2}{\pi} \int_0^\pi 4\frac{f}{b} [1 - (1 - \cos\theta_1)] \cos\theta_1 d\theta_1 = 4\frac{f}{b} \qquad (8.135)$$

$$A_n = \frac{2}{\pi} \int_0^\pi 4\frac{f}{b} (1 - 2\frac{x}{b}) \cos n\theta_1 d\theta_1$$

$$= \frac{2}{\pi} \int_0^\pi 4\frac{f}{b} [1 - (1 - \cos\theta_1)] \cos n\theta_1 d\theta_1 = 0, n = 2, 3, \ldots. \qquad (8.136)$$

Substituting into $\gamma(\theta)$,

$$\gamma(\theta) = 2V_\infty \alpha ctg\frac{\theta}{2} + 8V_\infty \frac{f}{b} \sin\theta$$

$$\gamma(x) = 2V_\infty \alpha \sqrt{\frac{b-x}{x}} + 16V_\infty \frac{f}{b} \sqrt{\frac{x}{b} \left(1 - \frac{x}{b} \right)} \qquad (8.137)$$

The leading-edge suction coefficient is

$$C_F = \frac{2}{V_\infty b} \int_0^b \gamma(x) \left(\alpha - \frac{dy_f}{dx} \right) dx$$

$$= \frac{2}{V_\infty b} \int_0^\pi \left[2V_\infty \alpha \frac{1 + \cos \theta}{\sin \theta} + 8V_\infty \frac{f}{b} \sin \theta \right]$$

$$\left(\alpha - 4\frac{f}{b} \cos \theta \right) \frac{b}{2} \sin \theta \mathrm{d}\theta = 2\pi \alpha^2 \qquad (8.138)$$

Zero-lift angle of attack is

$$\alpha_0 = \frac{1}{\pi} \int_0^\pi 4\frac{f}{b}\left(1 - 2\frac{x}{b} \right)(1 - \cos \theta_1)\mathrm{d}\theta_1 = -2\frac{f}{b} \qquad (8.139)$$

The zero-lift moment coefficient is

$$C_{m0} = \frac{\pi}{4}(A_2 - A_1) = \frac{1}{2} \int_0^\pi \frac{\mathrm{d}y_f}{\mathrm{d}x}(\cos 2\theta_1 - \cos \theta_1)\mathrm{d}\theta_1 = -\pi\frac{f}{b} \qquad (8.140)$$

The lift coefficient is

$$C_L = 2\pi(\alpha - \alpha_0) = 2\pi\left(\alpha + 2\frac{f}{b} \right) \qquad (8.141)$$

The distance between the pressure center and the leading-edge point is

$$\frac{x_p}{b} = -\frac{M_z}{L} = \frac{1}{4} - \frac{C_{m0}}{C_L}\frac{1}{4} = \frac{1}{4} - \frac{-\pi\frac{f}{b}}{\pi\left(\alpha + 2\frac{f}{b} \right)} = \frac{1}{4} + \frac{f}{(b\alpha + 2f)} \qquad (8.142)$$

The distance between pressure center and aerodynamic center is

$$\frac{\Delta x_p}{b} = \frac{f}{(b\alpha + 2f)} \qquad (8.143)$$

8.8.4 Solution of Thickness Problem

The flow around a symmetric airfoil with thickness distribution function y_c at zero angle of attack is called the thickness problem. For the thickness problem of airfoil flow, the layout area source method can be used to solve it. That is, the surface sources are continuously arranged on the airfoil surface. However, for thin airfoils, the source distribution on the chord line can be used to approximately replace the source distribution on the airfoil surface. The surface source is continuously arranged

Fig. 8.73 Distribution of area source function

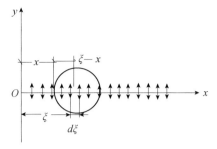

on the x-axis, and the strength is q (negative value is sink). q is determined according to the condition that the object surface is streamline, as shown in Fig. 8.73.

The boundary condition that the object surface is a streamline is

$$v'_w\big|_{ud} = v'(x, \pm 0) = \pm V_\infty \frac{dy_c}{dx} \tag{8.144}$$

and due to

$$v'(x, +0) = -v'(x, -0) = q/2 \tag{8.145}$$

$$q(\xi) = 2V_\infty \left(\frac{dy_c}{dx}\right)_{x=\xi} \tag{8.146}$$

Pressure on airfoil surface is

$$C_{pw} = -\frac{2u'_{wc}}{V_\infty} \tag{8.147}$$

$$u'_{wc} = u'(x, \pm 0) = \int_0^b \frac{q\,d\xi}{2\pi(x-\xi)} = \int_0^b \frac{V_\infty \left(\frac{dy_c}{dx}\right) d\xi}{\pi(x-\xi)}$$

$$C_{pw} = -2 \int_0^b \frac{\left(\frac{dy_c}{dx}\right) d\xi}{\pi(x-\xi)} \tag{8.148}$$

Example: a low-speed symmetrical thin airfoil with an angle of attack of zero is shown in Fig. 8.74, and the profile function is

$$y_c = 2c\frac{x}{b}\left(1 - \frac{x}{b}\right) \tag{8.149}$$

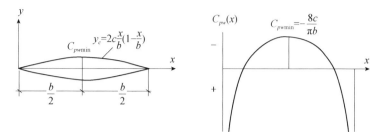

Fig. 8.74 Thickness problem of thin airfoil

where c is the maximum thickness of the airfoil. The pressure coefficient on the airfoil surface and the pressure coefficient at the maximum thickness point are solved by using the thickness problem.

Solution:

   Due to

$$\frac{dy_c}{dx} = 2\frac{c}{b}\left(1 - \frac{2x}{b}\right) \tag{8.150}$$

$$q(\xi) = 4\frac{V_\infty c}{b}\left(1 - \frac{2\xi}{b}\right) \tag{8.151}$$

Pressure on airfoil surface is

$$u'_{wc} = u'(x, \pm 0) = -\int_0^b \frac{q d\xi}{2\pi(\xi - x)} = \int_0^b \frac{V_\infty \frac{2c}{b}\left(1 - \frac{2x}{b}\right)d\xi}{\pi(x - \xi)}$$

$$= \frac{2V_\infty c}{b} \int_0^b \frac{\left(1 - \frac{2x}{b}\right)d\xi}{\pi(x - \xi)}$$

$$= \frac{2V_\infty c}{\pi b}\left[2 + \left(2\frac{x}{b} - 1\right)\ln\frac{b - x}{x}\right] \tag{8.152}$$

$$C_{pw} = -\frac{2u'_{wc}}{V_\infty} = -\frac{4c}{\pi b}\left[2 + \left(2\frac{x}{b} - 1\right)\ln\frac{b - x}{x}\right] \tag{8.153}$$

At $x = b/2$, the pressure coefficient is

$$C_{pw\,min} = -\frac{8c}{\pi b} \tag{8.154}$$

8.9 Theory of Thick Airfoil

The thin airfoil theory is only applicable to the flow around the thin airfoil at small angle of attack. If the relative thickness of the airfoil is more than 12%, or the angle of attack is large, the theoretical and experimental values of the thin airfoil are quite different, so the thick airfoil theory needs to be used.

8.9.1 Numerical Calculation Method of Flow Around Symmetrical Thick Airfoil Without Angle of Attack

The flow around an ideal incompressible two-dimensional symmetrical airfoil without lift is numerically simulated by the surface source method. It can also be solved by arranging plane dipoles on the axis of symmetry and superposition of incoming flow. Now consider the flow superimposed by the straight uniform flow and the dipole source arranged on the AB line segment on the x-axis, as shown in Fig. 8.75. It is assumed that the dipole strength is $\mu(x)$. At point P (x, y), the stream function is

$$\mathrm{d}\psi = -\frac{\mu(\xi)\mathrm{d}\xi\, y}{(x - \xi)^2 + y^2} \tag{8.155}$$

The superposition result of direct uniform current and dipole is

$$\psi = V_\infty y - \int_0^b \frac{\mu(\xi)\mathrm{d}\xi\, y}{(x - \xi)^2 + y^2} \tag{8.156}$$

If $y = 0$ is given as the object surface condition, the dipole distribution can be determined by the above equation. The stream function at any point outside the object plane is

Fig. 8.75 Dipole distribution on symmetry axis

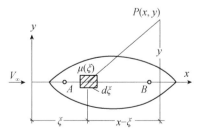

$$\psi_p = V_\infty y_p - \sum_{j=1}^{n} \frac{\mu(\xi_j)\Delta\xi y_p}{(x_p - \xi_j)^2 + y_p^2} \tag{8.157}$$

The velocity component on the object surface is

$$u_s = \frac{\partial\psi}{\partial y}, \quad v_s = -\frac{\partial\psi}{\partial x} \tag{8.158}$$

The surface pressure coefficient is

$$C_{ps} = 1 - \frac{u_s^2 + v_s^2}{V_\infty^2} \tag{8.159}$$

8.9.2   Numerical Calculation Method of Flow Around Arbitrary Thick Airfoil with Angle of Attack

For the flow around an airfoil with arbitrary shape, thickness and angle of attack, the surface vortex method can be used to calculate the pressure distribution, lift and moment characteristics on the airfoil. The idea of this method is that the wing surface is divided into n segments, the unknown vortex strength is arranged on each subsegment, the vortex strength of each point is respectively $\gamma_1, \gamma_2, \gamma_3, \ldots, \gamma_n$. The appropriate control points are taken on each vortex sheet, and the object surface boundary conditions are satisfied on these control points. The disturbance velocity potential function caused by the jth vortex sheet at the ith control point is

$$\mathrm{d}\varphi_{ij} = -\frac{\gamma_j \Delta s_j}{2\pi}\theta_{ij}$$
$$\theta_{ij} = tg^{-1}\frac{y_i - y_j}{x_i - x_j} \tag{8.160}$$

The total disturbance velocity potential function caused by all vortices on the wing to the i control point is

$$\varphi_i = -\frac{1}{2\pi}\sum_{j=1}^{n}\gamma_j \Delta s_j \theta_{ij} \tag{8.161}$$

At control point i, the velocity component of the normal disturbance is

$$v_{ni} = \frac{\partial\varphi_i}{\partial n_i} = -\frac{1}{2\pi}\sum_{j=1}^{n}\gamma_j \Delta s_j \frac{\partial\theta_{ij}}{\partial n_i} \tag{8.162}$$

Fig. 8.76 Surface vortex method for general thick airfoils

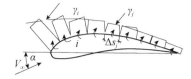

The boundary at point i is satisfied on the surface condition.

$$V_\infty \cos \beta_i - \frac{1}{2\pi} \sum_{j=1}^{n} \gamma_j \Delta s_j \frac{\partial \theta_{ij}}{\partial n_i} = 0 \tag{8.163}$$

Among them, β_i is the angle between the incoming flow and the external normal of the ith vortex plate. This equation is the key to the surface vortex method.

Applying the above equation to each control point, n linear equations can be established. In order to meet the trailing-edge conditions, it is necessary to make the control points of the first vortex on the lower wing and the last vortex on the upper wing as close to the trailing edge as possible.

$$\gamma_1(0) = \gamma_n(b) \tag{8.164}$$

After calculating the vortex strength at each point, the tangential velocity and pressure coefficient at the control point are (Fig. 8.76)

$$u_{si} = V_\infty \sin \beta_i + \frac{\gamma_i}{2} - \sum_{\substack{j=1 \\ j \neq i}}^{n} \frac{\gamma_j}{2\pi} \Delta s_j \frac{\partial \theta_{ij}}{\partial s_i}$$

$$C_{pi} = 1 - \left(\frac{u_{si}}{V_\infty}\right)^2 \tag{8.165}$$

8.10 Aerodynamic Characteristics of Practical Low-Speed Airfoils

8.10.1 Wing Pressure Distribution and Lift Characteristics

The pressure distribution on the airfoil surface not only determines the aerodynamic lift and moment but also the external load basis of aircraft structure design. Therefore, it has always been an important measurement subject of wind tunnel test. The lift and drag coefficients can be obtained by integrating the pressure distribution on the upper and lower airfoils.

$$C_l = \int_0^1 (C_{pl} - C_{pu}) \mathrm{d}x$$

$$m_z = -\int_0^1 x(C_{pl} - C_{pu}) \mathrm{d}x \tag{8.166}$$

The relationship between lift coefficient and angle of attack is the key curve to characterize the lift characteristics of airfoil. The experimental and calculation results show that the linear relationship between lift coefficient and angle of attack is

$$C_L = C_L^\alpha (\alpha - \alpha_0) \tag{8.167}$$

At the stall angle of attack, the lift coefficient reaches the maximum C_{Lmax}. Therefore, the four parameters to determine the lift characteristic curve are: lift line slope, zero lift angle of attack, maximum lift coefficient, and stall angle of attack. The details are as follows.

(1) The lift line slope is not related to Re number but mainly related to the shape of airfoil. The theoretical value for thin airfoil is 2π, and the theoretical value for thick airfoil is larger than 2π (the thickness is between 12 and 20%, which increases with the increase of thickness and trailing edge angle). Since the influence of viscosity is not included, the experimental value is less than the theoretical value. For the flat plate $C_L^\alpha = 0.9 \times 2\pi$, the lift line slope of NACA airfoil is close to the theoretical value. An empirical equation often used is

$$C_L^\alpha = 0.9 \times 2\pi \left(1 + 0.8 \frac{c}{b}\right) \tag{8.168}$$

(2) The zero-lift angle is mainly related to the camber of the airfoil. NACA four-digit airfoil is

$$\alpha_0 = -\frac{f}{b} \times 100 \tag{8.169}$$

(3) The maximum lift coefficient is mainly related to the boundary layer separation. It depends on the geometric parameters, Re number and surface finish of the airfoil. The commonly used low-speed airfoil is 1.3–1.7, which increases with the increase of Re number, which can be found in the airfoil data book.

8.10.2 Longitudinal Moment Characteristics of Airfoils

The longitudinal moment characteristic of airfoils is represented by C_m–C_L curve.

$$C_m = C_{m0} + C_m^{C_L} C_L \qquad (8.170)$$

For airfoils with positive camber, C_{m0} is a decimal less than zero and the slope of the torque curve is also negative. The thin wing theory can estimate these two values. C_{m0} is related to the camber of the airfoil, and the torque slope is -0.25.

8.10.3 Pressure Center Position and Focus (Aerodynamic Center) Position

The acting point of resultant force on the airfoil is the pressure center, and the chord position at low angle of attack is

$$\frac{x_p}{b} = -\frac{M_z}{L} = \frac{1}{4} - \frac{C_{m0}}{C_L} \qquad (8.171)$$

The smaller the angle of attack, the more backward the pressure center. The aerodynamic center position of the airfoil is the position where the moment coefficient remains unchanged, and the chord position is

$$\frac{x_p}{b} = \frac{x_a}{b} - \frac{C_{m0}}{C_L} \qquad (8.172)$$

The aerodynamic center reflects the action point of the lift increment caused by the variation of the airfoil with the angle of attack, located at the position before the pressure center.

8.10.4 Drag Characteristics and Polar Curve of Airfoil

Airfoil drag includes friction drag and pressure drag. The essence of airfoil drag is caused by air viscosity. The frictional drag is caused by the direct frictional shear stress on the surface, and the pressure drag is caused by the change of pressure distribution due to the existence of boundary layer on the surface. Generally, under the airfoil design of lift coefficient, the airfoil drag is the smallest, and its size can be obtained by properly modifying the plate friction drag coefficient. The minimum drag coefficient of the airfoil can be approximated as

$$C_{D\,min} = 0.925(2C_f)\eta_c$$

$$C_{D\,min} = 2C_f\left[1 + 2\frac{c}{b} + 6\left(\frac{c}{b}\right)^4\right] \tag{8.173}$$

where C_f is the plate drag coefficient with the same flow pattern as the airfoil. η_c is the thickness correction factor. The viscous pressure drag coefficient is

$$C_{D\,min} = kC_L^2 \tag{8.174}$$

where k is the viscous piezoresistive coefficient.

In aircraft design, the drag coefficient with zero-lift coefficient is often called zero-lift drag coefficient, $C_{D0} = 0.005 - 0.008$. Generally, when the lift is not zero, the drag coefficient is

$$C_D = C_{D0} + kC_L^2 \tag{8.175}$$

The lift-to-drag characteristic curve is called polar curve, as shown in Fig. 8.77, which was proposed by Otto Lilienthal (1834–1906), a German aviation pioneer. The lift-to-drag ratio of an airfoil represents the aerodynamic efficiency, which is defined as

$$K = \frac{C_L}{C_D} \tag{8.176}$$

For the airfoil with good performance, the maximum lift-to-drag ratio can reach more than 50–80.

Fig. 8.77 Lift-to-drag characteristic curve (polar curve)

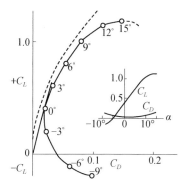

8.11 Exercises

A. Thinking questions

1. Please briefly describe the definition of airfoil thickness, camber and angle
    of attack and point to the thickness, camber and x value of the maximum
    camber of NACA4412 and NACA23012 airfoils.
2. The lift force affecting the airfoil is the incoming flow velocity V_∞ and
    density ρ_∞, chord length b, sound velocity a_∞, thickness C, curvature f,
    angle of attack α, sideslip angle β, and gravitational acceleration g. Please
    use the dimensional analysis method to give the dimensionless expression
    of the lift coefficient.
3. Explain the steady flow around a low-speed airfoil. What is the trailing-edge
    condition.
4. Explain the formation process of starting vortex and attached vortex.
5. Explain Kutta–Joukowski rule of lift circulation.
6. Try to point out the factors affecting the strength of attached vortex around
    steady airfoil and the reasons.
7. Try to establish the relationship between the strength of the attached vortex
    and the lift coefficient.
8. What is potential flow superposition for steady airfoil flow?
9. Describe the basic characteristics of surface source and surface vortex?
10. How to deal with the problems of thickness, camber, and angle of attack in
    the steady flow around a thin airfoil?
11. How to solve the lift problem of thin wing theory? Try to point out the
    variation trend of surface vortex intensity along the chord line.
12. Try to point the lift line slope, zero-lift angle of attack and zero-lift moment
    coefficient of flow around thin airfoil.
13. Try to point out the physical cause of leading-edge suction.
14. Briefly describe the solution idea of airfoil thickness problem.
15. What are the pressure center and pneumatic center and their relationship?
16. As the angle of attack increases, why does the moment around the
    aerodynamic center remain unchanged?
17. How does the lift coefficient of the airfoil change as the angle of attack
    increases? Why does the maximum lift coefficient of the airfoil appear?
18. Try to explain the physical parameters that the Re number affects the flow
    around the airfoil and the reasons.
19. For the steady flow around a flat plate airfoil without thickness at small
    angle of attack, try to prove that the concentrated vortex is arranged at 1/4
    chord and the boundary condition is satisfied at 3/4 chord.
20. What are the main parameters affecting the drag coefficient of airfoil steady
    flow?
21. Try to explain the physical causes of viscous pressure drag of steady airfoil
    flow.

22. Try to explain the main factors affecting the lift-to-drag ratio of an airfoil.
23. Try to point out the physical phenomenon of flow separation around airfoil. What are the trailing-edge separation and the leading-edge separation?
24. Briefly describe the effect of separation on airfoil lift and drag coefficient.

B. **Calculation questions**

1. NACA 2412 airfoil with chord length of 0.64 m in the air flow under standard sea-level conditions is considered. The free flow velocity is 70 m/s, and the lift per unit chord is 1254 N/m. Try to calculate the angle of attack and drag per unit chord.
2. NACA2415 airfoil is used as the flat wing of a low-speed aircraft. What is the $\overline{f}, \overline{x}_f$ and $\overline{c}$?
3. The vortices are distributed on the position of 1/4 chord of a plate at small angle of attack as shown in the fellow figure. Try to prove that if the boundary condition is satisfied at 3/4 chord point, then $C_l^\alpha = 2\pi \, \mathrm{rad}^{-1}$.

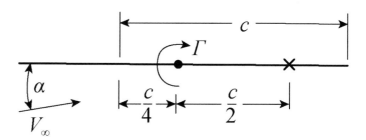

4. NACA 2412 airfoil with chord length of 0.64 m in the air flow under standard sea level conditions is considered. The free flow velocity is 70 m/s, and the lift per unit chord is 1254 N/m. Try to calculate the unit chord moment about the aerodynamic center.
5. Consider a thin flat plate with an angle of attack of 5°. Calculate: (a) the lift coefficient, (b) the moment coefficient about the leading edge, (c) the moment coefficient about the 1/4 chord point, and (d) the moment coefficient about the trailing edge.
6. For the flow around a flat plate airfoil at a small angle of attack, try to prove that $y(0)$ can have the following two forms of solutions.
   (1) (not met the trailing-edge condition)
   (2) (met the trailing-edge condition)
7. The arc equation of NACA2412 airfoil is

$$\begin{cases} y_f = \dfrac{1}{8}(0.80x - x^2), 0 \le x \le 0.4 \\ y_f = 0.0555(0.20 + 0.80x - x^2), 0.4 \le x \le 1.0 \end{cases}$$

As shown in the fellow figure. Try to calculate $C_l^\alpha, \alpha_0, \bar{x}_f$ and C_{m0} according to thin airfoil theory.

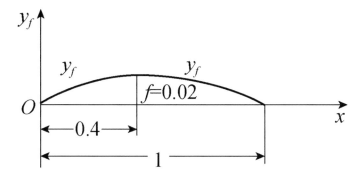

8. NACA 23012 is considered in the study at $\alpha = 4°$, while the $C_L = 0.55$ and $C_{m,c/4} = -0.005$. The zero-lift angle of attack is $-1.1°$. In addition, at $\alpha = -4°$, $C_{m,c/4} = -0.0125$. Based on the given information, calculate the aerodynamic center position of NACA 23012 airfoil.

9. Thin symmetrical airfoils with an angle of attack of $1.5°$ are considered. According to the calculation results of thin airfoil theory, try to calculate the lift coefficient and moment coefficient of the leading edge.

10. A curved airfoil with $c = 1$ and $y_f = kx(x-1)(x-2)$ where k is a constant and $\bar{f} = 2\%$. Try to calculate the C_L and C_m at $\alpha=3°$.

11. People often ask such a question: can the wing fly backwards? To answer this question, perform the following calculation. Consider airfoils with a zero-lift angle of $-3°$. The lift slope is 0.1 per degree.
    (a) Calculate the lift coefficient at an angle of attack of $5°$.
    (b) Now imagine the same wing upside down, but at the same $5°$ angle of attack as part (a). Calculate its lift coefficient.
    (c) At what angle of attack can the upside down airfoil produce the same lift as the original airfoil at an angle of attack of $5°$?

12. The front section of an airfoil is a flat plate, and the rear section is a flat plate flap with a downward deviation of $15°$, as shown in the fellow figure, try to calculate the C_L at angle of attack of $5°$.

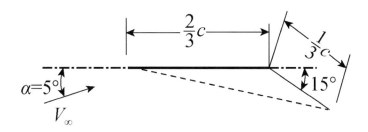

13. The airflow flows through a parabolic curved airfoil $\overline{f} \ll 1$. The vortices distributed on the bending plate are concentrated at $\overline{x} = 1/8$ and $\overline{x} = 5/8$ two points, and the vortex strength is Γ_1 and Γ_2 respectively, as shown in the fellow figure. The front $\overline{x} = 3/8$ and rear $\overline{x} = 7/8$ control points are taken to meet the wing boundary conditions. Try to prove the lift coefficient of the simplified model is $C_{l0} = 4\pi \overline{f}$.

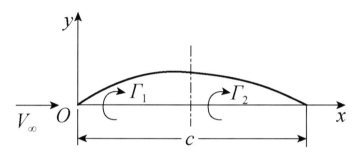

14. There is a flat elliptical airfoil with $c = 1, t \ll 1, y_t = \frac{t}{2}\sqrt{1 - (2x - 1)^2}$ as shown in the fellow figure. Try to apply the thickness problem of thin airfoil theory to calculate the lowest pressure strength coefficient $C_{\text{point}} = C_{p(x=1/2)}$ at the midpoint of chord.

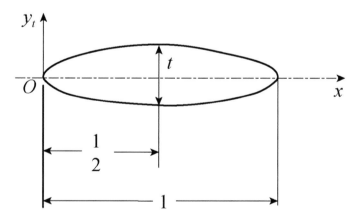

15. For NACA 2412 airfoil, the lift coefficient and moment coefficient of quarter chord at $-6°$ angle of attack are -0.39 and -0.045, respectively. At $4°$ angle of attack, these coefficients are 0.65 and -0.037, respectively. Try to calculate the position of the aerodynamic center.

16. There is a curved airfoil with $c = 1, y_f = 8.28\overline{f}\left(x^3 - \frac{15}{8}x^2 + \frac{7}{8}x\right)$ as shown in the fellow figure. Try to prove $C_{m0} = 0$ and $\alpha_0 = -2.07\overline{f}$ rad.

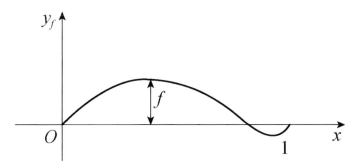

17. The low-speed air flow V_∞ flows through a thin symmetrical airfoil $\bar{y}_t = 4\left(\frac{t}{2}\right)\bar{x}(1 - \bar{x})$ at a small angle of attack as shown in the fellow figure. Try to solve the fellow questions using the method of angle of attack problem and thickness problem.
    (1) Functional relation expression of surface C_p and $\bar{x}$
    (2) $C_{p(\bar{x}=1/2)}$

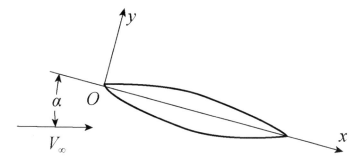

18. Consider NACA 23012 airfoil. The average arc of the airfoil is given by

$$\frac{z}{c} = 2.6595\left[\left(\frac{x}{c}\right)^3 - 0.6075\left(\frac{x}{c}\right)^2 + 0.1147\left(\frac{x}{c}\right)\right] \text{ for } 0 \le \frac{X}{c} \le 0.2025$$

$$\frac{z}{c} = 0.02208\left(1 - \frac{x}{c}\right) \text{ for } 0.2025 \le \frac{x}{c} \le 1.0$$

Calculate (a) lift at zero angle of attack, (b) lift coefficient at $\alpha = 4°$, (c) moment coefficient at quarter chord length, and (d) position of pressure center presented by x_{cp}/c.

Chapter 9
Aerodynamic Characteristics of Low Speed Wing Flow

This chapter introduces the basic concept and principle of flow around the low-speed wing. It includes wing geometric characteristics and parameters, aerodynamic coefficients of the wing, low-speed flow characteristics of high aspect ratio straight wing, vortex model of the low-speed wing, Prandtl's lifting-line theory, stall characteristics of high aspect ratio straight wing, low-speed aerodynamic characteristics of the swept-back wing, lifting-surface theory of wing, low-speed aerodynamic characteristics of low aspect ratio wing, engineering estimation of low-speed aerodynamic characteristics of wing, and aerodynamic characteristics of the control surface.

Learning points:

(1) Familiar with the basic concepts of low-speed wing flow, including flow characteristics, aerodynamic characteristics and coefficients, vortex model of low-speed wing flow, etc.;
(2) Master the derivation process and application of Prandtl's lifting-line theory for flow around a straight wing with a high aspect ratio;
(3) Familiar with stall characteristics of high aspect ratio straight wing, low-speed aerodynamic characteristics of the swept-back wing, lifting-surface theory of wing, and low-speed aerodynamic characteristics of low aspect ratio wing;
(4) Understand the engineering estimation of low-speed aerodynamic characteristics of the wing and the aerodynamic characteristics of the control surface.

9.1 Geometric Characteristics and Parameters of the Wing

9.1.1 Plane Shape of the Wing

The wing is the aerodynamic component of the aircraft to generate lift. In order to obtain a good aerodynamic shape, the wing is usually made into a 3D thin slender

© Science Press 2022
P. Liu, *Aerodynamics*, https://doi.org/10.1007/978-981-19-4586-1_9

structure, which is arranged on both sides of the fuselage (can be located above the fuselage, below the fuselage, and in the middle of the fuselage). There are various shapes of the wing, and the design can be optimized according to the different flight speeds and missions. The earliest wing shape is flat, such as the shape of the Chinese kite. The lift-to-drag ratio of the flat wing is the smallest, generally 2–3. Then there is the curved plate, whose lift-to-drag ratio can reach more than 5, and the lift-to-drag ratio generated by the later designed wing profile can reach more than 20. For example, the lift-to-drag ratio of the pure wing of a large airliner can reach about 30. Due to the influence of the wingtip, the lift-to-drag ratio of the 3D wing is smaller than that of the 2D airfoil. Because the fuselage mainly produces drag, the lift-to-drag ratio of the whole aircraft will be smaller if the fuselage and other resistance components are added. For example, the cruising lift-to-drag ratio of large passenger aircraft B747 is between 17 and 18, which is equivalent to lifting 1 kg of gravity only needs to overcome 55 g of resistance.

Although there are various wing shapes, they mainly include straight wing, delta wing, and swept-back wing (also swept-forward wing), as shown in Fig. 9.1. However, no matter what shape is adopted, the designer must make the aircraft have a good aerodynamic shape and make the structural weight as light as possible. The so-called good aerodynamic shape means that the wing has a high lift, low drag, and good handling stability. For the low-speed wing, in order to reduce induced drag, a straight wing with a high aspect ratio is often used. For high-subsonic transport aircraft and large passenger aircraft, in order to suppress and control shock waves, the swept-back supercritical wing is generally used. For supersonic fighters, in order to reduce wave drag, large swept-back delta wings are mostly used. For example, the U.S. tactical transport aircraft C-130 (propeller aircraft, as shown in Fig. 9.2) has a cruising speed of 540 km/h and a maximum take-off weight of 70.3 t, with an upper single rectangular straight wing. China's self-developed strategic transport aircraft Y-20 (as shown in Fig. 1.50) has a cruise speed of 800 km/h and a maximum take-off weight of 220 t. The wing adopts a supercritical wing with a high aspect ratio and a medium sweep. The large aircraft C919 (as shown in Fig. 1.48) developed by China is a single-aisle narrow-body 150-seat high-subsonic mainline passenger aircraft with a cruising speed of 850 km/h and a maximum take-off weight of 72.5 t. A supercritical wing with a high aspect ratio and medium sweep is adopted. China's self-developed fifth-generation fighter J20 (as shown in Fig. 1.44) has a maximum take-off weight of 37 t and a maximum flight speed of Mach 2.5, and the swept-back delta wing is used.

9.1.2 Characterization of the Wing Geometry

(1) Coordinate system

Take the body coordinate system as shown in Fig. 9.3. The x-axis is the longitudinal axis of the wing, along the chord line of the airfoil on the symmetry plane of the wing,

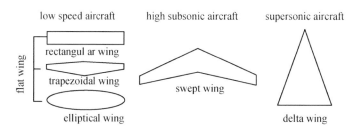

low speed aircraft   high subsonic aircraft   supersonic aircraft

flat wing

rectangul ar wing

trapezoidal wing

swept wing

elliptical wing

delta wing

Fig. 9.1 Plane shape of the wing

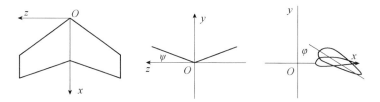

Fig. 9.2 American C-130 transport aircraft (cruising speed 540 km/h, propeller aircraft)

which is positive backward; the y-axis is the wing vertical axis, which is orthogonal to the x-axis in the wing symmetry plane, and is positive upward; and z-axis is the horizontal axis of the wing, which forms a right-handed coordinate system with the x- and y-axes, and it is positive to the left. The main geometric parameters for measuring the aerodynamic shape of the wing are expressed as follows:

(1) Wingspan length: Wingspan refers to the length between the left and right wingtips of the wing, denoted by l.
(2) Wing area: It refers to the projected area of the wing on the Oxz-plane, denoted by S.

Fig. 9.3 Wing coordinate

Fig. 9.4 Plane shape of the
trapezoid wing

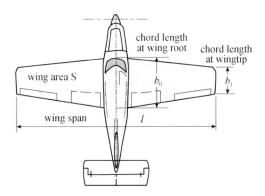

(3) Wing chord length: Wing chord length refers to the chord length of the wing parallel to the fuselage. In addition to the rectangular wing, the chord lengths of different parts of the wing are different, such as root chord length b_0 and tip chord length b_1, as shown in Fig. 9.4.

(4) The geometric mean chord length b_{pj} is defined as

$$b_{pj} = \frac{S}{l} \tag{9.1}$$

(5) Aspect ratio: The ratio of the wingspan length l to the average geometric chord length b_{pj} is called the aspect ratio, expressed by λ, and the calculation equation is

$$\lambda = \frac{l}{b_{pj}} \tag{9.2}$$

The aspect ratio can also be expressed as the ratio of the square of the wingspan length to the wing area, which is

$$\lambda = \frac{l^2}{S} \tag{9.3}$$

Studies have shown that with the increase of the aspect ratio, the lift of the wing increases, and the induced drag decreases, but the friction drag increases. High-speed aircraft generally use wings with a small aspect ratio.

(6) Taper ratio: Taper ratio is the ratio of the wing root chord length b_0 to the wingtip chord length b_1, expressed by η, which is

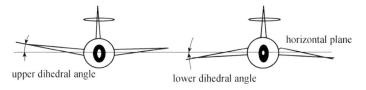

upper dihedral angle   lower dihedral angle

Fig. 9.5 Arrangements of the dihedral and cathedral angle

$$\eta = \frac{b_0}{b_1} \tag{9.4}$$

(7) Tip root ratio: It refers to the ratio of the wingtip chord length b_1 to the wing root chord length b_0, expressed by ξ, which is

$$\xi = \frac{b_1}{b_0} \tag{9.5}$$

(8) Dihedral angle: It refers to the angle between the chord plane of the wing and the xOz-plane (as shown in Fig. 9.5). When the wing is twisted, it refers to the angle between the twist axis and the xOz-plane. When the dihedral angle is negative, it becomes the cathedral angle. The low-speed wing adopts a certain dihedral angle to improve the lateral stability, generally $\psi = +7° - 3°$.

(9) Swept-back angle: It refers to the angle between the wing edge or specific line and the vertical line of the fuselage axis, as shown in Fig. 9.6. Different specific lines correspond to different swept-back angles including the following: the leading-edge swept-back angle represents the angle between the leading edge of the wing and the vertical line of the fuselage axis, represented by χ_0; the trailing edge swept-back angle represents the angle between the trailing edge of the wing and the vertical line of the fuselage axis, represented by χ_1; and the 1/4 chord line swept-back angle represents the angle between the 1/4 chord line of the wing and the vertical line of the fuselage axis, which is represented by $\chi_{0.25}$. If the wing is swept forward, the swept-back angle becomes negative and becomes a swept-forward angle.

(10) Geometric twist angle: The angle between the chord line of the wing profile parallel to the symmetry plane and the chord line of the wing root profile is

Fig. 9.6 Sweep angle of the wing

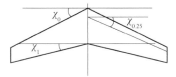

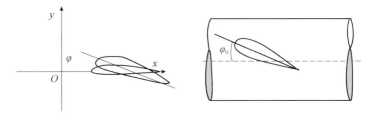

Fig. 9.7 Twist and installation angle of the wing

called the geometric twist angle of the wing φ; as shown in Fig. 9.7. If the local angle of attack of the wing profile is greater than the angle of attack of the wing root profile, the twist angle is positive. The twist in which the local angle of attack of the spanwise wing profile decreases from root to wingtip is called washout, and the twist angle is negative. On the contrary, it becomes washin. In addition to the geometric twist angle, there is also an aerodynamic twist angle, which refers to the angle between the zero-lift line of any wing profile parallel to the symmetry plane of the wing and the zero-lift line of the wing root profile.

(11)   Installation angle: The wing is installed on the fuselage. The angle between the wing root profile chord line and the fuselage axis is called the wing installation angle.

9.2 Aerodynamic Coefficient, Mean Aerodynamic Chord Length, and the Focus of the Wing

9.2.1 Aerodynamic Coefficient of the Wing

Take the wind axis coordinate Oxyz, as shown in Fig. 9.8, where the x-axis is backward along the incoming flow V_∞, and the y- and z-axes and the x-axis form a right-handed coordinate system. If the incoming flow V_∞ is parallel to the symmetry plane of the wing, it is called the longitudinal flow around the wing. The angle between V_∞ and the chord line of the wing profile (wing root profile) at the symmetry plane is defined as the angle of attack α of the wing. When flowing longitudinally, the aerodynamic forces acting on the wing include wing lift L (perpendicular to V_∞-direction), drag D (parallel to V_∞-direction), and longitudinal moment Mz (moment around the z-axis at a reference point).

The lift coefficient of the wing is

$$C_L = \frac{L}{\frac{1}{2}\rho V_\infty^2 S} \tag{9.6}$$

Fig. 9.8 Wind coordinate

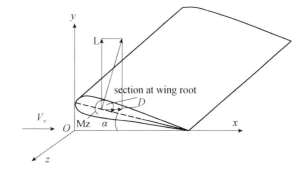

The drag coefficient of the wing is

$$C_D = \frac{D}{\frac{1}{2}\rho V_\infty^2 S} \tag{9.7}$$

The longitudinal moment coefficient of the wing is

$$C_m = \frac{M_z}{\frac{1}{2}\rho V_\infty^2 S b_A} \tag{9.8}$$

In the equation, b_A is the mean aerodynamic chord length of the wing.

9.2.2 Mean Aerodynamic Chord Length of the Wing

According to airfoil aerodynamics, the aerodynamic force acting on the airfoil can be expressed as the sum of lift, drag, and moment around the neutral point of the airfoil. The reference length of the moment is the chord length of the airfoil. Similarly, the longitudinal aerodynamic force acting on the wing can also be expressed as the sum of the lift and drag acting on the neutral point and the pitch moment around that point, but the reference length of the moment is the mean aerodynamic chord length b_A of the wing.

The mean aerodynamic chord length of the wing is the chord length of an imaginary rectangular wing. The area S of this imaginary wing is equal to the area of the actual wing, and its moment characteristic is also the same as the actual wing. The zero-lift pitch moment of the imaginary rectangular wing is

$$M'_{z0} = C_{m0}\frac{1}{2}\rho V_\infty^2 S b_A, \quad q_\infty = \frac{1}{2}\rho V_\infty^2 \tag{9.9}$$

In the above equation, C_{m0} is the zero-lift pitching moment coefficient of the imaginary wing, and also the zero-lift pitching moment coefficient of the actual

Fig. 9.9 Mean aerodynamic
chord length of the wing

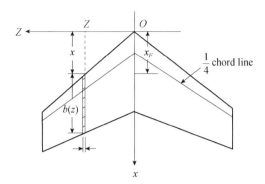

wing, and q_∞ is the incoming dynamic pressure. As shown in Fig. 9.9, the zero-lift
pitch moment of the actual wing element area $b(z)\,dz$ is

$$dM_{z0} = C'_{m0}q_\infty b(z)b(z)dz \tag{9.10}$$

In the equation, C'_{m0} is the zero-lift pitching moment coefficient of the airfoil.
Then the zero-lift pitch moment of the actual wing is

$$M_{z0} = 2q_\infty \int_0^{1/2} C'_{m0}b^2(z)dz \tag{9.11}$$

Assuming $C_{m0} = C'_{m0}$, then the above equation becomes

$$M_{z0} = 2q_\infty C'_{m0} \int_0^{1/2} b^2(z)dz \tag{9.12}$$

Assuming that the zero-lift pitching moment of the rectangular wing is equal to
the zero-lift pitching moment of the actual wing, then $M_{z0} = M'_{z0}$; from Eqs. (9.9)
and (9.12), it can be got that

$$b_A = \frac{2}{S} \int_0^{1/2} b^2(z)dz \tag{9.13}$$

For trapezoidal wings, the method shown in Fig. 9.10 can be used to determine
the mean aerodynamic chord length.

Fig. 9.10 Graphical method
to find the mean
aerodynamic chord length

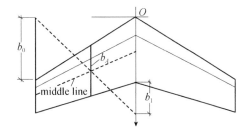

9.2.3 The Focus of the Wing

Because the wing is symmetrical, and the incoming flow is parallel to the symmetry
plane of the wing, the neutral point of the wing must be on the symmetry plane of the
wing (wing root profile). Suppose the wing neutral point is x_F from the wing apex,
as shown in Fig. 9.9, the moment of the total lift acting on the wing neutral point to
the oz-axis passing through the apex is

$$M_z = C_L q_\infty S x_F \tag{9.14}$$

Assuming that the neutral point of each profile of the wing is still at 1/4 of the
chord length of the profile like the airfoil, the lift acting on the element area $b(z)$
dz is $C'_L q_\infty b(z) \cdot dz$. The distance between the leading edge of the profile and the
oz-axis is x, and the distance between the neutral point of the profile oz is $x + \frac{1}{4}b(z)$;
therefore, the moment of the lift acting on the element to the neutral on the oz-axis is

$$C'_L q_\infty [x + \frac{1}{4}b(z)]b(z)dz \tag{9.15}$$

In the equation, $C'_L(z)$ is the lift coefficient of the local profile. Integrate the entire
wing and combine it with Eq. (9.14) to obtain

$$M_z = C_L q_\infty S x_F = 2 \int_0^{l/2} C'_L q_\infty \left(x + \frac{1}{4}b(z) \right) b(z)dz \tag{9.16}$$

Assuming $C_L \approx C'_L \approx$ constant, the neutral point position can be obtained as

$$x_F = \frac{2}{S} \int_0^{l/2} \left(x + \frac{1}{4}b(z) \right) b(z)dz \tag{9.17}$$

Substituting the b_A expression (9.13) into it, it can be got that

$$x_F = \frac{1}{4}b_A + \frac{2}{S}\int_0^{1/2} xb(z)\mathrm{d}z \tag{9.18}$$

Therefore, when the plane shape of the wing is given, the neutral point position x_F of the wing can be determined. Since the neutral point position of the profile was assumed to be at 1/4 of the chord length during the derivation process, this assumption is correct for straight wings with high aspect ratios, but there is a certain error for swept-back wings and wings with small aspect ratios. To more accurately determine the neutral point position of the swept-back wing, it is necessary to rely on experiments or numerical calculations.

The airfoil is equivalent to the wing whose span length tends to be infinite, that is, $\lambda = \infty$, and the flow is 2D. While the actual span length and the corresponding λ are both finite values, the flow must be 3D. This chapter mainly discusses the low-speed aerodynamic characteristics of high aspect ratio ($\lambda \geq 5$) straight wing $\left(\chi_{1/4} < 20°\right)$.

9.3   Low-Speed Aerodynamic Characteristics of Large Aspect Ratio Straight Wing

9.3.1   Flow State

In order to understand the flow around a straight wing with a high aspect ratio, a row of silk threads is evenly pasted along the span on the trailing edge of a straight wing with a high aspect ratio, and small cotton balls are tied at the end of the silk threads. Then place the wing in a low-speed wind tunnel. By changing the angle of attack of the wing and observing the flow phenomenon when the air flows around the wing, as shown in Fig. 9.11, the following characteristics can be seen:

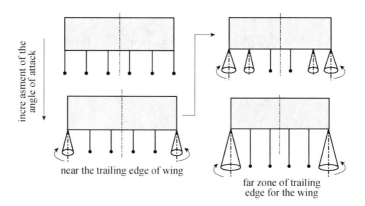

Fig. 9.11 Description of the phenomenon of flow around the wing

(1) When the angle of attack is zero, the flow around the wing is attached, and the small cotton balls on the silk threads downstream of the trailing edge of the wing are stationary;

(2) When the angle of attack is very small, the flow around the wing is attached flow. The small cotton balls at the downstream of the trailing edge do not rotate except for the opposite spinning at the tips of the wing;

(3) If the angle of attack increases, the spinning part of the cotton balls on the silk threads at the downstream of the wing trailing edge gradually develops from the wingtip to the wing root, and the spinning speed is also different. The cotton balls at the wingtip rotate fastest, while the cotton balls near the wing root rotate slower;

(4) No matter how the angle of attack changes, the cotton ball at the symmetrical part of the wing does not rotate;

(5) When the angle of attack is constant, if the silk threads are lengthened to more than the span length, it is found that only the cotton balls at the wingtips rotate, while the balls in other regions do not.

These phenomena clearly show that, compared with the 2D airfoil flow, except for the attached flow on both the wing and airfoil, the flow around the 3D wing has obvious vortex motion after leaving the trailing edge. The vortex-free flow in front of the wing bypasses the wing and then drags out the vortex, forming a free vortex layer at the trailing edge of the wing, and far away from the trailing edge, these free vortices will be self-induced to form concentrated wake vortices, which will move with the aircraft, as shown in Figs. 9.12 and 9.13.

Fig. 9.12 Smoke flow experiment of a rectangular straight wing

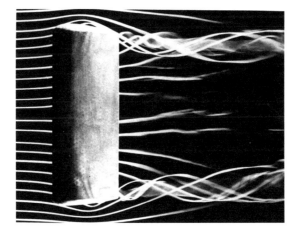

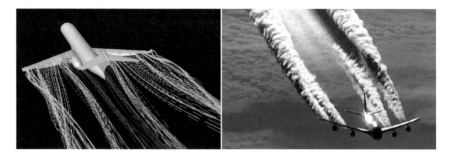

Fig. 9.13 Wake vortex of the wing

9.3.2   Vortex Structure of 3D Wing Flow at Low Speed

British aerodynamicist F. W. Lanchester (1868~1946), first proposed in 1891 to replace the wing with the vortex structure acting on it and proposed a method for estimating the lift of the wing. In 1894, he preceded the German physicist Kutta (1867~1944) and Russian physicist Joukowski (1847~1921) to explain the principle of wing lift generation; in 1915, the concepts of the attached vortex and free vortex were proposed for the calculation of wing lift with finite span as shown in Fig. 9.14. This concept was later developed by the world's fluid mechanics master German mechanist Ludwig Prandtl (1875~1953), and the famous lifting-line theory was proposed in 1918. This work made people realize the importance of the wingtip effect of a wing with a finite span on the overall aerodynamic performance of the wing and pointed out the essential relationship between wingtip vortices and induced drag. As we all know, for the steady flow of a 2D airfoil, the starting vortex is thrown far from the airfoil, only the attached vortex moves with the airfoil, the flow outside the airfoil is potential flow, and there is no vortex system. In contrast, the 3D wing flow is different. In addition to the attached vortex system on the wing, there is also a free vortex system on the trailing edge of the wing that moves together with the wing. These vortices constitute the overall characteristics of the flow around the wing. What kind of relationship exists between them and how they contribute to lift require a detailed analysis.

(1)   Starting vortex

When a wing is in the process of starting and accelerating, the vortices continuously fall off from the sharp trailing edge of the wing, resulting in the increasing vorticity on the wing surface, and the boundary layer gradually forms on the wing surface. When the wing reaches a stable speed, the boundary layer tends to be stable, the attached vorticity reaches the maximum, and the airflow smoothly leaves from the trailing edge of the wing. The vortices falling off the trailing edge of the wing gather together to form a vortex of the same size and opposite direction as the attached vortex of the wing, which moves to infinity far downstream of the wing with the airflow, which is called the starting vortex, as shown in Fig. 9.15.

Fig. 9.14 Wingtip vortex given by Lanchester

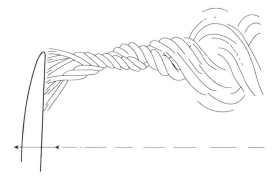

Fig. 9.15 Flow around attached vortex airfoil

(2) Attached vortex

The attached vortex moves with the wing. If the wing moves at a constant speed, the strength of the attached vortex will remain unchanged. For thin wing flow, if the influence of the thickness of the wing is ignored, the attached vortex system arranged on the wing chord can be used to replace the wing. The flow field induced by the attached vortex system is basically the same as the flow field generated by the wing not far from the wing, so the attached vortices can be used to calculate the aerodynamic force and moment. Changing the angle of attack of the wing will change the aerodynamic force, so changing the angle of attack actually changes the strength of the attached vortices around the wing. When calculating aerodynamic force with attached vortices, in the actual calculation, the attached vortices can be arranged on the chord plane, middle camber, and upper and lower wing surfaces of the wing according to the accuracy requirements as shown in Fig. 9.16.

(3) Free vortex sheet and wake vortex (concentrated vortex)

For the flow around a wing with a finite span, a large number of experiments have found that the wingtip effect of the wing is the main cause of the free vortex. This is because the airflow with higher pressure on the lower surface of the wing at a positive angle of attack will turn from the wingtip to the upper wing surface, so that the airflow on the upper wing surface flows from the wingtip to the wing root (the upper wing surface airflow pressure is higher at the wingtip than the wing root), and the airflow on the lower wing surface flows from the wing root to the wingtip (the lower wing surface airflow pressure is higher at the wing root than the wingtip). There is a reverse and large speed difference between the upper and lower wing surfaces in the spanwise direction. Once the airflow leaves the trailing edge of the wing, the spanwise speed difference between the upper and lower surfaces will cause

Fig. 9.16 Arrangement of
attached vortex

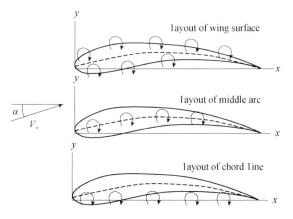

the trailing edge air to undergo spanwise shearing, resulting in the shear layer vortex, which is the free vortex sheet, as shown in Fig. 9.17. At the distance from the central axis z, take a small section of dz, the vortex intensity increment produced by the spanwise velocity difference of the upper and lower surface trailing edges

$$d\Gamma_w(z) = (w_d + w_u)dz \tag{9.19}$$

Obviously, the strength of the free vortex layer is determined by the velocity changes in the spanwise direction. Since spanwise velocity is larger at the wingtip and smaller at the wing root, the spanwise distribution of the free vortex strength is that the wingtip is larger and the wing root is smaller, and the total strength of the wake vortex is

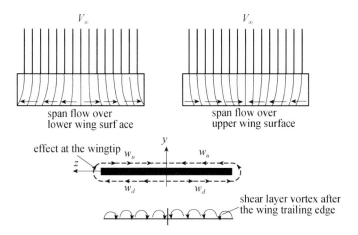

Fig. 9.17 The formation mechanism of the free vortex sheet on the trailing edge of a wing with finite span

$$\Gamma_w = \int\limits_0^{L/2} (w_d + w_u)\mathrm{d}z \tag{9.20}$$

Therefore, the free vortex sheet is caused by the spanwise shearing effect of the spanwise airflow of the upper and lower surfaces after leaving the trailing edge. The free vortex sheet leaving the trailing edge of the wing will induce each other due to different strengths. In the far region about 1 time wingspan length away from the trailing edge, these free vortex sheets will be rolled into two concentrated vortices with opposite direction and equal strength, that is, the wake vortex of the aircraft. The wake vortex is dragged out from the trailing edge of the wing, and the axis of the wake vortex is approximately parallel to the direction of the incoming flow, as shown in Fig. 9.18. Figure 9.19 shows the dissipation of the wake vortex.

(4) Horseshoe vortex

The airflow passes the wing as a whole. What is the relationship between the attached vortex, starting vortex, and free vortex sheet, and how do they affect the aerodynamic

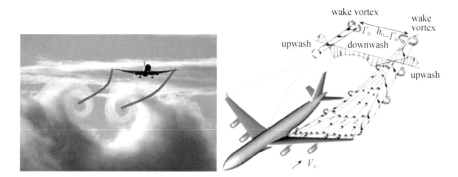

Fig. 9.18 Formation of wake vortex and downwash of airflow

Fig. 9.19 Wake vortex dispersion of large aircraft

Fig. 9.20 Vortex ring model

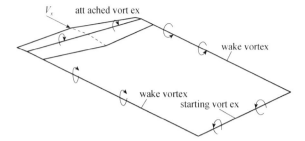

force? This involves the correct establishment of the physical concept of the aircraft vortex model. According to the vorticity conservation law of the German physicist Helmholtz (1821~1894), for an ideal barotropic fluid, the vorticity remains conserved when the mass force is potential. For an arbitrarily large contour around the wing, there is no vortex at the beginning and no vortex in the future. Therefore, only a closed vortex ring can be formed between the starting vortex, the attached vortex, and the free vortex (wake vortex) that appear around the wing in the contour. The head of this vortex ring is the attached vortex, the tail is the starting vortex, and the two sides are the wake vortex, as shown in Fig. 9.20. Because the starting vortex is far behind the wing, it has no effect on the aerodynamic force of the wing. Therefore, when performing the aerodynamic modeling, only the attached vortex and the wake vortices on both sides are considered. In form, it looks like a π-shaped horseshoe vortex. Since the area near the trailing edge of the wing is a free vortex sheet, the starting vortex can be ignored in actual modeling and replaced by a series of horseshoe vortices.

9.4   Vortex System Model of Low-Speed Wing Flow

9.4.1   Characteristics of Vortex Model

In order to analyze and estimate the aerodynamic characteristics of the wing, it is necessary to establish an adaptive vortex model, also called the aerodynamic model. For a 3D wing flow problem, if the flow is decomposed into the longitudinal flow (chordal) and spanwise flow (as shown in Fig. 9.21), the resultant aerodynamic force should be the combination of these two components. For the aerodynamic force of the longitudinal flow component, the analysis is given in the airfoil flow; for the effect of the spanwise flow, the key point is how to account for the resultant induced effect of the free surface vortex. Obviously, if this part of the influence is ignored, it will return to the flow of the 2D airfoil. The actual situation is only valid for the flow around the infinite wingspan. For the flow around a wing with a finite wingspan, the influence of the spanwise flow must be considered. In the establishment of the vortex system model, the free vortex system must be considered.

Fig. 9.21 Decomposition of
flow around a 3D wing

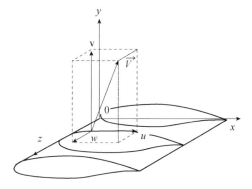

According to the thin airfoil theory, the lift of an airfoil (infinite span wing) is determined by the angle of attack and camber of the airfoil. Distributed vortices can be arranged on the upper and lower middle cambers of the airfoil profile, or chord line to replace the airfoil, and this vortex system is called the attached vortex. The total lift of the airfoil is proportional to the total strength Γ of the attached vortex. According to the Kutta–Joukowski lift circulation theorem, we know

$$\Gamma = \int_0^b \gamma \, dx \quad L = \rho V_\infty \Gamma \tag{9.21}$$

In the flow around a 3D wing, in addition to the distributed vortex system that can be arranged on the wing surface, the middle camber surface, or the chord plane, it is also necessary to consider the free vortex sheet caused by the spanwise flow. Such vortex system model includes the combination of the attached vortex and free vortex. Of course, the analysis of aerodynamic forces needs to be based on this combined vortex model.

Due to the wingtip effect of the flow around the 3D wing, in terms of the pressure difference between the upper and lower wing surfaces, the lift of each profile of the wing is from zero at the wingtip (the pressures on the upper and lower wing surface are equal) to the maximum at the wing root (the maximum pressure difference). Therefore, when the attached vortex strength is arranged in the spanwise direction, the vorticity distribution curve should also be the curve with the maximum in the middle and zero at the two wingtips. As shown in Fig. 9.22, the vortex intensity distribution is different for different wing plane shapes.

This shows that the main differences between the finite span straight wing and the infinite span wing are as follows: (1) the strength of the attached vortex varies along the spanwise direction (longitudinal flow), and $\Gamma_{z=0} = \Gamma_{max}$, $\Gamma_{z=\pm\frac{l}{2}} = 0$; (2) the trailing edge of the wing will drag out the free vortex sheet (spanwise flow). These two points are also the main reason why the flow around a 3D wing is different from that around a 2D airfoil.

Fig. 9.22 Vorticity
distribution curves of
different plane shapes

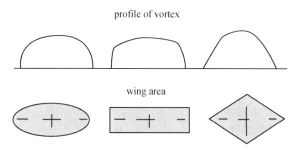

9.4.2 Aerodynamic Model of the Superposition of Straight Uniform Flow and a Single Π-Shaped Horseshoe Vortex

For the sake of simplicity, early people used a single Π-shaped horseshoe vortex model to replace the wing. That is, the equal strength attached vortex and free vortex models (concentrated wake vortex) are used, as shown in Fig. 9.23.

Using the Biot–Savart law, the attached vortex does not induce velocity along its axis, and the downwash velocity induced by the free vortex on the attached vortex line is

$$w = \frac{\Gamma}{4\pi(L/2+z)} + \frac{\Gamma}{4\pi(L/2-z)} = \frac{\Gamma}{4\pi(L^2/4-z^2)} \tag{9.22}$$

It can be seen from the above equation that when z-$L/2$, the downwash speed tends to be infinite, which is unrealistic, as shown in Fig. 9.24. It can be seen that the model has a large deviation from the actual wing flow. The main manifestations are as follows: (1) The wingtip induces infinite downwash speed, which is unrealistic; (2) The vortex strength does not change in the spanwise direction, which is not in accordance with reality.

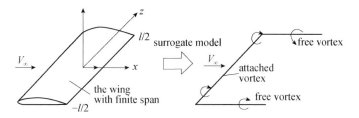

Fig. 9.23 Straight uniform flow and single Π-shaped horseshoe vortex model

Fig. 9.24 The induced
velocity field of a single
horseshoe vortex

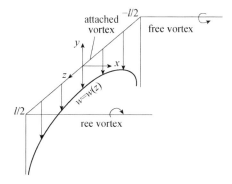

9.4.3   Aerodynamic Model of the Superposition of Straight Uniform Flow, Attached Vortex Sheet and Free Vortex Sheet

For a straight wing with a high aspect ratio, the curling and bending of the free vortex sheet mainly occur far away from the wing (approximately one wingspan length from the trailing edge of the wing). For the sake of simplicity, it is assumed that the free vortex sheet neither rolls up nor dissipates and extends to infinity in the direction of the incoming flow. Therefore, the flow around a straight wing with a high aspect ratio can use the aerodynamic model of straight uniform flow superimposed attached vortex sheet and the free vortex surface, and this model is also called the lift surface theoretical model; if the strength distribution of the vortex sheet can be obtained theoretically, the aerodynamic force and moment of the wing can be calculated. The combined model of the attached vortex sheet and free vortex sheet is composed of countless Π-shaped horseshoe vortices. The superposition of the Π-shaped horseshoe vortex system and the straight uniform flow is a reasonable and practical aerodynamic model for a straight wing with a high aspect ratio. This is because of the following reasons: (1) The model conforms that the strength is constant along a vortex line and the vortex line cannot be interrupted in the flow field, the basic vortex model is a horseshoe vortex; (2) The part of Π-shaped horseshoe vortex vertical to the inflow is attached vortex system. The number of vortex lines passing through each profile along the spanwise direction represents different strengths. The middle profile has the most vortex lines and circulation, while no vortex line passes through the wingtip profile, and the circulation is zero. The spanwise distribution of circulation and lift is simulated; (3) The Π-shaped horseshoe vortex system flows in parallel to the inflow and drags to infinitely downstream, which simulates the free vortex sheet. Since the strength of the free vortex drawn out between two adjacent profiles in the spanwise direction is equal to the circulation difference of the attached vortices on the two profiles, the relationship between the strength of the free vortex line in the spanwise direction and the circulation of the attached vortex on the wing is established as shown in Fig. 9.25.

Fig. 9.25 Combined model
of attached vortex sheet and
free vortex sheet

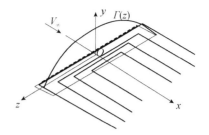

9.4.4 Aerodynamic Model of the Superposition of Straight Uniform Flow, Attached Vortex Line and Free Vortex Sheet

For a straight wing with a high aspect ratio, since the wing chord length is much smaller than the wingspan length, in 1918, the famous German mechanist Ludwig Prandtl (1875~1953) simplified the aerodynamic model of straight uniform flow superimposed attached vortex sheet and free vortex sheet. It was proposed that the attached vortex sheet on the wing should be replaced by an attached vortex line. The vortex line is a combination of a series of vortex lines in the chord direction. The strength of the vortex on the vortex line changes along the span. This vortex line is called the attached vortex line. The strength of the vortex at each point on this vortex line determines the lift of each profile. The simplified aerodynamic model is the straight uniform flow superimposed attached vortex line and free vortex sheet, which is called the lifting-line model. Because the lift increment point of the low-speed airfoil is located at the 1/4 chord, the attached vortex line can be placed on the line connecting the 1/4 chord point of each profile in the spanwise direction, which is called the lifting line.

9.5 Prandtl's Lifting-Line Theory

The calculation method of wing aerodynamic force based on the lifting-line model is called the lifting-line theory.

9.5.1 Profile Hypothesis

The wing profile on a wing with a finite span is different from a 2D airfoil, and the difference reflects the 3D effect of wing flow. For flow around a straight wing with a high aspect ratio and small angle of attack, the change of spanwise velocity component and flow parameters in each profile is much smaller than that in the other two directions. Therefore, the flow on each profile can be approximately regarded as 2D, and the 2D flows on different profiles in the spanwise direction are different from each other due to the influence of the free vortex. This kind of flow is 2D viewed from a local profile and 3D viewed from the profile of the entire wing, so it is called the profile hypothesis. The profile hypothesis is actually a quasi-2D flow hypothesis. The larger the λ value of the wing, the closer this hypothesis is to reality. When $\lambda \to \infty$, this hypothesis is accurate.

9.5.2 Downwash Speed, Downwash Angle, Lift, and Induced Drag

For a straight wing with a high aspect ratio, Prandtl's lifting-line theory uses a variable strength attached vortex line $\Gamma(z)$ at the 1/4 chord line and a free vortex sheet dragged out from the attached vortex line to replace the wing. Take x-axis in direction of incoming flow, y-axis upward, and z-axis and xOy-plane satisfy the right-handed spiral law. The free vortex sheet coincides with the xOz-plane, and each vortex line is dragged to $+ \infty$ along the x-axis as shown in Fig. 9.26.

The main difference between the spanwise profile of a high aspect ratio straight wing and a 2D airfoil is that the free vortex sheet induces a downward (positive lift) speed at the spanwise profile, which is called downwash speed. Since the wing has been replaced by an attached vortex line with variable strength $\Gamma(z)$ in the spanwise direction—lifting line. Therefore, the induced downwash speed of the free vortex on the wing can also be replaced by the induced downwash speed on the attached

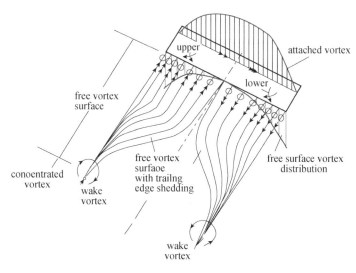

Fig. 9.26 Lifting-line model

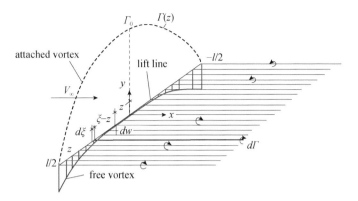

Fig. 9.27 Lifting-line model and its downwash speed

vortex line, as shown in Fig. 9.27. Suppose the intensity of the attached vortex at the spanwise position ζ is $\Gamma(\zeta)$, and the vortex intensity at $\zeta + d\zeta$ is $\Gamma(\zeta) + \frac{d\Gamma}{d\zeta} d\zeta$, according to the conservation law of vorticity, the strength of the free vortex dragged out by the $d\zeta$ element is $\frac{d\Gamma}{d\zeta} d\zeta$. The downwash speed of this free vortex line at any point z on the attached vortex line is

$$dw = \frac{|d\Gamma|}{4\pi(\zeta - z)} = \frac{-\frac{d\Gamma}{d\zeta}}{4\pi(\zeta - z)} d\zeta \tag{9.23}$$

The induced downwash speed of the entire vortex system at point z is

$$w(z) = \int_{l/2}^{l/2} \frac{-\frac{d\Gamma}{d\zeta}}{4\pi(\zeta - z)} d\zeta \tag{9.24}$$

Because of the downwash speed, the actual effective speed V_e on each profile of the wingspan is the vector sum of the incoming flow velocity V_∞ at infinity and the downwash speed, and the effective angle of attack α_e is also smaller than the geometric angle of attack α by α_i, α_i is called the downwash angle, as shown in Fig. 9.28. According to the velocity triangle, we can get

$$\alpha_i = tg^{-1}\left(\frac{w(z)}{V_\infty}\right) \approx \frac{w(z)}{V_\infty}$$

$$\alpha_e = \alpha - \alpha_i$$

$$V_e = \frac{V_\infty}{\cos\alpha_i} \approx V_\infty \tag{9.25}$$

Since the downwash speed is much smaller than the incoming flow speed, we can get

Fig. 9.28 The effect of downwash speed on wing profile flow

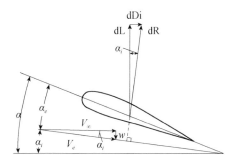

$$\alpha_i = tg^{-1}\left(\frac{w(z)}{V_\infty}\right) \approx \frac{w(z)}{V_\infty} = \frac{1}{V_\infty}\int\limits_{l/2}^{l/2}\frac{-\frac{d\Gamma}{d\zeta}}{4\pi(\zeta - z)}d\zeta \tag{9.26}$$

When calculating the lift acting on the element of the wing, it is necessary to introduce the assumption of "profile flow". It is assumed that the aerodynamic force on each profile of the wing with a finite wingspan is the same as the 2D airfoil with the same shape and the angle of attack α_e at effective velocity V_e. Therefore, acting at the point $P(z)$, the force R on the wing element dz is determined by the Kutta–Joukowski lift circulation theorem, namely

$$dR = \rho V_e \Gamma(z)dz \approx \rho V_\infty \Gamma(z)dz \tag{9.27}$$

The direction of dR is perpendicular to the effective velocity V_e, and its components in the vertical and parallel V_∞-directions are lift dL and drag dD_i, expressed as

$$dL = dR\cos\alpha_i \approx dR \approx \rho V_\infty \Gamma(z)dz$$
$$dD_i = dR\sin\alpha_i \approx dL\alpha_i \tag{9.28}$$

Integrate along the entire wingspan to obtain the lift and induced drag of the entire wing as

$$L = \int\limits_{l/2}^{l/2} \rho V_\infty \Gamma(z)dz$$

$$D_i = \int\limits_{l/2}^{l/2} \rho V_\infty \Gamma(z)\alpha_i(z)dz \tag{9.29}$$

In the equation, the drag term D_i does not exist in the flow around an ideal 2D airfoil, which is the result of the induction by the free vortex dragged out by the trailing edge of the wing with a finite span. In other words, it is the drag formed in the direction of the incoming flow due to the decrease of the effective angle of attack of the profile by the downwash angle, so it is called induced drag. This induced drag has nothing to do with the viscosity of the fluid and is the drag price that a wing with a finite span must pay to generate lift. From the energy point of view, the rotational kinetic energy required by the fluid elements on the free vortex sheet dragged out by the trailing edge of the wing must be provided by the aircraft with an additional thrust to overcome the induced drag in order to maintain a lifted flight. The additional power paid by the propulsion system is

$$\Delta P_w = D_i V_\infty = \rho V_\infty^2 \int_{l/2}^{l/2} \Gamma(z) \alpha_i(z) \mathrm{d}z \tag{9.30}$$

9.5.3 Differential–Integral Equation on the Intensity of the Attached Vortex

It can be seen from the above analysis that solving the problem of lift and drag of a straight wing with a high aspect ratio comes down to determining the distribution of the attached vortex strength along the wingspan direction $\Gamma(z)$. The following derivation determines the connection equation of $\Gamma(z)$. According to the airfoil theory, the lift dL acting on the wing element dz is

$$\mathrm{d}L = C_L' \frac{1}{2} \rho V_\infty^2(z) b(z) \mathrm{d}z, \quad \mathrm{d}L = \rho V_\infty \Gamma(z) \mathrm{d}z \tag{9.31}$$

Based on the profile flow hypothesis, the profile lift coefficient can be expressed as

$$C_L'(z) = C_{L\infty}^\alpha [\alpha_e - \alpha_{0\infty}] = C_{L\infty}^\alpha [\alpha - \alpha_i - \alpha_{0\infty}] = C_{L\infty}^\alpha [\alpha_a - \alpha_i] \tag{9.32}$$

In the equation, $C_{y\infty}^\alpha(z)$ and $\alpha_{0\infty}(z)$ is the lift line slope and zero-lift angle of attack of the 2D airfoil, then, $\alpha_a = \alpha - \alpha_{0\infty}$ is called the absolute angle of attack. From Eqs. (9.26), (9.31), and (9.32), we can get

$$\Gamma(z) = \frac{1}{2} V_\infty b(z) C_{L\infty}^\alpha \left[\alpha_a + \frac{1}{V_\infty} \int_{l/2}^{l/2} \frac{\frac{\mathrm{d}\Gamma}{\mathrm{d}\zeta}}{4\pi(\zeta - z)} \mathrm{d}\zeta \right] \tag{9.33}$$

This equation is the differential–integral equation for determining the attached vortex intensity $\Gamma(z)$ under the condition of a given angle of attack and wing geometry (airfoil). The exact solution of this equation can be obtained only in a few special cases, and the elliptical circulation distribution is one of the most important. If the circulation distribution of the wing $\Gamma(z)$ is elliptical, then

$$\frac{\Gamma(z)}{\Gamma_0} = \sqrt{1 - \left(\frac{2z}{l}\right)^2} \tag{9.34}$$

Among them, Γ_0 is the maximum circulation value on the symmetry plane of the wing. From the circulation distribution function, the downwash velocity and angle at point z can be obtained

$$w(z) = \int_{l/2}^{1/2} \frac{-\frac{d\Gamma}{d\zeta}}{4\pi(\zeta - z)} d\zeta = \frac{\Gamma_0}{2l} \tag{9.35}$$

$$\alpha_i = \frac{1}{V_\infty} \int_{l/2}^{1/2} \frac{-\frac{d\Gamma}{d\zeta}}{4\pi(\zeta - z)} d\zeta = \frac{\Gamma_0}{2l V_\infty} \tag{9.36}$$

The above results show that for a wing with elliptical distribution, the downwash speed and downwash angle are constant along the wingspan. If the wing has no twist, neither geometric twist nor aerodynamic twist, then the geometric angle of attack α, the zero-lift angle of attack $\alpha_{0\infty}$, and the profile lift line slope $C_{L\infty}^\alpha$ are also constant along the wingspan, so along the wingspan there is

$$\alpha_e - \alpha_{0\infty} = \alpha - \alpha_i - \alpha_{0\infty} = \alpha_a - \alpha_i = Const.$$
$$C_L' = C_{L\infty}^\alpha[\alpha - \alpha_i - \alpha_{0\infty}] = Const. \tag{9.37}$$
$$C_{Di}' = C_L'\alpha_i = Const.$$

For the entire wing

$$C_L = \frac{L}{\frac{1}{2}\rho V_\infty^2 S} = \frac{\int_{-\frac{l}{2}}^{\frac{l}{2}} C_L' \frac{1}{2}\rho V_\infty^2 b(z) dz}{\frac{1}{2}\rho V_\infty^2 S} = C_L' \frac{\int_{-\frac{l}{2}}^{\frac{l}{2}} b(z) dz}{S} = C_L'$$

$$C_{Di} = \frac{D_i}{\frac{1}{2}\rho V_\infty^2 S} = \frac{\int_{-\frac{l}{2}}^{\frac{l}{2}} C_{Di}' \frac{1}{2}\rho V_\infty^2 b(z) dz}{\frac{1}{2}\rho V_\infty^2 S} = C_{Di}' \frac{\int_{-\frac{l}{2}}^{\frac{l}{2}} b(z) dz}{S} = C_{Di}' \tag{9.38}$$

The above two equations show that the lift coefficient and induced drag coefficient of the elliptical circulation distribution non-twisted straight wing are equal to the lift coefficient and induced drag coefficient of each profile along the wingspan direction. The expression of the aerodynamic coefficient of a straight wing with an elliptical circulation distribution is obtained below, which is

$$L = \rho V_\infty \int_{-\frac{l}{2}}^{\frac{l}{2}} \Gamma(z) dz, \quad \frac{\Gamma(z)}{\Gamma_0} = \sqrt{1 - \left(\frac{2z}{l}\right)^2} \tag{9.39}$$

$$C_L = \frac{2}{V_\infty S} \int_{-l/2}^{l/2} \Gamma(z) dz = \frac{\Gamma_0 \pi l}{2 V_\infty S}, \quad \Gamma_0 = C_L \frac{2 V_\infty S}{\pi l} \tag{9.40}$$

$$\alpha_i = \frac{1}{V_\infty} \int_{l/2}^{l/2} \frac{-\frac{d\Gamma}{d\zeta}}{4\pi(\zeta - z)} d\zeta = \frac{\Gamma_0}{2l V_\infty} = \frac{C_L}{\pi \lambda} \tag{9.41}$$

due to

$$C_L = C_L' = C_{L\infty}^\alpha [\alpha_a - \alpha_i] = C_{L\infty}^\alpha \left[\alpha_a - \frac{C_L}{\pi \lambda}\right] \tag{9.42}$$

Thus, the lift coefficient is

$$C_L = \frac{C_{L\infty}^\alpha}{1 + \frac{C_{L\infty}^\alpha}{\pi \lambda}} [\alpha - \alpha_{0\infty}] \tag{9.43}$$

The induced drag coefficient is

$$C_{Di} = \frac{C_L^2}{\pi \lambda} \tag{9.44}$$

The above two equations illustrate that the main differences in aerodynamic characteristics between a straight wing with an elliptical circulation distribution and an infinite span wing (airfoil) are as follows:

(1) The lift line slope of a wing with a finite span is smaller than that of a wing with an infinite span, and it decreases with the decrease of the aspect ratio λ;
(2) There is induced drag in a wing with a finite span, and the induced drag coefficient is proportional to the square of the lift coefficient and inversely proportional to the aspect ratio λ. When the C_L value is given, increasing λ can decrease the C_{Di} value. Increasing the aspect ratio can increase the wing's lift line slope value.

The plane shape of the wing with the elliptical circulation distribution is given below. The lift dL acting on the wing element dz is

$$dL = C_L'(z) \frac{1}{2} \rho V_\infty^2 b(z) dz, \quad dL = \rho V_\infty \Gamma(z) dz \tag{9.45}$$

$$\Gamma(z) = \frac{1}{2} V_\infty b(z) C_L' \tag{9.46}$$

Since $z = 0$, $b(z) = b_0$, $\Gamma(z) = \Gamma_0$, substituting in the above equations to get

$$\Gamma_0 = \frac{1}{2} V_\infty b_0 C_L'$$ (9.47)

Because $C_L'(z) = C_L'(0) = $ constant, so

$$\frac{\Gamma(z)}{\Gamma_0} = \frac{b(z)}{b_0} = \sqrt{1 - \left(\frac{2z}{l}\right)^2}$$ (9.48)

The above equation shows that the spanwise distribution of chord length of the wing with elliptical circulation distribution is also elliptical, and this kind of wing is called an elliptical wing, as shown in Figs. 9.29 and 9.30. The downwash speed and downwash angle of an elliptical wing are constant and do not vary with spanwise position z, as shown in Fig. 9.31.

Fig. 9.29 Elliptical wing

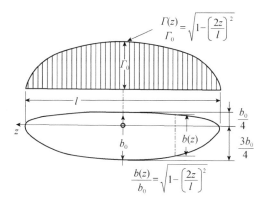

Fig. 9.30 Elliptical wing fighter

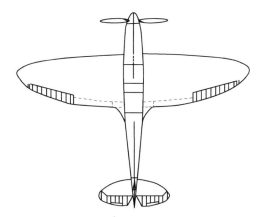

Fig. 9.31 The downwash
velocity of the elliptical wing

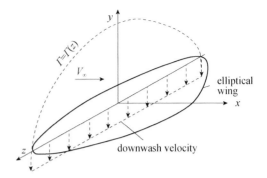

9.5.4 Aerodynamic Characteristics of a Straight Wing with Large Aspect Ratio in General Plane Shape

The circulation distribution of an elliptical wing is elliptical, which is the simplest analytical solution in the lifting-line theory. Although the lifting-line theory can prove that the elliptical wing is the shape with the best lift-to-drag ratio characteristic under the same aspect ratio, due to the convenience of structure and manufacturing technique, rectangular wing and trapezoidal wing are also commonly used in industry. Using the lifting-line theory to solve these non-elliptical $\Gamma(z)$ at a given angle of attack, the trigonometric series method is often used.

(1) Trigonometric series solutions of basic differential–integral equations

$$\Gamma(z) = \frac{1}{2} V_\infty C_{L\infty}^\alpha b(z) \left[\alpha_a(z) + \frac{1}{4\pi V_\infty} \int_{-\frac{l}{2}}^{\frac{l}{2}} \frac{\frac{d\Gamma}{d\zeta} d\zeta}{\zeta - z} \right] \qquad (9.49)$$

Perform variable substitution, as shown in Fig. 9.32, let

$$z = -\frac{l}{2} \cos\theta, \quad \zeta = -\frac{l}{2} \cos\theta_1 \qquad (9.50)$$

Thus

Fig. 9.32 Coordinate
transformation

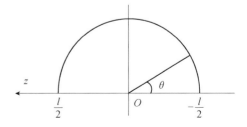

$$\Gamma(\theta) = \frac{1}{2} V_\infty C_{L\infty}^\alpha b(\theta) \left[\alpha_a(\theta) + \frac{1}{2\pi V_\infty l} \int_{-\frac{l}{2}}^{\frac{l}{2}} \frac{\frac{d\Gamma}{d\theta_1} d\theta_1}{\cos\theta - \cos\theta_1} \right] \tag{9.51}$$

Then expand the above equation into the following trigonometric series, which is

$$\Gamma(\theta) = 2l V_\infty \sum_{n=1}^{\infty} A_n \sin(n\theta) \tag{9.52}$$

Since the wingtip circulation is zero, $\Gamma(0) = \Gamma(\pi) = 0$, the above equation can only take the sine term.

In addition, the circulation distribution on the wing is symmetrical, $\Gamma(\theta) = \Gamma(\pi - \theta)$, so A_n is 0 when n is an even number, $A_2 = A_4 = A_6 = \ldots = A_{2n} = \ldots = 0$. It can be obtained from the above two equations

$$\mu\alpha_a(\theta) \sin\theta = \sum_{n=1}^{\infty} A_n \sin(n\theta)(\mu n + \sin\theta) \tag{9.53}$$

Among them, $\mu = \frac{C_{L\infty}^\alpha(\theta)b(\theta)}{4L}$. As long as the number of items n is kept enough and the corresponding coefficient A_n is selected, the actual circulation distribution can be approximated. Therefore, the final problem becomes to solve for A1, A3, A5, … under the given wing chord, airfoil, and absolute angle of attack distribution. In fact, it is only necessary to keep the first few series when solving. Taking the four terms of the trigonometric series can approximate the actual circulation distribution. Take four θ (corresponding to the four profiles of the right half of the wing), take $\theta_1 = 22.5°$, $\theta_2 = 45°$, $\theta_3 = 67.5°$, and $\theta_4 = 90°$ into Eq. (9.53), then four algebraic equations of A1, A3, A5, and A7 can be obtained.

The circulation distribution of an elliptical wing is a special case in the expression of the circulation trigonometric series. When only one item is taken in the expression of the circulation trigonometric series, $\Gamma(\theta) = 2l V_\infty A_1 \sin\theta$, the variable θ is restored to z, then

$$\Gamma(z) = 2l V_\infty A_1 \sqrt{1 - \left(\frac{2z}{l}\right)^2} \tag{9.54}$$

When $z = 0$, $\Gamma = \Gamma_0$, we can get

$$A_1 = \frac{\Gamma_0}{2V_\infty l} \tag{9.55}$$

So, there is $\frac{\Gamma(z)}{\Gamma_0} = \sqrt{1 - \left(\frac{2z}{l}\right)^2}$.

9.5.5   Influence of Plane Shape on Spanwise Circulation Distribution of Wing

Using the trigonometric series method, the circulation distribution along the span-wise direction of wings with different plane shapes can be obtained. With $\Gamma(z)$, the distribution of the lift coefficient of the wing profile along spanwise can be obtained. Figure 9.33 shows the spanwise circulation distribution of four typical plane shape non-twisted wings with aspect ratio $\lambda = 6$. Figure 9.34 shows the profile lift coefficient distribution of non-twisted trapezoidal wing with different taper ratios and constant aspect ratio $\lambda = 6$.

Fig. 9.33 $\lambda = 6$, circulation distributions along the span direction of four typical plane shape non-twisted wings

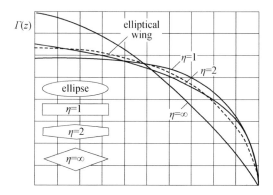

Fig. 9.34 $\lambda = 6$, profile lift coefficient distributions of non-twisted trapezoidal wings with different taper ratios

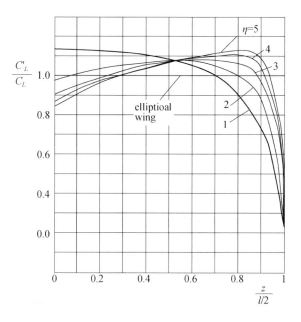

From the above results, the following conclusions can be drawn:

(1) The maximum profile lift coefficient of the rectangular wing is at the wing root profile;
(2) For trapezoidal wings with a large taper ratio $\eta > 2$, the maximum profile lift coefficient occurs near the wingtip, and as the taper ratio increases, the maximum profile lift coefficient moves closer to the wingtip;
(3) $\eta = 2 \sim 3$, the circulation distribution and profile lift coefficient distribution of the trapezoidal wing are closest to the elliptical wing.

The distribution law of the profile lift coefficient of the trapezoidal wing with a relatively high taper ratio is detrimental to the stall characteristic, so the taper ratio must be selected appropriately; otherwise, appropriate measures (such as geometric twist or aerodynamic twist) need to be taken to improve the stall characteristic.

9.5.6 Aerodynamic Characteristics of General Non-Twisted Straight Wing

(1) Lift coefficient

Integrate the circulation distribution function to get the lift, which is

$$\Gamma(\theta) = 2l V_\infty \sum_{n=1}^{\infty} A_n \sin(n\theta)$$

$$L = \rho V_\infty \int_{-\frac{l}{2}}^{\frac{l}{2}} \Gamma(z)dz$$

The lift coefficient is

$$C_L = \frac{2}{V_\infty S} \int_{-\frac{l}{2}}^{\frac{l}{2}} \Gamma(z)dz = \frac{4l}{S} \int_{-\frac{l}{2}}^{\frac{l}{2}} \sum_{n=1}^{\infty} A_n \sin(n\theta) \cdot \frac{l}{2} \sin\theta d\theta$$

$$= \pi \lambda A_1 \tag{9.56}$$

The above equation shows that the lift coefficient C_L of a wing with a finite span is only related to the first coefficient A_1 in the trigonometric series expansion that represents the circulation. The remaining coefficients do not affect the magnitude of the total lift, but only affect the circulation distribution along the spanwise, thus only affects the profile lift coefficient distribution along the spanwise. Here, we discuss a straight wing with neither geometric twist nor aerodynamic twist α_a, and $C_{L\infty}^\alpha$ is constant along the span, which is

$$C'_L(z) = C^\alpha_{L\infty}[\alpha_a - \Delta\alpha_i(z)] \tag{9.57}$$

$$C_L = \frac{\int_{-l/2}^{l/2} C'_L \frac{1}{2}\rho V^2_\infty b(z)dz}{\frac{1}{2}\rho V^2_\infty S} = \frac{\int_{-l/2}^{l/2} C'_L b(z)dz}{S} = C^\alpha_{L\infty}\left[\alpha_a - \frac{1}{S}\int_{-l/2}^{l/2}\alpha_i b(z)dz\right] \tag{9.58}$$

Substituting the circulation distribution of the trigonometric series $\Gamma(\theta) = 2lV_\infty \sum_{n=1}^{\infty} A_n \sin(n\theta)$ into Eq. (9.26), we get

$$\Delta\alpha_i(z) = \frac{1}{4\pi V_\infty}\int_{-\frac{l}{2}}^{\frac{l}{2}}\frac{\frac{d\Gamma}{d\zeta}d\zeta}{z-\zeta} = \frac{1}{\pi}\sum_{n=1}^{\infty}A_n n\int_0^\pi\frac{\cos n\theta_1}{\cos\theta_1 - \cos\theta}d\theta_1 = \sum_{n=1}^{\infty}\frac{A_n n \sin n\theta}{\sin\theta}$$

$$\Delta\alpha_i(z) = A_1\left(1 + \sum_{n=2}^{\infty}\frac{A_n n \sin n\theta}{A_1 \sin\theta}\right) \tag{9.59}$$

The wing lift coefficient is

$$C_L = C^\alpha_{L\infty}\left[\alpha_a - \frac{1}{S}\int_{-\frac{l}{2}}^{\frac{l}{2}}\Delta\alpha_i(z)b(z)dz\right]$$

$$= C^\alpha_{L\infty}\left[\alpha_a - A_1(1 + \frac{L}{2S}\int_0^\pi\frac{\sum_{n=2}^{\infty}nA_n\sin n\theta}{A_1}b(\theta)d\theta)\right]$$

$$= C^\alpha_{L\infty}[\alpha_a - A_1(1 + \tau)]$$

$$\tau = \frac{l}{2S}\int_0^\pi\frac{\sum_{n=2}^{\infty}nA_n\sin n\theta}{A_1}b(\theta)d\theta \tag{9.60}$$

In the equation, τ is a small positive value related to the plane shape of the wing, and its expression is

$$\tau = \frac{l}{2S}\int_0^\pi\frac{\sum_{n=2}^{\infty}nA_n\sin(n\theta)}{A_1}b(\theta)d\theta \tag{9.61}$$

Also, because $C_L = \pi\lambda A_1$, we can get

$$C_L = \frac{C_{L\infty}^{\alpha}}{1 + \frac{C_{L\infty}^{\alpha}}{\pi\lambda}(1+\tau)}\alpha_a = C_L^{\alpha}(\alpha - \alpha_{0\infty}), \quad C_L^{\alpha} = \frac{C_{L\infty}^{\alpha}}{1 + \frac{C_{L\infty}^{\alpha}}{\pi\lambda}(1+\tau)} \tag{9.62}$$

(2) Induced drag coefficient

$$C_{Di} = \frac{2}{V_\infty S}\int_{-l/2}^{l/2}\Gamma(z)\alpha_i(z)dz \tag{9.63}$$

Substituting $\Gamma(\theta) = 2lV_\infty \sum_{m=1}^{\infty} A_m \sin(m\theta)$ into Eq. (9.26), we get

$$\alpha_i = \frac{1}{V_\infty}\int_{l/2}^{l/2}\frac{-\frac{d\Gamma}{d\zeta}}{4\pi(\zeta - z)}d\zeta = \sum_{n=1}^{\infty}\frac{A_n n \sin n\theta}{\sin\theta} \tag{9.64}$$

Substituting into Eq. (9.63), we get

$$\begin{aligned}C_{Di} &= \frac{2}{V_\infty S}\int_{-\frac{l}{2}}^{\frac{l}{2}}\Gamma(z)\Delta\alpha_i(z)dz \\ &= \frac{2l^2}{S}\int_0^\pi \sum_{m=1}^{\infty}A_m\sin(m\theta)\sum_{n=1}^{\infty}A_n n\sin n\theta d\theta \\ &= 2\lambda\sum_{m=1}^{\infty}\sum_{n=1}^{\infty}A_m A_n n\int_0^\pi \sin m\theta \sin n\theta d\theta = \pi\lambda\sum_{n=1}^{\infty}nA_n^2\end{aligned} \tag{9.65}$$

After sorting out

$$C_{Di} = \pi\lambda\sum_{n=1}^{\infty}nA_n^2 == \pi\lambda A_1^2\left(1 + \sum_{n=2}^{\infty}\frac{nA_n^2}{A_1^2}\right) = \frac{C_L^2}{\pi\lambda}(1+\delta) \tag{9.66}$$

Among them, the parameter δ is

$$\delta = \sum_{n=2}^{\infty}\frac{nA_n^2}{A_1^2} = \frac{3A_3^2}{A_1^2} + \frac{5A_5^2}{A_1^2} + \frac{7A_7^2}{A_1^2} + \cdots \tag{9.67}$$

It is a small positive number related to the shape of the plane. Because it is always positive, the induced drag is always positive, which means that as long as the lift of a 3D finite span wing is not zero, induced drag is inevitable. In a physical sense, the induced drag is related to the energy consumed by the free vortex system behind the wing. For an elliptical wing, because $A_2 = A_3 = \ldots = A_n = 0$, now

$$\tau = \frac{l}{2S} \int_0^\pi \frac{\sum\limits_{n=2}^\infty n A_n \sin(n\theta)}{A_1} b(\theta) \mathrm{d}\theta = 0$$

$$\delta = \sum_{n=2}^\infty \frac{n A_n^2}{A_1^2} = \frac{3A_3^2}{A_1^2} + \frac{5A_5^2}{A_1^2} + \frac{7A_7^2}{A_1^2} + \cdots = 0$$

For non-elliptical wing, $\tau > 0$, $\delta > 0$. This indicates that under the same aspect ratio, the lift line slope of the elliptical wing is the largest, the induced drag coefficient is the smallest, and the lift-to-drag ratio is the largest. Therefore, the elliptical wing is the plane shape with the best lift-to-drag ratio. The aerodynamic characteristics of a high aspect ratio straight wing in any plane shape can be obtained by correcting τ and δ on the basis of the calculation equation of the elliptical wing, which is

$$C_L = C_L^\alpha (\alpha - \alpha_{0\infty}), \quad C_L^\alpha = \frac{C_{L\infty}^\alpha}{1 + \frac{C_{L\infty}^\alpha}{\pi\lambda}(1+\tau)}, \quad C_{Di} = \frac{C_L^2}{\pi\lambda}(1+\delta) \qquad (9.68)$$

Among them, τ and δ are usually referred to as the correction coefficient of the general wing to the aerodynamic force of the elliptical wing, which indicates the degree of deviation of other plane-shaped wings from the best plane-shaped wings. τ and δ mainly depend on the plane shape and aspect ratio of the wing and can be calculated by the trigonometric series method. Figure 9.35 shows the lift coefficient of the rectangular wing with different aspect ratios changes with the angle of attack. Table 9.1 shows the τ and δ values of several common plane shapes with aspect ratio $\lambda = 6$.

In summary, we can get the following:

(1) Trapezoidal wings with taper ratio $\eta = 2 \sim 3$ are widely used in the wings of low-speed aircraft, and the correction coefficients of lift coefficient and induced drag coefficient are the smallest, which is the closest to elliptical wing;

(2) The induced drag coefficient C_{Di} is proportional to the square of the lift coefficient and inversely proportional to λ. In order to obtain a large lift-to-drag ratio at low subsonic speed, it is best to use a large aspect ratio λ. But in fact, due to structural considerations, the aspect ratio used is generally between 6 and 10.

(3) In theory, the elliptical wing is the best plane shape with the best aerodynamic performance, but the structural design and manufacturing are inconvenient. In practice, trapezoidal wings with a taper ratio $\eta = 2 \sim 3$ are commonly used. The circulation distribution of the trapezoidal wing with $\eta = 2 \sim 3$ is closest to the circulation distribution of the elliptical wing.

Fig. 9.35 The variation curve of lift coefficient with angle of attack of Prandtl's classic rectangular wing

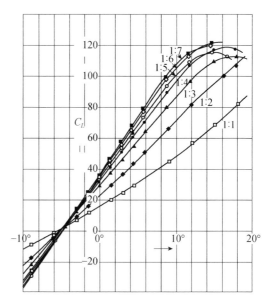

Table 9.1 Values of correction coefficients for wings with different aspect ratios

Planar shape	Root-tip ratio	τ	δ
Ellipse	/	0	0
Rectangle	1	0.17	0.049
Trapezoidal	4/3	0.10	0.026
Trapezoidal	2.5	0.01	0.01
Diamond	∞	0.17	0.141

9.5.7 Effect of Aspect Ratio on the Aerodynamic Characteristics of the Wing

There are two straight wings with a high aspect ratio which are composed of the same airfoil, but with different aspect ratios, λ_1 and λ_2, respectively. From the Eq. (9.68), we can get

$$\alpha_a = \alpha - \alpha_{0\infty} = \frac{C_L}{C_{L\infty}^{\alpha}} + \frac{C_L}{\pi\lambda}(1+\tau) \tag{9.69}$$

In the equation, $\frac{C_L}{C_{L\infty}^{\alpha}} = \alpha_{\alpha\infty}$ is the absolute angle of attack for the infinite span wing to get the same magnitude lift coefficient, thus

$$\alpha_a = (\alpha_a)_\infty + \frac{C_L}{\pi\lambda}(1+\tau) \tag{9.70}$$

Fig. 9.36 Variation curves
of lift coefficient versus
angle of attack for different
aspect ratios

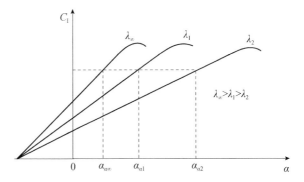

The above equation shows that the absolute angle of attack required to achieve the same C_L value of a 3D wing is greater than that of an infinite span wing, that is, the lift line slope of a 3D wing is smaller than that of an infinite span wing, and the lift line slope decrease as the aspect ratio decreases. Under the same C_L, the absolute angle of attack relationship for the two aspect ratio wings is

$$(\alpha_a)_{\lambda_2} = (\alpha_a)_{\lambda_1} + \frac{C_L}{\pi}\left(\frac{1+\tau_2}{\lambda_2} - \frac{1+\tau_1}{\lambda_1}\right) \qquad (9.71)$$

The above equation can convert the C_L-α curve of the λ_1 wing to the λ_2 wing, as shown in Fig. 9.36. From Eq. (9.68), the conversion equation for the drag coefficient of different aspect ratios under the same C_L is

$$(C_{Di})_{\lambda_2} = (C_{Di})_{\lambda_1} + \frac{C_L^2}{\pi}\left(\frac{1+\delta_2}{\lambda_2} - \frac{1+\delta_1}{\lambda_1}\right) \qquad (9.72)$$

If the lift coefficient and induced drag coefficient curve of the wing with aspect ratio λ_1 are known, the above conversion methods can be used to obtain the lift coefficient and drag coefficient of the wing with aspect ratio λ_2, as shown in Fig. 9.37. Figure 9.38 shows the lift coefficient and drag coefficient curve of Prandtl's classic rectangular wing. This conversion method has been experimentally proved to be satisfactory for wings with high aspect ratios.

9.5.8 Application Range of Lifting-Line Theory

The lifting-line theory is an approximate potential flow theory for solving the flow of a straight wing with a high aspect ratio. After knowing the plane shape of the wing and the aerodynamic data of the airfoil, the circulation distribution, the profile lift coefficient distribution, and the lift coefficient of the entire wing, the lift line slope and the induced drag coefficient can be obtained. Its outstanding advantage is that

Fig. 9.37 Variation curves of lift coefficient versus induced drag coefficient for different aspect ratios

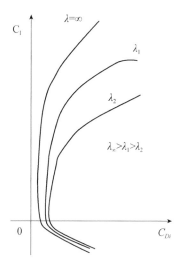

Fig. 9.38 Variation curves of lift coefficient versus drag coefficient of Prandtl's classical rectangular wing

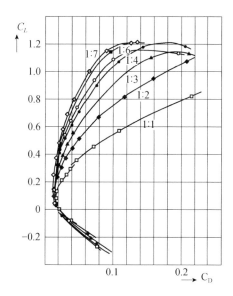

the influence of the wing plane parameters on the aerodynamic characteristics of the wing can be clearly given. The theory can be briefly summarized as follows:

(1) The wing is represented by a lifting line which is arranged at the 1/4 chord line;
(2) The strength of the attached vortex (the circulation of attached vortex) on the lifting line varies along the span;
(3) The variable strength attached vortex line drags out of the free vortex sheet extending downstream;

(4) The disturbance speed caused by the vortex system can be regarded as a small disturbance;
(5) The free vortex sheet is parallel to the flight direction;
(6) The flow around the wing profile can be solved by the Kutta–Joukowski lift circulation theorem.

The lifting-line theory provides theoretical guidance for the selection of aerodynamic design parameters and performance calculations for straight wings with high aspect ratios. The adaptation conditions of the lifting-line theory are as follows:

(1) The angle of attack cannot be too large ($\alpha < 10°$). The lifting-line theory does not consider the viscosity of air, but at high angles of attack, there is a clear flow separation;
(2) The aspect ratio cannot be too small ($\lambda \geq 5$);
(3) The sweep angle cannot be too large ($\chi \leq 20°$).

When the span is relatively small or the sweep angle is large, the lifting-line model and profile hypothesis are no longer suitable. The potential flow aerodynamic characteristics of swept-back wings and wings with small aspect ratios should be calculated using lifting-surface theory or other methods.

According to Eq. (9.43), for a straight elliptical wing with a high aspect ratio ($\lambda \geq 5$), the lift line slope is

$$\frac{dC_L}{d\alpha} = \frac{C_{L\infty}^\alpha}{1 + \frac{C_{L\infty}^\alpha}{\pi\lambda}} \tag{9.73}$$

For straight wings with small aspect ratios ($\lambda \leq 4$), German aerodynamicist H. B. Helmbold (1942) gave the correction equation for the lift line slope based on Eq. (9.73)

$$\frac{dC_L}{d\alpha} = \frac{C_{L\infty}^\alpha}{\sqrt{1 + \left(\frac{C_{L\infty}^\alpha}{\pi\lambda}\right)^2} + \frac{C_{L\infty}^\alpha}{\pi\lambda}} \tag{9.74}$$

For the swept-back wing, the correction equation for the lift line slope given by the German aerodynamicist Kuchemann (1978) is

$$\frac{dC_L}{d\alpha} = \frac{C_{L\infty}^\alpha \cos\chi}{\sqrt{1 + \left(\frac{C_{L\infty}^\alpha \cos\chi}{\pi\lambda}\right)^2} + \frac{C_{L\infty}^\alpha \cos\chi}{\pi\lambda}} \tag{9.75}$$

Among them, χ is the sweep angle of the center line. Figure 9.39 shows the comparison of Eqs. (9.73), (9.74), and (9.128).

Fig. 9.39 Variation curves of the lift line slopes of straight wings with different aspect ratios

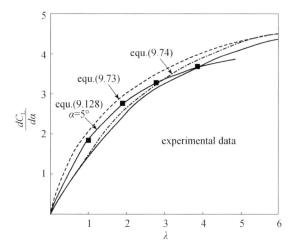

9.6 Stall Characteristics of a Straight Wing with a Large Aspect Ratio

At low angles of attack, the wing's lift coefficient C_L has a linear relationship with the angle of attack α. But when α continues to increase to a certain value, the $CL \sim \alpha$ curve begins to deviate from the linear relationship. At this time, the boundary layer near the trailing edge of the wing surface begins to locally separate, but it has not spread over the entire wing surface. Therefore, when continues to increase the angle of attack, the lift coefficient C_L still increases, but it is lower than the potential flow solution. Then, as the separation zone gradually expands, the lift coefficient increases more and more slowly. Finally, when the separation zone reaches a certain range, the lift coefficient C_L reaches the maximum value C_{Lmax}, and then as α continues to increase, C_L begins to decrease. This is the stall phenomenon of the wing, and the angle of attack corresponding to the maximum lift coefficient is the stall angle of attack. Experiments have found that there are many factors that affect the stall characteristics of the wing, such as the airfoil used, the Reynolds number, the Mach number, the plane shape, twist angle, thickness, and camber. The following only discusses the low-speed flow stall characteristics of non-twisted elliptical, rectangular, and trapezoidal wings.

9.6.1 Stall Characteristics of an Elliptical Wing

According to the lifting-line theory, for an elliptical wing, the downwash speed is constant along the span (the downwash angle is constant), so the effective angle of attack of each wing profile along the span does not change. In this way, if an elliptical wing is designed with the same airfoil, as α increases, each spanwise profile of the

Fig. 9.40 Stall
characteristics of an elliptical
wing

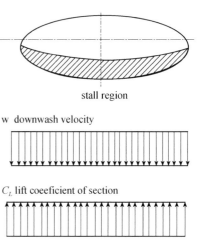

wing separates at the same time, reaching C_{Lmax} and stalling at the same time. The stall characteristics are good, as shown in Fig. 9.40.

9.6.2   Stall Characteristics of a Rectangular Wing

Since the downwash speed of a rectangular wing increases from the wing root to the wingtip (the same is true for the change in the downwash angle), the effective angle of attack of the wing root profile will be greater than that of the wingtip profile, and the corresponding profile lift coefficient will be greater than that of the wingtip. Therefore, the separation first occurs at the wing root, and then the separation zone gradually expands to both ends of the wing, the stall is gradual, as shown in Fig. 9.41.

Fig. 9.41 Stall
characteristics of a
rectangular wing

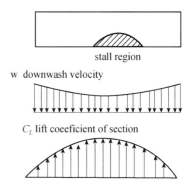

9.6.3   Stall Characteristics of a Trapezoidal Wing

Compared with the rectangular wing, the stall characteristic of the trapezoidal straight wing is just the opposite. The downwash speed decreases from the wing root to the wingtip. Therefore, the effective angle of attack of the wing profile increases toward the wingtip, and this trend becomes more obvious as the taper ratio increases. Therefore, the separation first occurs near the wingtip, which not only reduces the maximum lift coefficient of the wing, but also affects the efficiency of control surfaces such as ailerons as shown in Fig. 9.42.

It can be seen that the elliptical wing not only has good lift-to-drag ratio characteristics at small and medium angles of attack, but also has good stall characteristics at high angles of attack. The lift-to-drag characteristics of the rectangular wing at small and medium angles of attack are not as good as the elliptical wing, and the C_{Lmax} value at high angles of attack is also small, but the first separation at the wing root will not cause the deterioration of the aileron characteristics, which can give the pilot a warning that it is about to stall. The trapezoidal wing has a lift-to-drag ratio characteristic close to an elliptical wing at small and medium angles of attack, and its structural weight is lighter, so it is widely used. However, the separation first occurs near the wingtip, causing the wingtip to stall first. Therefore, in terms of stall characteristics, the trapezoidal straight wing is the worst among the above three types of wings. In particular, the separation first occurs at the wingtip and causes the reduction of aileron efficiency which may lead to the flight safety problems. However, as pointed out above, the plane shape of the trapezoidal wing is closest to the ellipse, and the lift-to-drag ratio characteristics are the best. Therefore, the trapezoidal wing is often used, but measures must be taken to improve the wingtip stall characteristics.

The commonly used improvement methods are as follows:

(1) Adopt geometric twist, the angle of attack of the wingtip region is reduced by using external wash twist to avoid the premature stall of the wingtip, the twist angle $\phi = -2° \sim -4°$;

Fig. 9.42 Stall characteristics of a trapezoidal wing

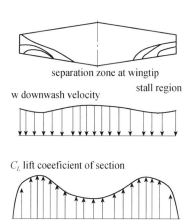

separation zone at wingtip

stall region

w downwash velocity

C_L lift coeeficient of section

(2) Adopt aerodynamic twist, the airfoil with a high stall angle of attack is selected near the wingtip;

(3) Use slat at the outer part of the wing to make the airflow with higher pressure flow from the lower wing surface to the upper surface through the gap, thus accelerating the airflow on the upper wing surface and delaying the separation of the boundary layer on the outer part of the wing.

9.6.4 Common Methods of Controlling Wing Separation

In order to slow down the spanwise flow and restrain the wing separation, the method of generating streamwise vortices is commonly used in aircraft design.

(1) Vortex generator

Arrange a row of small fins perpendicular to the wing surface along the span at the appropriate position on the upper wing surface. The height of these small fins is equivalent to the height of the airflow boundary layer at the location, about 2–3 cm. The fins have a certain deflection angle relative to the local airflow and will produce a series of streamwise vortices during flight. These small vortices increase the mixing of the high-speed mainstream and the low-speed boundary layer flow, so that the boundary layer obtains additional kinetic energy, thereby delaying the airflow separation and stalling of the wing (as shown in Fig. 9.43). An obvious shortcoming is that these small fins will increase the drag when there is no need to restrain separation in the cruise state.

(2) Fence

Fence is generally arranged on the leading edge of the aircraft wing which is a small wing-like knife. The airflow generates streamwise vortices through the fence, which prevents the boundary layer from flowing to the outer wing to slow down the spanwise flow and restrain the wingtip separation as shown in Fig. 9.44.

 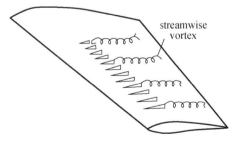

Fig. 9.43 Vortex generators and their streamwise vortices

Fig. 9.44 Fence and its streamwise vortices

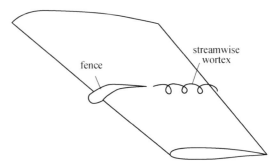

(3)   Leading-edge serrations

The leading-edge serrations are often arranged in the wingtip area, and the airflow generates streamwise vortices through the leading-edge serrations to slow down the spanwise flow and restrain wingtip separation, as shown in Fig. 9.45.

(4)   Leading-edge strake

The leading-edge strake refers to a long and narrow fin with a high swept-back that is located at the leading edge of the wing root and extends forward. The airflow generates streamwise vortices through the leading-edge strip, which can delay the separation in the root zone of the wing, increase the lift of the wing, and improve the stall characteristics of the wing as shown in Fig. 9.46.

Fig. 9.45 Leading-edge serrations and their streamwise vortices

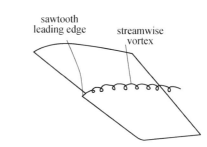

Fig. 9.46 Leading-edge strake and its streamwise vortices

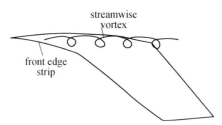

9.7 Low-Speed Aerodynamic Characteristics of a Swept-Back Wing

9.7.1 Flow Around a Swept-Back Wing

For low-speed aircraft, in order to obtain a good aerodynamic shape, a straight wing with a high aspect ratio (with or without twist) is often used in the design. However, when the speed of the aircraft increases to transonic or low supersonic, if the swept-back wing with sweep angle 35~65°is adopted, it can delay the appearance of shock wave and weaken shock wave drag. Therefore, the aircraft with high-subsonic speed and above uses all kinds of swept wings with various aspect ratios and plane shapes. The swept-back wing was first proposed by German aerodynamics Adolph Busemann (1901–1986, Prandtl's colleague, as shown in Fig. 9.47) at the Volta conference in 1935. Swept-back wing refers to the wing with both leading edge and trailing edge that are swept backward. The indicator of the degree of wing sweep is the sweep angle, which is the angle between the leading edge of the wing and the horizontal line. The aerodynamic characteristics of a swept-back wing are that it can increase the critical Mach number of the wing, delay the appearance of the shock wave, and reduce the shock wave drag in supersonic flight. When the air flows around the swept-back wing, the aerodynamic force of the wing is determined by the velocity component perpendicular to the leading edge of the wing, rather than the incoming flow velocity of the straight wing, so the swept-back wing will generate a shock wave at higher flight speed, which delays the generation of shock wave on the wing surface. Even if a shock wave appears, it also helps to weaken the shock wave intensity and reduce the shock wave drag.

Fig. 9.47 German aerodynamicist Adolph Busemann (1901~1986)

Now a swept-back wing of equal chord length is placed in the wind tunnel. When the incoming flow V_∞ flows around the wing with a small positive angle of attack, it can be found that the upper surface of the wing has an S-shaped streamline as shown in Fig. 9.48. In order to analyze this characteristic of the flow around the swept-back wing, we first discuss the flow around the oblique wing with infinite span. Suppose the sweep angle of the oblique wing with infinite span is χ, and the flow velocity V_∞ can be decomposed into the following: one is normal velocity $V_n = V_\infty \cos \chi$ which is perpendicular to the leading edge, and the other is spanwise velocity $V_t = V_\infty \sin \chi$ which is parallel to the leading edge as shown in Fig. 9.49. Taking the coordinate system $x'Oz'$, the streamline that passes any point on the upper wing surface is approximated as

$$\frac{dx'}{dz'} = \frac{u'}{w'} \approx \frac{u'}{V_\infty \sin \chi} \tag{9.76}$$

Fig. 9.48 S-shaped streamlines around the upper wing surface of the swept-back wing

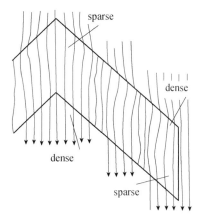

Fig. 9.49 Decomposition of leading-edge velocity

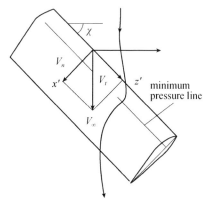

Among them, u' is the velocity component along the x'-direction (perpendicular to the leading edge), and w' is the velocity component along the z'-direction (parallel to the leading edge). When $u' = V_\infty \cos \chi$, the directions of upstream and downstream streamlines are consistent with the incoming flow. As the airflow approaches the leading edge of the wing, u' gradually decreases and the streamline deviates outward; when the normal velocity is zero, $u' = 0$, $w' = V_\infty \sin \chi$, the streamline deviates the most outward. As the airflow accelerates and bypasses the leading edge, u' increases rapidly, and the streamline starts to deflect in the opposite direction. When the minimum pressure point is reached, $u' = u'_{max}$, $w' = V_\infty \sin \chi$, and the streamline deviates inward most. After the minimum pressure point, u' decreases, and the streamline deflection continues to decrease and finally approaches to inflow direction, and the streamline that bypasses the upper wing surface presents an S shape. The swept-back wing can be considered to be composed of two symmetrical oblique wings. The flow around the middle part of the half-span of the swept-back wing is very close to that of an oblique wing with an infinite span. The analysis conclusion of the oblique wing with infinite span can be used to qualitatively analyze the influence of the sweep angle on the flow around the wing.

9.7.2 Load Distribution Characteristics of a Swept-Back Wing

Swept-back wing with a limited wingspan has a wing root and wingtip region. Due to the influence of S-shaped streamlines, "wing root effect" and "wingtip effect" will be caused, which will cause the aerodynamic characteristics of the swept-back wings with infinite wingspan and oblique wings are different. For the wing root effect, it can be seen from Fig. 9.48 that in the front section of the upper surface of the wing root, the streamlines deviate from the symmetry plane, and the flow tube expands and becomes thicker, while in the rear section, the streamlines deflect inward, and the stream tube shrinks and becomes thinner. At low speed or subsonic speed, due to the flow tube in the front section becomes thicker, the flow velocity decreases and the pressure rises (the suction force becomes smaller); while in the rear section, the flow tube becomes thinner, the flow velocity increases, and the pressure decreases (the suction force becomes larger). As for the wingtip effect, the density of the streamlines is just the opposite of the wing root area, where the suction force at the front section of the wing profile becomes larger, and the suction force at the rear section becomes smaller. Therefore, the pressure distribution in the wing root and wingtip area will be different from that in the middle area as shown in Fig. 9.50.

The wing root effect and wingtip effect of the swept-back wing cause changes in the pressure distribution of the wing chord, which is more obvious in the front section of the upper wing surface. Because the front section of the upper surface contributes a lot to the lift, the wing root effect reduces the lift coefficient of the wing root part, while the wingtip effect increases the lift coefficient of the wingtip

Fig. 9.50 Wing root and wingtip effect

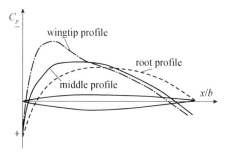

Fig. 9.51 The profile lift coefficient distribution along the span of the swept-back wing

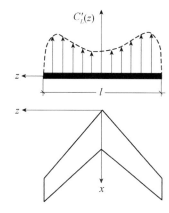

part, resulting in the distribution of the lift coefficient of the swept-back wing profile along the spanwise direction as shown in Fig. 9.51.

From the perspective of vorticity induction, there is a downwash at the wing root due to the attached vortices of the swept-back wing, as shown in Fig. 9.52. Compared with the pressure distribution on a straight wing, the suction peak of the root area of the swept-back wing is significantly smaller than that of a straight wing, as shown in Fig. 9.53. Furthermore, there is a favorable pressure gradient along the span of the swept-back wing, while the straight wing has an adverse pressure gradient along the span. Therefore, the spanwise speed of the swept-back wing is much greater than that of the straight wing, so the wing tip of the swept-back wing is prone to stall.

9.7.3 Aerodynamic Characteristics of an Oblique Wing with Infinite Span

As mentioned earlier, for an oblique wing with an infinite span, the pressure distribution is only related to the normal velocity V_n. In other words, when the incoming flow passes through an oblique wing with infinite span at a speed V_∞, the aerodynamic force on the wing is equal to the flow passes through the infinite span straight wing

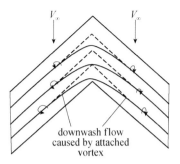

Fig. 9.52 Root vortex system of swept-back wing

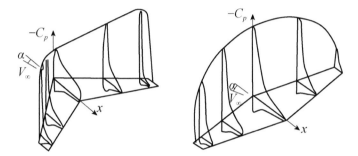

Fig. 9.53 Pressure distribution of swept-back wing and straight wing

with the oblique wing normalized at the normal velocity V_n, as shown in Fig. 9.54. Therefore, the aerodynamic force of an oblique wing with infinite span can be calculated by means of the airflow passing through the normal 2D airfoil at a normal velocity V_n, as shown in Fig. 9.55. From a geometric point of view, the relationship between the chord length b_n and the angle of attack α_n of the normal wing and the b and α of the oblique wing is

$$b_n = b \cos \chi \tag{9.77}$$

$$\sin \alpha_n = \frac{h}{b_n} = \frac{b \sin \alpha}{b \cos \chi} = \frac{\sin \alpha}{\cos \chi} \tag{9.78}$$

when α is small, $\sin \alpha \approx \alpha$, $\sin \alpha_n \approx \alpha_n$, substitute to the above equation

$$\alpha_n \approx \frac{\alpha}{\cos \chi} \tag{9.79}$$

It indicates that the chord length of the normal wing is shorter than that of the oblique wing, and the angle of attack of the normal wing is larger than that of the

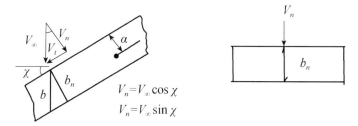

Fig. 9.54 Chord length of oblique wing and normal wing

Fig. 9.55 Angle of attack of oblique wing and normal wing

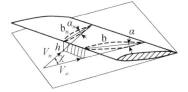

oblique wing. According to the definition, the pressure at a certain point on the wing is p, and its pressure coefficient C_P

$$C_p = \frac{p - p_\infty}{\frac{1}{2}\rho_\infty V_\infty^2} = \frac{p - p_\infty}{\frac{1}{2}\rho_\infty V_\infty^2 \cos^2 \chi} \cos^2 \chi = C_{pn} \cos^2 \chi \qquad (9.80)$$

In the equation, the footnote n indicates the normal wing, as shown in Fig. 9.56. For the dz_n of the normal wing, assume the lift acting on the normal wing as L, and the lift coefficient C_{Ln} as

$$C_{Ln} = \frac{L}{\frac{1}{2}\rho V_n^2 b_n dz_n} \qquad (9.81)$$

Due to $b_n = b \cos \chi$, $dz_n = dz / \cos \chi$, $V_n = V_\infty \cos \chi$, substitute to Eq. (9.81), we get

Fig. 9.56 The area relationship between oblique wing and normal wing

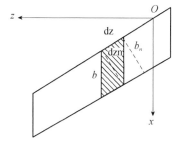

$$C_{Ln} = \frac{L}{\frac{1}{2}\rho V_\infty^2 \cos^2 \chi\, b \cos \chi\, dz / \cos \chi} = \frac{L}{\frac{1}{2}\rho V_\infty^2 b\, dz \cos^2 \chi} = \frac{C_L}{\cos^2 \chi} \qquad (9.82)$$

Or written as

$$C_L = \frac{L}{\frac{1}{2}\rho V_\infty^2 b\, dz} = \frac{L \cos^2 \chi}{\frac{1}{2}\rho V_\infty^2 \cos^2 \chi\, b \cos \chi\, dz / \cos \chi} = C_{Ln} \cos^2 \chi \qquad (9.83)$$

Suppose the drag acting on the unit wingspan of a normal wing is D_n, then the drag coefficient C_{Dn} in the V_n-direction is

$$C_{Dn} = \frac{D_n}{\frac{1}{2}\rho V_n^2 b_n\, dz_n} \qquad (9.84)$$

The drag in the V_∞-direction acting on the corresponding section of the oblique wing is $D = D_n \cos \chi$, so the drag coefficient C_D is

$$C_D = \frac{D_n \cos \chi}{\frac{1}{2}\rho V_\infty^2 b\, dz} = \frac{D_n \cos^3 \chi}{\frac{1}{2}\rho V_\infty^2 \cos^2 \chi\, b \cos \chi\, dz / \cos \chi} = C_{Dn} \cos^3 \chi \qquad (9.85)$$

The lift line slope of the oblique wing is

$$C_L^\alpha = \frac{dC_L}{d\alpha} = \frac{d(C_{Ln} \cos^2 \chi)}{d(\alpha_n \cos \chi)} = \frac{dC_{Ln}}{d\alpha_n} \cos \chi = (C_L^\alpha)_n \cos \chi$$

It can be seen from the above results that the pressure coefficient, lift coefficient, lift line slope, and drag coefficient of the oblique wing are smaller than those of the corresponding normal wing. In addition, no matter for low speed or high speed, the simple sweep theory relationship between the infinite oblique wing and the normal wing holds.

9.8   Lifting-Surface Theory of Wing

For a wing with a small sweep angle and a relatively large aspect ratio, the aerodynamic characteristics obtained by the lifting-line theory are in good agreement with the experimental results. However, for wings with larger sweep angles or smaller aspect ratios, neither the lifting-line theory nor the profile hypothesis can correctly describe the actual flow field. The calculated aerodynamic characteristics deviate greatly from the experiment, and the lifting-surface theory needs to be used instead.

Fig. 9.57 Flow around 3D
thin airfoil

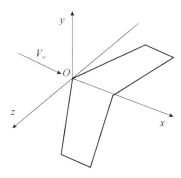

9.8.1 Aerodynamic Model of Lifting-Surface

Figure 9.57 represents the flow around a slightly curved thin wing with an incoming flow V_∞ and a small angle of attack α. Take the wind axis Oxyz. The upper and lower surfaces of the wing are very close to the Oxz-plane, and their projections on the Oxz-plane are basic planes. To solve the angle of the attack-curvature problem of wings with large sweep angles or small and medium aspect ratios although can still use the Π-shaped horseshoe vortex as the basic solution to superimpose the straight uniform flow, the use of one attached vortex line instead of the attached vortex surface of the wing should be abandoned. It is necessary to replace the wing with an attached vortex surface. At this time, the vortex density is the surface vortex density $\gamma(\xi, \zeta)$, the unit is the speed. The lifting-surface model is a combination of straight and uniform flow with attached vortex surface and free vortex surface.

The lifting-surface theory is aimed at the flow around a slightly curved thin wing at a small angle of attack. The attached vortex surface on the wing and the free vortex surface dragged backward are assumed to be in the Oxz-plane.

9.8.2 Integral Equation of Vortex Surface Intensity $\gamma(\xi, \zeta)$

As shown in Fig. 9.58, take any element area $d\xi d\zeta$ on the projection of the wing on the Oxz-plane, the strength of the attached vortex AB is $\gamma d\xi$, and the strength of free vortices AC and BD that extend downstream from the two corner points of the attached vortex AB is also $\gamma d\xi$, the two are in the opposite direction, and they are dragged to infinity in the direction of the incoming flow. According to the Biot–Savart law, the induced speed of arbitrary line segment EF to P (as shown in Fig. 9.58) is

$$v_P = \frac{\Gamma}{4\pi h}(\sin \gamma_2 - \sin \gamma_1) \tag{9.86}$$

where h is the distance from point P to line EF.

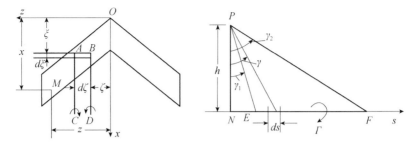

Fig. 9.58 The layout of the vortex sheet and the induced velocity of the straight vortex

Now, using the basic Eq. (9.86), the induced velocity of horseshoe vortex on microsegment CA-AB-BD is derived as follows. For any point $M(x, z)$ on the surface, the induced velocity is

$$dv_M = dv_{CA} + dv_{AB} + dv_{BD} \qquad (9.87)$$

The induced velocity of the attached vortex AB is

$$
\begin{aligned}
dv_{AB} &= \frac{\gamma d\xi}{4\pi(x-\xi)}\left[\frac{z-\zeta}{\sqrt{(x-\xi)^2+(z-\zeta)^2}} - \frac{z-\zeta-d\zeta}{\sqrt{(x-\xi)^2+(z-\zeta-d\zeta)^2}}\right] \\
&= -\frac{\gamma d\xi}{4\pi(x-\xi)}\frac{\partial}{\partial\zeta}\left[\frac{z-\zeta}{\sqrt{(x-\xi)^2+(z-\zeta)^2}}\right]d\zeta \\
&= \frac{\gamma d\xi}{4\pi}\frac{x-\xi}{\left[(x-\xi)^2+(z-\zeta)^2\right]^{3/2}}d\zeta \qquad (9.88)
\end{aligned}
$$

The induced velocity of free vortex CA to point M (negative sign means upwash) is

$$dv_{CA} = \frac{-\gamma d\xi}{4\pi(z-\zeta-d\zeta)}\left[1 + \frac{x-\xi}{\sqrt{(x-\xi)^2+(z-\zeta-d\zeta)^2}}\right] \qquad (9.89)$$

The induced velocity of free vortex BD to point M is

$$dv_{BD} = \frac{\gamma d\xi}{4\pi(z-\zeta)}\left[1 + \frac{x-\xi}{\sqrt{(x-\xi)^2+(z-\zeta)^2}}\right] \qquad (9.90)$$

Substituting the above equations into Eq. (9.87), the induced velocity on $M(x, z)$ by the horseshoe vortex CABD is obtained

$$dv_M = \frac{-\gamma d\xi}{4\pi(z-\zeta-d\zeta)}\left[1 + \frac{x-\xi}{\sqrt{(x-\xi)^2 + (z-\zeta-d\zeta)^2}}\right]$$
$$+ \frac{\gamma d\xi}{4\pi}\frac{x-\xi}{\left[(x-\xi)^2 + (z-\zeta)^2\right]^{3/2}}d\zeta$$
$$+ \frac{\gamma d\xi}{4\pi(z-\zeta)}\left[1 + \frac{x-\xi}{\sqrt{(x-\xi)^2 + (z-\zeta)^2}}\right] \qquad (9.91)$$

After sorting and simplification, we get

$$dv_M = \frac{\gamma d\xi}{4\pi}\left[\frac{1}{z-\zeta} - \frac{1}{z-\zeta-d\zeta}\right]$$
$$+ \frac{\gamma d\xi(x-\xi)}{4\pi}\left[\begin{array}{c}\dfrac{1}{(z-\zeta)\sqrt{(x-\xi)^2 + (z-\zeta)^2}} \\[2mm] -\dfrac{1}{(z-\zeta-d\zeta)\sqrt{(x-\xi)^2 + (z-\zeta-d\zeta)^2}}\end{array}\right]$$
$$+ \frac{\gamma d\xi}{4\pi}\frac{x-\xi}{\left[(x-\xi)^2 + (z-\zeta)^2\right]^{3/2}}d\zeta \qquad (9.92)$$

$$dv_M = -\frac{\gamma d\xi}{4\pi}\frac{\partial}{\partial\zeta}\left[\frac{1}{z-\zeta}\right]d\zeta$$
$$- \frac{\gamma d\xi(x-\xi)}{4\pi}\frac{\partial}{\partial\zeta}\left[\frac{1}{(z-\zeta)\sqrt{(x-\xi)^2 + (z-\zeta)^2}}\right]d\zeta$$
$$+ \frac{\gamma d\xi}{4\pi}\frac{x-\xi}{\left[(x-\xi)^2 + (z-\zeta)^2\right]^{3/2}}d\zeta \qquad (9.93)$$

$$dv_M = -\frac{\gamma d\xi}{4\pi}\frac{d\zeta}{(z-\zeta)^2}$$
$$- \frac{\gamma}{4\pi}\frac{x-\xi}{(z-\zeta)^2\sqrt{(x-\xi)^2 + (z-\zeta)^2}}d\xi d\zeta$$
$$= -\frac{1}{4\pi}\frac{\gamma d\xi d\zeta}{(z-\zeta)^2}\left[1 + \frac{x-\xi}{(z-\zeta)^2\sqrt{(x-\xi)^2 + (z-\zeta)^2}}\right] \qquad (9.94)$$

Integrate on the wing surface xOz, and the induced velocity of all attached vortices and free vortex sheets on the wing surface at point M is

$$v_M(x,z) = -\frac{1}{4\pi}\iint\limits_S \frac{\gamma(\xi,\zeta)}{(z-\zeta)^2}\left[1 + \frac{x-\xi}{(z-\zeta)^2\sqrt{(x-\xi)^2 + (z-\zeta)^2}}\right]d\xi d\zeta$$
$$(9.95)$$

Now examine the surface condition of the wing. The equation for the middle camber of the wing is

$$y = y(x, z) \tag{9.96}$$

The normal vector is

$$\vec{n} = \frac{\partial y}{\partial x}\vec{i} - \vec{j} + \frac{\partial y}{\partial z}\vec{k} \tag{9.97}$$

The boundary condition (non-penetration condition) of the potential flow surface is

$$\vec{V}_s \cdot \vec{n} = u_s \frac{\partial y}{\partial x} - v_s + w_s \frac{\partial y}{\partial z} = 0 \tag{9.98}$$

For a slightly curved thin wing at a small angle of attack, assuming that the wing's disturbance to the flow is small, the above equations can be linearized. In the wind axis coordinate, the velocity component at any point in the flow field is

$$\begin{cases} u = V_\infty + u' \\ v = v' \\ w = w' \end{cases}$$

In the equation, u', v', w', respectively, represent the disturbance velocity components. Under small disturbance conditions, these quantities are all first-order small quantities of y, $\frac{\partial y}{\partial x}$. Now take the wing surface boundary condition, which is approximately satisfied on the $y = 0$ plane, then there is

$$(V_\infty + u')_s \frac{\partial y}{\partial x} - v'_s + w'_s \frac{\partial y}{\partial z} = 0 \tag{9.99}$$

$$v'_s = v'(x, 0, z) + \frac{\partial v'}{\partial y}dy \approx v'(x, 0, z) \tag{9.100}$$

After simplification, the surface boundary condition is obtained as

$$v_s(x, 0, z) = V_\infty \frac{\partial y}{\partial x} \tag{9.101}$$

Because $V_s(x, 0, z) = -V_M$, use Eq. (9.95) to obtain the basic equation of vortex sheet intensity

$$\frac{\partial y}{\partial x} = \frac{1}{4\pi V_\infty} \iint_S \left[\frac{\gamma(\xi, \zeta)}{(z - \zeta)^2} \left(1 + \frac{x - \xi}{\sqrt{(x - \xi)^2 + (z - \zeta)^2}} \right) \right] d\xi d\zeta \tag{9.102}$$

In principle, as long as $\gamma(\xi, \zeta)$ can be solved by a mathematical method, the aerodynamic characteristics of the wing can be obtained.

9.8.3 A Numerical Method

(1) The vortex lattice method

It is very difficult to mathematically obtain the analytical solution of the integral equation of the vortex sheet intensity, so people resort to the numerical method. Falkner approximates the continuous variation of the circulation along the wingspan as a stepped circulation distribution. In the chord direction, four discrete attached vortices are used to replace the chordwise continuous vortex lines. Free vortices are dragged out at the two ends of each attached vortex which extend to infinity along the incoming flow direction. Weissinger further simplified the above discrete model, concentrated the attached vortices on the 1/4 chord point of the wing, and took the points on the 3/4 chord line to satisfy the boundary conditions (called control points). Both of these simplified models have been widely used. Due to the rapid development of computers, discrete horseshoe vortices distributed in spanwise and chordwise directions can be arranged in multiple layers. A commonly used numerical method is called the vortex lattice method.

The specific method is to divide the wing into several columns parallel to the x-axis along the spanwise on the Oxz projection surface and then divide it into several rows along the chord line of equal percentage, and the entire projection surface is divided into a finite number of tiny surface elements, which are called lattices. A horseshoe vortex is arranged on each lattice, its attached vortex line coincides with the 1/4 chord line of the lattice element, and two free vortex lines extend from the two end points of the 1/4 chord line to infinity downstream along the x-axis, as shown in Fig. 9.59. The vortex lattice method not only arranges discrete horseshoe vortices along the spanwise direction, but also arranges discrete horseshoe vortices along the chordwise direction. The entire wing is replaced by a limited number of discrete horseshoe vortices. The strength of each horseshoe vortex is constant, but the strength of vortices on different lattices is different. The lattice of the vortex distribution is called the vortex lattice, and the corresponding aerodynamic model is called the vortex lattice model.

The midpoint of the 3/4 chord line of each vortex lattice is taken as the control point. At these points, the induced velocity caused by all discrete horseshoe vortices is calculated, and the non-penetration condition on the wing surface is satisfied. The reason for choosing the 3/4 chord line point as the control point is derived from the 2D airfoil. For a 2D flat wing (as shown in Fig. 9.60), if a vortex of strength Γ is placed at its 1/4 chord point to replace the airfoil, it can be proved that the 3/4 chord point satisfies the non-penetration surface condition. In the thin wing theory, the solution to the angle of attack-curvature problem is $A_0 = \alpha$, $A_1 = A_2 = \cdots A_n = \cdots = 0$, then

Fig. 9.59 Vortex lattice
method

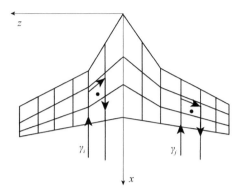

Fig. 9.60 Control points of
the flat airfoil

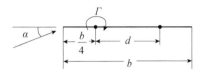

$$\Gamma = \pi V_\infty b (A_0 + \frac{A_1}{2}) = \pi V_\infty b \alpha$$

The surface boundary condition is satisfied at the control point

$$-\frac{\Gamma}{2\pi d} + V_\infty \sin \alpha = 0 \tag{9.103}$$

d is solved as

$$d = \frac{\Gamma}{2\pi V_\infty \sin \alpha} = \frac{\pi V_\infty b \alpha}{2\pi V_\infty \alpha} = \frac{b}{2} \tag{9.104}$$

The above equation shows that 0.5 b after the 1/4 chord point, that is, at the 3/4 chord point, the non-penetration condition of the object surface is satisfied, so this point is selected as the control point of the vortex lattice.

The control point position of the 3D wing is an extended conclusion drawn from the 2D airfoil, taking the 3/4 chord point as the control point. The calculation results show that the trailing edge condition is automatically met when the control point is selected in this way.

(2) Numerical solution process of vortex lattice method

1. Divide panel and arrange horseshoe vortex and control point

Assume there are N panels and N control points. Horseshoe vortices on N vortex lattices are represented by $\gamma_1, \gamma_2, ...\gamma_n$, respectively. The vortex intensity γ_j on the jth vortex lattice induces a velocity of v_{ij} on the control point of the ith vortex lattice. According to the Biot–Savart law (three-element vortex), we get

$$v_{ij} = V_\infty C_{ij} \gamma_j \tag{9.105}$$

Among them, C_{ij} is the influence coefficient. Summing j, the induced velocity of all vortex lattices to point i is

$$v_i = V_\infty \sum_{j=1}^{N} C_{ij} \gamma_j \tag{9.106}$$

2. Linear algebraic equation to determine the intensity of the vortex lattice

On the object surface, the boundary condition at the ith control point is

$$v_i = V_\infty \sum_{j=1}^{N} C_{ij} \gamma_j = V_\infty \frac{\partial y}{\partial x}\bigg|_i = V_\infty \beta_i \tag{9.107}$$

From this, the linear algebraic equation about γ_j is

$$\sum_{j=1}^{N} C_{ij} \gamma_j = \beta_i, \beta_i = \frac{\partial y}{\partial x}\bigg|_i, (i = 1, 2, \cdots, N) \tag{9.108}$$

3. Calculate aerodynamic force and aerodynamic coefficient

First find out the induced velocity V_j of all vortices to the midpoint of the attached vortex line on the jth vortex panel, and then calculate the aerodynamic force ΔL_j on the panel according to the Kutta–Joukowski lift circulation theorem, which is

$$\Delta L_j = \rho(\vec{V}_\infty + \vec{V}_j) \times \Delta \vec{\Gamma}_j, \quad \Delta \Gamma_j = \gamma_j \Delta \xi \Delta \zeta \tag{9.109}$$

Then sum up to get the total aerodynamic force and aerodynamic coefficient.

9.9 Low-Speed Aerodynamic Characteristics of a Wing with a Small Aspect Ratio

9.9.1 Vortex Lift

Generally, a wing with an aspect ratio $\lambda < 3$ is called a small aspect ratio wing. Because the small aspect ratio wing has the characteristic of low wave drag when flying at supersonic speed, this kind of wing is often used in tactical missiles and supersonic fighters. Its basic shapes are triangle, chamfered triangle, double triangle, and so on. Usually use sharp-edged non-bending symmetrical thin wings. It is found

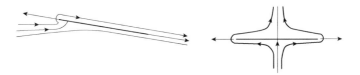

Fig. 9.61 Leading-edge suction and side-edge suction at low angle of attack

in practice that for thin wings with rounded corners and small aspect ratios, the flow at a small angle of attack is attached, and there is a leading-edge suction at the leading edge, as shown in Fig. 9.61. However, when the angle of attack is greater than 3–4°, the airflow around the leading edge will change from an attached flow to a separated flow. The high-pressure airflow on the lower wing surface bypasses the side edge and flows to the upper surface, which will inevitably cause separation at the side edge and form a detached vortex on the upper surface, as shown in Fig. 9.62. The appearance of these detached vortices will generate greater negative pressure on the upper surface, resulting in greater lift. The vortex-induced lift caused by the separation of the leading edge is often called vortex lift. Figure 9.63 shows the leading-edge vortices in experiment and practice. Figure 9.64 shows the surface pressure distribution and lift characteristic curve of a small aspect ratio wing. Figure 9.65 gives the complex vortex system of the canard layout aircraft at a high angle of attack.

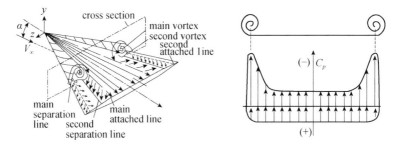

Fig. 9.62 Leading-edge separation vortex and vortex lift

Fig. 9.63 Leading-edge vortex in experiment and practice

Fig. 9.64 Pressure distributions and lift characteristic curves on the upper wing surface of a small aspect ratio wing

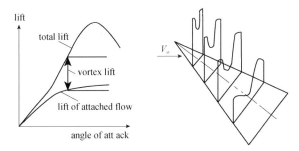

Fig. 9.65 The complex vortex system of a canard layout aircraft at a high angle of attack

9.9.2 Leading-Edge Suction Analogy

In 1966, American aerodynamicist Pohlhamus E.C. proposed the leading-edge suction analogy method. The basic idea of this method is to decompose the total lift of the wing with a detached vortex into two parts: potential lift and vortex lift. The normal force generated by the vortex on the wing surface is equal to the suction force generated by passing the leading edge, and the direction rotates 90° upwards. Physically, this analogy actually assumes that when the airflow separates at the leading edge and reattaches to the upper surface of the wing, the force required to maintain the flow balance around the separation vortex is equal to the suction of potential flow which keeps the attached flow around the leading edge. For a sharp-edged delta wing with a small aspect ratio, at a larger angle of attack, due to the presence of a detached vortex dragging downward on the wing surface, the lift characteristic curve has obvious nonlinear characteristics. The method of attached flow with a large aspect ratio is not suitable, and the leading-edge suction analogy is especially proposed for this kind of small aspect ratio wing. For the lift coefficient, there are

$$C_L = C_{Lp} + C_{Lv} \qquad (9.110)$$

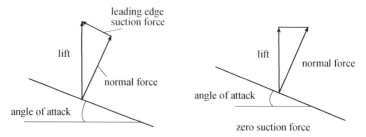

Fig. 9.66 The aerodynamic forces of flows at low and high angles of attack (with and without leading-edge separation)

Among them, C_{Lp} is the potential lift coefficient, and C_{Lv} is the vortex lift coefficient. For a sharp-edged delta wing with a small aspect ratio, the linearized small disturbance potential flow lift at a small angle of attack is different from the potential flow lift L_p at a larger angle of attack. The former does not separate when the airflow bypasses the wing, and there is a leading-edge suction peak, its potential flow lift includes the contribution of normal force and leading-edge suction; the latter will separate when the airflow bypasses the wing, the leading-edge suction is lost, but the separation flow reattaches on the upper surface, and its potential flow lift is only the projected part of the normal force which is perpendicular to the incoming flow direction, as shown in Fig. 9.66.

9.9.3 Potential Flow Solution of a Small Aspect Ratio Wing

Assuming that the flow pasts the wing with a small aspect ratio is attached, take the $oxyz$ coordinate system as shown in Fig. 9.67, the velocity component of the flow is

$$u = V_\infty \cos \alpha + u'$$
$$v = V_\infty \sin \alpha + v'$$
$$w = w' \tag{9.111}$$

where u', v', w' are the disturbance velocity components, and the corresponding velocity potential function is φ' which is

$$u' = \frac{\partial \varphi'}{\partial x}, \quad v' = \frac{\partial \varphi'}{\partial y}, \quad w' = \frac{\partial \varphi'}{\partial z} \tag{9.112}$$

For a slender delta wing, assume φ' is determined by the 2D potential flow of wing profile in a spanwise direction which satisfies the governing equation

Fig. 9.67 Cross potential flow of delta wing

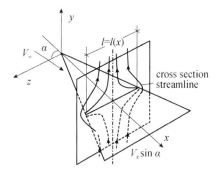

$$\frac{\partial^2 \varphi'}{\partial y^2} + \frac{\partial^2 \varphi'}{\partial z^2} = 0 \tag{9.113}$$

For a straight wing with no thickness, the flow around the profile is equivalent to the potential flow of the transverse flow $V_\infty \sin \alpha$ around the same length plate, and its velocity potential function is

$$\varphi' = \pm V_\infty \sin \alpha \sqrt{(l/2)^2 - z^2} \tag{9.114}$$

where $l = l(x)$ is the span length at x, and positive and negative represent the velocity potential function of the upper and lower wing surfaces. Using Eq. (9.114), we can get

$$u' = \frac{\partial \varphi'}{\partial x} = \pm \frac{V_\infty \sin \alpha}{2} \frac{l}{\sqrt{l^2 - 4z^2}} \frac{dl}{dx} \tag{9.115}$$

According to Bernoulli's equation, the pressure on the wing surface is

$$p = p_\infty + \frac{1}{2}\rho V_\infty^2 - \frac{1}{2}\rho\left(V_\infty + u' + v' + w'\right)^2 \approx p_\infty + \frac{1}{2}\rho V_\infty^2 - \rho V_\infty u' \tag{9.116}$$

Substituting Eq. (9.115) into Eq. (9.116), we get

$$\Delta p = p_d - p_u \approx \frac{\rho V_\infty^2 \sin \alpha}{2} \frac{2l}{\sqrt{l^2 - 4z^2}} \frac{dl}{dx} \tag{9.117}$$

Integrating the above equation and projecting it perpendicular to the flow direction, the potential flow lift is obtained as

$$L_p = \frac{\rho V_\infty^2 \sin \alpha \cos \alpha}{2} \int_0^{b_0} l(x) \frac{dl}{dx} \cdot \int_{-1}^{1} \frac{d\zeta}{\sqrt{1 - \zeta^2}} dx \tag{9.118}$$

where $\zeta = 2z/l$. Due to

$$\int_{-1}^{1} \frac{d\zeta}{\sqrt{1 - \zeta^2}} = \pi \tag{9.119}$$

For the delta wing, we have

$$l(x) = \frac{2x}{tg\chi} \tag{9.120}$$

$$L_p = \frac{\rho V_\infty^2 \sin\alpha \cos\alpha}{2} \int_0^{b_0} l(x)\frac{dl}{dx} \int_{-1}^{1} \frac{d\zeta}{\sqrt{1 - \zeta^2}} dx = \pi \frac{\rho V_\infty^2 \sin\alpha \cos\alpha}{2} \frac{2b_0^2}{tg^2\chi} \tag{9.121}$$

The lift coefficient is

$$C_{Lp} = \frac{L_p}{\frac{\rho V_\infty^2}{2} \frac{b_0^2}{tg\chi}} = \frac{2\pi}{tg\chi} \sin\alpha \cos\alpha \tag{9.122}$$

Let

$$C_{Lp} = \frac{2\pi}{tg\chi \cos\alpha} \sin\alpha \cos^2\alpha = K_p \sin\alpha \cos^2\alpha \tag{9.123}$$

In the equation, K_p is the coefficient. For small angles of attack

$$C_{Lp} = K_p \alpha \tag{9.124}$$

Indicating K_p is the lift line slope of the potential flow.

9.9.4 Vortex Lift Coefficient C_{Lv}

The vortex lift increment is compared with the leading-edge suction force, that is, it is assumed that the normal force increment generated by the main vortex on the wing surface is equal to the suction at the leading edge when the flow is attached, as shown in Fig. 9.68. Combine with Eq. (9.121), at small angles of attack, the leading-edge suction

$$F_s = L\sin\alpha \propto \frac{1}{2}\rho V_\infty^2 S \sin\alpha \sin\alpha = \frac{1}{2}\rho V_\infty^2 S \sin^2\alpha \tag{9.125}$$

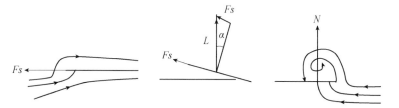

Fig. 9.68 Leading-edge suction analogy

Among them, S is the wing area. The vortex lift is

$$L_v = N \cos \alpha = F_s \cos \alpha \propto \frac{1}{2} \rho V_\infty^2 S \sin^2 \alpha \cos \alpha \qquad (9.126)$$

The vortex lift coefficient is

$$C_{Lv} = \frac{L_v}{\frac{1}{2} \rho V_\infty^2 S} = K_v \sin^2 \alpha \cos \alpha \qquad (9.127)$$

This method is suitable for aspect ratio 0.5–1.0.

9.9.5 Determination of K_p and K_v

Combining Eqs. (9.123) and (9.127), the total lift coefficient is

$$C_L = C_{Lp} + C_{Lv} = K_p \sin \alpha \cos^2 \alpha + K_v \sin^2 \alpha \cos \alpha \qquad (9.128)$$

Because the sharp-edged delta wing forms a detached vortex at a high angle of attack, the leading-edge suction is lost, so the lift-induced drag coefficient of the wing is

$$C_{Di} = C_L tg\alpha = K_p \sin^2 \alpha \cos \alpha + K_v \sin^3 \alpha \qquad (9.129)$$

For the sharp-edged delta wing, the values of K_p and K_v can be determined by the vortex sheet method (as shown in Fig. 9.69). Approximate values are

$$K_p = -0.8223 - 0.0612\lambda + 2.2142\lambda^{0.5}$$
$$K_v = 3.0941 + 0.1181\lambda - 0.0069\lambda^2 \qquad (9.130)$$

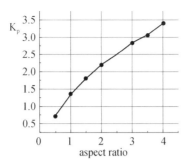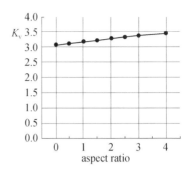

Fig. 9.69 Approximate values of K_p and K_v

9.10   Engineering Calculation Method for Low-Speed Aerodynamic Characteristics of a Wing

(1)   Spanwise distribution of profile lift coefficient

The spanwise distribution of the profile lift coefficient is an important basis for calculating the lift, moment, and induced drag of the wing. Generally, the profile lift coefficient of a straight wing with a high aspect ratio is obtained by the lifting-line theory, and a wing with a small aspect ratio or a swept-back wing is obtained by the lifting-surface theory. For a straight wing with a high aspect ratio, let

$$C'_L(z) = C'_{Lb}(z) + C'_{La}(z) \tag{9.131}$$

Among them, the first term is the basic lift coefficient, which represents the distribution of the profile lift coefficient when the entire wing lift coefficient is zero. When the wing is twisted, it depends on the local absolute angle of attack at zero lift. When the wing is not twisted, this term is zero. If the wing has no aerodynamic twist but a geometric twist, the basic lift coefficient is

$$C'_{Lb}(z) = \frac{C^\alpha_{L\infty}}{2}(\alpha_{0a} + \varphi(z)) \tag{9.132}$$

where α_{0a} is the absolute zero-lift angle of attack of the wing due to the geometric twist of the wing, that is, the angle between the zero-lift line of the wing and the zero-lift line of the middle profile.

$$\int_{-L/2}^{L/2} C'_{Lb}(z)b(z)dz = \int_{-L/2}^{L/2} \frac{C^\alpha_{L\infty}}{2}(\alpha_{0a} + \varphi(z))b(z)dz = 0$$

$$\alpha_{0a} = -\frac{1}{S} \int_{-L/2}^{L/2} \varphi(z)b(z)dz \tag{9.133}$$

For the trapezoidal wing, there is

$$\alpha_{0a} = -\frac{\varphi_1}{3}\frac{\eta+2}{\eta+1} \tag{9.134}$$

where φ_1 is the twist angle of the wingtip profile.

The second term is the distribution of the additional lift coefficient, which depends on the additional absolute angle of attack of the wing (that is, the angle between the inflow and the zero-lift line of the wing), and has nothing to do with the twist of the wing. According to the experimental results of a large number of untwisted wings, it can be considered that the magnitude of the additional lift at a certain spanwise position is proportional to the chord length and arithmetic mean of the actual chord length of the elliptical wing with the same total area and span length. Namely

$$L'_a = \frac{1}{2}\left[b(z) + b_0\sqrt{1 - (\frac{2z}{L})^2}\right]q_\infty C_L$$

$$C'_{La} = \frac{L'_a}{q_\infty b(z)} = \frac{1}{2}\left[1 + \frac{b_0}{b(z)}\sqrt{1 - (\frac{2z}{L})^2}\right]C_L, \quad b_0 = \frac{4S}{\pi L} \tag{9.135}$$

The general calculation equation is

$$C'_L(z) = C'_{Lb}(z) + C_{La}(z)C_L \tag{9.136}$$

(2) Lifting-line characteristics

The lifting line of the wing is expressed by the zero-lift angle of attack α_0, the lift line slope C_{la}, and the maximum lift coefficient C_{lmax}, which is

$$C_L = C_L^\alpha(\alpha - \alpha_0) \tag{9.137}$$

1. Zero-lift angle of attack α_0

The angle of attack at which the lift of the wing is zero (measured from the middle profile chord line) is called the zero-lift angle of attack. For a twist-free wing, $\alpha_0 = \alpha_{0\infty}$. For a high aspect ratio trapezoidal wing with a linear geometric twist, there are

$$\alpha_0 = \alpha_{0\infty} - \frac{\varphi_1}{3}\frac{\eta+2}{\eta+1} \tag{9.138}$$

2. Lift line slope of the wing

The lift line slope of the wing with arbitrary plane shape is calculated by lifting-surface theory. A common equation is

$$\frac{C_L^\alpha}{\lambda} = \frac{2\pi}{2 + \sqrt{4 + \left(\frac{\lambda}{\frac{C_{L\infty}^\alpha}{2\pi}\cos\chi_{1/2}}\right)^2}}$$
(9.139)

3. Maximum lift coefficient

$$C_{L\max} = k_s \frac{C_{L\max 0} + C_{L\max 1}}{2}$$
(9.140)

for $\eta = 1$ (rectangular wing), $k_s = 0.88$; for $\eta > 1$ (trapezoidal wing), $k_s = 0.95$. For swept-back wing, there is

$$C_{L\max} = (C_{L\max})_0 \cos\chi_{1/4}$$
(9.141)

(3) Drag characteristics

The drag coefficient of the wing can be expressed as the sum of the zero-lift drag coefficient and the lift-induced drag coefficient, which is

$$C_d = C_{d0} + C_{dL}$$
(9.142)

1. Zero-lift drag coefficient

When the wing has no camber or the camber is not large, there is

$$C_{d0} = C_{d\min} = \frac{2}{S} \int_0^{L/2} C'_{d\min}(z)b(z)dz$$
(9.143)

In the equation, $C_{D\min}$ is the minimum drag coefficient of the wing; $C'_{D\min}(z)$ is the minimum drag coefficient of the wing profile. In engineering calculation, the parameters at the geometric mean chord length can be used for calculation, which is

$$C_{d0} = (C'_{d\min})_{bav} = 0.925(2C_F\eta_c)_{bav}$$
(9.144)

$$C_{d0} = (C'_{d\min})_{bav} = (2C_F)_{bav}(` + 0.1C + 0.4C^2)$$
(9.145)

Among them, C_F is the frictional drag coefficient of the plate with the same chord length, η_c is the thickness correction coefficient, and C is the relative thickness.

2. Induced drag coefficient

The lift-induced drag coefficient is the sum of the induced drag coefficient and the viscous pressure drag coefficient.

The lift-induced drag coefficient is expressed as

$$C_{dL} = C_{di} + C_{dn} \tag{9.146}$$

The induced drag coefficient is expressed as

$$C_{di} = \frac{C_L^2}{\pi \lambda}(1 + \delta) \tag{9.147}$$

The viscous pressure drag coefficient

$$C_{dn} = kC_L^2 \tag{9.148}$$

In the equation, k is the correction coefficient, which is determined by the experiment. Approximately

$$k = \frac{0.025\lambda - \delta}{\pi \lambda} \tag{9.149}$$

The lift-induced drag factor and effective aspect ratio are

$$C_{dL} = C_{di} + C_{dn} = AC_L^2, A = \frac{1}{\pi \lambda}((1 + \delta) + \pi \lambda k) \tag{9.150}$$

The effective aspect ratio is

$$C_{dL} = \frac{C_L^2}{\pi \lambda_e}, \lambda_e = \frac{\lambda}{1 + \delta + k\pi \lambda} < \lambda \tag{9.151}$$

9.11 Aerodynamic Characteristics of Control Surfaces

9.11.1 Moment and Tails

If an aircraft wants to fly smoothly, in addition to the balance between lift and gravity, engine thrust and drag, there is also a balance of moment around the center of gravity, as shown in Fig. 9.70, lift acts on the neutral point of the wing, and the center of gravity of the aircraft is in front of the lift. The lift is not at the same point as the gravity of the aircraft, and this will produce a nose-down moment around the center of gravity of the aircraft. If there is no empennage behind the fuselage, the aircraft cannot fly stably.

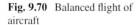

Fig. 9.70 Balanced flight of aircraft

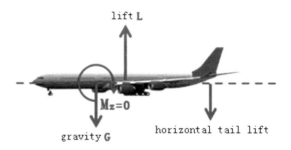

In order to make the aircraft to not bow its head during flight, there must be a moment to raise the aircraft's head, so that an empennage is placed on the tail of the fuselage to generate downward force (negative lift), just like the measuring weight of a scale, making the aircraft raise its head. When the aircraft is flying, the empennage produces a small negative lift. The upward lift and downward gravity are balanced with the negative lift, and the total moment of the aircraft's center of gravity is zero, the aircraft can fly smoothly. When designing an airplane, it is necessary to reasonably select the relative position and area of the empennage. Although the moment arm is short with a large empennage area (shorter fuselage), the negative lift is large. Large negative lift, on the one hand, will reduce the lift of the wing too much, resulting in insufficient total lift of the aircraft; on the other hand, a large empennage area will produce a large empennage drag. On the contrary, the empennage area is too small, and the moment arm is too large, which causes the fuselage to be too long and not easy to take off. Therefore, the aircraft designer must match the relative position and area of the wing and empennage reasonably, so that the aircraft can maintain a better balance of moments in various attitudes during flight. Because at small angles of attack, the lift has a linear relationship with the angle of attack. When the angle of attack is negative, the airflow around the airfoil will produce a negative lift, so the empennage is usually installed at a negative angle of attack.

9.11.2 Horizontal Tail Design

The basic requirements for the aerodynamic design of a horizontal tail are as follows:

(1) The layout design of the horizontal tail must be coordinated with the overall layout, as shown in Fig. 9.71, and the common layout of the horizontal tail is installed at the rear of the fuselage; T-tail form or crossed with the vertical tail; V-shaped or butterfly-shaped. As a general principle, it is best not to place the horizontal tail directly in the slipstream area of the propeller. Although some aircraft actually put the horizontal tail in the slipstream to improve the elevator efficiency during take-off, the slipstream usually causes the empennage buffeting, leading to fuselage noise and premature structural fatigue. In addition, rapid power changes can cause large trim changes;

Fig. 9.71 Low horizontal tail, high horizontal tail, and V-shaped empennage

(2) In order to reduce the structural weight and drag of the aircraft as much as possible, the wetted area of the empennage should be reduced as much as possible, so the moment of the horizontal tail should be as large as possible relative to the rear center of gravity of the aircraft. For the layout of the engine compartment hanging on the rear fuselage, placing the horizontal tail on the vertical tail and increasing the vertical tail sweep angle to increase the moment arm is an effective measure for keeping the horizontal tail area from being too large for this type of layout.

(3) The layout and parameter design of the horizontal tail should be combined with other parts of the aircraft to achieve the desired trim lift-to-drag ratio and the maximum lift coefficient of the flaps in each posture of stowed, take-off, and landing;

(4) The selection plane shape and parameters should meet the requirements of stability, operability, and flight quality. These requirements are given in the national military standards and civil aviation regulations. In the process of aircraft design, analysis and calculation are gradually carried out from simple to detailed, so that the horizontal tail can meet various requirements;

(5) The control efficiency of the horizontal tail and elevator must make the aircraft have good stall characteristics. For the T-tail layout, the aircraft must have the ability to recover from a deep stall;

(6) The structural layout of the horizontal tail must be coordinated with other aircraft components;

(7) The profile of the horizontal tail. The horizontal tail needs to provide lift in both the up and down directions, so most horizontal tails use symmetrical airfoils or modifications of symmetrical airfoils. The typical airfoils are NACA0009-NACA0018. If it is found that the front center of gravity trim when the flaps are lowered or the horizontal tail design when the front wheels are raised during take-off is in a critical state, the reverse camber airfoil can be used to reduce the area of the horizontal tail.

9.11.3 Vertical Tail Design

Most of the design requirements for the horizontal tail are applicable to the vertical tail. The layout of the vertical tail can be selected from the following options: installed

Fig. 9.72 Spoiler and
aileron

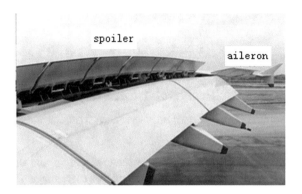

at the rear of the fuselage; twin-tail boom installation; single vertical tail or double vertical tail; V-shaped or butterfly-shaped. The vertical tail is located at the rear of the fuselage in the propeller slipstream area, which can improve the efficiency of the rudder during the take-off. In order to reduce the area of the vertical tail, the product of the lift line slope corresponding to the vertical tail position and the tail moment arm is maximized as much as possible. The plane shape design of vertical tail not only meets the stability and maneuverability, but also some aircraft manufacturers take the vertical tail shape as their trademark from the aesthetic requirements.

For the vertical tail airfoil configuration, the vertical tail adopts a symmetrical airfoil, so as to ensure that the vertical tail does not produce a yaw moment when the aircraft has no sideslip. Measures to improve the efficiency of the vertical tail and rudder include the following: (1) The influence of horizontal tail parameters on horizontal tail efficiency is also applicable to the vertical tail; (2) Reducing the gap between the leading edge of the rudder and the basic vertical tail can significantly improve the efficiency of the rudder. The use of a chordwise double-hinged rudder can greatly improve the lift coefficient of the vertical tail when the rudder is at its maximum deflection.

The main lateral control surfaces on the wing are ailerons and spoilers (as shown in Fig. 9.72). These control surfaces should be able to achieve rapid inclination changes, maintain wing level during no pitch maneuver under specified crosswind values, and provide trim and maneuverability in case of one engine failure or steady sideslip.

9.11.4   Requirements of the Lateral Control Surface for Aircraft Static Balance

(1)   During take-off, when the engine suddenly stops and the rudder is loosened, the lateral control efficiency of the aileron and spoiler should be sufficient to complete the sideslip coordination. For small speeds and small flying weights, the lateral control efficiency should be sufficient for static control, and there

should be remaining control efficiency to control the overshoot of the sideslip angle;

(2) In the normal flight envelope, the lateral control function formed by the aileron and the spoiler should exceed that of the rudder, especially in the landing configuration of low speed and weight, the lateral control function should be sufficient to overcome the sideslip of full rudder deflection, and there should be margin to control the disturbance related to the gust;

(3) In the take-off and landing configuration, the lateral control should be sufficient to control the failure of a single leading-edge device unless the probability of failure is shown to be extremely low;

(4) Requirements of aircraft roll response on the lateral control surface. The lateral control should be sufficient to achieve the rapid change of the aircraft's roll angle.

9.11.5 Aerodynamic Requirements for Aileron Configuration

(1) On the premise of meeting the lateral control requirements of the aircraft, the aileron span should be as small as possible in order to make the flap span as long as possible;

(2) On a high-speed wing, in order to prevent the deflection load of the aileron on the thin wing from causing the wing to twist in the opposite direction (aileron reversal) and reduce the aileron efficiency, high- and low-speed ailerons can be set. The outer aileron can only be used at low speed, while the high-speed aileron set in the wing root area can be used at both high and low speed. A suitable position is after the (inboard) engine, because if the flap is set at this position, it may be affected by the jet;

(3) In order to reduce the aileron area, the lateral control of high-speed aircraft usually adopts the joint control of the spoiler and aileron. Most of the lateral control is provided by the spoiler on the aircraft with deflector/aileron.

9.11.6 Basic Requirements for Spoiler Configuration

(1) The spoiler has many functions. It is called a flight spoiler for lateral control, deceleration plate (brake plate) for increasing drag in the air, and ground spoiler for reducing the lift of the ground motion;

(2) The spoiler is generally arranged on the upper surface of the main wing before the trailing edge flap, immediately behind the rear beam, and often accounts for most of the flap span;

(3) The spoiler generally has inner, middle, and outer parts, and its function depends on the needs of the aircraft type. The inner spoiler is generally used to make the ground running brake more effective and reduce the lift. The middle and outer

spoilers are often used as lateral control and speed brake. More use of the outer spoiler can prevent the inner and middle spoilers from disturbing the tail flow and preventing buffet;

(4) During lateral control, the spoiler often forms a linkage relationship with the aileron, and its law is determined according to the aircraft type characteristics and the requirements of the lateral control efficiency in each flight stage;

(5) The deflection of the outer spoiler decreases with the increase of dynamic pressure, which can reduce the power required by the hydraulic actuator of the spoiler and the rigidity requirements of the spoiler itself;

(6) Lateral control is usually the combined control of aileron and spoiler. This kind of control form has the advantages of simple structure, light weight, and good maintainability and can meet the requirements of lateral control within a certain speed range. At high speed, the efficiency of the aileron is reduced due to the aeroelasticity, but it has little effect on the control efficiency of the spoiler. When the spoiler is used for deceleration and drag increases during landing, it increases the force on the tires and makes the brake deceleration more effective.

Exercises

1. There is a straight trapezoidal wing, $S = 35$ m^2, taper ratio $\eta = 4$, and wingtip chord length $b_1 = 1.5$ m, find the aspect ratio λ of the wing.

2. Given the sweep angle χ_0 of the delta wing and the aspect ratio λ, try to prove from the geometric relationship that $\lambda \tan \chi_0 = 4$.

3. Consider a finite wing with an aspect ratio of 8 and a taper ratio of 0.8. The airfoil profile is thin and symmetrical. Calculate the lift and induced drag coefficient of the wing at an angle of attack of 5°. Assume that $\delta = \tau$.

4. Given a trapezoidal swept-back wing, the wingtip chord length b_1, the wing root chord length b_0, the taper ratio $\eta = b_0/b_1$, and the spanwise length l, try to derive the calculation formulas of area, aspect ratio, the tangent of middle line sweep angle, and mean aerodynamic chord length of the wing from the geometric relationship.

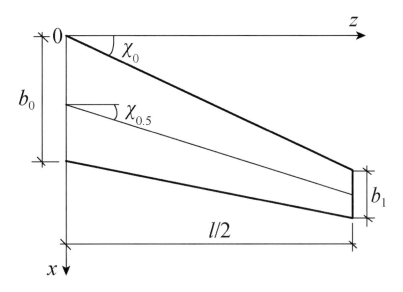

$$S = b_1 \frac{1+\eta}{2} l$$

$$\lambda = \frac{l^2}{S}, \frac{b_1}{l} = \frac{2}{\lambda(1+\eta)}, b_0 = \frac{2l\eta}{\lambda(1+\eta)}, tg\chi_0 - tg\chi_1 = \frac{4(\eta-1)}{\lambda(\eta+1)}$$

$$b_A = \frac{2}{S} \int_0^{l/2} b(z)^2 dz = \frac{4}{3} \frac{l}{\lambda} \left[1 - \frac{\eta}{(1+\eta)^2} \right]$$

5. Consider a rectangular wing with an aspect ratio of 6, an induced drag coefficient $\delta = 0.055$, and a zero-lift angle of attack of $-2°$. When the angle of attack is $3.4°$, the induced drag coefficient of the wing is 0.01. Calculate the induced drag coefficient of a similar wing (a rectangular wing with the same airfoil profile) with the same angle of attack and an aspect ratio of 10. Assume that the inducing factors δ and τ of drag and lift slope are, respectively, equal (i.e., $\delta = \tau$). Similarly, for $\lambda = 10$, $\delta = 0.105$.

6. Assume that the spanwise circulation distribution of a straight wing with a high aspect ratio is parabolic, $\Gamma(z) = \Gamma_0 \left[1 - \left(\frac{2z}{l} \right)^2 \right]$, as shown in the figure. If the total lift is equal to the elliptical circulation distribution wing, try to find the corresponding relationship of Γ_0 and w_i on the symmetry plane of the two circulation distributions.

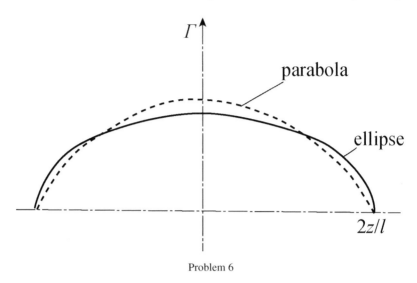

Problem 6

7. The measured lift slope of NACA 23,012 airfoil is 0.1080 per degree, and $\alpha_{L=0} = -1.3$. Consider the finite wing using this airfoil, $\lambda = 8$, taper ratio $= 0.8$. Assume that $\delta = \tau$. Calculate the lift and induced drag coefficient of the wing with the geometric angle of attack $7°$.

8. Given that the spanwise circulation distribution of a high aspect ratio wing is $\Gamma(z) = \Gamma_0 \left[1 - \left(\frac{2z}{l}\right)^2\right]^{3/2}$, try to use the lifting-line theory to solve

   (1) Downwash speed w_i at $z = l/4$
   (2) Downwash speed w_i at $z = l/2$

   Tip: use integral

   $$\int_0^\pi \frac{\cos(n\theta)}{\cos\theta - \cos\theta_1} d\theta = \pi \frac{\sin(n\theta_1)}{\sin\theta_1}$$

9. If the wing is replaced by a Ⅱ-shaped horseshoe vortex line, the span length of the attached vortex is L, as shown in Fig. 6.6 of the exercises, try to prove

   (1) The downwash angle at position a after the attached vortex at the middle of the wing is

   $$\alpha_i = \frac{C_L}{2\pi\lambda}\left[1 + \frac{\sqrt{a^2 + (l/2)^2}}{a}\right]$$

   where C_L is the lift coefficient and λ is the aspect ratio.

(2) If the airfoil has $C_{l\infty}^a = 2\pi$, assuming that the aspect ratio correction adopts the correction of the elliptical wing, the rate of change of the downwash angle at a position the back of the middle wing to the angle of attack is

$$\frac{\mathrm{d}\alpha_i}{\mathrm{d}\alpha} = \frac{1}{\lambda + 2}\left[\frac{\sqrt{a^2 + (l/2)^2}}{a} + 1\right]$$

And calculate the value of $\frac{\mathrm{d}\alpha}{\mathrm{d}\alpha}$; when $\lambda = 8$, $a = 0.4l$.

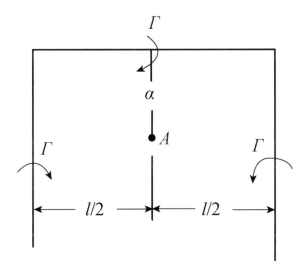

Problem 9

10. A light single-engine general aircraft has a wing area of 15 square meters, a wingspan of 9.7 m, and a maximum gross weight of 1111 kg. The wing adopts NACA 65–415 airfoil, the lift coefficient is 0.1033 per degree, and $\alpha_{L=0} = -3°$. Assume $\tau = 0.12$. If the aircraft cruises at 120 mph at standard sea level at its maximum gross weight. Calculate the geometric angle of attack of the wing in straight and level flight.

11. An airplane with a weight of $G = 14700$ N, Cruising $(Y = G)$ at $h = 3000$ m with $V\infty = 300$ km/h, wing area $S = 17\,\mathrm{m}^2$, $\lambda = 6.2$, NACA23012 airfoil $(a_{0\infty} = -4°, C_l^\alpha = 0.108/(°)$, non-twisted elliptical plane shape. Try to calculate C_L, α, and C_{Dv}.

12. There is a monoplane with a weight of $G = 7.38 \times 10^4$ N, the wing is an elliptical plane shape, the wingspan length is $L = 15.23$ m, and it is flying straight at the sea level with a speed of 90 m/s, try to calculate the induced drag D_v and the value of Γ_0 at the wing root profile.

13. Try to prove that if a horseshoe vortex line with a span length of l and a strength of the circulation Γ_0 of the wing root profile of the original wing is used to simulate the total lift of the elliptical wing with a span length of l, it can be obtained

$$\frac{l'}{l} = \frac{\pi}{4}$$

14. A curved airfoil, $a_{0\infty} = -4°$, $C_l^\alpha = 2\pi/rad$. If this airfoil is placed on an elliptical wing with $\lambda = 5$ and no twist, try to find the C_L at $\alpha = 8°$.

15. For a flat delta wing with $\lambda = 3$, known $C_{L\infty}^\alpha = 2\pi$, try to use the engineering calculation method to find the values of C_L^α and $\frac{x_F}{b_A}$ at small α.

16. For a rectangular wing, $\lambda = 6, l = 12$ m, wing load $G/S = 900$ N/m². Try to calculate the induced drag and the ratio of induced drag to total lift when the aircraft is flying at sea level with $V\infty = 150$km/h.

17. The C_L-α curve of a non-twisted straight wing with $\lambda = 9$, $\eta = 2.5$ under a certain Reynolds number is shown in the figure. From the figure, it can be obtained that $a_{0\infty} = -1.5°$, $C_l^\alpha = \frac{0.084}{(°)}$, $C_{lmx} = 1.22$. If other parameters remain the same, but λ is reduced to 5, find the a_0 and C_l^α at this time and draw the C_L-α curve of the wing when $\lambda = 5$.

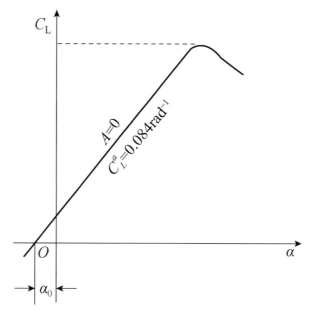

Problem 17

18. Consider a finite wing with an aspect ratio of 6. Suppose the lift is distributed in an elliptical shape. The lift slope of the airfoil profile is 0.1/degree. Calculate and compare the lift coefficients of (a) straight wing and (b) swept-back wing, the half-chord sweep angle is 45°.

19. A civil aircraft uses a trapezoidal swept-back wing with $\lambda = 8$, $\eta = 2$, $\chi_0 = 45^0$. Use the engineering calculation method to calculate the C_L^α and $\frac{x_F}{b_A}$ of the wing.

20. When the angle of attack is small, the polar curve of the wing can be expressed as a parabola $C_D = C_{D0} + \lambda C_L^2$, try to prove

$$K_{max} = \left(\frac{C_L}{C_D}\right)_{max} = \frac{1}{2\sqrt{\lambda C_{D0}}}, \quad (C_L)_{K_{max}} = \sqrt{\frac{C_{D0}}{\lambda}}$$

Chapter 10
Aerodynamic Characteristics of Low-Speed Fuselage and Wing-Body Configuration

This chapter introduces the low-speed aerodynamic characteristics of the aircraft wing-body configuration, including the aerodynamic characteristics of the low-speed fuselage, the theory and application of the flow around the slender axis-symmetric body, the engineering estimation method of the low-speed aerodynamic characteristics of the wing-body configuration and the numerical calculation of the flow around the aircraft wing, etc.

Learning points:

Familiar with the basic concepts and aerodynamic characteristics of the low-speed wing-body configuration, the theory and application of the slender axis-symmetric body flow, the engineering estimation method of the low-speed aerodynamic characteristics of the wing-body configuration, etc.

10.1 Overview of Aerodynamic Characteristics of Low-Speed Fuselage

10.1.1 Introduction

The aircraft fuselage is not only a component used to carry personnel, cargo, weapons, and airborne equipment but also a connector of the wing, tail, landing gear, and other components. In light aircraft and fighter aircraft, the engine is often installed in the fuselage. The drag of the fuselage in flight accounts for about 30–40% of the whole aircraft drag. Therefore, the slender and streamlined fuselage plays an important role in reducing aircraft drag and improving flight performance.

© Science Press 2022
P. Liu, *Aerodynamics*, https://doi.org/10.1007/978-981-19-4586-1_10

Since the pilots, passengers, cargo, and airborne equipment are all concentrated on the fuselage, most of the requirements related to the use of the aircraft (such as the visual field of the pilot, the environmental requirements of the cockpit, the loading and unloading of cargo and weapons, the inspection and maintenance of system equipment, etc.) have a direct impact on the shape and structure of the fuselage. Similarly, when the bird is soaring, it also pursues a low-drag body shape. At this time, the bird puts its legs under the stomach, and after that, the legs are covered with feathers to form a slender cone to reduce air drag. The shape of the seagull's body is a cone with a large aspect ratio when it is soaring. Aspect ratio refers to the ratio of body length to maximum diameter. Generally, when the aircraft is between 6 and 13, the air drag is small. Imitating the body shape of a bird, the fuselage of a human-built aircraft is also a conical body with a large aspect ratio to obtain the minimum drag.

For transport aircraft, the fuselage is divided into five segments: nose, front fuselage, middle fuselage, rear fuselage, and tail cone according to the processed structural surface (as shown in Fig. 10.1). The basic tasks and structural requirements of each segment are different. The basic requirements for the overall aerodynamic design of the fuselage shape are:

(1) Under a given dynamic pressure, drag mainly depends on the shape of the aircraft and the wet area. To meet the same passenger and cargo requirements, the cross-sectional area of the fuselage should be as small as possible to reduce the windward drag of the fuselage. This is the criterion for the design of the fuselage section.

(2) The aspect ratio of the actual fuselage can be determined according to the overall layout. For subsonic aircraft with a cruising Mach number lower than 0.85, the design should be as close as possible to the slender streamlined shape, only if the commercial load is not affected.

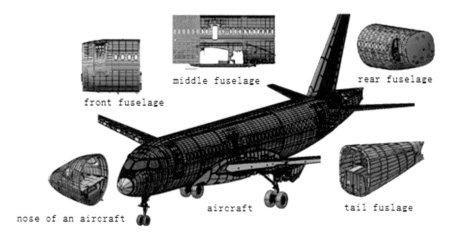

Fig. 10.1 Structure of the fuselage

(3) The shape of the front fuselage is mainly based on the smooth transition requirements of the nose radome, cockpit, and skylight glass, and fully considers the overall layout requirements of vision, cockpit, and electronic equipment cabin, to avoid separation and increase drag aerodynamically. The aspect ratio of the front fuselage (the ratio of length to diameter of the front fuselage) is usually 2.0–2.5.

(4) The aspect ratio of the rear fuselage (the ratio of length to diameter of the rear fuselage) is usually 3.0–3.5, if the shrinkage is too fast, it is easy to cause boundary layer separation. To obtain the required front wheel lift angle during take-off or landing (the rear fuselage does not scratch the ground), the rear fuselage is often slightly tilted up. From the point of view of drag, the tail tilt angle (which determines the scratching angle of the aircraft) should not exceed 6°–7°. If the tail tilt angle is too large, especially for the rear fuselage of a cargo aircraft. Due to the influence of wing downwash, landing gear fairing drum and the flow around the rear fuselage, there may be large adverse interference, such as the formation of unstable vortex system at the rear of the fuselage, which causes lateral oscillation, especially in low-speed and large flap deflection angle. The tail tilt angle of large aircraft fuselage produces great drag during cruise.

10.1.2 Geometric Parameters of Axis-Symmetric Body

The fuselage of an airplane is mostly made into an axis-symmetric body. The so-called axis-symmetric body refers to the volume formed by a smooth (or broken line) generatrix rotating around an axis, such as cone, cylinder, sphere, etc. The axis is the rotation axis. Any section of the body perpendicular to the rotation axis is circular, and any plane passing through the rotation axis is called the meridian plane. The boundary shape of the axis-symmetric body on any meridian plane is the same. The intersecting line between the axis-symmetric body and the meridian plane is the generatrix. The trajectory of a point rotating around the axis is called the weft circle. In order to reduce the drag of fuselage, the fuselage shape is closely related to the flight speed of the aircraft, involving the development of the boundary layer (related to friction drag and pressure drag), boundary layer separation, shock wave shape and control (related to wave drag), etc. Generally, for low-speed and subsonic aircraft, the fuselage shape is round nose and pointed tail (streamlined). The fuselage shape of supersonic aircraft is pointed nose (to reduce wave drag), as shown in Fig. 10.2.

As shown in Fig. 10.3, the axis-symmetric body is divided into head, middle, and tail along the axial direction. The length of the axis-symmetric body is expressed by

Fig. 10.2 Shapes of the fuselage

Fig. 10.3 Shape and geometric parameters of the axis-symmetric body

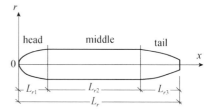

L_r, $R(x)$ is the radius of the axis-symmetric body along the axial direction, R_{max} is the maximum diameter of the axis-symmetric body, S_{max} is the maximum cross-sectional area of the body, R_d is the radius of the bottom circle of the axis-symmetric body, L_{r1}, L_{r2} and L_{r3} are the length of the head, middle and tail of the axis-symmetric body, respectively, $\lambda = \frac{L_r}{2R_{max}}$ is the aspect ratio of the body, $\lambda_2 = \frac{L_{r2}}{2R_{max}}$ is the aspect ratio of the head, $\lambda_2 = \frac{L_{r2}}{2R_{max}}$ is the aspect ratio of the middle, $\lambda_3 = \frac{L_{r3}}{2R_{max}}$ is the aspect ratio of the tail.

10.2   Theory and Application of Slender Body

In view of the characteristics of the axis-symmetric body, when the air flows along the symmetry axis of the axis-symmetric body, the flow around the body is the same on any meridian plane, which is called axisymmetric flow. Its characteristics are: (1) the air flows in the plane passing through the axis of rotation; (2) all the flow properties passing through the meridian plane are the same. The low and subsonic axisymmetric flow of the axis-symmetric body is similar to the flow around a symmetrical airfoil, which can be superimposed by sources arranged on the axis. For supersonic axisymmetric flow, shock waves and expansion waves will appear in the flow field. When the incoming flow has an angle of attack, the flow around the axis-symmetric body is asymmetric, the boundary layer on the leeward side is thickened, and the boundary layer on the windward side is thinned. When the angle of attack reaches a certain value, the boundary layer on the upper surface of the axis-symmetric body separates, and vortices appear, as shown in Fig. 10.4.

Fig. 10.4 Boundary layer separation and vortices of a slender body at a high angle of attack

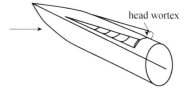

10.2.1 Linearized Potential Flow Equation in Cylindrical Coordinate System

Generally, the aircraft adopts the axis-symmetric fuselage that is relatively slender, and the angle of attack of the aircraft is relatively small. Therefore, the disturbance of the fuselage to the flow field is a small disturbance. The entire flow field can be regarded as an irrotational flow, and the disturbance velocity potential function satisfies the linearized potential flow equation. Taking the cylindrical coordinate system shown in Fig. 10.5, at any point $P(x, r, \theta)$ in the flow field, the disturbance velocity components are, respectively, v_x, v_r, v_θ. In the case of ideal potential flow, there is a disturbance velocity potential function φ, and the velocity components are

$$v_x = \frac{\partial \varphi}{\partial x}, \quad v_r = \frac{\partial \varphi}{\partial r}, \quad v_\theta = \frac{1}{r}\frac{\partial \varphi}{\partial \theta} \tag{10.1}$$

The velocity components of the incoming flow are

$$\begin{aligned}
V_{x\infty} &= V_\infty \cos \alpha \\
V_{r\infty} &= V_\infty \sin \alpha \cos \theta \\
V_{\theta\infty} &= -V_\infty \sin \alpha \sin \theta
\end{aligned} \tag{10.2}$$

At small angle of attack, the incoming velocity components can be simplified as

$$\begin{aligned}
V_{x\infty} &\approx V_\infty \\
V_{r\infty} &\approx V_\infty \alpha \cos \theta \\
V_{\theta\infty} &\approx -V_\infty \alpha \sin \theta
\end{aligned} \tag{10.3}$$

At small angle of attack, the velocity component at any point in the flow field can be written as the sum of the undisturbed velocity component and the disturbed velocity component.

$$V_x = V_{x\infty} + v_x \approx V_\infty + v_x$$

Fig. 10.5 Cylindrical coordinate system

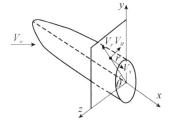

$$V_r = V_{r\infty} + v_r \approx V_\infty \alpha \cos \theta + v_r$$
$$V_\theta = V_{\theta\infty} + v_\theta \approx -V_\infty \alpha \sin \theta + v_\theta \qquad (10.4)$$

Under small disturbances, ignoring the second-order small quantities, the energy equation is

$$a^2 = a_\infty^2 - \frac{\kappa - 1}{2}(V^2 - V_\infty^2) \approx a_\infty^2 - (\kappa - 1)V_\infty v_x \qquad (10.5)$$

When the Mach number is not close to 1, and it is not a very large supersonic flow, omit the small quantities higher than the first order, the disturbance velocity potential function equation is obtained as

$$(1 - M_\infty)^2 \frac{\partial^2 \varphi}{\partial x^2} + \frac{\partial^2 \varphi}{\partial r^2} + \frac{1}{r^2}\frac{\partial^2 \varphi}{\partial \theta^2} + \frac{1}{r}\frac{\partial \varphi}{\partial r} = 0 \qquad (10.6)$$

The application condition of the above equation is: the axis-symmetric body must be a slender body, the slope of any point on the spiral generatrix is less than 1, which means that the apex of the axis-symmetric body must be sharp, and the incoming Mach number is not too close to 1, nor very high. The boundary condition that the disturbance velocity potential function needs to meet is that the airflow non-penetration condition is satisfied on the surface of the axis-symmetric body. that is

$$\left(\frac{V_r}{V_x}\right)_{r=R} = \left[\frac{\frac{\partial \varphi}{\partial r} + V_\infty \alpha \cos \theta}{V_\infty + \frac{\partial \varphi}{\partial x}}\right]_{r=R} = \frac{\mathrm{d}R(x)}{\mathrm{d}x} \qquad (10.7)$$

For the flow around the fuselage with an angle of attack (the boundary layer does not separate), since the equation is linear, the disturbance velocity potential function can be decomposed into the axial potential flow and the lateral potential flow velocity potential function, as shown in Fig. 10.6, which is

$$\varphi = \varphi_1 + \varphi_2$$
$$\varphi_1 \; generated \; by \; incoming \; flow \; V_\infty \cos \alpha \approx V_\infty.$$
$$\varphi_2 \; generated \; by \; incoming \; flow \; V_\infty \sin \alpha \approx V_\infty \alpha. \qquad (10.8)$$

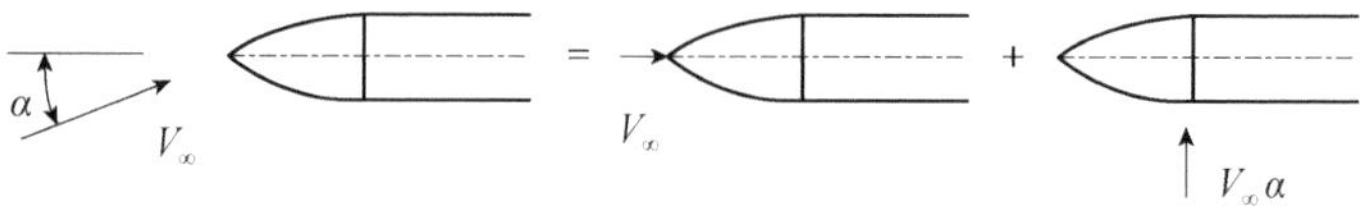

Fig. 10.6 Decomposition of small disturbance velocity potential function

For the flow around a subsonic slender fuselage, the governing equation of the axial velocity potential function is

$$(1 - M_\infty)^2 \frac{\partial^2 \varphi_1}{\partial x^2} + \frac{\partial^2 \varphi_1}{\partial r^2} + \frac{1}{r} \frac{\partial \varphi_1}{\partial r} = 0 \tag{10.9}$$

The governing equation of the lateral velocity potential function is

$$(1 - M_\infty)^2 \frac{\partial^2 \varphi_2}{\partial x^2} + \frac{\partial^2 \varphi_2}{\partial r^2} + \frac{1}{r^2} \frac{\partial^2 \varphi_2}{\partial \theta^2} + \frac{1}{r} \frac{\partial \varphi_2}{\partial r} = 0 \tag{10.10}$$

It can be seen from the above two equations that the relationship between (axisymmetric) and φ_2 (non-axisymmetric) is

$$\varphi_2 = \frac{\partial \varphi_1}{\partial r} \cos \theta \tag{10.11}$$

Substitute the surface condition of the axis-symmetric body, there is

$$\left(V_\infty + \frac{\partial \varphi_1}{\partial x} + \frac{\partial \varphi_2}{\partial x} \right)_{r=R} \frac{dR(x)}{dx} = \left(V_\infty \alpha \cos \theta + \frac{\partial \varphi_1}{\partial r} + \frac{\partial \varphi_2}{\partial r} \right)_{r=R} \tag{10.12}$$

When $\alpha = 0$, the surface condition of axial flow is

$$\left(V_\infty + \frac{\partial \varphi_1}{\partial x} \right)_{r=R} \frac{dR(x)}{dx} = \left(\frac{\partial \varphi_1}{\partial r} \right)_{r=R} \tag{10.13}$$

Ignore the second-order small quantities, we have

$$\left(\frac{\partial \varphi_1}{\partial r} \right)_{r=R} = V_\infty \frac{dR(x)}{dx} \tag{10.14}$$

The surface boundary condition for lateral flow (decomposition of Eq. (10.12)) is

$$\left(\frac{\partial \varphi_2}{\partial x} \right)_{r=R} \frac{dR(x)}{dx} = \left(V_\infty \alpha \cos \theta + \frac{\partial \varphi_2}{\partial r} \right)_{r=R} \tag{10.15}$$

Ignore the second-order small quantities, we have

$$\left(\frac{\partial \varphi_2}{\partial r} \right)_{r=R} = -V_\infty \alpha \cos \theta \tag{10.16}$$

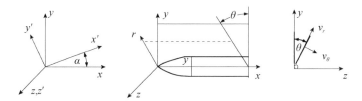

Fig. 10.7 Coordinate system transformation relationship

The expression of the pressure coefficient of the slender axis-symmetric body in the wind axis coordinate system is

$$C_p = -\frac{2}{V_\infty}\frac{\partial \varphi'}{\partial x} - \frac{1}{V_\infty^2}\left[\left(\frac{\partial \varphi'}{\partial y}\right)^2 + \left(\frac{\partial \varphi'}{\partial z}\right)^2\right] \tag{10.17}$$

Use the relationship between the wind coordinate system and the body coordinate system, as shown in Fig. 10.7.

The relationship between the wind and the body coordinate system is

$$x = x'\cos\alpha - y'\sin\alpha \approx x' - y'\alpha$$
$$y = y'\cos\alpha + x'\sin\alpha \approx y' + x'\alpha$$
$$z = z' \tag{10.18}$$

Therefore, we have

$$\varphi_x' = \frac{\partial \varphi}{\partial x}\frac{\partial x}{\partial x'} + \frac{\partial \varphi}{\partial y}\frac{\partial y}{\partial x'} + \frac{\partial \varphi}{\partial z}\frac{\partial z}{\partial x'} = \varphi_x + \varphi_y\alpha \tag{10.19}$$

And

$$\varphi_y = v_y = v_r\cos\theta - v_\theta\sin\theta = \varphi_r\cos\theta - \frac{1}{r}\varphi_\theta\sin\theta$$
$$\varphi_r = \varphi_r', \ \varphi_\theta = \varphi_\theta' \tag{10.20}$$

Therefore, for the body coordinate system, the expression of the pressure coefficient is

$$C_p = -\frac{2\varphi_x}{V_\infty} - \frac{1}{V_\infty}\left[(V_\infty\alpha\cos\theta + \varphi_r)^2 + \left(V_\infty\alpha\sin\theta - \frac{1}{r}\varphi_\theta\right)^2 - V_\infty^2\alpha^2\right] \tag{10.21}$$

$$C_p = -\frac{2}{V_\infty}\left(\frac{\partial \varphi_1}{\partial x} + \frac{\partial \varphi_2}{\partial x}\right)$$

$$-\frac{1}{V_\infty^2}\left[\left(V_\infty \alpha \cos\theta + \frac{\partial \varphi_1}{\partial r} + \frac{\partial \varphi_2}{\partial r}\right)^2 + \left(V_\infty \alpha \sin\theta - \frac{1}{r}\frac{\partial \varphi_2}{\partial \theta}\right)^2 - V_\infty^2 \alpha^2\right]$$

$$(10.22)$$

It can be seen from the above equations that when there is a flow with an angle of attack, the pressure coefficient at any point in the flow field is generally not equal to the sum of the pressure coefficient generated by the axial flow and the pressure coefficient generated by the transverse flow. Only when the pressure coefficient on the surface of the axis-symmetric body is calculated, the pressure coefficient has superposition. Suppose the length of the slender body is l, the bottom area is $S(l)$, if the tail area is zero, use the largest cross-sectional area. The incoming flow velocity V_∞, the angle of attack α, the normal force is N, and the axial force is A. Therefore, the calculation equations of lift and drag of subsonic flow are as follows:

$$\begin{aligned} F_L &= N\cos\alpha - A\sin\alpha \\ F_d &= N\sin\alpha + A\cos\alpha \end{aligned}$$

$$(10.23)$$

For a slender body with a small angle of attack, the normal force and axial force can be obtained from the potential flow theory, which is

$$\begin{aligned} N &= \tfrac{1}{2}\rho_\infty V_\infty^2 S(L_r)(2\alpha) \\ A &= -\tfrac{1}{2}\rho_\infty V_\infty^2 S(L_r)(\alpha^2) \end{aligned}$$

$$(10.24)$$

In this way, the lift and drag coefficients of the slender body are

$$\begin{aligned} C_L &= \frac{N\cos\alpha - A\sin\alpha}{\tfrac{1}{2}\rho_\infty V_\infty^2 S(L_r)} \approx 2\alpha \\ C_d &= \frac{N\sin\alpha + A\cos\alpha}{\tfrac{1}{2}\rho_\infty V_\infty^2 S(L_r)} \approx \alpha^2 \end{aligned}$$

$$(10.25)$$

If the maximum area S_{max} is taken as the characteristic area, the lift and drag coefficients become Fig. 10.8.

$$C_L \approx 2\alpha\frac{S(L_r)}{S_{\text{max}}}, \quad C_D \approx \alpha^2\frac{S(L_r)}{S_{\text{max}}}$$

$$(10.26)$$

Fig. 10.8 Aerodynamic
forces of a slender body

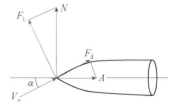

10.2.2   Cross-Flow Theory at High Angles of Attack

If a separation vortex above the leeward surface appears in the slender body flow, it
will have a great influence on the surface pressure distribution. Therefore, the separation of cross-flow needs to be considered when calculating the normal force and the
axial force. Due to the axis-symmetric body is slender, assuming that the surface pressure of the lateral flow is the same as the pressure of the incoming flow $V_\infty \sin \alpha$ passes
the cross-section cylinder, the resulting normal force is approximately regarded as
the drag of the flow around a cylinder.

As shown in Fig. 10.9, the additional normal force on the per unit length of the
axis-symmetric body can be expressed as

$$N = 2\alpha \left(\frac{1}{2}\rho_\infty V_\infty^2 \right) S(L_r) + C_x \left(\frac{1}{2}\rho_\infty V_\infty^2 \right)\alpha^2 S(L_r) \qquad (10.27)$$

The lift coefficient and drag coefficient are

$$\begin{aligned}
C_L &= \frac{N \cos \alpha - A \sin \alpha}{\frac{1}{2}\rho_\infty V_\infty^2 S(L_r)} \approx 2\alpha + C_x \alpha^2 \\
C_d &= \frac{N \sin \alpha + A \cos \alpha}{\frac{1}{2}\rho_\infty V_\infty^2 S(L_r)} \approx \alpha^2 + C_x \alpha^3
\end{aligned} \qquad (10.28)$$

In the equation, C_x is the drag coefficient of the cross-flow separation on 2D cylinder.
The comparison between the above theoretical results and the experimental data

Fig. 10.9 Cross-flow
separation and vortices

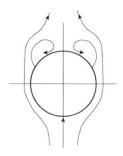

finds that the viscosity influence is greater, the lift coefficient is much larger than the calculation result of the potential flow theory, and the pressure center is also moved back a lot, the results calculated by the cross-flow theory are more consistent with the experimental results.

10.3 Engineering Estimation Method for Aerodynamic Characteristics of Wing-Body Assembly

The aircraft flies under the wing-body configuration, so the aerodynamic forces of each component will be affected by each other, which is different from the aerodynamic forces of individual components. This is mainly due to the changes in the boundary state of each component. The lift of a wing-body configuration cannot simply be obtained by adding the lift of a single wing to the lift of a single fuselage. To be precise, once the wing and the fuselage are combined, the flow field flowing through the fuselage will change the flow field flowing through the wing, and vice versa. This phenomenon is called wing-body interference, and there is an interference term in aerodynamic force. In addition to the interference of lift, there is also a drag interference of wing-body configuration. As in the case of lift, due to the influence of the wing-body interference, the drag of the aircraft cannot be obtained by simply adding the drag of each part.

(1) Interference of fuselage to wing

   Due to the influence of fuselage cross-flow, an upwash flow will be generated on the wing, so the effective angle of attack of the wing near the fuselage will be increased, resulting in additional lift, as shown in Fig. 10.10.

(2) Interference of wing to fuselage

   If the lift effect of the wing is replaced by vortices arranged along the wing chord plane, these vortices will produce induced velocities in the vertical direction, which will be combined with the incoming velocity, and the distribution of angle of attack along the fuselage axis will be changed. Under the induction of attached vortex, the flow near fuselage head washes up and the effective angle

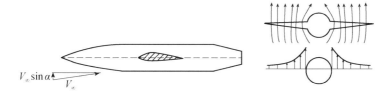

$V_\infty \sin \alpha$ V_∞

Fig. 10.10 Wing and fuselage flow

Fig. 10.11 Interference of
the wing on the fuselage

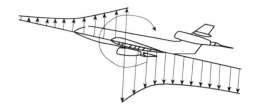

of attack increases; under the downwash induction of wing free vortex sheet, the
angle of attack in the middle of fuselage decreases; At the junction of the wing
and fuselage, the air flows along the wing, when there is no wing installation
angle, the angle of attack is zero; otherwise, it is the flow with an angle of
attack. Because the wing produces lift, the fuselage also produces additional
lift, as shown in Fig. 10.11.

From the flow phenomenon and interference of the wing-body configuration, the
lift L_t generated by the wing-body configuration with single middle wing and no
installation angle can be written as

$$L_t = L_{ws} + L_{sw} \tag{10.29}$$

Among them, L_{ws} is the wing lift generated by the presence of the fuselage; L_{sw} is
the fuselage lift generated by the presence of wing. Define the characteristic area of
the wing (the area of the wing alone or the exposed area of the wing) as S_0, then the
lift coefficient of the wing-body configuration is

$$C_{Lt} = \frac{L_{ws} + L_{sw}}{\frac{1}{2}\rho_\infty V_\infty^2 S_0} \tag{10.30}$$

10.4   Numerical Calculation of Wing Flow

Computational fluid dynamics is a subject developed in the 1960s. It constitutes the
three major branches of modern fluid dynamics with theoretical fluid dynamics and
experimental fluid dynamics. Computational fluid dynamics is playing an increas-
ingly important role in the industry. Especially for aircraft design, computational
fluid dynamics has become a universal and important analysis tool. Take the CRM
model as an example. As shown in Fig. 10.12, a supercritical wing with a wing span
of 58.8 m, wing area is 383.7 m^2, quarter-chord sweep angle of the wing is 35°, wing
aspect ratio is 9, and wing thickness of 9.5% from wingtip to 15.5% of wing root, the
thickness gradually decreases from wing root to wingtip. Wing twist angle of $-3.8°$

Fig. 10.12 Supercritical wing

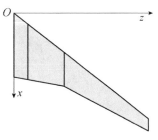

from wingtip to 6.8° of wing root, the wing root is twisted positively, the wingtip is twisted negatively, washout. Figure 10.13 shows the mesh near the fuselage and wing for numerical calculation. Figure 10.14 shows the pressure distribution when the angle of attack is 2.7°, and the Mach number is 0.8. Figure 10.15 shows the streamlines around the wing when the angle of attack is 2.7° and the Mach number is 0.8.

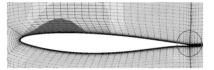

Fig. 10.13 Mesh near fuselage and wing for numerical calculation

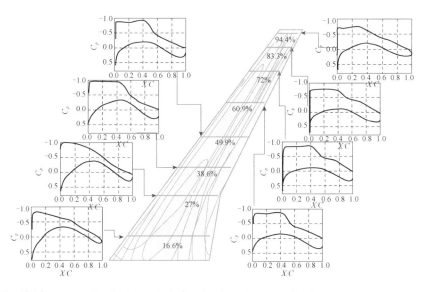

Fig. 10.14 Pressure distribution at 2.7° of angle of attack and Mach 0.8

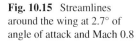
Fig. 10.15 Streamlines around the wing at 2.7° of angle of attack and Mach 0.8

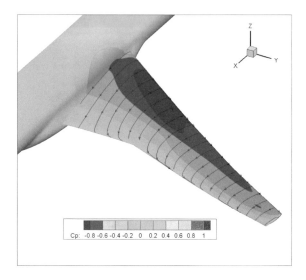

Exercises

1. Explain the general principles of aerodynamic design for slender fuselage of transport aircraft.
2. Give the linearization assumption and governing equation of slender axis-symmetric body.
3. Explain the basic idea of cross-flow theory of slender body flow at high angle of attack.
4. Explain the physical concept of wing-body interference in the wing-body configuration flow.
5. Briefly describe the separation phenomenon of slender axis-symmetric body flow at high angle of attack.

Chapter 11
Aerodynamic Characteristics of Subsonic Thin Airfoil and Wing

This chapter introduces the flow characteristics of thin subsonic airfoils and wings, including the flow around the subsonic compressible airfoil, the derivation of the ideal steady compressible velocity potential function equation, the theory of small disturbance linearization, the theoretical solution of the two-dimensional subsonic streamline around the corrugated wall, the compressibility correction method, and its application (the Prandtl-Glauert compressibility correction). The results show that the influence of Mach number on the aerodynamic characteristics of the wing is discussed.

Learning points:

(1) Learn the flow around the thin airfoil with subsonic velocity and the derivation of the ideal steady compressible velocity potential function equation;
(2) Learn the theory and solution process of small disturbance linearization and the method of compressibility correction and its application (the compressibility correction of Prandtl-Glauert, Kármán-QianXuesen, and Wright compressibility);
(3) Understand the flow and aerodynamic characteristics of a thin subsonic wing and the influence of Mach number on the aerodynamic characteristics of the wing.

11.1 Subsonic Compressible Flow Around an Airfoil

It has been known that when the Mach number is less than 0.3, the compressibility of the air can be neglected and treated as if it were an incompressible flow. When the Mach number is greater than 0.3, the effect of compressibility must be taken into account, otherwise large errors will occur. If there is subsonic flow everywhere in the flow field, the flow field is called the subsonic flow field. There is no essential difference between the flow over a subsonic airfoil and the flow over a low-speed

© Science Press 2022
P. Liu, *Aerodynamics*, https://doi.org/10.1007/978-981-19-4586-1_11

Fig. 11.1 Streamlines of flow around an airfoil in incompressible and a subsonic flow

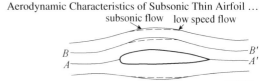

airfoil in incompressible flow, except that the expansion of the subsonic compressible flow at the contraction of the upper and lower tubes of the airfoil is larger than that of the flow over a low-speed incompressible flow, that is, the vertical disturbance caused by compressibility is larger than the low-speed incompressible flow, the impact of the scope of the disturbance is also larger. As shown in Fig. 11.1, the dotted line is the incompressible flow at low speed and the solid line is the compressible flow at subsonic speed.

This can be clearly illustrated by using the one-dimensional isentropic flow theory. As shown in Fig. 11.1, the flow tube analysis is between AA' and BB'. From the continuous equation and momentum equation of one-dimensional isentropic flow, it is known that

$$\mathrm{d}m = \mathrm{d}(\rho A V) = 0, \quad \frac{\mathrm{d}A}{A} + \frac{\mathrm{d}V}{V} + \frac{\mathrm{d}\rho}{\rho} = 0$$

$$\mathrm{d}p = -\rho V \mathrm{d}V \tag{11.1}$$

From the acoustic equation, yeild

$$a^2 = \frac{\mathrm{d}p}{\mathrm{d}\rho} \tag{11.2}$$

According to Eqs. (11.1) and (11.2), we can solve

$$\frac{\mathrm{d}A}{A} = -(1 - Ma^2)\frac{\mathrm{d}V}{V} \tag{11.3}$$

For low-speed incompressible flows, the formula (11.3) is simplified to

$$\frac{\mathrm{d}A}{A} = -\frac{\mathrm{d}V}{V} \tag{11.4}$$

From the equations above, for the same relative velocity increment $\mathrm{d}V/V$, the relative decrease of flow tube cross section $\mathrm{d}A$ in subsonic compressible flow is smaller than that inlow-velocity incompressible flow. Therefore, the local flow tube is larger than the non-compressible flow tube. This is because the density of the compressible flow decreases with the increase of velocity, so the cross-sectional area decreases less than that of the non-compressible flow in order to maintain mass conservation, that is, the flow tube is larger than the incompressible case. In a subsonic compressible flow, the relative increments of pressure and density are

$$\frac{\mathrm{d}p}{\rho a^2} = -Ma^2 \frac{\mathrm{d}V}{V}$$

$$\frac{\mathrm{d}\rho}{\rho} = -Ma^2 \frac{\mathrm{d}V}{V} \qquad (11.5)$$

It is shown that while the increment of relative velocity $\frac{\mathrm{d}V}{V}$ is kept constant, in the negative pressure region on the upper surface of the subsonic airfoil flow, the compressible flow will cause the density to decrease (the rate of change of the relative density is negative) and the flow to expand in order to keep the mass conservation in the flow tube. Compared with the incompressible flow, the compressible flow tube has to be expanded, so the streamline must be expanded vertically to increase the flow tube area.

11.2 Velocity Potential Function Equation of Ideal Steady Compressible Flow

For ideal fluid, steady, and incompressible potential flows, the velocity potential function satisfies Laplace's equation. But for the ideal fluid, steady, isentropic compressible potential flow, the governing equations for the velocity potential function no longer satisfy the incompressibility condition because the density is not a constant, so the governing equations for the velocity potential function no longer satisfy Laplace's equation condition. For ideal fluid, steady, isentropic compressible potential flow, the continuous equation is

$$\frac{\partial(\rho u)}{\partial x} + \frac{\partial(\rho v)}{\partial y} + \frac{\partial(\rho w)}{\partial z} = 0 \qquad (11.6)$$

If mass forces are not taken into account, the Euler equations of fluid motion are

$$u\frac{\partial u}{\partial x} + v\frac{\partial u}{\partial y} + w\frac{\partial u}{\partial z} = -\frac{1}{\rho}\frac{\partial p}{\partial x}$$

$$u\frac{\partial v}{\partial x} + v\frac{\partial v}{\partial y} + w\frac{\partial v}{\partial z} = -\frac{1}{\rho}\frac{\partial p}{\partial y}$$

$$u\frac{\partial w}{\partial x} + v\frac{\partial w}{\partial y} + w\frac{\partial w}{\partial z} = -\frac{1}{\rho}\frac{\partial p}{\partial z} \qquad (11.7)$$

In isentropic flow, density is just a function of pressure, it's a barotropic fluid. From $\rho = \rho(p)$ and wave equation $a^2 = \frac{\mathrm{d}p}{\mathrm{d}\rho}$, we can get

$$\frac{\partial\rho}{\partial x} = \frac{\mathrm{d}\rho}{\mathrm{d}p}\frac{\partial p}{\partial x} = \frac{1}{a^2}\frac{\partial p}{\partial x}\frac{\partial\rho}{\partial y} = \frac{1}{a^2}\frac{\partial p}{\partial y}\frac{\partial\rho}{\partial z} = \frac{1}{a^2}\frac{\partial p}{\partial z} \qquad (11.8)$$

By substituting it into the continuous Eq. (11.6), we get

$$u\frac{\partial \rho}{\partial x} + v\frac{\partial \rho}{\partial y} + w\frac{\partial \rho}{\partial z} + \rho(\frac{\partial u}{\partial x} + \frac{\partial v}{\partial y} + \frac{\partial w}{\partial z}) = 0$$

$$\frac{1}{a^2}\left(\frac{u}{\rho}\frac{\partial p}{\partial x} + \frac{v}{\rho}\frac{\partial p}{\partial y} + \frac{w}{\rho}\frac{\partial p}{\partial z}\right) + \frac{\partial u}{\partial x} + \frac{\partial v}{\partial y} + \frac{\partial w}{\partial z} = 0 \qquad (11.9)$$

By replacing the partial pressure derivative in the Euler equations with the density derivative and substituting it into the continuous equation, the equation containing only the velocity and the speed of sound is

$$\frac{-u}{a^2}\left(u\frac{\partial u}{\partial x} + v\frac{\partial u}{\partial y} + w\frac{\partial u}{\partial z}\right) - \frac{v}{a^2}\left(u\frac{\partial v}{\partial x} + v\frac{\partial v}{\partial y} + w\frac{\partial v}{\partial z}\right)$$

$$- \frac{w}{a^2}\left(u\frac{\partial w}{\partial x} + v\frac{\partial w}{\partial y} + w\frac{\partial w}{\partial z}\right) + \frac{\partial u}{\partial x} + \frac{\partial v}{\partial y} + \frac{\partial w}{\partial z} = 0 \qquad (11.10)$$

After arranged

$$(1 - \frac{u^2}{a^2})\frac{\partial u}{\partial x} + (1 - \frac{v^2}{a^2})\frac{\partial v}{\partial y} + (1 - \frac{w^2}{a^2})\frac{\partial w}{\partial z} -$$

$$\frac{uv}{a^2}\left(\frac{\partial u}{\partial y} + \frac{\partial v}{\partial x}\right) - \frac{vw}{a^2}\left(\frac{\partial v}{\partial z} + \frac{\partial w}{\partial y}\right) - \frac{uw}{a^2}\left(\frac{\partial u}{\partial z} + \frac{\partial w}{\partial x}\right) = 0 \qquad (11.11)$$

For potential flow, there is a potential function of velocity, and

$$u = \frac{\partial \varphi}{\partial x}, v = \frac{\partial \varphi}{\partial y}, w = \frac{\partial \varphi}{\partial z}$$

Substitute in (11.11)

$$(1 - \frac{u^2}{a^2})\frac{\partial^2 \varphi}{\partial x^2} + (1 - \frac{v^2}{a^2})\frac{\partial^2 \varphi}{\partial y^2} + (1 - \frac{w^2}{a^2})\frac{\partial^2 \varphi}{\partial z^2} -$$

$$2\frac{uv}{a^2}\frac{\partial^2 \varphi}{\partial x \partial y} - 2\frac{vw}{a^2}\frac{\partial^2 \varphi}{\partial y \partial z} - 2\frac{uw}{a^2}\frac{\partial^2 \varphi}{\partial x \partial z} = 0 \qquad (11.12)$$

Or

$$(a^2 - u^2)\frac{\partial^2 \varphi}{\partial x^2} + (a^2 - v^2)\frac{\partial^2 \varphi}{\partial y^2} + (a^2 - w^2)\frac{\partial^2 \varphi}{\partial z^2}$$

$$- 2uv\frac{\partial^2 \varphi}{\partial x \partial y} - 2vw\frac{\partial^2 \varphi}{\partial y \partial z} - 2wu\frac{\partial^2 \varphi}{\partial z \partial x} = 0 \qquad (11.13)$$

The Eq. (11.12) or (11.13) is called the ideal steady compressible flow full velocity potential function equation, also known as the full velocity potential equation (or full velocity potential equation), which is a second order nonlinear partial differential equation about the velocity potential function. For incompressible flows, where the speed of sound approaches infinity. Consider it in the full potential equation is replaced with Laplace's equation (second order linear partial differential equation). From the energy equation of isentropic flow, we can know

$$\frac{V^2}{2} + \int \frac{dp}{\rho} = C \tag{11.14}$$

Take the isentropic relation $p = C\rho^\gamma$, get

$$\frac{V^2}{2} + \frac{\gamma}{\gamma - 1} \frac{p}{\rho} = C \tag{11.15}$$

Among them $\frac{V^2}{2} = \frac{u^2 + v^2 + w^2}{2}$. According to the ideal gas law and the acoustic equation $\frac{p}{\rho} = RT$, $a^2 = \gamma RT$, substitute them into formula (11.15)

$$\frac{V^2}{2} + \frac{\gamma}{\gamma - 1} RT = C, \quad \frac{V^2}{2} + \frac{a^2}{\gamma - 1} = C \tag{11.16}$$

For the velocity of the incoming flow V_∞ and the pressure p_∞, we get

$$\frac{V^2}{2} + \frac{a^2}{\gamma - 1} = \frac{V_\infty^2}{2} + \frac{a_\infty^2}{\gamma - 1}$$

$$a = \sqrt{a_\infty^2 + \frac{\gamma - 1}{2}(V_\infty^2 - V^2)} \tag{11.17}$$

of velocity. Therefore, the full velocity potential equation contains only one unknown function, namely the velocity potential function. For axisymmetric flows, the equation for the velocity potential function is

$$\left(1 - \frac{u_x^2}{a^2}\right)\frac{\partial^2 \varphi}{\partial x^2} - 2\frac{u_x v_r}{a^2}\frac{\partial^2 \varphi}{\partial x \partial r} + \left(1 - \frac{v_r^2}{a^2}\right)\frac{\partial^2 \varphi}{\partial r^2} + \frac{v_r}{r} = 0 \tag{11.18}$$

For an ideal, steady, isentropic compressible flow, it is necessary to solve the full-velocity potential equation with specific boundary conditions. Because of the nonlinearity of the equation, it is difficult to find an exact solution to the problem of flow around an actual body. The approximate solution of small disturbance linearization and numerical method can be used.

11.3 Small Perturbation Linearization Theory

11.3.1 Small Disturbance Approximation

In order to reduce the drag, the relative thickness and camber of the wing are small and the angle of attack is small when the aircraft flies at high speed. As shown in Fig. 11.2, the total disturbance for an infinite stream is small except in a few places and satisfies the small disturbance condition.

The X axis is consistent with the unperturbed straight flow, that is, in the wind axis coordinate system, the velocity of each point in the flow field is u, v, w. This can be divided into two parts, one is the forward flow velocity V_∞, and the other is because of the existence of the object, the disturbance velocity of the flow field is u', v', w', there is

$$u = V_\infty + u'$$
$$v = v'$$
$$w = w' \tag{11.19}$$

If the disturbance fraction is small in comparison with the velocity of flow, i.e.,

$$\frac{u'}{V_\infty} \ll 1, \quad \frac{v'}{V_\infty} << 1, \quad \frac{w'}{V_\infty} \ll 1 \tag{11.20}$$

This flow is called a small disturbance flow. Under the condition of small disturbance, the full velocity potential equation can be simplified as a linearized equation of the disturbed velocity potential. Substituted the formula above into the full velocity potential equation, and the sound velocity a is given by the energy equation

$$a^2 = a_\infty^2 + \frac{\gamma - 1}{2}\left(V_\infty^2 - V^2\right)$$
$$a^2 = a_\infty^2 - \frac{\gamma - 1}{2}\left(2V_\infty u' + u'^2 + v'^2 + w'^2\right) \tag{11.21}$$

Utilization formula (11.11)

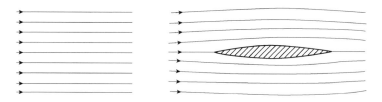

Fig. 11.2 Uniform flow over thin airfoils (uniform and perturbed flow fields)

$$(a^2 - u^2)\frac{\partial u}{\partial x} + (a^2 - v^2)\frac{\partial v}{\partial y} + (a^2 - w^2)\frac{\partial w}{\partial z}$$

$$= uv\left(\frac{\partial u}{\partial y} + \frac{\partial v}{\partial x}\right) + vw\left(\frac{\partial v}{\partial z} + \frac{\partial w}{\partial y}\right) + uw\left(\frac{\partial u}{\partial z} + \frac{\partial w}{\partial x}\right) \tag{11.22}$$

Using the formula (11.21), carrying out the simplified formula (11.22), we get

$$\frac{a^2 - u^2}{V_\infty^2} = \frac{1}{V_\infty^2}\left(a_\infty^2 - \frac{\gamma - 1}{2}(2V_\infty u' + u'^2 + v'^2 + w'^2) - (V_\infty + u')^2\right)$$

$$= \frac{1}{Ma_\infty^2} - \left((\gamma + 1)\frac{u'}{V_\infty} + \frac{\gamma + 1}{2}(\frac{u'}{V_\infty})^2 + \frac{\gamma - 1}{2}(\frac{v'^2 + w'^2}{V_\infty^2})\right) - 1 \tag{11.23}$$

$$\frac{a^2 - v^2}{V_\infty^2} = \frac{1}{V_\infty^2}\left(a_\infty^2 - \frac{\gamma - 1}{2}(2V_\infty u' + u'^2 + v'^2 + w'^2) - v'^2\right)$$

$$= \frac{1}{Ma_\infty^2} - \left((\gamma - 1)\frac{u'}{V_\infty} + \frac{\gamma + 1}{2}(\frac{v'}{V_\infty})^2 + \frac{\gamma - 1}{2}(\frac{u'^2 + w'^2}{V_\infty^2})\right) \tag{11.24}$$

$$\frac{a^2 - w^2}{V_\infty^2} = \frac{1}{V_\infty^2}\left(a_\infty^2 - \frac{\gamma - 1}{2}(2V_\infty u' + u'^2 + v'^2 + w'^2) - w'^2\right)$$

$$= \frac{1}{Ma_\infty^2} - \left((\gamma - 1)\frac{u'}{V_\infty} + \frac{\gamma + 1}{2}(\frac{w'}{V_\infty})^2 + \frac{\gamma - 1}{2}(\frac{u'^2 + v'^2}{V_\infty^2})\right) \tag{11.25}$$

After arranged, we get

$$(1 - Ma_\infty^2)\frac{\partial u'}{\partial x} + \frac{\partial v'}{\partial y} + \frac{\partial w'}{\partial z}$$

$$= Ma_\infty^2\left[(\gamma + 1)\frac{u'}{V_\infty} + \frac{\gamma + 1}{2}\left(\frac{u'}{V_\infty}\right)^2 + \frac{\gamma - 1}{2}\left(\frac{v'^2 + w'^2}{V_\infty^2}\right)\right]\frac{\partial u'}{\partial x}$$

$$+ Ma_\infty^2\left[(\gamma - 1)\frac{u'}{V_\infty} + \frac{\gamma + 1}{2}\left(\frac{v'}{V_\infty}\right)^2 + \frac{\gamma - 1}{2}\left(\frac{u'^2 + w'^2}{V_\infty^2}\right)\right]\frac{\partial v'}{\partial y}$$

$$+ Ma_\infty^2\left[(\gamma - 1)\frac{u'}{V_\infty} + \frac{\gamma + 1}{2}\left(\frac{w'}{V_\infty}\right)^2 + \frac{\gamma - 1}{2}\left(\frac{v'^2 + u'^2}{V_\infty^2}\right)\right]\frac{\partial w'}{\partial z}$$

$$+ Ma_\infty^2\frac{v'}{V_\infty}\left(1 + \frac{u'}{V_\infty}\right)\left(\frac{\partial u'}{\partial y} + \frac{\partial v'}{\partial x}\right) + Ma_\infty^2\frac{v'}{V_\infty}\frac{w'}{V_\infty}\left(\frac{\partial v'}{\partial z} + \frac{\partial w'}{\partial y}\right)$$

$$+ Ma_\infty^2\frac{w'}{V_\infty}\left(1 + \frac{u'}{V_\infty}\right)\left(\frac{\partial u'}{\partial z} + \frac{\partial w'}{\partial x}\right) \tag{11.26}$$

On the left side is a linear operator with constant coefficients, and on the right is a nonlinear term. If we use the small perturbation hypothesis and ignore the third-order small quantity, we have

$$(1 - Ma_\infty^2)\frac{\partial u'}{\partial x} + \frac{\partial v'}{\partial y} + \frac{\partial w'}{\partial z}$$

$$= Ma_\infty^2(\gamma + 1)\frac{u'}{V_\infty}\frac{\partial u'}{\partial x} + Ma_\infty^2(\gamma - 1)\frac{u'}{V_\infty}\frac{\partial v'}{\partial y} + Ma_\infty^2(\gamma - 1)\frac{u'}{V_\infty}\frac{\partial w'}{\partial z}$$

$$Ma_\infty^2\frac{v'}{V_\infty}\left(\frac{\partial u'}{\partial y} + \frac{\partial v'}{\partial x}\right) + Ma_\infty^2\frac{w'}{V_\infty}\left(\frac{\partial u'}{\partial z} + \frac{\partial w'}{\partial x}\right) \qquad (11.27)$$

For the convenience of representation, the signal of perturbation velocity "'" is removed to obtain

$$(1 - Ma_\infty^2)\frac{\partial u}{\partial x} + \frac{\partial v}{\partial y} + \frac{\partial w}{\partial z}$$

$$= Ma_\infty^2(\gamma + 1)\frac{u}{V_\infty}\frac{\partial u}{\partial x} + Ma_\infty^2(\gamma - 1)\frac{u}{V_\infty}(\frac{\partial v}{\partial y} + \frac{\partial w}{\partial z})$$

$$Ma_\infty^2\frac{v}{V_\infty}\left(\frac{\partial u}{\partial y} + \frac{\partial v}{\partial x}\right) + Ma_\infty^2\frac{w}{V_\infty}\left(\frac{\partial u}{\partial z} + \frac{\partial w}{\partial x}\right) \qquad (11.28)$$

11.3.2   Linearization Equation of Perturbed Velocity Potential Function

Except for small perturbation, if the flow is non-transonic meanwhile, that is, M_∞ far less than 1, so $|1 - M_\infty^2|$ is not small; or non-hypersonic, that is, M_∞ is not very large. At this point, on the left side of the formula is an order of magnitude, and on the right side for the second-order small quantity to be omitted, get

$$(1 - Ma_\infty^2)\frac{\partial u}{\partial x} + \frac{\partial v}{\partial y} + \frac{\partial w}{\partial z} = 0 \qquad (11.29)$$

For irrotational flows, there is a perturbed velocity potential function and the equation becomes

$$(1 - Ma_\infty^2)\frac{\partial^2 \varphi}{\partial x^2} + \frac{\partial^2 \varphi}{\partial y^2} + \frac{\partial^2 \varphi}{\partial z^2} = 0 \qquad (11.30)$$

The equation is a linear second order partial differential equation, so it's called a small perturbation linear equation of the full velocity potential function. When < 1, $\beta = \sqrt{1 - Ma_\infty^2}$, get

$$\beta^2 \frac{\partial^2 \varphi}{\partial x^2} + \frac{\partial^2 \varphi}{\partial y^2} + \frac{\partial^2 \varphi}{\partial z^2} = 0 \tag{11.31}$$

When $M_\infty > 1$, $B = \sqrt{Ma_\infty^2 - 1}$, make the expression (11.30) become

$$B^2 \frac{\partial^2 \varphi}{\partial x^2} - \frac{\partial^2 \varphi}{\partial y^2} - \frac{\partial^2 \varphi}{\partial z^2} = 0 \tag{11.32}$$

It can be seen that the small perturbation linearized equation of the perturbed velocity potential function is an elliptic equation at subsonic speed and a hyperbolic equation at supersonic speed.

11.3.3 Pressure Coefficient Linearization

According to the definition of the pressure coefficient, the pressure is proportional to the square of the velocity. If the small disturbance is linearized, the pressure coefficient can also be expressed as a linear relationship of the velocity.

$$
\begin{aligned}
C_p &= \frac{p - p_\infty}{\frac{1}{2}\rho_\infty V_\infty^2} = \frac{2(p/p_\infty - 1)}{V_\infty^2} \frac{p_\infty}{\rho_\infty} \\
&= \frac{2(p/p_\infty - 1)}{\gamma V_\infty^2} \gamma R T_\infty = \frac{2(p/p_\infty - 1)}{\gamma V_\infty^2} a_\infty^2 \\
&= \frac{2(p/p_\infty - 1)}{\gamma Ma_\infty^2}
\end{aligned}
\tag{11.33}
$$

Using the energy equation and the isentropic relation, there are

$$\frac{V_\infty^2}{2} + \frac{\gamma}{\gamma - 1}\frac{p_\infty}{\rho_\infty} = \frac{V^2}{2} + \frac{\gamma}{\gamma - 1}\frac{p}{\rho} \quad p = C\rho^\gamma \tag{11.34}$$

$$\frac{\gamma}{\gamma - 1}\frac{p_\infty}{\rho_\infty}\left(\frac{p}{p_\infty}\frac{\rho_\infty}{\rho} - 1\right) = \frac{V_\infty^2}{2}\left(1 - \frac{V^2}{V_\infty^2}\right)$$

$$\left(\frac{p}{p_\infty}\frac{\rho_\infty}{\rho} - 1\right) = \frac{\gamma - 1}{2}Ma_\infty^2\left(1 - \frac{V^2}{V_\infty^2}\right) \tag{11.35}$$

$$\frac{p}{p_\infty} = \left[1 + \frac{\gamma - 1}{2}Ma_\infty^2\left(1 - \frac{V^2}{V_\infty^2}\right)\right]^{\frac{\gamma}{\gamma - 1}} \tag{11.36}$$

The pressure coefficient is

$$C_p = \frac{2}{\gamma Ma_\infty^2}\left\{\left[1 + \frac{\gamma-1}{2}Ma_\infty^2\left(1 - \frac{V^2}{V_\infty^2}\right)\right]^{\frac{\gamma}{\gamma-1}} - 1\right\} \qquad (11.37)$$

In the case of small disturbance, the above formula is expanded by binomial, omitting the small quantity of the disturbance velocity more than third order, and getting

$$C_p = -\left[\frac{2u}{V_\infty} + \frac{u^2}{V_\infty^2} + \frac{v^2 + w^2}{V_\infty^2}\right] \qquad (11.38)$$

For a flat object, such as an airplane wing, take only one approximation

$$C_p = -\frac{2u}{V_\infty} \qquad (11.39)$$

The formula above is the same with incompressible pressure coefficiency linearized, which shows that the pressure coefficient is only determined by the x-direction perturbation velocity component.

11.3.4   Linearization of Boundary Conditions

The boundary conditions include those at far field and on the object surface. The surface boundary condition of the ideal fluid is that the normal velocity of the fluid on the surface is zero, also called the non-penetrating condition. A simple linearized surface boundary condition could be obtained under the condition of small disturbance. Let the middle arc equation of the object surface be as shown in Fig. 11.3, $y = f(x, z)$. The condition of the object surface not penetrating requires the normal velocity component on the object surface to be zero, i.e.,

$$\vec{V} \cdot \vec{n} = 0 \qquad (11.40)$$

$$(V_\infty + u)\frac{\partial f}{\partial x} - v + w\frac{\partial f}{\partial z} = 0 \qquad (11.41)$$

Under the condition of small disturbance, the thickness and curvature of the object are very small, that is, $\frac{\partial f}{\partial x}$, $\frac{\partial f}{\partial z}$ are small quantities. Retaining the first order small quantity in the upper formula

$$V_\infty \frac{\partial f}{\partial x} - v = 0 \qquad (11.42)$$

Fig. 11.3 Boundary
conditions on the object
surface

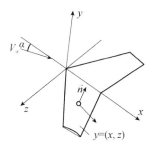

$$v|_s = V_\infty \frac{\partial y}{\partial x} \tag{11.43}$$

The velocity above is the velocity on the surface. If the object is flat and the angle of attack is small, we can use the xoz plane instead of it.

$$(v)_s = v(x, y_s) \cong v(x, 0) \tag{11.44}$$

And the boundary condition can be written as

$$v(x, 0) = \left(\frac{\partial \phi}{\partial y}\right)_{y=0} = V_\infty (\frac{\partial y}{\partial x})_S \tag{11.45}$$

For the plane flow problem, the upper form becomes

$$v(x, 0) = V_\infty \left(\frac{dy}{dx}\right)_s \tag{11.46}$$

In the equation, $(dy/dx)_s$ is the slope of the object surface.

For a boundary condition at infinity, the perturbation velocity always approaches zero for a body such as a wing that flows uniformly in a direct current. However, for the flow around a finite-span wing with lift, there will be a wake vortex system. The vortex system would extend downstream to infinity in ideal flow. At the downstream reaches of infinity, the perturbation velocity in this localized region of vortices does not tend to zero. But it does not need to be satisfied by any other condition, because the vortex system which satisfies the boundary condition on the wing, naturally extends down to infinity.

11.4　Theoretical Linearization Solution of Two-Dimensional Subsonic Flow Around the Corrugated Wall

The linearized equation of the perturbed velocity potential function (11.31) is used to solve a two-dimensional flow around an infinitely long (x-direction) corrugated wall. By this example, the effect of compressibility on the flow and pressure distribution can be explained qualitatively and quantitatively. As shown in Fig. 11.4, let the wave surface be a sine curve with a wavelength of l and an amplitude of d. Put the x axis on the average of the peaks and troughs. A uniform flow flows from the left side of the wall in a direction parallel to the x-axis. The equation for the corrugated wall is

$$y_s = d \sin \frac{2\pi x}{l} \tag{11.47}$$

The corresponding solution problem is

$$\begin{cases} \beta^2 \frac{\partial^2 \phi}{\partial x^2} + \frac{\partial^2 \phi}{\partial y^2} = 0 \\ \left(\frac{\partial \phi}{\partial y}\right)_{y=0} = V_\infty \frac{dy_s}{dx} \end{cases} \tag{11.48}$$

By using the method of separation of variables, assuming the disturbance velocity potential function $\phi = F(x)G(y)$, the above formula can be substituted

$$\beta^2 G \frac{d^2 F}{dx^2} + F \frac{d^2 G}{dy^2} = 0 \quad \frac{1}{F} \frac{d^2 F}{dx^2} = -\frac{1}{\beta^2 G} \frac{d^2 G}{dy^2} = -k^2$$
$$\frac{d^2 F}{dx^2} + k^2 F = 0 \qquad \frac{d^2 G}{dy^2} = \beta^2 k^2 G \tag{11.49}$$

get a solution

$$\varphi = (C_1 \cos kx + C_2 \sin kx)\left(C_3 e^{k\beta y} + C_4 e^{-k\beta y}\right) \tag{11.50}$$

The velocity in the y direction is

Fig. 11.4 Pressure coefficient distribution of compressible flow around the corrugated wall

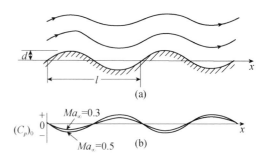

$$v = \frac{\partial \varphi}{\partial y} = (C_1 \cos kx + C_2 \sin kx) k \beta \left(C_3 e^{k\beta y} - C_4 e^{-k\beta y} \right) \tag{11.51}$$

While $y \to \infty$, the perturbation velocity v should go to zero, and thus $C3 = 0$. Approximation of v at $y = 0$ by surface velocity v. The slope of the corrugated wall is

$$\frac{dy_s}{dx} = \frac{2\pi d}{l} \cos \frac{2\pi x}{l} \tag{11.52}$$

$$-C_4 \beta k \left(C_1 \cos kx + C_2 \sin kx \right) = V_\infty \frac{2\pi d}{l} \cos \frac{2\pi x}{l} \tag{11.53}$$

Solve for the above formula, and get

$$C_2 = 0 \quad C_1 C_4 = -V_\infty \frac{d}{\beta} \quad k = \frac{2\pi}{l} \tag{11.54}$$

Therefore, the perturbation velocity potential function of two-dimensional subsonic flow around the wave wall is

$$\phi(x, y) = C_1 C_4 e^{-\beta ky} \cos kx$$

$$\phi(x, y) = -\frac{V_\infty d}{\beta} e^{-\beta \frac{2\pi}{l} y} \cos \frac{2\pi x}{l} \tag{11.55}$$

Velocity component is

$$u = \frac{\partial \phi}{\partial x} = -C_1 C_4 k e^{-\beta ky} \sin kx = \frac{V_\infty}{\beta} \frac{2\pi d}{l} e^{-\beta \frac{2\pi y}{l}} \sin \frac{2\pi x}{l}$$

$$v = \frac{\partial \phi}{\partial y} = -C_1 C_4 \beta k e^{-\beta ky} \cos kx = V_\infty \frac{2\pi d}{l} e^{-\beta \frac{2\pi y}{l}} \cos \frac{2\pi x}{l} \tag{11.56}$$

The streamline equation is

$$\frac{dy}{v} = \frac{dx}{V_\infty + u} \approx \frac{dx}{V_\infty} \frac{dy}{dx} = \frac{v}{V_\infty}$$

$$\frac{dy}{dx} = \frac{2\pi d}{l} e^{-\beta \frac{2\pi y}{l}} \cos \frac{2\pi x}{l}$$

$$y = de^{-\beta \frac{2\pi h}{l}} \sin \frac{2\pi x}{l} \tag{11.57}$$

The pressure coefficient on the wall is

$$C_{ps} = -\frac{2u(x, 0)}{V_\infty} = -\frac{1}{\beta} \frac{4\pi d}{l} \sin \frac{2\pi x}{l} \tag{11.58}$$

The undulations of the streamline are in the same phase as the wall. The further away from the wall, the smaller the fluctuation. However at the same altitude, for different incoming Mach numbers, the larger the Ma_∞ is, the less β is, and the larger $e^{-\beta ky}$ make the larger the amplitude. In another word, the larger Ma_∞ is, the larger the disturbance at the same point, or the smaller the attenuation. The $(c_p)_s$ distribution along X on the wall is shown in Fig. 11.4. The fluctuation of the pressure is also sinusoidal, but it is a minus sign of the difference between the fluctuation of the wall. There is no essential difference between the flow of subsonic and incompressible flow, because when $Ma_\infty \to 0$, $\beta \to 1$, the flow field changes from subsonic to incompressible. The absolute value of $(c_p)_s$ is increased with the increase of Ma_∞, and its magnification factor is $1/\beta$. The two curves $= 0.3$ and 0.5 are shown in the figure. The pressure coefficient from Eq. (11.58) is the pressure coefficient of the flow around the incompressible flow, as $C_{ps}(0)$. However, at any other Ma_∞ the wall pressure coefficient is expressed as $C_{ps}(Ma_\infty)$. From (11.58)

$$C_{ps}(Ma_\infty) = \frac{1}{\beta}\left[-\frac{4\pi d}{l}\sin\frac{2\pi x}{l} \right] = \frac{1}{\beta}C_{ps}(0) \qquad (11.59)$$

11.5 Prandtl-Glauert Compressibility Correction of Two-Dimensional Subsonic Flow

For the subsonic compressible flow around an object, the perturbation velocity potential function satisfies the linearized equation and the linearized boundary condition under the small disturbance condition. So, the pressure along the surface could be solved and aerodynamic further. For the same object, there is no essential difference between the subsonic flow field and the incompressible flow field, but there is some difference in quantity. If the low-speed aerodynamic characteristics are known, can the subsonic aerodynamic characteristics be obtained by some relations? The answer is yes.

The subsonic streamline equation is elliptical. Compared to Laplace's equation of incompressible flow, except that the coefficient of the first term is not 1.0, but the constant factor B2. This can be done by converting the linearized equations to Laplace type equations and by converting the boundary conditions and the pressure coefficient accordingly. In this way, the problem of solving the linearized equation to satisfy the boundary condition becomes the problem of solving Laplace's equation to satisfy the boundary condition.

11.5.1 Transformation of Linearized Equations

(1) Affine transformation

There is the following affine transformation

$$
\begin{aligned}
X &= x, Y = \beta y, Z = \beta z \\
\Phi &= k\varphi \\
V_\infty &= V_{\infty c}
\end{aligned}
\tag{11.60}
$$

In which, $V_{\infty c}$ represents the flow velocity in compressible flow, V_∞ represents the flow velocity in the incompressible field. The vertical x and the other two directions Y, and Z with a different scale in the equations above. This kind of keeping the vertical scale unchanged, only the other two directions of the scale to be enlarged or reduced by the transformation called an affine transformation. The objects in the two flow fields are not geometrically similar, but affine similar after this transformation. In which X, Y, Z, and Φ are the incompressible coordinate and perturbation velocity potential functions. By substituting them into the linearized equations, we could get

$$
\frac{\partial^2 \varphi}{\partial x^2} = \frac{1}{k}\frac{\partial^2 \Phi}{\partial X^2}, \ \frac{\partial^2 \varphi}{\partial y^2} = \frac{\beta^2}{k}\frac{\partial^2 \Phi}{\partial Y^2}, \ \frac{\partial^2 \varphi}{\partial z^2} = \frac{\beta^2}{k}\frac{\partial^2 \Phi}{\partial Z^2}
\tag{11.61}
$$

Substitute to (11.31) and get

$$
\frac{\partial^2 \Phi}{\partial X^2} + \frac{\partial^2 \Phi}{\partial Y^2} + \frac{\partial^2 \Phi}{\partial Z^2} = 0
\tag{11.62}
$$

The transformed Eq. (11.62) is completely Laplace's equation with incompressible flow. For flow pass plane, yield

$$
\frac{\partial^2 \Phi}{\partial X^2} + \frac{\partial^2 \Phi}{\partial Y^2} = 0
\tag{11.63}
$$

(2) Transformation of boundary conditions

For the far front boundary condition, the perturbation velocity must be zero, which is still satisfied after affine transformation. For the boundary condition of two-dimensional object surface, while the affine transformation is substituted

$$
\begin{aligned}
v(x,0) &= \left(\frac{\partial \varphi}{\partial y}\right)_{y=0} = \frac{\beta}{k}\left(\frac{\partial \Phi}{\partial Y}\right)_{Y=0} \\
V_\infty \left(\frac{dy}{dx}\right)_s &= \frac{V_\infty}{\beta}\left(\frac{dY}{dX}\right)_s
\end{aligned}
\tag{11.64}
$$

The utility Eq. (11.46) then becomes

$$\frac{\beta}{k}\left(\frac{\partial \Phi}{\partial Y}\right)_{Y=0} = \frac{V_\infty}{\beta}\left(\frac{dY}{dX}\right)_s, \left(\frac{\partial \Phi}{\partial Y}\right)_{Y=0} = k\frac{V_\infty}{\beta^2}\left(\frac{dY}{dX}\right)_s \tag{11.65}$$

If $k = \beta^2$, the formula above became

$$\left.\frac{\partial \Phi}{\partial Y}\right|_{Y=0} = V_\infty(\frac{dY}{dX})_s \tag{11.66}$$

Thus, the boundary condition with the same form as the incompressible flow is obtained. In this way, the problem of solving the linearized boundary condition of the equations for the potential function of the compressible flow is transformed into the problem of solving Laplace's equation with the same boundary condition in incompressible flow. Which means the subsonic flow around thin airfoils has been transformed into the incompressible flow around thin airfoils.

(3) Transformation of definite solution problem

(1) Affine transformation

$$X = x, Y = \beta y, Z = \beta z$$
$$\Phi = \beta^2 \varphi \tag{11.67}$$
$$V_\infty = V_{\infty c}$$

(2) Subsonic compressible flows

$$\begin{cases} \beta^2 \dfrac{\partial^2 \varphi}{\partial x^2} + \dfrac{\partial^2 \varphi}{\partial y^2} = 0 \\ v(x, 0) = V_\infty \dfrac{dy_s}{dx} \end{cases} \tag{11.68}$$

(3) The incompressible flow around the object after affine transformation

$$\begin{cases} \dfrac{\partial^2 \Phi}{\partial X^2} + \dfrac{\partial^2 \Phi}{\partial Y^2} = 0 \\ \left.\dfrac{\partial \Phi}{\partial Y}\right|_{Y=0} = V_\infty \dfrac{dY_s}{dX}, Y_s = \beta y_s \end{cases} \tag{11.69}$$

(4) Transformation of the flow around a compressible thin airfoil

Now, let's deal with the geometry relationship between the flow over thin airfoils at subsonic speed and the corresponding incompressible thin airfoils. According to the

Fig. 11.5 Relationship between compressible and incompressible flows over airfoils

compressible flow   incompressible flow

affine transformation, the size of the airfoil in the incompressible flow is constant in the X-direction, and the y direction is equal to the size of the subsonic airfoil in the y direction product β (reduce it). Therefore, the relative thickness and the relative camber of the incompressible airfoil are all β times the corresponding value of the compressible flow airfoil. Similarly, the angle of attack should also be β times. It can be seen that the flow around the corresponding incompressible airfoil is thinner, less camber, and smaller angle of attack than that of the original airfoil, as shown in Fig. 11.5.

$$c' = \beta c$$
$$f' = \beta c$$
$$\alpha' = \beta \alpha \tag{11.70}$$

11.5.2 Compressibility correction based on linearization theory

(1) The transformation of pressure coefficients at corresponding points on airfoils

By substituting the affine transformation relation (11.66) into the linearized formula of pressure coefficient (11.39), we obtain

$$C_{ps}(Ma_\infty) = -\frac{2u}{V_\infty} = -\frac{2}{V_\infty}\frac{\partial\varphi}{\partial x} = \frac{1}{\beta^2}\left(-\frac{2}{V_\infty}\frac{\partial\Phi}{\partial X}\right) = \frac{1}{\beta^2}C_p(0)$$
$$C_p(0) = -\frac{2}{V_\infty}\frac{\partial\Phi}{\partial X} \tag{11.71}$$

$C_p(0)$ is the pressure coefficient of the incompressible potential flow. The euqation above can be rewritten as

$$C_p(Ma_\infty, \alpha, c, f) = \frac{1}{\beta^2}C_p(0, \beta\alpha, \beta c, \beta f) \tag{11.72}$$

That is the pressure coefficient at a point in the compressible flow field is equal to the pressure coefficient at the corresponding point of the affine transformation times

$1/\beta^2$. This transformation law is called the Gothert Law (Gothert, German aerodynamicist, 1939). This method shows that: In order to obtain the aerodynamic characteristics of subsonic airfoils, it is necessary to calculate the flow field of different airfoils at various angles of attack in incompressible flow. Which makes the application very inconvenient. Can we establish the relationship between the pressure coefficients of compressible flow and incompressible flow for the same airfoil at the same angle of attack? The answer is undoubted too.

According to the thin-wing theory, the disturbances in flow around an incompressible airfoil with small disturbances can be considered as a linear superposition of the disturbances caused by the thickness, camber, and angle of attack of the airfoil. According to this principle, the solution of dimensionless potential flow is the same when the thickness, camber, and angle of attack of airfoil are enlarged or decreased in the same proportion in the incompressible flow field. If the shape of affine transformation is magnified $1/\beta$ times, the disturbance velocity will also be magnified $1/\beta$ times. Since the linear pressure coefficient will be proportional to it, so it will be magnified $1/\beta$ times, then

$$C_p(0, \alpha, c, f) = \frac{1}{\beta} C_p(0, \beta\alpha, \beta c, \beta f) \qquad (11.73)$$

Taking Eq. (11.73) into the transformation of the pressure coefficient (11.72) we can obtain

$$C_p(Ma_\infty, \alpha, c, f) = \frac{1}{\beta} C_p(0, \alpha, c, f) \qquad (11.74)$$

That is to say, the relationship between the pressure coefficient at the corresponding point of incompressible flow and compressible flow with exactly the same airfoil and angle of attack is that the CP value of incompressible flow multiplied by $1/\beta$ is the CP value of subsonic compressible flow. The conversion rule is called the Prandtl- Glauert Rules. This law was derived by H. Glauert in 1027 was been used by Prandtl firstly in 1922. $1/\beta$ is called the compressibility factor for subsonic flows.

(2) Subsonic aerodynamic characteristics of airfoils

The lift coefficient CL of the airfoil is derived from the integral of the pressure coefficient at each point on the airfoil. The difference between moment coefficient Cm, and the lift coefficient is only the force arm in the X-direction. So the same airfoil is at the same angle of attack, CL and Cm at the subsonic velocity are $1/\beta$ times the values of CL and CM for the incompressible flow, i.e.,

$$C_L(Ma_\infty, \alpha, c, f) = \frac{1}{\beta} C_L(0, \alpha, c, f) \qquad (11.75)$$

Due to the same angle of attack of the corresponding airfoils

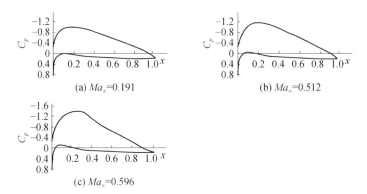

Fig. 11.6 Distribution of pressure coefficients of NACA4415 under different Mach numbers

$$C_m = (Ma_\infty, \alpha, c, f) = \frac{1}{\beta} C_m(0, \alpha, c, f) \tag{11.76}$$

$$C_L^\alpha(Ma_\infty, \alpha, c, f) = \frac{1}{\beta} C_L^\alpha(0, \alpha, c, f) \tag{11.77}$$

Figure 11.6 shows the CP distributions of NACA4415 airfoil at the same angle of attack and three incoming Mach Number, $Ma_\infty = 0.191, 0.512, 0.596$, respectively. These three curves are the result of the experiment. According to Prandtl-Glauert law, these curves can be converted to each other by a factor of $1/\beta$. From the experimental results, it is true that the pressure coefficient distribution increases with the increase of Mach number and the suction peak increases also.

11.6 Karman-Qian Compressibility Correction

11.6.1 Characteristics of Karman-Qian Compressibility Correction

It was found that the modified results of **Prandtl-Glauert** were quite different from the experimental data when the Mach Number was between 0.5 and 0.7. In 1939, Von Karman and Qian Xuesen proposed a new compressibility correction called the Karman-Qian formula.

$$C_p(Ma_\infty, \alpha, c, f) = \frac{C_p(0, \alpha, c, f)}{\sqrt{1 - Ma_\infty^2} + \frac{Ma_\infty^2}{\sqrt{1 - Ma_\infty^2 + 1}} \frac{C_p(0, \alpha, c, f)}{2}} \tag{11.78}$$

The correction of the formula is no longer a constant $1/\beta$ but is related to the local pressure coefficient $(C_P)_{0,\alpha,\bar{c},\bar{f}}$. If it is a negative pressure point, as the wind behind

Fig. 11.7 Comparison of
three pressure coefficient
curves of NACA4412 airfoil

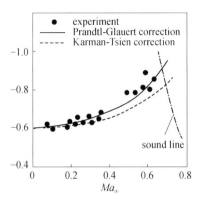

the side of the airfoil has a larger correction ratio than $1/\beta$; if it is a positive pressure point, the airfoil's forward wind side has a smaller correction ratio than $1/\beta$, this method has a higher correction accuracy. Figure 11.7 shows three sets of pressure coefficient comparison curves for NACA4412 airfoil: one is experimental data from a two dimensional subsonic wind tunnel; the second is modified by Karman-Qian formula; and the third one is modified by Prandtl-Glauert Law. The angle of attack of the airfoil is $2°$, and the static pressure hole is 30% chord length from the leading edge, local flow velocity reached the sound speed velocity. As you can see from the diagram, Karman-Qian correction could be applied to local velocity up to the speed of sound, while Prandtl-Glauert correction has been shown to be insufficient when M_∞ is not very large.

11.6.2 Governing Equations for Perfectly Compressible Planar Flows

For ideal, steady, and planar subsonic flows, Karman and Qian derived a modified subsonic compressibility formula for the point pressure by using the transformation of the velocity potential function equation and the flow function equation between the physical plane and the velocity plane.

By using formula (11.13), the full-velocity potential equation for ideal, steady, and plane subsonic flows is obtained

$$\left(1 - \frac{u^2}{a^2}\right)\frac{\partial^2\varphi}{\partial x^2} + \left(1 - \frac{v^2}{a^2}\right)\frac{\partial^2\varphi}{\partial y^2} - 2\frac{uv}{a^2}\frac{\partial^2\varphi}{\partial x\partial y} = 0 \qquad (11.79)$$

For the plane flow, the continuous equation is used

$$\frac{\partial\rho u}{\partial x} + \frac{\partial\rho v}{\partial y} = 0 \qquad (11.80)$$

If density ρ is included in the definition of stream function, the stream function is defined as

$$\rho u = \frac{\partial \psi}{\partial y}, \quad \rho v = -\frac{\partial \psi}{\partial x} \tag{11.81}$$

Using the formula (11.81), we get

$$\rho \frac{\partial u}{\partial x} = \frac{\partial^2 \psi}{\partial x \partial y} - u\frac{\partial \rho}{\partial x}, \quad \rho\frac{\partial u}{\partial y} = \frac{\partial^2 \psi}{\partial y^2} - u\frac{\partial \rho}{\partial y}$$

$$\rho \frac{\partial v}{\partial x} = -\frac{\partial^2 \psi}{\partial x^2} - v\frac{\partial \rho}{\partial x}, \quad \rho\frac{\partial v}{\partial y} = -\frac{\partial^2 \psi}{\partial x \partial y} - v\frac{\partial \rho}{\partial y} \tag{11.82}$$

For the ideal plane potential flow, the rotation angle is zero

$$\frac{\partial v}{\partial x} - \frac{\partial u}{\partial y} = 0 \tag{11.83}$$

By substituting Eq. (11.81) into Eq. (11.83), we get the equation of the stream function

$$-\rho\left(\frac{\partial v}{\partial x} - \frac{\partial u}{\partial y}\right) = -\frac{\partial \rho v}{\partial x} + \frac{\partial \rho u}{\partial y} + v\frac{\partial \rho}{\partial x} - u\frac{\partial \rho}{\partial y} = 0$$

$$\frac{\partial^2 \psi}{\partial x^2} + \frac{\partial^2 \psi}{\partial y^2} + v\frac{\partial \rho}{\partial x} - u\frac{\partial \rho}{\partial y} = 0 \tag{11.84}$$

For the ideal barotropic fluid and the isentropic acoustic equation, there are

$$\frac{dp}{d\rho} = a^2, \quad \frac{\partial \rho}{\partial x} = \frac{\partial \rho}{\partial p}\frac{\partial p}{\partial x} = \frac{1}{a^2}\frac{\partial p}{\partial x}, \quad \frac{\partial \rho}{\partial y} = \frac{1}{a^2}\frac{\partial p}{\partial y} \tag{11.85}$$

Using Euler Eqs. (11.7) and substitution to (11.84), we have

$$\frac{\partial^2 \psi}{\partial x^2} + \frac{\partial^2 \psi}{\partial y^2} + v\frac{\partial \rho}{\partial x} - u\frac{\partial \rho}{\partial y} = 0$$

$$\frac{\partial^2 \psi}{\partial x^2} + \frac{\partial^2 \psi}{\partial y^2} - \frac{\rho v}{a^2}\left(u\frac{\partial u}{\partial x} + v\frac{\partial u}{\partial y}\right) + \frac{\rho u}{a^2}\left(u\frac{\partial v}{\partial x} + v\frac{\partial v}{\partial y}\right) = 0 \tag{11.86}$$

By the formula (11.82), then

$$\frac{\partial^2 \psi}{\partial x^2} + \frac{\partial^2 \psi}{\partial y^2} - \frac{\rho v}{a^2}\left(u\frac{\partial u}{\partial x} + v\frac{\partial u}{\partial y}\right) + \frac{\rho u}{a^2}\left(u\frac{\partial v}{\partial x} + v\frac{\partial v}{\partial y}\right) = 0$$

$$\frac{\partial^2 \psi}{\partial x^2} + \frac{\partial^2 \psi}{\partial y^2} + \frac{1}{a^2}\left[-vu\frac{\partial^2 \psi}{\partial x \partial y} + vu^2\frac{\partial \rho}{\partial x} - v^2\frac{\partial^2 \psi}{\partial y^2} + v^2 u\frac{\partial \rho}{\partial y}\right]$$

Fig. 11.8 The physical plane and the velocity plane

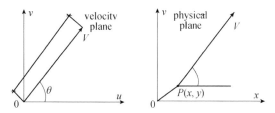

$$+ \frac{1}{a^2}\left[-u^2\frac{\partial^2\psi}{\partial x^2} - u^2 v\frac{\partial\rho}{\partial x} - uv\frac{\partial^2\psi}{\partial x\partial y} - uv^2\frac{\partial\rho}{\partial y}\right] \tag{11.87}$$

After simplification

$$\left(1 - \frac{u^2}{a^2}\right)\frac{\partial^2\psi}{\partial x^2} + \left(1 - \frac{v^2}{a^2}\right)\frac{\partial^2\psi}{\partial y^2} - 2\frac{uv}{a^2}\frac{\partial^2\psi}{\partial x\partial y} = 0 \tag{11.88}$$

This is the stream function equation for a plane compressible flow.

Because of the nonlinear characteristics of the Eqs. (11.79) and (11.88) in the physical plane (x, Y), in order to solve the equations transformed to the velocity plane (v, θ), the velocity graph method could be used. The equations are linear after being transformed and then they could be solved in the velocity plane. As shown in Fig. 11.8.

11.6.3   Transformation in Velocity Plane

In the physical plane, according to the definition of the velocity potential function φ and the stream function ψ, we can get

$$d\varphi = \frac{\partial\varphi}{\partial x}dx + \frac{\partial\varphi}{\partial y}dy = udx + vdy = V(\cos\theta dx + \sin\theta dy)$$

$$d\psi = \frac{\partial\psi}{\partial x}dx + \frac{\partial\psi}{\partial y}dy = \frac{\rho}{\rho_0}(udy - vdx) = \frac{\rho}{\rho_0}V(-\sin\theta dx + \cos\theta dy) \tag{11.89}$$

Here the stream function is defined by dividing by ρ_0 (as a constant), which is the stagnation point density. Solve for dx and dy, then

$$dx = \frac{\cos\theta}{V}d\varphi - \frac{\rho_0}{\rho}\frac{\sin\theta}{V}d\psi$$

$$dy = \frac{\sin\theta}{V}d\varphi + \frac{\rho_0}{\rho}\frac{\cos\theta}{V}d\psi \tag{11.90}$$

With (V, θ) as an independent variable on the velocity plane, the stream function and the velocity potential function can be expressed as

$$d\varphi = \frac{\partial \varphi}{\partial V} dV + \frac{\partial \varphi}{\partial \theta} d\theta$$
$$d\psi = \frac{\partial \psi}{\partial V} dV + \frac{\partial \psi}{\partial \theta} d\theta \tag{11.91}$$

Substituted (11.91) in (11.90), we have

$$dx = \left(\frac{\cos\theta}{V} \frac{\partial\varphi}{\partial V} - \frac{\rho_0 \sin\theta}{\rho} \frac{1}{V} \frac{\partial\psi}{\partial V} \right) dV + \left(\frac{\cos\theta}{V} \frac{\partial\varphi}{\partial\theta} - \frac{\rho_0 \sin\theta}{\rho} \frac{1}{V} \frac{\partial\psi}{\partial\theta} \right) d\theta$$
$$dy = \left(\frac{\sin\theta}{V} \frac{\partial\varphi}{\partial V} + \frac{\rho_0 \cos\theta}{\rho} \frac{1}{V} \frac{\partial\psi}{\partial V} \right) dV + \left(\frac{\sin\theta}{V} \frac{\partial\varphi}{\partial\theta} + \frac{\rho_0 \cos\theta}{\rho} \frac{1}{V} \frac{\partial\psi}{\partial\theta} \right) d\theta \tag{11.92}$$

If we use the complex variable $(z = x + iy)$, it could be

$$dz = dx + i\,dy = \frac{\partial z}{\partial V} dV + \frac{\partial z}{\partial\theta} d\theta \tag{11.93}$$

Substituted (11.92) in equations above

$$dz = e^{i\theta} \left(\frac{1}{V} \frac{\partial\varphi}{\partial V} + i\frac{\rho_0}{\rho} \frac{1}{V} \frac{\partial\psi}{\partial V} \right) dV + e^{i\theta} \left(\frac{1}{V} \frac{\partial\varphi}{\partial\theta} + i\frac{\rho_0}{\rho} \frac{1}{V} \frac{\partial\psi}{\partial\theta} \right) d\theta \tag{11.94}$$

Compare the previous two formulas and we could get

$$\frac{\partial z}{\partial V} = e^{i\theta} \left(\frac{1}{V} \frac{\partial\varphi}{\partial V} + i\frac{\rho_0}{\rho} \frac{1}{V} \frac{\partial\psi}{\partial V} \right)$$
$$\frac{\partial z}{\partial\theta} = e^{i\theta} \left(\frac{1}{V} \frac{\partial\varphi}{\partial\theta} + i\frac{\rho_0}{\rho} \frac{1}{V} \frac{\partial\psi}{\partial\theta} \right) \tag{11.95}$$

Make a partial derivative of θ in the first formula above. And the partial derivative of V in the second one. Then

$$\frac{\partial^2 z}{\partial\theta\partial V} = e^{i\theta} \left[\left(\frac{1}{V} \frac{\partial^2\varphi}{\partial\theta\partial V} - \frac{\rho_0}{\rho} \frac{1}{V} \frac{\partial\psi}{\partial V} \right) + i\left(\frac{\rho_0}{\rho} \frac{1}{V} \frac{\partial\psi}{\partial\theta\partial V} + \frac{1}{V} \frac{\partial\varphi}{\partial V} \right) \right]$$
$$\frac{\partial^2 z}{\partial V\partial\theta} = e^{i\theta} \left[\left(\frac{1}{V} \frac{\partial^2\varphi}{\partial\theta\partial V} - \frac{1}{V^2} \frac{\partial\varphi}{\partial\theta} \right) + i\left(\frac{\rho_0}{\rho} \frac{1}{V} \frac{\partial\psi}{\partial V\partial\theta} + \frac{\partial}{\partial V}\left(\frac{\rho_0}{\rho} \frac{1}{V} \right) \frac{\partial\psi}{\partial\theta} \right) \right] \tag{11.96}$$

In the comparison of (11.94), the first term is equal to the second term, except for the derivative part. By separated imaginary part and real part, it becomes

$$\frac{\rho_0}{\rho}\frac{\partial\psi}{\partial V} = \frac{1}{V}\frac{\partial\varphi}{\partial\theta}$$

$$V\frac{\partial}{\partial V}\left(\frac{\rho_0}{\rho}\frac{1}{V}\right)\frac{\partial\psi}{\partial\theta} = \frac{\partial\varphi}{\partial V} \tag{11.97}$$

Using the isentropic relation $\frac{\partial p}{\partial\rho} = a^2$ and the Euler equations $\frac{\partial p}{\partial V} = -\rho V$, we get

$$\frac{\partial\rho}{\partial V} = -\frac{\rho V}{a^2} \tag{11.98}$$

Using formula (11.98), formula (11.97) is obtained

$$\frac{\partial}{\partial V}\left(\frac{\rho_0}{\rho}\frac{1}{V}\right) = -\frac{\rho_0}{\rho^2 V}\frac{\partial\rho}{\partial V} - \frac{\rho_0}{\rho V^2} = -\frac{\rho_0}{\rho V^2}(1 - Ma^2) \tag{11.99}$$

Then Eq. (11.97) changes to

$$\frac{\rho_0}{\rho}\frac{\partial\psi}{\partial V} = \frac{1}{V}\frac{\partial\varphi}{\partial\theta}$$

$$-\frac{\rho_0}{\rho}(1 - Ma^2)\frac{1}{V}\frac{\partial\psi}{\partial\theta} = \frac{\partial\varphi}{\partial V} \tag{11.100}$$

The formula (11.100) is called the fundamental equation for the Molenbok-Chaplechin transformation. By the crossing partial derivative and eliminating the potential function, the equation of the stream function is

$$\frac{\partial}{\partial V}\left(V\frac{\rho_0}{\rho}\frac{\partial\psi}{\partial V}\right) + \frac{\rho_0}{\rho}(1 - Ma^2)\frac{1}{V}\frac{\partial^2\psi}{\partial\theta^2} = 0$$

$$V^2\frac{\partial^2\psi}{\partial V^2} + (1 + Ma^2)V\frac{\partial\psi}{\partial V} + (1 - Ma^2)\frac{\partial^2\psi}{\partial\theta^2} = 0 \tag{11.101}$$

Similarly, by the crossing partial derivative and eliminating the stream function, the equation of the velocity potential function is

$$V^2(1 - Ma^2)\frac{\partial^2\varphi}{\partial V^2} + V(1 + \gamma Ma^4)\frac{\partial\varphi}{\partial V} + (1 - Ma^2)^2\frac{\partial^2\varphi}{\partial\theta^2} = 0 \tag{11.102}$$

If we find the stream function satisfying Eq. (11.101) on the velocity plane, we can find the corresponding flow in the physical plane. It could be done by: (1) deriving from the stream function $\psi(V, \theta)$, we could get $\frac{\partial\psi}{\partial V}, \frac{\partial\psi}{\partial\theta}$ (2) deriving from the transformation relation (11.100), the partial derivative of the potential function is $\frac{\partial\varphi}{\partial V}, \frac{\partial\varphi}{\partial\theta}$; (3) by integrating the convection function and the potential function, the streamline coordinates x and y on the physical plane are obtained.

11.6.4 Relation Between Compressible and Incompressible Flow Velocity Planes

For incompressible flow, substitution $Ma \to 0, \rho_0 = \rho$ to (11.100), yield

$$W \frac{\partial \psi}{\partial W} = \frac{\partial \varphi}{\partial \theta}$$
$$-\frac{1}{W} \frac{\partial \psi}{\partial \theta} = \frac{\partial \varphi}{\partial W} \tag{11.103}$$

where in the incompressible flow, the velocity is W and the directivity is θ. If you take

$$W \frac{\partial}{\partial W} = \frac{\partial}{\partial (\ln W)} \tag{11.104}$$

Let $Q = \ln W$, the Bernhard-Riemann condition appears as

$$\frac{\partial \psi}{\partial Q} = \frac{\partial \varphi}{\partial \theta}$$
$$-\frac{\partial \psi}{\partial \theta} = \frac{\partial \varphi}{\partial Q} \tag{11.105}$$

For compressible flows, the expression (11.100) can also be changed into a form similar to the expression (11.103). By using

$$d\Pi = \sqrt{1 - Ma^2} \frac{dV}{V} \tag{11.106}$$

Then, the relationship between compressible and incompressible flows can be established, i.e.,

$$Ma \to 0, \rho_0 = \rho, \ V \to W, \Pi \to Q \tag{11.107}$$

Take (11.106) into (11.100), then

$$\frac{\partial \varphi}{\partial \Pi} = -\frac{\rho_0}{\rho} \sqrt{1 - Ma^2} \frac{\partial \psi}{\partial \theta}$$
$$\frac{\partial \varphi}{\partial \theta} = \frac{\rho_0}{\rho} \sqrt{1 - Ma^2} \frac{\partial \psi}{\partial \Pi} \tag{11.108}$$

Chaplygin found that in the subsonic range, where the Mach Number is not too close to 1, the formula below is close to 1, i.e.,

$$\frac{\rho_0}{\rho}\sqrt{1-Ma^2} \approx 1.0 \tag{11.109}$$

This is because

$$\frac{\rho_0}{\rho}\sqrt{1-Ma^2} = \left(1+\frac{\gamma-1}{2}Ma^2\right)^{\frac{1}{\gamma-1}}(1-Ma^2)^{\frac{1}{2}}$$

$$\approx \left(1+\frac{Ma^2}{2}+(2-\gamma)\frac{Ma^4}{8}+\dots\right)$$

$$\left(1-\frac{Ma^2}{2}-\frac{Ma^4}{8}+\right) \approx 1-\frac{\gamma+1}{8}Ma^4+\dots \tag{11.110}$$

When $\gamma = 1.4$, $Ma = 0.5$, the error is less than 2% if the value of the above formula is 1.0. Take it into (11.108), then

$$\frac{\partial\varphi}{\partial\Pi} = -\frac{\partial\psi}{\partial\theta}$$

$$\frac{\partial\varphi}{\partial\theta} = \frac{\partial\psi}{\partial\Pi} \tag{11.111}$$

The above formula is also reduced to the Bernhard Riemann condition. This shows that a similar relationship can be established between the compressible flow problem and the Incompressible flow problem by transformation. Using the formula (11.110) together with $\frac{\rho_0}{\rho}\sqrt{1-Ma^2} \approx 1.0$, it could be shown that $\gamma = -1.0$.

Although, gamma $= -1.0$ doesn't exist actually. However, by the isentropic relation $p = C\rho^\gamma$, it can be regarded as the use tangent of $\gamma = -1.0$ to approximate the curvilinear relation in the relationship between P and $\frac{1}{\rho}$ because

$$\frac{dp}{d\rho} = C\gamma\rho^{\gamma-1}, d\left(\frac{1}{\rho}\right) = -\frac{d\rho}{\rho^2} \tag{11.112}$$

So,

$$\frac{dp}{d\left(\frac{1}{\rho}\right)} = -C\gamma\rho^{\gamma+1} \tag{11.113}$$

In the case of $r = -1$, derived the above formula

$$\frac{dp}{d\left(\frac{1}{\rho}\right)} = -C(-1)\rho^{-1+1} = C \tag{11.114}$$

By using the formula (11.114), instead of the isentropic relation

$$p = \frac{C}{\rho} + B \tag{11.115}$$

where C and B are constants. If the in flow condition (V_∞, p_∞) is used as a tangent point. Since $\frac{\mathrm{d}p}{\mathrm{d}\rho} = a^2$ and according to the formula (11.114), there are

$$C = \frac{\mathrm{d}p}{\mathrm{d}(\frac{1}{\rho})} = -\rho^2 \frac{\mathrm{d}p}{\mathrm{d}\rho} = -\rho^2 a^2 = -\rho_\infty^2 a_\infty^2 \tag{11.116}$$

Using the tangent equation, there are (Fig. 11.9)

$$p - p_\infty = \rho_\infty^2 a_\infty^2 \left(\frac{1}{\rho_\infty} - \frac{1}{\rho} \right) \tag{11.117}$$

Since the expression for compressible flow (11.107) and the non-compressible flow (11.101) are formally identical, the relationship between the velocity of compressible flow and incompressible flow can be expressed as follows

$$\mathrm{d}\Pi = \mathrm{d}Q, \quad \frac{\mathrm{d}W}{W} = \sqrt{1 - Ma^2}\, \frac{\mathrm{d}V}{V} \tag{11.118}$$

The two flows can be described by the same equation. The problem is solved by first finding the flow function $\psi(Q, \theta)$ on the incompressible flow, then getting the flow function $\psi(\Pi, \theta)$ on the velocity surface of the compressible flow which is then converted to the physical plane as $\psi(x, y)$, as shown in Fig. 11.10.

Integral of (11.118), get

Fig. 11.9 Tangent approximation gas

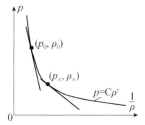

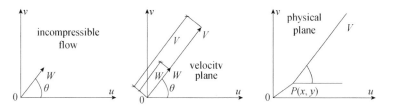

Fig. 11.10 Transformation between incompressible and compressible flows

$$\ln W = \int \sqrt{1 - Ma^2} \frac{dV}{V} = \int \frac{\rho}{\rho_0} \frac{dV}{V} = \int \frac{\rho}{\rho_0} \frac{dV}{V} \tag{11.119}$$

Consider

$$\frac{\rho_0}{\rho} = \left(1 + \frac{\gamma - 1}{2} Ma^2\right)^{\frac{1}{\gamma-1}}, \quad \frac{a_0^2}{a^2} = 1 + \frac{\gamma - 1}{2} Ma^2 \tag{11.120}$$

From that,

$$\frac{\rho_0}{\rho} = \left(1 - \frac{\gamma - 1}{2} \frac{V^2}{a_0^2}\right)^{\frac{-1}{\gamma-1}} \tag{11.121}$$

Let $\gamma = -1$ in the equation above, then

$$\frac{\rho_0}{\rho} = \left(1 - \frac{\gamma - 1}{2} \frac{V^2}{a_0^2}\right)^{\frac{-1}{\gamma-1}} = \sqrt{1 + \frac{V^2}{a_0^2}} \tag{11.122}$$

Substitution to (11.119)

$$\ln W = \int \frac{dV}{V\sqrt{1 + \frac{V^2}{a_0^2}}} \tag{11.123}$$

Integral the above formula, get

$$\ln W = \int \frac{dV}{V\sqrt{1 + \frac{V^2}{a_0^2}}} = \ln \frac{CV}{1 + \sqrt{1 + \frac{V^2}{a_0^2}}} \tag{11.124}$$

$$W = \frac{CV}{1 + \sqrt{1 + \frac{V^2}{a_0^2}}} \tag{11.125}$$

In the formula, C is an integral constant. While $Ma \to 0, a_0 \to \infty, V \to W, C = 2$. Thus obtained

$$V = \frac{4a_0^2 W}{4a_0^2 - W^2} \tag{11.126}$$

Substitution to (11.122) and

$$\frac{\rho_0}{\rho} = \sqrt{1 + \frac{V^2}{a_0^2}} = \sqrt{1 + \frac{(4a_0^2 W)^2}{a_0^2(4a_0^2 - W^2)^2}} = \frac{4a_0^2 + W^2}{4a_0^2 - W^2} \tag{11.127}$$

For compressible flow V_∞, incompressible flow can be assumed as W_∞ and the substitute to (11.126) to get

$$W_\infty = \frac{2a_0 V_\infty}{a_0 + \sqrt{a_0^2 + V_\infty^2}}, \quad V_\infty = \frac{4a_0^2 W_\infty}{4a_0^2 - W_\infty^2} \tag{11.128}$$

Using the formula (11.117), the pressure coefficient of compressible flow is

$$C_p = \frac{p - p_\infty}{\frac{1}{2}\rho_\infty V_\infty^2} = \frac{2a_\infty^2}{V_\infty^2}\left(1 - \frac{\rho_\infty}{\rho}\right) \tag{11.129}$$

The expression $\frac{\rho_\infty}{\rho}$ in the equation above is

$$\frac{\rho_\infty}{\rho} = \frac{\rho_\infty}{\rho_0}\frac{\rho_0}{\rho} = \frac{4a_0^2 - W_\infty^2}{4a_0^2 + W_\infty^2} \cdot \frac{4a_0^2 + W^2}{4a_0^2 - W^2} = \frac{1 - E}{1 + E} \cdot \frac{1 + E\frac{W^2}{W_\infty^2}}{1 - E\frac{W^2}{W_\infty^2}} \tag{11.130}$$

where E stands for

$$E = \frac{W_\infty^2}{4a_0^2} = \frac{V_\infty^2}{\left(a_0 + \sqrt{a_0^2 + V_\infty^2}\right)^2} \tag{11.131}$$

Since $\gamma = -1$, substitute to (11.120), we can get

$$\frac{a_0^2}{a_\infty^2} = 1 - Ma_\infty^2 \tag{11.132}$$

Substitute to (11.131)

$$E = \frac{W_\infty^2}{4a_0^2} = \frac{Ma_\infty^2}{\left(\frac{a_0}{a_\infty} + \sqrt{\left(\frac{a_0}{a_\infty}\right)^2 + Ma_\infty^2}\right)^2} = \frac{Ma_\infty^2}{\left(\sqrt{1 - Ma_\infty^2} + 1\right)^2} \tag{11.133}$$

Taking it into (11.129), we have

$$C_p = \frac{4E}{Ma_\infty^2(1 + E)} \cdot \frac{1 - \frac{W^2}{W_\infty^2}}{(1 - E) + E(1 - \frac{W^2}{W_\infty^2})} \tag{11.134}$$

For incompressible flow, yield

$$C_p(0) = 1 - \frac{W^2}{W_\infty^2} \qquad (11.135)$$

Substitute (11.133) and (11.135) into (11.134), and we get

$$C_p(Ma_\infty) = \frac{C_p(0)}{\sqrt{1 - Ma_\infty^2} + \frac{Ma_\infty^2}{1+\sqrt{1-Ma_\infty^2}}\frac{C_p(0)}{2}} \qquad (11.136)$$

Finally, this is the **Karman-Qian** correction formula.

11.7   Laitone Compressibility Correction Method

In 1951, E.V. **Laitone** from the University of California, USA, replace the incoming Mach number in the **Prandtl-Glauert** law with the local Mach number and proposed a new compressibility correction method. In which, the incoming Mach number in the Eq. (11.74) is replaced by the local Mach number

$$C_p(Ma_\infty, \alpha, c, f) = \frac{1}{\sqrt{1 - Ma^2}} C_p(0, \alpha, c, f) \qquad (11.137)$$

In this formula, Ma is the local Mach number, not the incoming Mach number. As defined by the pressure coefficient

$$C_p(Ma) = \frac{p - p_\infty}{\frac{1}{2}\rho_\infty V_\infty^2} = \frac{2p_\infty}{\rho_\infty V_\infty^2}\left(\frac{p}{p_\infty} - 1\right) = \frac{2}{\gamma Ma_\infty^2}\left(\frac{p}{p_\infty} - 1\right) \qquad (11.138)$$

Using compressible isentropic flow Eq. (11.117)

$$\frac{p_0}{p} = \left(\frac{T_0}{T}\right)^{\frac{\gamma}{\gamma-1}} = \left(1 + \frac{\gamma - 1}{2}Ma^2\right)^{\frac{\gamma}{\gamma-1}}$$

It could be known,

$$\frac{p}{p_\infty} = \left[\frac{1 + \frac{\gamma-1}{2}Ma_\infty^2}{1 + \frac{\gamma-1}{2}Ma^2}\right]^{\frac{\gamma}{\gamma-1}} \qquad (11.139)$$

Take the formula (11.139) into the formula (11.138) to find the local Mach number

$$Ma^2 = \left(Ma_\infty^2 + \frac{2}{\gamma - 1}\right)\left[1 + \frac{\gamma Ma_\infty^2}{2}C_p(Ma)\right]^{-\frac{\gamma}{\gamma-1}} - \frac{2}{\gamma - 1} \qquad (11.140)$$

Taking into account the small perturbations of the first order approximation

Fig. 11.11 Wright's
compressibility correction

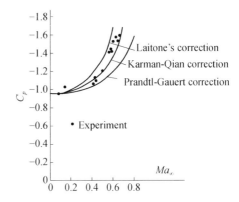

$$Ma^2 = Ma_\infty^2 - \left[1 + \frac{\gamma - 1}{2} Ma_\infty^2\right] Ma_\infty^2 C_p(0) \tag{11.141}$$

Substituted (11.141) into (11.137), it is

$$C_p(Ma) = \frac{C_p(0)}{\sqrt{1 - Ma_\infty^2} + \frac{Ma_\infty^2}{\sqrt{1-Ma_\infty^2}}(1 + \frac{\gamma-1}{2} Ma_\infty^2)\frac{C_p(0)}{2}} \tag{11.142}$$

In the derivation of this correction relation, the standard isentropic relation $p = C\rho^\gamma$ is used. The comparison with **Prandtl-Glauert** correction (11.74) and **Karman-Qian** correction (11.136) is shown in Fig. 11.11, which shows that Laitone's correction is higher qualified than **Karman-Qian** correction.

11.8 Aerodynamic Characteristics of Subsonic Thin Wing

By extending the two dimensional **Prandtl-Glauert** law to three dimensions, the corresponding relationship between the flow over a subsonic wing and an incompressible wing can be obtained.

11.8.1 Compressibility Correction of Sweep Wing With Infinite Span

As shown in Fig. 11.12, for an infinite wingspan swept wing, the incoming Mach number can be decomposed into a normal Mach number and a tangential Mach number. Because it is an ideal gas, the tangential Mach number does not affect the pressure distribution. For normal Mach numbers, the potential function of the linearization theory is

Fig. 11.12 Compressibility
correction of a sweep wing
with infinite span

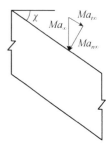

$$(1 - Ma_\infty^2 \cos^2 \chi)\frac{\partial^2 \varphi}{\partial x^2} + \frac{\partial^2 \varphi}{\partial y^2} + \frac{\partial^2 \varphi}{\partial z^2} = 0 \tag{11.143}$$

Thus, the compressibility correction factor caused by the normal Mach number is

$$\frac{1}{\beta} = \frac{1}{\sqrt{1 - Ma_\infty^2 \cos^2 \chi}} \tag{11.144}$$

11.8.2   Transformation Between Planform Shapes of Wings

For a flat object such as an aircraft wing, the first order approximation of the pressure
coefficient is

$$C_p = -\frac{2u}{V_\infty}$$

The pressure coefficient is determined only by the disturbance velocity in the
X-direction, which is the same as the pressure coefficient linearization formula of
incompressible flow. For the wings, according to the affine transformation relation,
the x-direction is invariable, and the z direction shrinks to $Z = \beta z$. Therefore the plane
geometry parameter between the corresponding wings has the following relations.

Root-tip ratio (chord length remains constant)

$$\eta' = \eta \tag{11.145}$$

Aspect ratio (length reduced):

$$\lambda' = \beta \lambda \tag{11.146}$$

Sweep angle (increase in sweep angle):

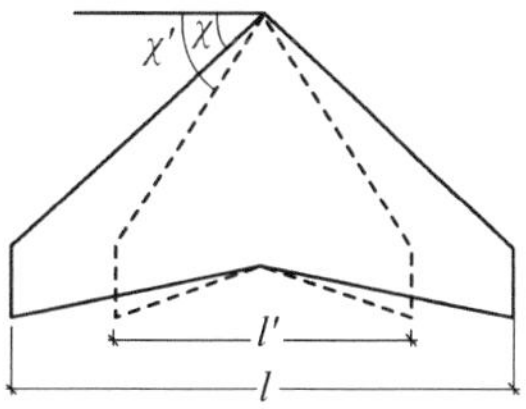

Fig. 11.13 Relationship between the subsonic velocity and the corresponding planform shape of an incompressible wing

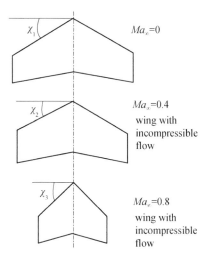

Fig. 11.14 Compressibility correction for flow around a flat object at different numbers

$$tg\chi' = \frac{1}{\beta}tg\chi \qquad (11.147)$$

It can be seen that the wing in the incompressible flow has a smaller aspect ratio, a larger sweep angle, and an invariable root-tip ratio, as shown in Fig. 11.13. Figure 11.14 shows the shape of a flat object flowing around it in different Mach numbers.

11.8.3 Prandtl-Glauert Law

By affine transformation, compared with the corresponding incompressible flow wing, the planform shape of the incompressible flow wing after transformation is changed: the aspect ratio becomes $\beta\lambda$, the tangent of sweep angle becomes $tg\chi' = \frac{1}{\beta}tg\chi$, and the airfoil shape and angle of attack are invariable. By extending Prandtl-Glauert to three dimensional, the pressure coefficient of the wing in compressible flow is equal to $1/\beta$ times the pressure coefficient at the corresponding point on the wing corresponding to the incompressible flow, i.e.,

(1) Pressure coefficient

$$C_p(Ma_\infty, \alpha, c, f, \lambda, tg\chi, \eta) = \frac{1}{\beta} C_p(0, \alpha, c, f, \beta\lambda, \frac{1}{\beta}tg\chi, \eta) \qquad (11.148)$$

Or

$$C_p(Ma_\infty, \alpha, \lambda, tg\chi, \eta) = \frac{1}{\beta} C_p(0, \alpha, \beta\lambda, \frac{1}{\beta}tg\chi, \eta) \qquad (11.149)$$

According to the Prandtl-Glauert law, the aerodynamic characteristics of an airfoil in subsonic compressible flow can be derived from the aerodynamic characteristics of the corresponding airfoil in incompressible flow.

(2) Lift coefficient and moment coefficient

$$C_L(Ma_\infty, \alpha, \lambda, tg\chi, \eta) = \frac{1}{\beta} C_L(0, \alpha, \beta\lambda, \frac{1}{\beta}tg\chi, \eta)$$
$$m_z(Ma_\infty, \alpha, \lambda, tg\chi, \eta) = \frac{1}{\beta} m_z(0, \alpha, \beta\lambda, \frac{1}{\beta}tg\chi, \eta) \qquad (11.150)$$

(3) Slope of the lift line

$$C_L^\alpha(Ma_\infty, \alpha, \lambda, tg\chi, \eta) = \frac{1}{\beta} C_L^\alpha(0, \alpha, \beta\lambda, \frac{1}{\beta}tg\chi, \eta) \qquad (11.151)$$

$$\frac{C_L^\alpha}{\lambda}(Ma_\infty, \alpha, \lambda, tg\chi, \eta) = \frac{1}{\beta\lambda} C_L^\alpha(0, \alpha, \beta\lambda, \frac{1}{\beta}tg\chi, \eta) \qquad (11.152)$$

The right-hand term of the above formula is a function of the plane geometric parameters $\beta\lambda$, $\lambda \tan\chi$, η, which are called the affine combination parameters. The equation above can be written as follows

$$\frac{C_L^\alpha}{\lambda} = Y(\beta\lambda, \frac{\lambda tg\chi}{\beta\lambda}, \eta) \qquad (11.153)$$

In the formula, Y is a function of the affine combination of parameters. It can be seen that if $\beta\lambda$, $\lambda \tan\chi$, η is the same, $\frac{C_L^\alpha}{\lambda}$ of subsonic wing will be the same too. Therefore, if the arrange computational or experimental values of airfoils of different planform shapes without torsion three affine combinations $\beta\lambda$, $\lambda \tan\chi$, η, the curves obtained will provide the values of $\frac{C_L^\alpha}{\lambda}$ for airfoils in the subsonic flow patterns of arbitrary plane shapes. It could be done by fixing one parameter at the beginning as

$\lambda \tan \chi$ and making a set of curves. As drawn on the left side (subsonic on the left and supersonic on the right), the effect of η can be neglected at this time. Then we could turn to another parameter as $\lambda \tan \chi$ and get another group of curves. Then a set of curves for calculating the slope of the wing lift line in compressible flow is obtained. In Fig. 11.15, the x axis is, the y axis is $\frac{C_L^{\alpha}}{\lambda}$, the parameter $\lambda c^{1/3}$ (relative thickness); $\chi_{0.5}$ represents the sweep of the midline of the wing.

(4) Pressure center

If the pressure center of the compressible flow wing is $(x_p)_{Ma, \lambda, \tan \chi, \eta}$ in x-direction from the leading edge of the wing, and the corresponding value of the incompressible wing is $(x_p)_{0, \alpha, \beta\lambda, \frac{\lambda tg\chi}{\beta\lambda}, \eta}$, then the moment coefficient at the leading edge point is

$$m_z(Ma_\infty, \alpha, \lambda, tg\chi, \eta) = C_L(Ma_\infty, \alpha, \lambda, tg\chi, \eta)\left(\frac{x_p}{b_A}\right)(Ma_\infty, \alpha, \lambda, tg\chi, \eta)$$

$$m_z(0, \alpha, \beta\lambda, \frac{\lambda tg\chi}{\beta\lambda}, \eta) = C_L(0, \alpha, \beta\lambda, \frac{\lambda tg\chi}{\beta\lambda}, \eta)\left(\frac{x_p}{b_A}\right)(0, \alpha, \beta\lambda, \frac{\lambda tg\chi}{\beta\lambda}, \eta)$$

$$\tag{11.154}$$

$$m_z(Ma_\infty, \alpha, \lambda, tg\chi, \eta) = \frac{1}{\beta}m_z(0, \alpha, \beta\lambda, \frac{\lambda tg\chi}{\beta\lambda}, \eta) \tag{11.155}$$

In which, the mean aerodynamic chord of a subsonic wing is $(b_A)_{Ma, \lambda, \tan \chi, \eta}$ and an incompressible wing is $(b_A)_{0, \beta\lambda, \frac{1}{\beta} \tan \chi, \eta}$, respectively.

By Prandtl-Glauert Law

$$\left(\frac{x_p}{b_A}\right)(Ma_\infty, \alpha, \lambda, tg\chi, \eta) = \left(\frac{x_p}{b_A}\right)(0, \alpha, \beta\lambda, \frac{\lambda tg\chi}{\beta\lambda}, \eta) \tag{11.156}$$

Because it is no torsion symmetric airfoil, the pressure center is the focus. The relationship between the relative position of the wing focus in compressible flow and that in incompressible flow is

$$\left(\frac{x_F}{b_A}\right)(Ma_\infty, \alpha, \lambda, tg\chi, \eta) = \left(\frac{x_F}{b_A}\right)(0, \alpha, \beta\lambda, \frac{\lambda tg\chi}{\beta\lambda}, \eta) \tag{11.157}$$

Therefore, the relative pressure center and focus of a compressible flow wing are a function of the affine combination parameters. The data curves can be arranged by affine combination parameters in the same way as the lift characteristics, as shown in Fig. 11.16. The relative pressure center and focus of an arbitrary planform shape airfoil without torsion in subsonic compressible flow can be obtained for design.

The experimental results show that when the angle of attack continues to increase, the pressure center of the wing will move backward. It is practically always approximate that there exists the following linear relationship in the range of $5° < \alpha < 20°$:

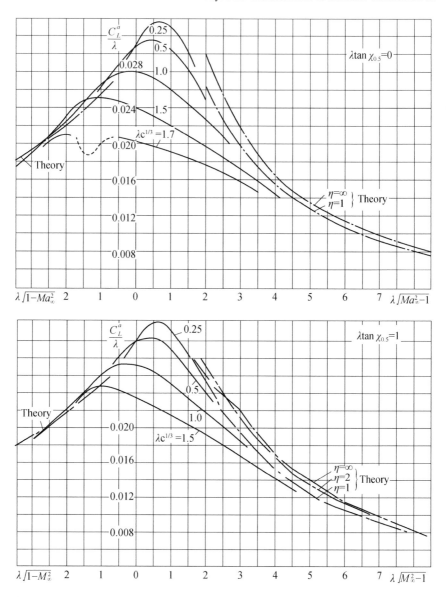

Fig. 11.15 Slope of lift coefficient for wing with different swept-back angles at subsonic speed

$$\left(\frac{x_p}{b_A}\right) = \left(\frac{x_p}{b_A}\right)_{\alpha=5^0} + \frac{\alpha - 5}{15}\Delta\left(\frac{x_p}{b}\right) \tag{11.158}$$

In the supersonic flow still exists a similar flow curve to the subsonic. The similar parameters form is not changed, just needs to change $\sqrt{1 - Ma^2}$ to $\sqrt{Ma^2 - 1}$.

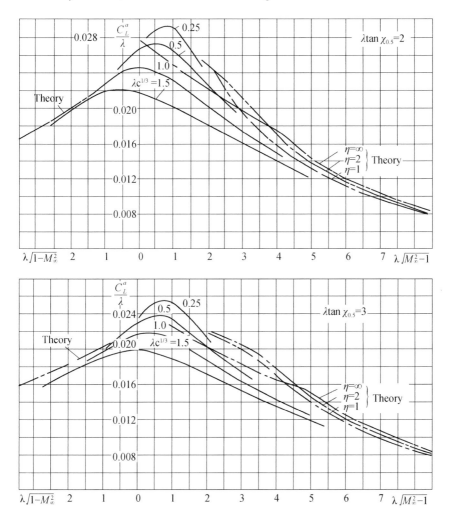

Fig. 11.15 (continued)

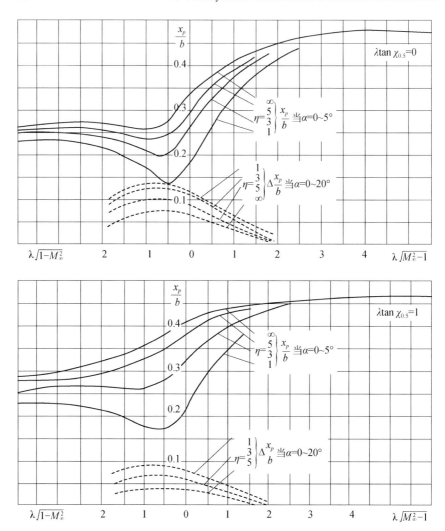

Fig. 11.16 Subsonic speed similitude ratio wing pressure center position curve

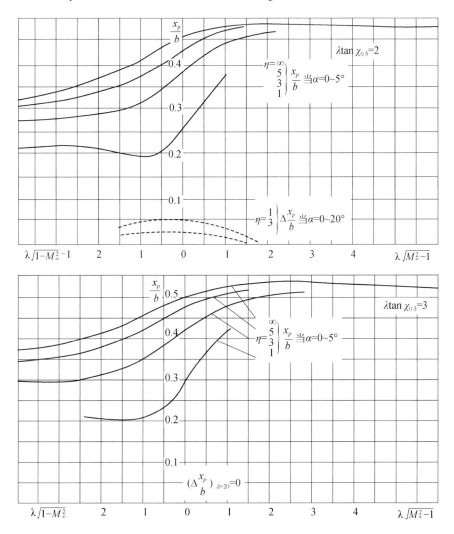

Fig. 11.16 (continued)

11.9 Effect of Mach Number of Incoming Flow on Aerodynamic Characteristics of Airfoil

11.9.1 Effect of Mach Number on Wing Lift Characteristics

In the subsonic range, the lift-line slope of a wing of the same planform shape increases with increasing Ma. Since with the same angle of attack, the absolute value of the negative pressure coefficient on the upper surface and the positive pressure coefficient on the lower surface of the wing increase with C_L^α increased. In the subsonic range, the maximum lift coefficient of the wing $C_{L\max}$ is related to the shape of the airfoil and generally decreases as Ma increases, as shown in Fig. 11.17. This is because the absolute value of the pressure coefficient on the airfoil surface increases with the same proportion $\frac{1}{\beta}$ with the increase of the Ma.

So the pressure at the minimum pressure point on the airfoil decreases the most which makes the inverse pressure gradient at the rear of the airfoil increases. As a result, the airfoil separates stall at a small angle of attack, and the lift coefficient of the wing decreases.

11.9.2 Effect of Mach Number on the Position of the Pressure Center of the Wing

According to the linearization theory, the center of pressure is the focus for thin wings with symmetric airfoils without torsion. It can be seen from the curve of the pressure center position of the subsonic velocity similarity ratio that with invariable $\lambda tg\chi_{0.5}$ and η the pressure center position moves forward and backward with the increase of Ma.

According to Prandtl-Glauert law, the pressure center of the wing in the subsonic flow is the same as the wing in the incompressible flow which the aspect ratio

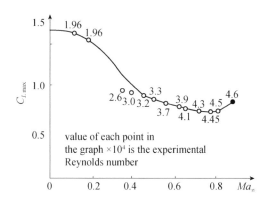

Fig. 11.17 The effect of mach number on the maximum lift coefficient

decreases to $\beta\lambda$, and the sweep angle increases to $\frac{1}{\beta}tg\chi$, i.e., with the increasing of Ma for the correspond arifoil in imcompressible flow, the aspect ratio become smaller, and swept angle increased. The low-speed experiments show that the smaller the aspect ratio is, the more forward the pressure center position is. And the larger the sweep angle is, the more backward the pressure center position is. The effect of these two factors is opposite, so the position of the pressure center depends on the combination of them together. In general, for swept wing with the larger η and $\lambda tg\chi_{0.5}$ sweep angle usually plays a major role, so with the increase of Ma the pressure center back shifts. $\overline{x}_p$ become larger. In contrast, slightly forward, $\overline{x}_p$ become smaller.

11.9.3   Effect of Mach Number on Drag Characteristics of Airfoil

As in the case of low speed, the drag coefficient of the wing in subsonic flow is still composed of the shape drag coefficient and the induced drag coefficient. The shape drag coefficient is

$$C_{dp} = (2C_f)_{Ma_\infty=0}\eta_C\eta_M \qquad (11.159)$$

In which $C_{f\,Ma_\infty=0}$ is the drag coefficient of low speed flat plate, which is related to Reynold Number and turning point. η_c is the correction factor for wing thickness. And the correction factor for compressibility η_M will change with the change of Mach number and transition position $\overline{X}_T = \frac{X_T}{b}$; however, it is always less than 1 and decreases with the Ma increase, as shown in Fig. 11.18. The friction coefficient decreases with the increase of Ma because of temperature in boundary layer will increase at this condition and density will decrease then. So the friction coefficient decreases and the viscosity coefficient increases very less to make an affection.

Fig. 11.18 Effect of mach number on friction coefficient

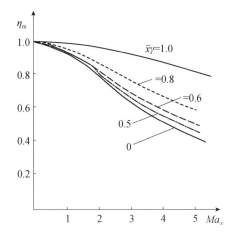

The compressibility of the induced drag coefficient is negligible when the Mach number is below the critical Mach number (at which the maximum flow velocity on the surface of the object reaches the sound velocity), it could still be treated as incompressible flow with an error of no more than 5%.

For moderate to high aspect ratios, the induced drag coefficient of a subsonic wing is

$$C_{di} = AC_L^2 \quad A = \frac{1+\delta}{\pi\lambda} \tag{11.160}$$

The effect of Mach number on the induced drag factor A is not significant, where $1 + \delta$ can still be determined in accordance with the low velocity incompressible flow. For small aspect ratios wing, there will be additional induce drag except the self vortices to induce drag when the angle of attack is small. That is induced drag caused by leading edge vortices and the side edge vortices separated.

11.10   Exercises

A.   Thinking questions

1. What is subsonic flow? Why is the subsonic flow more curved perpendicular to the airfoil than the incompressible flow?
2. Derive a differential equations that describes the motion of an ideal compressible fluid.
3. Why the governing equation of ideal compressible potential flow is nonlinear?
4. What is the linearized equation for the velocity potential function of compressible potential flow? Please describe the linearization conditions.
5. Under the condition of linearization, please write out the formulation of the solution for the steady compressible potential flow? And A linearized expression of the wall pressure coefficient?
6. How to use affine transformation to solve the linearization of velocity potential function of steady compressible plane potential flow?
7. How to use affine transformation to transform the equation below to Laplace's equation. Required $Y = y, Z = z$.

$$\beta^2\frac{\partial^2\varphi}{\partial x^2} + \frac{\partial^2\varphi}{\partial y^2} + \frac{\partial^2\varphi}{\partial z^2} = 0$$

8. Please demonstrate the physical reason that

$$C_p(Ma_\infty, \alpha, c, f) = \frac{1}{\beta^2}C_p(0, \beta\alpha, \beta c, \beta f)$$

can be written as

$$C_p(Ma_\infty, \alpha, c, f) = \frac{1}{\beta} C_p(0, \alpha, c, f)$$

9. Explain the difference between **Karman-Qian** formula and **Prandtl-Glauert law**

$$C_p(Ma_\infty, \alpha, c, f) = \frac{C_p(0, \alpha, c, f)}{\sqrt{1 - Ma_\infty^2} + \frac{Ma_\infty^2}{\sqrt{1-Ma_\infty^2}+1} \cdot \frac{C_p(0,\alpha,c,f)}{2}}$$

10. How does the lift coefficient of an airfoil with ideal steady compressible flow vary with the Mach number of the incoming flow and why?

11. How does the drag coefficient of an airfoil with ideal steady compressible flow vary with the Mach number and lift-drag ratio of the incoming flow?

B. Calculation

1. At a given point on the airfoil surface with very low speeds, the pressure coefficient is 0.3. If the Mach number of the free stream is 0.6, then CP calculated at this point is?

2. The theoretical lift coefficient for a symmetric thin airfoil in the Incompressible flow is $c_l = 2\pi\alpha$. The lift coefficient when M∞ = 0.7 is?

3. The pressure coefficient distribution at Re = 3.65×10^6 on the airfoil measured in the low speed wind tunnel is shown below. Based on these information, estimat that the Critical Mach number of the NACA0012 airfoil at zero angle of attack.

4. This is a two-dimensional parallel channel with a corrugated plate (sine curve), which the distance between the upper and lower walls is h, $\beta=\sqrt{1 - Ma_\infty^2}$

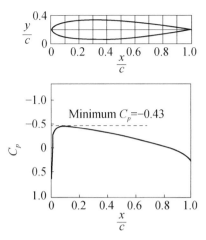

Question 3

Prove that

$$\phi = \frac{V_\infty}{\beta} \frac{d}{1 - e^{-4\pi\beta y/l}} \sin\left(\frac{2\pi}{l}x\right) e^{-2\pi\beta y/l}(1 + e^{4\pi\beta(y-h)/l})$$

is a solution to the subsonic linearization equation.

5. Consider a subsonic compressible flow in the Cartesian coordinate system, where the velocity potential is determined by

$$\phi(x, y) = V_\infty x + \frac{70}{\sqrt{1 - M_\infty^2}} e^{-2\pi\sqrt{1 - M_\infty^2}\, y} \sin 2\pi x$$

If the free flow properties are given by $V\infty = \frac{213\,\text{m}}{s}$, $p\infty = 101325$ Pa, and $T\infty = 15^\circ$ C , ask M, P, and T properties at position (0.061 m, 0.061 m).

6. $\phi = \frac{V_\infty}{\beta} h \sin\left(\frac{2\pi}{l}x\right) e^{-2\pi\beta y/l}$.

Suppose the formula satisfying the subsonic small disturbance linearization equation, with $\beta = \sqrt{1 - Ma_\infty^2}$. Try to write the differential equation of the streamline and solve the streamline equation by integral. Assuming that h is very small, find an approximate expression for a streamline near the x axis and the streamline shape when y is very large.

7. At low speed Incompressible flow, the pressure coefficient at a given point on the wing is 0.54. When the free flow Mach number is 0.58, by use of
    (1) Prandtl-Glauert law
    (2) Karman-Qian law
    (3) Laitone law.
    What is the pressure coefficient?

8. At a very low Mach number, the pressure coefficient at certain point of 2D airfoil is 0.5. What is the pressure coefficient at this point when the summation $Ma_\infty = 0.5$ and 0.8 according to the linearization theory?

9. A two dimensional airfoil is placed in the air so that its lowest pressure point appears on the lower surface. The pressure coefficient at this point is 0.782 when far field Mach number is 0.3. Try to use **Prandtl-Glauert law to** find the Critical Mach number of the airfoil.

10. For a given airfoil, Critical Mach number is 0.8. When $M\infty = 0.8$, find the minimum value of p/p_∞.

11. The experiment of airfoil NACA006 was carried out in subsonic wind tunnel, and the slope of lift curve d at $a = 0$ was measured as

Ma_∞	0.3	0.4	0.5	0.6	0.7	0.8
$\frac{dC_L}{d\alpha}$ /rad	0.595	0.620	0.654	0.710	0.801	0.963

Try to plot experimental curve and compare with the results from **Prandtl-Glauert law.**

12. Figure below shows four cases of flow over the same airfoil, where M_∞ gradually increases from 0.3 to $M_{cr} = 0.61$.

Point A on the airfoil is the minimum pressure point on the airfoil (and hence the maximum M). Assume that the minimum pressure (Max Mach number) continues to occur at the same point of M_∞ increase. In figure (a), for $M_\infty = 0.3$, the local Mach number at point a is arbitrarily chosen to be $M_A = 0.435$. This arbitrariness is reasonable because no airfoil shape is specified, regardless of shape, maximum Mach 0.435 occurs at point A on the airfoil surface. However, once the parameters in (a) are given, then (b), (c), and (d) are not arbitrary. Indeed, M_a is the only unknown function of M_∞ in the figure. Taking all of this as background information, starting from the data shown in Figure a, M_A is calculated when $M_\infty = 0.61$. Obviously, from figure d, the result should be $M_A = 1.0$ because $M_\infty = 0.61$ is considered Critical Mach number. Which means the Critical Mach number of this airfoil is 0.61. Tip: assume the conditions under the application region of **Prandtl-Glauert law.**

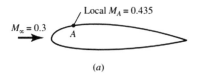

(a)

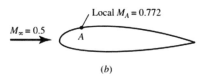

(b)

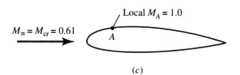

(c)

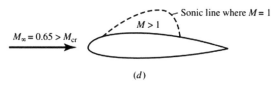

(d)

Question 12

13. Consider the flow pass a cylinder. Subsonic compressible flows on cylinder and sphere are similar in but differ in magnitude from them in Incompressible flow. In fact, because of the blunt shape of these objects, their Critical Mach number are relatively low. Specifically: for cylinders: $M_{cr} = 0.404$; for spheres: $M_{cr} = 0.57$.

Explain why the M_{cr} of a sphere is higher than that of a cylinder in a physical basis.

Chapter 12
Aerodynamic Characteristics of Supersonic Thin Airfoil and Wing

In this chapter, phenomena and aerodynamic characteristics of thin airfoil and wing at supersonic flow are introduced, including a linearized supersonic theory of thin airfoil and aerodynamic force characteristics, aerodynamic characteristics of an oblique wing with an infinite wingspan at supersonic flow, phenomena and aerodynamic characteristics of a thin wing at supersonic flow, lift characteristics of rectangular flat wing at supersonic flow, and characteristic line theory of supersonic flow.

Learning points:

(1) On top of the linearized supersonic theory of thin airfoil and aerodynamic characteristics;
(2) Be familiar with aerodynamic characteristics of an oblique wing with an infinite wingspan at supersonic flow, and phenomena and aerodynamic characteristics of a thin wing at supersonic flow;
(3) Know about lift characteristics of the rectangular flat wing at supersonic flow, and the characteristic line theory of supersonic flow.

12.1 Phenomena of the Thin Airfoil at Supersonic Flow

12.1.1 Shock Wave Drag of Thin Airfoil at Supersonic Flow

Different from subsonic flow, airfoil at supersonic flow involve with shock wave, expansion wave, and especially, shock wave drag, which is one of the main difference between the aerodynamic characteristics of supersonic and subsonic flow. Assuming gas ideal, leaving out flow viscosity, flow around the airfoil is a small perturbance problem with thin thickness, small bending, and small angle of attack. Airfoil has different effects on disturbance propagation of flow with subsonic and supersonic speed, as shown in Fig. 12.1.

© Science Press 2022

P. Liu, *Aerodynamics*, https://doi.org/10.1007/978-981-19-4586-1_12

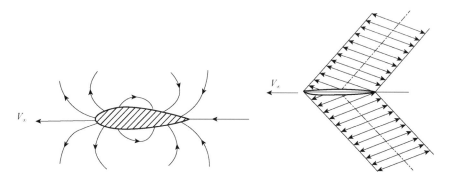

Fig. 12.1 Effects on disturbance propagation from an airfoil at subsonic and supersonic flow

For subsonic flow, disturbance from the airfoil propagates to the whole flow field. For supersonic flow around an airfoil, in contrast, disturbance propagation is restricted in the front Mach cone, the front part of the flow is compressed, and the back part of the flow is expanded. If we move with an airfoil, the phenomenon of flow being strong compressed when a pass from the airfoil nose can be seen, called a shock wave. Compared with the subsonic flow, airfoil at supersonic flow need to overcome additional drag (called shock wave drag), obviously, shock wave drag has a large correlation with the strength of shock wave attracted from disturbance of airfoil on flow. Shock wave strength has a direct relation with the nose shape of the airfoil, so shock wave drag is in contact with the dullness of the airfoil nose closely. As shown in Fig. 12.2, detached shock wave caused by supersonic flow around bluff object has strong shock wave strength and big shock wave drag; while attached shock wave caused by supersonic flow around sharp object (oblique shock wave) has weak strength and small drag.

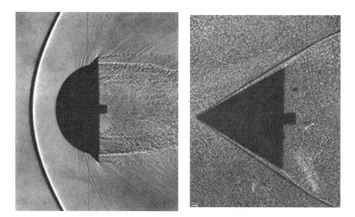

Fig. 12.2 Shock wave strength of supersonic flow around objects with different nose shapes

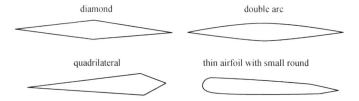

Fig. 12.3 Supersonic airfoils

Therefore, supersonic airfoil, unlike low speed or subsonic airfoil with a round nose and sharp tail (to reduce shape drag), adopts shapes with both sharp nose and tail (to reduce both shock wave and shape drag), like rhomboid, quadrilateral, double-cambered, etc. However, for supersonic aircraft, taking off, and landing period cannot be avoided. Airfoil with a sharp nose at low speed flow will occur flow separation at a small angle of attack, deteriorating the aerodynamic characteristics of the airfoil. In order to solve the low speed issue of supersonic aircraft, for the airfoil of low supersonic aircraft, thin airfoil with small round nose and sharp symmetric tail is usually used, as shown in Fig. 12.3.

12.1.2 Supersonic Flow Around Double-Cambered Airfoil

When supersonic flow passes double-cambered airfoil at a small angle of attack as Fig. 12.4, for $\alpha < \theta$ (Fig. 12.4a), both upper and lower sides of leading edge are compressed, forming oblique shock wave with different strengths, where shock wave of lower side of airfoil is stronger, and which of the upper side is weak; for $\alpha > \theta$ (Fig. 12.4b), expansion wave appears on the upper side of airfoil, and oblique shock wave on the lower side.

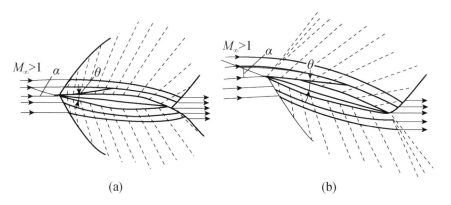

Fig. 12.4 Supersonic flow around double-cambered airfoil

For the case of the angle of attack smaller than half vertex angle, flow pass through oblique shock wave of leading edge, join together in trailing edge though series of expansion waves over both upper and lower side of airfoil, form two oblique shock waves because of inconformity of flow direction and pressure at trailing edge, and joined flow behind shock wave has the same direction and equal pressure (similar to incoming flow). For the case of the angle of attack bigger than half vertex angle, expansion wave occurs on the upper surface of airfoil leading edge, oblique shock wave appear on the lower side of leading edge, join together in trailing edge though series of expansion waves over both side of airfoil, and pass through an oblique shock wave and a series of expansion wave, letting flow get same direction and pressure (similar to incoming flow).

Affected by shock wave (strong compression waves) and expansion waves, pressure over the airfoil increases behind the shock wave and reduces behind expansion waves. Shock wave drag and lift are connected with pressure over the airfoil surface. Flow over airfoil surface is compressed by passing shock wave then flow along surface develops by passing a series of expansion wave. Due to pressure of forepart greater than backpart, the resultant force of pressure around airfoil surface has a backward component on flow direction, namely shock wave drag. When airfoil is at a small positive angle of attack, compression angle between tangent line of upper side of leading edge and streamline is smaller than which of lower side, so shock wave on upper surface is weaker than lower surface, its Mach number behind wave is larger than the other side, pressure behind wave is weaker than lower side. As a result, pressure of upper surface of airfoil is smaller than lower surface, resultant force of pressure has a component on vertical direction of incoming flow, that is lift.

12.2　Linearized Supersonic Theory

12.2.1　Fundamental Solution of Linearized Theory

In order to reduce shock wave drag, thickness of supersonic airfoil is usually small, camber is very small or zero, and angle of attack while in flight is also slight. Hence shock wave produced is weak, the entropy increased when flow pass shock wave can be neglected as first approximation. For ideal gas flow, assuming it is isentropic and potential, it can be solved with linearized potential flow equation.

For ideal gas, two-dimensional flow around supersonic thin airfoil, perturbation velocity potential function satisfies the linearized potential equation

$$B^2 \frac{\partial^2 \varphi}{\partial x^2} - \frac{\partial^2 \varphi}{\partial y^2} = 0, \quad \text{where}: B = \sqrt{M^2 - 1} \qquad (12.1)$$

wherein, x is along the direction of incoming flow, y is perpendicular to the direction of incoming flow, and φ is the velocity potential function of turbulence around airfoil.

The equation is linear hyperbolic partial differential equations of second order, which can be solved with characteristic line method or traveling wave method in mathematical equations. To follow the wave peak and introduce the combined variable of traveling wave, namely

$$\xi = x - By, \quad \eta = x + By \tag{12.2}$$

Thus, there are

$$\frac{\partial \varphi}{\partial x} = \frac{\partial \varphi}{\partial \xi}\frac{\partial \xi}{\partial x} + \frac{\partial \varphi}{\partial \eta}\frac{\partial \eta}{\partial x} = \frac{\partial \varphi}{\partial \xi} + \frac{\partial \varphi}{\partial \eta} \tag{12.3}$$

$$\frac{\partial^2 \varphi}{\partial x^2} = \frac{\partial^2 \varphi}{\partial \xi^2} + 2\frac{\partial^2 \varphi}{\partial \xi \partial \eta} + \frac{\partial^2 \varphi}{\partial \eta^2} \tag{12.4}$$

Similarly, there is

$$\frac{\partial^2 \varphi}{\partial y^2} = B^2\left(\frac{\partial^2 \varphi}{\partial \xi^2} - 2\frac{\partial^2 \varphi}{\partial \xi \partial \eta} + \frac{\partial^2 \varphi}{\partial \eta^2}\right) \tag{12.5}$$

Substitute Eqs. (12.4) and (12.5) into Eq. (12.1), and get

$$\frac{\partial^2 \varphi(\xi, \eta)}{\partial \xi \partial \eta} = 0 \tag{12.6}$$

Integrate upper formula along ξ, and get

$$\frac{\partial \varphi(\xi, \eta)}{\partial \eta} = f^*(\eta) \tag{12.7}$$

In the formula, f^* is a function of independent variable η, then integrate Eq. (12.7) along η and get

$$\varphi(\xi, \eta) = \int f^*(\eta)\mathrm{d}\eta + f_1(\xi) = f_1(\xi) + f_2(\eta) \tag{12.8}$$

wherein, $f_2(\eta) = \int f^*(\eta)\mathrm{d}\eta$ is a negative waveform function, $f_1(\xi)$ is a positive waveform function. Substitute transform Formula (12.2) to (12.8), the general solution of the linearized equation is obtained as

$$\phi(\xi, \eta) = f_1(x - By) + f_2(x + By) \tag{12.9}$$

In Fig. 12.5, perturbation in supersonic flow field at point $A(x_0, 0)$ propagate within the range of Mach cone originated from point A, the forward Mach line is AC^+ and the backward Mach line is AC^-. Perturbation generated at A and $t = 0$

Fig. 12.5 The supersonic
Mach cone

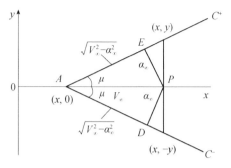

moments propagate along AC^+, the distance propagated along x axis at moment t is Vt. Then, propagation time t satisfies

$$t = \frac{x - x_0}{V}, \quad x = x_0 + Vt \tag{12.10}$$

At the same time, the wavefront of the perturbation wave is a sphere with a radius of at, which is

$$R = at = a\frac{x - x_0}{V}, \quad x = x_0 + R\mathrm{Ma} \tag{12.11}$$

According to the geometric relationship, the included Angle between the Mach line and the X axis is the Mach Angle, which is

$$\sin \mu = \frac{R}{x - x_0} = \frac{R}{R\mathrm{Ma}} = \frac{1}{\mathrm{Ma}} \tag{12.12}$$

Slope of the Mach line is

$$tg\mu = \frac{y}{x - x_0} = \frac{R}{\sqrt{(x - x_0)^2 - R^2}} = \frac{1}{\sqrt{\mathrm{Ma}^2 - 1}} = \frac{1}{B} \tag{12.13}$$

Therefore, in the $y > 0$ region, the equation of the Mach front is

$$x = x_0 + By, \quad x_0 = x - By, \quad \xi = x - By \tag{12.14}$$

Similarly, along with AC^-, in the $y < 0$ area, the equation of the AC^- Mach front is

$$tg\mu = \frac{-y}{x - x_0} = \frac{R}{\sqrt{(x - x_0)^2 - R^2}} = \frac{1}{\sqrt{\mathrm{Ma}^2 - 1}} = \frac{1}{B} \tag{12.15}$$

In the region $y < 0$, the equation of the Mach front is

Fig. 12.6 The supersonic
flow around airfoil at 0 angle
of attack

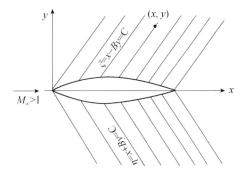

$$x = x_0 - By, \quad x_0 = x + By, \quad \eta = x + By \qquad (12.16)$$

It follows that $\xi = x - By$ is the forward (y positive) Mach front, is $f_1(\xi)$ the
waveform function of forward Mach wave front, which remain the waveform during
propagation; $\eta = x + By$ is the negative (y negative) Mach front, $f_2(\eta)$ is the waveform function of the backward Mach wave front, whose waveform does not change
during propagation. The physical meaning of Eq. (12.8) is that the velocity potential
function at any point in the supersonic flow field is the result of the superposition of
the positive wave function and the negative wave function.

As shown in Fig. 12.6, for a symmetrical airfoil flying flat at zero Angle of attack,
flight Mach number is $\mathrm{Ma}_\infty > 1$, in the plane of $Y > 0$, At any point $P(x, y)$, the
perturbation velocity potential function caused by the airfoil perturbation (forward
Mach wave) is

$$\varphi(x, y) = f_1(\xi) = f_1(x - By) \qquad (12.17)$$

Formula (12.17) is the solution of the perturbation velocity potential function
caused by the airfoil in the upper half plane. The component of the perturbation
velocity along the x- and y-directions is

$$
\begin{aligned}
u &= \frac{\partial \varphi(x, y)}{\partial x} = \frac{\partial f_1}{\partial \xi} \frac{\partial \xi}{\partial x} = \frac{df_1}{d\xi} = f_1'(x - By) \\
v &= \frac{\partial \varphi(x, y)}{\partial y} = \frac{\partial f_1}{\partial \xi} \frac{\partial \xi}{\partial y} = -B \frac{df_1}{d\xi} = -B f_1'(x - By)
\end{aligned}
\qquad (12.18)
$$

It can be seen that in the upper half plane, the disturbance velocities u and v are
constant along the Mach line $\xi = x - By$, indicating that in the Linearized theory,
the wave forms on the airfoil are unattenuated no matter shock wave, compression
wave or expansion wave.

In the plane $y < 0$, the perturbation velocity potential function of any point $P(x,$
$-y)$ caused by the airfoil perturbation (backward Mach wave) is

$$\varphi(x, y) = f_2(\eta) = f_2(x + By) \qquad (12.19)$$

Formula (12.19) is the solution of the perturbation velocity potential function caused by the airfoil in the lower half plane. The component of the perturbation velocity along the x- and y-direction is

$$
\begin{aligned}
u &= \frac{\partial\varphi(x,y)}{\partial x} = \frac{\partial f_2}{\partial \eta}\frac{\partial \eta}{\partial x} = \frac{\mathrm{d} f_2}{\mathrm{d}\eta} = f_2'(x+By) \\
v &= \frac{\partial\varphi(x,y)}{\partial y} = \frac{\partial f_2}{\partial \eta}\frac{\partial \eta}{\partial y} = B\frac{\mathrm{d} f_2}{\mathrm{d}\eta} = Bf_2'(x+By)
\end{aligned}
\tag{12.20}
$$

It can be seen that in the lower half plane, the disturbance velocities u and v are constant along the Mach line $\xi = x + By$, indicating that in the Linearized theory, the wave forms on the airfoil are unattenuated no matter shock waves, compression waves or expansion waves.

12.2.2   Supersonic Flow Over Corrugated Wall

When supersonic airflow bypasses a two-dimensional corrugated wall surface, the curve of the corrugated wall surface is set as

$$
y_s = d \sin \frac{2\pi x}{l}
\tag{12.21}
$$

where, l is the wave wavelength of the wall surface, and d is the wave amplitude, $\frac{d}{l} \ll 1$. According to the wall boundary condition, when $y = 0$

$$
\frac{\mathrm{d} y_s}{\mathrm{d} x} = \frac{2\pi d}{l}\cos\frac{2\pi x}{l}
\tag{12.22}
$$

The wall boundary condition of potential flow is

$$
v(x,0) = V_\infty \frac{\mathrm{d} y_s}{\mathrm{d} x} = V_\infty \frac{2\pi d}{l}\cos\frac{2\pi x}{l}
\tag{12.23}
$$

Using Eq. (12.18), it can be obtained

$$
v = -Bf_1'(x) = V_\infty \frac{2\pi d}{l}\cos\frac{2\pi x}{l}
\tag{12.24}
$$

Integrate the above equation, get

$$
\varphi(x,y) = -\frac{V_\infty d}{B}\sin\frac{2\pi(x-By)}{l}
\tag{12.25}
$$

The perturbation velocity component at any point in the flow field is

$$
\begin{aligned}
u &= \frac{\partial \varphi(x, y)}{\partial x} = \frac{\partial f_1}{\partial \xi} \frac{\partial \xi}{\partial x} = -\frac{V_\infty d}{B} \frac{2\pi}{l} \cos \frac{2\pi(x - By)}{l} \\
v &= \frac{\partial \varphi(x, y)}{\partial y} = \frac{\partial f_1}{\partial \xi} \frac{\partial \xi}{\partial y} = V_\infty \frac{2\pi d}{l} \cos \frac{2\pi(x - By)}{l}
\end{aligned}
\tag{12.26}
$$

The pressure coefficient is

$$
C_p = -\frac{2u}{V_\infty} = +\frac{4\pi}{B} \frac{d}{l} \cos \frac{2\pi(x - By)}{l}
\tag{12.27}
$$

The equation of the streamline is

$$
\begin{aligned}
\frac{dy}{v} &= \frac{dx}{V_\infty + u} \approx \frac{dx}{V_\infty}, \frac{dy}{dx} = \frac{v}{V_\infty} = \frac{2\pi d}{l} \cos \frac{2\pi(x - By)}{l} \\
y &= d \sin \frac{2\pi(x - Bh)}{l}
\end{aligned}
\tag{12.28}
$$

According to the Linearized theory, neither the perturbation velocity nor the amplitude of the streamline for the supersonic airflow passing around the corrugated wall decreases with the distance from the wall. On the wall at $y = 0$, the pressure coefficient is

$$
C_{ps} = -\frac{2u}{V_\infty} = +\frac{4\pi}{B} \frac{d}{l} \cos \frac{2\pi x}{l}
\tag{12.29}
$$

The pressure coefficient of supersonic flow over the wall and the corrugated wall have a phase difference value of $\pi/2$, rather than π in the subsonic case. Presented in Fig. 12.7.

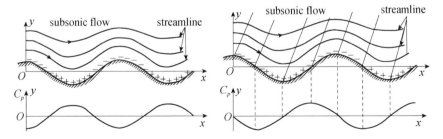

Fig. 12.7 Subsonic and supersonic flow over corrugated wall

12.3 Linearized Theory and Loading Coefficient of Thin Airfoil at Supersonic Flow

12.3.1 Linearized Theory of Thin Airfoil at Supersonic Flow

Let the slope of the upper surface of airfoil be $\frac{dy_u}{dx}$, the linear boundary condition of flow around airfoil is

$$v\left(x, 0^+\right) = V_\infty \frac{dy_u}{dx} \tag{12.30}$$

According to Formula (12.18), the velocity component at $y = 0$ is

$$
\begin{aligned}
u\left(x, 0^+\right) &= \frac{\partial\varphi\left(x, 0^+\right)}{\partial x} = \frac{\partial f_1}{\partial \xi}\frac{\partial \xi}{\partial x} = \frac{d f_1}{d\xi} = f_1'(x) \\
v\left(x, 0^+\right) &= \frac{\partial\varphi\left(x, 0^+\right)}{\partial y} = \frac{\partial f_1}{\partial \xi}\frac{\partial \xi}{\partial y} = -B\frac{d f_1}{d\xi} = -Bf_1'(x)
\end{aligned}
\tag{12.31}
$$

Resulting

$$f_1'(x) = u\left(x, 0^+\right) = -\frac{V_\infty}{B}\frac{dy_u}{dx} \tag{12.32}$$

Substituting the above equation into the linear pressure coefficient formula, the pressure coefficient of the upper surface can be obtained as

$$C_{pu} = -\frac{2u\left(x, 0^+\right)}{V_\infty} = \frac{2}{B}\frac{dy_u}{dx} \tag{12.33}$$

For flows in the lower half plane, the perturbation velocity potential function is shown in Formula (12.17), it can be obtained at $y = 0^-$ by the velocity component Formula (12.18) that

$$
\begin{aligned}
u\left(x, 0^-\right) &= \frac{\partial\varphi\left(x, 0^-\right)}{\partial x} = f_2'(x) \\
v\left(x, 0^-\right) &= \frac{\partial\varphi\left(x, 0^-\right)}{\partial y} = Bf_2'(x)
\end{aligned}
\tag{12.34}
$$

According to the boundary conditions

$$\frac{u\left(x, 0^-\right)}{V_\infty} = \frac{1}{B}\frac{dy_d}{dx} \tag{12.35}$$

Thus, the pressure coefficient of the lower surface is

Fig. 12.8 The changes of
flow get though a Mach wave

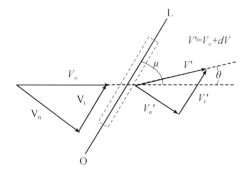

$$C_{pd} = -\frac{2u(x, 0^-)}{V_\infty} = -\frac{2}{B}\frac{dy_d}{dx} \tag{12.36}$$

wherein, 0^+ and 0^- are the upper and lower parts in the $y = 0$ plane and approximately represent the upper and lower surfaces of the airfoil, respectively. The above results can also be obtained by unified analyzing the supersonic flow through a small compression angle. The relation between the pressure coefficient and the turning angle is derived by using the principle that the tangential velocity of the weak oblique shock wave or the Mach compression wave is constant when the airflow passes through the small compression angle θ. In Fig. 12.8, because of $V_t' = V_t$, it is obtained that

$$V_\infty \cos \mu = (V_\infty + dV) \cos(\mu - \theta) \tag{12.37}$$

Expand the above equation, the compression angle θ is a small quantity, take the first order approximation, get

$$\frac{dV}{V_\infty} = -\frac{\sin \mu}{\cos \mu}\theta \tag{12.38}$$

And

$$\sin \mu = \frac{1}{Ma_\infty}, \cos \mu = \sqrt{1 - \sin^2 \mu} = \frac{\sqrt{Ma_\infty^2 - 1}}{Ma_\infty} \tag{12.39}$$

Substituting the above equation into Eq. (12.38), and get

$$\frac{dV}{V_\infty} = -\frac{\theta}{\sqrt{Ma_\infty^2 - 1}} = -\frac{\theta}{B} \tag{12.40}$$

where, $\mathrm{Ma}_\infty (= V_\infty / a_\infty)$ is the Mach number of incoming flow, and θ is the compression angle causing a small disturbance. If θ is an expansion Angle, the above equation can be signed by+. Since the compression Angle is small, the Mach compression wave is approximately isentropic, so the velocity and pressure relationship before and after the Mach wave satisfies (Euler equation)

$$dp = -\rho_\infty V_\infty dV, \quad \frac{dp}{p_\infty} = -\gamma \mathrm{Ma}_\infty^2 \frac{dV}{V_\infty} \tag{12.41}$$

Substitute the relation between the velocity and the folding Angle into, and get

$$\frac{dp}{p_\infty} = \frac{\gamma \mathrm{Ma}_\infty^2}{\sqrt{\mathrm{Ma}_\infty^2 - 1}} \theta \tag{12.42}$$

The pressure coefficient is

$$C_\mathrm{p} = \frac{(p_\infty + dp) - p_\infty}{\frac{1}{2}\rho_\infty V_\infty^2} = \frac{2dp}{\gamma \mathrm{Ma}_\infty^2 p_\infty} = -2\frac{dV}{V_\infty} = \frac{2\theta}{\sqrt{\mathrm{Ma}_\infty^2 - 1}} = \frac{2\theta}{B} \tag{12.43}$$

When θ is a compression Angle, C_p is positive. When θ is an expansion Angle, C_p is negative. Compared with Formula (12.33), when θ is very small, there is $\theta \approx \frac{dy_u}{dx}$. This is the first order small approximation formula for the wall pressure coefficient. The second-order approximate formula for the wall pressure coefficient is

$$C_\mathrm{p\,wall} = \frac{p - p_\infty}{\frac{1}{2}\rho_\infty V_\infty^2} = \frac{2\theta}{\sqrt{\mathrm{Ma}_\infty^2 - 1}} + \frac{(\gamma + 1)\mathrm{Ma}_\infty^4 - 4(\mathrm{Ma}_\infty^2 - 1)}{2(\mathrm{Ma}_\infty^2 - 1)^2}\theta^2 + \cdots$$

$$\tag{12.44}$$

When the wall angle θ is a small value, it can be regard as the slope $\frac{dy}{dx}$ of the angle between the tangent line at a point on the airfoil and the incoming flow along the x-axis.

A comparison between the linear theoretical pressure coefficient calculation formula and the experiment is shown in Fig. 12.9. Double-cambered airfoil is selected (leading edge half-vertical Angle is 11.3°, airfoil thickness is 10%, angle of attack is $-10°$, and inlet Mach number is 2.13). Figure 12.9 shows the wind tunnel test results of the upper and lower airfoil surfaces and a comparison of the first and second approximations. It can be seen that the first-order approximation theory is not compressed enough in the front half of the upper surface due to the impact of strong shock waves on the nose, and the second-order approximation theory agrees well. Under the influence of the tail shock wave, the first-order approximation theory expands more than the second stage approximation theory, and the second-order approximation theory agrees well.

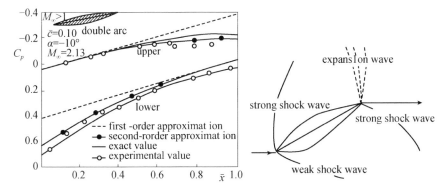

Fig. 12.9 Surface pressure coefficient distribution of double-cambered airfoil in supersonic flow

Figure 12.10 shows the pressure coefficient distribution of upper and lower surface of the double-cambered airfoil selected at different angles of attack (leading edge half-vertical Angle is 11.3°, airfoil thickness is 10%). Angle of attack is 0°, 4°, 8°, and inlet Mach number is 2.13. The increase of the actual pressure coefficient in the second half of the lower surface is, on the one hand, caused by the existence of the boundary layer, and the high pressure propagates upstream through the subsonic region of the boundary layer after the tail shock wave; on the other hand, the boundary layer is thickened or even separated due to the interference between the tail shock wave and the boundary layer, which reduces the actual expansion angle and forms a λ-shaped shock wave, thus increasing the pressure coefficient. The first-order approximation theory does not consider the above situation, so the expansion is excessive.

12.3.2 The Relationship Between Pressure Coefficient and Mach Number in Supersonic and Subsonic Flow

In compressible flow, the pressure coefficient at any point varies with the Mach number is different for subsonic and supersonic flow fields. In the subsonic flow field, according to Prandtl-Glauert's rule, the relationship between pressure coefficient and Mach number [Formula (11.74)] is

$$C_p \propto \frac{1}{\beta} = \frac{1}{\sqrt{1 - Ma_\infty^2}} \tag{12.45}$$

In supersonic flow field, the relation between pressure coefficient and Mach number [Formula (12.43)] is

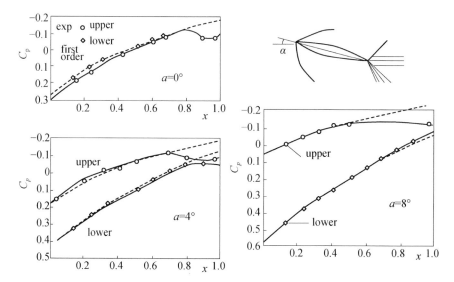

Fig. 12.10 Pressure distribution of double-cambered airfoil in supersonic flow at different angles of attack (less than half vertex angle)

$$C_p \propto \frac{1}{B} = \frac{1}{\sqrt{Ma_\infty^2 - 1}} \tag{12.46}$$

Thus, for subsonic flows, the pressure coefficient increases with the increase of the incoming Mach number; for ultrasonic flow, the pressure coefficient decreases with the increase of the incoming Mach number. As is shown in Fig. 12.11, when the Mach number approaches 1.0, the pressure coefficient approaches infinity, indicating that the Linearized theory is not suitable for the transonic region.

Fig. 12.11 Pressure coefficient as a function of Mach number

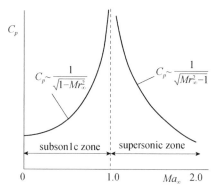

12.3.3 Loading Coefficient of Thin Airfoil at Supersonic Flow

It is shown by the linearized theory of thin airfoil at supersonic flow or the first-order approximation, that the pressure coefficient has a linear relationship with the slope of the airfoil, and the influence of the geometrical factors of the airfoil can be decomposed into three parts: Angle of attack, camber, and thickness. The pressure coefficient of the airfoil can be obtained by superposition of these three parts, which is

$$C_p = C_{p\alpha} + C_{pf} + C_{pc} \tag{12.47}$$

where, subscript α represents the flow around a panel with an angle of attack of α; f represents the flow around a curved panel with an angle of attack of zero and a camber of f in the middle arc; c represents the flow around a symmetric airfoil with an Angle of attack and a camber of zero and a thickness of c, as shown in Fig. 12.12.

Therefore, the pressure coefficients on the upper and lower surfaces can be written as

$$C_{pu}(x, 0^+) = (C_{pu})_\alpha + (C_{pu})_f + (C_{pu})_c \tag{12.48}$$

$$C_{pd}(x, 0^-) = (C_{pd})_\alpha + (C_{pd})_f + (C_{pd})_c \tag{12.49}$$

Using Formulas (12.33) and (12.36), it can be obtained

$$C_{pu}(x, 0^+) = \frac{2}{B}\left[\left(\frac{dy_u}{dx}\right)_\alpha + \left(\frac{dy_u}{dx}\right)_f + \left(\frac{dy_u}{dx}\right)_c\right] \tag{12.50}$$

$$C_{pd}(x, 0^-) = -\frac{2}{B}\left[\left(\frac{dy_d}{dx}\right)_\alpha + \left(\frac{dy_d}{dx}\right)_f + \left(\frac{dy_d}{dx}\right)_c\right] \tag{12.51}$$

(1) Part of angle of attack

Because the slope of the upper and lower surfaces is the same $\left(\frac{dy}{dx}\right)_\alpha = -\alpha$, so that the upper surface expands, the lower surface compresses, so there is

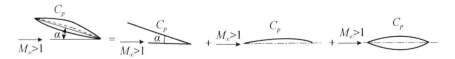

Fig. 12.12 The linear decomposition of flow around a supersonic airfoil

$$C_{pu}(x, 0^+) = -\frac{2}{B}\alpha, \quad C_{pd}(x, 0^-) = \frac{2}{B}\alpha \tag{12.52}$$

The load coefficient is

$$\Delta C_{p\alpha} = (C_{pd} - C_{pu})_\alpha = \frac{4}{B}\alpha = \frac{4\alpha}{\sqrt{Ma_\infty^2 - 1}} \tag{12.53}$$

(2) Camber part

Since the upper and lower surfaces have the same slope, when $\left(\frac{dy}{dx}\right)_f$ is positive, the upper surface is compressive, the lower surface is expansive; when $\left(\frac{dy}{dx}\right)_f$ is negative, the upper surface is expansive, the lower surface is compressive, so there is

$$(C_{pu})_f = \frac{2}{B}\left(\frac{dy}{dx}\right)_f, \quad (C_{pd})_f = -\frac{2}{B}\left(\frac{dy}{dx}\right)_f \tag{12.54}$$

The load coefficient is

$$\Delta C_{pf} = (C_{pd} - C_{pu})_f = -\frac{4}{B}\left(\frac{dy}{dx}\right)_f \tag{12.55}$$

(3) Thickness part

When the slope of the upper surface $\left(\frac{dy_u}{dx}\right)_c$ is positive, the upper surface is compressive; when the slope is negative, the upper surface is expansive; on the contrary, when $\left(\frac{dy_d}{dx}\right)_c$ is positive, the lower surface is expansive, and when is negative, the lower surface is compressive. Therefore, there is

$$(C_{pu})_c = \frac{2}{B}\left(\frac{dy_u}{dx}\right)_c, \quad (C_{pd})_c = -\frac{2}{B}\left(\frac{dy_d}{dx}\right)_c \tag{12.56}$$

Since the slopes of the upper and lower wings have equal value and opposite direction, $\left(\frac{dy_u}{dx}\right)_c = -\left(\frac{dy_d}{dx}\right)_c$, so the load coefficient is

$$\Delta C_{pc} = (C_{pd} - C_{pu})_c = 0 \tag{12.57}$$

Therefore, the pressure coefficient at any point on the upper and lower wings of the thin airfoil is

$$C_{pu}(x, 0^+) = \frac{2}{B}\left[-\alpha + \left(\frac{dy}{dx}\right)_f + \left(\frac{dy_u}{dx}\right)_c\right] \tag{12.58}$$

$$C_{pd}(x, 0^-) = -\frac{2}{B}\left[-\alpha + \left(\frac{dy}{dx}\right)_f + \left(\frac{dy_d}{dx}\right)_c\right] \qquad (12.59)$$

The load coefficient at any point on the upper and lower surfaces of the thin airfoil can be expressed as

$$\Delta C_p = \left(C_{pd} - C_{pu}\right)_\alpha + \left(C_{pd} - C_{pu}\right)_f + \left(C_{pd} - C_{pu}\right)_\alpha$$
$$= \frac{4}{B}\alpha - \frac{4}{B}\left(\frac{dy}{dx}\right)_f \qquad (12.60)$$

The pressure coefficient distributions of the airfoil angle of attack, camber, and thickness are given in the above equation as shown in Fig. 12.13. In this figure, the load coefficient distribution around the supersonic flat panel airfoil is shown on the left, from which we can see the difference of the load coefficient distribution between the subsonic and supersonic flat panels. The subsonic panel: the leading edge load is large, because the leading edge flows around the lower surface at a high velocity, and the trailing edge load is zero (to ensure that the trailing edge meets the Kutta-Joukowski trailing edge condition with equal pressure); while for plat in supersonic flow: the pressure coefficients on the upper and lower surfaces are equal, and the load coefficients are constant because the supersonic panel flow around the upper and lower surfaces does not affect each other. For supersonic flow around symmetric airfoil (thickness) problem: the first half compress and the second half expand, only generate drag without lift. For supersonic flow around bending plate problem: The front half of the upper surface is compressive and the rear half is expansive; The front half of the lower surface expands and the back half compacts, producing no lift but only drag, which is completely different from subsonic flow around it.

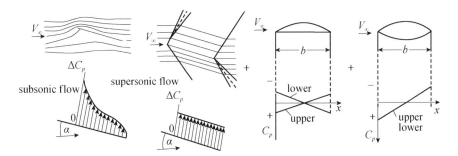

Fig. 12.13 Decomposition of angle of attack, camber, and thickness supersonic flow around thin airfoil

12.4   Aerodynamic Force Characteristics of Thin Airfoil at Supersonic Flow

Based on the linearized theory, lift coefficient, shock wave drag coefficient and the pitching moment coefficient to the leading edge of supersonic flow around thin airfoil can be expressed as superposition of contribution of angle of attack, camber and thickness, just like the pressure coefficient.

12.4.1   Lift Coefficient of Thin Airfoil at Supersonic Flow

According to the definition of lift coefficient of airfoil

$$C_{\mathrm{L}} = \frac{L}{\frac{1}{2}\rho_\infty V_\infty^2 S} = \frac{L}{q_\infty \cdot b \cdot 1} \tag{12.61}$$

where L is the lift force received by the airfoil of unit length, $q_\infty = \frac{1}{2}\rho_\infty V_\infty^2$ is the dynamic pressure of incoming flow, and b is the chord length of the airfoil. According to the linearized theory, the lift force can be written as

$$L = L_\alpha + L_f + L_c \tag{12.62}$$

(1)   Part of angle of attack (flow around panel)

Since the pressure distribution along the chordal direction is constant and the upper and lower surfaces are perpendicular to the panel, the normal force N_α perpendicular to the panel is

$$N_\alpha = \left(C_{\mathrm{pd}} - C_{\mathrm{pu}}\right)_\alpha q_\infty b \tag{12.63}$$

Substituting the panel load factor Eq. (12.53) into the above equation, there is

$$N_\alpha = \frac{4\alpha}{B} \cdot q_\infty b \tag{12.64}$$

As shown in Fig. 12.14, the lift force perpendicular to the incoming flow is

$$L_\alpha = N_\alpha \cos\alpha \approx N_\alpha = \frac{4\alpha}{B} q_\infty b \tag{12.65}$$

(2)   Camber part

As shown in Fig. 12.15, the lift force acting on the micro-segment ds is

Fig. 12.14 Force of part of angle of attack (flow around panel)

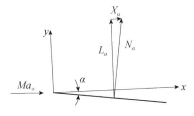

Fig. 12.15 Force of camber part (flow around bending plate at 0 angle of attack)

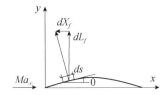

$$dL_f = \left(C_{pd} - C_{pu}\right)_f q_\infty ds \ \cos\theta \qquad (12.66)$$

Because of $dx = ds \ \cos\theta$, there is

$$L_f = \int_0^b \left(C_{pd} - C_{pu}\right)_f q_\infty dx \qquad (12.67)$$

By substituting Eq. (12.55) into the above equation, we can get

$$L_f = -\int_0^b \frac{4\left(\frac{dy}{dx}\right)_f}{B} q_\infty dx = -\frac{4q_\infty}{B} \int_0^0 dy_f = 0 \qquad (12.68)$$

This result shows that airfoil camber does not generate lift in supersonic flow under the condition of linearized small perturbation, which is different from that in low or subsonic flow.

(3) Thickness part

As shown in Fig. 12.16, as the upper and lower surface is symmetrical, dL_{cu} and dL_{cd} cancel out with each other at the corresponding point, so

$$L_c = \int_0^b dL_{cu} + \int_0^b dL_{cd} = 0 \qquad (12.69)$$

It can be concluded that, in the supersonic linearized theory, the airfoil thickness and camber do not generate lift force, and it is only generated by angle of attack of the panel. Namely

Fig. 12.16 Force of thickness part (flow around symmetrical airfoil at 0 angle of attack)

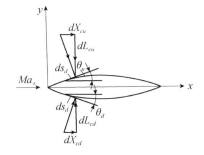

$$C_L = (C_L)_\alpha = \frac{4\alpha}{B} \tag{12.70}$$

12.4.2 Shock Wave Drag Coefficient of Thin Airfoil at Supersonic Flow

According to the airfoil shock wave drag coefficient definition

$$C_{\mathrm{Dw}} = \frac{X}{\frac{1}{2}\rho_\infty V_\infty^2 b} = \frac{X}{q_\infty b} \tag{12.71}$$

where X is the shock wave drag suffered by the airfoil of unit length, $q_\infty = \frac{1}{2}\rho_\infty V_\infty^2$ is the dynamic pressure of incoming flow, and b is the chord length of the airfoil. According to the linearized theory, the drag force can be written as

$$X = X_\alpha + X_f + X_c \tag{12.72}$$

(1) Part of angle of attack (flow around panel)

As shown in Fig. 12.14, the component of normal force in the direction of incoming flow can be obtained by using Eq. (12.64)

$$X_\alpha = N_a \sin \alpha = \frac{4\alpha^2}{B} q_\infty b \tag{12.73}$$

(2) Camber part (flow around bending plate at 0 angle of attack)

As shown in Fig. 12.15, the component of the normal force acting on the micro-segment ds in the direction of incoming flow is

$$dX_f = -\left(C_{\mathrm{pd}} - C_{\mathrm{pu}}\right)_f q_\infty \mathrm{d}s \, \sin \theta \tag{12.74}$$

Since

$$tg\theta = \frac{\sin\theta}{\cos\theta} = \left(\frac{dy}{dx}\right)_f, \quad ds\cos\theta = dx \tag{12.75}$$

Substituted into Formula (12.74) will get

$$dX_f = -\left(C_{pd} - C_{pu}\right)_f \left(\frac{dy}{dx}\right)_f q_\infty dx \tag{12.76}$$

By substituting Eq. (12.55) and integrating with x, we can get

$$X_f = \frac{4q_\infty}{B} \int\limits_0^b \left(\frac{dy}{dx}\right)_f^2 dx \tag{12.77}$$

(3) Thickness part

As shown in Fig. 12.16, as the contributions of the upper and lower surface to the drag are in the same direction, the drag generated by the upper and lower surface to the micro-segment at the corresponding point is twice that of the drag of the upper surface's micro point. Namely

$$dX_c = 2\left(C_{pu}\right)_c q_\infty ds_u \sin\theta_u \tag{12.78}$$

Since

$$tg\theta_u = \frac{\sin\theta_u}{\cos\theta_u} = \left(\frac{dy}{dx}\right)_{cu}, \quad ds_u \cos\theta_u = dx \tag{12.79}$$

Substituted into Formula (12.78) will get

$$dX_c = 2\left(C_{pu}\right)_c \left(\frac{dy_u}{dx}\right)_c q_\infty dx \tag{12.80}$$

According to Formula (12.56), $\left(C_{pu}\right)_c$ can be written as

$$\left(C_{pu}\right)_c = \frac{2}{B}\left(\frac{dy_u}{dx}\right)_c \tag{12.81}$$

Integrating in the x-direction and get

$$X_c = \frac{4q_\infty}{B} \int\limits_0^b \left(\frac{dy_u}{dx}\right)_c^2 dx \tag{12.82}$$

Substitute Eqs. (12.73), (12.77) and (12.82) into Eq. (12.72). The total shock wave drag is

$$X = X_\alpha + X_f + X_c$$
$$= \frac{4\alpha^2}{B} q_\infty b + \frac{4q_\infty}{B} \left[\int_0^b \left(\frac{dy}{dx} \right)_f^2 dx + \int_0^b \left(\frac{dy_u}{dx} \right)_c^2 dx \right] \tag{12.83}$$

The total shock wave drag coefficient is

$$C_{Dw} = \frac{X}{q_\infty b} = \frac{4\alpha^2}{B} + \frac{4}{Bb} \left[\int_0^b \left(\frac{dy}{dx} \right)_f^2 dx + \int_0^b \left(\frac{dy_u}{dx} \right)_c^2 dx \right] \tag{12.84}$$

Or

$$C_{Dw} = \frac{X}{q_\infty b} = C_L \alpha + \frac{4}{Bb} \left[\int_0^b \left(\frac{dy}{dx} \right)_f^2 dx + \int_0^b \left(\frac{dy_u}{dx} \right)_c^2 dx \right] \tag{12.85}$$

The above equation shows that the shock drag coefficient of flow around a supersonic thin airfoil is composed of two parts, one part is related to lift force, the other part is related to camber and thickness. The shock wave drag coefficient independent of lift force and related to camber and thickness is called zero-lift shock wave drag coefficient. Namely

$$C_{Dw0} = \frac{4}{Bb} \left[\int_0^b \left(\frac{dy}{dx} \right)_f^2 dx + \int_0^b \left(\frac{dy_u}{dx} \right)_c^2 dx \right] \tag{12.86}$$

In conclusion, since camber has no contribution to the lift of airfoil at supersonic flow, in order to reduce the zero-lift resistance, the supersonic airfoil is generally a symmetric airfoil without camber and with a small thickness. In order to reduce the flight drag, the flight angle of attack is also small, so there is $C_L \sim \alpha$, $C_{Dw} \sim \alpha^2$. If the angle of attack is large, the lift-drag ratio of flow around a supersonic airfoil decreases very fast.

Example 1 Symmetric rhombus airfoil, thickness c, chord length B, use linearized theory to find the lift coefficient and the shock wave drag coefficient.

Solution According to the linearized theory of thin airfoil at supersonic flow, the lift coefficient of a rhombus airfoil is

$$C_L = \frac{4\alpha}{B} = \frac{4\alpha}{\sqrt{M_\infty^2 - 1}}$$

Fig. 12.17 Rhombus airfoil

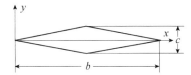

The slope of the lift line is

$$C_L^\alpha = \frac{dC_L}{d\alpha} = \frac{4}{\sqrt{M_\infty^2 - 1}}$$

It can be concluded that the slope of lift line of supersonic airfoil decreases with the increase of inlet Mach number. The shock wave drag coefficient is

$$C_{Dw} = \frac{X}{q_\infty b} = \frac{4\alpha^2}{B} + \frac{4}{Bb}\left[\int_0^b \left(\frac{dy}{dx}\right)_f^2 dx + \int_0^b \left(\frac{dy_u}{dx}\right)_c^2 dx\right]$$

Using the geometric relationship as shown in Fig. 12.17, there are

$$C_{Dw} = \frac{X}{q_\infty b} = \frac{4\alpha^2}{B} + \frac{4}{Bb}\left[\int_0^{b/2} \left(\frac{c}{b}\right)^2 dx + \int_{b/2}^b \left(-\frac{c}{b}\right)^2 dx\right] = \frac{4\alpha^2}{B} + \frac{4}{B}\left(\frac{c}{b}\right)^2 \tag{12.87}$$

The zero-lift shock wave drag coefficient is

$$C_{Dw0} = \frac{4}{B}\left(\frac{c}{b}\right)^2 \tag{12.88}$$

Example 2 A symmetrical double-chambered thin airfoil is provided, as shown in Fig. 12.18. The thickness is C, the chord length is B, and the arc equation of the upper surface of the thin airfoil is.

$$x^2 + (y_u + a)^2 = R^2 \tag{12.89}$$

Use linearized theory to obtain lift and shock wave drag coefficient of the double-chambered airfoil.

Solution According to the linearized theory of thin airfoil at supersonic flow, the lift coefficient of a double-chambered airfoil is

$$C_L = \frac{4\alpha}{B} = \frac{4\alpha}{\sqrt{M_\infty^2 - 1}}$$

Fig. 12.18 The
double-cambered thin airfoil

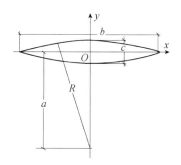

The slope of the lift line is

$$C_L^\alpha = \frac{dC_L}{d\alpha} = \frac{4}{\sqrt{M_\infty^2 - 1}}$$

The shock wave drag coefficient is

$$C_{\mathrm{Dw}} = \frac{X}{q_\infty b} = \frac{4\alpha^2}{B} + \frac{4}{Bb} \left[\int_{-b/2}^{b/2} \left(\frac{dy_u}{dx}\right)^2 dx \right]_c$$

Based on the geometry shown in Fig. 12.18 and taking into account the airfoil is thin, $\frac{y_u}{a} \ll 1$, just take the first-order approximation, use Eq. (12.89), and then get

$$\frac{dy_u}{dx} = -\frac{a}{x} \tag{12.90}$$

Therefore,

$$C_{\mathrm{Dw}} = \frac{X}{q_\infty b} = \frac{4\alpha^2}{B} + \frac{4}{Bb} \left[\int_{-b/2}^{b/2} \left(-\frac{x}{a}\right)^2 dx \right]_c = \frac{4\alpha^2}{B} + \frac{4}{12B} \frac{b^2}{a^2} \tag{12.91}$$

According to the geometrical relationship

$$\left(\frac{b}{2}\right)^2 + a^2 = R^2 = \left(a + \frac{c}{2}\right)^2 \tag{12.92}$$

Ignore the small value $\left(\frac{c}{2}\right)^2$, and get

$$a = \frac{b^2}{4c} \tag{12.93}$$

Substituting Eq. (12.93) into Eq. (12.91), it can be obtained that

$$C_{Dw} = \frac{4\alpha^2}{B} + \frac{4}{12B}\frac{b^2}{a^2} = \frac{4}{B}\left[\alpha^2 + \frac{4}{3}\left(\frac{c}{b}\right)^2\right] = \frac{4}{\sqrt{Ma_\infty^2 - 1}}\left[\alpha^2 + \frac{4}{3}\left(\frac{c}{b}\right)^2\right]$$

(12.94)

The zero-lift shock wave drag coefficient is

$$C_{Dw0} = \frac{4}{B}\frac{4}{3}\left(\frac{c}{b}\right)^2$$

(12.95)

12.4.3 Pitching Moment Coefficient of Thin Airfoil at Supersonic Flow

The pitching moment coefficient of aerodynamic force around a thin airfoil at supersonic flow on the leading edge of the airfoil is defined as

$$C_m = \frac{M_z}{q_\infty b^2}$$

(12.96)

where, M_z is the pitching moment to the leading edge of the airfoil, and the head-up is specified as positive.

(1) Part of angle of attack (flow around panel)

Since the pressure distribution along the panel is constant and the action point of lift is located at the middle of the panel (as shown in Fig. 12.13), the moment to the leading edge is

$$(M_z)_\alpha = C_L q_\infty b\left(-\frac{b}{2}\right) = -\frac{1}{2}C_L q_\infty b^2$$

(12.97)

(2) Camber part (flow around bending plate at 0 angle of attack)

As shown in Fig. 12.15, when the distance between micro-segment ds and the leading edge is x, the moment exerted on the leading edge by the micro-segment is

$$(dM_z)_f = -dL_f x = \frac{4\left(\frac{dy}{dx}\right)_f}{B}q_\infty x dx$$

(12.98)

$$(M_z)_f = \int_0^b \frac{4}{B} q_\infty \left(\frac{dy}{dx} \right)_f x dx \tag{12.99}$$

Noticing $y_f \big|_0^b = 0$, integrate the above equation by part and get

$$(M_z)_f = -\int_0^b \frac{4}{B} q_\infty y_f dx \tag{12.100}$$

When the curve equation in airfoil camber $y = y_f(x)$ is known, the camber moment can be obtained by integrating the above equation.

(3) Thickness part (flow around symmetric airfoil at 0 angle of attack)

As shown in Fig. 12.16, because the pressure on the upper and lower surface is symmetrical, the dL_u and dL_d at the corresponding point cancel out with each other, so the contribution of the airfoil thickness to the leading edge moment is zero.

The total moment is

$$M_z = -\frac{1}{2} C_L q_\infty b^2 - \int_0^b \frac{4}{B} q_\infty y_f dx \tag{12.101}$$

The moment coefficient to the leading edge point is

$$C_m = \frac{M_z}{q_\infty b^2} = \frac{-\frac{1}{2} C_L q_\infty b^2 - \int_0^b \frac{4}{B} q_\infty y_f dx}{q_\infty b^2} = -\frac{1}{2} C_L - \frac{4}{Bb^2} \int_0^b y_f dx \tag{12.102}$$

According to the linearized theory, airfoil camber and thickness do not generate lift, and airfoil thickness does not generate leading edge moments as well, so the camber moment coefficient is also called the zero-lift moment coefficient. Suppose the distance between the pressure center of the airfoil and the leading edge is x_p, there is

$$\frac{x_p}{b} = \frac{1}{2} + \frac{4}{C_L Bb^2} \int_0^b y_f dx \tag{12.103}$$

The center of pressure is related to camber, and when camber is zero, the center of pressure is at the midpoint. According to the definition of focus

$$\frac{x_F}{b} = -\frac{\partial C_m}{\partial C_L} = \frac{1}{2} \tag{12.104}$$

The above equation indicates that the focus of a thin airfoil at supersonic flow is located at the midpoint of the wing chord. Because the focus is the point at which the lift increment occurs, and the lift depends only on the angle of attack. Because the load is evenly distributed on the panel, the focus is located at the midpoint of the wing chord. Compared with the low-speed airfoil, the focus of the low-speed airfoil is located at the 1/4 chord point from the leading edge, while the focus of the supersonic thin airfoil is located at the 1/2 chord point from the leading edge, that is, from low speed flow to the supersonic flow, the focus of the thin airfoil moves significantly back, which has a great influence on the stability and maneuverability of the aircraft.

12.4.4 Comparison of Linearized Theory and Experimental Results of Supersonic Thin Airfoil

For supersonic flow around a symmetrical thin double-cambered airfoil, aerodynamic forces obtained from the linearized theory and experimental results are compared as shown in Fig. 12.19. It can be seen that:

(1) The slope of the lift line obtained by the supersonic linearized theory of thin airfoil is 2.5% higher than the experimental value. The reason is that the increase of trailing pressure and decrease of lift due to the boundary layer on the upper surface and its interference with the shock wave at the trailing edge are not taken into account in the linearized theory;

(2) The shock wave drag coefficient obtained by the linearized theory is slightly less than the experimental value, and is almost constant in the whole angle of

Fig. 12.19 Comparison between theoretical and experimental results of lift and drag coefficient of flow over supersonic thin airfoil

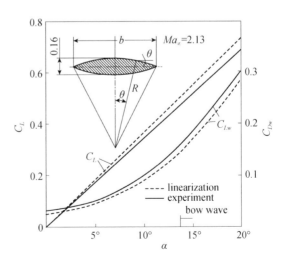

Fig. 12.20 Comparison between theoretical and experimental results of leading edge moment coefficient of flow over supersonic thin airfoil

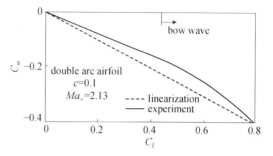

attack range, which is approximately equal to the friction drag caused by the viscosity and pressure drag not included in the theory;

(3) The moment coefficient obtained from the supersonic linearized theory is compared with the experimental results, as shown in Fig. 12.20. It can be seen that the result of moment coefficient of linearized theory is slightly lower than the experimental result, because the actual pressure near the trailing edge of the upper surface is higher than the result of linearized theory, and the moment arm is larger, resulting in the linearized theoretical value of is lower than the experimental result.

12.5 Aerodynamic Characteristics of Oblique Wing with Infinite Wingspan at Supersonic Flow

As shown in Fig. 12.21, for an oblique wing with infinite wingspan, the sweep angle of the oblique wing is χ, and the inlet Mach number can be decomposed into the normal component perpendicular to the leading edge and the tangential component parallel to the leading edge. Namely

$$\text{Ma}_{\infty n} = \text{Ma}_\infty \cos \chi, \quad \text{Ma}_{\infty t} = \text{Ma}_\infty \sin \chi \qquad (12.105)$$

If not consider air viscosity, the tangential component of the impact on the aerodynamic characteristics of the wing does not produce, infinite wingspan oblique wing aerodynamic characteristics depend on the normal component of Mach number, and only when the $\text{Ma}_{\infty n} > 1$ flow around oblique wing is the supersonic, even if $\text{Ma}_\infty > 1$, flow around infinite oblique wing still for subsonic or transonic characteristics, shock wave drag is almost zero. The supersonic aerodynamic characteristics of an infinite oblique wing with $\text{Ma}_{\infty n} > 1$ are given below.

According to the derivation of oblique wing in Chap. 9, the pressure coefficient, lift coefficient and shock wave drag coefficient between oblique wing and positive wing with infinite wingspan exist as follows. Namely

The relationship between pressure coefficient of oblique wing and straight wing

Fig. 12.21 Oblique wing
with infinite wingspan

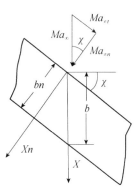

$$C_p = \left(C_p\right)_n \cos^2 \chi \tag{12.106}$$

The relationship between lift coefficient of oblique wing and straight wing

$$C_L = (C_L)_n \cos^2 \chi \tag{12.107}$$

The relationship between shock wave drag coefficient of oblique wing and straight wing

$$C_{Dw} = (C_{Dw})_n \cos^3 \chi \tag{12.108}$$

According to the geometric relation, the relation between the chord length, thickness and angle of attack of oblique wing and straight wing is

$$b_n = b \cos \chi \tag{12.109}$$

$$\frac{dy}{dx_n} = \frac{dy}{dx} / \cos \chi \tag{12.110}$$

$$\alpha_n = \alpha / \cos \chi \tag{12.111}$$

(1) The formula of pressure coefficient and load coefficient of infinite oblique wing

According to the pressure coefficient formula of upper and lower surfaces of airfoil at supersonic flow, the Mach number is written as the normal Mach number $Ma_{\infty n}$, the angle of attack as the normal angle of attack, and the surface derivative as the normal derivative, so that the normal pressure coefficient of upper and lower surfaces is

$$\begin{cases} (C_{pu})_n = \dfrac{2}{\sqrt{Ma_\infty^2 \cos^2 \chi - 1}} \left[-\alpha_n + \left(\dfrac{dy}{dx_n}\right)_f + \left(\dfrac{dy_u}{dx_n}\right)_c \right] \\[4mm] (C_{pd})_n = \dfrac{2}{\sqrt{Ma_\infty^2 \cos^2 \chi - 1}} \left[+\alpha_n - \left(\dfrac{dy}{dx_n}\right)_f - \left(\dfrac{dy_d}{dx_n}\right)_c \right] \end{cases} \quad (12.112)$$

Replace the normal derivative with the normal angle of attack, and get

$$\begin{cases} (C_{pu})_n = \dfrac{2}{\cos \chi \sqrt{Ma_\infty^2 \cos^2 \chi - 1}} \left[-\alpha + \left(\dfrac{dy}{dx}\right)_f + \left(\dfrac{dy_u}{dx}\right)_c \right] \\[4mm] (C_{pd})_n = \dfrac{2}{\cos \chi \sqrt{Ma_\infty^2 \cos^2 \chi - 1}} \left[+\alpha - \left(\dfrac{dy}{dx}\right)_f - \left(\dfrac{dy_d}{dx}\right)_c \right] \end{cases} \quad (12.113)$$

Note that for the thickness problem, due to the symmetry of the wing, at the same x, there is

$$\left(\frac{dy_u}{dx}\right)_c = -\left(\frac{dy_d}{dx}\right)_c \quad (12.114)$$

The normal load coefficient is

$$(\Delta C_p)_n = (C_{pu} - C_{pd})_n = \frac{4}{\cos \chi \sqrt{Ma_\infty^2 \cos^2 \chi - 1}} \left[\alpha - \left(\frac{dy}{dx}\right)_f \right] \quad (12.115)$$

Using Eq. (12.106), the pressure and load coefficient of the oblique wing can be obtained as

$$\begin{cases} C_{pu} = (C_{pu})_n \cos^2 \chi = \dfrac{2\cos \chi}{\sqrt{Ma_\infty^2 \cos^2 \chi - 1}} \left[-\alpha + \left(\dfrac{dy}{dx}\right)_f + \left(\dfrac{dy_u}{dx}\right)_c \right] \\[4mm] C_{pd} = (C_{pd})_n \cos^2 \chi = \dfrac{2\cos \chi}{\sqrt{Ma_\infty^2 \cos^2 \chi - 1}} \left[+\alpha - \left(\dfrac{dy}{dx}\right)_f - \left(\dfrac{dy_d}{dx}\right)_c \right] \end{cases} \quad (12.116)$$

According to Eq. (12.115), the load coefficient of oblique wing can be obtained as

$$\Delta C_p = (\Delta C_p)_n \cos^2 \chi = \frac{4\cos \chi}{\sqrt{Ma_\infty^2 \cos^2 \chi - 1}} \left[\alpha - \left(\frac{dy}{dx}\right)_f \right] \quad (12.117)$$

(2) The lift coefficient of infinite oblique wing

Using the lift coefficient of a thin airfoil at supersonic flow, the Mach number is written as the normal Mach number $\text{Ma}_{\infty n}$, and the angle of attack is written as the normal angle of attack. The normal lift coefficient is obtained

$$C_{Ln} = \frac{4\alpha_n}{\sqrt{\text{Ma}_{\infty}^2 \cos^2 \chi - 1}} = \frac{4\alpha}{\cos \chi \sqrt{\text{Ma}_{\infty}^2 \cos^2 \chi - 1}} \qquad (12.118)$$

Using Eq. (12.107), the lift coefficient of infinite oblique wing can be obtained as

$$C_L = \frac{4\alpha \cos \chi}{\sqrt{\text{Ma}_{\infty}^2 \cos^2 \chi - 1}} \qquad (12.119)$$

(3) The shock wave drag coefficient of infinite oblique wing

The normal shock wave drag coefficient is

$$C_{Dwn} = \frac{4}{\sqrt{\text{Ma}_{\infty}^2 \cos^2 \chi - 1}} \left[\alpha_n^2 + \frac{1}{b_n} \int_0^{b_n} \left(\frac{dy}{dx_n} \right)_f^2 dx_n + \frac{1}{b_n} \int_0^{b_n} \left(\frac{dy_u}{dx_n} \right)_c^2 dx_n \right] \qquad (12.120)$$

Substitute the normal relation, and get

$$C_{Dwn} = \frac{4}{\cos^2 \chi \sqrt{\text{Ma}_{\infty}^2 \cos^2 \chi - 1}} \left[\alpha^2 + \frac{1}{b} \int_0^b \left(\frac{dy}{dx} \right)_f^2 dx + \frac{1}{b} \int_0^b \left(\frac{dy_u}{dx} \right)_c^2 dx \right] \qquad (12.121)$$

Using Eq. (12.108), the shock wave drag coefficient of infinite oblique wing can be obtained as

$$C_{Dw} = C_{Dwn} \cos^3 \chi$$

$$= \frac{4 \cos \chi}{\sqrt{\text{Ma}_{\infty}^2 \cos^2 \chi - 1}} \left[\alpha^2 + \frac{1}{b} \int_0^b \left(\frac{dy}{dx} \right)_f^2 dx + \frac{1}{b} \int_0^b \left(\frac{dy_u}{dx} \right)_c^2 dx \right] \qquad (12.122)$$

If the surface derivative remains normal derivative and no substitution is made in the above shock wave drag coefficient, the shock wave drag coefficient can be expressed as

$$
\begin{aligned}
C_{\mathrm{Dw}} &= \frac{4\cos^3\chi}{\sqrt{\mathrm{Ma}_\infty^2\cos^2\chi - 1}}\left[\alpha_n^2 + \frac{1}{b_n}\int_0^{b_n}\left(\frac{\mathrm{d}y}{\mathrm{d}x_n}\right)_f^2\mathrm{d}x_n + \frac{1}{b_n}\int_0^{b_n}\left(\frac{\mathrm{d}y_u}{\mathrm{d}x_n}\right)_c^2\mathrm{d}x_n\right] \\
&= \frac{4\alpha^2\cos\chi}{\sqrt{\mathrm{Ma}_\infty^2\cos^2\chi - 1}} + \frac{4\cos^3\chi}{\sqrt{\mathrm{Ma}_\infty^2\cos^2\chi - 1}}I
\end{aligned}
$$

$$
C_{\mathrm{Dw0}} = \frac{4\cos^3\chi}{\sqrt{\mathrm{Ma}_\infty^2\cos^2\chi - 1}}I, \quad I = \frac{1}{b_n}\int_0^{b_n}\left(\frac{\mathrm{d}y}{\mathrm{d}x_n}\right)_f^2\mathrm{d}x_n + \frac{1}{b_n}\int_0^{b_n}\left(\frac{\mathrm{d}y_u}{\mathrm{d}x_n}\right)_c^2\mathrm{d}x_n
$$

$$(12.123)$$

In the above equation, the second term on the right of the drag formula is the zero-lift shock drag coefficient of infinite oblique wing (expressed by the normal derivative of the wing surface). Make

$$
\frac{\mathrm{d}}{\mathrm{d}\chi}\left(\frac{C_{\mathrm{Dw0}}}{4I}\right) = \frac{\mathrm{d}}{\mathrm{d}\chi}\left(\frac{\cos^3\chi}{\sqrt{\mathrm{Ma}_\infty^2\cos^2\chi - 1}}\right) = 0
$$

The sweep back angle where the zero-lift shock wave drag coefficient obtain extreme value (minimum) is

$$
\cos\chi = \frac{1}{\mathrm{Ma}}\sqrt{\frac{3}{2}} \tag{12.124}
$$

Substituted into Eq. (12.123), the minimum zero-lift shock wave drag coefficient can be obtained as

$$
C_{\mathrm{Dw0\,min}} = 4I\frac{\left(\frac{1}{\mathrm{Ma}_\infty}\sqrt{\frac{3}{2}}\right)^3}{\sqrt{\mathrm{Ma}_\infty^2\left(\frac{1}{\mathrm{Ma}_\infty}\sqrt{\frac{3}{2}}\right)^2 - 1}} = 4I\sqrt{2}\left(\frac{1}{\mathrm{Ma}_\infty}\sqrt{\frac{3}{2}}\right)^3 = 2I\frac{3^{1.5}}{\mathrm{Ma}_\infty^3}
$$

$$(12.125)$$

According to the above aerodynamic characteristics formula of infinite oblique wing at supersonic flow, the slope of lift line and the zero-lift shock wave drag coefficient follow the sweep angle can be calculated, as shown in Fig. 12.22.

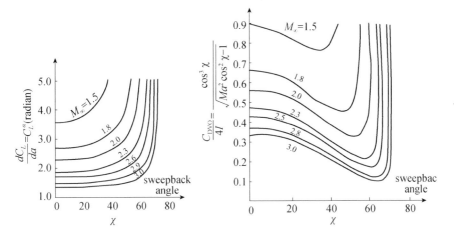

Fig. 12.22 Characteristic curve of sweep wing with thin airfoil in supersonic flow

12.6 Conceptual Framework of Thin Wing at Supersonic Flow

12.6.1 The Concept of Front and Rear Mach Cone

In order to better understand the aerodynamic characteristics of supersonic flow around a thin wing, it is necessary to explain several basic concepts of supersonic thin wing. In supersonic flow field, two Mach cones with axes parallel to the flow direction can be formed from any point P. The cone upstream of point P is called the front Mach cone, and the cone downstream of point P is called the rear Mach cone, as shown in Fig. 12.23.

Fig. 12.23 Front and rear supersonic Mach cone

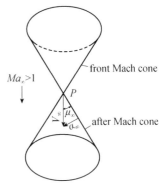

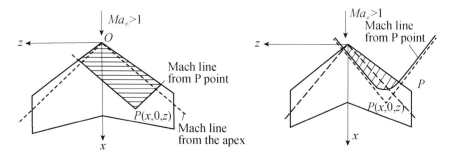

Fig. 12.24 The dependence region of point P on the wing plate

The half-apex angle of the Mach cone is

$$\mu_\infty = tg^{-1} \frac{1}{\sqrt{Ma_\infty^2 - 1}} \tag{12.126}$$

The region bounded by the front Mach cone is called the dependence region of point P, in which all disturbance sources affect the solution of P. The region enclosed by the rear Mach cone is called the influence zone of point P, and all spatial points within the Mach cone are affected by the P source.

As shown in Fig. 12.24, for example, point P is located on the wing plate, and the coordinate of point P is $(x, 0, z)$, which is only affected by the part of the wing located in the front Mach cone. However, when point P is located above the wing, the coordinate of point P is (x, y, z), and its dependence region is the area of the intersection line between the Mach cone and the wing surface.

12.6.2 Leading Edge, Trailing Edge and Side Edge

According to the propagation characteristics of the disturbance wave in the supersonic flow field, different boundaries have different effects on the aerodynamic characteristics. Therefore, the boundaries of a wing should be divided into leading edge, trailing edge and side edge. The boundary at which the wing first intersects the line parallel to the incoming flow direction is the leading edge, the boundary at which the second intersects the trailing edge, and the boundary parallel to the incoming flow is the side edge. Whether the leading edge, trailing edge, or side edge is also naturally related to the direction of incoming flow relative to the wing, as shown in Fig. 12.25. If the normal component of velocity of the incoming flow relative to the leading (trailing) edge is less than the sound velocity ($Ma_{\infty n} < 1$), the leading (trailing) edge is called the subsonic leading (trailing) edge, as shown in Fig. 12.26a; otherwise, if $Ma_{\infty n} > 1$, the leading (trailing) edge is called the supersonic leading (trailing) edge, as shown in Fig. 12.26b; if $Ma_{\infty n} = 1$, it is called sonic leading (trailing) edge. The geometric relationship between supersonic leading edge and subsonic leading edge is shown in Fig. 12.26. When the incoming Mach line is behind the leading edge, it is the supersonic leading edge, and in front it is the subsonic leading edge: According to the above geometric relationship, the parameter K is introduced to represent the ratio of the tangent of the leading edge half angle to the tangent of the leading edge Mach angle, which is

$$K = \frac{tg\left(\frac{\pi}{2} - \chi\right)}{tg\,\mu_\infty} = \frac{ctg\,\chi}{\frac{1}{\sqrt{Ma_\infty^2 - 1}}} = \frac{\sqrt{Ma_\infty^2 - 1}}{tg\,\chi} = \frac{B}{tg\,\chi} \tag{12.127}$$

wherein, $B = \sqrt{Ma_\infty^2 - 1}$. Obviously, when $K > 1$, the leading (trailing) edge is supersonic; $K < 1$, the leading (trailing) edge is subsonic; and $K = 1$, the leading (trailing) edge is sonic.: To sum up, there are three methods to distinguish the supersonic leading (trailing) edge:

(1) $Ma_{\infty n} > 1$ or $V_{\infty n} > a_\infty$;
(2) Geometrically, the Mach line is located behind the leading (trailing) edge;
(3) $K > 1$, is supersonic leading (trailing) edge.

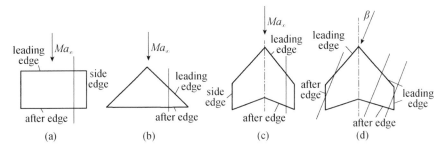

Fig. 12.25 The names of edges of wing at supersonic flow

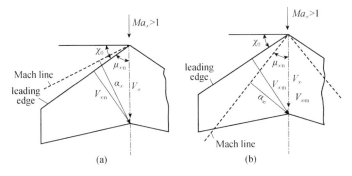

Fig. 12.26 The normal flow characteristic of leading edge of wing at supersonic flow

12.6.3 Two-Dimensional Flow Region and Three-Dimensional Flow Region

According to the propagation properties of the disturbance waves, there are different disturbance affected regions in supersonic flow around a wing. Regions affected only by a single leading edge are called two-dimensional flow regions (in the flow region, the dependent region of each point contain only one leading edge), such as the shaded regions shown in Fig. 12.27. The rest of the non-shaded region, whose dependence zone contains two leading edges (or one leading edge and one trailing or side edge), is called three-dimensional region. In the two-dimensional flow region, the wing can be viewed as an infinite wingspan straight wing or an infinite wingspan oblique wing. The perturbation characteristic is that it is only affected by the normal airfoil perpendicular to the leading edge, and is independent of the plane shape of the wing. For a flat wing, the pressure coefficient of the upper and lower surfaces of the two-dimensional flow region is

$$
\begin{cases}
C_{pu} = -\dfrac{2\alpha \cos \chi}{\sqrt{Ma_\infty^2 \cos^2 \chi - 1}} \\[4mm]
C_{pd} = \dfrac{2\alpha \cos \chi}{\sqrt{Ma_\infty^2 \cos^2 \chi - 1}}
\end{cases}
\tag{12.128}
$$

If the wing with airfoil, the pressure coefficient of the upper and lower wing surfaces in this region can be obtained by Eq. (12.116). By using Eq. (12.127), the above equation can be written as

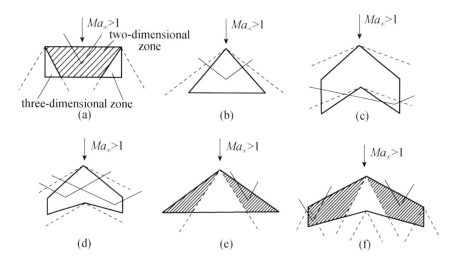

Fig. 12.27 Two-dimensional and three-dimensional flow regions of wings at supersonic flow

$$\begin{cases} C_{pu} = -\dfrac{2\alpha K}{B\sqrt{K^2 - 1}} \\ C_{pd} = \dfrac{2\alpha K}{B\sqrt{K^2 - 1}} \end{cases} \qquad (12.129)$$

In the three-dimensional region, the flow parameters are related to both the airfoil and the plane shape of the wing. For the situation shown in Fig. 12.27, it can be seen that Fig. 12.27b is subsonic leading edge, Fig. 12.27c is subsonic leading and trailing edge, Fig. 12.27d is subsonic leading edge and supersonic trailing edge, Fig. 12.27e is supersonic leading edge, and Fig. 12.27f is supersonic leading and trailing edge.

12.7 Aerodynamic Characteristics of Thin Wing with Finite Wingspan at Supersonic Flow

The aerodynamic characteristics of the supersonic flow around a thin wing with finite wingspan are closely related to the properties of the leading and trailing edges. The same swept wing has different aerodynamic characteristics under different inlet Mach numbers. There may be subsonic leading (trailing) edges, subsonic leading edge and supersonic trailing edge, and supersonic leading and trailing edge, etc. Taking a swept flat wing as an example, the characteristics of normal flow, that is, the characteristics of flow around different leading and trailing edge conditions, are illustrated.

(1) Flow around subsonic leading edge and subsonic trailing edge

As shown in Fig. 12.28, the flows around the upper and lower wings interact with each other through the leading edge. As a result, the section perpendicular to the leading edge shows subsonic flow characteristics around the leading edge. For the subsonic trailing edge, the section perpendicular to the trailing edge is required to satisfy Kutta-Joukowski conditions. Figure 12.29 shows the pressure coefficient distribution around the subsonic leading edge and the subsonic trailing edge. The pressure coefficient tends to negative infinity at the subsonic leading edge and is zero at the subsonic trailing edge.

(2) Flow around subsonic leading edge and supersonic trailing edge

As shown in Fig. 12.30, the characteristics of flow around subsonic leading edge are similar to the flow around the leading edge in Fig. 12.28. The trailing edge is the supersonic trailing edge. At this time, the normal air flow on the upper wing reaches the supersonic zone, and a shock wave occurs at the trailing edge to deflect the air flow and achieve the condition of incoming flow. A series of expansion waves will appear to accelerate and satisfy the inlet flow conditions in the flow around the trailing edge. The pressure coefficient distribution at subsonic leading edge is the same as

Fig. 12.28 The flow around subsonic leading edge and subsonic trailing edge (normal flow)

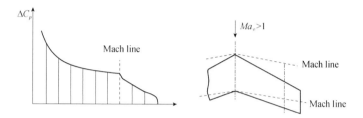

Fig. 12.29 The pressure coefficient distribution of subsonic leading edge and subsonic trailing edge (normal flow)

shown in Fig. 12.29, and the pressure coefficient at subsonic leading edge tends to negative infinite. However, a shock wave appears at supersonic trailing edge, and a finite pressure coefficient appears on the upper wing surface, which is determined by the normal Mach number of the trailing edge. See Fig. 12.31.

(3)   Flow around supersonic leading edge and supersonic trailing edge

As shown in Fig. 12.32, are the supersonic leading edge and the supersonic trailing edge. For the supersonic flow around the leading edge, expansion waves appear on the upper wing surface of the leading edge nose, and oblique shock wave appears on the lower wing surface; for supersonic flow around the trailing edge, oblique shock wave appears on the trailing edge and expansion waves appear on the lower wing surface. As shown in Fig. 12.33, expansion waves appear on the upper wing surface of the supersonic leading edge, oblique shock wave appears on the lower wing surface,

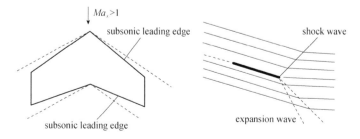

Fig. 12.30 The flow around subsonic leading edge and supersonic trailing edge (normal flow)

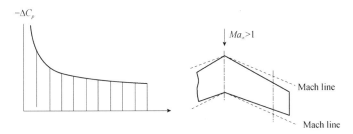

Fig. 12.31 The pressure coefficient distribution of subsonic leading edge and supersonic trailing edge (normal flow)

and the pressure coefficient of the upper wing surface shows a finite value; at the supersonic trailing edge, oblique shock wave appears on the upper wing surface, expansion waves appear on the lower wing surface, and the pressure coefficient of the upper wing surface shows a finite value.

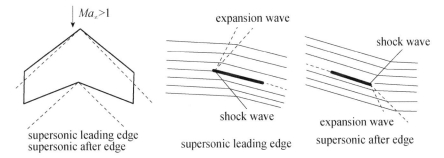

Fig. 12.32 The flow around supersonic leading edge and supersonic trailing edge (normal flow)

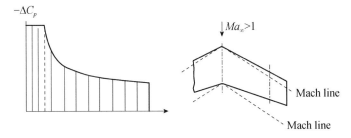

Fig. 12.33 The pressure coefficient distribution of supersonic leading edge and supersonic trailing edge (normal flow)

12.8 Lift Characteristics of Rectangular Flat Wing at Supersonic Flow

12.8.1 Conical Flow in the Three Dimensional Region of Supersonic Leading Edge

As shown in Fig. 12.34, for the supersonic flow around the leading (trailing) edge, the intersection point of the Mach rays emitted from the vertex o and o' with the trailing edge of the wing are A and F respectively, where AF is greater than or equal to zero. There are two-dimensional and three-dimensional regions on the wing plane. As shown in Fig. 12.34, the shaded area is a two-dimensional area and the non-shaded areas are three-dimensional areas. For the two-dimensional region of a swept wing and the two-dimensional region of a flat wing, when the angle of attack is α, the pressure coefficients on the upper and lower wing surfaces can be obtained by Eq. (12.129).

However, in the three-dimensional region, due to the influence of corners and side edges, the high pressure on the lower wing surface offsets part of the low pressure on the upper wing surface through the side edges, making the pressure in this region higher than the negative pressure in the two-dimensional region. In the supersonic region, according to the characteristics of the disturbance propagating along the Mach line, the conical coordinates are taken as

$$t' = B\frac{z'}{x'} \tag{12.130}$$

According to the characteristics of Mach rays, the pressure variation in the three-dimensional region is equal to the negative pressure intensity in the two-dimensional region at one end, so that the side edge of the end is the incoming pressure. In this region, the pressure change is obtained from the conical flow theory (refer to relevant data) as

Fig. 12.34 Supersonic flow around leading (trailing) edge of a panel (normal flow)

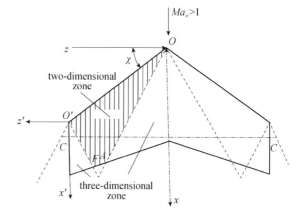

$$\begin{cases} C_{pu} = -\dfrac{2\alpha K}{\pi B \sqrt{K^2 - 1}} \cos^{-1}\left(1 + 2t' \dfrac{K+1}{K-t'}\right) \\ C_{pd} = \dfrac{2\alpha K}{\pi B \sqrt{K^2 - 1}} \cos^{-1}\left(1 + 2t' \dfrac{K+1}{K-t'}\right) \end{cases} \tag{12.131}$$

12.8.2 Three-Dimensional Region of Supersonic Flow Around Rectangular Flat Wing

For the supersonic flow over a rectangular flat plane wing, the Mach ray emitted from the tip of the wing has a three-dimensional conical flow region in the wing tip area, and the pressure distribution is determined by the conical flow. As shown in Fig. 12.35, since the leading edge swept angle of the rectangular wing is zero, then

$$K = \frac{B}{tg\chi} = \frac{\sqrt{Ma_\infty^2 - 1}}{tg\chi} \to \infty \tag{12.132}$$

It can be obtained from Eq. (12.131) that

$$\begin{cases} C_{pu} = -\dfrac{2\alpha}{\pi B} \cos^{-1}\left(1 + 2t'\right) \\ C_{pd} = \dfrac{2\alpha}{\pi B} \cos^{-1}\left(1 + 2t'\right) \end{cases} \tag{12.133}$$

Or, there are

Fig. 12.35 Supersonic flow around rectangular flat wing

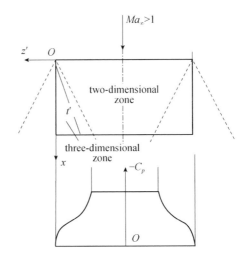

$$\begin{cases} C_{pu} = -\dfrac{4\alpha}{\pi B} \sin^{-1}\left(\sqrt{-t'}\right) \\ C_{pd} = \dfrac{4\alpha}{\pi B} \sin^{-1}\left(\sqrt{-t'}\right) \end{cases} \qquad (12.134)$$

wherein, $t' = B\frac{z'}{x'}$. It can be proved that the average pressure in the three-dimensional region of the rectangular flat wing tip is half that in the two-dimensional region.

12.8.3 Lift Characteristics of Supersonic Flow Around Rectangular Flat Wing

According to the pressure distribution characteristics of the rectangular flat wing, the lift coefficient (small angle of attack) of the wing is obtained by integrating the load coefficient of the three-dimensional region of the wing surface as

$$C_L = \frac{1}{S} \iint_S \Delta C_p dS \qquad (12.135)$$

where, S is the area of the rectangular flat wing. The wing surface can be divided into three-dimensional region and two-dimensional region. Suppose the area of three-dimensional region is $2S_1$ (the area of the trigonometric region at the tip of S_1) and the area of two-dimensional region is S_2, then

$$S = 2S_1 + S_2 \qquad (12.136)$$

Substitute it into Eq. (12.135) and get

$$C_L = \frac{1}{S} \iint_{2S_1+S_2} \Delta C_p dS = \frac{1}{S}\left[2\iint_{S_1} \Delta C_p dS + \iint_{S_2} \Delta C_p dS \right] = C_{L1} + C_{L2} \qquad (12.137)$$

As shown in Fig. 12.35, according to Eq. (12.134), the load factor in the three-dimensional region is

$$\Delta C_p = C_{pd} - C_{pu} = \frac{8\alpha}{\pi B} \sin^{-1}\left(\sqrt{-t'}\right) \qquad (12.138)$$

As shown in Fig. 12.36, the geometric relationship of trigonometric region is

$$S = \frac{btg\theta}{2}b, \quad t' = B\frac{z'}{x'} = Btg\theta, \quad S = \frac{b^2}{2}\frac{t'}{B} \qquad (12.139)$$

Fig. 12.36 A diagram of three-dimensional region of a rectangular flat wing

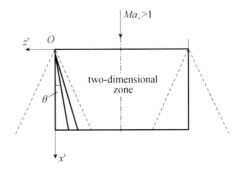

Since $dS = \frac{b^2}{2}\frac{dt}{B}$, substituted into Eq. (12.137), the contribution of the three-dimensional region of rectangular flat wing to the average lift coefficient can be obtained as

$$C_{L1} = \frac{2}{S}\iint\limits_{S_1}\Delta C_p dS = \frac{8\alpha b^2}{S\pi B^2}\int\limits_{-1}^{0}\sin^{-1}\sqrt{-t'}dt' = \frac{2\alpha}{B^2\lambda} \qquad (12.140)$$

where, $\lambda = \frac{S}{b^2}$. For the two-dimensional region in the middle of the wing, the area S_2 determined by

$$\Delta l = btg\mu_\infty = b\frac{\sin\mu_\infty}{\cos\mu_\infty} = b\frac{\sin\mu_\infty}{\sqrt{1-\sin^2\mu_\infty}} = \frac{b}{\sqrt{Ma_\infty^2-1}} = \frac{b}{B} \qquad (12.141)$$

So, S_2 is

$$S_2 = b\frac{l+l-2\Delta l}{2} = b\left(1-\frac{b}{B}\right) \qquad (12.142)$$

In the two-dimensional zone, the load coefficient is $\Delta C_p = \frac{4\alpha}{B}$, and is substituted into the second part of Eq. (12.137) and then get

$$C_{L2} = \frac{1}{S}\iint\limits_{S_2}\Delta C_p dS = \frac{4\alpha}{B}\left(1-\frac{1}{\lambda B}\right) \qquad (12.143)$$

Finally, the lift coefficient of rectangular flat wing is

$$C_L = C_{L1} + C_{L2} = \frac{2\alpha}{B^2\lambda} + \frac{4\alpha}{B}\left(1-\frac{1}{\lambda B}\right) = \frac{4\alpha}{B}\left(1-\frac{1}{2\lambda B}\right) \qquad (12.144)$$

$$BC_L^\alpha = \frac{dC_L}{d\alpha} = 4\left(1-\frac{1}{2\lambda B}\right) \qquad (12.145)$$

Fig. 12.37 Slope curve of lift line of supersonic flow around rectangular flat wing

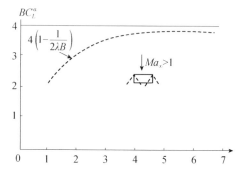

According to Eq. (12.145), when the aspect ratio of rectangular airfoil tends to infinity, the lift coefficient tends to the lift coefficient of two-dimensional airfoil, $\frac{4\alpha}{B}$. Figure 12.37 shows the relationship between the aspect ratio of rectangular flat wing and the slope of lift line (Eq. (12.145)).

12.9   Characteristic Line Theory of Supersonic Flow

In the steady supersonic flow field, the disturbance propagates in the form of wave. According to whether the disturbance wave causes compression or expansion of air flow, the disturbance wave can be divided into compression wave and expansion wave. Shock wave is a kind of strong disturbance compression wave. Its typical characteristic is that it causes sudden changes in flow parameters at the shock surface, leading to discontinuity in velocity and pressure. For weak disturbance waves (weak compression waves, expansion waves), although there is no discontinuity in the flow parameters of Mach line after the boundary of the wave, their partial derivatives are discontinuous.

According to Eq. (11.12), for the two-dimensional supersonic flow, it is assumed to be an ideal, steady and potential flow, the governing equation of its full velocity potential function (not the disturbance velocity potential function) is

$$\left(1 - \frac{u^2}{a^2}\right)\frac{\partial^2\varphi}{\partial x^2} - \frac{2uv}{a^2}\frac{\partial^2\varphi}{\partial x\partial y} + \left(1 - \frac{v^2}{a^2}\right)\frac{\partial^2\varphi}{\partial y^2} = 0 \qquad (12.146)$$

Using the relationship between the full velocity potential function and the velocity component

$$u = \frac{\partial\varphi}{\partial x}, v = \frac{\partial\varphi}{\partial y} \qquad (12.147)$$

It can be obtained that

$$du = \frac{\partial u}{\partial x}dx + \frac{\partial u}{\partial y}dy = \frac{\partial^2 \varphi}{\partial x^2}dx + \frac{\partial^2 \varphi}{\partial x \partial y}dy \tag{12.148}$$

$$dv = \frac{\partial v}{\partial x}dx + \frac{\partial v}{\partial y}dy = \frac{\partial^2 \varphi}{\partial x \partial y}dx + \frac{\partial^2 \varphi}{\partial y^2}dy \tag{12.149}$$

In the supersonic flow field, the partial derivative $\left(\frac{\partial u}{\partial x}, \frac{\partial v}{\partial y}, \frac{\partial p}{\partial x}, \ldots\right)$ of the flow parameters at the path of the disturbance wave will be uncertain, and the lines with uncertainty in the flow field are called characteristic lines (wave front of disturbance wave, Mach wave line). Since

$$\frac{\partial u}{\partial x} = \frac{\partial^2 \varphi}{\partial x^2}, \quad \frac{\partial v}{\partial x} = \frac{\partial u}{\partial y} = \frac{\partial^2 \varphi}{\partial x \partial y}, \quad \frac{\partial v}{\partial y} = \frac{\partial^2 \varphi}{\partial y^2} \tag{12.150}$$

According to Eqs. (12.146), (12.148) and (12.149), the determinant of coefficient matrix and permutation matrix of the following equations are all zero at the disturbance wave front. Namely

$$\begin{vmatrix} 1-\frac{u^2}{a^2} & 1-\frac{u^2}{a^2} & 1-\frac{v^2}{a^2} \\ dx & dy & 0 \\ 0 & dx & dy \end{vmatrix} \left\{ \begin{matrix} \frac{\partial^2 \varphi}{\partial x^2} \\ \frac{\partial^2 \varphi}{\partial x \partial y} \\ \frac{\partial^2 \varphi}{\partial y^2} \end{matrix} \right\} = \left\{ \begin{matrix} 0 \\ du \\ dv \end{matrix} \right\} \tag{12.151}$$

(1) Characteristic line equation

According to the factor determinant of Eq. (12.151) is zero, get

$$\begin{vmatrix} 1-\frac{u^2}{a^2} & -\frac{2uv}{a^2} & 1-\frac{v^2}{a^2} \\ dx & dy & 0 \\ 0 & dx & dy \end{vmatrix} = 0 \tag{12.152}$$

By expanding the above equation, the characteristic line equation can be obtained as

$$\left(1-\frac{u^2}{a^2}\right)dy^2 + \frac{2uv}{a^2}dxdy + \left(1-\frac{v^2}{a^2}\right)dx^2 = 0 \tag{12.153}$$

$$\left(1-\frac{u^2}{a^2}\right)\frac{dy^2}{dx^2} + \frac{2uv}{a^2}\frac{dy}{dx} + \left(1-\frac{v^2}{a^2}\right) = 0 \tag{12.154}$$

Solve it and get

$$\frac{dy}{dx} = \frac{-\frac{uv}{a^2} \pm \sqrt{\frac{u^2+v^2}{a^2}-1}}{1-\frac{u^2}{a^2}} \tag{12.155}$$

Substitute $u = V \cos \theta, \quad v = V \sin \theta$ into above equation

$$\frac{dy}{dx} = \frac{-Ma^2 \cos \theta \sin \theta \pm \sqrt{Ma^2 - 1}}{1 - Ma^2 \cos^2 \theta} \tag{12.156}$$

Then substitute $\sin \mu = \frac{1}{Ma}$ into above equation, and get

$$\frac{dy}{dx} = \frac{-\cos \theta \sin \theta \pm \sin \mu \cos \mu}{\sin^2 \mu - \cos^2 \theta} \tag{12.157}$$

Multiply the numerator and the denominator by $\cos \theta \cos \mu \pm \sin \theta \sin \mu$, and use trig identity

$$\sin^2 \mu - \cos^2 \theta = \sin^2 \theta - \cos^2 \mu \tag{12.158}$$

It can be obtained from Eq. (12.157) that

$$\frac{dy}{dx} = tg(\theta \mp \mu) \tag{12.159}$$

This equation tells the slope of the characteristic line on the physical plane. Wherein, the characteristic line represented by the negative sign is the right-extended characteristic line, which is represented by $C-$; The characteristic line represented by the plus sign is the left extension characteristic line, denoted by $C+$. This left/right extension refers to the left/right side that the observer is pointing in the direction of the velocity vector, as shown in Fig. 12.38.

(2)   The variation law of function derivative along the characteristic line

Using the permutation coefficient determinant of Formula (12.151) is zero, it can be obtained that

Fig. 12.38 The characteristic line through point A

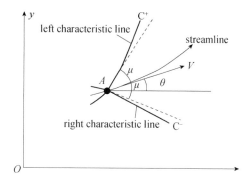

$$\begin{vmatrix} 1 - \frac{u^2}{a^2} & 0 & 1 - \frac{v^2}{a^2} \\ dx & du & 0 \\ 0 & dv & dy \end{vmatrix} = 0 \qquad (12.160)$$

Expand it and get

$$\left(1 - \frac{u^2}{a^2}\right) du \, dy + \left(1 - \frac{v^2}{a^2}\right) dv \, dx = 0 \qquad (12.161)$$

The result of solving is

$$\frac{dv}{du} = -\frac{1 - \frac{u^2}{a^2}}{1 - \frac{v^2}{a^2}} \frac{dy}{dx} \qquad (12.162)$$

Substituting Eq. (12.155) into the above equation, there is

$$\frac{dv}{du} = -\frac{1 - \frac{u^2}{a^2} - \frac{uv}{a^2} \pm \sqrt{\frac{u^2 + v^2}{a^2} - 1}}{1 - \frac{v^2}{a^2}} = -\frac{-\frac{uv}{a^2} \pm \sqrt{\frac{u^2 + v^2}{a^2} - 1}}{1 - \frac{v^2}{a^2}} \qquad (12.163)$$

By differentiating $u = V \cos\theta$, $v = V \sin\theta$, it can be get that

$$du = d(V \cos\theta) = \cos\theta \, dV - V \sin\theta \, d\theta$$
$$dv = d(V \sin\theta) = \sin\theta \, dV + V \cos\theta \, d\theta \qquad (12.164)$$

And substituted into Eq. (12.163), then

$$\frac{d(V \sin\theta)}{d(V \cos\theta)} = \frac{\sin\theta \, dV + V \cos\theta \, d\theta}{\cos\theta \, dV - V \sin\theta \, d\theta} = \frac{\mathrm{Ma}^2 \cos\theta \sin\theta \mp \sqrt{\mathrm{Ma}^2 - 1}}{1 - \mathrm{Ma}^2 \sin^2\theta} \qquad (12.165)$$

$$d\theta = \sqrt{\mathrm{Ma}^2 - 1} \frac{-tg\theta \sqrt{\mathrm{Ma}^2 - 1} \mp 1}{1 \mp tg\theta \sqrt{\mathrm{Ma}^2 - 1}} \frac{dV}{V} = \mp \sqrt{\mathrm{Ma}^2 - 1} \frac{dV}{V} \qquad (12.166)$$

(3) Equation set of characteristic lines

By combining the two families of characteristic line Eq. (12.159) and the relation of flow parameters along the characteristic line (12.166), the solution of the second-order nonlinear partial differential Eq. (12.146) can be transformed into the solution of a set of ordinary differential equations. Namely

Along the positive $C+$ characteristic line

$$\begin{cases} \dfrac{dy}{dx} = tg(\theta + \mu) \\ d\theta = \sqrt{Ma^2 - 1}\dfrac{dV}{V} \end{cases} \tag{12.167}$$

Along the negative $C-$ characteristic line

$$\begin{cases} \dfrac{dy}{dx} = tg(\theta - \mu) \\ d\theta = -\sqrt{Ma^2 - 1}\dfrac{dV}{V} \end{cases} \tag{12.168}$$

The Prandtl–Mayer Angle [see Eq. (7.175)] can be obtained by integrating the relationship of the characteristic lines

$$\delta(Ma) = \sqrt{\frac{\gamma + 1}{\gamma - 1}} \arctan\sqrt{\frac{\gamma - 1}{\gamma + 1}(Ma^2 - 1)} - \arctan\sqrt{Ma^2 - 1} \tag{12.169}$$

Then Eqs. (12.167) and (12.168) change into
Along the positive $C+$ characteristic line

$$\begin{cases} \dfrac{dy}{dx} = tg(\theta + \mu) \\ \theta - \delta(Ma) = 0 \end{cases} \tag{12.170}$$

Along the negative $C-$ characteristic line

$$\begin{cases} \dfrac{dy}{dx} = tg(\theta - \mu) \\ \theta + \delta(Ma) = 0 \end{cases} \tag{12.171}$$

Exercises

A. Thinking questions

1. What is supersonic flow around airfoil?
2. What are the main characteristics of supersonic flow around airfoil compared with subsonic flow around airfoil?
3. Under the linearized condition, please write down the formulation of the definite solution problem of the velocity potential function of supersonic flow? The linearized expression of the wall pressure coefficient?

4. Please write down the general solution form and its physical meaning of the definite solution problem of the velocity potential function of supersonic flow?

5. For small perturbation supersonic flow around a thin airfoil, write the linearized decomposition of the surface pressure coefficient?

6. Please explain the main differences and pressure distribution of supersonic and subsonic flow around a plate?

7. Draw a diagram of supersonic flow around bending plate at 0 angle of attack and the pressure distribution on the upper and lower surface?

8. Write down the expressions of the lift coefficient and drag coefficient of supersonic flow around a thin airfoil, and state the physical meaning of each term?

9. Why does the lift coefficient of supersonic flow around airfoil decrease with the increase of inlet Mach number?

10. Write down the expressions of shock wave drag supersonic flow around a thin airfoil, and explain the physical meaning of each term?

11. Explain the relationship between minimum zero-lift shock wave drag and Mach number.

B. Calculation questions

1. Using the results of linearization theory, calculate lift and shock wave drag of infinite thin plate in the free flow of Mach number of 2.6 at angle of attack of (a) $\alpha = 5°$ (b) $\alpha = 15°$ (c) $\alpha = 30°$.

2. An infinite wingspan wing with a symmetrical rhombus profile, as illustrated in the attached figure, to move left in sea level air at $Ma = 2$. Relative thickness $t = 0.15$, $a = 2°$, use the theory of shock wave and expansion wave, try to find the pressure at point B on the airfoil shown in the figure.

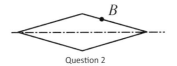

Question 2

3. Assume the angle of attack of the wing in above question is zero, try to find the pressure at point B by using shock wave—expansion wave theory and linearized theory respectively.

4. The Lockheed F-104 supersonic fighter, as shown in the figure, is the first fighter designed for sustained flight at Mach 2. The F-104 embodies good supersonic aircraft design. The airfoil thickness is 3.4%, and the plane area of the wing is 18.21 m². The F-104, for example, flies steadily level at Mach 2 at an altitude of 11 km. The aircraft has a combat weight of 9400 kg. Assume that all of the aircraft's lift comes from the wings (ignoring the fuselage and tail lift). What is the angle of attack of the wing relative to the free flow when the flight Mach number is 2 at the altitude of 11 km?

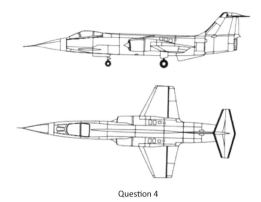

Question 4

5. Calculate the pressure of upper and lower surfaces of a infinite wingspan thin plate in the free flow of Mach number is 2.6, at angle of attack of (a) $\alpha = 5°$ (b) $\alpha = 15°$ (c) $\alpha = 30°$ (use the form of $p/p\infty$).

6. A two-dimensional plate flies at altitude of 6 km and $\text{Ma}_\infty = 2$, angle of attack is 10°. Try to find the pressure difference between the upper and the lower surface, using shock wave—expansion wave theory and linearized theory.

7. The airfoil shown in figure moves at $\text{Ma}_\infty = 2$, $\alpha = 2°$, the ratio of thickness and chord length is 0.1, and the maximum thickness takes place at 30% chord length behind leading edge. Use linearized theory to find
    (1) Pitching moment coefficient around the focus
    (2) Where is the center of pressure
    (3) How much is the shock wave drag coefficient
    (4) Zero-lift angle of attack

Question 7

8. The figure shows that the incoming flow with velocity V_∞ around a two-dimensional cylinder with radius R along the y-axis (see figure). Find the maximum and minimum values of the y-direction disturbance velocity component v' in the plane $y = -R$, and the corresponding z-coordinate. Let the flow be an incompressible potential flow without viscosity

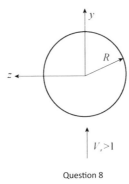

Question 8

9. The figure shows a combination of a slender cylindrical fuselage and a small aspect ratio triangular flat wing with a sweep Angle of 60°, $(l/2)/R = 2.5$. When the fuselage angle of attack is zero and the wing mounting angle is 5°, calculate the lift coefficient of the wing-body assembly with the fuselage cross section area as the characteristic area $(C_{l\text{wing - body}})_{\varphi,0}$

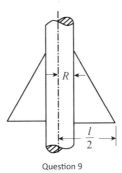

Question 9

10. According to the linearized supersonic theory, the equation $c_d = \dfrac{4\alpha^2}{\sqrt{M_\infty^2 - 1}}$ predicts that the plate c_d decreases with increasing M_∞, right? Does this mean that the drag itself decreases as M_∞ increase? To answer this question, derive an equation that takes the drag as a function of M_∞.

11. There is a wing with plane shape shown in figure, try to find the Ma_∞ range of supersonic leading edge and subsonic trailing edge.

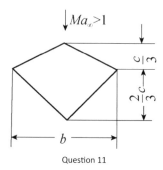

Question 11

12. A symmetrical double-cambered airfoil is placed in a supersonic flow, and the relative position of its focus (also its pressure center) is calculated by using the second-order approximate pressure coefficient formula

$$\overline{x}_{focus} = \frac{1}{2} + \frac{C_2}{C_1} I_5$$

where I_5 is the negative value of a sectional area of a dimensional symmetric double-cambered airfoil.

13. There is a triangular-shaped wing with a leading edge sweep angle χ_0 of $45°$ is now flying at speed of $V_\infty = 450$ m/s. Try to consider how the leading edge characteristic of the wing change when flying at 5500 m and 11,000 m above sea level.

14. Try to prove that the lift force in region I of the rectangular flat wing is equal to half of the lift force generated by the two-dimensional value of the supersonic velocity in this region.

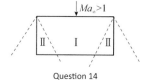

Question 14

15. Consider a plate at an angle of attack in an inviscous supersonic flow. According to linearized theory, what is the maximum lift-drag ratio, and at what angle of attack does it occur?

16. Try to find the ratio of the lift force of a slender cylindrical fuselage and triangular flat wing with small aspect ratio in the case of "$\varphi, 0$" to the lift force of which at the angle of attack $a = \varphi$ in the case of "a, a".

17. As shown in the figure, there is a long thin wing-body assembly with pointed tips. The body of the wing-body connection area is a cylindrical section. Try to find the ratio of cylindrical section R and half span length of the

wing $l/2$ when the ratio of the lift of the assembly $Y_{\text{wing - body}}$ and the lift of single wing Y_{wing} is minimum in "a, a" condition.

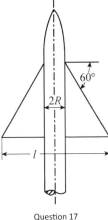

Question 17

18. There are two flat triangular wings with aspect ratio $A = 3.0$ and 1.0, flight Mach numbers are $\text{Ma}_\infty = \sqrt{2}$, and a small angle of attack α. Try to calculate the lift coefficients of the two wings and the errors of slender body theory, according to slender body theory and linearized theory respectively.

19. Consider a plate with an angle of attack in a viscous supersonic flow; that is, there is both surface friction drag and shock wave drag on the plate. Lift and shock wave drag coefficients use linear theory. C_f represents the total surface friction drag coefficient and assumes that it does not change with the angle of attack. Deduce the relationship between the maximum lift-drag ratio and C_f and the free-flow Mach number M_∞.

20. The cross-section area distribution along the x-direction of the body axis of a thin wing-body assembly is given as

$$\frac{\mathrm{d}S(x)}{\mathrm{d}x} = \frac{64(V)}{\pi L^2}\left(\frac{x}{L} - \frac{x^2}{L^2}\right)^{1/2} \cdot \left(1 - 2\frac{x}{L}\right)$$

where V is the volume of the combination, $S(x)$ is the cross-sectional area of the combination at x, L is the length of the combination, Try to find the zero-lift shock wave drag of the combination at $\text{Ma}_\infty = 1$.

Chapter 13
Aerodynamic Characteristics
of Transonic Thin Airfoil and Wing

This chapter introduces the aerodynamic characteristics of transonic thin airfoil and wing. It includes the concept of critical Mach number of transonic airfoil flow, transonic thin airfoil flow phenomenon and aerodynamic characteristics, transonic small perturbation potential flow equation and similarity rule, the influence of wing geometric parameters on transonic critical Mach number, and aerodynamic characteristics of supercritical airfoil flow, high subsonic flow over a swept wing with a high aspect ratio and transonic area rule.

Learning points:

(1) Familiar with the concept of critical Mach number of transonic airfoil flow, transonic thin airfoil flow phenomenon and aerodynamic characteristics, and transonic small perturbation potential flow equation and similarity rule;
(2) The influence of wing geometry parameters on the critical Mach number of transonic flow, aerodynamic characteristics of supercritical airfoil, and high subsonic flow over a swept wing with a high aspect ratio and transonic area rule are investigated.

13.1 Critical Mach Number of Transonic Airfoil Flow

13.1.1 Problem of Transonic Flow

When air flows around an object, in addition to subsonic and supersonic flow, if there is a local supersonic region in the subsonic flow field or a local subsonic region in the supersonic flow field, such flow is called transonic flow. Since shock waves are often used to realize the transition from supersonic to subsonic, local shock waves are often included in the transonic flow field. The transonic flow around the thin wing mainly occurs when the incoming Mach number is close to 1. In the supersonic flow around the blunt body, the transonic flow also appears in the region behind the shock wave.

© Science Press 2022

P. Liu, *Aerodynamics*, https://doi.org/10.1007/978-981-19-4586-1_13

The transonic flow field is far more complex than the supersonic and supersonic flow fields, because the flow field is mixed (the coupling of flow and wave) and there are local shock waves that needs further study in theory and experimental technology.

13.1.2 Critical Mach Number

When the incoming Mach number Ma_∞ over the object (or wing) is at subsonic speed, the velocity of each point on the surface of the object is different. Speed of some areas are larger than the incoming velocity, and some areas are smaller than the incoming velocity. When the incoming Mach number reaches a certain value ($Ma_\infty < 1$), the maximum point velocity (minimum pressure point) on the surface is just equal to the local sound velocity ($Ma = 1$). At this time, the corresponding incoming Mach number is called critical Mach number (or lower critical Mach number) $M_{\infty c}$, and the corresponding pressure at $Ma = 1$ is called critical pressure, which is expressed by P_C. For the thin airfoil, the pressure is related to the thickness, curvature, and angle of attack of the airfoil, so the critical Mach number of the airfoil is also related to these parameters. For the wing, the critical Mach number is also related to its plane shape. If the incoming Mach number Ma_∞ continues to increase ($Ma_\infty > M_{\infty c}$), a local supersonic region and shock wave will appear on the airfoil surface, and the aerodynamic characteristics will change dramatically. It is obvious that this change will begin when the stream Mach number exceeds the critical Mach number, so it is very important to determine $M_{\infty c}$.

According to the pressure ratio formula of isentropic flow, the relationship between Ma, P at a certain point on the airfoil surface and Ma_∞, P_∞ of the incoming flow is as follows:

$$\frac{p}{p_\infty} = \left[\frac{1 + \frac{\gamma-1}{2}Ma_\infty^2}{1 + \frac{\gamma-1}{2}Ma^2} \right]^{\frac{\gamma}{\gamma-1}} \tag{13.1}$$

When $Ma_\infty = Ma_{\infty c}$, $Ma = 1$, and $P = P_c$, Eq. (13.1) can be changed into Eq. (13.2).

$$\frac{p_c}{p_\infty} = \left[\frac{1 + \frac{\gamma-1}{2}Ma_{\infty c}^2}{\frac{\gamma+1}{2}} \right]^{\frac{\gamma}{\gamma-1}} \tag{13.2}$$

By definition, the critical pressure coefficient is as follows:

$$C_{pc} = \frac{p_c - p_\infty}{\frac{1}{2}\rho_\infty V_{\infty c}^2} = \frac{2}{\gamma Ma_{\infty c}^2}\left(\frac{p_c}{p_\infty} - 1 \right) \tag{13.3}$$

Substituting Eq. (13.2) into Eq. (13.3).

$$C_{\mathrm{pc}} = \frac{p_c - p_\infty}{\frac{1}{2}\rho_\infty V_{\infty c}^2} = \frac{2}{\gamma \mathrm{Ma}_{\infty c}^2}\left\{\left[\frac{2}{\gamma + 1}\left(1 + \frac{\gamma - 1}{2}\mathrm{Ma}_{\infty c}^2\right)\right]^{\frac{\gamma}{\gamma - 1}} - 1\right\} \quad (13.4)$$

Equation (13.4) shows the relationship between the critical pressure coefficient C_{pc} and the critical Mach number over the object in the isentropic flow field. When subsonic air flow a airfoil, the relationship between critical Mach number and critical pressure coefficient is shown in Fig. 13.1. It can be seen that the smaller the critical Mach number is, the larger the negative value of the pressure coefficient is. For a given airfoil, with the increase of Ma_∞, the lowest pressure point of the airfoil first reaches the critical state, as shown in Fig. 13.2. The variation of pressure coefficient cpmin with Mach number Ma_∞ at the lowest pressure point of the airfoil can be calculated according to the Prandtl–Glauert compressibility correction rule.

$$\left(C_{p\,\min}\right)_{\mathrm{Ma}_\infty} = \frac{\left(C_{p\,\min}\right)_0}{\sqrt{1 - \mathrm{Ma}_\infty^2}} \quad (13.5)$$

It can also be modified by Kármán-Qian Xuesen formula.

$$\left(C_{p\,\min}\right)_{\mathrm{Ma}_\infty} = \frac{\left(C_{p\,\min}\right)_0}{\sqrt{1 - \mathrm{Ma}_\infty^2} + \frac{\mathrm{Ma}_\infty^2}{1+\sqrt{1-\mathrm{Ma}_\infty^2}}\frac{\left(C_{p\,\min}\right)_0}{2}} \quad (13.6)$$

Among them, $\left(C_{p\,\min}\right)_0$ is the minimum pressure coefficient of an airfoil in incompressible flow, which can be obtained by low-speed flow simulation or by experiment. The relationship between the minimum pressure coefficient and the incoming Mach number calculated according to Kármán-Qian Xuesen Formula (13.6) is given for symmetrical airfoils with different thicknesses as shown in Fig. 13.1. The $C_{p\,\min}$ and Ma_∞ corresponding to the intersection of the two curves are the critical pressure

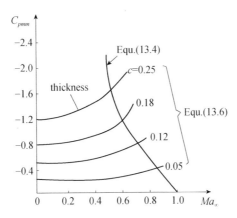

Fig. 13.1 Curve of critical Mach number with relative thickness

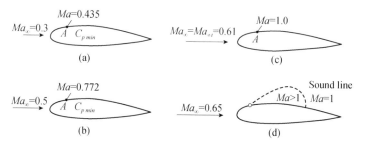

Fig. 13.2 Mach number variation of the lowest pressure point on the upper surface of an airfoil at different incoming Mach numbers

coefficient C_{PC} and the critical Mach number $Ma_{\infty C}$ of the airfoil. The figure shows that the critical Mach number of the airfoil decreases with the increase of the relative thickness of the airfoil.

Taking NACA0012 airfoil as an example, in incompressible flow, the minimum pressure coefficient is $C_{p\,min} = -0.43$, and the critical Mach number of the airfoil is calculated. According to the Prandtl–Glauert compressibility correction (13.5), the minimum pressure coefficient in compressible turbulent flow is

$$\left(C_{p\,min}\right)_{Ma_{\infty}} = \frac{\left(C_{p\,min}\right)_0}{\sqrt{1 - Ma_{\infty}^2}} = \frac{-0.43}{\sqrt{1 - Ma_{\infty}^2}} \tag{13.7}$$

At the critical point, let Eq. (13.7) be equal to Eq. (13.4)

$$\frac{-0.43}{\sqrt{1 - Ma_{\infty c}^2}} = \frac{2}{\gamma Ma_{\infty c}^2}\left\{\left[\frac{2}{\gamma + 1}\left(1 + \frac{\gamma - 1}{2}Ma_{\infty c}^2\right)\right]^{\frac{\gamma}{\gamma - 1}} - 1\right\} \tag{13.8}$$

The result of the solution is the critical Mach number for NACA0012 airfoil equaling $Ma_{\infty c} = 0.74$.

13.2   Transonic Flow Over a Thin Airfoil

The local shock wave system and pressure distribution of a thin airfoil at the same low angle of attack and different Mach numbers are discussed. When the inflow Ma_{∞} is less than the critical Mach number, the whole flow over the airfoil is subsonic. When the inflow Mach number increases to the critical Mach number (the critical Mach number of the airfoil is 0.57), some complex phenomena such as supersonic flow region and shock waves begin to appear in some regions of the airfoil, which seriously affect the aerodynamic force of the airfoil.

Fig. 13.3 Transonic flow over a thin airfoil at different Mach numbers

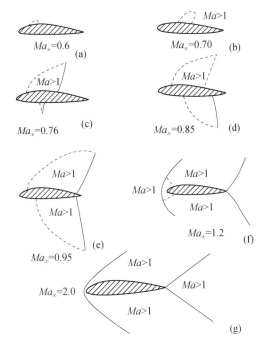

(1) When Ma_∞ gradually increases and slightly exceeds the critical Mach number, the upper wing first reaches the sound velocity at a certain point, and there is a small range of the supersonic region. The dotted line is the boundary of subsonic velocity, which is called the sonic line. Because the supersonic region is small, the airflow can smoothly transition from subsonic velocity to supersonic velocity without shock wave, and the pressure distribution has no sudden jump, as shown in Figs. 13.3a and 13.4b, $Ma_\infty = 0.6$.

(2) When the incoming Ma_∞ continues to increase, the supersonic region of the upper wing expands. Due to the pressure condition, the supersonic region ends with a local shock wave, and the pressure suddenly increases after the shock wave, and the velocity is no longer a smooth transition, as shown in Figs. 13.3b and 13.4c, $Ma_\infty = 0.7$.

(3) As the incoming Ma_∞ continues to increase, the supersonic region of the upper wing continues to expand, the shock wave position continues to move backward, and the shock wave also appears on the lower surface, and moves to the trailing edge faster than the lower wing, as shown in Figs. 13.3c, d, e, $Ma_\infty = 0.76$, 0.85, 0.95 and 13.4d, e, f, $Ma_\infty = 0.76$, 0.8, 0.88. At these Mach numbers, most of the upper and lower surfaces of the airfoil are supersonic. When the wake shock moves to the trailing edge of the upper and lower wings, there is no sudden jump in the pressure distribution of the upper and lower wings.

(4) When $Ma_\infty > 1$, the bow shock wave appears in front of the airfoil, and with the increase of Ma_∞, the bow shock wave gradually approaches the leading edge

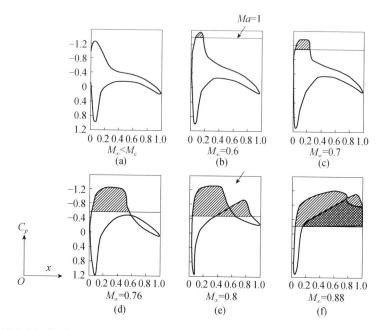

Fig. 13.4 Distribution of pressure coefficients on the upper and lower surfaces of a thin airfoil in transonic flow at different Mach numbers

of the airfoil, as shown in Fig. 13.3f. Because there is a positive shock zone in the detached shock, there is a subsonic zone behind the shock in a certain range near the leading edge, and then the flow along the wing accelerates to reach the supersonic zone. At the trailing edge of the airfoil, the flow decelerates to a Mach number close to the incoming flow through the trailing edge shock wave.

(5) When Ma_∞ continues to increase, the leading edge shock is detached and the shock appendage forms the head oblique shock. The whole flow field is a single supersonic flow field, as shown in Fig. 13.3g. When the oblique shock wave is formed by the front edge shock appendage, the corresponding incoming flow Ma_∞ is called the upper critical Mach number. Generally, the upper critical Mach number is between 1.2 and 1.4, which is related to the shape of the airfoil head. In the field of aerodynamics, the flow between the upper critical Mach number and the lower critical Mach number is called transonic flow. In transonic flow, in addition to the appearance of the shock wave, the interaction between the wing shock wave and boundary layer is also a complex flow phenomenon. In transonic flow, the interaction characteristics between shock wave and laminar boundary layer or turbulent boundary layer (see Sect. 7.11 for detailed analysis). The interaction between shock wave and boundary layer results in the thickening of the boundary layer, which is easy to cause the separation of the boundary layer (called shock-induced separation), so as to reduce the lift of airfoil (so-called shock stall) and increase the drag.

13.3 Aerodynamic Characteristics of Transonic Thin Airfoil Flow and Its Influence by Geometric Parameters

13.3.1 Relationship Between Lift Characteristics and Incoming Mach Number

As shown in Fig. 13.5, it is the curve of lift coefficient of a transonic airfoil with the Mach number of the incoming flow. It can be seen that the lift coefficient C_L changes according to the subsonic rule and supersonic rule before A point and after E point. The lift coefficient C_L increases with the increase of Mach number Ma_∞ in subsonic flow, while decreases in the case of supersonic flow. The lift coefficient continues to increase due to the expansion of the supersonic region of the upper wing and the decrease of the pressure when the Mach number Ma_∞ of the incoming flow is between A point and B point. After the incoming Mach number increases to point B, the shock wave on the upper wing continues to move backward, and the intensity increases. The negative pressure gradient in the boundary layer increases sharply which leads to the separation of the upper surface boundary layer and the sudden drop in the lift coefficient. This phenomenon caused by the interference of the shock wave and boundary layer is called shock stall. With the increase of incoming Mach number, the supersonic region and shock wave appears on the lower wing, and the shock wave on the lower wing moves to the trailing edge faster than that on the upper wing, which reduces the pressure on the lower wing and causes the lift coefficient to continue to drop to point C. With the increase of Mach number, the shock wave on the upper wing moves to the trailing edge, the separation point of the boundary layer also moves back, the pressure on the upper wing continues to decrease, and the lift coefficient rises to D again. After point D, bow-shaped detached shock wave appears in front of the airfoil. Before the detached shock wave is not attached to the airfoil, the pressure distribution on the upper and lower airfoil does not change with Mach number, but the flow pressure increases with the increase of Mach number, so the lift coefficient still decreases with the increase of Mach number.

It can be seen from the above that in the transonic range, the lift coefficient of the airfoil will change with the complex physical phenomena. In this process, there will be complex phenomena, such as the appearance of shock waves on the lower airfoil, the movement of the shock wave, the interference between shock wave and boundary layer, and the shock wave-induced separation.

Fig. 13.5 Curve of lift coefficient versus Mach number for transonic airfoil flow

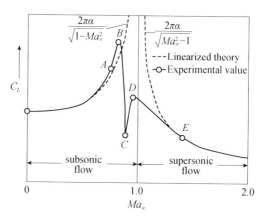

13.3.2 Relationship Between Drag Characteristics and Incoming Mach Number (Drag Divergence Mach Number)

As shown in Fig. 13.6, when the incoming Mach number Ma_∞ is less than the critical Mach number $Ma_{\infty C}$, the flow over the airfoil is subsonic, and the airfoil drag is mainly composed of the pressure drag and friction drag caused by the flow viscosity, so the drag coefficient has little change with the incoming Mach number Ma_∞. When the incoming Mach number Ma_∞ exceeds $Ma_{\infty C}$, the flow over the airfoil is transonic. With the increase of the incoming Mach number Ma_∞, the supersonic region on the airfoil gradually expands, and a shock wave appears, which results in shock wave drag. When the shock wave passes over the top of the airfoil, the strength of the shock wave increases rapidly, which leads to the sharp increase of the shock drag coefficient and the divergence of the drag. The Mach number of a shock wave passing over the apex is called drag divergence Mach number M_{DD}. The drag divergence Mach number can also be defined by the incoming Mach number corresponding to the point on the $C_d \sim Ma_\infty$ curve where the drag coefficient increases sharply (in aircraft aerodynamic design, the incoming Mach number corresponding to the derivative of drag coefficient with respect to the incoming Mach number is equal to 0.1). With the increase of Ma_∞, the shock wave continues to move backward, the supersonic wavefront continues to expand and accelerate, the shock wave intensity continues to increase, and the drag coefficient continues to increase. When the inlet Mach number Ma_∞ is close to 1, the shock waves on the upper and lower airfoils move to the trailing edge. At this time, the drag coefficient of the shock wave formed by the suction generated by the negative pressure in the rear section of the airfoil reaches the maximum, and the so-called sound barrier appears. Then, although the incoming Mach number Ma_∞ continues to increase, the drag coefficient decreases gradually because the pressure distribution on the wing surface is basically unchanged, but the incoming pressure continues to increase with the increase of M_∞.

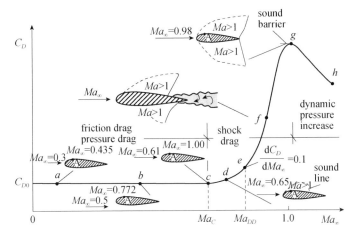

Fig. 13.6 Variation of drag coefficient with incoming Mach number (drag divergence Mach number)

13.3.3 Relationship Between Pitching Moment Characteristics and Incoming Mach Number

The change of pitching moment characteristics of a transonic airfoil with Mach number Ma_∞ is closely related to the change of relative position of the pressure center with Ma_∞. In subsonic flow, the pressure center of the airfoil changes slightly at different Ma_∞, but it doesn't change much, and it floats about 1/4 of the chord length.

When Ma_∞ exceeds $Ma_{\infty C}$, the local supersonic region appears on the upper wing and increases with the Ma_∞, and the low-pressure region expands backward which causes the pressure center to move backward and the low head torque to increase. When Ma_∞ continues to increase, local supersonic and local shock waves also appear on the lower wing, and the local shock wave on the lower wing moves backward faster than that on the upper wing. The local supersonic region of low pressure expands backward faster, so the suction on the rear section of the lower wing increases rapidly, which makes the pressure center move forward and causes the lift up torque. It can be seen that, in the transonic range, due to the movement of the shock wave on the airfoil surface, the position of the pressure center moves back and forth sharply, resulting in a great change in the longitudinal moment of the airfoil as shown in Fig. 13.7.

Fig. 13.7 Curve of moment coefficient for transonic airfoil flow

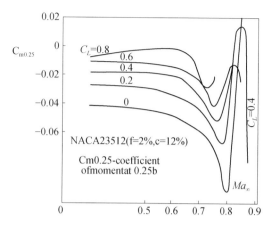

13.3.4 Influence of Airfoil Geometric Parameters on Transonic Aerodynamic Characteristics

The critical Mach number of the airfoil is related to the minimum pressure coefficient $(C_{p\,min})_0$. As the absolute value of the minimum pressure coefficient increases with the increase of the relative thickness c and the relative camber f, the critical Mach number of the airfoil decreases with the increase of c, f, and the lift coefficient (angle of attack).

(1) At the same lift coefficient, the critical Mach number decreases with the increase of airfoil thickness, as shown in Fig. 13.8.
(2) At the same lift coefficient, the critical Mach number decreases with the increase of airfoil camber, as shown in Fig. 13.9.
(3) At low C_L (small angle of attack), the larger the relative thickness is, the smaller the critical Mach number is, as shown in Fig. 13.10.

Fig. 13.8 Relationship between airfoil thickness and critical Mach number

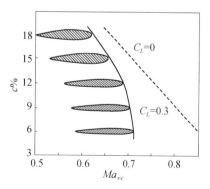

Fig. 13.9 Relationship between airfoil camber and critical Mach number

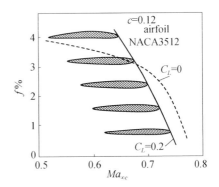

Fig. 13.10 Relationship between lift coefficient and critical Mach number of airfoil

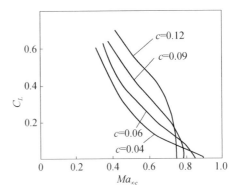

13.4 Transonic Small Perturbation Potential Flow Equation and Similarity Rule

If there is a local supersonic region in the subsonic flow field or a local subsonic region in the supersonic flow field, this kind of flow is called transonic flow. Because the transition from supersonic to subsonic is usually realized by the shock wave, the transonic flow field often contains local shock wave. The results show that the complex flow over a thin wing mainly occurs when the Mach number Ma_∞ is close to 1. When the blunt body moves at supersonic speed, the transonic flow also appears after the shock wave of the head. When the incoming Mach number Ma_∞ is close to 1, $Ma_\infty - 1$ is a small quantity. At this time, when the equation of small perturbation potential flow is simplified, the influence of this should be considered. After simplification, the second-order small quantity is retained, and the nonlinear part of the equation appears. By retaining the first term on the right in Eq. (11.28), we obtain Eq. (11.9)

$$\left(1 - Ma_\infty^2\right)\frac{\partial u}{\partial x} + \frac{\partial v}{\partial y} + \frac{\partial w}{\partial z} = Ma_\infty^2(\gamma+1)\frac{u}{V_\infty}\frac{\partial u}{\partial x} \qquad (13.9)$$

or Eq. (13.10).

$$\left(1 - \text{Ma}_\infty^2\right)\frac{\partial^2 \varphi}{\partial x^2} + \frac{\partial^2 \varphi}{\partial y^2} + \frac{\partial^2 \varphi}{\partial z^2} = \text{Ma}_\infty^2 (\gamma + 1)\frac{1}{V_\infty}\frac{\partial \varphi}{\partial x}\frac{\partial^2 \varphi}{\partial x^2} \tag{13.10}$$

If the first term is greater than zero, it is an elliptic equation; If the first term is less than zero, it is a hyperbolic equation. Move the right term of the equation to the left and change it to Eq. (13.11).

$$\left[1 - \text{Ma}_\infty^2 - \frac{\text{Ma}_\infty^2(\gamma + 1)}{V_\infty}\frac{\partial \varphi}{\partial x}\right]\frac{\partial^2 \varphi}{\partial x^2} + \frac{\partial^2 \varphi}{\partial y^2} + \frac{\partial^2 \varphi}{\partial z^2} = 0 \tag{13.11}$$

When studying the rule of the transonic airflow over a thin wing, the main factors that affect the aerodynamic characteristics of the thin wing can be qualitatively recognized. It is assumed that l is the wing length, c_{jm} is the average relative thickness of the wing, and b_{jm} is the average geometric chord length of the wing. Let

$$B = \sqrt{\left|1 - \text{Ma}_\infty^2\right|}, \quad K = (\gamma + 1)\text{Ma}_\infty^2 \tag{13.12}$$

the similarity rule of pressure coefficient is

$$C_{\text{p}} = \frac{C_{\text{jm}}^{2/3}}{K^{1/3}} f_p\left(B\lambda, \lambda\left(K C_{\text{jm}}\right)^{1/3}, \lambda tg\chi, \frac{x}{b_{\text{jm}}}, \frac{z}{l}\right) \tag{13.13}$$

and the similarity rule of lift coefficient is

$$C_{\text{L}} = \frac{C_{\text{jm}}^{2/3}}{K^{1/3}} f_L\left(B\lambda, \lambda\left(K C_{\text{jm}}\right)^{1/3}, \lambda tg\chi, \frac{x}{b_{\text{jm}}}, \frac{z}{l}\right) \tag{13.14}$$

The similarity rule of zero-lift wave drag coefficient is

$$\frac{C_{D0}}{\lambda c_{\text{jm}}^2} = f_D\left(B\lambda, \lambda\left(K C_{\text{jm}}\right)^{1/3}, \lambda tg\chi, \frac{x}{b_{\text{jm}}}, \frac{z}{l}\right) \tag{13.15}$$

13.5 Influence of Wing Geometry Parameters on Critical Mach Number of Transonic Flow

For sweep forward or backward wing, the pressure of the wing is only related to the normal Mach number perpendicular to the leading edge and the flow around the profile. The relationship between pressure P at any point on the wing surface and the pressure P_∞ of the incoming flow is determined by the normal Mach number.

$$\frac{p}{p_\infty} = \left[\frac{1 + \frac{\gamma-1}{2}\mathrm{Ma}_\infty^2 \cos^2 \chi}{1 + \frac{\gamma-1}{2}\mathrm{Ma}_n^2} \right]^{\frac{\gamma}{\gamma-1}} \tag{13.16}$$

where Ma_n is the normal Mach number corresponding to the flow over the airfoil normal section. The corresponding pressure coefficient is

$$\begin{aligned}
C_p &= \frac{2}{\gamma \mathrm{Ma}_\infty^2}\left(\frac{p}{p_\infty} - 1 \right) \\
&= \frac{2}{\gamma \mathrm{Ma}_\infty^2}\left\{ \left[\frac{1 + \frac{\gamma-1}{2}\mathrm{Ma}_\infty^2 \cos^2 \chi}{1 + \frac{\gamma-1}{2}\mathrm{Ma}_n^2} \right]^{\frac{\gamma}{\gamma-1}} - 1 \right\}
\end{aligned} \tag{13.17}$$

When $\mathrm{Ma}_n = 1$, the critical pressure coefficient of the sweep forward or backward wing is

$$C_{pc} = \frac{p_c - p_\infty}{\frac{1}{2}\rho_\infty V_{\infty c}^2} = \frac{2}{\gamma \mathrm{Ma}_{\infty c}^2}\left\{ \left[\frac{2}{\gamma+1}\left(1 + \frac{\gamma-1}{2}\mathrm{Ma}_{\infty c}^2 \cos^2 \chi \right) \right]^{\frac{\gamma}{\gamma-1}} - 1 \right\} \tag{13.18}$$

Using the above equation, the relationship between $\mathrm{Ma}_{\infty C}$ and the critical pressure coefficient can be established, as shown in Fig. 13.11. The figure shows that for a given critical pressure coefficient, increasing the sweep angle will increase the critical Mach number of the wing. Given the inlet Mach number Ma_∞ and lift coefficient, if the minimum pressure coefficient of the normal section of the forward-sweep or backward-sweep wing is less than the critical pressure coefficient determined by Eq. (13.18), then the flow over the wing with infinite span is subsonic flow, that is, the normal Mach number Ma_n on the wing surface is less than 1. Therefore, increasing the wing sweep angle is an effective measure to increase the critical Mach number of the wing.

The effect of aspect ratio on the critical Mach number of the wing is obvious. The smaller the aspect ratio is, the higher the critical Mach number of the wing is, because the interaction between the upper and lower surfaces of the wing will increase with the decrease of the aspect ratio, thus reducing the maximum velocity of the airfoil flow. In other words, the smaller the aspect ratio is, the smaller the maximum velocity increment of each section of the wing is than that of the wing with a larger aspect ratio under the same incoming Mach number and C_L.

The influence of swept angle and aspect ratio on the critical Mach number of the wing is estimated as follows

$$(\mathrm{Ma}_{\infty c})_{\mathrm{wing}} = (\mathrm{Ma}_{\infty c})_{\mathrm{airfoil}} + (\Delta \mathrm{Ma}_{\infty c})_\chi + (\Delta \mathrm{Ma}_{\infty c})_\lambda \tag{13.19}$$

Fig. 13.11 Critical pressure
coefficient of normal section
of sweep forward or
backward wing

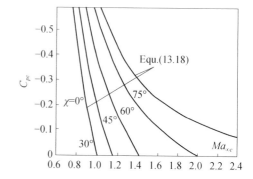

Fig. 13.12 Effects of aspect
ratio and sweep angle on
critical Mach number

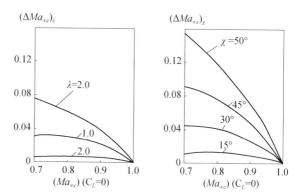

The correction considering aspect ratio and sweep angle can be found in Fig. 13.12. It can be seen that the larger the sweep angle and the smaller the aspect ratio, the higher the critical Mach number of the wing.

13.6 Aerodynamic Characteristics of Supercritical Airfoil Flow

13.6.1 Basic Concepts of Supercritical Airfoil

The concept of supercritical airfoil (as shown in Fig. 13.13) was proposed by Richard T. Whitcomb (1921–2009, as shown in Fig. 1.41), director of the wind tunnel laboratory of NASA Langley Research Center, in 1967, in order to increase the MA_{DD} number of resistance divergence of subsonic transport aircraft. Its profile is shown in Fig. 13.14. It was first used on A300 (Airbus, 1972) and A320 in the 1980s. It is the core technology of wing design of large transport aircraft (turbofan engine) (called supercritical wing). Whitcomb's other two innovations are: the area rule for reducing the zero-lift wave drag in transonic and supersonic flight (proposed in 1955) and the

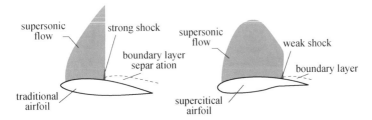

Fig. 13.13 Transonic airfoil flow (normal airfoil and supercritical airfoil)

Fig. 13.14 Comparison of laminar flow airfoil and supercritical airfoil

winglet for reducing the wing-induced drag (proposed in the early 1970s). These three achievements are widely used in aircraft design.

As shown in Fig. 13.14, compared with the laminar flow airfoil (peaked airfoil), the profile characteristics of a supercritical airfoil are that the head is blunt and round, the upper surface is flat, the lower surface is concave near the trailing edge, and the trailing edge is thin and curved downward. When the flow bypasses the leading edge of the laminar airfoil, the flow acceleration is faster (the sharper the leading edge, the higher the angle of attack, the greater the acceleration), which is slower than that of the supercritical airfoil bypassing the blunt circle. Similarly, on the upper surface of the airfoil, the upper surface of the laminar airfoil bulges upward, the positive pressure gradient is large and the acceleration is fast, while the upper surface of the supercritical airfoil is flat and the flow acceleration is slow. When the flight speed is large enough (the incoming Mach number is between 0.8 and 0.9), a local supersonic flow area will appear on the upper surface of the airfoil. At this time, the flow over the airfoil belongs to transonic flow. At the downstream boundary of the supersonic flow area, the shock wave will connect with the subsonic flow field, as shown in Fig. 13.15. For a laminar airfoil, when the incoming flow Ma_∞ exceeds the critical $Ma_{\infty}c$, the flow on the upper airfoil appears supersonic region. The transition from supersonic flow to subsonic flow must go through a shock wave. Due to the large convex height of the upper airfoil and the rapid expansion of the flow, the Mach number in front of the shock wave is large, which leads to the strong shock wave and the forward position. Similarly, the negative pressure gradient of the boundary layer behind the shock wave is also large, which leads to the obvious thickening or separation of the boundary layer behind the shock wave and the sharp increase of the shock wave resistance, so the drag divergence Mach number is low. From the point of view of shock wave formation and its influence on the boundary layer, the key is to reduce the shock wave intensity, which needs to reduce the Mach number in front of the shock wave. Therefore, improving the shape of the upper airfoil and reducing the protrusion height become the decisive factors. For this reason, the supercritical

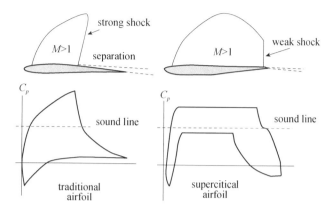

Fig. 13.15 Comparison of transonic flow over normal airfoil and supercritical airfoil

airfoil adopts a flat upper airfoil, which can minimize the expansion and acceleration process of the supersonic flow, so that the upper airfoil flow will maintain a certain distance with a small Mach number supersonic flow, and then transits to the subsonic outflow region through a small shock wave. Compared with a laminar airfoil, the acceleration expansion process of a supercritical airfoil is greatly slowed down, and the Mach number in front of the shock wave is small, which leads to the low shock wave strength and backward position. Similarly, the negative pressure gradient of the boundary layer behind the shock wave is also small, which leads to the small interference between the shock wave and the boundary layer, the thickness of the boundary layer behind the shock wave is thin, it is not easy to separate, and the shock wave drag is weakened, which effectively increases the drag divergence Mach number.

13.6.2　Expansion Mechanism of Supersonic Flow Over Supercritical Airfoil

Because the upper surface of a supercritical airfoil is flat, the expansion angle of supersonic flow is reduced, so the process of accelerating expansion is slowed down. As shown in Fig. 13.16, in the supersonic region, the flow expands from point 1 to point 2, and the expansion angle of the flow from Mach 1 to Mach 2 is

$$d\theta = \theta_1 - \theta_2 \approx \frac{ds}{R} = \frac{\sqrt{1 + \left(\frac{dy_u}{dx}\right)^2}}{R} dx, \quad ds = \sqrt{dx^2 + dy_u^2} \qquad (13.20)$$

where the radius of curvature of the upper wing is R.

Fig. 13.16 Accelerated expansion process of supersonic flow

$$R = \frac{\left[1 + \left(\frac{dy_u}{dx}\right)^2\right]^{3/2}}{\left|\frac{d^2 y_u}{dx^2}\right|}$$

(13.21)

After passing the expansion angle, the velocity increment is

$$dV = \frac{V}{\sqrt{Ma^2 - 1}} d\theta = \frac{V}{\sqrt{Ma^2 - 1}} \frac{\sqrt{1 + \left(\frac{dy_u}{dx}\right)^2}}{R} dx$$

$$= \frac{V}{\sqrt{Ma^2 - 1}} \frac{\left|\frac{d^2 y_u}{dx^2}\right|}{\left[1 + \left(\frac{dy_u}{dx}\right)^2\right]} dx$$

(13.22)

The pressure increment (negative value) is

$$dC_p = -\frac{2}{\sqrt{Ma^2 - 1}} d\theta = -\frac{2}{\sqrt{Ma^2 - 1}} \frac{\sqrt{1 + \left(\frac{dy_u}{dx}\right)^2}}{R} dx$$

$$= -\frac{2}{\sqrt{Ma^2 - 1}} \frac{\left|\frac{d^2 y_u}{dx^2}\right|}{\left[1 + \left(\frac{dy_u}{dx}\right)^2\right]} dx$$

(13.23)

The energy equation of isentropic flow is

$$\frac{a^2}{\gamma - 1} + \frac{V^2}{2} = C$$

For the above differential, we obtain

$$\frac{dV}{V} = \frac{\sqrt{Ma^2 - 1}}{1 + \frac{\gamma - 1}{2} Ma^2} \frac{dMa}{Ma}$$

(13.24)

Substituting Eq. (13.22), we get

$$\frac{\left|\frac{d^2 y_u}{dx^2}\right|}{1 + \left(\frac{dy_u}{dx}\right)^2} dx = \frac{\sqrt{Ma^2 - 1}}{1 + \frac{\gamma-1}{2} Ma^2} \frac{dMa}{Ma} \qquad (13.25)$$

Starting from the upper wing facing the boundary point ($Ma = 1$), the integral is obtained

$$tg^{-1}\left(\frac{dy_u}{dx}\right) = \sqrt{\frac{\gamma + 1}{\gamma - 1}} tg^{-1}\sqrt{\frac{\gamma - 1}{\gamma + 1}(Ma^2 - 1)} - tg^{-1}\sqrt{Ma^2 - 1} \qquad (13.26)$$

Given the variation of Ma along the airfoil, the airfoil contour can be determined by Eq. (13.26). It can be seen from Eqs. (13.22) and (13.23) that to reduce the increment of velocity and negative pressure, it is necessary to increase the curvature radius of the upper wing or reduce the second derivative of the upper wing coordinate curve. Comparing the change of curvature radius along the X direction between the ordinary laminar flow airfoil and the supercritical airfoil, it can be clearly seen that the curvature radius of the supercritical airfoil is much larger than that of the laminar flow airfoil, so the accelerated expansion process of the supercritical airfoil is obviously smaller than that of the laminar flow airfoil. This is the reason why supercritical airfoil weakens shock wave strength.

13.6.3 Aerodynamic Characteristics of Supercritical Airfoil Flow

As shown in Fig. 13.17, the main aerodynamic characteristics of supercritical airfoil flow are as follows:

(1) The curvature of the upper wing is relatively small and flat, so that after the incoming Mach number exceeds, the critical Mach number, the supersonic velocity flow along the upper surface is very small from the chord length of 5% from the leading edge. Thus, the supersonic Mach number before the shock wave is lower, the shock intensity is weak, the extension range is not large, the negative pressure gradient after the wave is small, and the boundary layer is not easy to separate, Thus, the phenomenon of drag divergence is eased;

(2) In order to compensate the acceleration of the front section of the upper wing of supercritical airfoil and the negative pressure of the leading edge are insufficient, the lower wing surface near the trailing edge can be concave to form a negative pressure gradient, reduce the velocity of the lower wing back edge area and increase the pressure. Thus increasing the contribution of the lower wing back section to the lift and making up for the insufficient lift caused by the flat upper wing can be realized. This effect of increasing lift from the concave of the lower wing surface of the rear section is called the post loading effect; The concave shape of the lower wing is actually increasing the curvature of the airfoil;

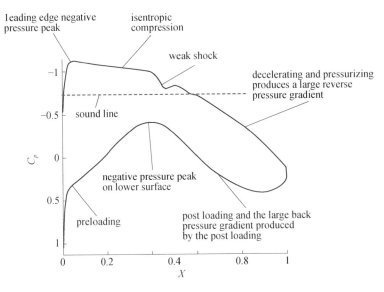

Fig. 13.17 Pressure distribution characteristics of supercritical airfoil

(3) Compared with laminar flow airfoil, the concave layout of the lower wing can make up for the lack of lift, but it obviously increases the low head moment of the airfoil, which increases the burden of the tail;

(4) Compared with the common peak airfoil, supercritical airfoil can increase the Mach number of drag divergence by about 0.05–0.12, or increase the maximum relative thickness of the airfoil by 2–5%. The thickness of the airfoil can increase the aspect to chord ratio of the wing by 2.5–3.0, or reduce the sweep angle by about 5–10° under the condition of keeping the Mach number of drag divergence unchanged.

13.7 High-Subsonic Flow Over a Swept Wing with a High Aspect Ratio

Supercritical wing is widely used in modern transport aircraft and fighters because of its good transonic aerodynamic effect, high cruise Mach number, and large relative thickness (which is beneficial to wing structure and fuel tank volume). Since 1980s, almost all transonic aircraft have adopted this kind of wing. In particular, modern large transport aircraft are equipped with turbofan engines, wing crane or tail crane. In order to improve the economic efficiency, the cruising speed is 800–970 km/h, flying $Ma_\infty = 0.75$–0.90, which is called high subsonic flight. In this flight speed range, the aircraft achieves the unity of speed and efficiency, that is, the speed is not low and the energy consumption is low. In high subsonic aircraft, there will be supersonic flow in the local region of the upper wing surface, and a shock wave will

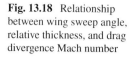

Fig. 13.18 Relationship between wing sweep angle, relative thickness, and drag divergence Mach number

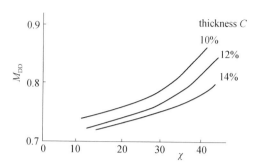

appear when the wing is in transonic flow. In order to control the shock wave, reduce the shock wave drag and improve the drag divergence Mach number, in addition to the swept wing and the thin airfoil (as shown in Fig. 13.18, the swept thin airfoil has a large drag divergence Mach number), the supercritical airfoil design is also an effective measure. Generally, the sweep angle of large transport aircraft is between 250 and 350. Compared with the flat wing, the swept wing can increase the drag divergence Mach number, but it can obviously increase the speed of the aircraft with little drag increase. However, the lift of the swept wing decreases, the drag increases, and the lift drag ratio decreases. Therefore, the swept wing should not be used at low speed.

In order to further improve the drag divergence Ma number of subsonic transport aircraft, modern large transport aircraft adopt supercritical airfoil design wings (called supercritical wings). Compared with the common airfoil, the leading edge of this airfoil is thicker and blunt, the upper airfoil is flat, and the lower airfoil is concave and curved downward near the trailing edge. In this way, the velocity of the flow around the leading edge increases a little, so that the drag divergence Mach number is increased. Under the given flight Mach number, the supercritical wing can reduce the aircraft sweep angle, reduce the weight of the aircraft, and increase the thickness of the wing (increase the volume of the fuel tank). In addition, the backward swept wing, because of the increase of the flow velocity, causes the tip to stall easily, so it is necessary to adopt the torsional wing in the design. In this way, for the wings of high subsonic transport aircraft, most of them adopt swept back, twisted, and variable thickness supercritical wings.

With the maturity of the theory and wide application of supercritical airfoil, a large number of data have been accumulated in the design of supercritical airfoil and wing, and they are constantly improved. There are three generations of supercritical airfoil families released by NASA (as shown in Figs. 13.19 and 13.20). Figure 13.21 is the second generation airfoil family of NACA. Figure 13.22 shows the typical supercritical airfoil of large passenger aircraft (A340, A350, A380) developed by Airbus, and A350 also uses a supercritical airfoil. Boeing B777 and B787 also adopt supercritical wing, and C919 developed in China also adopts supercritical wing design.

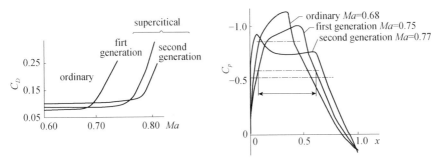

Fig. 13.19 Drag divergence characteristics and pressure distribution of supercritical airfoil and normal airfoil

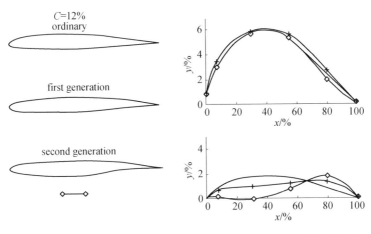

Fig. 13.20 Comparison of shape, thickness, and camber between the normal airfoil and supercritical airfoil

13.8 Transonic Area Rule

13.8.1 The Concept of Area Rule

In transonic flight, the zero-lift drag increases significantly near the sound speed and then decreases with the increase of Mach number until a certain supersonic speed, which is called the sound barrier. It is found that the time and magnitude of the drag rise are closely related to the distribution of the cross-sectional area along the longitudinal axis. In order to reduce the zero-lift wave drag near Mach number $Ma_\infty = 1$, in addition to adopting swept wing, thin airfoil, and supercritical airfoil, Richard T. Whitcomb proposed a drag reduction limit law in 1955, which is called area rule, for transonic and supersonic vehicles. Assuming that the longitudinal axis of the aircraft is the x-axis, the variation of the cross-sectional area $A(x)$ of the aircraft with

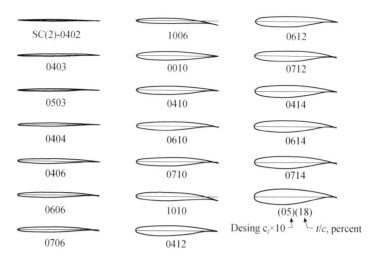

Fig. 13.21 NACA second generation supercritical airfoil family

Fig. 13.22 Typical wing profiles of Airbus

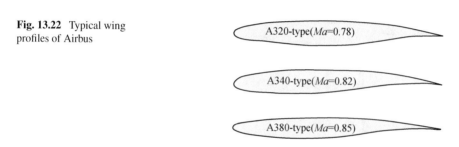

X plays an important role in reducing the wave resistance when Ma_∞ is close to 1. As shown in Fig. 13.23, for a certain aircraft delta wing layout without considering the area rule, the change curve of axial x cross section $A(x)$ has a great relationship with the layout of fuselage, wing, and tail. For example, when the wing and tail are encountered, the cross section $A(x)$ will change suddenly, and $A(x)$ will increase along the course. In order to reduce the drag coefficient of the sound barrier, Whitcomb proposes to reduce the increase of the cross-sectional area $A(x)$ by reducing the cross-sectional area of the fuselage, and adopts the shape of the bee waist fuselage, so that the change of the cross-sectional area $A(x)$ is a smooth curve similar to that of a slender body of revolution, as shown in Fig. 13.24. As shown in Fig. 13.25, the increase in drag coefficient in the sound barrier region of the two layouts is obvious. The layout considering the area rule obviously decreases near Mach number 1, which is about half of the drag coefficient without considering the area rule. The increase of shock wave drag mainly comes from the sudden change and increase of cross-sectional area, so the modified fuselage section becomes the area rule. It can be seen that in order to reduce the sound barrier resistance and improve the aerodynamic performance of the aircraft when the Mach number is near 1 in transonic flight, the area rule is an effective method to design the fuselage. In practical design, in order to simplify, a simple equivalent body of revolution wave drag is often used to replace the wave resistance of complex aircraft. Area rule is widely used in the design of transonic and supersonic aircraft.

The wind tunnel test shows that the drag of the body with a wing rises earlier (the corresponding Mach number can be less than 0.9), and the rise is several times that of the body without a wing. If the cross-sectional area of an aircraft is converted into an equivalent body of revolution, and the change of the cross-sectional area of the equivalent body of revolution along the longitudinal axis is gentle and smooth

Fig. 13.23 Change of section area without considering area rule

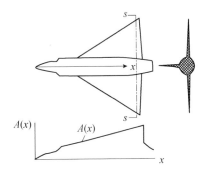

Fig. 13.24 Change of section area considering area rule

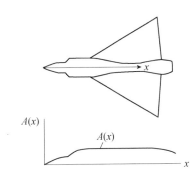

Fig. 13.25 Influence of area rule on drag coefficient of sound barrier

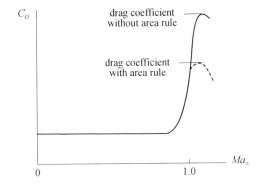

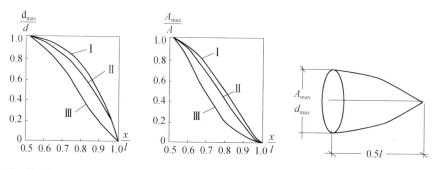

Fig. 13.26 Sears–Haack body of revolution

(without mutation), the wave drag rise of the aircraft in the transonic range is delayed and the rise is reduced, which is the transonic area rule. In order to meet the requirement of no mutation in cross-sectional area, the cross section of the fuselage at the organic wing can be reduced in design, and even the fuselage before the wing can be enlarged. In view of the fact that the shape of the cross-sectional area of the aircraft has little influence on the wave resistance, what kind of longitudinal distribution law of the cross-sectional area should be adopted in the design to minimize the sound barrier resistance? The body of revolution satisfying this characteristic is called Sears–Haack (SH), and the curve of the rear half of the aircraft is shown in Fig. 13.26. It is found that the drag reduction effect of the area rule is significant in the transonic region, but with the increase of Mach number, the drag reduction effect of the area rule gradually weakens. When the Mach number is 1.8–2.0, the effect of the area rule is almost zero.

In Fig. 13.26, type I curve is the curve of the body of revolution with the minimum zero-lift wave drag when the length and volume are given; Type II curve shows the curve of the body of revolution with the minimum zero-lift wave drag when the length and diameter are given; The curve III represents the curve of the body of revolution with the smallest zero-lift wave drag when the diameter and volume are given. In aircraft design, according to the requirements of length (carrier-based aircraft), volume (fuel tank), and diameter (cockpit), it is necessary to optimize the compromise. Figure 13.27 shows the variation curve of zero-lift wave drag for the combined configuration of the delta wing and fuse large with and without considering the area rule. It can be seen that the area rule can significantly reduce the drag coefficient of the sound barrier. The effect of the transonic area rule for the supercritical airfoil is shown in Fig. 13.28.

13.8.2 Slender Waist Fuselage

In practical application, it is usually assumed that the wing and tail remain unchanged, but the distribution of the total cross-sectional area of the aircraft is changed by

Fig. 13.27 Variation of zero-lift wave drag considering area rule correction

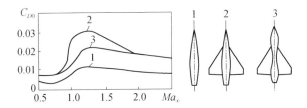

Fig. 13.28 Effect of transonic area rule on supercritical airfoil

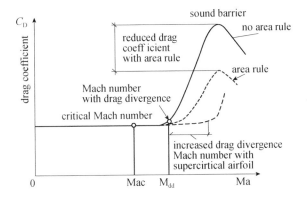

modifying the fuselage, so the called bee waist fuselage structure is proposed. This is an effective method proposed by NASA Whitcomb to reduce the zero-lift wave drag by modifying the fuselage.

A certain type of bomber, cruising Mach number 0.75, adopts a cantilever single wing, double beam box structure. The sweep angle of the focus line is 35°, the lower reflection angle of the chord plane is 3°, and the installation angle is 1°. The whole wing is composed of the central wing, left and right outer wings. There are inner and outer flaps and ailerons on the full span of the wing trailing edge. The flap is of backward slotted type with a maximum deflection angle of 35°. The whole metal semi hard shell fuselage structure is the bee waist streamline fuselage. It is found that the zero-lift wave drag can be reduced by 25–30% by using the area rule. As shown in Fig. 13.29.

Fig. 13.29 The bee waist structure of aircraft (area rule)

Exercises

A. Thinking questions

1. What is transonic airfoil flow?
2. Compared with subsonic airfoil flow, what are the main characteristics of transonic airfoil flow?
3. What are the lower critical Mach number and the upper critical Mach number?
4. Write the expression of critical pressure. How to calculate the lower critical Mach number of transonic airfoil flow?
5. Try to sketch the trend of critical Mach number with the angle of attack, curvature, and thickness.
6. Try to sketch the trend of the lift coefficient of an airfoil with the increase of Mach number in the transonic region (Mach number 0.8–1.6)?
7. What is a shock stall? How does the lift coefficient change with Mach number when a shock stall occurs?
8. What conditions is the shock wave drag of transonic airfoil the largest?
9. In transonic airfoil flow, how to control the strength and stability of shock wave on the upper airfoil?
10. What are the main characteristics of the supercritical airfoil flow? Why can the flat upper wing suppress shock and reduce shock intensity?
11. In transonic airfoil flow, how can the lift loss due to the flatness of the upper airfoil be compensated by the change of the lower airfoil?
12. Why can variable camber suppress the shock wave on the upper wing of transonic flow?
13. Please point out the main measures to increase the drag divergence Mach number?

B. Calculation questions

1. When Ma_∞ of an airfoil increases to 0.8, the velocity of the maximum velocity point on the airfoil reaches the sound velocity. What is the pressure coefficient of the airfoil at the maximum speed point at low speed? It is assumed that the Prandtl–Glauert rule is available.

2. There is a rectangular wing with aspect ratio λ of 3.5, its profile is NACA0006 airfoil, and it flies in a steady straight line at $Ma_\infty = 0.85$ at the altitude of 12 km. Try to calculate the relative thickness t of the airfoil corresponding to the affine-related wing in incompressible flow after an affine transformation $\bar{t}$ and aspect ratio $\bar{\lambda}$.

3. A group of affine airfoils is tested in a low-speed wind tunnel under the condition of $\alpha = 0°$ and the relationship between the minimum pressure coefficient and the relative thickness is as follows:

c	0.05	0.10	0.15	0.20
C_{pmin}	−0.1357	−0.1786	−0.2286	−0.300

Try to find the critical Mach number of the airfoil when $t = 0.10$.

4. Try to calculate the lift line slope of a rectangular wing with aspect ratio $\lambda = 10$ in the condition of $Ma_\infty = 0.6$ and compare it with the C_l^a of the same wing in incompressible flow.

5. The data obtained from the wind tunnel test of a rectangular thin wing of $\lambda = 3$ are as follows:

Ma_∞	0.40	0.70	0.80	0.90	0.95
$\frac{dC_L}{d\alpha}$ /rad	0.0600	0.0660	0.0690	0.0750	0.0825

On the basis of the above experimental data, try to calculate the lift line slope value C_l^a of the rectangular wing of $\lambda = 4$ with the same airfoil at $Ma_\infty = 0.80$.

6. Try to calculate the lift coefficient of a thin rectangular wing of $\lambda = 5$ with $Ma_\infty = 0.85$ using Prandtl–Glauert rule and the affine combination parameter chart method.

7. For the existing transonic flow $= 0.95$, a rectangular airfoil with relative thickness $t = 0.08$ and aspect ratio $\lambda = 4$. If the flow is similar try to calculate the aspect ratio and the relative thickness of the wing at the condition of $Ma_\infty = 1.07$

8. In the process of the space shuttle entering the earth's atmosphere, the maximum stagnation point heating occurs at the trajectory point corresponding to the altitude of 68.9 km, where $\rho_\infty = 1.075 \times 10^{-4}$ kg/m³, with a flight speed of 6.61 km/s. At this point of entering orbit, the shuttle is at an angle of attack of 40.2°, showing an effective nose radius at the

stagnation point of 1.29 m. If the wall temperature $T_W = 1110$ K, calculate the stagnation point heating rate.

9. For the two-dimensional transonic flow, in the range of small disturbance theory, the flow can be considered as irrotational under the following conditions:

$$\frac{\partial u}{\partial y} - \frac{\partial v}{\partial x} = 0$$

Try to prove that if the above differential equations are transformed by u and v as independent variables instead of x and y, that is

$$x = X(u, v)$$
$$y = Y(u, v)$$

The nonlinear equations in xy plane can be changed into linear equations in velocity plane

$$\left. \begin{array}{l} \left(1 - \text{Ma}_\infty^2\right)\frac{\partial y}{\partial v} + \frac{\partial x}{\partial u} = (\gamma + 1)\text{Ma}_\infty^2 \frac{u}{V_\infty}\frac{\partial y}{\partial v} \\ \frac{\partial x}{\partial u} - \frac{\partial y}{\partial u} = 0 \end{array} \right\}$$

where Ma_∞ is the incoming Mach number and V is the incoming velocity.

Chapter 14
High Lift Devices and Their Aerodynamic Performances

This chapter introduces the high lift device and its aerodynamic characteristics for large aircraft, including the development history of the high lift device, the basic type of the high lift device, the support and driving mechanism of the high lift device, the aerodynamic principle of the high lift device, the aerodynamic noise of the high lift device, the wind tunnel experiment and calculation method of the high lift device, and the combined control technology of the trailing-edge hinge flap and the lower deflection of the spoiler.

Learning points:

(1) Understand the development history of high lift device, basic type of high lift device, support and driving mechanism of high lift device;
(2) Learn the aerodynamic principle of high lift device, aeroacoustics of high lift device and wind tunnel test and calculation method of high lift device;
(3) Learn the technical characteristics of the combined control of hinged flap with deflection of spoilers.

14.1 Development of High Lift Devices

High lift device plays an important role in takeoff, landing, climbing performance, and controlling the best attitude of the aircraft, and is an important part related to the safety of the aircraft. The design of modern aircraft high lift devices belongs to the optimization design problem of multidisciplinary, multi-objective, and multi-technical synthesis. In the aspect of aerodynamics, the requirements of takeoff, landing, and climbing of the aircraft shall be met. In terms of structure, it is required that the components are less, and lighter. They are simply connected with enough strength and rigidity. In terms of operation, easy maintenance, reliable, cost-effective, and the requirements of damage tolerance are met. Early low-speed transport aircraft did not need high lift devices because they had low wing loads and the ratio of their

© Science Press 2022
P. Liu, *Aerodynamics*, https://doi.org/10.1007/978-981-19-4586-1_14

cruise speed to speed of takeoff or landing was not greater than 2:1. With the development of high-power engines, the cruise speed of transport aircraft is constantly increasing, and the wing load of large aircraft is also increasing, generally 450–650 kg/m². Therefore, practical high lift devices are emerged to keep the takeoff and landing speed within a reasonable safety range. At that time, the type of high lift device was mainly the trailing-edge flap. With the use of high-thrust turbofan engines, the cruise speed of the aircraft can reach the high subsonic speed zone, and the cruise Mach number is between 0.78 and 0.85, so as to meet the economic requirements of the engine. But in order to make the aircraft meet the requirements of safe speed during takeoff and landing, it is necessary to have a powerful, safe, and reliable high lift device on the aircraft. The trailing-edge high lift devices of the wing evolved from a plain flap to a single-slotted, double-slotted, or even triple-slotted Fowler flap. The leading-edge high lift device of the wing evolved from the fixed to the simple Kruger flap, from the fixed flap with the slot to the two or three positions slat and the variable curvature Kruger flap. The complexity of the high lift device may be reflected to the maximum extent on the Boeing 747–400 aircraft (as shown in Fig. 14.1, it has a round head Kruger flap for the inner wing, variable curvature Kruger flaps for mid and outboard wing with trailing-edge high lift devices of triple-slotted Fowler flaps). Since B747-400, the trend of high lift devices development has changed to use the simple device and active flow control to obtain lightweight, low noise, high lift, and low-cost devices. Generally speaking, the design of modern high-efficiency high lift devices mainly includes: general layout and mechanism design, aerodynamic design and optimization, flow control and noise reduction technology, structural design, etc.

At present, the success of Boeing and Airbus in the development of large transport aircraft provides rich experience and data for the design and manufacture of high lift devices. At present, the basic trend is that in order to avoid following the complex high lift devices mechanism of Boeing 747, Airbus has carried out bold and innovative designs from simplistic thinking. By using the multi-objective and comprehensive design concepts of aerodynamics, mechanism, structure, strength, maintenance, and

Fig. 14.1 The high lift devices of Boeing 747-400

economy, the simplified mechanism has been successful on A320, especially on the premise of meeting the aerodynamic requirements, the support and drive system was reformed boldly. In the design of A380 high lift devices (as shown in Fig. 14.2, it has a leading-edge drooped nose for the inboard wing and slats for the mid and outboard wing with single-slotted Fowler flaps), Airbus broke the defect of separate design of aerodynamics and mechanism and put forward the integrated design concept of aerodynamics, mechanism, and drive system. At present, B787 and A350 large commercial aircraft adopt the configuration of the hinged flap with deflection of spoilers for the trailing-edge high lift devices. As shown in Fig. 14.3, the leading edge of the inboard wing and the outboard wing of B787 are both slats (in order to improve the lift-to-drag ratio and reduce the aerodynamic noise, the leading-edge slats of the inboard wing are sealed during takeoff), and the trailing-edge is equipped with a new type of simple hinge flaps with deflection of spoilers. As shown in Fig. 14.4, the leading edge of the inboard wing of A350 is drooped nose with that of the outboard wing is the slat, and the type of trailing-edge high lift devices is the same as B787.

14.2 Basic Types of High Lift Devices

According to the position of the high lift device, it can be divided into leading-edge and trailing-edge high lift devices. The specific types are as follows.

14.2.1 Trailing-Edge High Lift Devices

(1) Plain flap

Fig. 14.2 The high lift devices of Airbus A380

Fig. 14.3 The high lift devices of Boeing B787

Fig. 14.4 The high lift devices of Airbus A350

As shown in Fig. 14.5, the trailing edge of the original airfoil is made to be movable and then bent down to form a flap. The characteristics of this flap are as follows: by bending down the flap, the camber of the wing is increased, the zero-lift angle of the airfoil is increased, and the lift coefficient is increased. This kind of flap is prone to separation in the trailing edge and its wake is unstable which may cause separation at a medium deflection angle resulting in little increase in lift and a large increase in drag.

(2) Split flap

As shown in Fig. 14.6, the flaps are arranged on the lower surface of the trailing edge of the original airfoil to form a moving surface. Principle: the flap is deflected

Fig. 14.5 Simple flap

Fig. 14.6 Split flap

Fig. 14.7 Single-slotted flap

downward to increase the camber properly. More importantly, a low-pressure area is formed between the flap and the fixed-wing surface, resulting in the increase of negative pressure on the upper surface of the trailing edge of the airfoil, thus increasing the lift and the maximum lift coefficient. It is characterized by simple structure, high lift, and high drag.

(3) Single-slotted flap

As shown in Fig. 14.7, the hinge axis of the flap is slightly lower than the chord line of the basic wing, and a gap is formed through the downward deflection of the flap. When the high-speed air flows through the gap, the boundary layer flow in the upper wing area of the flap is changed, the separation of the flap surface is delayed, and the maximum lift coefficient of the overall airfoil is greatly improved. The deflection angle of the flap can reach 40°. The effect is very obvious: the curvature increases, the chord length slightly increases, and the gap flow delays the separation of the trailing edge.

(4) Fowler flap

As shown in Fig. 14.8, this kind of flap is formed by turning the trailing edge of the airfoil into a moving surface and retreating at the same time when the flap deflects, which is proposed by Fowler. The principle is: in addition to the aerodynamic characteristics of the single-slotted flap, the chord length of the airfoil is greatly increased, the effective area of the airfoil is increased, and the lift is increased more than that of the single-slotted flap. It is a kind of trailing-edge high lift device widely used in modern large civil aircraft.

(5) Double-slotted Flap

As shown in Fig. 14.9, there are two main types of flaps. One is vane\main double-slotted flap, the other is main\aft double-slotted flap. The function of the deflector is to form a double-slot channel through the deflector, which can double control the boundary layer of the main wing and the flap, and play a greater role in the delayed flow separation. In the latter form, a flap cabin is made at the trailing edge of the main flap, followed by a conventional flap. Compared with the former, this kind of double slot is more obvious in controlling the flow and increasing the effective area.

Fig. 14.8 Fowler flap

(a) vane\main double-slotted flap   (b) main\aft double-slotted flap

Fig. 14.9 Double-slotted flap

Fig. 14.10 Triple-slotted
flap

Its function is to increase the camber of the wing and the effective area of the wing. High-speed slot flow is used to control the separation of an airfoil and improve the maximum lift coefficient. Because of the obvious lift-increasing effect of the double-slotted flap, it is widely used in modern large commercial aircraft. For the vane\main double-slotted flap, the single slot mode is usually adopted during takeoff. At this time, the lift is not the maximum, but the lift-to-drag is relatively large. When landing, the double slots are all opened, the deflection angle is the largest, the lift is the largest, and the resistance is the largest, which is beneficial to reduce the grounding speed and the running distance.

(6) Triple-slotted flap

As shown in Fig. 14.10, the flap consists of the main wing, vane flap, main flap, and aft flap. It is equivalent to the double slot formed by adding a guide vane on the basis of the latter one. This kind of flap is more effective in increasing the curvature, increasing the effective area, and controlling the slot flow, so the lift effect is better. The disadvantage is a complex structure. The range of application is far less than that of the single-slotted and double-slotted flap.

14.2.2 Leading-Edge High Lift Devices

(1) Plain flap

As shown in Fig. 14.11, when the leading-edge flap is deflected downward, there is a transition section, called the joint section, besides the flap and main wing. With the increase of head camber, the peak of leading-edge suction increases, the critical angle of attack increases, and the maximum lift coefficient increases.

(2) Drooped nose

As shown in Fig. 14.12, the mechanism is the same as that of the plain flap. However, the leading-edge drooped nose is fixed and cannot change shape with the change of flight state. Therefore, it has to be modified by other flight states (such as high speed).

Fig. 14.11 Plain flap

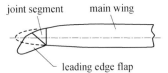

joint segment   main wing

leading edge flap

Fig. 14.12 Drooped nose

Fig. 14.13 Slat

(3) Slat

As shown in Fig. 14.13, the leading-edge slat is an auxiliary airfoil extending forward
to the airfoil. It is used to help the airflow smoothly (without separation) around the
leading edge in the high lift state. In order to obtain good aerodynamic performance, the leading-edge slat must have a retractable mechanism. The leading-edge
slat deflection produces large lift increment, high stall angle of attack, and good
stall characteristics. The slat has two deflection angles and three positions for cruise,
takeoff, and landing.

(4) Krueger Flap

As shown in Fig. 14.14, according to the form of its motion mechanism, it can be
divided into "upper skin extended Kruger flap" and "rotating along the leading-edge
Kruger flap". Compared with the leading-edge slat, it has the advantages of high lift-
to-drag ratio and low aerodynamic noise (suitable for aircraft takeoff configuration),
but it has the disadvantages of small stall angle of attack and small lift increment.

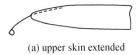

(a) upper skin extended

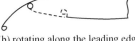

(b) rotating along the leading edge

Fig. 14.14 Krueger flap

14.3   Supporting and Driving Mechanism of High Lift Devices

The high lift device is the moving surface of the aircraft, which must be guided to the corresponding predetermined position through the supporting and retracting mechanism. Different types of leading-edge and trailing-edge high lift devices lead to different forms of supporting and driving mechanisms as listed below:

(1)   Mechanism of drooped nose

As shown in Fig. 14.15, this kind of mechanism (called hinged mechanism) is for seamless and downward deflection drooped nose. It uses a hinged mechanism to realize the downward deflection of the drooped nose at a certain angle. Because the curvature of the wing leading edge is increased, the peak suction is reduced, the negative pressure gradient of the upper surface is reduced, and the stall angle of attack is increased, so the maximum lift coefficient is increased. The disadvantage is that the curvature radius of the leading edge of the wing is too small, which is easy to cause airflow separation. It has not been widely used in the past commercial transport aircraft. However, because it can delay the laminar transition, it has been used in the latest A380 and A350.

(2)   Mechanism of Kruger flap

Kruger flap is a leading-edge high lift device widely used in high-performance large aircraft, as shown in Fig. 14.16. It mainly includes a simple Kruger flap, round head Kruger flap, and variable curvature Kruger flap. The simple Kruger flap consists of a flat plate that can rotate around the hinge. The circular head Kruger flap is a folding head that is added to the simple Kruger flap. The difference between the variable curvature Kruger flap and the round head Kruger flap is that the surface of the main Kruger flap is made of flexible glass fiber, which can change the profile curve of the airfoil. After the Kruger flap is opened, it not only increases the wing area, but also increases the wing curvature, which has a good effect on increasing lift. The increasing lift effect of these three types is obvious, but the mechanism complexity increases accordingly.

(3)   Mechanism of slat

Three-position slat (Fig. 14.17) is most used in the high lift devices of commercial aircraft. It is composed of cruise, takeoff, and landing positions. In takeoff position,

Fig. 14.15 Mechanism of the drooped nose

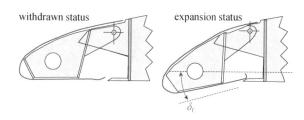

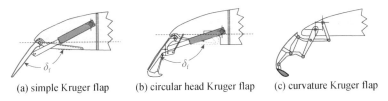

(a) simple Kruger flap   (b) circular head Kruger flap   (c) curvature Kruger flap

Fig. 14.16 Three kinds of guidance mechanisms for Kruger flap

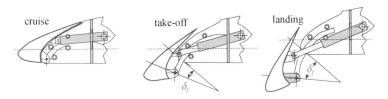

Fig. 14.17 Mechanism of slat

the upper surface of the slat tail and the leading edge of the main wing can be sealed or unsealed. In the case of sealing, the lift-to-drag ratio is high and noise is low (such as Boeing B787). But, Airbus A380 is in the case of unsealed. At the landing position, the slat is fully extended forward to 20° to 38° and forms a slot with the leading edge of the wing.

(4) Mechanism of trailing-edge flap

At present, the trailing-edge high lift device of large civil aircraft is mainly the Fowler flap (as shown in Fig. 14.18). According to the number of slots, it can be divided into single-slotted flap, double-slotted flap, and triple-slotted flap. There are three types of double-slotted flaps: vane\main double-slotted flap, main\aft double-slotted flap. The single-slotted flap is widely used. Although the lift effect is not as good as the double-slotted and triple-slotted flaps, the mechanism is simple and the weight is light. The double-slotted flap with fins (deflectors) forms a double-slot channel mainly through the sub fins and flaps, which double controls the boundary layer of the main wing and flaps and delays the flow separation. The results show that the double-slotted flap with the main wing and aft wing is more effective in controlling the slot flow and increasing the effective area. The triple-slotted flap is composed of vane flap, main flap, and aft flap. The triple-slotted flap can greatly increase the effective area of the wing and improve the lift effect, but the mechanism is complex and the weight is heavy.

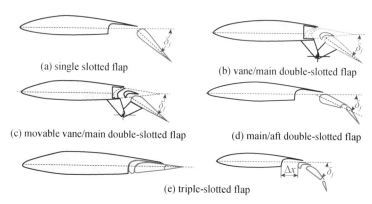

(a) single slotted flap

(b) vane/main double-slotted flap

(c) movable vane/main double-slotted flap

(d) main/aft double-slotted flap

(e) triple-slotted flap

Fig. 14.18 Type of trailing-edge flap

14.4 Aerodynamic Principles of High Lift Devices

When the high lift device is opened, the aerodynamic principle of bypassing these configurations is extremely complex, which mainly involves the aerodynamic mechanism of multi-stage wing flow. For two-dimensional multi-element airfoil flow, the possible flow phenomena include: boundary layer transition, shock/boundary layer interaction, wake/boundary layer mixing, boundary layer separation, laminar separation bubble, separated concave corner flow, large curvature of streamline, and so on. Almost all the complex problems of viscous fluid mechanics are involved, which brings great difficulty to the design and optimization of high lift devices as shown in Figs. 14.19 and 14.20. Figures 14.21 and 14.22 show the flow field and pressure coefficient distribution of the three-element wing. The flow characteristics of the multi-element airfoil are as follows:

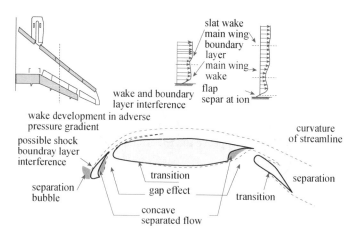

Fig. 14.19 Complexity of flow around a multi-element airfoil

Fig. 14.20 Wake shedding from discontinuities at trailing-edge high lift device

Fig. 14.21 Flow field around slat and trailing-edge Fowler flap

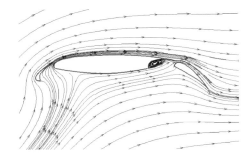

Fig. 14.22 Pressure distribution coefficient of flow around slat and trailing-edge Fowler flap

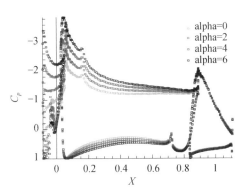

(1) For the multi-element airfoil with a large flap deflection angle, even if the Mach number of the incoming flow is not high, there may be a limited supersonic flow region on the upper surface of the leading edge, and there may be a transonic region or shock boundary layer interference;

(2) The wake of the upstream wing often mixes with the boundary layer on the surface of the downstream wing, and the resultant shear layer is a mixed boundary layer;

(3) The viscous flow region on the upper surface of the trailing-edge flap is relatively thick, especially in landing configuration. Even under normal flight conditions (i.e., before the maximum lift coefficient is reached far away), the relatively thick viscous flow region will lead to flow separation;

(4) When the trailing-edge flap deflects, separation bubbles will form at the trailing
     edge of the main wing. When the leading-edge high lift device (slat or Kruger
     flap) is deflected, the same flow state may occur at a low angle of attack.

The main reasons for the increase in lift of multi-stage airfoil are as follows:

(1) Increasing the camber effect of the wing

When the camber of the wing is increased, that is to say, the circulation is increased,
a large bow moment will be produced, especially when landing and approaching, the
horizontal stabilizer or the upper part of the elevator trailing edge is needed for trim.

(2) Increasing the effective area of the wing

Most of the high lift devices increase the wing area by increasing the basic chord
length of the wing. When the high lift device is not opened, the wing area is taken as
the reference area. When the high lift device is opened, the effective area of the wing
increases, and the lift increases. In this way, if the reference area is not changed, the
lift coefficient at zero angle of attack is increased, so the maximum lift coefficient is
greatly increased.

(3) Improving the flow quality and delaying the separation

By improving the flow quality of the slot between the wings and the boundary layer
state on the wing surface, the ability of the boundary layer to bear the negative pressure
gradient is enhanced, the separation is delayed, the stall angle of attack is increased,
and the maximum lift coefficient is increased. The lift curve of a multi-element
airfoil with angle of attack is shown in Fig. 14.23. For different types of multi-
element airfoils, the lift curve is shown in Fig. 14.24, the lift-to-drag characteristic
curve is shown in Fig. 14.25, and the maximum lift coefficient values of different
types of multi-element airfoils are shown in Fig. 14.26.

   In 1975, A.M.O. Smith, a famous American aircraft designer, made a detailed
exposition of the aerodynamics principle of multi-element wing in his famous paper.
Generally speaking, there are five main effects on the lift coefficient increment of a
multi-element airfoil:

(1) Slat effect: the velocity generated by the front wing circulation near the leading
     edge of the rear wing is opposite to the original velocity direction of the leading

Fig. 14.23 Effect of slat and
trailing-edge flap on lift
coefficient

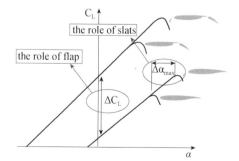

Fig. 14.24 Lift curve of multi-element airfoils

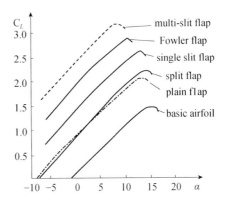

Fig. 14.25 Lift to drag curve of multi-element airfoils

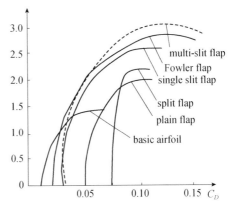

Fig. 14.26 Maximum lift coefficients for different types of multi-element airfoils

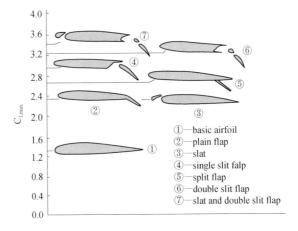

Fig. 14.27 Mixed boundary
layer of a multi-element
airfoil

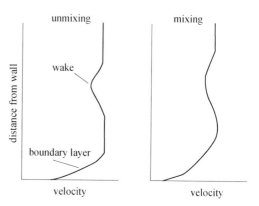

edge of the rear wing, thus suppressing the suction peak on the upper surface
of the rear wing;

(2) Circulation effect: the velocity near the leading edge of the rear wing increases
the local angle of attack of the flow around the trailing edge of the front wing,
which induces and enhances the circulation of the front wing;

(3) Tipping effect: the velocity near the leading edge of the rear wing makes the
boundary layer of the trailing edge of the front wing leave the front wing at a
higher speed, which reduces the negative pressure gradient of the upper wing
surface of the front wing, thus inhibiting the possible separation;

(4) Pressure build-up effect without the wall: the front wing dumps at a higher speed,
causing the boundary layer to decelerate and recover pressure after leaving the
front wing. In this process, there is no contact with the wall, which is more
efficient than the pressure build-up when contacting with the wall, as shown in
Fig. 14.27.

(5) New boundary layer effect: each wing will form a new thinner boundary layer
at its leading edge, so it is more resistant to the negative pressure gradient, so it
is more difficult to separate.

As shown in Fig. 14.28, the lift coefficient of a typical three-element airfoil varies
with the angle of attack. The lift coefficient of the main wing reaches the maximum
at stall. The stall of the three-element airfoil may be related to the wake flow in the
space of a negative pressure gradient. For a swept wing, the flow around a three-
dimensional high lift device is more complex, which is affected by the boundary
layer rotation and position beside the various flow phenomena of a two-dimensional
multi-element airfoil.

Generally, the flap span length accounts for 60–70% of the wingspan length.
Generally, the chord length of flap is selected according to the following ratio: b_j/b
$= 0.25$ for split flap; Simple flap $b_j/b = 0.30$; Fowler flap $b_j/b = 0.30$–0.4. The range
of maximum deflection angle: split flap $\delta_{jmax} = 55$–$60°$; Simple flap $\delta_{jmax} = 40$–$50°$;
Multi-slot backward flap $\delta_{jmax} = 50$–$60°$. In the preliminary design stage, the lift
enhancement effect of different high lift devices on the straight wing trailing edge is
as follows: ordinary slotted flap, $\Delta C_{Lmax} = 0.8$–1.0, $\alpha = 13$–$14°$; Single-slotted flap:

Fig. 14.28 Lift characteristics of a typical three-element airfoil

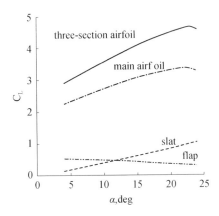

$\Delta C_{Lmax} = 1.4$–1.5, $\alpha = 12°$; The results show that for the multi-slotted flap, $\Delta C_{Lmax} = 1.6$–1.8, $\alpha = 12$–$13°$. For the wing with a sweep angle of $35°$, the chord length of the multi-slotted flap is 0.30–0.35, the relative span is 60%, and ΔC_{Lmax} is 0.9–1.0. For the leading-edge high lift device, the lift coefficient increment is about 0.1–0.2 with a fixed slotted flap; For slat, the lift coefficient increment is about 0.5–0.7. For the Kruger flap, the lift coefficient increment is about 0.3–0.4. The maximum lift increment of the high lift device belongs to a multidisciplinary and multi-objective optimization problem, as shown in Figs. 14.29 and 14.30. It generally depends on: the type of high lift device and its span; wing root tip ratio η; aspect ratio of wing λ; wing sweep angle χ; chord length b_j of slat or flap; the type and relative thickness of airfoil t/b; deflection angle of lifting surface δ_j; the amount of backward flap; the gap form and slot width between flap and wing.

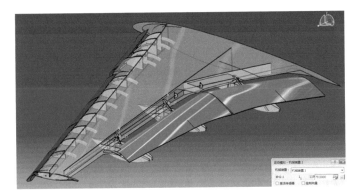

Fig. 14.29. 3D multi-element wing

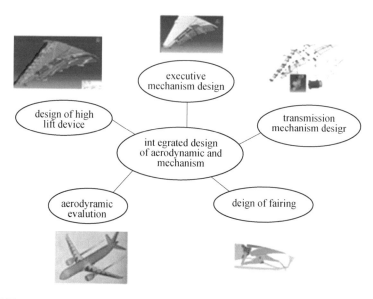

Fig. 14.30 Integrated design concept of aero-mechanism of high lift devices

14.5   Aeroacoustics of High Lift Devices

Generally speaking, aircraft noise mainly includes propulsion system noise and body noise. As shown in Fig. 14.31, body noise is mainly composed of landing gear noise and high lift device noise. The use of turbofan engines with a large bypass ratio not only greatly reduces the fuel consumption of aircraft, but also reduces the noise of the propulsion system to a great extent. Coupled with the application of noise reduction technologies such as anechoic nacelle and V-shaped petal nozzle, the proportion of aircraft body noise in the overall noise is increasing, and the high lift device noise accounts for the main part of the body noise, as shown in Fig. 14.32.

The results show that the main noise sources of slat are located in the trailing edge of the slat, the shear layer inside the slot, and the flow reattachment zone, in which the turbulent kinetic energy is the largest and the pressure fluctuation is the strongest. There are strong vortices on the side edge of the trailing-edge flap, including high-frequency small-scale unstable vortices and low-frequency large-scale vortices. These two vortices form the side edge noise source of the flap. In the design of the high lift device, it is necessary to fully understand the causes and mechanisms of the noise and seek the corresponding noise reduction methods. At present, there are several noise reduction methods for leading-edge slat: (1) micro perforation at the trailing edge of slat, (2) the cover of the slot of slat, (3) Slot filling, (4) reduction of the thickness of the trailing edge, (5) Caudal margin serrate, (6) leading edge droop, and (7) a sound liner at the lower surface of the slat and main wing. The passive control method is mainly used to control the side edge noise of the trailing-edge flap. One is to install porous materials; The other is the fence installed

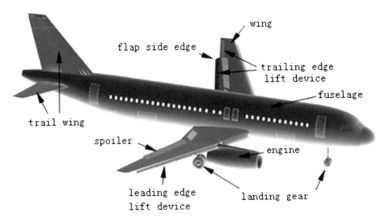

Fig. 14.31 Distribution of main noise sources of aircraft

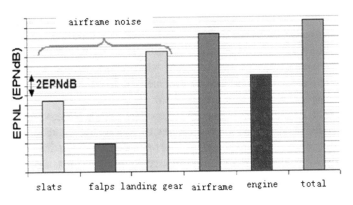

Fig. 14.32 Noise ratio of aircraft components

on the side edge of the flap, which is similar to the structure of the winglet. The flap fence can weaken the instability of the side edge shear layer and avoid interference between the side edge vortex structure and the sharp side edge. More importantly, the fence delays the fusion of the vortex structure, which makes the fence structure produce strong interference to the flow field. It is found that the flap side fence can reduce noise by 5–7 dB in approach and landing configurations; The third is the continuous profile method. The original purpose of this method is to eliminate the flap side edge and make the flap and the main wing connected smoothly and continuously, so as to fundamentally eliminate the flap side edge noise.

14.6   Method of Wind Tunnel and Numerical Simulation for High Lift Devices

The design of high lift devices used to rely on wind tunnel experiments. Despite the rapid development of CFD technology, the wind tunnel test is still an important means to evaluate the aerodynamic performance of high lift devices and determine the final type selection. The physical flow phenomena of the high lift device are very sensitive to the experimental Reynolds number, including the transition of the attachment line, the re-laminar fluidization of the turbulent boundary layer, the mutual interference and mixing of viscous wakes, the separated flow, and other complex flows are all related to the Reynolds number (as shown in Fig. 14.33). Therefore, the design of high lift devices generally needs to be carried out in a high Reynolds number wind tunnel. At present, there are three kinds of high Reynolds number wind tunnels for studying high lift devices: atmospheric wind tunnel, pressurized wind tunnel, and cryogenic wind tunnel. The size of the atmospheric wind tunnel is relatively large, so it can be used for a high Reynolds number test of the whole aircraft model. The Reynolds number is increased by increasing the wind speed, but the Mach number is also increased correspondingly, which leads to the increase of the influence of viscosity and compression effect. In a supercharged wind tunnel, increasing the Reynolds number by increasing the pressure can eliminate the adverse effects brought by the increase of Mach number, but at the same time, the larger pressure may change the slot parameters of the high lift device and cause the change of aerodynamic performance. In order to overcome the disadvantages of the supercharged wind tunnel, the low-temperature wind tunnel can be used to increase the experimental Reynolds number to the flight level. The world's aviation powers attach great importance to the construction of wind tunnels, and have invested huge human and material resources.

Another method of high lift device design is CFD (Computational Fluid Dynamics) technology. CFD has developed rapidly in recent years and has been widely used in the world. The flow phenomenon of the multi-element wing is complex. The difficulty of CFD technology in the design of high lift devices is whether it can accurately capture the complex physical phenomenon. At present,

Fig. 14.33 Effect of Reynolds number on maximum lift coefficient

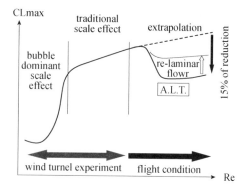

CFD methods are basically based on Reynolds averaged Navier–Stokes equations (RANS). RANS method shows a good ability to capture the complex flow characteristics of high lift configuration. The biggest uncertainty of the RANS method comes from the choice of turbulence mode. By comparison, the early algebraic model (BL) and the classical K-ε two-equation model have been proved to be used to calculate the flow around a multi-element airfoil with separation. At present, the one equation model (SA) and two-equation model (SST-k-w) developed since 1990s are mainly used to improve the calculation results of separated flow. Under the current conditions, due to the limitations of wind tunnel test and CFD technology, the design of a high lift device cannot be completed by any single method, but needs an efficient combination of the two. CFD technology is mainly used in the selection and optimization design of high lift devices, which can evaluate the performance changes brought by the modified design in a short period, greatly reducing the research and development time. A wind tunnel test is used to check the accuracy of the CFD method and design finalization. Any aircraft designed by the CFD method needs to accept the finalization of the wind tunnel test.

14.7 Technology of Hinged Flap with Deflection of Spoilers

The combined control technology of trailing-edge hinged flap and spoiler downward deflection has been applied in B787 and A350. The basic idea is: when the trailing-edge hinged flap deflects downward, the computer controls the spoiler to the most favorable position at the same time, so that the slot parameters can reach the best state of flow quality, and the flap can deflect a larger angle, thus greatly improving the aerodynamic performance of the single-slotted flap, as shown in Fig. 14.34. Figure 14.35 shows the lift coefficient comparison between the adaptive drooped hinge flap (ADHF) and the ordinary single-slotted Fowler flap. It can be seen that when the two flaps deflect at the same angle, the lift coefficient of the adaptive downward hinge flap is better than that of the ordinary single-slotted flap in most angles of attack, but the maximum lift coefficient is lower than that of the ordinary single-slotted flap. If the two flaps are deflected to their maximum angles (ADHF can deflect 45° and single-slotted flap can deflect 35° as shown in the figure), the lift coefficient of the adaptive downward deflection hinge flap is significantly higher than that of the ordinary single-slotted flap at all angles of attack, so the spoiler downward deflection can better improve the aerodynamic performance of the ordinary single-slotted flap. In addition, in the past, the trailing-edge flap only works in the process of takeoff and landing, but now the trailing-edge flap can also change the trailing-edge curvature of the wing by slightly deflecting the flap and the spoiler at the same time. Usually, the wing shape and lift coefficient remain unchanged during the cruise. However, the fuel is constantly consumed, resulting in the weight of the aircraft being constantly reduced, and the lift required is reduced, generally by increasing the cruise altitude to solve this contradiction. Variable camber technology solves this problem by changing the camber of the trailing edge and adjusting the lift of the

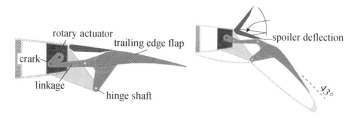

Fig. 14.34 Technology of trailing-edge hinged flap and spoiler downward deflection adopted by A350 and 787

Fig. 14.35 Comparison of ADHF and single-slotted Fowler flap

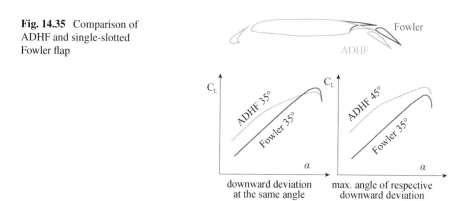

wing. This measure can reduce drag, improve the lift-to-drag ratio, and improve fuel efficiency. B787 and A350 also adopt variable curvature technology.

Exercises

1. Please explain the main characteristics of flow around multi-element airfoil and the physical mechanism of increasing lift?
2. The physical reason for the increase of lift coefficient of leading-edge slat? What is the effect of slot flow on lift increase?
3. Explain the effect of the downward deflection of the trailing-edge flap on the increase of lift?
4. What is the main difference between a drooped nose and slat flow?
5. The main physical causes of aerodynamic noise produced by multi-element wing?
6. Sketch the technical characteristics of the combined control of hinged flap with deflection of spoilers.
7. What is the development trend of high lift devices in the future?

Appendix A

Appendix Vector Operation and Control Equations in Orthogonal Curvilinear Coordinate System

1. Vector and vector operation

(1) Definition of Vector

$$\vec{A} = A_x \vec{i} + A_y \vec{j} + A_z \vec{k}$$
$$\vec{B} = B_x \vec{i} + B_y \vec{j} + B_z \vec{k}$$
$$\vec{C} = C_x \vec{i} + C_y \vec{j} + C_z \vec{k}$$

(2) Definition of Vector Operation

$$\vec{A} \cdot \vec{B} = A_x B_x + A_y B_y + A_z B_z$$

$$\vec{A} \times \vec{B} = \begin{vmatrix} \vec{i} & \vec{j} & \vec{k} \\ A_x & A_y & A_z \\ B_x & B_y & B_z \end{vmatrix} = (A_y B_z - A_z B_y)\vec{i}$$
$$+ (A_z B_x - A_x B_z)\vec{j} + (A_x B_y - A_y B_x)\vec{k}$$
$$(\vec{A} \times \vec{B}) \cdot \vec{C} = \vec{A} \cdot (\vec{B} \times \vec{C}) = (\vec{A} \times \vec{B}) \cdot \vec{C} = \vec{B} \cdot (\vec{C} \times \vec{A})$$
$$\vec{A} \times (\vec{B} \times \vec{C}) = (\vec{A} \cdot \vec{C})\vec{B} \cdot (\vec{A} \cdot \vec{B})\vec{C}$$

2. Formula of vector field calculation

(1) Hamilton operator

Hamiltonian operator $\triangledown$ is an operator with dual operations of differentiation and vector. Hamiltonian operator is defined in the rectangular coordinate system as

© Science Press 2022
P. Liu, *Aerodynamics*, https://doi.org/10.1007/978-981-19-4586-1

$$\nabla = \frac{\partial}{\partial x}\vec{i} + \frac{\partial}{\partial y}\vec{j} + \frac{\partial}{\partial z}\vec{k}$$

(2) Gradient of scalar and vector fields

Gradient of scalar field φ

$$\nabla\varphi = \frac{\partial\varphi}{\partial x}\vec{i} + \frac{\partial\varphi}{\partial y}\vec{j} + \frac{\partial\varphi}{\partial z}\vec{k}$$

Gradient of Vector field A

$$\nabla\vec{A} = \vec{i}\frac{\partial\vec{A}}{\partial x} + \vec{j}\frac{\partial\vec{A}}{\partial y} + \vec{k}\frac{\partial\vec{A}}{\partial z}$$

(3) Divergence of a vector field

$$\nabla \cdot \vec{A} = \frac{\partial A_x}{\partial x} + \frac{\partial A_y}{\partial y} + \frac{\partial A_z}{\partial z}$$

(4) Curl of a vector field

$$\text{rot}\,\vec{A} = \nabla \times \vec{A} = \vec{i}\left(\frac{\partial A_z}{\partial y} - \frac{\partial A_y}{\partial z}\right) + \vec{j}\left(\frac{\partial A_x}{\partial z} - \frac{\partial A_z}{\partial x}\right) + \vec{k}\left(\frac{\partial A_y}{\partial x} - \frac{\partial A_x}{\partial y}\right)$$

(5) Hamiltonian operator formula

$$\nabla(\varphi + \psi) = \nabla\varphi + \nabla\psi$$
$$\nabla(\varphi\psi) = \psi\nabla\varphi + \varphi\nabla\psi$$
$$\nabla F(\varphi) = \frac{dF}{d\varphi}\nabla\varphi$$
$$\nabla \cdot (\vec{A} + \vec{B}) = \nabla \cdot \vec{A} + \nabla \cdot \vec{B}$$
$$\nabla \cdot (\varphi\vec{A}) = \varphi\nabla \cdot \vec{A} + \vec{A} \cdot \nabla\varphi$$
$$\nabla \times (\vec{A} \times \vec{B}) = (\vec{B} \cdot \nabla)\vec{A} - (\vec{A} \cdot \nabla)\vec{B} + \vec{A}(\nabla \cdot \vec{B}) - \vec{B}(\nabla \cdot \vec{A})$$
$$(\vec{A} \cdot \nabla)\vec{A} = \nabla\left(\frac{A^2}{2}\right) - \vec{A} \times (\nabla \times \vec{A})$$
$$\nabla \cdot (\nabla\varphi) = \nabla^2\varphi = \Delta\varphi, \quad \Delta = \frac{\partial^2}{\partial x^2} + \frac{\partial^2}{\partial y^2} + \frac{\partial^2}{\partial z^2}$$
$$\nabla \times (\nabla\varphi) = 0$$
$$\nabla \cdot (\nabla \times \vec{A}) = 0$$
$$\nabla \times (\nabla \times \vec{A}) = \nabla(\nabla \cdot \vec{A}) - \nabla^2\vec{A}$$

$$\nabla \cdot (\varphi \nabla \psi) = \varphi \nabla^2 \psi + \nabla \varphi \cdot \nabla \psi$$

$$\nabla^2 (\varphi \psi) = \psi \nabla^2 \varphi + \varphi \nabla^2 \psi + 2 \nabla \varphi \cdot \nabla \psi$$

3. The generalized Gauss theorem and Stokes theorem

(1) The generalized Gauss theorem

If S is a closed surface of the space volume τ, the relationship between the volume and area of the physical quantity $\vec{A}$ or φ is

$$\iiint_\tau \nabla \cdot \vec{A} d\tau = \oiint_S \vec{n} \cdot \vec{A} dS$$

$$\iiint_\tau \nabla \varphi d\tau = \oiint_S \vec{n} \varphi dS$$

$$\iiint_\tau \nabla \times \vec{A} d\tau = \oiint_S \vec{n} \times \vec{A} dS$$

where $\vec{n}$ is the unit vector of the outer normal direction of the integral surface.

(2) Integral formula of the Hamiltonian operator

$$\iiint_\tau \nabla \varphi d\tau = \oiint_S \vec{n} \varphi dS$$

$$\iiint_\tau \nabla \cdot \vec{A} d\tau = \oiint_S \vec{n} \cdot \vec{A} dS$$

$$\iiint_\tau (\vec{V} \cdot \nabla) \vec{A} d\tau = \oiint_S (\vec{V} \cdot \vec{n}) \vec{A} dS - \iiint_\tau \vec{A} (\nabla \cdot \vec{V}) d\tau$$

$$\iiint_\tau \nabla^2 \varphi d\tau = \oiint_S \vec{n} \cdot \nabla \varphi dS = \oiint_S \frac{\partial \varphi}{\partial n} dS$$

$$\iiint_\tau \nabla^2 \vec{A} d\tau = \oiint_S \vec{n} \cdot \nabla \vec{A} dS = \oiint_S \frac{\partial \vec{A}}{\partial n} dS$$

$$\iiint_\tau \varphi \nabla^2 \psi + \nabla \varphi \cdot \nabla \psi) d\tau = \oiint_S \varphi \frac{\partial \psi}{\partial n} dS$$

$$\iiint_\tau \varphi \nabla^2 \psi - \psi \nabla^2 \varphi) d\tau = \oiint_S \left(\varphi \frac{\partial \psi}{\partial n} - \psi \frac{\partial \varphi}{\partial n} \right) dS$$

(3) Stokes theorem

If L is the boundary closed curve of the surface S and it is a shrinkable curve, the relationship between the line integral and the area fraction of the vector

$\vec{A}$ on the curve L and the surface S is

$$\oint_L \vec{A} \cdot d\vec{r} = \iint_S (\nabla \times \vec{A}) \cdot d\vec{S}$$

4. Orthogonal curvilinear coordinate system

(1) Orthogonal curvilinear coordinate system

The space curve coordinate system consists of the intersection lines of three sets of space surfaces. If the intersection lines of the three sets of space surfaces are perpendicular to each other, an orthogonal curve coordinate system (as shown in Fig. A.1) is formed. The space surface's equation in the rectangular coordinate system (x, y, z) is written as

$$\vec{r} = \vec{r}(q_1, q_2, q_3)$$

where q_1, q_2, and q_3 represent curve coordinates. The expression in the curve coordinate system on any micro-element line segment is

$$d\vec{r} = h_1 dq_1 \vec{e}_1 + h_2 dq_2 \vec{e}_2 + h_3 dq_3 \vec{e}_3$$

In the above formula, h_1, h_2, and h_3 are Lame coefficients, respectively; $\vec{e}_1, \vec{e}_2, \vec{e}_3$ are the unit vectors (coordinate base vectors) on the curve coordinates.

(2) Lame coefficients

$$h_1 = \left| \frac{\partial \vec{r}}{\partial q_1} \right| = \sqrt{\left(\frac{\partial x}{\partial q_1} \right)^2 + \left(\frac{\partial y}{\partial q_1} \right)^2 + \left(\frac{\partial z}{\partial q_1} \right)^2}$$

$$h_2 = \left| \frac{\partial \vec{r}}{\partial q_2} \right| = \sqrt{\left(\frac{\partial x}{\partial q_2} \right)^2 + \left(\frac{\partial y}{\partial q_2} \right)^2 + \left(\frac{\partial z}{\partial q_2} \right)^2}$$

$$h_3 = \left| \frac{\partial \vec{r}}{\partial q_3} \right| = \sqrt{\left(\frac{\partial x}{\partial q_3} \right)^2 + \left(\frac{\partial y}{\partial q_3} \right)^2 + \left(\frac{\partial z}{\partial q_3} \right)^2}$$

(3) Coordinate base vector

$$\vec{e}_1 = \frac{\frac{\partial x}{\partial q_1}\vec{i} + \frac{\partial y}{\partial q_1}\vec{j} + \frac{\partial z}{\partial q_1}\vec{k}}{h_1}$$

$$\vec{e}_2 = \frac{\frac{\partial x}{\partial q_2}\vec{i} + \frac{\partial y}{\partial q_2}\vec{j} + \frac{\partial z}{\partial q_2}\vec{k}}{h_2}$$

$$\vec{e}_3 = \frac{\frac{\partial x}{\partial q_3}\vec{i} + \frac{\partial y}{\partial q_3}\vec{j} + \frac{\partial z}{\partial q_3}\vec{k}}{h_3}$$

(4) Coordinate basis vector derivative

In the curvilinear coordinate system, the partial derivative of the unit vector with respect to the coordinate is

$$\frac{\partial \vec{e}_i}{\partial q_j} = \frac{1}{h}\frac{\partial h_j}{\partial q_i}\vec{e}_j \quad (i \neq j, i, j \text{ not sum up})$$

$$\frac{\partial \vec{e}_1}{\partial q_1} = -\left(\frac{1}{h_2}\frac{\partial h_1}{\partial q_2}\vec{e}_2 + \frac{1}{h_3}\frac{\partial h_1}{\partial q_3}\vec{e}_3\right)$$

$$\frac{\partial \vec{e}_2}{\partial q_2} = -\left(\frac{1}{h_1}\frac{\partial h_2}{\partial q_1}\vec{e}_1 + \frac{1}{h_3}\frac{\partial h_2}{\partial q_3}\vec{e}_3\right)$$

$$\frac{\partial \vec{e}_3}{\partial q_3} = -\left(\frac{1}{h_2}\frac{\partial h_3}{\partial q_2}\vec{e}_2 + \frac{1}{h_1}\frac{\partial h_3}{\partial q_1}\vec{e}_1\right)$$

(5) Expressions of gradient, divergence, and curl

Scalar gradient

$$\nabla\varphi = \frac{1}{h_1}\frac{\partial \varphi}{\partial q_1}\vec{e}_1 + \frac{1}{h_2}\frac{\partial \varphi}{\partial q_2}\vec{e}_2 + \frac{1}{h_3}\frac{\partial \varphi}{\partial q_3}\vec{e}_3$$

Vector divergence

$$\nabla \cdot \vec{A} = \frac{1}{h_1 h_2 h_3}\left[\frac{\partial (A_1 h_2 h_3)}{\partial q_1} + \frac{\partial (A_2 h_3 h_1)}{\partial q_2} + \frac{\partial (A_3 h_1 h_2)}{\partial q_3}\right]$$

Vector curl

$$\nabla \times \vec{A} = \frac{1}{h_1 h_2 h_3}\begin{vmatrix} h_1\vec{e}_1 & h_2\vec{e}_2 & h_3\vec{e}_3 \\ \frac{\partial}{\partial q_1} & \frac{\partial}{\partial q_2} & \frac{\partial}{\partial q_3} \\ h_1 A_1 & h_2 A_2 & h_3 A_3 \end{vmatrix}$$

Laplace operator

$$\Delta\varphi = \frac{1}{h_1 h_2 h_3}\left[\frac{\partial}{\partial q_1}\left(\frac{h_2 h_3}{h_1}\frac{\partial \varphi}{\partial q_1}\right) + \frac{\partial}{\partial q_2}\left(\frac{h_3 h_1}{h_2}\frac{\partial \varphi}{\partial q_2}\right) + \frac{\partial}{\partial q_3}\left(\frac{h_1 h_2}{h_3}\frac{\partial \varphi}{\partial q_3}\right)\right]$$

(6) In the cylindrical coordinate system, the expressions such as scalar gradient are

In the cylindrical coordinate system $q_1 = r, q_2 = \theta, q_3 = z$, the Lame coefficient is $h_1 = 1, h_2 = r, h_3 = 1$, then there is

Scalar gradient

$$\nabla\varphi = \frac{\partial\varphi}{\partial r}\vec{e}_r + \frac{1}{r}\frac{\partial\varphi}{\partial\theta}\vec{e}_\theta + \frac{\partial\varphi}{\partial z}\vec{e}_z$$

Vector divergence

$$\nabla\cdot\vec{A} = \frac{1}{r}\left[\frac{\partial(r A_r)}{\partial r} + \frac{\partial A_\theta}{\partial\theta} + \frac{\partial(r A_z)}{\partial z}\right]$$

Vector curl

$$\nabla\times\vec{A} = \frac{1}{r}\begin{vmatrix} \vec{e}_r & r\vec{e}_\theta & \vec{e}_z \\ \frac{\partial}{\partial r} & \frac{\partial}{\partial\theta} & \frac{\partial}{\partial z} \\ A_r & r A_\theta & A_z \end{vmatrix}$$

Laplace operator

$$\nabla^2\varphi = \frac{1}{r}\frac{\partial}{\partial r}\left(r\frac{\partial\varphi}{\partial r}\right) + \frac{1}{r^2}\frac{\partial^2\varphi}{\partial\theta^2} + \frac{\partial^2\varphi}{\partial z^2}$$

(7) In the spherical coordinate system, the expressions such as scalar gradient are

In the spherical coordinate system $q_1 = r, q_2 = \theta, q_3 = \gamma$, the Lame coefficient is $h_1 = 1, h_2 = r, h_3 = r\sin\theta$, then there is

Scalar gradient

$$\nabla\varphi = \frac{\partial\varphi}{\partial r}\vec{e}_r + \frac{1}{r}\frac{\partial\varphi}{\partial\theta}\vec{e}_\theta + \frac{1}{r\sin\theta}\frac{\partial\varphi}{\partial\gamma}\vec{e}_\gamma$$

Vector divergence

$$\nabla\cdot\vec{A} = \frac{1}{r^2}\frac{\partial(r^2 A_r)}{\partial r} + \frac{1}{r\sin\theta}\frac{\partial(\sin\theta A_\theta)}{\partial\theta} + \frac{1}{r\sin\theta}\frac{\partial A_\gamma}{\partial\gamma}$$

Vector curl

Fig. A.1 Orthogonal curvilinear coordinate system

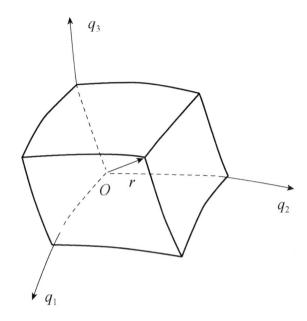

$$\nabla \times \vec{A} = \frac{1}{r^2 \sin \theta} \begin{vmatrix} \vec{e}_r & r\vec{e}_\theta & r \sin \theta \vec{e}_\gamma \\ \dfrac{\partial}{\partial r} & \dfrac{\partial}{\partial \theta} & \dfrac{\partial}{\partial \gamma} \\ A_r & r A_\theta & r \sin \theta A_\gamma \end{vmatrix}$$

Laplace operator

$$\nabla^2 \varphi = \frac{1}{r^2} \frac{\partial}{\partial r} \left(r^2 \frac{\partial \varphi}{\partial r} \right) + \frac{1}{r^2 \sin \theta} \frac{\partial}{\partial \theta} \left(\sin \theta \frac{\partial \varphi}{\partial \theta} \right) + \frac{1}{r^2 \sin \theta} \frac{\partial^2 \varphi}{\partial \gamma^2}$$

(1) **The fluid deformation rate matrix and the Newtonian fluid constitutive relationship**

The deformation rate matrix of the fluid is

$$[\varepsilon] = \begin{bmatrix} \varepsilon_{11} & \varepsilon_{12} & \varepsilon_{13} \\ \varepsilon_{21} & \varepsilon_{22} & \varepsilon_{23} \\ \varepsilon_{31} & \varepsilon_{32} & \varepsilon_{33} \end{bmatrix} \quad (x = 1, y = 2, z = 3)$$

The stress matrix is

$$[\tau] = \begin{bmatrix} \tau_{11} & \tau_{12} & \tau_{13} \\ \tau_{21} & \tau_{22} & \tau_{23} \\ \tau_{31} & \tau_{32} & \tau_{33} \end{bmatrix} (x = 1, y = 2, z = 3)$$

The Generalized Newton's Internal Friction Theorem

$$[\tau] = 2\mu[\varepsilon] - \left(p + \frac{2}{3}\mu\nabla \cdot \vec{V} \right)[I]$$

(2) Deformation rate matrix in orthogonal curvilinear coordinate system

Velocity vector

$$\vec{V} = V_1\vec{e}_1 + V_2\vec{e}_2 + V_3\vec{e}_3$$

$$\varepsilon_{11} = \frac{1}{h_1}\frac{\partial V_1}{\partial q_1} + \frac{V_2}{h_1 h_2}\frac{\partial h_1}{\partial q_2} + \frac{V_3}{h_1 h_3}\frac{\partial h_1}{\partial q_3}$$

$$\varepsilon_{22} = \frac{1}{h_2}\frac{\partial V_2}{\partial q_2} + \frac{V_3}{h_2 h_3}\frac{\partial h_2}{\partial q_3} + \frac{V_1}{h_1 h_2}\frac{\partial h_2}{\partial q_1}$$

$$\varepsilon_{33} = \frac{1}{h_3}\frac{\partial V_3}{\partial q_3} + \frac{V_1}{h_1 h_3}\frac{\partial h_3}{\partial q_1} + \frac{V_2}{h_2 h_3}\frac{\partial h_3}{\partial q_2}$$

$$\varepsilon_{12} = \varepsilon_{21} = \frac{1}{2}\left[\frac{h_2}{h_1}\frac{\partial}{\partial q_1}\left(\frac{V_2}{h_2}\right) + \frac{h_1}{h_2}\frac{\partial}{\partial q_2}\left(\frac{V_1}{h_1}\right) \right]$$

$$\varepsilon_{23} = \varepsilon_{32} = \frac{1}{2}\left[\frac{h_3}{h_2}\frac{\partial}{\partial q_2}\left(\frac{V_3}{h_3}\right) + \frac{h_2}{h_3}\frac{\partial}{\partial q_3}\left(\frac{V_2}{h_2}\right) \right]$$

$$\varepsilon_{31} = \varepsilon_{13} = \frac{1}{2}\left[\frac{h_1}{h_3}\frac{\partial}{\partial q_3}\left(\frac{V_1}{h_1}\right) + \frac{h_3}{h_1}\frac{\partial}{\partial q_1}\left(\frac{V_3}{h_3}\right) \right]$$

(3) Deformation rate matrix in rectangular coordinate system

Velocity vector

$$\vec{V} = V_x\vec{i} + V_y\vec{j} + V_z\vec{k}$$

$$\varepsilon_{xx} = \frac{\partial V_x}{\partial x}, \varepsilon_{yy} = \frac{\partial V_y}{\partial y}, \varepsilon_{zz} = \frac{\partial V_z}{\partial z}$$

$$\varepsilon_{xy} = \varepsilon_{yx} = \frac{1}{2}\left[\frac{\partial V_y}{\partial x} + \frac{\partial V_x}{\partial y} \right]$$

$$\varepsilon_{yz} = \varepsilon_{yz} = \frac{1}{2}\left[\frac{\partial V_z}{\partial y} + \frac{\partial V_y}{\partial z} \right]$$

$$\varepsilon_{zx} = \varepsilon_{xz} = \frac{1}{2}\left[\frac{\partial V_x}{\partial z} + \frac{\partial V_z}{\partial x} \right]$$

(4) Deformation rate matrix in cylindrical coordinate system

Velocity vector

$$\vec{V} = V_r\vec{e}_r + V_\theta\vec{e}_\theta + V_z\vec{e}_z$$

$$\varepsilon_{rr} = \frac{\partial V_r}{\partial r}, \quad \varepsilon_{\theta\theta} = \frac{1}{r}\frac{\partial V_\theta}{\partial\theta} + \frac{V_r}{r}, \quad \varepsilon_{zz} = \frac{\partial V_z}{\partial z}$$

$$\varepsilon_{r\theta} = \varepsilon_{\theta r} = \frac{1}{2}\left[\frac{1}{r}\frac{\partial V_r}{\partial\theta} + r\frac{\partial}{\partial r}\left(\frac{V_\theta}{r}\right)\right]$$

$$\varepsilon_{z\theta} = \varepsilon_{\theta z} = \frac{1}{2}\left[r\frac{\partial}{\partial z}\left(\frac{V_\theta}{r}\right) + \frac{1}{r}\frac{\partial V_z}{\partial\theta}\right]$$

$$\varepsilon_{zr} = \varepsilon_{rz} = \frac{1}{2}\left[\frac{\partial V_z}{\partial r} + \frac{\partial V_r}{\partial z}\right]$$

(5) Deformation rate matrix in spherical coordinate system

Velocity vector

$$\vec{V} = V_r\vec{e}_r + V_\theta\vec{e}_\theta + V_\gamma\vec{e}_\gamma$$

$$\varepsilon_{rr} = \frac{\partial V_r}{\partial r}, \quad \varepsilon_{\theta\theta} = \frac{1}{r}\frac{\partial V_\theta}{\partial\theta} + \frac{V_r}{r}, \quad \varepsilon_{\gamma\gamma} = \frac{1}{r\sin\theta}\frac{\partial V_\gamma}{\partial\gamma} + \frac{V_r}{r} + \frac{V_\theta\cos\theta}{r\sin\theta}$$

$$\varepsilon_{r\theta} = \varepsilon_{\theta r} = \frac{1}{2}\left[\frac{1}{r}\frac{\partial V_r}{\partial\theta} + r\frac{\partial}{\partial r}\left(\frac{V_\theta}{r}\right)\right]$$

$$\varepsilon_{\gamma\theta} = \varepsilon_{\theta\gamma} = \frac{1}{2}\left[\frac{1}{r\sin\theta}\frac{\partial V_\theta}{\partial\gamma} + \frac{1}{r}\frac{\partial V_\gamma}{\partial\theta} - \frac{V_\gamma\cos\theta}{r\sin\theta}\right]$$

$$\varepsilon_{\gamma r} = \varepsilon_{r\gamma} = \frac{1}{2}\left[\frac{\partial V_\gamma}{\partial r} + \frac{1}{r\sin\theta}\frac{\partial V_r}{\partial\gamma} - \frac{V_\gamma}{r}\right]$$

6. The differential equations of the motion of viscous fluids (Navier-Stokes equations)

(1) Vector form

Continuous differential equation

$$\frac{\partial\rho}{\partial t} + \nabla\cdot(\rho\vec{A}) = 0$$

Differential equation of motion

$$\frac{d\vec{V}}{dt} = \vec{f} + \frac{1}{\rho}\nabla\cdot[\tau]$$

The component form is

$$\begin{cases} \dfrac{\mathrm{d}V_x}{\mathrm{d}t} = f_x + \dfrac{1}{\rho}\left(\dfrac{\partial \tau_{xx}}{\partial x} + \dfrac{\partial \tau_{yx}}{\partial y} + \dfrac{\partial \tau_{zx}}{\partial z}\right) \\[2mm] \dfrac{\mathrm{d}V_y}{\mathrm{d}t} = f_y + \dfrac{1}{\rho}\left(\dfrac{\partial \tau_{xy}}{\partial x} + \dfrac{\partial \tau_{yy}}{\partial y} + \dfrac{\partial \tau_{zy}}{\partial z}\right) \\[2mm] \dfrac{\mathrm{d}V_z}{\mathrm{d}t} = f_z + \dfrac{1}{\rho}\left(\dfrac{\partial \tau_{xz}}{\partial x} + \dfrac{\partial \tau_{yz}}{\partial y} + \dfrac{\partial \tau_{zz}}{\partial z}\right) \end{cases}$$

Energy equation

$$\rho \frac{\mathrm{d}h}{\partial t} = \frac{\mathrm{d}p}{\mathrm{d}t} + \nabla \cdot (\kappa \nabla T) + \rho q + \Phi$$

For incompressible fluid flow

$$\rho \frac{\mathrm{d}e}{\partial t} = \nabla \cdot (\kappa \nabla T) + \rho q + \Phi$$

(2) **Differential equations of rectangular coordinate system**

For compressible fluid movement
Continuous differential equation

$$\frac{\partial \rho}{\partial t} + \frac{\partial (\rho u)}{\partial x} + \frac{\partial (\rho v)}{\partial y} + \frac{\partial (\rho w)}{\partial z} = 0$$

Differential equations of motion

$$\frac{\mathrm{d}u}{\mathrm{d}t} = f_x - \frac{1}{\rho}\frac{\partial p}{\partial x} + \frac{1}{\rho}\frac{\partial}{\partial x}\left(2\mu\frac{\partial u}{\partial x} - \frac{2}{3}\mu\nabla \cdot \vec{V}\right)$$
$$+ \frac{1}{\rho}\frac{\partial}{\partial y}\left[\mu\left(\frac{\partial v}{\partial x} + \frac{\partial u}{\partial y}\right)\right] + \frac{1}{\rho}\frac{\partial}{\partial z}\left[\mu\left(\frac{\partial w}{\partial x} + \frac{\partial u}{\partial z}\right)\right]$$
$$\frac{\mathrm{d}v}{\mathrm{d}t} = f_y - \frac{1}{\rho}\frac{\partial p}{\partial y} + \frac{1}{\rho}\frac{\partial}{\partial x}\left[\mu\left(\frac{\partial v}{\partial x} + \frac{\partial u}{\partial y}\right)\right]$$
$$+ \frac{1}{\rho}\frac{\partial}{\partial y}\left(2\mu\frac{\partial v}{\partial y} - \frac{2}{3}\mu\nabla \cdot \vec{V}\right) + \frac{1}{\rho}\frac{\partial}{\partial z}\left[\mu\left(\frac{\partial w}{\partial y} + \frac{\partial v}{\partial z}\right)\right]$$
$$\frac{\mathrm{d}w}{\mathrm{d}t} = f_z - \frac{1}{\rho}\frac{\partial p}{\partial z} + \frac{1}{\rho}\frac{\partial}{\partial x}\left[\mu\left(\frac{\partial w}{\partial x} + \frac{\partial u}{\partial z}\right)\right]$$
$$+ \frac{1}{\rho}\frac{\partial}{\partial y}\left[\mu\left(\frac{\partial w}{\partial y} + \frac{\partial v}{\partial z}\right)\right] + \frac{1}{\rho}\frac{\partial}{\partial z}\left(2\mu\frac{\partial w}{\partial z} - \frac{2}{3}\mu\nabla \cdot \vec{V}\right)$$

Energy equation

$$\rho C_p \frac{\mathrm{d}T}{\mathrm{d}t} = \frac{\mathrm{d}p}{\mathrm{d}t} + \frac{\partial}{\partial x}\left(\kappa\frac{\partial T}{\partial x}\right) + \frac{\partial}{\partial y}\left(\kappa\frac{\partial T}{\partial y}\right) + \frac{\partial}{\partial z}\left(\kappa\frac{\partial T}{\partial z}\right) + \rho q + \phi$$

$$\phi = -\frac{2}{3}\mu\left(\frac{\partial u}{\partial x} + \frac{\partial v}{\partial y} + \frac{\partial w}{\partial z}\right)^2$$

$$+ \mu\left[2\left(\frac{\partial u}{\partial x}\right)^2 + 2\left(\frac{\partial v}{\partial y}\right)^2 + 2\left(\frac{\partial w}{\partial z}\right)^2\right]$$

$$+ \mu\left[\left(\frac{\partial v}{\partial x} + \frac{\partial u}{\partial y}\right)^2 + \left(\frac{\partial w}{\partial y} + \frac{\partial v}{\partial z}\right)^2 + \left(\frac{\partial u}{\partial z} + \frac{\partial w}{\partial x}\right)^2\right]$$

For incompressible fluid flow.
Continuous differential equation

$$\frac{\partial u}{\partial x} + \frac{\partial v}{\partial y} + \frac{\partial w}{\partial z} = 0$$

Differential equations of motion

$$\frac{du}{dt} = f_x - \frac{1}{\rho}\frac{\partial p}{\partial x} + \frac{\mu}{\rho}\left(\frac{\partial^2 u}{\partial x^2} + \frac{\partial^2 u}{\partial y^2} + \frac{\partial^2 u}{\partial z^2}\right)$$

$$\frac{dv}{dt} = f_y - \frac{1}{\rho}\frac{\partial p}{\partial y} + \frac{\mu}{\rho}\left(\frac{\partial^2 v}{\partial x^2} + \frac{\partial^2 v}{\partial y^2} + \frac{\partial^2 v}{\partial z^2}\right)$$

$$\frac{dw}{dt} = f_z - \frac{1}{\rho}\frac{\partial p}{\partial z} + \frac{\mu}{\rho}\left(\frac{\partial^2 w}{\partial x^2} + \frac{\partial^2 w}{\partial y^2} + \frac{\partial^2 w}{\partial z^2}\right)$$

Energy equation

$$\rho C_v\frac{dT}{dt} = \mu\left[2\left(\frac{\partial u}{\partial x}\right)^2 + 2\left(\frac{\partial v}{\partial y}\right)^2 + 2\left(\frac{\partial w}{\partial z}\right)^2\right]$$

$$+ \mu\left[\left(\frac{\partial v}{\partial x} + \frac{\partial u}{\partial y}\right)^2 + \left(\frac{\partial w}{\partial y} + \frac{\partial v}{\partial z}\right)^2 + \left(\frac{\partial u}{\partial z} + \frac{\partial w}{\partial x}\right)^2\right]$$

$$+ \kappa\left(\frac{\partial^2 T}{\partial x^2} + \frac{\partial^2 T}{\partial y^2} + \frac{\partial^2 T}{\partial z^2}\right) + \rho q$$

(3) Differential Equations of Viscous Fluid Motion in Cylindrical Coordinate System

For compressible fluid movement
Continuous differential equation

$$\frac{\partial \rho}{\partial t} + \frac{1}{r}\frac{\partial (r\rho V_r)}{\partial r} + \frac{1}{r}\frac{\partial (\rho V_\theta)}{\partial \theta} + \frac{\partial (\rho V_z)}{\partial z} = 0$$

Differential equations of motion

$$\frac{\partial V_r}{\partial t} + V_r \frac{\partial V_r}{\partial r} + \frac{V_\theta}{r} \frac{\partial V_r}{\partial \theta} + V_z \frac{\partial V_r}{\partial z} - \frac{V_\theta^2}{r}$$

$$= f_r + \frac{1}{\rho r} \left[\frac{\partial (r \tau_{rr})}{\partial r} + \frac{\partial \tau_{\theta r}}{\partial \theta} + \frac{\partial (r \tau_{zr})}{\partial z} \right] - \frac{\tau_{\theta\theta}}{\rho r}$$

$$\frac{\partial V_\theta}{\partial t} + V_r \frac{\partial V_\theta}{\partial r} + \frac{V_\theta}{r} \frac{\partial V_\theta}{\partial \theta} + V_z \frac{\partial V_\theta}{\partial z} + \frac{V_r V_\theta}{r}$$

$$= f_\theta + \frac{1}{\rho r} \left[\frac{\partial (r \tau_{r\theta})}{\partial r} + \frac{\partial \tau_{\theta\theta}}{\partial \theta} + \frac{\partial (r \tau_{z\theta})}{\partial z} \right] + \frac{\tau_{r\theta}}{\rho r}$$

$$\frac{\partial V_z}{\partial t} + V_r \frac{\partial V_z}{\partial r} + \frac{V_\theta}{r} \frac{\partial V_z}{\partial \theta} + V_z \frac{\partial V_z}{\partial z}$$

$$= f_z + \frac{1}{\rho r} \left[\frac{\partial (r \tau_{rz})}{\partial r} + \frac{\partial \tau_{\theta z}}{\partial \theta} + \frac{\partial (r \tau_{zz})}{\partial z} \right]$$

Energy equation

$$\rho \left(\frac{\partial e}{\partial t} + V_r \frac{\partial e}{\partial r} + \frac{V_\theta}{r} \frac{\partial e}{\partial \theta} + V_z \frac{\partial e}{\partial z} \right)$$

$$= \tau_{rr} \frac{\partial V_r}{\partial r} + \tau_{\theta\theta} \left(\frac{1}{r} \frac{\partial V_\theta}{\partial \theta} + \frac{V_r}{r} \right) + \tau_{zz} \frac{\partial V_z}{\partial z}$$

$$+ \tau_{r\theta} \left(\frac{\partial V_\theta}{\partial r} + \frac{1}{r} \frac{\partial V_r}{\partial \theta} - \frac{V_\theta}{r} \right) + \tau_{\theta z} \left(\frac{1}{r} \frac{\partial V_z}{\partial \theta} + \frac{\partial V_\theta}{\partial z} \right)$$

$$+ \tau_{rz} \left(\frac{\partial V_r}{\partial z} + \frac{\partial V_z}{\partial r} \right) + \frac{1}{r} \left[\frac{\partial}{\partial r} \left(\kappa r \frac{\partial T}{\partial r} \right) + \frac{\partial}{\partial \theta} \left(\frac{k}{r} \frac{\partial T}{\partial \theta} \right) + \frac{\partial}{\partial z} \left(\kappa r \frac{\partial T}{\partial z} \right) \right] + \rho q$$

The constitutive relationship between stress and deformation rate is

$$\tau_{rr} = (-p - \frac{2}{3} \mu \nabla \cdot \vec{V}) + 2\mu \frac{\partial V_r}{\partial r}$$

$$\tau_{\theta\theta} = (-p - \frac{2}{3} \mu \nabla \cdot \vec{V}) + 2\mu \left(\frac{1}{r} \frac{\partial V_\theta}{\partial \theta} + \frac{V_r}{r} \right)$$

$$\tau_{zz} = (-p - \frac{2}{3} \mu \nabla \cdot \vec{V}) + 2\mu \frac{\partial V_z}{\partial z}$$

$$\tau_{r\theta} = 2\mu \varepsilon_{\theta r} = \mu \left[\frac{1}{r} \frac{\partial V_r}{\partial \theta} + r \frac{\partial}{\partial r} \left(\frac{V_\theta}{r} \right) \right]$$

$$\tau_{\theta z} = 2\mu \varepsilon_{\theta z} = \mu \left[\frac{1}{r} \frac{\partial V_z}{\partial \theta} + \frac{\partial V_\theta}{\partial z} \right]$$

$$\tau_{zr} = 2\mu \varepsilon_{zr} = \mu \left[\frac{\partial V_r}{\partial z} + \frac{\partial V_z}{\partial r} \right]$$

For incompressible fluid flow
Continuous differential equation

Appendix A

$$\frac{1}{r}\frac{\partial(rV_r)}{\partial r} + \frac{1}{r}\frac{\partial V_\theta}{\partial \theta} + \frac{\partial V_z}{\partial z} = 0$$

Differential equations of motion

$$\frac{\partial V_r}{\partial t} + V_r\frac{\partial V_r}{\partial r} + \frac{V_\theta}{r}\frac{\partial V_r}{\partial \theta} + V_z\frac{\partial V_r}{\partial z} - \frac{V_\theta^2}{r}$$

$$= f_r - \frac{1}{\rho}\frac{\partial p}{\partial r} + \frac{\mu}{\rho}\left[\frac{\partial^2 V_r}{\partial r^2} + \frac{1}{r}\frac{\partial V_r}{\partial r} + \frac{1}{r^2}\frac{\partial^2 V_r}{\partial \theta^2} + \frac{\partial^2 V_r}{\partial z^2} - \frac{2}{r^2}\frac{\partial V_\theta}{\partial \theta} - \frac{V_r}{r^2}\right]$$

$$\frac{\partial V_\theta}{\partial t} + V_r\frac{\partial V_\theta}{\partial r} + \frac{V_\theta}{r}\frac{\partial V_\theta}{\partial \theta} + V_z\frac{\partial V_\theta}{\partial z} + \frac{V_r V_\theta}{r}$$

$$= f_\theta - \frac{1}{\rho r}\frac{\partial p}{\partial \theta} + \frac{\mu}{\rho}\left[\frac{\partial^2 V_\theta}{\partial r^2} + \frac{1}{r}\frac{\partial V_\theta}{\partial r} + \frac{1}{r^2}\frac{\partial^2 V_\theta}{\partial \theta^2} + \frac{\partial^2 V_\theta}{\partial z^2} - \frac{V_\theta}{r^2} + \frac{2}{r^2}\frac{\partial V_r}{\partial \theta}\right]$$

$$\frac{\partial V_z}{\partial t} + V_r\frac{\partial V_z}{\partial r} + \frac{V_\theta}{r}\frac{\partial V_z}{\partial \theta} + V_z\frac{\partial V_z}{\partial z}$$

$$= f_z - \frac{1}{\rho}\frac{\partial p}{\partial z} + \frac{\mu}{\rho}\left[\frac{\partial^2 V_z}{\partial r^2} + \frac{1}{r}\frac{\partial V_z}{\partial r} + \frac{1}{r^2}\frac{\partial^2 V_z}{\partial \theta^2} + \frac{\partial^2 V_z}{\partial z^2}\right]$$

Energy equation

$$\rho C_v\left(\frac{\partial T}{\partial t} + V_r\frac{\partial T}{\partial r} + \frac{V_\theta}{r}\frac{\partial T}{\partial \theta} + V_z\frac{\partial T}{\partial z}\right)$$

$$= \kappa\left[\frac{\partial^2 T}{\partial r^2} + \frac{1}{r}\frac{\partial T}{\partial r} + \frac{1}{r^2}\frac{\partial^2 T}{\partial \theta^2} + \frac{\partial^2 T}{\partial z^2}\right] + q + \phi$$

$$\phi = 2\mu\left[\left(\frac{\partial V_r}{\partial r}\right)^2 + \left(\frac{1}{r}\frac{\partial V_\theta}{\partial \theta} + \frac{V_r}{r}\right)^2 + \left(\frac{\partial V_z}{\partial z}\right)^2\right]$$

$$+\mu\left[\left(\frac{1}{r}\frac{\partial V_r}{\partial \theta} + \frac{\partial V_\theta}{\partial r} - \frac{V_\theta}{r}\right)^2 + \left(\frac{1}{r}\frac{\partial V_z}{\partial \theta} + \frac{\partial V_\theta}{\partial z}\right)^2 + \left(\frac{\partial V_r}{\partial z} + \frac{\partial V_z}{\partial r}\right)^2\right]$$

(4) Differential Equations of Viscous Fluid Motion in Spherical Coordinate System

For compressible fluid movement
Continuous differential equation

$$\frac{\partial \rho}{\partial t} + \frac{1}{r^2}\frac{\partial(r^2\rho V_r)}{\partial r} + \frac{1}{r\sin\theta}\frac{\partial(\rho\sin\theta V_\theta)}{\partial \theta} + \frac{1}{r\sin\theta}\frac{\partial(\rho V_\gamma)}{\partial \gamma} = 0$$

Differential equations of motion

$$\frac{\partial V_r}{\partial t} + V_r\frac{\partial V_r}{\partial r} + \frac{V_\theta}{r}\frac{\partial V_r}{\partial \theta} + \frac{V_\gamma}{r\sin\theta}\frac{\partial V_r}{\partial \gamma} - \frac{V_\theta^2 + V_\gamma^2}{r}$$

$$= f_r + \frac{1}{\rho r^2 \sin\theta}\left[\frac{\partial(r^2\sin\theta\,\tau_{rr})}{\partial r} + \frac{\partial(r\sin\theta\,\tau_{\theta r})}{\partial\theta} + \frac{\partial(r\tau_{\gamma r})}{\partial\gamma}\right] - \frac{\tau_{\theta\theta}+\tau_{\gamma\gamma}}{\rho r}$$

$$\frac{\partial V_\theta}{\partial t} + V_r\frac{\partial V_\theta}{\partial r} + \frac{V_\theta}{r}\frac{\partial V_\theta}{\partial\theta} + \frac{V_\gamma}{r\sin\theta}\frac{\partial V_\theta}{\partial\gamma} + \frac{V_r V_\theta}{r} - \frac{V_\gamma^2\cos\theta}{r\sin\theta}$$

$$= f_\theta + \frac{1}{\rho r^2 \sin\theta}\left[\frac{\partial(r^2\sin\theta\,\tau_{r\theta})}{\partial r} + \frac{\partial(r\sin\theta\,\tau_{\theta\theta})}{\partial\theta} + \frac{\partial(r\tau_{\gamma\theta})}{\partial\gamma}\right] + \frac{\tau_{r\theta}}{\rho r} - \frac{\tau_{\gamma\gamma}\cos\theta}{r\sin\theta}$$

$$\frac{\partial V_\gamma}{\partial t} + V_r\frac{\partial V_\gamma}{\partial r} + \frac{V_\theta}{r}\frac{\partial V_\gamma}{\partial\theta} + \frac{V_\gamma}{r\sin\theta}\frac{\partial V_\gamma}{\partial\gamma} + \frac{V_r V_\gamma}{r} + \frac{V_\theta V_\gamma\cos\theta}{r\sin\theta}$$

$$= f_\gamma + \frac{1}{\rho r^2 \sin\theta}\left[\frac{\partial(r^2\sin\theta\,\tau_{r\gamma})}{\partial r} + \frac{\partial(r\sin\theta\,\tau_{\gamma\theta})}{\partial\theta} + \frac{\partial(r\tau_{\gamma\gamma})}{\partial\gamma}\right] + \frac{\tau_{r\gamma}}{\rho r} + \frac{\tau_{\theta\gamma}\cos\theta}{r\sin\theta}$$

Energy equation

$$\rho\left(\frac{\partial e}{\partial t} + V_r\frac{\partial e}{\partial r} + \frac{V_\theta}{r}\frac{\partial e}{\partial\theta} + \frac{V_\gamma}{r\sin\theta}\frac{\partial e}{\partial\gamma}\right)$$

$$= \tau_{rr}\frac{\partial V_r}{\partial r} + \tau_{\theta\theta}\left(\frac{1}{r}\frac{\partial V_\theta}{\partial\theta} + \frac{V_r}{r}\right) + \tau_{\gamma\gamma}\left(\frac{1}{r\sin\theta}\frac{\partial V_\gamma}{\partial\gamma} + \frac{V_r}{r} + \frac{V_\theta\cos\theta}{r\sin\theta}\right)$$

$$+\tau_{r\theta}\left(\frac{1}{r}\frac{\partial V_r}{\partial\theta} + \frac{\partial V_\theta}{\partial r} - \frac{V_\theta}{r}\right) + \tau_{\theta\gamma}\left(\frac{1}{r\sin\theta}\frac{\partial V_\theta}{\partial\gamma} + \frac{1}{r}\frac{\partial V_\gamma}{\partial\theta} - \frac{V_\gamma\cos\theta}{r\sin\theta}\right)$$

$$+ \tau_{r\gamma}\left(\frac{\partial V_\gamma}{\partial r} + \frac{1}{r\sin\theta}\frac{\partial V_r}{\partial\gamma} - \frac{V_\gamma}{r}\right)$$

$$+ \frac{1}{r^2\sin\theta}\left[\frac{\partial}{\partial r}\left(\kappa r^2\sin\theta\frac{\partial T}{\partial r}\right) + \frac{\partial}{\partial\theta}\left(\sin\theta k\frac{\partial T}{\partial\theta}\right) + \frac{\partial}{\partial\gamma}\left(\frac{k}{\sin\theta}\frac{\partial T}{\partial\gamma}\right)\right] + \rho q$$

The constitutive relationship between stress and deformation rate is

$$\tau_{rr} = (-p - \frac{2}{3}\mu\nabla\cdot\vec{V}) + 2\mu\frac{\partial V_r}{\partial r}$$

$$\tau_{\theta\theta} = (-p - \frac{2}{3}\mu\nabla\cdot\vec{V}) + 2\mu\left(\frac{1}{r}\frac{\partial V_\theta}{\partial\theta} + \frac{V_r}{r}\right)$$

$$\tau_{\gamma\gamma} = (-p - \frac{2}{3}\mu\nabla\cdot\vec{V}) + 2\mu\left(\frac{1}{r\sin\theta}\frac{\partial V_\gamma}{\partial\gamma} + \frac{V_r}{r} + \frac{V_\theta\cos\theta}{r\sin\theta}\right)$$

$$\tau_{r\theta} = 2\mu\varepsilon_{\theta r} = \mu\left[\frac{1}{r}\frac{\partial V_r}{\partial\theta} + r\frac{\partial}{\partial r}\left(\frac{V_\theta}{r}\right)\right]$$

$$\tau_{\theta\gamma} = 2\mu\varepsilon_{\theta\gamma} = \mu\left[\frac{1}{r\sin\theta}\frac{\partial V_\theta}{\partial\gamma} + \frac{1}{r}\frac{\partial V_\gamma}{\partial\theta} - \frac{V_\gamma\cos\theta}{r\sin\theta}\right]$$

$$\tau_{\gamma r} = 2\mu\varepsilon_{\gamma r} = \mu\left[\frac{\partial V_\gamma}{\partial r} + \frac{1}{r\sin\theta}\frac{\partial V_r}{\partial\gamma} - \frac{V_\gamma}{r}\right]$$

For incompressible fluid flow
Continuous differential equation

$$\frac{1}{r^2}\frac{\partial(r^2 V_r)}{\partial r} + \frac{1}{r\sin\theta}\frac{\partial(\sin\theta\, V_\theta)}{\partial\theta} + \frac{1}{r\sin\theta}\frac{\partial(V_\gamma)}{\partial\gamma} = 0$$

Differential equations of motion

$$\frac{\partial V_r}{\partial t} + V_r\frac{\partial V_r}{\partial r} + \frac{V_\theta}{r}\frac{\partial V_r}{\partial\theta} + \frac{V_\gamma}{r\sin\theta}\frac{\partial V_r}{\partial\gamma} - \frac{V_\theta^2 + V_\gamma^2}{r}$$

$$= f_r - \frac{1}{\rho}\frac{\partial p}{\partial r} + \frac{\mu}{\rho}\left[\begin{array}{l}\dfrac{\partial^2 V_r}{\partial r^2} + \dfrac{2}{r}\dfrac{\partial V_r}{\partial r} + \dfrac{1}{r^2}\dfrac{\partial^2 V_r}{\partial\theta^2} + \dfrac{ctg\theta}{r^2}\dfrac{\partial V_r}{\partial\theta} \\[2mm] + \dfrac{1}{r^2\sin^2\theta}\dfrac{\partial^2 V_r}{\partial\gamma^2} - \dfrac{2}{r^2}\dfrac{\partial V_\theta}{\partial\theta} - \dfrac{2}{r^2\sin\theta}\dfrac{\partial V_\gamma}{\partial\gamma} - \dfrac{2}{r^2}V_r - \dfrac{2V_\theta ctg\theta}{r^2}\end{array}\right]$$

$$\frac{\partial V_\theta}{\partial t} + V_r\frac{\partial V_\theta}{\partial r} + \frac{V_\theta}{r}\frac{\partial V_\theta}{\partial\theta} + \frac{V_\gamma}{r\sin\theta}\frac{\partial V_\theta}{\partial\gamma} + \frac{V_r V_\theta}{r} - \frac{V_\gamma^2\cos\theta}{r\,\sin\theta}$$

$$= f_\theta - \frac{1}{\rho r}\frac{\partial p}{\partial\theta} + \frac{\mu}{\rho}\left[\begin{array}{l}\dfrac{\partial^2 V_\theta}{\partial r^2} + \dfrac{2}{r}\dfrac{\partial V_\theta}{\partial r} + \dfrac{1}{r^2}\dfrac{\partial^2 V_\theta}{\partial\theta^2} + \dfrac{ctg\theta}{r^2}\dfrac{\partial V_\theta}{\partial\theta} \\[2mm] + \dfrac{1}{r^2\sin^2\theta}\dfrac{\partial^2 V_\theta}{\partial\gamma^2} + \dfrac{2}{r^2}\dfrac{\partial V_r}{\partial\theta} - \dfrac{2\cos\theta}{r^2\sin^2\theta}\dfrac{\partial V_\gamma}{\partial\gamma} - \dfrac{V_\theta}{r^2\sin^2\theta}\end{array}\right]$$

$$\frac{\partial V_\gamma}{\partial t} + V_r\frac{\partial V_\gamma}{\partial r} + \frac{V_\theta}{r}\frac{\partial V_\gamma}{\partial\theta} + \frac{V_\gamma}{r\sin\theta}\frac{\partial V_\gamma}{\partial\gamma} + \frac{V_r V_\gamma}{r} + \frac{V_\theta V_\gamma\cos\theta}{r\,\sin\theta}$$

$$= f_\gamma - \frac{1}{\rho r\sin\theta}\frac{\partial p}{\partial\gamma} + \frac{\mu}{\rho}\left[\begin{array}{l}\dfrac{\partial^2 V_\gamma}{\partial r^2} + \dfrac{2}{r}\dfrac{\partial V_\gamma}{\partial r} + \dfrac{1}{r^2}\dfrac{\partial^2 V_\gamma}{\partial\theta^2} + \dfrac{ctg\theta}{r^2}\dfrac{\partial V_\gamma}{\partial\theta} \\[2mm] + \dfrac{1}{r^2\sin^2\theta}\dfrac{\partial^2 V_\gamma}{\partial\gamma^2} + \dfrac{2}{r^2\sin\theta}\dfrac{\partial V_r}{\partial\gamma} + \dfrac{2\cos\theta}{r^2\sin^2\theta}\dfrac{\partial V_\theta}{\partial\gamma} - \dfrac{V_\gamma}{r^2\sin^2\theta}\end{array}\right]$$

Energy equation

$$\rho C_v\left(\frac{\partial T}{\partial t} + V_r\frac{\partial T}{\partial r} + \frac{V_\theta}{r}\frac{\partial T}{\partial\theta} + \frac{V_\gamma}{r\sin\theta}\frac{\partial T}{\partial\gamma}\right)$$

$$= k\left[\frac{\partial^2 T}{\partial r^2} + \frac{2}{r}\frac{\partial T}{\partial r} + \frac{1}{r^2}\frac{\partial^2 T}{\partial\theta^2} + \frac{ctg\theta}{r^2}\frac{\partial T}{\partial\theta} + \frac{1}{r^2\sin^2\theta}\frac{\partial^2 T}{\partial\gamma^2}\right] + \rho q + \phi$$

$$\phi = 2\mu\left[\left(\frac{\partial V_r}{\partial r}\right)^2 + \left(\frac{1}{r}\frac{\partial V_\theta}{\partial\theta} + \frac{V_r}{r}\right)^2 + \left(\frac{1}{r\sin\theta}\frac{\partial V_\gamma}{\partial\gamma} + \frac{V_r}{r} + \frac{V_\theta\cos\theta}{r\,\sin\theta}\right)^2\right]$$

$$+ \mu\left(\frac{1}{r}\frac{\partial V_r}{\partial\theta} + \frac{\partial V_\theta}{\partial r} - \frac{V_\theta}{r}\right)^2 + \mu\left(\frac{1}{r\sin\theta}\frac{\partial V_\theta}{\partial\gamma} + \frac{1}{r}\frac{\partial V_\gamma}{\partial\theta} - \frac{V_\gamma\cos\theta}{r\,\sin\theta}\right)^2$$

$$+ \mu\left(\frac{\partial V_\gamma}{\partial r} + \frac{1}{r\sin\theta}\frac{\partial V_r}{\partial\gamma} - \frac{V_\gamma}{r}\right)^2$$

Appendix B

Appendix-Atmospheric Parameter Table

See Tables B.1, B.2, B.3, B.4, B.5, B.6 and B.7 .

Table B.1 Density and gravity of common fluids

Fluid	t (°C)	Density ρ (kg/m³)	Gravity γ (N/m³)
Air	0	1.293	12.68
Air	15	1.225	12.01
Oxygen	0	1.429	14.02
Hydrogen	0	0.0899	0.881
Nitrogen	0	1.251	12.28
Carbon monoxide	0	1.251	12.27
Carbon dioxide	0	1.976	19.4
Distilled water	4	1000	9806
Seawater	15	1020–1030	10,000–10,100
Gasoline	15	700–750	6860–7350
Oil	15	880–890	8630–8730
Lubricating oil	15	890–920	8730–9030
Alcohol	15	790–800	7750–7840
HG	0	13,600	13,400

© Science Press 2022
P. Liu, *Aerodynamics*, https://doi.org/10.1007/978-981-19-4586-1

Table B.2 The density, dynamics, and kinematic viscosity coefficient of air and water

Air $p = 1.0132 \times 10^5$ N/m^2

Temperature (Air) °C	P (Kg/m^3)	$\mu \times 10^6$ Kg/m/s (Pa.s)	$\nu \times 10^6$ m^2/s
−20	1.39	15.6	11.2
−10	1.35	16.2	12.0
0	1.29	16.8	13.0
10	1.25	17.3	13.9
*15	1.23	17.8	14.4
20	1.21	18.0	14.9
40	1.12	19.1	17.1
60	1.06	20.3	19.2
80	0.99	21.5	21.7
100	0.94	22.8	24.3
Temperature(Water)°C	ρ (Kg/m^3)	$\mu \times 10^6$ Kg/m/s (Pa.s)	$\nu \times 10^6$ m^2/s
−20			
−10			
0	1000	1787	1.80
10	1000	1307	1.31
[a]15	999	1054	1.16
20	997	1002	1.01
40	992	635	0.66
60	983	467	0.48
80	972	355	0.37
100	959	282	0.30

[a]Standard status

Appendix B

Table B.3 Physical property parameters of standard atmosphere

H (km)	t (°C)	a (m/s)	$p \times 10^{-4}$ (N/m^2) (Pa)	P kg/m^3	$\mu \times 10^5$ (Pa.s)
0	15.0	340	10.132	1.225	1.780
1	8.5	336	8.987	1.112	1.749
2	2.0	332	7.948	1.007	1.717
3	−4.5	329	7.010	0.909	1.684
4	−11.0	325	6.163	0.820	1.652
5	−17.5	320	5.400	0.737	1.619
6	−24.0	316	4.717	0.660	1.586
7	−30.5	312	4.104	0.589	1.552
8	−37.0	308	3.558	0.526	1.517
9	−43.5	304	3.073	0.467	1.482
10	−50.0	299	2.642	0.413	1.447
11	−56.5	295	2.261	0.364	1.418
12	−56.5	295	1.932	0.311	1.418
13	−56.5	295	1.650	0.265	1.418
14	−56.5	295	1.409	0.227	1.418
15	−56.5	295	1.203	0.194	1.418
16	−56.5	295	1.027	0.163	1.418
17	−56.5	295	0.785	0.141	1.418
18	−56.5	295	0.749	0.121	1.418
19	−56.5	295	0.640	0.103	1.418
20	−56.5	295	0.546	0.088	1.418
30	−56.5	295	0.117	0.019	1.418
45	40.0	355	0.017	0.002	1.912
60	70.8	372	0.003	3.9×10^{-4}	2.047
75	−10.0	325	0.0006	8×10^{-5}	1.667

Table B.4 The flow parameter of Subsonic flow ($\gamma = 1.4$) in accordance with Mach number

M	p/p_o	ρ/ρ_o	T/T_0	a/a_0	$A*/A$	λ
0.00	1.0000	1.0000	1.0000	1.0000	0.0000	0.00000
0.01	0.9999	1.0000	1.0000	1.0000	0.01728	0.01096
0.02	0.9997	0.9998	0.9999	1.0000	0.03455	0.02191
0.03	0.9994	0.9996	0.9998	0.9999	0.05181	0.03286
0.04	0.9989	0.9992	0.9997	0.9998	0.06905	0.04381
0.05	0.9983	0.9988	0.9995	0.9998	0.08627	0.05476
0.06	0.9975	0.9982	0.9993	0.9996	0.1035	0.06570
0.07	0.9966	0.9976	0.9990	0.9995	0.1206	0.07664
0.08	0.9955	0.9968	0.9987	0.9994	0.1377	0.08758
0.09	0.9944	0.9960	0.9984	0.9992	0.1548	0.09851
0.10	0.9930	0.9950	0.9980	0.9990	0.1718	0.1094
0.11	0.9916	0.9940	0.9976	0.9988	0.1887	0.1204
0.12	0.9900	0.9928	0.9971	9986	0.2056	0.1313
0.13	0.9883	0.9916	0.9966	0.9983	0.2224	0.1422
0.14	0.9564	0.9903	0.9961	0.9980	0.2391	0.1531
0.15	0.9844	0.9888	0.9955	0.9978	0.2557	0.1640
0.16	0.9823	0.9873	0.9949	0.9974	0.2723	0.1748
0.17	0.9800	0.9857	0.9943	0.9971	0.2887	0.1857
0.18	0.9776	0.9840	0.9936	0.9968	0.3051	0.1965
0.19	0.9751	0.9822	0.9928	0.9964	0.3213	0.2074
0.20	0.9725	0.9803	0.9921	0.9960	0.3374	0.2182
0.21	0.9697	0.9783	0.9913	0.9956	0.3534	0.2290
0.22	0.9668	0.9762	0.9904	0.9952	0.3693	0.2399
0.23	0.9638	0.9740	0.9895	0.9948	0.3851	0.2506
0.24	0.9607	0.9718	0.9886	0.9943	0.4007	0.2614
0.25	0.9575	0.9694	0.9877	0.9938	0.4162	0.2722
0.26	0.9541	0.9670	0.9867	0.9933	0.4315	0.2829
0.27	0.9506	0.9645	0.9856	0.9928	0.4467	0.2936
0.28	0.9470	0.9619	0.9846	0.9923	0.4618	0.3044
0.29	0.9433	0.9592	0.9835	0.9917	0.4767	0.3150
0.30	0.9395	0.9564	0.9823	0.9911	0.4914	0.3257
0.31	0.9355	0.9535	0.9811	0.9905	0.5059	0.3364
0.32	0.9315	0.9506	0.9799	0.9899	0.5203	0.3470
0.33	0.9274	0.9476	0.9787	0.9893	0.5345	0.3576
0.34	0.9231	0.9445	0.9774	0.9886	0.5486	0.3682
0.35	0.9188	0.9413	0.9761	0.9880	0.5624	0.3788

(continued)

Table B.4 (continued)

M	p/p_o	ρ/ρ_o	T/T_0	a/a_0	A^*/A	λ
0.36	0.9143	0.9380	0.9747	0.9873	0.5761	0.3894
0.37	0.9098	0.9347	0.9733	0.9866	0.5896	0.3999
0.38	0.9052	0.9313	0.9719	0.9859	0.6029	0.4104
0.39	0.9004	0.9278	0.9705	0.9851	0.6160	0.4209
0.40	0.8956	0.9243	0.9690	0.9844	0.6289	0.4313
0.41	0.8907	0.9207	0.9675	0.9836	0.6416	0.4418
0.42	0.8857	0.9170	0.9659	0.9828	0.6541	0.4522
0.43	0.8807	0.9132	0.9643	0.9820	0.6663	0.4626
0.44	0.8755	0.9094	0.9627	0.9812	0.6784	0.4730
0.45	0.8703	0.9055	0.9611	0.9803	0.6903	0.4833
0.46	0.8650	0.9016	0.9594	0.9795	0.7019	0.4936
0.47	0.8596	0.8976	0.9577	0.9786	0.7134	0.5039
0.48	0.8541	0.8935	0.9560	0.9777	0.7246	0.5141
0.49	0.8486	0.8894	0.9542	0.9768	0.7356	0.5243
0.50	0.8430	0.8852	0.9524	0.9759	0.7464	0.5345
0.51	0.8374	0.8809	0.9506	0.9750	0.7569	0.5447
0.52	0.8317	0.8766	0.9487	0.9740	0.7672	0.5548
0.53	0.8259	0.8723	0.9469	0.9730	0.7773	0.5649
0.54	0.8201	0.8679	0.9449	0.9721	0.7872	0.5750
0.55	0.8142	0.8634	0.9430	0.9711	0.7968	0.5751
0.56	0.8082	0.8589	0.9410	0.9701	0.8063	0.5951
0.57	0.8022	0.8544	0.9390	0.9690	0.8155	0.6051
0.58	0.7962	0.8598	0.9370	0.9680	0.8244	0.6150
0.59	0.7901	0.8451	0.9349	0.9669	0.8331	0.6249
0.60	0.7840	0.8405	0.9328	0.9658	0.8416	0.6348
0.61	0.7778	0.8357	0.9307	0.9647	0.8499	0.6447
0.62	0.7716	0.8310	0.9286	0.9636	0.8579	0.6545
0.63	0.7654	0.8262	0.9265	0.9625	0.8657	0.6643
0.64	0.7591	0.8213	0.9243	0.9614	0.8732	0.6740
0.65	0.7528	0.8164	0.9221	0.9603	0.8806	0.6837
0.66	0.7465	0.8115	0.9199	0.9591	0.8877	0.6934
0.67	0.7401	0.8066	0.9176	0.9579	0.8945	0.7031
0.68	0.7338	0.8016	0.9153	0.9567	0.9012	0.7127
0.69	0.7274	0.7966	0.9113	0.9555	0.9076	0.7223
0.70	0.7209	0.7916	0.9107	0.9543	0.9138	0.7318
0.71	0.7145	0.7865	0.9084	0.9531	0.9197	0.7413

(continued)

Table B.4 (continued)

M	p/p_o	ρ/ρ_o	T/T_0	a/a_0	A^*/A	λ
0.72	0.7080	0.7814	0.9061	0.9519	0.925	0.7508
0.73	0.7016	0.7763	0.9037	0.9506	0.9309	0.7602
0.74	0.6951	0.7712	0.9013	0.9494	0.9362	0.7696
0.75	0.6886	0.7660	0.8989	0.9481	0.9412	0.7789
0.76	0.6821	0.7609	0.8964	0.9468	0.9461	0.7883
0.77	0.6756	0.7557	0.8940	0.9455	0.9507	0.7975
0.78	0.6690	0.7505	0.8915	0.9442	0.9551	0.8068
0.79	0.6625	0.7452	0.8890	0.9429	0.9592	0.8160
0.80	0.6560	0.7400	0.8865	0.9416	0.9632	0.8251
0.81	0.6495	0.7347	0.8840	0.9402	0.9669	0.8343
0.82	0.6430	0.7295	0.8815	0.9389	0.9704	0.8433
0.83	0.6365	0.7242	0.8789	0.9375	0.9737	0.8524
0.84	0.6300	0.7189	0.8763	0.9361	0.9769	0.8614
0.85	0.6235	0.7136	0.8737	0.9347	0.9797	0.8704
0.86	0.6170	0.7083	0.8711	0.9333	0.9824	0.8793
0.87	0.6106	0.7030	0.8685	0.9319	0.9849	0.8882
0.88	0.6041	0.6977	0.8659	0.9305	0.9872	0.8970
0.89	0.5977	0.6924	0.8632	0.9291	0.9893	0.9058
0.90	0.5913	0.6870	0.8606	0.9277	0.9912	9146
0.91	0.5849	0.6817	0.8579	0.9262	0.9929	0.9233
0.92	0.5785	0.6764	0.8552	0.9248	0.9944	0.9320
0.93	0.5721	0.6711	0.8525	0.9233	0.9958	0.9407
0.94	0.5658	0.6658	0.8798	0.9218	0.9969	0.9493
0.95	0.5595	0.6604	0.8471	0.9204	0.9979	0.9578
0.96	0.5532	0.6551	0.8444	0.9189	0.9986	0.9663
0.97	0.5469	0.6498	0.8416	0.9174	0.9992	0.9748
0.98	0.5407	0.6445	0.8389	0.9159	0.9997	0.9833
0.99	0.5345	0.6392	0.8361	0.9144	0.9999	0.9917
0.100	0.5283	0.6339	0.8333	0.9129	1.0000	1.0000

The original data were acquired from NACA TM1428

Table B.5 The flow parameter of Supersonic flow ($\gamma = 1.4$) in accordance with mach number

M	$\frac{p}{p_0}$	$\frac{\rho}{\rho_0}$	$\frac{T}{T_0}$	$\frac{a}{a_0}$	$\frac{A^*}{A}$	$\delta \ (°)$	λ
1.00	0.5283	0.6339	0.8333	0.9192	1.0000	0	1.0000
1.01	0.5221	0.6287	0.8306	0.9113	0.9999	0.04473	1.0083
1.02	0.5160	6234	0.8278	0.9098	0.9997	0.1257	1.0166
1.03	0.5099	0.6181	0.8250	0.9083	0.9993	0.2294	1.0248
1.04	0.5039	0.6129	0.8222	0.9067	0.9987	0.3510	1.0330
1.05	0.4979	0.6077	0.8193	0.9052	0.9980	0.4874	1.0411
1.06	0.4919	0.6024	0.8165	0.9036	0.9971	0.6367	1.0492
1.07	0.4860	0.5972	0.8137	0.9020	0.9961	0.7973	1.0573
1.08	0.4800	0.5920	0.8108	0.9005	0.9949	0.9680	1.0653
1.09	0.4742	0.5869	0.8080	0.8989	0.9936	1.148	1.0733
1.10	0.4684	0.5817	0.8052	0.8973	0.9921	1.336	1.0812
1.11	0.4626	0.5766	0.8023	0.8957	0.9905	1.532	1.0891
1.12	0.4568	0.5714	0.7994	0.8941	0.9888	1.735	1.0970
1.13	0.4511	0.5663	0.7966	0.8925	0.9870	1.944	1.1048
1.14	0.4455	0.5612	0.7937	0.8909	0.9850	2.160	1.1126
1.15	0.4398	0.5562	0.7908	0.8893	0.9828	2.381	1.1203
1.16	0.4343	0.5511	0.7879	0.8877	0.9806	2.607	1.1280
1.17	0.4287	0.5461	0.7851	0.8860	0.9782	2.839	1.1356
1.18	0.4232	0.5411	0.7822	0.8844	0.9758	3.074	1.1432
1.19	0.4178	0.5361	0.7793	0.8828	0.9732	3.314	1.1508
1.20	0.4124	0.5311	0.7764	0.8811	0.9705	3.558	1.1582
1.21	0.4070	0.5262	0.7735	0.8795	0.9676	3.806	1.1658
1.22	0.4017	0.5213	0.7706	0.8778	0.9647	4.057	1.1732
1.23	0.3964	0.5164	0.7677	0.8762	0.9617	4.312	1.1806
1.24	0.3912	0.5115	0.7648	0.8745	0.9586	4.569	1.1879
1.25	0.3861	0.5067	0.7619	0.8729	0.9553	4.830	1.1952
1.26	0.38009	0.5019	0.7590	0.8712	0.9520	5.093	1.2025
1.27	0.3759	0.4971	0.7561	0.8695	0.9486	5.359	1.2097
1.28	0.3708	0.4923	0.7532	0.8679	0.9451	5.627	1.2169
1.29	0.3658	0.4876	0.78503	0.8662	0.9415	5.898	1.2240
1.30	0.3609	0.4829	0.7474	0.8645	0.9378	6.170	1.2311
1.31	0.3560	0.4728	0.7445	0.8628	0.9341	6.445	1.2382
1.32	0.3512	0.4736	0.7416	0.8611	0.9302	6.721	1.2452
1.33	0.3464	0.4690	0.7387	0.8595	0.9263	7.000	1.2522
1.34	0.3417	0.4644	0.7358	0.8578	0.9223	7.279	1.2591
1.35	0.3370	0.4598	0.7329	0.8561	0.9182	4.561	1.2660

(continued)

Table B.5 (continued)

M	$\frac{p}{p_0}$	$\frac{\rho}{\rho_0}$	$\frac{T}{T_0}$	$\frac{a}{a_0}$	$\frac{A^*}{A}$	δ (°)	λ
1.36	0.3323	0.4553	0.7300	0.8544	0.9141	7.844	1.2729
1.37	0.3277	0.4508	0.7271	0.8527	0.9099	8.128	1.2797
1.38	0.3232	0.4463	0.7242	0.8510	0.9056	8.413	1.2865
1.39	0.3187	0.4418	0.7213	0.8493	0.9013	8.699	1.2932
1.40	0.3142	0.4374	0.7184	0.8476	0.8969	8.987	1.2999
1.41	0.3098	0.4330	0.7155	0.8459	0.8925	9.276	1.3065
1.42	0.3055	0.4287	0.7126	0.8442	0.8880	9.565	1.3131
1.43	0.3012	0.4244	0.7097	0.8425	0.8834	9.855	1.3197
1.44	0.2969	0.4201	0.7069	0.8407	0.8788	10.15	1.3262
1.45	0.2927	0.4158	0.7040	0.8390	0.8742	10.44	1.3327
1.46	0.2886	0.4116	0.7011	0.8373	0.8695	10.73	1.3329
1.47	0.2845	0.4074	0.6982	0.8356	0.8647	11.02	1.3456
1.48	0.2804	0.4032	0.6954	0.8339	0.8599	11.32	1.3520
1.49	0.2764	0.3991	0.6925	0.8322	0.8551	11.61	1.3583
1.50	0.2724	0.3950	0.6897	0.8305	0.8502	11.91	1.3646
1.51	0.2685	0.3909	0.6868	0.8287	0.8453	12.20	1.3708
1.52	0.2646	0.3869	0.6840	0.8270	0.8404	12.49	1.3770
1.53	0.2608	0.3928	0.6811	0.8253	0.8354	12.79	1.3832
1.54	0.2570	0.3789	0.6783	0.8236	0.8304	13.09	1.3894
1.55	0.2533	0.3750	0.6754	0.8219	0.8254	13.38	1.3955
1.56	0.2496	0.3710	0.6726	0.8201	0.8203	13.38	1.4016
1.57	0.2459	0.3672	0.6698	0.8184	0.8152	13.97	1.4076
1.58	0.2423	0.3633	0.6670	0.8167	0.8101	14.27	1.4135
1.59	0.2388	0.3595	0.6642	0.8150	0.8050	14.56	1.4195
1.60	0.2353	0.3557	66.14	0.8133	0.7998	14.86	1.4254
1.61	0.2318	0.3520	0.6586	0.8115	0.7947	15.16	1.313
1.62	0.2284	0.3483	0.6558	0.8098	0.7895	15.45	1.4371
1.63	0.2250	0.3446	0.6530	0.8081	0.7843	15.75	1.4429
1.64	0.2217	0.3409	0.6502	0.8064	0.7791	16.04	1.4487
1.65	0.2184	0.3373	0.6475	0.8046	0.7739	16.34	1.4511
1.66	0.2151	0.3337	0.6447	0.8029	0.7686	16.63	1.4601
1.67	0.2119	0.3302	0.6419	0.8012	0.7634	16.93	1.4657
1.68	0.2088	0.3266	0.6392	0.7995	0.7581	17.22	1.4713
1.69	0.2057	0.3232	0.6364	0.7978	0.7529	17.52	1.4769
1.70	0.2026	0.3197	0.6337	0.7961	0.7476	17.81	1.4825
1.71	0.1996	0.3163	0.6310	0.7943	0.7423	18.10	1.4880

(continued)

Table B.5 (continued)

M	$\frac{p}{p_0}$	$\frac{\rho}{\rho_0}$	$\frac{T}{T_0}$	$\frac{a}{a_0}$	$\frac{A^*}{A}$	$\delta\ (°)$	λ
1.72	0.1966	0.3129	0.6283	0.7926	0.7371	18.40	1.4935
1.73	0.1936	0.3095	0.6256	0.7909	0.7318	18.69	1.4989
1.74	0.1907	0.3062	0.6229	0.7892	0.7265	18.98	1.5043
1.75	0.1878	0.3029	0.6202	0.7875	0.7212	19.27	1.5097
1.76	0.1850	0.2996	0.6175	0.7858	0.7160	19.56	1.5150
1.77	0.1822	0.2964	0.6148	0.7841	0.7107	19.86	1.5203
1.78	0.1794	0.2932	0.6121	0.7824	0.7054	20.15	1.5256
1.79	0.1767	0.2900	0.6095	0.7807	0.7002	20.44	1.5308
1.80	0.1740	0.2868	0.6068	0.7790	0.6949	20.73	1.5360
1.81	0.1714	0.2837	0.6041	0.7773	0.6897	21.01	1.5412
1.82	0.1688	0.2806	0.6015	0.7756	0.6845	21.30	1.5463
1.83	0.1662	0.2776	0.5989	0.7739	0.6792	21.59	1.5514
1.84	0.1637	0.2745	0.5963	0.7722	0.6740	21.88	1.5564
1.85	0.1612	0.2715	0.5963	0.7705	0.6688	22.16	1.5614
1.86	0.1587	0.2686	0.5910	0.7688	0.6636	22.45	1.5664
1.87	0.1563	0.2656	0.5884	0.7671	0.6584	22.73	1.5714
1.88	0.1539	0.2627	0.5859	0.7654	0.6533	23.02	1.5763
1.89	0.1516	0.2598	0.5833	0.7637	0.6481	23.30	1.5812
1.90	0.1492	0.2570	0.5807	0.7620	0.6430	23.59	1.5861
1.91	0.1470	0.2542	0.5782	0.7604	0.6379	23.87	1.5909
1.92	0.1447	0.2514	0.5756	0.7587	0.6328	24.15	1.5957
1.93	0.1425	0.2486	0.5731	0.7570	0.6277	24.43	1.6005
1.94	0.1403	0.2459	0.5705	0.7553	0.6226	24.71	1.6052
1.95	0.1381	0.2432	0.5680	0.7537	0.6175	24.99	1.6099
1.96	0.1360	0.2405	0.5655	0.7520	0.6125	25.27	1.6146
1.97	0.1339	0.2378	0.5630	0.7503	0.6075	25.55	1.6193
1.98	0.1318	0.2352	0.5605	0.7487	0.6025	25.83	1.6239
1.99	0.1298	0.2326	0.5580	0.7470	0.5975	26.10	1.6285
2.00	0.1278	0.2300	0.5556	0.7454	0.5926	26.38	1.6330
2.01	0.1258	0.2275	0.5531	0.7437	0.5877	26.66	1.6375
2.02	0.1239	0.2250	0.5506	0.7420	0.5828	26.93	1.6420
2.03	0.1220	0.2225	0.5482	0.7404	0.5779	27.20	1.6465
2.04	0.1201	0.2200	5458	0.7388	0.5730	27.48	1.6509
2.05	0.1182	0.2176	0.5433	0.7371	0.5682	27.75	1.6553
2.06	0.1164	0.2152	0.5409	0.7355	0.5634	28.02	1.6597
2.07	0.1146	0.2128	0.5385	0.7338	0.5586	28.29	1.6640

(continued)

Table B.5 (continued)

M	$\frac{p}{p_0}$	$\frac{\rho}{\rho_0}$	$\frac{T}{T_0}$	$\frac{a}{a_0}$	$\frac{A^*}{A}$	δ (°)	λ
2.08	0.1128	0.2104	0.5361	0.7322	0.5538	28.56	1.6683
2.09	0.1111	0.2081	0.5337	0.7306	0.5491	28.83	1.6726
2.10	0.1094	0.2058	0.5313	0.7289	0.5444	29.10	1.6769
2.11	0.1077	0.2035	0.5290	7273	0.5379	29.36	1.6811
2.12	0.1060	0.2013	0.5266	0.7257	0.5350	29.63	1.6853
2.13	0.1043	0.19 90	0.5243	0.7241	0.5304	29.90	1.6895
2.14	0.1027	0.1968	0.5219	7225	0.5258	30.16	1.6936
2.15	0.1011	0.1946	0.5196	0.7208	0.5212	30.43	1.6977
2.16	0.09956	0.1925	0.5173	0.7192	0.5167	30.69	1.7018
2.17	0.09802	0.1903	0.5150	0.7176	0.5122	30.95	1.7059
2.18	0.09650	0.1882	0.5127	0.7160	0.5077	31.21	1.7099
2.19	0.09500	0.1861	0.5104	0.7144	0.5032	31.47	1.7139
2.20	0.09352	0.1841	0.5081	0.7128	0.4988	31.73	1.7179
2.21	0.09207	0.1820	0.5059	0.7112	0.4944	31.99	1.7219
2.22	0.09064	0.1800	0.5036	0.7097	0.4900	32.25	1.7258
2.23	0.08923	0.1780	0.5014	0.7081	0.4856	32.51	1.7297
2.24	0.08785	0.1760	0.4991	0.7065	0.4813	32.76	1.7336
2.25	0.08648	0.1740	0.4969	0.7049	0.4770	33.02	1.7374
2.26	0.08514	0.1721	0.4947	0.7033	0.4727	33.27	1.7412
2.27	0.08382	0.1702	0.4925	0.7018	0.4685	33.53	1.7450
2.28	0.8252	0.1683	0. 4903	0.7002	0.4643	33.78	1.7488
2.29	0.08123	0.1664	0.4881	0.6986	0.4601	34.03	1.7526
2.30	0.07997	0.1646	0.4859	0.6971	0.4560	34.28	1.7563
2.31	0.07873	0.1628	0.4837	0.6955	0.4519	34.53	1.7600
2.32	0.07751	0.1609	0.4816	0.6940	0.4478	34.78	1.7637
2.33	0.07631	0.1592	0.4794	0.6924	0.4437	35.03	1.7673
2.34	0.07512	0.1574	0.4773	0.6909	0.4397	35.28	1.7709
2.35	0.07396	0.1556	0.4752	0.6893	0.4357	35.53	1.7745
2.36	0.07281	0.1539	0.4731	0.6878	0.4317	35.77	1.7781
2.37	0.07168	0.1522	0.4709	0.6863	0.4278	36.02	1.7781
2.38	0.07057	0.1505	0.4688	0.6847	0.4239	36.26	1.7852
2.39	0.06948	0.1488	0.4668	0.6832	0.4200	36.50	1.7887
2.40	0.06840	0.1472	0.4647	0.6817	0.4161	36.75	1.7922
2.41	0.06734	0.1456	0.4626	0.6802	0.4123	36.99	1.7957
2.42	0.06630	0.1439	0.4606	0.6786	0.4085	37.23	1.7991
2.43	0.06527	0.1424	0.4585	0.6771	0.4048	37.47	1.8025

(continued)

Table B.5 (continued)

M	$\frac{p}{p_0}$	$\frac{\rho}{\rho_0}$	$\frac{T}{T_0}$	$\frac{a}{a_0}$	$\frac{A^*}{A}$	δ (°)	λ
2.44	0.06426	0.1408	0.4565	0.6756	0.4010	37.71	1.8059
2.45	0.06327	0.1392	0.4544	0.6741	0.3973	37.95	1.8093
2.46	0.06229	0.1377	0.4524	0.6726	0.3937	38.18	1.8126
2.47	0.06133	0.1362	0.4504	0.6711	0.3900	38.42	1.8159
2.48	0.06038	0.1347	0.4484	0.6696	0.3864	38.66	1.8192
2.49	0.05945	0.1332	0.4464	0.6681	0.3828	38.89	1.8225
2.50	0.05853	0.1317	0.4444	0.6667	0.3793	39.12	1.8258
2.51	0.05762	0.1302	0.4425	0.6652	0.3757	39.36	1.8290
2.52	0.05674	0.1288	0.4405	0.6637	0.3722	39.59	1.8322
2.53	0.05586	0.1274	0.4386	0.6622	0.3688	39.28	1.8354
2.54	0.05500	0.1260	0.4366	0.6608	0.3653	40.05	1.8386
2.55	0.05415	0.1246	0.4347	0.6593	0.3619	40.28	1.8417
2.56	0.05332	0.1232	0.1328	0.6579	0.3585	40.51	1.8448
2.57	0.05250	0.1218	0.4309	0.6564	0.3552	40.75	1.8479
2.58	0.05169	0.1205	0.4289	0.6549	0.3519	40.96	1.8510
2.59	0.05090	0.1192	0.4271	0.6535	0.3486	41.19	1.8541
2.60	0.05012	0.1179	0.4252	0.6521	0.3453	41.41	1.8572
2.61	0.04935	0.1166	0.4233	0.6506	0.3421	41.64	1.8602
2.62	0.04859	0.1153	0.4214	0.6492	0.3389	41.86	1.8632
2.63	0.04784	0.1140	0.4196	0.6477	0.3357	42.09	1.8662
2.64	0.04711	0.1128	0.4177	0.6463	0.3325	42.31	1.8692
2.65	0.04639	0.1115	0.4159	0.6449	0.3294	42.53	1.8721
2.66	0.04568	0.1103	0.4141	0.6435	0.3263	42.75	1.8750
2.67	0.04498	0.1091	0.4122	0.6421	0.6232	42.97	1.8779
2.68	0.04429	0.1079	0.4104	0.6406	0.3202	43.19	1.8808
2.69	0.04362	0.1067	0.4086	0.6392	0.3172	43.40	1.8837
2.70	0.04295	0.1056	0.4086	0.6378	0.3142	43.62	1.8865
2.71	0.04229	0.1044	0.4051	0.6364	0.3112	43.84	1.8894
2.72	0.04165	0.1033	0.4033	0.6350	0.3083	44.05	1.8922
2.73	0.04102	0.1022	0.4015	0.6337	0.3054	44.27	1.8950
2.74	0.04039	0.1010	0.3998	0.6323	0.3025	44.48	1.8978
2.75	0.03978	0.09994	0.3980	0.6309	0.2996	44.69	1.9005
2.76	0.03917	0.09885	0.3963	0.6295	0.2968	44.91	1.9032
2.77	0.03858	0.09778	0.3945	0.6281	0.2940	45.12	1.9060
2.78	0.03799	0.09671	0.3928	0.6268	0.2912	45.33	1.9081
2.79	0.03742	0.09566	0.3911	0.6254	0.2884	45.54	1.9114

(continued)

Table B.5 (continued)

M	$\frac{p}{p_0}$	$\frac{\rho}{\rho_0}$	$\frac{T}{T_0}$	$\frac{a}{a_0}$	$\frac{A^*}{A}$	δ (°)	λ
2.80	0.03685	0.09463	0.3894	0.6240	0.2857	45.75	1.9140
2.81	0.03629	0.09360	0.3877	0.6227	0.2830	45.95	1.9167
2.82	0.03574	0.09259	0.3860	0.6213	0.2803	46.16	1.9193
2.83	0.03520	0.09158	0.3844	0.6200	0.2777	46.37	1.9220
2.84	0.03467	0.09059	0.3827	0.6186	0.2750	46.57	1.9246
2.85	0.03415	0.08962	0.3810	0.6173	0.2724	46.78	1.9271
2.86	0.03363	0.08865	0.3794	0.6159	0.2698	46.98	1.9297
2.87	0.03312	0.08769	0.3777	0.6146	0.2673	47.19	1.9322
2.88	0.03263	0.08675	0.3761	0.6133	0.2648	47.39	1.9348
2.89	0.03213	0.08581	0.3745	0.6119	0.2622	47.59	1.9373
2.90	0.03165	0.08489	0.3729	0.6106	0.2598	47.79	1.9398
2.91	0.03118	0.08398	0.3712	0.6093	0.2573	47.99	1.9424
2.92	0.03071	0.08307	0.3696	0.6080	0.2549	48.19	1.9448
2.93	0.03025	0.08218	0.3681	0.6067	0.2524	48.39	1.9472
2.94	0.02980	0.08130	0.3665	0.6054	0.2500	48.59	1.9497
2.95	0.02935	0.08043	0.3649	0.6041	0.2477	48.78	1.9521
2.96	0.02891	0.07957	0.3633	0.6028	0.2453	48.98	1.9545
2.97	0.02848	0.07852	0.3618	0.6015	0.2430	49.18	1.9569
2.98	0.02805	0.07788	0.3602	0.6002	0.2407	49.37	1.9593
2.99	0.02764	0.07705	0.3587	0.5989	0.2384	49.56	1.9616
3.00	0.02722	0.07623	0.3571	0.5976	0.2362	49.76	1.9640
3.01	0.02682	0.07541	0.3556	0.5963	0.2339	49.95	
3.02	0.02642	0.07461	0.3541	0.5951	0.2317	50.14	
3.03	0.02603	07,382	0.3526	0.5938	0.2295	50.33	
3.04	0.02564	0.07303	0.3511	0.5925	0.2273	50.52	
3.05	0.02526	0.07226	0.3496	0.5913	0.2252	50.71	
3.06	0.02489	0.07149	0.3481	0.5900	0.2230	50.90	
3.07	0.02452	0.07074	0.3466	0.5887	0.2209	51.09	
3.08	0.02416	0.06999	0.3452	0.5875	0.2188	51.28	
3.09	0.02380	0.06925	0.3437	0.5862	0.2168	51.46	
3.10	0.02345	0.06852	0.3422	0.5850	0.2147	51.65	1.8866
3.11	0.02310	0.06779	0.3408	0.5838	0.2127	51.84	
3.12	0.02276	0.06708	0.3393	0.5825	0.2107	52.02	
3.13	0.02243	0.06637	0.3379	0.5813	0.2087	52.20	
3.14	0.02210	0.06568	0.3365	0.5801	0.2067	52.39	
3.15	0.02177	0.06499	0.3351	0.5788	0.2048	52.57	

(continued)

Table B.5 (continued)

M	$\frac{p}{p_0}$	$\frac{\rho}{\rho_0}$	$\frac{T}{T_0}$	$\frac{a}{a_0}$	$\frac{A^*}{A}$	δ (°)	λ
3.16	0.02146	0.06430	0.3337	0.5776	0.2028	52.75	
3.17	0.02114	0.06363	0.3323	0.5764	0.2009	52.93	
3.18	0.02083	0.06269	0.3309	0.5752	0.1990	53.11	
3.19	0.02053	0.06231	0.3295	0.5740	0.1971	53.29	
3.20	0.02023	0.06165	0.3281	0.5728	0.1953	53.47	2.0079
3.21	0.01993	0.06101	0.3267	0.5716	0.1934	53.65	
3.22	0.01964	0.06037	0.3253	0.5704	0.1916	53.83	
3.23	0.01936	0.05975	0.3240	5692	0.1898	54.00	
3.24	0.01908	0.05912	0.3226	0.5680	0.1880	54.18	
3.25	0.01880	0.05851	0.3213	0.5668	0.1863	54.35	
3.26	0.01853	0.05990	0.3199	0.5656	0.1845	54.53	
3.27	0.01826	0.05730	0.3186	0.5645	0.1828	54.71	
3.28	0.01799	0.05671	0.3173	0.5633	0.1810	54.88	
3.29	0.01773	0.05612	0.3160	0.5621	0.1793	5.05	
3.30	0.01748	0.05554	0.3147	0.5609	0.1777	55.22	2.0279
3.31	0.01722	0.05497	0.3134	0.5598	0.1760	55.39	
3.32	0.01698	0.05440	0.3121	0.5586	0.1743	55.56	
3.33	0.01673	0.05384	0.308	0.5575	0.1727	55.73	
3.34	0.01649	0.05329	0.3095	0.5563	0.1711	55.90	
3.35	0.01625	0.05274	0.3082	0.5552	0.1695	56.07	
3.36	0.01602	0.05220	0.3069	0.5540	0.1679	56.24	
3.37	0.01579	0.05166	0.3057	0.5529	0.1663	56.41	
3.38	0.01557	0.05113	0.3044	0.5517	0.1648	56.58	
3.39	0.01534	0.05061	0.3032	0.5506	0.1632	56.75	
3.40	0.01513	0.05009	0.3019	0.5595	0.1617	56.91	2.0466
3.41	0.01491	0.04958	0.3007	0.5484	0.1602	57.07	
3.42	0.01470	0.04908	0.2995	0.5472	0.1587	57.24	
3.43	0.01449	0.04858	0.2982	0.5461	0.1572	57.40	
3.44	0.01428	0.04808	0.2970	0.5450	0.1558	57.56	
3.45	0.01408	0.04759	0.2958	0.5439	0.1543	57.73	
3.46	0.01388	0.04711	0.2946	0.5428	0.1529	57.89	
3.47	0.01368	0.04663	0.2934	0.5417	0.1515	58.05	
3.48	0.01349	0.04616	0.2922	0.5406	0.1501	58.21	
3.49	0.01330	0.04569	0.2910	0.5395	1487	58.37	
3.50	0.01311	0.04523	0.2899	0.5384	0.1473	58.53	2.0642
3.60	0.01138	0.04089	0.2784	0.5276	0.1342	60.09	2.0808

(continued)

Table B.5 (continued)

M	$\frac{p}{p_0}$	$\frac{\rho}{\rho_0}$	$\frac{T}{T_0}$	$\frac{a}{a_0}$	$\frac{A^*}{A}$	$\delta\ (°)$	λ
3.70	9.903×10^{-3}	0.03702	0.2675	0.5172	0.1224	61.60	2.0964
3.80	8.629×10^{-3}	0.03355	0.2572	0.5072	0.1117	63.04	2.1111
3.90	7.532×10^{-3}	0.03044	0.2474	0.4974	0.1021	64.44	2.1250
4.00	6.586×10^{-3}	0.02766	0.2381	0.4880	0.09239	64.78	2.1381
4.10	5.769×10^{-3}	0.02516	0.2293	0.4788	0.08536	67.08	2.1505
4.20	5.062×10^{-3}	0.02292	0.2208	0.4699	0.07818	68.33	1.1622
4.30	4.449×10^{-3}	0.02090	0.2129	0.4614	0.07166	69.54	2.1732
4.40	3.918×10^{-3}	0.01909	0.2053	0.4531	0.06575	70.71	2.1837
4.50	3.455×10^{-3}	0.01745	0.1980	0.4450	0.00038	71.83	2.1936
4.60	3.053×10^{-3}	0.01597	0.1911	0.4372	0.05550	72.92	2.2030
4.70	2.701×10^{-3}	0.01464	0.1846	0.4296	0.05107	73.97	20.2119
4.80	2.394×10^{-3}	0.01343	0.1783	0.4223	0.04703	74.99	2.2204
4.90	2.126×10^{-3}	0.01233	0.1734	0.4152	0.04335	75.97	2.2284
5.00	1.890×10^{-3}	0.01134	0.1667	0.4082	0.04000	76.92	2.2361
6.00	6.334×10^{-4}	5.194×10^{-3}	0.1220	0.3492	0.01880×10^{-3}	84.96	2.2953
7.00	2.416×10^{-4}	2.609×10^{-3}	0.09259	0.3043	9.602×10^{-3}	90.97	2.3333
8.00	1.024×10^{-4}	1.414×10^{-3}	0.07246	0.2692	5.260×10^{-3}	95.62	2.3591
9.00	4.739×10^{-5}	8.150×10^{-4}	0.05814	0.2411	3.056×10^{-3}	99.32	2.3772
10.00	2.356×10^{-5}	4.948×10^{-4}	0.04762	0.2182	1.866×10^{-3}	102.3	2.3904
100.00	2.790×10^{-12}	5.583×10^{-9}	4.998×10^{-4}	0.02236	2.157×10^{-8}	127.6	
∞	0	0	0	0	0	130.5	2.4495

Table B.6 Prandtl–Meyer expansion flow ($\gamma = 1.4$)

$\delta°$	M	$\mu°$	$\delta°$	M	$\mu°$
0.0	1.000	90.000	22.5	1.862	32.488
0.5	1.051	72.099	23.0	1.879	32.148
1.0	1.082	67.574	23.5	1.897	31.814
1.5	1.108	64.451	24.0	1.915	31.486
2.0	1.133	61.997	24.5	1.932	31.164
2.5	1.155	59.950	25.0	1.950	30.847
3.0	1.177	58.180	25.5	1.968	30.536
3.5	1.198	56.614	26.0	1.986	30.229
4.0	1.218	55.205	26.5	2.004	29.928
4.5	1.237	53.920	27.0	2.023	29.632
5.0	1.256	52.738	27.5	2.041	29.340
5.5	1.275	51.642	28.0	2.059	39.052
6.0	1.294	50.619	28.5	2.078	28.769
6.5	1.312	49.658	29.0	2.096	28.491
7.0	1.330	48.753	29.5	2.115	28.216
7.5	1.348	47.896	30.0	2.134	27.945
8.0	1.366	47.082	30.5	2.153	27.678
8.5	1.383	46.306	31.0	2.172	27.415
9.0	1.400	45.566	31.5	1.191	27.155
9.5	1.418	44.857	32.0	2.210	26.899
10.0	1.435	44.177	32.5	2.230	26.646
10.5	1.452	43.523	33.0	2.246	26.397
11.0	1.469	42.894	33.5	2.269	26.151
11.5	1.486	42.287	34.0	2.289	25.908
12.0	1.503	41.701	34.5	2.309	25.668
12.5	1.520	41.134	35.0	2.329	25.430
13.0	1.537	40.585	35.5	2.349	25.196
13.5	1.554	40.053	36.0	2.369	24.965
14.0	1.571	39.537	36.5	2.390	24.736
14.5	1.588	39.035	37.0	2.410	24.510
15.0	1.605	38.547	37.5	2.431	24.287
15.5	1.622	38.073	38.0	2.452	24.066
16.0	1.639	37.611	38.5	2.473	23.847
16.5	1.655	37.160	39.0	2.495	23.631
17.0	1.672	36.721	39.5	2.516	23.418
17.5	1.689	36.293	40.0	2.538	23.206

(continued)

Table B.6 (continued)

$\delta°$	M	$\mu°$	$\delta°$	M	$\mu°$
18.0	1.706	35.874	40.5	2.560	22.997
18.5	1.724	35.465	41.0	2.583	22.790
19.0	1.741	35.065	41.5	2.604	22.585
19.5	1.758	34.673	42.0	2.626	22.382
20.0	1.775	34.290	42.5	2.649	22.182
20.5	1.792	33.915	43.0	2.671	21.983
21.0	1.810	33.548	43.5	2.694	21.786
21.5	1.827	33.188	44.0	2.718	21.591
22.0	1.844	32.834	44.5	2.741	21.398
45.0	2.764	21.207	67.5	4.133	14.002
45.5	2.788	21.017	68.0	4.173	13.865
46.0	2.812	20.830	68.5	4.214	13.729
46.5	2.836	20.644	69.0	4.255	13.593
47.0	2.861	20.459	69.5	4.297	13.459
47.5	2.886	20.277	70.0	4.339	13.325
48.0	2.910	20.096	70.5	4.382	13.191
48.5	2.936	19.916	71.0	4.426	13.059
49.0	2.961	19.738	71.5	4.470	12.927
49.5	2.987	15.561	72.0	4.515	12.795
50.0	3.013	19.386	72.5	4.561	12.665
50.5	3.039	19.213	73.0	4.608	12.535
51.0	3.065	19.041	73.5	4.655	12.406
51.5	3.092	18.870	74.0	4.703	12.277
52.0	3.119	18.701	74.5	4.752	12.149
52.5	3.146	18.532	75.0	4.801	12.021
53.0	3.174	18.366	75.5	4.852	11.894
53.5	3.202	18.200	76.0	4.903	11.768
54.0	3.230	18.036	76.5	4.955	11.642
54.5	3.258	17.873	77.0	4.009	11.517
55.0	3.287	17.711	77.5	5.063	11.392
55.5	3.316	17.551	78.0	5.118	11.268
56.0	3.346	17.391	78.5	5.175	11.145
56.5	3.375	17.233	79.0	5.231	11.022
57.0	3.106	17.076	79.5	5.289	10.899
57.5	3.436	16.920	80.0	5.348	10.777
58.0	3.467	16.765	80.5	5.408	10.656

(continued)

Table B.6 (continued)

$\delta°$	M	$\mu°$	$\delta°$	M	$\mu°$
58.5	3.498	16.611	81.0	5.470	10.535
59.0	3.530	16.458	81.5	5.532	10.414
59.5	3.562	16.306	82.0	5.596	10.294
60.0	3.594	16.1585	82.5	5.661	10.175
60.5	3.627	16.006	83.0	5.727	10.056
61.0	3.660	15.856	83.5	5.795	9.937
61.5	3.694	15.708	84.0	5.864	9.819
62.0	3.728	15.561	84.5	5.864	9.819
62.5	3.762	15.415	85.0	6.006	9.584
63.0	3.797	15.270	85.5	6.080	9.467
63.5	3.832	15.126	86.0	6.155	9.350
64.0	3.868	14.983	86.5	6.232	9.234
64.5	3.904	14.830	87.0	6.310	9.119
65.0	3.941	14.698	87.5	6.390	9.003
65.5	3.979	14.557	88.0	6.472	8.888
66.0	4.016	14.417	88.5	6.556	8.774
66.5	4.055	14.278	89.0	6.642	8.660
67.0	4.094	14.140	89.5	6.729	8.546
90.0	6.819	8.433	97.5	8.480	6.772
90.5	6.911	8.320	98.0	8.618	6.664
91.0	7.005	8.207	98.5	8.759	6.556
91.5	7.102	8.095	99.0	8.905	6.448
92.0	7.201	7.983	99.5	9.055	6.340
92.5	7.302	7.871	100.0	9.210	6.233
93.0	7.406	7.760	100.5	9.371	6.126
93.5	7.513	7.649	101.0	9.536	6.019
94.0	7.623	7.538	101.5	9.708	5.913
94.5	7.735	7.428	102.0	9.885	5.806
95.0	7.851	7.318			
95.5	7.970	7.208			
96.0	8.092	7.099			
96.5	8.218	6.989			
97.0	8.347	6.881			

Table B.7 The parameter of Shockwave flow ($r = 1.4$)

M_{1n}	p_2/p_1	ρ_2/ρ_1	T_2/T_1	a_2/a_1	p_{20}/p_{10}	M_2 (Only for positive shock)
1.00	1.000	1.000	1.000	1.000	1.0000	1.0000
1.01	1.023	1.017	1.007	1.003	1.0000	0.9901
1.02	1.047	1.033	1.013	1.007	1.0000	0.9805
1.03	1.071	1.050	1.020	1.010	1.0000	0.9712
1.04	1.095	1.067	1.026	1.013	0.9999	0.9620
1.05	1.120	1.084	1.033	1.016	0.9999	0.9531
1.06	1.144	1.101	1.039	1.019	0.9998	0.9444
1.07	1.169	1.118	1.046	1.023	0.9999	0.9360
1.08	1.194	1.135	1.052	1.026	0.9994	0.9277
1.09	1.219	1.152	1.059	1.029	0.9992	0.9196
1.10	1.245	1.169	1.065	1.032	0.9989	0.9118
1.11	1.271	1.186	1.071	1.035	0.9986	0.9041
1.12	1.297	1.203	1.078	1.038	0.9982	0.8966
1.13	1.323	1.221	1.084	1.041	0.9978	0.8892
1.14	1.350	1.238	1.090	1.044	0.9973	0.8820
1.15	1.376	1.255	1.097	1.047	0.9967	0.8750
1.16	1.403	1.272	1.103	1.050	0.9961	0.8682
1.17	1.430	1.290	1.109	1.053	0.9953	0.8615
1.18	1.458	1.307	1.115	1.056	0.9946	0.8549
1.19	1.485	1.324	1.122	1.059	0.9937	0.8485
1.20	1.513	1.342	1.128	1.062	0.9928	0.8422
1.21	1.541	1.359	1.134	1.065	0.9918	0.8360
1.22	1.570	1.376	1.141	1.068	0.9907	0.8300
1.23	1.598	1.394	1.147	1.071	0.9896	0.8241
1.24	1.627	1.411	1.153	1.074	0.9884	0.8183
1.25	1.656	1.429	1.159	1.077	0.9871	0.8126
1.26	1.686	1.446	1.166	1.080	0.9857	0.8071
1.27	1.715	1.463	1.172	1.083	0.9842	0.8016
1.28	1.745	1.481	1.178	1.085	0.9827	0.7963
1.29	1.775	1.498	1.185	1.088	0.9811	0.7911
1.30	1.805	1.516	1.191	1.091	0.9794	0.7860
1.31	1.835	1.533	1.197	1.094	0.9776	0.7809
1.32	1.866	1.551	1.204	1.097	0.9758	0.7760
1.33	1.897	1.568	1.210	1.100	0.9738	0.7712
1.34	1.928	1.585	1.216	1.103	0.9718	0.7664

(continued)

Table B.7 (continued)

M_{1n}	p_2/p_1	ρ_2/ρ_1	T_2/T_1	a_2/a_1	p_{20}/p_{10}	M_2 (Only for positive shock)
1.35	1.960	1.603	1.223	1.106	0.9697	0.7618
1.36	1.991	1.620	1.229	1.109	0.9676	0.7572
1.37	2.023	1.638	1.235	1.111	0.9653	0.7527
1.38	2.055	1.655	1.242	1.114	0.9630	0.7483
1.39	2.087	1.672	1.248	1.117	0.9606	0.7440
1.40	2.120	1.690	1.255	1.120	0.9582	0.7379
1.41	2.153	1.707	1.261	1.123	0.9557	0.7355
1.42	2.186	1.724	1.268	1.126	0.9531	0.7314
1.43	2.219	1.742	1.274	1.129	0.9504	0.7274
1.44	2.253	1.759	1.281	1.132	0.9476	0.7235
1.45	2.286	1.776	1.287	1.135	0.9448	0.7196
1.46	2.320	1.793	1.294	1.137	0.9420	0.7157
1.47	2.354	1.811	1.300	1.140	0.9390	0.7120
1.48	2.389	1.828	1.307	1.143	0.9360	0.7083
1.49	2.423	1.845	1.314	1.146	0.9329	0.7047
1.50	2.458	1.862	1.320	1.149	0.9298	0.7011
1.51	2.493	1.879	1.327	1.152	0.9266	0.6967
1.52	2.529	1.896	1.334	1.155	0.9233	0.6941
1.53	2.564	1.913	1.340	1.158	0.9200	0.6907
1.54	2.600	1.930	1.347	1.161	0.9166	0.6874
1.55	2.636	1.947	1.354	1.164	0.9132	0.6841
1.56	2.673	1.964	1.361	1.166	0.9097	0.6809
1.57	2.709	1.981	1.367	1.169	0.9061	0.6777
1.58	2.746	1.998	1.374	1.172	0.9026	0.6746
1.59	1.783	2.015	1.381	1.175	0.8989	0.6715
1.60	2.280	2.032	1.388	1.178	0.8952	0.6684
1.61	2.857	2.049	1.395	1.181	0.8914	0.6655
1.62	2.895	2.065	1.402	1.184	0.8877	0.6625
1.63	2.933	2.082	1.409	1.187	0.8838	0.6596
1.64	2.971	2.099	1.416	1.190	0.8799	0.6568
1.65	3.010	2.115	1.423	1.193	0.8760	0.6540
1.66	3.048	2.132	1.430	1.196	0.8720	0.6512
1.67	3.087	2.148	1.437	1.199	0.8680	0.6485
1.68	3.126	2.165	1.444	1.202	0.8640	0.6458
1.69	3.165	2.181	1.451	1.205	0.8599	0.6431

(continued)

Table B.7 (continued)

M_{1n}	p_2/p_1	ρ_2/ρ_1	T_2/T_1	a_2/a_1	p_{20}/p_{10}	M_2 (Only for positive shock)
1.70	3.205	2.198	1.458	1.208	0.8557	0.6405
1.71	2.245	2.214	1.466	1.211	0.8516	0.6380
1.72	3.285	2.230	1.473	1.214	0.8474	0.6355
1.73	2.325	2.247	1.480	1.217	0.8431	0.6330
1.74	3.366	2.263	1.487	1.220	0.8389	0.6305
1.75	3.406	2.279	1.495	1.223	0.8346	0.6281
1.76	3.447	2.295	1.502	1.226	0.8302	0.6257
1.77	3.488	2.311	1.509	1.229	0.8259	0.6234
1.78	3.530	2.327	1.517	1.232	0.8215	0.6210
1.79	3.571	2.343	1.524	1.235	0.8171	0.6188
1.80	3.613	2.359	1.532	1.238	0.8127	0.6165
1.81	3.655	3.375	1.539	1.241	0.8082	0.6143
1.82	3.698	2.391	1.547	1.244	0.8038	0.6121
1.83	3.740	2.407	1.554	1.247	0.7993	0.6099
1.84	3.783	2.422	1.562	1.250	0.7948	0.6078
1.85	3.826	2.483	1.569	1.253	0.7902	0.6057
1.86	3.870	2.454	1.577	1.256	0.7857	0.6036
1.87	3.913	2.469	1.585	1.259	0.7811	0.6016
1.88	3.957	2.485	1.592	1.262	0.7765	0.5996
1.89	4.001	2.500	1.600	1.265	0.7720	0.5976
1.90	4.045	2.516	1.608	1.268	0.7674	0.5956
1.91	4.089	2.531	1.616	1.271	0.7628	0.5937
1.92	4.134	2.546	1.624	1.274	0.7581	0.5918
1.93	4.179	2.562	1.631	1.277	0.7535	0.5899
1.94	4.224	2.577	1.639	1.280	0.7488	0.5889
1.95	4.270	2.592	1.647	1.283	0.7442	0.5862
1.96	4.315	2.607	1.655	1.287	0.7395	0.5844
1.97	4.361	2.622	1.663	1.290	0.7349	0.5826
1.98	4.407	2.637	1.671	1.293	0.7302	0.5808
1.99	4.453	2.652	1.679	1.296	0.7255	5791
2.00	4.500	2.667	1.688	1.299	0.7209	0.5773
2.01	4.547	2.681	1.696	1.302	0.7162	0.5757
2.02	4.594	2.696	1.704	1.305	0.7115	0.5740
2.03	4.641	2.711	1.712	1.308	0.7069	0.5723
2.04	4.689	2.725	1.720	1.312	0.7022	0.5707

(continued)

Table B.7 (continued)

M_{1n}	p_2/p_1	ρ_2/ρ_1	T_2/T_1	a_2/a_1	p_{20}/p_{10}	M_2 (Only for positive shock)
2.05	4.736	2.740	1.729	1.315	0.6975	0.5691
2.06	4.784	2.755	1.737	1.318	0.6928	0.5675
2.07	4.832	2.769	1.745	1.321	0.6928	0.5659
2.08	4.881	2.783	1.754	1.324	0.6835	0.5643
2.09	4.929	2.798	1.762	1.327	0.6789	0.5628
2.10	4.978	2.812	1.770	1.331	0.6742	0.5613
2.11	5.027	2.826	1.779	1.334	0.6696	0.5598
2.12	5.077	2.840	1.787	1.337	0.6649	0.5583
2.13	5.126	2.854	1.796	1.340	0.6603	0.5568
2.14	5.176	2.868	1.805	1.343	0.6557	0.5554
2.15	5.226	2.882	1.813	1.347	0.6511	0.5540
2.16	5.277	.2896	1.822	1.350	0.6464	0.5525
2.17	5.327	2.910	1.831	1.353	0.6419	0.5511
2.18	5.378	2.924	1.839	1.356	0.6373	0.5498
2.19	5.429	2.938	1.848	1.359	0.6327	0.5484
2.20	5.480	2.951	1.857	1.363	0.6281	0.5471
2.21	5.531	2.965	1.866	1.366	0.6263	0.5457
2.22	5.583	2.978	1.875	1.369	0.6191	0.5444
2.23	5.635	2.992	1.883	1.372	0.6145	0.5431
2.24	5.687	3.005	1.892	1.376	0.6100	0.5418
2.25	5.740	3.019	1.901	1.379	0.6055	0.5406
2.26	5.792	3.032	1.910	1.382	0.6011	0.5393
2.27	5.845	3.045	1.919	1.385	0.5966	0.5381
2.28	5.898	3.058	1.929	1.389	0.5921	0.5368
2.29	5.951	3.071	1.938	1.392	0.5877	0.5356
2.30	6.005	3.085	1.947	1.395	0.5833	0.5344
2.31	6.059	3.098	1.956	1.399	0.5789	0.5332
2.32	6.113	3.110	1.965	1.402	0.5745	0.5321
2.33	6.167	3.123	1.947	1.405	0.5702	0.5309
2.34	6.222	3.136	1.984	1.408	0.5658	0.5297
2.35	6.276	3.149	1.993	1.412	0.5615	0.5286
2.36	6.331	3.162	2.002	1.415	0.5572	0.5275
2.37	6.386	3.174	2.012	1.418	0.5529	0.5264
2.38	6.442	3.187	2.021	1.422	0.5486	0.5253
2.39	6.497	3.199	2.031	1.425	0.5444	0.5242

(continued)

Table B.7 (continued)

M_{1n}	p_2/p_1	ρ_2/ρ_1	T_2/T_1	a_2/a_1	p_{20}/p_{10}	M_2 (Only for positive shock)
2.40	6.553	3.212	2.040	1.428	0.5401	0.5231
2.41	6.609	3.224	2.050	1.432	0.5359	0.5221
2.42	6.666	3.237	2.059	1.435	0.5317	0.5210
2.43	6.722	3.249	2.069	1.438	0.5276	0.5200
2.44	6.779	3.261	2.079	1.442	0.5234	0.5189
2.45	6.836	3.273	2.088	1.445	0.5193	0.5179
2.46	6.894	3.285	2.098	1.449	0.5152	0.5169
2.47	6.951	3.298	2.108	1.452	0.5111	0.5159
2.48	7.009	3.310	2.118	1.445	0.5071	0.5149
2.49	7.067	3.321	2.128	1.459	0.5030	0.5140
2.50	7.125	3.333	2.138	1.462	0.4990	0.5130
2.51	7.183	3.345	2.147	1.465	0.4950	0.5120
2.52	7.242	3.357	2.157	1.469	0.4911	0.5111
2.53	7.301	3.369	2.167	1.472	0.4871	0.5102
2.54	7.360	3.380	2.177	1.476	0.4832	0.5092
2.55	7.420	3.392	2.187	1.479	0.4793	0.5083
2.56	7.479	3.403	2.198	1.482	0.4754	0.5074
2.57	7.539	3.415	2.208	1.486	0.4715	0.5065
2.58	7.599	3.426	2.218	1.489	0.4677	0.5056
2.59	7.659	3.438	2.228	1.493	0.4639	0.5047
2.60	7.720	3.449	2.238	1.496	0.4601	0.5039
2.61	7.781	3.460	2.249	1.500	0.4564	0.5030
2.62	7.842	3.471	2.259	1.503	0.4526	0.5022
2.63	7.903	3.483	2.269	1.506	0.4489	0.5013
2.64	7.965	3.494	2.280	1.510	0.4452	0.5005
2.65	8.026	3.505	2.290	1.513	0.4416	0.4996
2.66	8.088	3.516	2.301	1.517	0.4379	0.4988
2.67	8.150	3.527	2.311	1.520	0.4343	0.4980
2.68	8.213	3.537	2.322	1.524	0.4307	0.4972
2.69	8.275	3.548	2.332	1.527	0.4271	0.4964
2.70	8.338	3.559	2.343	1.531	0.4236	0.4956
2.71	8.401	3.570	2.354	1.534	0.4201	0.4949
2.72	8.465	3.680	2.364	1.538	0.4166	0.4941
2.73	8.528	3.591	2.375	1.541	0.4131	0.4933
2.74	8.592	3.601	2.386	1.545	0.4097	0.4926

(continued)

Table B.7 (continued)

M_{1n}	p_2/p_1	ρ_2/ρ_1	T_2/T_1	a_2/a_1	p_{20}/p_{10}	M_2 (Only for positive shock)
2.75	8.656	3.612	2.397	1.548	0.4062	0.4918
2.76	8.721	3.622	2.407	1.552	0.4028	0.4911
2.77	8.785	3.633	2.418	1.555	0.3994	0.4903
2.78	8.850	3.643	2.429	1.559	0.3961	0.4896
2.79	8.915	3.653	2.440	1.562	0.3928	0.4889
2.80	8.980	3.664	2.451	1.566	0.3895	0.4882
2.81	9.045	3.674	2.462	1.569	0.3862	0.4875
2.82	9.111	3.684	2.473	1.573	0.3829	0.4868
2.83	9.177	3.694	2.484	1.576	0.3797	0.4861
2.84	9.243	3.704	2.496	1.580	0.3765	0.4854
2.85	9.310	3.714	2.507	1.583	0.3733	0.4847
2.86	9.376	3.724	2.518	1.587	0.3701	0.4840
2.87	9.443	3.734	2.529	1.590	0.3670	0.4833
2.88	9.510	3.743	2.540	1.594	0.3639	0.4827
2.89	9.577	3.753	2.552	1.597	0.3608	0.4820
2.90	9.645	3.763	2.563	1.601	0.3577	0.4814
2.91	9.713	3.773	2.575	1.605	0.3547	0.4807
2.92	9.781	3.782	2.586	1.608	0.3517	0.4801
2.93	9.849	3.792	2.598	1.612	0.3487	0.4795
2.94	9.918	3.801	2.609	1.615	0.3457	0.4788
2.95	9.986	3.811	2.621	1.619	0.3428	0.4782
2.96	10.06	3.820	2.632	1.622	0.3398	0.4776
2.97	10.12	3.829	2.644	1.626	0.3369	0.4770
2.98	10.19	3.839	2.656	1.630	0.3340	0.4764
2.99	10.26	3.848	2.667	1.633	0.3312	0.4758
3.00	10.33	3.857	2.679	1.637	0.3283	0.4752
3.10	11.05	3.947	2.799	1.673	0.3012	0.4695
3.20	11.78	4.031	2.922	1.709	0.2762	0.4643
3.30	12.54	4.112	3.049	1.746	0.2533	0.4596
3.40	13.32	4.188	3.180	1.783	0.2322	0.4552
3.50	14.13	4.261	3.315	1.821	0.2129	0.4512
3.60	14.95	4.330	3.454	1.858	0.1953	0.4474
3.70	15.80	4.395	3.596	1.896	0.1792	0.4439
3.80	16.68	4.457	3.743	1.935	0.1645	0.4407
3.90	17.58	4.516	3.893	1.973	0.1510	0.4377

(continued)

Table B.7 (continued)

M_{1n}	p_2/p_1	ρ_2/ρ_1	T_2/T_1	a_2/a_1	p_{20}/p_{10}	M_2 (Only for positive shock)
4.00	18.50	4.517	4.047	2.012		0.4350
5.00	29.00	5.000	5.800	2.408	0.1388	0.4152
6.00	41.83	5.268	7.941	2.818	0.06172	0.4042
7.00	57.00	5.444	10.47	3.236	0.02965	0.3947
8.00	74.50	5.565	13.39	3.659	0.01535	0.3929
9.00	94.33	5.651	16.69	4.086	8.488×10^{-3}	0.3898
10.00	116.5	5.714	20.39	4.515	4.964×10^{-3}	0.3876
100.00	11666.5	5.997	1945.4	44.11	3.045×10^{-3}	0.3781
∞	∞	6	∞	∞	3.593×10^{-3}	0.3780
					0	

All the data were acquired from NACA TN1428

Bibliography

1. Prandtl L. Fuhere durch die stromungslehre [M]. Translated by GUO Yunhuai, LU Shijia. Beijing: Science Press, 1987.
2. Batchelor, G.K., An Introduction to Fluid Dynamics [M]. Translated by SHEN Qin, JIA Fu, Beijing: Science Press, 1997.
3. LIU Peiqing. Fluid Mechanics Biography[M]. Beijing: Science Press, 2017.
4. John D. Anderson, Jr., Fundamentals of Aerodynamics [M]. Third Edition, International Edition. Mechanical Engineering Series, New York: McGaw-Hill, 2001.
5. Frank M. White. Fluid Mechanics [M] (Seventh Edition). Mcgraw-Hill series in mechanical engineering, New York, 2011.
6. E. L. Houghton P. W. Carpenter Steven H. Collicott and Daniel T. Valentine. Aerodynamics for Engineering Students [M]. Amsterdam Boston Heidelberg, London, 2013.
7. Barnes W. McCormick. Aerodynamics, Aeronautics, and Flight Mechanics [M]. John Wiley and Sons, Inc. Canada.1995.
8. WU Wangyi. Fluid Mechanics [M]. Beijing: Peking University Press, 1983.
9. ZHOU Guangxi, YAN Zongyi, XU Shixiong, Zhang Keben eds. Fluid Mechanics (Volumes 1 and 2) [M]. Beijing: Higher Education Press, 2000.
10. J. F. Doulas, J. M. Gasiorek and J. A. Swaffield. Fluid Mechanics [M], 3rd Edition. Beijing: World Book Inc., 2000.
11. Victor L. Streeter and E. Benjamin Wylie. Fluid Mechanics [M]. McGraw-Hill Book Company, New York, 1979.
12. P. D. McCormack and Lawrence Crane. Physical Fluid Dynamics [M]. Academic Press, New York and London, 1973.
13. XU Weide eds. Fluid Mechanics [M]. Beijing: National Defense Industry Press, 1979.
14. CHEN Zaixin, LIU Fuchang, BAO Guohua, Aerodynamics [M]. Beijing: Aviation Industry Press, 1993.
15. XU Huafang, Basics of Aerodynamics (Volumes 1 and 2) [M]. Beijing: Beijing University of Aeronautics and Astronautics Press, 1987.
16. YANG Bisheng and YU Shouqin eds, Aerodynamics of Aircraft Parts [M]. Beijing: National Defense Industry Press, 1981.
17. QIAN Yiji, Aerodynamics [M]. Beijing: Beihang Univerisity Press, 2004.
18. WU Ziniu. Aerodynamics (Volumes 1 and 2) [M]. Beijing: Tsinghua University Press, 2007.
19. LU Zhiliang. Aerodynamics [M]. Beijing: Beihang University Press, 2009.
20. ZENG Ming, LIU Wei, ZOU Jianjun. Aerodynamics [M]. Beijing: Science Press, 2017.
21. ZHAO Xueduan, LIAO Qidian eds. Viscous fluid mechanics [M]. Beijing: Machinery Industry Press, 1983.
22. CHEN Maozhang eds. Fundamentals of Viscous Fluid Dynamics [M]. Beijing: Higher Education Press, 2002.

© Science Press 2022
P. Liu, *Aerodynamics*, https://doi.org/10.1007/978-981-19-4586-1

23. XU Wenxi and XU Wencan eds. Viscous fluid mechanics[M]. Beijing: Beijing Institute of Technology Press, 1989.
24. ZHANG Hanxin. Structural analysis of separation flow and vortex motion [M]. Beijing: National Defense Industry Press, 2005.
25. Batchelor G H. The theory of homogeneous turbulence [M]. New York: Cambridge University Press, 1953.
26. Schlichting H. Boundary layer theory [M]. New York: McGraw Hill Book Company, 1979.
27. Launder B E and Spalding D B. Mathematical models of turbulence [M]. London: Academic Press, 1972.
28. Stanisic M M. The Mathematical theory of turbulence [M]. New York: Springer-Verlag, 1984.
29. Michel A. Saad. Compressible Fluid Flow [M]. Prentice-Hall International, Inc., London.1985.
30. Hinze J O. Turbulence (Volumes 1 and 2) [M]. Huang Yongnian, Yan Dachun. Beijing: Science Press, 1987.
31. Frisch U. Turbulnce [M]. New York: Cambridge University Press, 1995.
32. SHI Xungang. Turbulence [M]. Tianjin: Tianjin University Press, 1994.
33. Zhang Zhaoshun, Cui Guixiang, Xu Chunxiao. Turbulence Theory and Simulation [M]. Beijing: Tsinghua University Press, 2005
34. Ed Obert. Aerodynamic Design of Transport Aircraft [M]. Delft University Press, 2009.
35. JIe-Zhi Wu, Hui-Yang Ma, Ming-De Zhou. Vortical Flows [M], Springer Heidelberg New York Dordrecht London, 2015.
36. LI Zhoufu. Manual of Wind Tunnel Test. Beijing: Aviation Industry Press, 2015.
37. YAN Dachun. Experimental Fluid Mechanics. Beijing: Higher Education Press, 1992.
38. Yun Qilin. Experimental Aerodynamics. Beijing: National Defense Industry Press, 1991.
39. WANG Tiecheng. Aerodynamics Experimental Technology. Beijing: Aviation Industry Press, 1995.
40. LI Suxun. Complex flow dominated by shock waves and boundary layer [M]. Beijing: Science Press, 2007.
41. LIU Peiqing. Free Turbulent Jet Theory [M]. Beijing: Beihang University Press, 2008.
42. LIU Peiqing. Air propeller theory and its application [M]. Beijing: Beihang University Press, 2008.
43. Sheldon I. Green. Fluid Vortices [M]. Kluwer Academic Publishers, The Netherlands, 1995.
44. ZHANG Zixiong and DONG Zengnan eds. Viscous fluid mechanics [M]. Beijing: Tsinghua University Press, 1998.
45. DONG Zengnan and ZHANG Zixiong. Non-viscous fluid mechanics [M]. Beijing: Tsinghua University Press, 2003.
46. ZHANG Zhaoshun and CUI Guixiang. Fluid Mechanics [M]. Beijing: Tsinghua University Press, 1999.
47. ZHANG Changgao. Hydrodynamics [M]. Beijing: Higher Education Press, 1993.
48. Tsien, S. H., Superaerodynamics, Mechanics of Rarefied Gases, J. Aero. Sci, 1946, 13: 653–664.
49. L. M. Milne-Thomson. Theoretical aerodynamics [M]. London: Macmillan and Co, 1948.
50. N. A. Cumpsty, Compressor Aerodynamics [M]. Krieger Publishing Company, Malabar, Florida, 1998.
51. A. A. Townsend, The Structure of Turbulent Shear Flow [M]. The University Press, Cambridge, 1956.
52. John D. Anderson Jr., Introduction to Flight [M]. McGraw-Hill Education, New York, 2016.
53. John D. Anderson Jr., Modern Compressible Flow [M]. McGraw-Hill Education, New York, 2003.
54. YANG Yunjun, GONG Anlong, BAI Peng, Hypersonic aerodynamic design and evaluation method [M]. Beijing: China Aerospace Development Publishing House, 2019.
55. Robert W. Fox and Alan T. McDonald, Introduction to Fluid Mechanics [M]. John Wiley and Sons,Inc., New York, 2001.
56. Doug McLean, Understanding Aerodynamics [M]. John Wiley and Sons Ltd., India, 2014.
57. Daniel P. Raymer, Aircraft Design: A Conceptual Approach [M]. American Institute of Aeronautics and Astronautics, Inc., Virginia, 2006.

Printed in the United States
by Baker & Taylor Publisher Services